Corrosion Handbook

Corrosive Agents and Their Interaction with Materials

Edited by G. Kreysa and M. Schütze

Volume 2
Hydrochloric Acid, Nitric Acid

Corrosion Handbook

Corrosive Agents and Their Interaction with Materials

Second, Completely Revised and Extended Edition

Volume 1
Sodium Hydroxide, Mixed Acids

Volume 2
Hydrochloric Acid, Nitric Acid

Volume 3
Hypochlorites, Phosphoric Acid

Volume 4
Drinking Water, Waste Water (Urban), Waste Water (Industrial)

Volume 5
Carbonic Acid, Chlorine Dioxide, Seawater

Volume 6
Atmosphere, Industrial Waste Gases

Volume 7
Sodium Chloride

Volume 8
Chlorinated Hydrocarbons – Chloromethanes,
Chlorinated Hydrocarbons – Chloroethanes, Alkanols

Volume 9
Potassium Hydroxide, Ammonium and Ammonium Hydroxide

Volume 10
Sulfur Dioxide, Sodium Sulfate

Volume 11
Sulfuric Acid

Volume 12
Ferrous Chlorides ($FeCl_2$, $FeCl_3$)

Volume 13
Index

Corrosion Handbook

Corrosive Agents and Their Interaction with Materials

Second, Completely Revised and Extended Edition

Edited by Gerhard Kreysa and Michael Schütze

Volume 2
Hydrochloric Acid, Nitric Acid

WILEY-VCH

WILEY-VCH Verlag GmbH & Co. KGaA

Editors

Prof. Dr. Dr.-Ing. E.h. Dr.h.c. Gerhard Kreysa
Chief Executive of DECHEMA
DECHEMA e. V.
Society for Chemical Engineering and Biotechnology
Theodor-Heuss-Allee 25
60486 Frankfurt (Main)
Germany

Prof. Dr.-Ing. Michael Schütze
Director Materials
Karl Winnacker Institute of DECHEMA e. V.
Society for Chemical Engineering and Biotechnology
Theodor-Heuss-Allee 25
60486 Frankfurt (Main)
Germany

First Edition 1987
Second, Completely Revised and Extended Edition 2004

Technical Editors:
Dr. Roman Bender, Birgit Czack

Cover Illustration
Source: Karl Winnacker Institute of DECHEMA e. V., Frankfurt (Main), Germany

Warranty Disclaimer

This book has been compiled from literature data with the greatest possible care and attention. The statements made only provide general descriptions and information.

Even for the correct selection of materials and correct processing, corrosive attack cannot be excluded in a corrosion system as it may be caused by previously unknown critical conditions and influencing factors or subsequently modified operating conditions.

No guarantee can be given for the chemical stability of the plant or equipment. Therefore, the given information and recommendations do not include any statements, from which warranty claims can be derived with respect to DECHEMA e. V. or its employees or the authors.

The DECHEMA e. V. is liable to the customer, irrespective of the legal grounds, for intentional or grossly negligent damage caused by their legal representatives or vicarious agents.

For a case of slight negligence, liability is limited to the infringement of essential contractual obligations (cardinal obligations). DECHEMA e. V. is not liable in the case of slight negligence for collateral damage or consequential damage as well as for damage that results from interruptions in the operations or delays which may arise from the deployment of the Corrosion Handbook.

■ This book was carefully produced. Nevertheless, editors, authors and publisher do not warrant the information contained therein to be free of errors. Readers are advised to keep in mind that statements, data, illustrations, procedural details or other items may inadvertently be inaccurate.

Library of Congress Card No.: Applied for.

British Library Cataloguing-in-Publication Data:
A catalogue record for this book is available from the British Library.

Bibliographic information published by Die Deutsche Bibliothek
Die Deutsche Bibliothek lists this publication in the Deutsche Nationalbibliografie; detailed bibliographic data is available in the Internet at <http://dnb.ddb.de>.

© 2005 DECHEMA e. V., Society for Chemical Engineering and Biotechnology, 60486 Frankfurt (Main), Germany

All rights reserved (including those of translation into other languages). No part of this book may be reproduced in any form – nor transmitted or translated into machine language without written permission from the publishers. Registered names, trademarks, etc. used in this book, even when not specifically marked as such, are not to be considered unprotected by law.

Printed in the Federal Republic of Germany
Printed on acid-free paper

Composition   Kühn & Weyh, Satz und Medien, Freiburg
Printing   betz-druck GmbH, Darmstadt
Bookbinding   Großbuchbinderei J. Schäffer GmbH & Co. KG, Grünstadt

ISBN   3-527-31118-1

Contents

List of Contributors *VII*

Preface *IX*

How to use the CORROSION HANDBOOK *XI*

Hydrochloric Acid *1*
Author: A. Bäumel, P. Drodten, E. Heitz/Editor: R. Bender

A Metallic materials *12*
   Aluminium and aluminium alloys, copper and copper alloys, iron, iron-based alloys and steels, nickel and nickel alloys, titanium and titanium alloys, zinc cadmium and their alloys

B Non-metallic inorganic materials *178*
   Carbon and graphite, binders for building materials, glass, quartz ware and quartz glass, enamel, oxide ceramic materials, metal ceramic materials

C Organic materials/plastics *206*
   Thermoplastics, thermosetting plastics, elastomers, duroplasts

D Materials with special properties *239*
   Coatings and linings, seals and packings, composite materials

E Materials recommendations *252*

Bibliography *265*

Nitric Acid *283*
Author: K. Hauffe / Editor: R. Bender

A   Metallic materials   *291*
    Aluminium and aluminium alloys, copper and copper alloys, iron, iron-based alloys and steels, nickel and nickel alloys, titanium and titanium alloys, zinc cadmium and their alloys

B   Non-metallic inorganic materials   *460*
    Carbon and graphite, binders for building materials, glass, quartz ware and quartz glass, enamel, oxide ceramic materials, metal ceramic materials

C   Organic materials/plastics   *475*
    Thermoplastics, thermosetting plastics, elastomers, duroplasts

D   Materials with special properties   *500*
    Coatings and linings, seals and packings, composite materials

Bibliography   *510*

Key to materials compositions   *568*

Index of materials   *595*

Subject index   *613*

List of Contributors

Editors

Prof. Dr. Dr.-Ing. E.h. Dr.h.c. Gerhard Kreysa
Chief executive of DECHEMA e.V.
DECHEMA e.V.
Society for Chemical Engineering and Biotechnology
Theodor-Heuss-Allee 25
60486 Frankfurt (Main)
Germany

Prof. Dr.-Ing. Michael Schütze
Director Materials
Karl Winnacker Institute of DECHEMA e.V.
Society for Chemical Engineering and Biotechnology
Theodor-Heuss-Allee 25
60486 Frankfurt (Main)
Germany

Authors

Prof. Dr. Anton Bäumel
formerly head of the Staatliche Materialprüfungsanstalt
Darmstadt University of Technology
Grafenstraße 2
64283 Darmstadt
Germany

Dr. Peter Drodten
formerly ThyssenKrupp Stahl AG
Kaiser-Wilhelm-Straße 100
47166 Duisburg
Germany

Prof. Dr. Karl Hauffe
formerly Institute of Physical Chemistry
University of Göttingen
Tammannstraße 6
37077 Göttingen
Germany

Prof. Dr. Ewald Heitz
formerly Karl Winnacker Institute of DECHEMA e.V.
Society for Chemical Engineering and Biotechnology
Theodor-Heuss-Allee 25
60486 Frankfurt (Main)
Germany

Coordination and Supervision

Dr. Roman Bender
DECHEMA e.V.
Society for Chemical Engineering and Biotechnology
Theodor-Heuss-Allee 25
60486 Frankfurt (Main)
Germany

Preface

Practically all industries face the problem of corrosion – from the micro-scale of components for the electronics industries to the macro-scale of those for the chemical and construction industries. This explains why the overall costs of corrosion still amount to about 2 to 4% of the gross national product of industrialized countries despite the fact that zillions of dollars have been spent on corrosion research during the last few decades.

Much of this research was necessary due to the development of new technologies, materials and products, but it is no secret that a considerable number of failures in technology nowadays could, to a significant extent, be avoided if existing knowledge were used properly. This fact is particularly true in the field of corrosion and corrosion protection. Here, a wealth of information exists, but unfortunately in most cases it is scattered over many different information sources. However, as far back as 1953, an initiative was launched in Germany to compile an information system from the existing knowledge of corrosion and to complement this information with commentaries and interpretations by corrosion experts. The information system, entitled "DECHEMA-WERKSTOFF-TABELLE" (DECHEMA Corrosion Data Sheets), grew rapidly in size and content during the following years and soon became an indispensable tool for all engineers and scientists dealing with corrosion problems. This tool is still a living system today: it is continuously revised and up-dated by corrosion experts and thus represents a unique source of information. Currently, it comprises more than 8,000 pages with approximately 110,000 corrosion systems (i.e., all relevant commercial materials and media), based on the evaluation of over 100,000 scientific and technical articles which are referenced in the database.

Last century, an increasing demand for an English version of the DECHEMA-WERKSTOFF-TABELLE arose in the 80s; accordingly the DECHEMA Corrosion Handbook was published in 1987. This was a slightly condensed version of the German edition and comprised 12 volumes. Before long, this handbook had spread all over the world and become a standard tool in countless laboratories outside Germany.

Now that almost 20 years have passed since the first DECHEMA Corrosion Handbook was prepared for publication, it seems timely to publish a completely revised edition which takes into account the advances that have been made in the meantime. A large-scale research programme, which was funded by the German Federal

Ministry of Research and Development over a period of two decades and ended only a few years ago, played an important role in the discovery of much of this new knowledge. In addition, the international state-of-the-art has developed significantly and this has also been taken into account in the new edition, which is now called "Corrosion Handbook".

The general character of the handbook remains unchanged. The chapters are arranged by the agents leading to individual corrosion reactions, and a vast number of materials are presented in terms of their behaviour in these agents. The key information consists of quantitative data on corrosion rates coupled with commentaries on the background and mechanisms of corrosion behind these data, together with the dependencies on secondary parameters, such as flow-rate, pH, temperature, etc. This information is complemented by more detailed annotations where necessary, and by an immense number of references listed at the end of each chapter.

An important feature of this handbook is that the data was compiled for industrial use. Therefore, particularly for those working in industrial laboratories or for industrial clients, the book will be an invaluable source of rapid information for day-to-day problem solving. The handbook will have fulfilled its task if it helps to avoid the failures and problems caused by corrosion simply by providing a comprehensive source of information summarizing the present state-of-the-art. Last but not least, in cases where this knowledge is applied, there is a good chance of decreasing the costs of corrosion significantly.

Finally the editors would like to express their appreciation to Birgit Czack and Dr. Roman Bender for their admirable commitment and meticulous editing of a work that is encyclopedic in scope.

They are also indebted to Karin Sora and Dr. Barbara Boeck of Wiley-VCH for their valuable assistance during all stages of the preparation of this book.

Gerhard Kreysa and Michael Schütze

How to use the CORROSION HANDBOOK

The CORROSION HANDBOOK, abbreviated to CHB in the following, provides information on the chemical resistance and the corrosion behavior of materials in approximately 1000 different attacking chemical media and mixtures of materials.

The user is given information on the range of applications and corrosion protection measures for metallic, non-metallic inorganic, and organic materials, including plastics.

Research results and operating experience reported by experts allow recommendations to be made for the selection of materials and to provide assistance in the assessment of damage.

The objective is to offer a comprehensive and concise description of the behavior of the different materials in contact with a particular medium.

Every chapter is a self-contained work that is subdivided according to four groups of materials A-D:

- A Metallic materials
- B Non-metallic inorganic materials
- C Organic materials and plastics
- D Materials with special properties

These material groups are each subdivided according to their chemical formula, the metals are classed according to different alloy groups. These groups are shown in the uniformly designed overview table at the start of each chapter.

Material recommendations are given for each of the four groups of materials. In more recent editions, these are summarized in the section

- E Material recommendations

The information on resistance is given as text, tables, and figures. The literature used by the author is cited at the corresponding point. There is an index of materials as well as a subject index at the end of the book so that the user can quickly find the information given for a particular keyword.

The CORROSION HANDBOOK is thus a guide that leads the reader to materials that have already been used in certain cases, that can be used or that are not suitable owing to their lack of resistance.

The resistance is coded with three evaluation symbols in order to compress the information. Uniform corrosion is evaluated according to the following criteria:

Symbol	Meaning	Area-related mass loss rate[1]		Corrosion rate
		g/(m²h)	g/(m²d)	mm/a
+	resistant	≤ 0.1	≤ 2.4	≤ 0.1[2]
⊕	fairly resistant	< 1.0	< 24.0	< 1.0
−	not resistant	> 1.0	> 24.0	> 1.0

[1] for Al, Mg, and its alloys, 1/3 of the value must be used
[2] the values for Ta, Ti, and Zr are too high (possible embrittlement due to hydrogen absorption in the event of corrosion! Therefore, corrosion rate = 0.01 mm/a, see the individual cases)

The evaluation of the corrosion resistance of metallic materials is given

- for uniform corrosion or local penetration rate, in: mm/a
- or if the density of the material is not known, in: g/(m²h) or g/(m²d).

Pitting corrosion, crevice corrosion, and stress corrosion cracking or non-uniform attack are particularly highlighted.

The following equations are used to convert mass loss rates, x, into the corrosion rate, y:

from x_1 into g/(m²h)   from x_2 into g/(m²d)   where

$$\frac{x_1 \cdot 365 \cdot 24}{\rho \cdot 1000} = y \, (mm/a) \qquad \frac{x_2 \cdot 365}{\rho \cdot 1000} = y \, (mm/a)$$

x_1: value in g/(m²h)
x_2: value in g/(m²d)
ρ: density of material in g/cm³
y: value in (mm/a)
d: days
h: hours

In those media in which uniform corrosion can be expected, if possible, isocorrosion curves (corrosion rate = 0.1 mm/a) or resistance ranges for non-metallic materials are given. The evaluation criteria for non-metallic inorganic materials are stated in the individual cases; depending on the material and medium, they may also be given as corrosion rates (mm/a).

The suitability of organic materials is generally evaluated by comparing property characteristics (e.g. mass, tensile strength, elasticity module or ultimate elongation) and other changes (e.g. cracking) after exposure to the medium with respect to these characteristics in the initial state before exposure. The extent of changes in the properties after exposure to the medium is decisive for the evaluation of the resistance to chemicals or the durability of the materials. The criteria listed below for the evaluation of the chemical resistance apply to thermoplastics used to manufacture pipes and are based on results from immersion tests with an immersion time of 112 days (see ISO 4433 Part 1 to 4). In principle, they are also applicable to other organic materials; however, they should be adapted to the individual material, because, as the following table shows, the evaluation criteria are not consistent, even within a group of thermoplasts, but depend on the type of thermoplastic material.

Symbol	Meaning	Permissible limiting value[1]			
		of the mass change[2] %	of the tensile strength[3] %	of the elasticity module[3] %	of the ultimate elongation[3] %
+	resistant/ durable	PE, PP, PB: −2 to 10	PE, PP, PB, PVC, PVDF: ≥ 80	PE, PP, PB: ≥ 38	PE, PP, PB: ≥ 50 to 200
		PVC, PVDF: −0.8 to 3.6		PVC: ≥ 83	PVC, PVDF: 50 to 125
				PVDF: ≥ 43	
⊕	limited resistance/ limited durability	PE, PB, PB: > 10 to 15 or < −2 to −5	PE, PB, PB, PVC, PVDF: < 80 to 46	PE, PB, PB: < 38 to 31	PE, PB, PB: < 50 to 30 or > 200 to 300
		PVC, PVDF: < −0.8 to −2 or > 3.6 to 10		PVC: < 83 to 46	PVC, PVDF: < 50 to 30 or > 125 to 150
				PVDF: < 43 to 30	
−	not resistant/ not durable	PE, PP, PB: < −5 or > 15	PE, PP, PB, PVC, PVDF: < 46	PE, PP, PB: <31	PE, PP, PB: < 30 or > 300
		PVC, PVDF: < −2 or > 10		PVC: < 46	PVC, PVDF: < 30 or > 150
				PVDF: < 30	

[1] The data applies to the values determined in the initial state without exposure to the medium which correspond to 100 %
[2] Relative mass change according to DIN EN ISO 175
[3] Tensile strength, elasticity module, and ultimate elongation according to DIN EN ISO 527-1
Scope of validity for PVC: PVC-U, PVC-HI, and PVC-C; for PE: PE-HD, PE-MD, PE-LD, and PE-X

Unless stated otherwise, the data was measured at atmospheric pressure and room temperature.

The resistance data should not be accepted by the user without question, and the materials for a particular purpose should not be regarded as the only ones that are suitable. To avoid wrong conclusions being drawn, it must be always taken into account that the expected material behavior depends on a variety of factors that are often difficult to recognize individually and which may not have been taken deliberately into account in the investigations upon which the data is based. Under certain circumstances, even slight deviations in the chemical composition of the medium, in the pressure, in the temperature or, for example, in the flow rate are sufficient to have a significant effect on the behavior of the materials. Furthermore, impurities in the medium or mixed media can result in a considerable increase in corrosion.

The composition or the pretreatment of the material itself can also be of decisive importance for its behavior. In this respect, welding should be mentioned. The suitability of the component's design with respect to corrosion is a further point which must be taken into account. In case of doubt, the corrosion resistance should be investigated under operating conditions to decide on the suitability of the selected materials.

Hydrochloric Acid

Author: A. Bäumel †, P. Drodten, E. Heitz / Editor: R. Bender

		Page
Introduction		4
A	**Metallic materials**	12
A 1	Silver and silver alloys	12
A 2	Aluminium	15
A 3	Aluminium alloys	18
A 4	Gold and gold alloys	40
A 5	Cobalt alloys	41
A 6	Chromium and chromium alloys	42
A 7	Copper	46
A 8	Copper-aluminium alloys	48
A 9	Copper-nickel alloys	51
A 10	Copper-tin alloys (bronzes)	54
A 11	Copper-tin-zinc alloys (red brass)	54
A 12	Copper-zinc alloys (brass)	54
A 13	Other copper alloys	56
A 14	Unalloyed and low-alloy steels/cast steel	56
A 15	Unalloyed cast iron and low-alloy cast iron	66
A 16	High-alloy cast iron	67
A 16.1	Silicon cast iron	67
A 16.2	Austenitic cast iron (and others)	67
A 17	Ferritic chromium steels with < 13 % Cr	67
A 18	Ferritic chromium steels with ≥ 13 % Cr	72
A 19	Ferritic/pearlitic-martensitic steels	74
A 20	Ferritic-austenitic steels/duplex steels	77
A 21	Austenitic CrNi steels	79
A 22	Austenitic CrNiMo(N) and CrNiMoCu(N) steels	96
A 23	Special iron-based alloys	97
A 24	Magnesium and magnesium alloys	98
A 25	Molybdenum and molybdenum alloys	98
A 26	Nickel	98
A 27	Nickel-chromium alloys	99
A 28	Nickel-chromium-iron alloys (without Mo)	99
A 29	Nickel-chromium-molybdenum alloys	99
A 30	Nickel-copper alloys	117

		Page
A 31	Nickel-molybdenum alloys	117
A 33	Lead and lead alloys	129
A 34	Platinum and platinum alloys	129
A 35	Platinum metals (Ir, Os, Pd, Rh, Ru) and their alloys	129
A 36	Tin and tin alloys	129
A 37	Tantalum, niobium and their alloys	130
A 38	Titanium and titanium alloys	143
A 39	Zinc, cadmium and their alloys	159
A 40	Zirconium and zirconium alloys	166
A 41	Other metals and their alloys	170
B	**Non-metallic inorganic materials**	178
B 3	Carbon and graphite	178
B 4	Binders for building materials (e. g. concrete, mortar)	181
B 5	Acid-resistant building materials and binders (putties)	181
B 6	Glass	184
B 8	Enamel	189
B 12	Oxide ceramic materials	197
B 13	Metal-ceramic materials (carbides, nitrides)	198
C	**Organic materials/plastics**	206
	General	206
	Type of material	206
	Abbreviations / Chemical structure	208
	Basic processes with respect to the action of chemicals	211
	Testing of the chemical resistance	216
	Resistance tables	218
	Resistance of individual material groups and materials	221
	Thermoplastics	221
	Vinyl polymers	224
	Styrene polymers (PS), polyacrylates	230
	High-temperature thermoplasts	230
	Further thermoplastic polycondensates	231
	Duroplasts	232
	Elastomers	236

Corrosion Handbook: Hydrochloric Acid, Nitric Acid
Edited by: G. Kreysa, M. Schütze
Copyright © 2005 DECHEMA e.V.
ISBN: 3-527-31118-1

		Page			Page
D	**Materials with special properties**	239	E	**Materials recommendations**	252
D 1	Coatings and linings	239			
D 2	Seals and packings	246	**Bibliography**		266
D 3	Composite materials	248			

Warranty disclaimer

This chapter of the DECHEMA Corrosion Handbook has been compiled from literature data with the greatest possible care and attention. The statements made in this chapter only provide general descriptions and information.

Even for the correct selection of materials and correct processing, corrosive attack cannot be excluded in a corrosion system as it may be caused by previously unknown critical conditions and influencing factors or subsequently modified operating conditions.

No guarantee can be given for the chemical stability of the plant or equipment. Therefore, the given information and recommendations do not include any statements, from which warranty claims can be derived with respect to DECHEMA e.V. or its employees or the authors.

DECHEMA is liable – regardless of the actual reason – for all damage caused by their official representatives and their employees on purpose or by gross negligence.

If there is only light negligence DECHEMA is only liable for infringing essential duties that arise from the use of the DECHEMA Corrosion Handbook ("cardinal duties"). DECHEMA can not be made liable for any indirect or subsequent damage and for any damage arising from production stops or delays, which are caused by using the DECHEMA Corrosion Handbook.

Introduction

Index of the Introduction

V 1	General	4	V 4	Utilisation	8
V 2	Physical and chemical properties	4	V 5	Corrosive properties	9
V 3	Production and formation	7	V 6	Corrosion-relevant additives and impurities	10

Introduction

V 1 General

Hydrochloric acid is a strong non-oxidising inorganic acid and is regarded as being one of the most important basic chemicals. It represents a solution of hydrogen chloride – which has the chemical formula HCl – in water. It is also known as muriatic acid. It is the second most frequent acid after sulphuric acid. It is used in a range of industrial sectors and in many everyday applications.

Free hydrogen chloride gas is ubiquitous in the stratosphere; however, only in low concentrations of approx. 10^{-11} vol%. This concentration can be significantly higher if the atmosphere contains volcanic vapors. Apart from anthropogenic pollutant emissions, the only natural sources of free hydrochloric acid are human and animal organisms.

Physiologically, hydrochloric acid is found in the gastric juice of humans and higher animals in concentrations of 0.1 to 0.5 mass%. Inhalation of hydrochloric acid vapors leads to pneumonia and to cauterisation of the pulmonary alveoli. The eyes, the mouth, the oesophagus and the skin can also be cauterised if they come into contact with hydrochloric acid. A high dosage can lead to unconsciousness, cardiac insufficiency and ultimately to death. Concentrations of 1 – 5 ppm in the air can be detected by the odour; strong irritation occurs above 10 ppm. The TLV value is 7 mg HCl/m^3 [1, 2]. When handling hydrochloric acid in concentrations above 10 mass%, the statutory directive regulating protective measures against hazardous materials (Hazardous Materials Ordinance) must be observed [3].

V 2 Physical and chemical properties

Hydrogen chloride is very soluble in water: at 293 K (20 °C), a saturated aqueous solution has a concentration of 40.4 %. Concentrated hydrochloric acid is marketed with concentrations of approx. 30 % and 36 – 38 % (formerly referred to as "fuming hydrochloric acid"). Dilute hydrochloric acid is usually approx. 7 %. Aqueous hydrochloric acid forms an azeotrope at a concentration of 20 % that boils at 381.5 K (108.5 °C). The characteristic feature of the azeotropic state is that the vapor and the liquid have the same concentration. If a hydrochloric acid solution below this concentration is heated, then water evaporates preferentially. However, if the concentration is higher, then hydrogen chloride evaporates preferentially until the azeotropic concentration is reached.

Pure hydrogen chloride gas has the following critical data:

- T_{crit} = 324.6 K (51.6 °C)
- p_{crit} = 83.9 bar
- d_{crit} = 0.421 g/cm^3.

The boiling and freezing curve of aqueous hydrochloric acid at a total pressure of 1 bar is shown in Figure 1. The boiling maximum is at 381.5 K (108.5 °C), the boiling point of 35 % hydrochloric acid is at a relatively low temperature of 336 K (65 °C) and its freezing point is at – 244 K (29 °C).

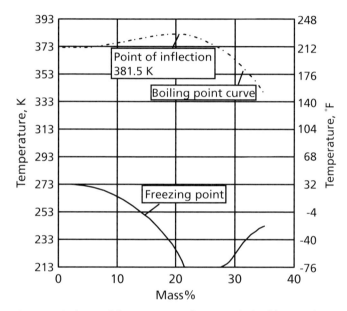

Figure 1: Boiling and freezing curve of aqueous hydrochloric acid at a total pressure of 1 bar [4; p. 7]

The pressure-temperature-concentration diagram of the HCl – H$_2$O system is shown in Figure 2. The vapor pressure-temperature curves of different concentrations of hydrochloric acid can be read off this diagram. For example, 35 % hydrochloric acid at 373 K (100 °C) has a vapor pressure of 4 bar (point B). This means that stability data of materials in concentrated hydrochloric acid at higher temperatures must be determined under pressure. Inversely, a HCl-H$_2$O vapor mixture with 0.45 volume% water vapor cooled in a condenser at 1 bar to 303K (30 °C) produces a condensate with 40 % hydrochloric acid (point C).

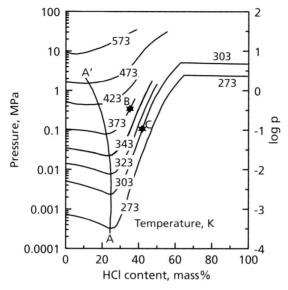

Figure 2: Pressure-temperature-concentration diagram of the system HCl-H$_2$O [1]

It can be seen from the diagram that the pressure of the HCl-H$_2$O vapor mixture of 37% hydrochloric acid is larger by several orders than that of 25% acid. This affects the permeation ability of HCl in organic materials.

The curve A-A' connects the points of the azeotropic concentrations at various temperatures.

At high pressures, the HCl-H$_2$O system has a miscibility gap in the liquid phase, i.e. highly concentrated aqueous hydrochloric acid exists alongside liquid hydrogen chloride [1]. From the practical point of view, such systems are of little importance at present. For reasons of clarity, the equilibrium curves of the two-phase liquid are not shown in Figure 2.

The partial pressure of water vapor over concentrated hydrochloric acid is very low. For example, the proportion of water vapor at 303 K (30 °C) and 5 bar is only 800 ppm (volume proportion), at 273 K (0 °C) the tolerable water content is < 100 ppm [1]. This means that above these low concentrations, concentrated hydrochloric acid condenses and this can lead to corrosion damage.

In aqueous hydrochloric acid, HCl is completely dissociated into hydrated protons (H$_3$O$^+$) and chloride ions (Cl$^-$) and thus forms solutions with a high electrolytic conductivity.

As a strong acid, hydrochloric acid undergoes many reactions with reducing and oxidising compounds. Of particular importance are reduction processes with the participation of metals that generate hydrogen from the protons, as well as oxidation processes that produce chlorine from the chloride ions. The reaction of hydrochloric acid with metal oxides and metal salts of other (weaker) acids produces metal chlorides.

V 3 Production and formation

In the following information on the current state-of-the-art of hydrochloric acid production, in addition to a description of the principles, information is also given concerning common usages of materials that are then described in more detail in the later sections.

The simplest method of producing hydrogen chloride is **synthesis from the elements** chlorine and hydrogen. This is usually carried out by combusting chlorine in a jet of hydrogen using a quartz burner. The burner is installed in a cylindrical reactor lined with refractory bricks. Another technique employs water-cooled graphite reactors. Lately, reactors made of unalloyed steel have become particularly attractive because they can be operated under pressure. In this case, particular care must be taken that no aqueous hydrochloric acid is able to condense. Mixtures of chlorine and hydrogen are explosive over large concentration ranges and can even be ignited by light.

The formation of HCl as a **by-product from the chlorination of organic compounds** is the greatest source, e.g. during manufacture of vinyl chloride for the plastic PVC. The gaseous hydrogen chloride thus formed can be processed further as follows [1]:

- direct utilisation as a gas after the chlorinated hydrocarbons have been separated by condensation
- isolation and purification by distillation
- absorption in water to form hydrochloric acid.

Of these three processes, the production of aqueous hydrochloric acid as an intermediate product poses the greatest problems for materials on account of its nature. If the process is carried out under atmospheric pressure or slightly increased pressure, then absorption columns made of phenol-formaldehyde resin with a fiberglass-reinforced polyester resin sheath are used. Stripper columns to separate HCl from chlorinated hydrocarbons are designed in the same way.

Recently, increased use has been made of steel columns that are lined with fluoropolymers (mainly PTFE), with success. Compared to graphite columns, they have the advantage that they can be operated at a higher pressure. On the other hand, the failure of the lining leads to a rapid destruction of the steel sheath. To avoid this, several small units are used that can be rapidly exchanged in the event of damage.

Heat exchangers are mainly made of graphite. Another possibility is the use of tantalum, whereby a possible hydrogen embrittlement (e.g. in the presence of fluoride ions as a contaminant) must be avoided.

Completely dry HCl gas does not attack unalloyed steel so that pipes and compressors made of this material can be used for further processing.

Another HCl source is the **incineration of waste** containing chlorinated hydrocarbons and chlorinated plastic waste to produce hydrogen chloride, water and carbon dioxide. The flue gases from the incineration, which contain HCl and often HF, at temperatures of up to 1273 K (1000 °C), are quenched with water in quencher towers to produce a very acidic condensate. Materials for the quencher tower include linings of silicon carbide and carbon bricks that, in addition to corrosion resistance, must also have sufficient erosion resistance.

The HCl product is separated, in principle, in the same way as for the combustion gases from the chlorination process, namely by absorption, but with the following differences [1]:

- water vapor is always present,
- for commercial reasons, the highest possible concentration of acid must be produced by the condensation,
- for further direct processing of gaseous HCl, the operating temperature must always lie above the condensation temperature of hydrochloric acid.
- in power stations fired with chloride-containing coal, dew point corrosion by sulphuric acid and hydrochloric acid in the electrostatic precipitators is a serious problem [5]. The dew point of hydrochloric acid lies between that of sulphuric acid and water and, particularly at lower temperatures, leads to strong corrosion. The use of linings with fluoroelastomers has overcome this problem in the critical dew point range from 373 to 408 K (100 to 135 °C).

In the 17th century, Glauber already carried out **reactions of sulphuric acid with sodium chloride** which led to the development of two technical processes, although their importance is no longer significant. In the so-called Mannheim process, a sodium chloride/sulphuric acid mixture is reacted in a refractory-lined muffle furnace by heating it to 873 K (600 °C). The liberated HCl gas has a concentration of 85 % and contains air, sulphuric acid mist and salt particles that have to be removed in complicated purification steps. In the Berlin process, the starting materials already react in fused sodium or potassium hydrogen sulphate at 573 K (300 °C) in cast iron retorts to form a purer HCl gas than that produced from hydrochloric acid.

A source of unwanted hydrochloric acid is the **hydrolysis of chlorinated hydrocarbons**; this can occur even in the presence of traces of water at elevated temperatures. The hydrocarbon solvent is cleaved into hydrochloric acid and an organic residue. This process can be suppressed by the addition of stabilisers to the solvent; however, they must always be present in sufficient quantities. In a series of processes in which chlorinated hydrocarbons participate there is a a risk of the unwanted formation of hydrochloric acid, e.g. cleaning of metal surfaces.

V 4   Utilisation

Hydrochloric acid and gaseous HCl are two of the most important industrial chemicals. Aqueous hydrochloric acid has many applications as a strong inorganic acid:

- preparation of chlorides,
- dissolution of minerals,
- pickling and etching of metals,
- regeneration of ion-exchange resins for the treatment of water,
- neutralisation of alkaline products or waste,

- additive in precipitation baths in the synthetic fiber industry,
- production of chlorine and hydrogen by electrolysis,
- digestion of animal and vegetable proteins in the food industry,
- pickling agent in tanning.

And, of course, it is used as a versatile reagent in the chemical laboratory.

Dry gaseous HCl is usually used directly by the manufacturer, e.g. if it arises as a by-product from the chlorination of organic compounds. It is also used in oxichlorination, hydrochlorination and in other less important processes [1]. There is an interesting proposal to use dry HCl, if necessary in a supercritical state, to increase the permeability of geological oil and gas reservoir rocks [6].

V 5 Corrosive properties

The corrosiveness of hydrochloric acid towards **metals** is based on the oxidising effect of the hydrated proton H_3O^+, which attacks metals that lie below hydrogen in the electrochemical series to form metal chlorides and liberate hydrogen. Furthermore, hydrochloric acid dissolves passivating surface films (oxides and other compounds) and leads to active corrosion. Finally, chloride ions form metal complexes that can lower the electrode potential of electropositive metals, e.g. copper, to below that of the hydrogen electrode so that corrosive attack can take place. Further information is given in Section A.

Non-metallic inorganic materials can undergo a series of purely chemical reactions. These include oxidation, dissolution and ion-exchange processes. In contrast to the behavior of metals, electrochemical reactions do not play a significant role in the corrosion mechanism.

Plastics are generally damaged by the action of hydrochloric acid due to hydrolytic degradation of labile chemical bonds in the polymer macromolecule. This attack can start at the surface or it occurs inside the plastic after preliminary diffusion of HCl and water into the polymer. The resulting degradation processes are swelling, with impairment of the mechanical properties, and hydrolytic degradation of the macromolecule. If the aqueous hydrochloric acid contains organic solvents as an impurity, then these effects are amplified. In the presence of oxidising agents as possible further components or impurities in hydrochloric acid, then oxidative degradation of the polymer occurs that can lead to its rapid destruction. Information on the destruction mechanisms are given in Section C.

In addition to aqueous hydrochloric acid, HCl solutions in non-aqueous solvents are used to a lesser extent. It is known that methanol and ethanol solutions containing hydrochloric acid have a high HCl solubility and are very aggressive to metallic materials. The solubility is low in hydrocarbons and halogenated hydrocarbons; however, because of the superimposition of swelling and permeation effects of the solvent and the hydrolytic action of the HCl, there is a high risk of damage to plastics. Table 1 gives a list of the solubility of HCl in various solvents.

Solvent	Temperature K (°C)	HCl dissolved g/kg solvent
Methanol	273 (0)	1092
	293 (20)	877
Ethanol	273 (0)	838
	293 (20)	681
Tetrahydrofuran	283 (10)	584
Dioxan	283 (10)	433
Dibutyl ether	283 (10)	250
Chloroform	288 (15)	8.5
Carbon tetrachloride	293 (20)	6.0
Benzene	293 (20)	20
n-Hexadecane	300 (27)	3.9
Water	273 (0)	825

Table 1: Solubility of HCl in various organic solvents and water at 1.013 bar HCl partial pressure and different temperatures [1]

V 6 Corrosion-relevant additives and impurities

Depending on the application, hydrochloric acid may contain **additives** that can decisively affect the corrosiveness.

In the production and further processing of metallic materials as well as the operation of technical equipment and plants, hydrochloric acid is a frequently employed pickling agent, e.g. to

- remove annealing or rolling scale after hot forming
- clean and activate steel surfaces before metallic coatings are applied by galvanic methods or hot-dip methods as well as inorganic or organic coatings
- remove deposits or corrosion products.

In these pickling processes, the acid is generally used to attack only the products to be removed from the surface and to activate, at most, the metal surface but not to remove a noticeable amount of material. To keep the attack by the acid as low as possible if longer treatment times are necessary or if the surface has already been cleaned, the addition of inhibitors to acidic pickling agents is customary. The choice of a suitable inhibitor depends not only on the pickling conditions, such as acid concentration, temperature and bath movement, but also on the actual metal being treated [7]. Further details are given under the corresponding material groups.

In technical pickling processes, good inhibitors must fulfil a series of requirements [7]:

- effective inhibition of metal dissolution
- no overpickling in the presence of higher salt contents
- no delay of the pickling process

- effective even at low concentrations
- also effective at higher temperatures
- thermal and chemical stability
- effective inhibition of hydrogen absorption by the metal
- good surfactant properties
- low foaming tendency.

During pickling of unalloyed and low-alloy steels in hydrochloric acid, if insufficient or no inhibitor is added or if an unsuitable inhibitor has been selected there is a risk not only that the surface is too strongly attacked but also the possibility that the hydrogen generated during the acid attack is able to diffuse into the material and thus damage it. A suitable inhibitor must therefore not only prevent the removal of material, but also exhibit an antipromotor effect for hydrogen absorption.

Therefore, for the characterisation and selection of an inhibitor and monitoring of its effectiveness during operational utilisation, investigations on the corrosion of the material are not enough, measurements of the hydrogen permeation should also be carried out. Corresponding information is given in [8].

Impurities (in HCl) in the hydrochloric acid are usually produced during the production process or from application processes. These include

- chlorinated hydrocarbons that have not been completely removed with the chlorination products after the manufacturing process and have remained in the hydrochloric acid, and which can lead to swelling and degradation of plastics.
- oxidising agents, such as chlorine, which is present in certain process steps in the chlor-alkali industry, in hydrochloric media and whose aggressiveness towards metals and plastics must be taken into account.
- oxygen in hydrochloric acid, which can also corrode those metals that should actually be stable because of their position in the electrochemical series.
- heavy metal ions from pickling processes (e.g. iron(III) ions), which, in addition to their effect on metals, can also oxidatively degrade plastics. In this respect, the yellowish colour of technical hydrochloric acid should be mentioned: this is caused by an iron hexachloro complex as an impurity.
- solids in flowing hydrochloric acid can cause abrasion and erosion corrosion of many materials in pumps and fittings. Solids as an impurity are often found in the mineral digestion industry and in food processing.

An increasingly important application field with particularly high requirements on the purity of hydrochloric acid are techniques using highly pure chemicals, e.g. those used to manufacture electronic components, especially chip manufacture. With regard to process engineering of purest chemicals, fluoropolymers play a dominant role in the handling of hydrochloric acid as essentially corrosion-inert materials.

A
Metallic materials

A 1 Silver and silver alloys

Figure 3 gives 4 potential-pH diagrams for the system silver/aqueous silver ion solution at 298 K (25 °C), 373 K (100 °C), 473 K (200 °C) and 573 K (300 °C) [9]. The lower of the inclined dotted lines gives the equilibrium potential of the hydrogen production in dependence on the pH value. The upper line limits the region of thermodynamic stability of silver, above which silver can go into solution as a silver ion in acidic solutions only at very high potentials, whereas a protective coating of silver oxides is formed at nearly all potentials in alkaline solutions.

Silver, and particularly silver alloys, exhibit a high corrosion resistance to hydrochloric acid in a relatively wide range of concentrations and temperatures. In hydrochloric acid up to 5 % and up to the boiling point, the corrosion rate is < 0.05 mm/a (< 2 mpy). However, at higher concentrations and with increasing temperatures as well as in the presence of oxidising agents the corrosion rate considerably increases. Both in non-aerated as well as in aerated 35 % hydrochloric acid the corrosion rate at 373 K (100 °C) reaches a value of 2.5 mm/a (98 mpy) [10]. An addition of 30 or more percent gold can increase the resistance by a factor of 20. A further improvement of the corrosion resistance is attained with the alloy Pallacid® (40 % Ag, 30 % Au, 30 % Pd). The corrosion rate in 20 % hydrochloric acid at 382 K (109 °C, boiling point), was determined to be 0.0025 mm/a (0.1 mpy) [11]. Silver and silver alloys are thus materials that can be used in the chemical industry – mostly as platings – under certain conditions [12].

For components that are used in heat transfer applications, the high thermal conductivity in combination with the high corrosion resistance of silver and silver alloys is an advantage in hydrochloric acid solutions. Although copper has a high thermal conductivity similar to that of silver, its corrosion resistance in hydrochloric acid solutions is lower (see A 7). The thermal conductivity of silver is approx. 25 times higher than the austenitic chromium-nickel and nickel-based alloys commonly used in the chemical industry.

Silver also has an excellent electrical conductivity, so that it can be used like gold and copper as a conductor in electronic assemblies. In addition to the high electrical conductivity, the corrosion behavior is a decisive factor for the contact behavior. Even at very low corrosion rates, expressed as mass loss per area and time, changes can

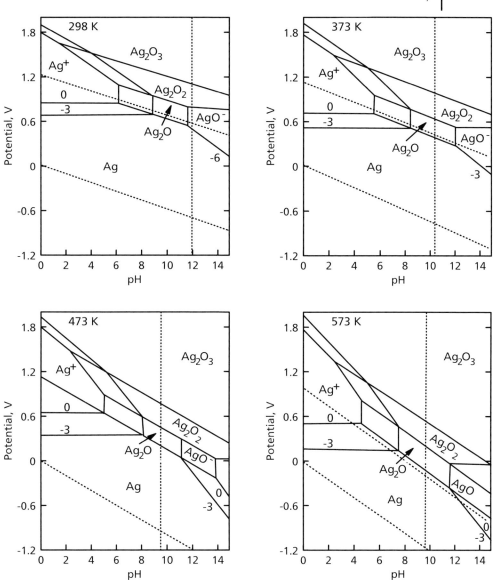

Figure 3: Potential – pH diagrams of the system silver/aqueous silver ion solution at different temperatures [9]

occur on the surface of conductor materials which greatly increase the contact resistance and thus affect the functioning of electronic equipment. This is particularly valid in regions with a high air temperature, high humidity and aggressive pollutants. These are essentially combustion products such as sulphur and chlorine compounds, which includes hydrochloric acid. They are produced by the combustion of chloride-containing coals or oil (heavy oil), as well as in chloride-containing combustion air in the vicinity of the sea and by the combustion of chlorinated hydrocarbons during waste incineration, in particular from PVC (polyvinyl chloride), which is the most frequently used chlorine-containing plastic.

With respect to this, the conditions in East Asia are particularly unfavorable. It is reported that approx. 20 % of the damages occurring in electronic equipment in aircraft is caused by corrosion [13]. Silver has a contact resistance of $R_c = 0.5$ mΩ in the initial non-corroded state. The R_c value should not exceed a value of 10 during utilisation, and this is generally the case. However, under certain conditions of exposure, over a period of several years R_c values were measured that were higher by five or six orders of magnitude. Such values are much too high for the correct electrical functioning capabilities to be maintained.

Because of this damage, experiments investigating the corrosion behavior of silver and copper were carried out indoors at various locations in Japan and South-east Asia (silver is rarely used in outdoor applications). Samples with dimensions 30 x 30 x 0.03 mm made of 99.9 % silver and 99.9 % copper (see A7) were given a uniform surface treatment and then glued onto acryl resin boards that were then exposed to the air in 23 rooms located in Bandung, Singapore, Bangkok and Tokyo in a vertical position for 1 to 6 months.

Various analysis methods were used to determine that the gaseous air pollutants were H_2S, SO_2, NO_2, HCl, Cl_2 and the solid pollutants were chloride salt particles. After exposure, the state of the sample's surface was studied with a light microscope, a microprobe, an Auger electron spectroscope and with an x-ray diffractometer. The corrosion rate was determined by electrolytic reduction in deaerated 0.1 mol/l KCl solution at 298 K (25 °C) against a platinum electrode. A deposit grew on the silver samples at all exposure locations. However, nearly all samples showed only very low pitting. C, N, S, Cl and O were uniformly distributed over the surface. However, in a hydrometallurgical laboratory in Bandung and in a chemical laboratory in Singapore, where there was a high chlorine and chloride loading, the silver samples exhibited pitting corrosion and a localised chloride deposit was found.

Initially, the polished samples are covered by an oxide layer (Ag_2O). Compounds containing S, Cl, N and C are formed on this during the period of exposure. While N and C compounds (the latter originate from volatile organic substances in the laboratory) are only present on the outer surface of the deposit, S and Cl compounds penetrate into the oxide layer. These are primarily silver sulphide Ag_2S and silver chloride AgCl. The high reaction tendency of silver with H_2S is more decisive for the corrosion rate than the reaction with the acidic impurities. Only if H_2S is virtually absent does silver react preferentially with hydrochloric acid to form AgCl. This reaction can also lead to pitting corrosion. The work showed that at a high ambient temperature and high humidity, even if the concentrations of the characteristic attacking agents are low in most environmental regions, it is still sufficient to form a corrosion deposit on silver which increases the electrical contact resistance to greater or lesser extents. Measured results of this were not reported. Protection by air-conditioning the rooms to avoid condensation as well as a regular servicing of the contact points is thus necessary to minimise functional impairment of electronic components. It is pointed out that the enrichment of aggressive materials can be minimised by filtering the outdoor air in air-conditioning units.

The reports [14] and [15] deal with corrosion of copper and silver in interior rooms. The reason for this is the use of these metals as printed circuits in electronic equipment. Corrosion lowers the electrical conductivity and may lead to malfunctions of such equipment. High humidity and contamination of the air in the room play a decisive role in this. In the case of silver, H_2S is the crucial factor. In a room with an increased level of H_2S in the air, a hardly soluble and firmly adhering AgS deposit is formed on the silver. If H_2S is not present in excess, then other contaminants, such as SO_2, SO_4, Cl_2 and HCl, become important.

Although it is only a minor corrosive attack, the coating on the surface disturbs the function of the component and thus leads to corrosion damage in the sense of the definitions in DIN EN ISO 8044 [16].

A 2 Aluminium

The reports [17] and [18] deal with the problem of filiform corrosion of coated aluminium because this type of damage has been increasing in the construction sector since the 1980s. The main focus was on the development of a short-term laboratory method that allows predictions of the long-term behavior of a coating system. Table 2 lists the 12 selected coating systems that were exposed for up to 5 years to atmospheric conditions with a differing aggressiveness in Hoek van Holland (industrial and marine atmosphere), in Düsseldorf (industrial atmosphere) and Stuttgart (mixed city-industrial atmosphere). The samples were given a cross-cut. Since filiform corrosion starts at so-called weak points in the coating (applied cuts, edges etc.), the affected area (in mm^2) is given with respect to the length of this weak point (in cm).

Parallel to the open-air weathering tests, laboratory tests were carried out using two different corrosion initiation methods at the bottom of the applied cut:

- NaCl test: Initiation by 24-hour exposure in a vapor chamber in the salt spray test (based on ISO 4623-1 (09/2000))
- HCl test: Initiation by 1-hour exposure in a vapor chamber over 32 – 34 % hydrochloric acid (based on DIN EN 3665 (08/1997))

System	Alloy	Pretreatment (amount applied) [g/m²]	Coating system*
1	AlMg 1 F 13	Yellow chromating (0.5 – 0.8)	1. Polyester powder, cross-linked with TGIC**
2	AlMg 1 F 13	Green chromating (1)	1. Polyester powder, cross-linked with TGIC
3	AlMg 1 F 13	Yellow chromating (0.5 – 0.8)	1. Polyester powder, cross-linked with TGIC (opaque) 1. Polyester powder, cross-linked with TGIC (clear coat)
4	AlMg 1 F 13	Yellow chromating (> 0.4)	1. Acrylate primer, containing chromate, wet 2. Polyvinylidene fluoride wet paint
5	AlMg 1 F 13	Yellow chromating (> 0.4)	1. Acrylate/polyvinylidene fluoride primer, containing chromate, wet 2. Polyvinylidene fluoride wet paint
6	AlMg 1 F 13	Yellow chromating (> 0.4)	1. 2K polyurethane primer, wet 2. 2K polyurethane/acryl wet paint
7	AlMg 1 F 13	Direct current anodisation in sulphuric acid (5 µm oxide layer)	1. Polyester powder, cross-linked with TGIC
8	AlMg 1.5	Green chromating (0.130)	1. Polyester wet paint
9	AlMn 1 Mg 0.5	Yellow chromating (0.145)	1. Polyester primer, wet 2. Polyvinylidene fluoride wet paint
10	AlMn 1 Mg 0.5	Green chromating (0.05)	1. Polyester primer, wet 2. Polyurethane fluoride wet paint
11	AlMg 1 F 13	Yellow chromating (> 0.4)	1. Epoxide primer, wet 2. Polyvinylidene fluoride powder
12	AlMg 1 F 13	Green chromating (0.05)	1. Polyester powder, cross-linked with TGIC

* Systems 8, 9, 10 are coil-coated, all others are individual coatings
** TGIC = Triglycidyl isocyanurate

Table 2: Coating systems [17]

In both methods, the samples were subsequently stored at 313 K (40 °C) and 80 % relative humidity. As expected, after 5 years of exposure to open-air weathering conditions, there were pronounced differences in the development of the filiform corrosion at the different locations (Figure 4). The greatest corrosion by far was exhibited by the samples weathered in Hoek van Holland (Figure 5). The salt loading as well as the high humidity in combination with the industrial atmosphere promoted filiform corrosion particularly strongly at this location.

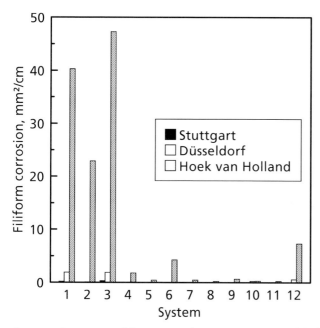

Figure 4: Comparison of the average values in Stuttgart, Düsseldorf and Hoek van Holland after 5 years of open-air weathering [17]

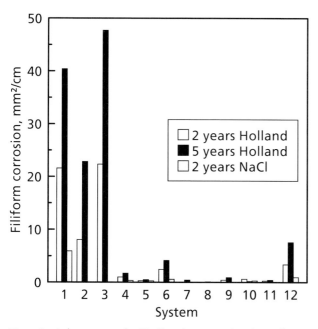

Figure 5: Laboratory results (NaCl test) compared to those of open-air weathering (Hoek van Holland) [17]

The laboratory results of the short-term tests on samples with HCl initiation correlate with the results of the samples with long-term weathering at Hoek van Holland. Particularly for systems with strong filiform corrosion, this is already the case in the 6-week laboratory test (Table 3). Therefore, this accelerated laboratory test method allows a prognosis to be made on the long-term behavior of coating systems on aluminium with regard to filiform corrosion.

Systems	Open-air weathering in Hoek van Holland Filiform corrosion [mm²/cm] Average value of 5 samples (standard deviation)				HCl laboratory test Filiform corrosion [mm²/cm] Average value of 3 samples (standard deviation)	
	2 years		5 years		6 weeks	
7	0		0.19	(0.06)	0.43	(0.05)
8	0		0.12	(0.05)	1.66	(0.18)
5	0.21	(0.22)	0.38	(0.24)	4.12	(1.91)
9	0.38	(0.06)	0.66	(0.30)	1.82	(0.13)
11	0.12	(0.09)	0.30	(0.24)	0.63	(0.23)
4	1.07	(0.54)	1.62	(0.73)	2.02	(0.54)
10	0.55	(0.17)	0.28	(0.05)	4.95	(0.50)
6	2.30	(2.35)	4.09	(7.29)	9.33	(3.26)
12	2.97	(0.64)	7.38	(1.94)	3.10	(0.33)
1	21.58	(7.18)	40.48	(24.56)	14.53	(1.09)
2	7.90	(2.44)	22.76	(6.98)	9.83	(2.02)
3	22.32	(8.86)	47.69	(14.45)	17.31	(1.44)

Table 3: Correlation between open-air weathering and the HCl laboratory test [17]

Filiform corrosion of the laboratory samples with NaCl initiation was noticeably lower than that exhibited by the open-air weathered samples in Hoek van Holland. This indicates that the acidic components in the atmosphere in Hoek van Holland in combination with the chloride present is the decisive factor with regard to filiform corrosion.

A 3 Aluminium alloys

Aluminium and its alloys are only corrosion-resistant in a relatively narrow range of approximately pH 7. As an amphoteric metal, aluminium dissolves in strongly acidic as well as in strongly alkaline media according to the reactions

Eq. 1 $\quad 2\,Al + 6\,HCl \rightarrow 2\,AlCl_3 + 3\,H_2$

Eq. 2 $\quad 2\,Al + 2\,NaOH + 6\,H_2O \rightarrow 2\,Na[Al(OH)_4] + 3\,H_2$

with liberation of hydrogen. The aluminium oxide layer, which gives aluminium its good resistance to the atmosphere and to many approximately neutral aqueous solutions in the pH range of approx. 4 to 8, cannot provide any protection in acidic and strongly alkaline solutions (Figure 6).

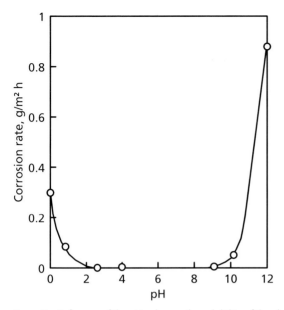

Figure 6: Influence of the pH value on the solubility of the aluminium oxide layer [19]

In the very acidic range, as is the case in hydrochloric acid solutions, attack mostly takes place in the form of pitting. Between areas covered with oxide and places that do not have a covering coating on the pure metal, which has a normal potential of $U_H = -1.6$ V and is thus very low in the electrochemical series, there are very effective local elements that can lead to pitting corrosion.

Because of its low resistance, aluminium cannot be used if it is in contact with hydrochloric acid. However, since the other properties are good, it was of great interest (as for iron) to develop inhibitors that greatly improve the corrosion resistance of aluminium to hydrochloric acid so that, under certain conditions, a cost-effective usage is possible.

Tests have also been carried out on alloying additives that increase the corrosion resistance of aluminium to acids in general, and to hydrochloric acid in particular. Corrosion damage to aluminium mainly occurs where the presence of hydrochloric acid, mostly in low concentrations, is unintended in the surrounding medium.

Effect of alloying on the corrosion behavior

In reference [20], pure aluminium 99.7 with 0.17 % Fe and 0.13 % Si was alloyed with mixed metals (50 – 52 % Ce, 10 – 12 % Nd, 15 – 17 % La and small amounts of other rare earths) in proportions of 0.25, 0.50 and 1.00 mass%. The degased melts were cast into 6 mm thick plates, which were rolled into sheets 3 mm thick and subsequently annealed for 1 h at 673 K (400 °C). The samples taken from these plates were given a fine mechanical polish and the current density/potential curves were recorded at 301 K (28 °C) in 0.1 M HCl, 0.05 M H_2SO_4 and 0.1 M HNO_3. The corrosion-current densities I_{corr} in $\mu A/cm^2$ determined from the Tafel lines (for an explanation of the electrochemical principles, see [21]) of these curves are given in Figure 7, and Table 4. In all three acids, the corrosion resistance increased with increasing content of mixed metals. This is mainly attributed to the formation of stable mixed metal oxides. Also in organic acids, e.g. citric and acetic acid, a similar improvement of the corrosion resistance of pure aluminium was found for lower initial values.

Alloy	Mixed metal additive mass%	Acid	$U_{R/SCE}$ mV	U_{corr} mV	I_{corr} $\mu A/cm^2$
A	0.25	HCl	−800	−800	30.0
B	0.5	HCl	−830	−840	20.0
C	1	HCl	−798	−840	16.0
Pure Al		HCl	−809	−780	60
A	0.25	H_2SO_4	−730	−735	13.0
B	0.5	H_2SO_4	−780	−760	6.5
C	1	H_2SO_4	−788	−760	5.0
Pure Al		H_2SO_4	−815	−820	18.0
A	0.25	HNO_3	−630	−635	11.0
B	0.5	HNO_3	−682	−675	8.5
C	1	HNO_3	−690	−700	6.0
Pure Al		HNO_3	−752	−725	30.0

Table 4: Influence of a mixed metal additive to pure aluminium on the potential values and corrosion current in 0.1 N inorganic acids [20]

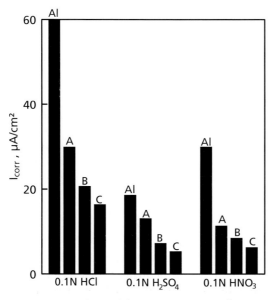

Figure 7: Dependence of the corrosion currents from pure aluminium and mixed metal-Al alloys (A, B and C) in 0.1 N inorganic acids [20]

AlLi alloys belong to the new generation of materials for aerospace applications. Because of their 10 % lower specific weight and a 10 % higher elasticity modulus compared to conventional aluminium alloys, weight reductions of approx. 15 % are possible. The great advantage of the AlLi materials compared to other new non-metallic aerospace materials is that the AlLi materials can replace conventional aluminium materials without complicated design modifications.

For utilisation in aerospace applications, the corrosion behavior under conditions of atmospheric attack is of great importance. The results from laboratory tests in neutral sodium chloride solutions and in open-air weathering tests from several test programs within the research and development program "Corrosion and corrosion protection" (FE-KKs) are summarised in reference [22]. The investigated materials are listed in Table 5.

Material	Manufacturer
AA 8090 A T851, T62 EN AW-8090 AlLiCuMg	Alcan
AA 8090 C T8, T62 EN AW-8090 AlLiCuMg	Alcan
AA 2091 CPHK T8 EN AW-2091 AlCuLiMg	Pechiney
AA 2090 T8 EN AW-2090 AlCuLi	Alcoa

Table 5: Investigated sheet materials of AlLi alloys [22]

The general corrosion properties of the investigated AlLi alloys are comparable with those of the conventional material AA 2024 T 4 (EN AW-2024). If the AlLi alloys are protected against corrosion with an anodised layer, which is common practice for the conventional alloys, then they are even superior to the conventional materials from a corrosion-chemical point of view. In spite of this, the AlLi materials are also not corrosion-resistant in acidic solutions with a pH < 4, i.e. in weak hydrochloric acid solutions.

Under the anodised layer there is an oxide layer on the aluminium surface generated by electrolytic oxidation. This layer improves the corrosion behavior and generally has a thickness of up to 30 µm, which is several times greater than that of the natural oxide layer (approx. 0.01 µm). The parts to be oxidised are connected as an anode to the positive pole of the current source and are subjected to a direct current in an electrolyte solution (e.g. in sulphuric acid). The terms anodic oxidation, anodising and anodised layer are derived from this. In Germany, the terms eloxal process, (electrolytic oxidation of aluminium), eloxation and eloxal layer are commonly used [23].

It is known that as alloys dissolve, the metal ions of the alloying components can influence the corrosion of the alloys. In another study [24], the corrosion behavior of AlLi alloy AA 8090–T851 (EN AW-8090) in HCl solutions of pH 2 with varying concentrations of metal ions of the alloying components was investigated and compared to that of pure aluminium. The mass loss rates were determined and the potentiodynamic current density/potential curves were recorded. The composition of the alloy is given in Table:

Alloying elements	Cu	Li	Mg	Fe + Si	Zr	Other	Al
Mass%	1.2 – 1.6	2.5	0.8 – 1.2	0.15	0.08 – 0.012	0.15	rest

Table 6: Composition of the investigated aluminium-lithium alloy

The material had been treated as follows: T 851 solution annealing, 1 – 2 % cold-rolled and artificially aged for stabilisation.

After exposure to the HCl solution for 7 days at ambient temperature, only the cations of Al, Li and Cu were detectable in the solution. Therefore, only the effect of these cations on the corrosion behavior was investigated. Table 7 shows the mass losses depending on the cation concentration. The results show that when the Cu^{2+} concentration of 10^{-3} M is exceeded there is an abrupt increase in the mass losses of pure aluminium, and particularly for the tested AlLi alloy. At a concentration of 10^{-2} M Cu^{2+} cations, the corrosion rate reached a value of 1.696 g/m² h. Compared to this, at a concentration of 10^{-2} M and even at 10^{-1} M, the Li^{+}- and Al^{3+} cations only increased the corrosion rate slightly.

Cation concentration (M)	Mass loss mg/cm²					
	Cu^{2+}		Li^{+}		Al^{3+}	
	AA 8090–T851		AA 8090-T851		AA 8090-T851	
measurement of the mass loss						
0	0.173	0.185	0.173	0.185	0.173	0.185
10^{-7}	0.170	0.159	–	–	–	–
10^{-5}	0.184	0.185	–	–	–	–
10^{-4}	0.232	0.242	0.168	0.185	0.170	0.172
5×10^{-4}	0.252	0.286	–	–	–	–
10^{-3}	0.286	0.326	0.195	0.182	0.175	0.190
5×10^{-3}	2.650	3.338	–	–	–	–
10^{-2}	3.271	4.068	0.229	0.230	0.188	0.303
10^{-1}	–	–	0.252	0.256	0.314	0.527
polarisation method						
0	0.070	0.082	–	–	–	–
10^{-5}	0.0886	0.104	–	–	–	–
10^{-3}	0.193	0.242	0.072	0.085	0.064	0.065
10^{-2}	2.070	2.131	0.082	0.086	0.058	0.074
10^{-1}	–	–	0.097	0.089	0.0051	0.0057

Table 7: Mass losses of both materials in a solution (pH 2.0) with increasing cation concentrations [24]

The promoting effect of copper ions is due to copper metal being deposited on the surface of aluminium and AlLi alloys above a certain concentration. The deposited copper particles are efficient cathodes with respect to the electrochemically much less noble aluminium and AlLi substrates. Therefore, numerous local elements are formed on the surface which increase the corrosion. The Al^{3+} and Li^+ cations, which are electrochemically the same as or similar to the substrate, are not deposited on the substrate.

For the copper-containing commercial AlLi alloy AA 8090–T851 (EN AW-8090), a selective dissolution of Al with an enrichment of Cu on the surface is to be expected; this would also promote corrosion. The negative effect on the corrosion behavior due to the deposition of copper is also known for other metallic materials, e.g. iron and low-alloy steels.

Since aluminium and its alloys only exhibit a low corrosion resistance in acidic media (see Figure 6), the addition of inhibitors to improve the corrosion resistance is particularly important. Dyes, such as those used to dye textiles, can have an inhibiting effect on metallic materials in acidic media. In reference [25], the inhibitive action of various dyes was investigated with the AlCu alloy AA 2017 (EN AW-2017) in 2.5 M HCl at 302 K (29 °C) with and without additional cathodic polarisation. The composition of alloy AA 2017 was as follows: Cu 3.88, Mn 0.87, Fe 0.43, Si 0.39, Mg < 0.32, Al rest. In the non-inhibited 2.5 M HCl at 302 K (29 °C), the immersion test for 30 min gave a mass loss rate of 77 g/m² h ~ 250 mm/a (9,843 mpy) for the alloy.

Table 8 summarises the results of the investigation. The concentration of all dyes was 0.002 M. The inhibition effect was very different and ranged from 11.1 to 90.5 % (column 3). The application of a cathodic current can provide 100 % protection against corrosion. Column 4 gives the current densities in mA/dm² that provide 100 % corrosion protection by the various dyes. In the pure acid, a current density of 481 mA/dm² is necessary for this. For dyes with a high inhibition effect, a lower current density f_{100} is usually necessary to provide complete corrosion protection. However, there are cases where, in spite of a high inhibition effect of the dye, a high current density is also necessary to attain complete corrosion protection.

In reference [25], the inhibitive action of various amines in 2.0 M trichloroacetic acid with and without a cathodic current was also investigated on EN AW-2017 (AlCu4SiMg) and the manganese-containing Al alloy EN AW-3003 (AlMn1Cu; Mn 1.3, Fe 0.54, Cu 0.15, Si 0.65, Zn ≤ 0.1 mass%, rest Al). These results are also listed in Table 8. At higher inhibitor concentrations of 0.1 and 0.2 M, all samples showed high efficiencies and the values of the cathodic current density necessary to give 100 % corrosion protection were correspondingly lower.

This investigation showed that additives that have a partial inhibitive effect on aluminium and Al alloys in acid media, these metals can be easily given complete corrosion protection by an additional cathodic current. In certain cases, this combined protection method could also be used to advantage with the system iron-based material/hydrochloric acid, which is used in practice.

Hydrochloric Acid

Inhibitor	Inhibitor concentration	Inhibition without current	Necessary current density [mA/dm²] for 100% protection with and without inhibitor	Protection in pure acid for f_{100}
	M	%	f_{100}	%
(a) for alloy AA 2017 in 2.5 M hydrochloric acid (30 min)				
No additive	–	–	481	100
Malachite green	0.002	88.5	32	4.5
Methyl violet 6B	0.002	78.2	32	4.5
Crystal violet	0.002	84.0	288	52.7
Fuchsine (basic)	0.002	52.7	300	61.3
Fuchsine acid	0.002	63.0	320	74.9
Alizarin red S	0.002	57.2	272	45.3
Acridine orange	0.002	38.3	176	24.3
Catechin violet	0.002	11.1	400	90.1
Rhodamine B	0.002	69.1	80	13.6
Dimethyl yellow	0.002	31.3	320	74.9
Methyl red	0.002	39.9	288	52.7
Eriochrome black T	0.002	90.5	48	7.0
(b) for alloy AA 3003 in 2.0 M trichloroacetic acid (30 min)				
No additive	–	–	228	100
Methylamine	0.1	72.0	98	33
Dimethylamine	0.1	70.0	98	33
n-Propylamine	0.1	80.0	81	28
Diethylamine	0.1	64.0	114	40
Triethylamine	0.1	45.9	146	64
(c) for alloy AA 2017 in 2.0 M trichloroacetic acid (15 min)				
No additive	–	–	230	100
Methylamine	0.2	85.3	69	25.0
Dimethylamine	0.2	91.5	31	17.6
Trimethylamine	0.2	91.7	15	12.1
Ethylamine	0.2	76.4	108	35.5
Diethylamine	0.2	68.0	100	32.5
n-Propylamine	0.2	92.0	31	17.6
n-Butylamine	0.2	86.4	61	23.0

AA 2017 (EN AW-2017)
AA 3003 (EN AW-3003)

Table 8: Necessary current density for the complete protection of aluminium alloys in inhibited acids [25]

The inhibition effect of amines on aluminium in hydrochloric acid was investigated in detail by determining the mass losses and by polarisation measurements in reference [26]. The material investigated had the composition (in mass%): Cd 0.06, Co 0.14, Cr 0.10, Cu 0.02, Mg 0.01, Mn 0.22, Ni 0.11, Pb 0.06, Zn 0.14, rest Al. The samples were polished and cleaned just before the tests were started. The test temperature for the determination of the mass losses was 302 ± 0.5 K (29 °C ± 0.5 °C), that for the polarisation measurements was 297 ± 0.5 K (24 °C ± 0.5 °C). The current density/potential curves were recorded to find the open-circuit potential in the potential range from U = –250 mV to + 250 mV with a ramping rate of 1 mV/s. Before the recordings of the curves were started, the samples were immersed for 20 min in the solution.

Figure 8 gives the mass loss of aluminium in pure hydrochloric acid depending on the time and concentration. At low concentrations, the dissolution is relatively low at the start of the tests and becomes increasingly greater until it reaches a linear dependence. This indicates that the acid initially reacts with the protective oxide layer which aluminium quickly forms in air. When this has dissolved, the accelerated reaction with aluminium starts. For a test duration of 160 min, the sample in the purified 1.5-molar HCl solution gave an extraordinarily high corrosion rate of 487 g/m^2 h, which corresponds to a corrosion rate of 1,580 mm/a (62,200 mpy). Therefore, it follows that aluminium can only be used in hydrochloric acid if it has been efficiently inhibited.

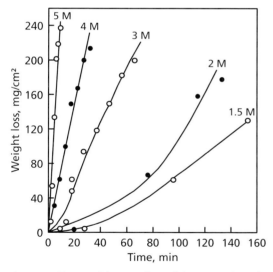

Figure 8: Change of the mass loss of aluminium depending on the time at various HCl concentrations. The numbers on the lines correspond to the individual HCl concentrations in mol/l [26]

Six different amines were used in the investigation. Figure 9 and Figure 10 show the inhibition effect with the same addition of 0.6 M inhibitor to 3 M and 5 M HCl solutions compared to the pure acid solutions depending on the exposure time. This showed that the inhibitor effect increased in the sequence methyl – (MA) < ethyl – (EA) < propyl – (PA) < butyl – (BA) < allyl (AA) < benzyl-amine (BzA).

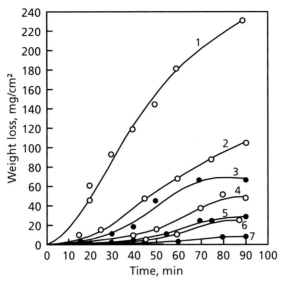

Figure 9: Change of the mass loss of aluminium depending on the time in 3 M HCl and with addition of 0.6 M inhibitor
1) pure acid; 2) acid + MA; 3) acid + EA; 4) acid + PA; 5) acid + BA; 6) acid + AA; 7) acid + BzA [26]

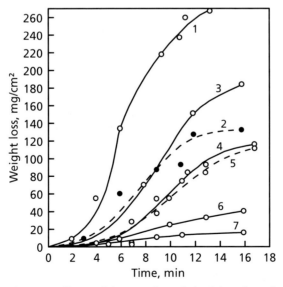

Figure 10: Change of the mass loss of aluminium depending on the time in 5 M HCl and with addition of 0.6 M inhibitor
1) pure acid; 2) acid + MA; 3) acid + EA; 4) acid + PA; 5) acid + BA; 6) acid + AA; 7) acid + BzA [26]

Amine	3 M HCl		5 M HCl	
	P		P	
	after 30 min	after 70 min	after 6 min	after 12 min
Methylamine	72.2	59.5	55.6	51.9
Ethylamine	84.4	67.5	74.1	42.3
Propylamine	94.4	81.0	88.9	66.5
Butylamine	97.8	87.0	88.9	70.0
Allylamine	96.9	90.0	91.9	88.1
Benzylamine	99.1	96.5	95.6	94.2

Table 9: Percentage inhibition (P) of aluminium corrosion by various amines in 3 M and 5 M HCl after different periods [26]

The chronological progression of the curves shows that, at least for strongly inhibiting amines, the mass loss tends to a more or less constant value. The percentage inhibition effect for an intermediate and a long testing time, given in Figure 9 and Figure 10, is summarised in Table 9. The best inhibitor, benzylamine, gave values between 94.2 and 99.1 %. For the longest test times of 90 min in 3 M HCl and 16 min in 5 M HCl, for which the best inhibitor benzylamine gave a constant efficiency, the removal rates were calculated to be 216 mm/a and 2,129 mm/a (8,503 and 83,800 mpy), respectively. This shows that, in spite of the high inhibition efficiency, the removal of aluminium in more highly concentrated HCl solution is still too high so that the use of aluminium is not cost-effective and is only worthwhile for short-term contact. Furthermore, Figure 13 also shows that a high inhibition efficiency by amines is only obtained at a relatively high inhibitor concentration of approximately > 0.5 M for the most efficient allyl- and benzylamines.

The electrochemical polarisation measurements confirmed the sequence of inhibition efficiency of the investigated amines found by determining the mass losses. From the electrochemical point of view, the investigated amines act by increasing the hydrogen overpotential on aluminium so that the corrosion rate is decreased.

In reference [27], the inhibition efficiencies of three heterocyclic hydrazone derivatives for aluminium in 2 M HCl solution at 298 K (25 °C) was investigated using three methods of determination, namely mass loss determination, as well as the generation of hydrogen and heat (thermometrics).

These compounds were:
I   N-(o-hydroxybenzylidene)-2-amino-1,2,4-triazole
II N-(o-hydroxybenzylidene)-5-amino-4-methylisotriazole
III N-(o-hydroxybenzylidene)-2-aminopyrazine

The last two methods are based on the fact that during dissolution of aluminium according to the equation

Eq. 3 $Al + 3 H^+ \rightarrow Al^{3+} + 3/2\ H_2$

H_2 is liberated, and heat is generated at a rate of 1049.2 kJ/mol H_2. As the inhibitor efficiency increases, the liberation of hydrogen decreases and the generated heat correspondingly decreases. This can be seen from the chronological progression of the temperature.

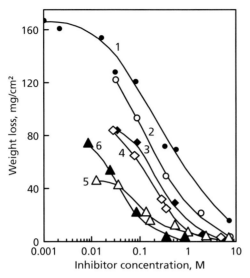

Figure 11: Influence of the inhibitor concentration on the mass losses of aluminium in 3 M HCl
1) pure acid; 2) acid + MA; 3) acid + EA; 4) acid + PA; 5) acid + BA; 6) acid + AA; 7) acid + BzA [26]

The result of the thermometric investigation on the influence of the addition of inhibitor I on the chronological dependence of the temperature progression is shown in Figure 12. As the inhibitor concentration increases, the temperature increase is delayed and the temperature maximum increasingly drops.

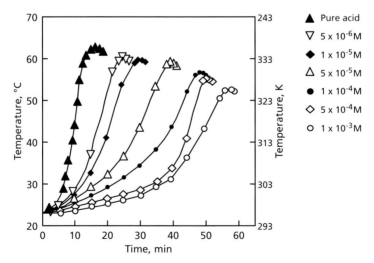

Figure 12: Time-dependence of the temperature profile for various concentrations of inhibitor I [27]

Figure 13, in contrast to Figure 12, gives the mass loss in dependence on the time for the same concentration of inhibitor I. It can be seen that the behavior is similar to that obtained by determination of the generated heat.

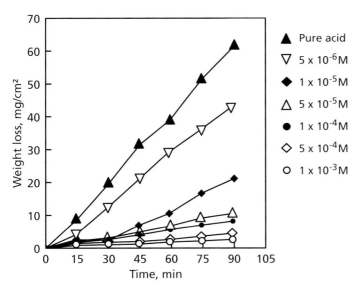

Figure 13: Mass loss in dependence on the time for various concentrations of inhibitor I [27]

Table 10 gives the effect of all three inhibitors for all six investigated concentrations in 2 M HCl solution at 298 K (25 °C) obtained by the three determination methods. According to all three methods, in the entire concentration range from 5×10^{-6} to 1×10^{-3} mol/l, the order of effectivity is I > II > III.

This result is also in agreement with the values of the activation energies that were obtained for the hydrogen liberation as a function of the temperature in 2 M HCl solution with and without inhibitors.

Concentration	Inhibition, %								
mol/l	Mass loss			Thermometrics			H_2 liberation		
	I	II	III	I	II	III	I	II	III
5×10^{-6}	29.6	25.1	21.5	20.2	20.1	19.5	37.1	31.4	16.1
1×10^{-5}	64.8	30.2	29.4	39.1	29.4	27.8	46.5	40.5	23.1
5×10^{-5}	81.8	48.3	35.2	63.5	40.3	39.9	54.5	53.8	30.2
1×10^{-4}	86.1	61.4	41.6	72.2	47.5	43.5	74.8	69.8	34.9
5×10^{-4}	92.1	74.7	45.3	74.3	60.4	62.6	79.1	76.9	47.5
1×10^{-3}	94.8	75.2	50.3	79.4	71.6	70.2	86.1	84.2	55.9

Table 10: Effect of the three inhibitors for six different concentrations in 2 M HCl at 298 K (25 °C) using the three methods of determination [27]

The value of the activation energy is a measure of the strength of the physical adsorption of the inhibitors onto the aluminium surface. Inhibitor I has the highest value.

The authors attributed the differing effectivity of the three inhibitors to differences in the molecular structure of the heterocyclic moiety in the compound. According to this hypothesis, compound I has two nitrogen centres and one active oxygen centre, compound II has one active sulphur and one active nitrogen centre and compound III has two active nitrogen centres that correspondingly decrease the effectivity.

The effectivity of carboxylic acids as corrosion inhibitors for aluminium in acidic and alkaline solutions was investigated in the study [28]. Because of the amphoteric corrosion behavior of aluminium, it is also strongly attacked in alkaline solutions (see also Corrosion Handbook volume 1 "Sodium hydroxide"), so that inhibition is necessary if aluminium is used e.g. in NaOH solutions.

The carboxylic acids used in the investigation were:

Monocarboxylic acids
(I) benzoic acid   (II) salicylic acid
(III) m-hydroxybenzoic acid   (IV) p-hydroxybenzoic acid
(V) anthranilic acid   (VI) m-chlorobenzoic acid

Polycarboxylic acids
(VII) terephthalic acid   (VIII) phthalic acid
(IX) trimellitic acid   (X) pyromellitic acid
(XI) 1.8-naphthalic acid   (XII) diphenic acid

Aliphatic carboxylic acids
(XIII) acetic acid   (XIV) lauryl acid

The main objective of the investigation was to show the relationship between the structure of the molecule and its inhibiting effect. To increase the solubility of the inhibitors, 50 % methanol was added to the 2 M HCl and 2 M NaOH solutions that were used in the investigations. The aluminium had the following composition:
99.535 Al, 0.19 Fe, 0.15 Si, 0.1 Mg, 0.02 Cu and 0.005 Mn. The inhibitor efficiency was determined by mass loss.

Figure 14 shows the mass losses depending on the time for the monocarboxylic acids, and Figure 19 gives those for polycarboxylic acids. Characteristic for both groups is the initially low dissolution. This is because the more corrosion-resistant oxide layer must be dissolved before the bare, less corrosion-resistant metal surface can be attacked.

The corresponding mass loss curves for the NaOH solution are given in Figure 16 and Figure 17. There is a strictly linear time-dependence of the mass loss, even at the beginning of the test. Thus, in the 2 M NaOH solution, the protective oxide layer is immediately dissolved; however, the overall inhibitor efficiency is higher and the scattering is significantly lower than for the 2 M HCl solution. The decreasing sequence of inhibitor efficiency is different to that found for the 2 M HCl solution.

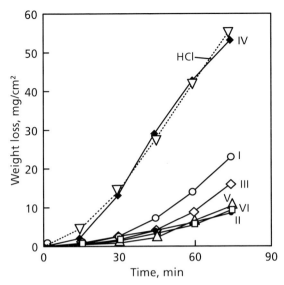

Figure 14: Time-dependent mass losses at constant concentration (1×10^{-3} M) of monocarboxylic acids in 2 M HCl at 303 K (30 °C, 50 vol% methanol) [28]

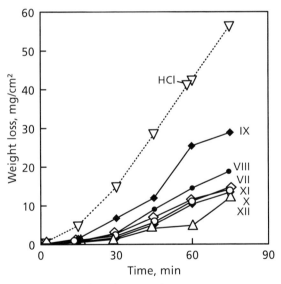

Figure 15: Time-dependent mass losses at constant concentration (1×10^{-3} M) of polycarboxylic acids in 2 M HCl at 303 K (30 °C, 50 vol% methanol) [28]

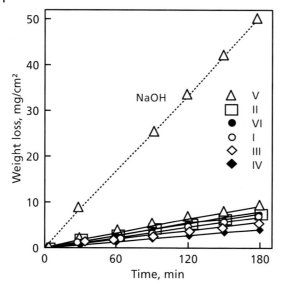

Figure 16: Time-dependent mass losses at constant concentration (1×10^{-3} M) of monocarboxylic acids in 2 M NaOH at 303 K (30 °C, 50 vol% methanol) [28]

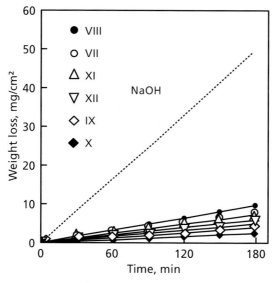

Figure 17: Time-dependent mass losses at constant concentration (1×10^{-3} M) of polycarboxylic acids in 2 M NaOH at 303 K (30 °C, 50 vol% methanol) [28]

Table 11 and Table 12 give the percentage decrease of the corrosion rate depending on the inhibitor concentration for a test period of 60 min. According to the graphical representation in Figure 18, compound IV (p-hydroxybenzoic acid) is an anomaly. It did not exhibit any inhibition effect at all in the HCl solution, although it was the best inhibitor in the NaOH solution.

	HCl					NaOH			
Concentration × 10⁵ M	1	5	10	50	100	1	5	10	50	100
Inhibitor									
I	16.9	21.5	40.7	56.1	64.6	77.2	79.0	83.2	85.6	86.8
II	65.3	66.9	69.2	70.0	80.7	74.2	76.6	79.0	81.4	84.4
III	3.8	15.3	23.0	57.6	69.2	71.2	73.0	77.8	79.6	80.8
IV	−16.9	−10.0	−7.6	−3.8	−0.7	75.4	78.4	82.0	91.6	92.8
V	38.4	53.8	55.3	61.5	63.5	71.8	78.0	76.0	79.6	82.0
VI	2.7	13.1	20.7	53.2	64.0	62.6	65.3	72.3	76.3	79.1

Table 11: Percentage decrease of the corrosion rate in dependence on the inhibitor concentration of the monocarboxylic acids in 2 M HCl and 2 M NaOH after 60 min (50 vol% methanol) [28]

	HCl					NaOH			
Concentration × 10⁵ M	1	5	10	50	100	1	5	10	50	100
Inhibitor									
VII	6.1	15.3	26.9	46.1	50.0	72.4	74.8	80.8	83.8	85.0
VIII	6.1	11.5	19.2	30.0	65.3	80.3	85.6	88.0	91.6	95.2
IX	5.3	10.0	28.4	39.2	46.1	79.0	82.0	85.6	86.2	89.2
X	15.3	25.3	33.0	29.2	61.5	88.8	89.9	91.9	92.9	94.9
XI	32.6	39.2	46.9	55.3	62.3	75.4	78.4	82.0	83.8	88.6
XII	27.6	45.3	63.8	75.3	85.3	73.0	77.2	80.2	86.2	90.4

Table 12: Percentage decrease of the corrosion rate in dependence on the inhibitor concentration of the polycarboxylic acids in 2 M HCl and 2 M NaOH after 60 min (50 vol% methanol) [28]

Table 13 summarises the results obtained with the two aliphatic carboxylic acids. The inhibition efficiency of acetic acid is significantly lower than that of lauryl acid. The authors attempted to rationalise the differing inhibition efficiencies of the investigated compounds on the basis of their molecular structure.

	Acetic acid			Lauryl acid	
Medium Inhibitor concentration 10⁵ M	HCl aqueous	HCl 50 vol% methanol	NaOH 50 vol% methanol	HCl 50 vol% methanol	NaOH 50 vol% methanol
1	3.1	9.7	13.4	32.1	71.7
5	7.7	12.1	17.9	58.0	73.5
10	9.2	17.0	20.8	62.5	75.9
50	13.6	21.2	23.8	71.5	77.9
100	17.3	23.6	49.2	83.7	78.9

Table 13: Inhibition efficiency (%) of acetic acid and lauryl acid in acidic and alkaline media [28]

The inhibition efficiency of the pyridine derivatives listed in Table 14 on the corrosion of aluminium in 2 M HCl solution at 298 K (25 °C) was investigated in reference [29].

Number	Designation	Active centres	Inhibition efficiency % at 10^{-3} mol/l
I	2-Aminopyridine	one N atom	43.5
II	2-Amino-5-chloropyridine	one N atom	36.9
III	2-Cyanopyridine	two N atoms	62.5
IV	3-Carboxylpyridine	one N and one O atom	64.1
V	4-Carboxylpyridine	one N and one O atom	70.2
VI	3-Nicotinamide	one N and one O atom	74.2

Table 14: Tested inhibitors and their efficiencies [29]

The aluminium material had the same composition as that used in references [28 and 30]: 99.535 Al, 0.19 Fe, 0.15 Si, 0.1 Mg, 0.02 Cu and 0.005 Mn. The test was evaluated using the methods of heat and hydrogen generation.

Table 15 summarises the results from the heat generation method in dependence on the inhibitor concentration. The inhibition efficiency increases with increasing concentration of the inhibitor; however at the highest concentration of 10^{-3} mol/l it was only 74 % for the most effective inhibitor VI. At this relatively high acid concentration, these pyridine derivatives are not sufficiently effective as inhibitors. In acid solutions of lower concentration, high efficiencies should be attained for the derivative concentrations investigated here.

Inhibitor concentration mol/l	Reduction of the reaction number (RN) %					
	I	II	III	IV	V	VI
1×10^{-6}	7.5	5.2	21.3	24.5	26.5	36.4
5×10^{-6}	13.4	9.3	28.9	33.1	35.2	45.5
10×10^{-6}	26.5	18.5	32.5	37.5	45.5	58.4
100×10^{-6}	38.5	30.9	47.3	52.2	59.5	67.5
500×10^{-6}	41.3	34.3	57.1	61.2	66.1	71.5
1000×10^{-6}	43.5	36.9	62.5	64.1	70.2	74.2

Table 15: Inhibition efficiency given as a percentage reduction of the reaction number RN [29]

The reaction number (RN [°C/min]) is defined as follows:

$$RN = \frac{T_m - T_i}{t}$$

T_i and T_m denote the initial and maximum temperature and t is the time from the start of the test until the maximum temperature is reached.

The aim of the investigation was to elucidate the fundamental relationship between the molecular structure and the inhibitive action. The decisive factor for this is the electron density of the active centres. Inhibitors I and II, each containing one active N centre, only have a low efficiency of approx. 40 % at the highest concentration of 10^{-3} mol/l. Inhibitor III, with two active N centres, shows a sudden increase in the efficiency of more than 60 %. This increases with inhibitors IV – VI, which each have one active N centre and one O centre, up to a maximum of 74.2 %.

Reference [31] deals with the inhibition efficiency of dimethyltin-dichloride (($CH_3)_2SnCl_2$) towards aluminium in 2 M HCl and 2 M NaOH solutions at 308, 318 and 333 K (35, 45 and 60 °C). The reason for this was the recent use in Japan of $(CH_3)_2SnCl_2$ instead of $SnCl_4$ to coat glass containing SnO_2 because $(CH_3)_2SnCl_2$ is easier to handle. The composition of the tested Al is given as 99 % Al, 0.2 % Fe, 0.2 % Cu, 0.2 % Si, 0.03 % Ti, 0.08 % Zn (the sample dimensions were: 0.1 × 10 × 30 mm). The hydrogen and heat generation methods were used. Figure 22 shows the dependence of the hydrogen liberation, which runs parallel to the dissolution of the metal, on the inhibitor concentration and the time. As for the heat generation method, the generation of hydrogen is initially delayed. After a certain incubation time, which increases with increasing inhibitor concentration, it reaches a linear dependence V = Kt. The incubation time is due to the more difficult dissolution of the oxide layer that is initially present. Table 16 summarises the corrosion rate, the incubation time and the inhibition efficiency in dependence on the inhibitor concentration in a 2 M HCl solution at 308 K (35 °C). At the highest concentration, an inhibitive action of 91.1 % is reached.

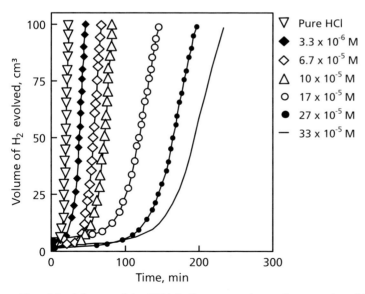

Figure 18: Influence of the $(CH_3)_2SnCl_2$ concentration on the generation of hydrogen depending on the time for aluminium in 2 M HCl at 308 K (35 °C) [31]

$(CH_3)_2SnCl_2$ concentration mol/l	0.0	3.3×10^{-5}	6.7×10^{-5}	10×10^{-5}	17×10^{-5}	27×10^{-5}	35×10^{-5}
Corrosion rate cm^3/min	13.88	6.87	5.29	4.70	1.94	1.57	1.24
Incubation period min	9.33	20	38.17	41.33	80	120	140
Inhibition efficiency %	0.0	50.5	61.9	66.1	86.0	88.7	91.1

Table 16: Influence of the $(CH_3)_2SnCl_2$ concentration on the corrosion rate of aluminium in 2 M HCl at 308 K (35 °C) [31]

Figure 19 shows the hydrogen liberation for the highest inhibitor concentration of 3.3×10^{-6} M at 308, 318 and 333 K (35, 45 and 60 °C). The incubation time strongly decreases with increasing temperature and the corrosion rate increases correspondingly, as can be seen in Table 17.

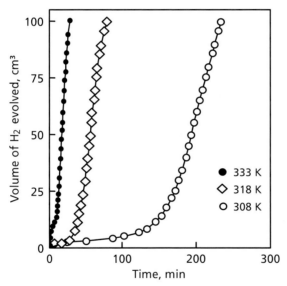

Figure 19: Influence of the temperature on the hydrogen liberation in dependence on the time for aluminium in 2 M HCl with 3.3×10^{-5} M $(CH_3)_2SnCl_2$ [31]

Temperature, K (°C)	308 (35)	318 (45)	333 (60)
Corrosion rate, cm^3/min	1.24	2.22	4.76
Incubation period, min	140.5	32.2	7.07

Table 17: Influence of the temperature on the corrosion of aluminium in 2 M HCl with 3.3×10^{-5} M $(CH_3)_2SnCl_2$ [31]

The influence of the $(CH_3)_2SnCl_2$ concentration on the hydrogen volume/time curve in a 2 M NaOH solution at 308 K (35 °C) is shown in Figure 20. As already found in reference [28], the curves from the starting point are linear. However, whereas the carboxylic acids used there as an inhibitor in the 2 M NaOH solution exhibited a much better efficiency than in the 2 M HCl solution, the reverse is true for $(CH_3)_2SnCl_2$. Even for an approx. 10-fold higher concentration there is only an insufficient inhibition efficiency of max. 52.54 % (Table 18). Similar to the test in the HCl solution, the inhibition efficiency strongly decreases with increasing temperature. $(CH_3)_2SnCl_2$ is not an effective inhibitor for aluminium in NaOH solutions.

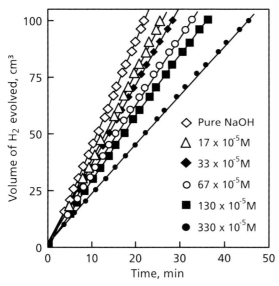

Figure 20: Influence of the $(CH_3)_2SnCl_2$ concentration on the generation of hydrogen depending on the time for aluminium in 2 M NaOH [31]

$(CH_3)_2SnCl_2$ concentration mol/l	0.0	17×10^{-5}	33×10^{-5}	67×10^{-5}	130×10^{-5}	330×10^{-5}
Corrosion rate cm³/min	4.54	3.85	3.46	3.00	2.72	2.15
Inhibition efficiency %	0.0	15.45	23.62	33.77	39.96	52.54

Table 18: Influence of the $(CH_3)_2SnCl_2$ concentration on the corrosion rate of aluminium in 2 M NaOH at 305 K (35 °C) [31]

In the study [32], the inhibition efficiency of 2-furancarboxyaldehyde-(2'-pyridylhydrazone) (inhibitor I), 2-pyrrolcarboxaldehyde-(2'-pyridylhydrazone) (inhibitor II) and 2-thiophencarboxaldehyde-(2'-pyridylhydrazone) (inhibitor III) towards aluminium in a 2 M HCl solution at 298 K (25 °C) was investigated using the methods of mass loss, of hydrogen liberation and by recording current density/potential curves.

The substrate was aluminium (99.5 %). The mass loss curves in Figure 21 show that all three compounds have a high efficiency for an addition of 10^{-4} mol/l.

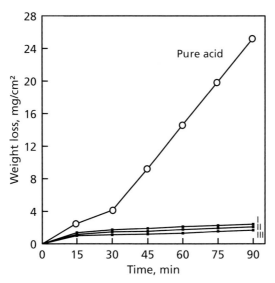

Figure 21: Time-dependent mass loss for the three inhibitors (10^{-4} M) at 298 K (25 °C) [32]

Inhibitor	Concentration mol/l	Inhibition %
HCl		–
(I)	10×10^{-6}	45.1
	5×10^{-5}	58.3
	10×10^{-5}	71.2
	5×10^{-4}	79.6
	10×10^{-4}	86.0
(II)	10×10^{-6}	34.9
	5×10^{-5}	57.4
	10×10^{-5}	60.2
	5×10^{-4}	68.0
	10×10^{-4}	74.9
(III)	10×10^{-6}	30.5
	5×10^{-5}	46.0
	10×10^{-5}	48.7
	5×10^{-4}	57.3
	10×10^{-4}	68.0

Table 19: Efficiency of the inhibitors on the corrosion of aluminium in 2 M HCl at 298 K (25 °C) [32]

Table 19 summarises the inhibition efficiency for all three compounds depending on their concentration. Thus the sequence of inhibition efficiency is I > II > III, as is already indicated in Figure 21.

The inhibition efficiency of four pyrazoline derivatives on aluminium (99.5 %) in 2 M HCl solutions at 303 K (30 °C) was investigated using the method of mass loss and by means of galvanostatic polarisation curves [33].

The following compounds were tested:
(I) 4-benzylidene-3-methyl-5-oxo-1-phenyl-2-pyrazoline
(II) diethyl-[a-(3-methyl-5-oxo-1-phenyl-2-pyrazolin-4-yl)benzyl]malonate
(III) ethyl-[a-(3-methyl-5-oxo-1-phenyl-2-pyrazolin-4-yl)benzyl]acetacetate
(IV) [3-(3-methyl-5-oxo-1-phenyl-2-pyrazolin-4-yl)benzyl)pentan-2,4-dione

The results are given as mass losses in Figure 22 and Table 20. Even for small additions, all four compounds are effective inhibitors. This result is also confirmed by the polarisation measurements. Although the differences are small, the following sequence of effectivity can be seen: II > III > IV > I.

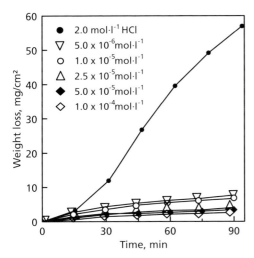

Figure 22: Time-dependent mass losses for inhibitor IV [33]

Inhibitor concentration mol/l	Inhibition %			
	I	II	III	IV
5.0×10^{-6}	84.30	91.80	90.90	84.50
1.0×10^{-5}	85.50	91.70	91.20	85.60
2.5×10^{-5}	91.60	92.70	91.70	91.70
5.0×10^{-5}	92.10	92.80	92.20	92.10
1.0×10^{-4}	92.20	93.60	93.40	92.40

Table 20: Inhibition efficiency determined from the mass losses after 60 minutes at 303 K (30 °C) [33]

In addition, the effect of the inorganic cations Ni^{2+}, Co^{2+} and Cu^{2+}, added as chlorides, was investigated. In combination with the inhibitors, there was a further stabilisation of the protective layer. The efficiency of these cations was $Ni^{2+} > Co^{2+} > Cu^{2+}$.

In reference [34], the inhibitive action of propargyl alcohol on pure aluminium and aluminium alloys was investigated with an electrochemical method (evaluation using the Tafel lines and polarisation resistance) in HCl, H_2SO_4, NaCl and NaCl + NaOH at 298 K (25 °C). Table 21 summarises the results of the investigated materials. As the chromium content increases and the iron/manganese ratio increases, the inhibition efficiency increases, while it decreases for an addition of copper, silicon and magnesium.

Sample	Mn	Fe	Cr	Cu	Mg	Si	Zn	Ti
1	0.28	–	0.24	0.28	–	< 0.1	–	–
2	–	–	0.27	–	–	< 0.1	–	–
3	0.36	–	–	–	–	< 0.1	–	–
4	1.53	0.41	–	–	–	< 0.1	–	–
5	1.38	0.43	–	0.19	–	< 0.1	–	–
6	1.46	0.33	–	–	–	< 0.1	–	–
7	1.82	0.60	–	–	–	< 0.1	–	–
8	0.22	0.26	–	–	0.15	< 0.1	–	–
9	0.24	0.18	–	–	0.41	< 0.1	–	–
10	0.11	0.71	–	–	–	< 0.1	–	–
DIN 1732	1.17	0.50	0.05	0.1	0.1	0.5	0.10	0.05
Pure Al	–	–	–	–	–	–	–	–

Table 21: Composition of the investigated aluminium alloys [34]

In 0.5 M H_2SO_4, the addition of 20 mmol/l propargyl alcohol increased the corrosion rate for all investigated materials; some of them were increased considerably by up to 82 %. In 1 M NaCl as well as in 1 M NaCl + 0.001 M NaOH, each with an addition of 20 mmol/l propargyl alcohol, the average inhibition is small and a max. of 68 % was obtained for an individual case.

A 4 Gold and gold alloys

In 5 % hydrochloric acid, even at elevated temperatures, the corrosion rate of gold lies below 0.01 mm/a (0.4 mpy). Reference [35] gives a general description of the behavior of metals that are used in electronic equipment to attack by HCl, which can be produced in the event of chlorine-containing plastics (PVC) catching fire. According to this, gold is essentially resistant to HCl under these conditions.

Gold is attacked by mixtures of hydrochloric acid and nitric acid and also in the presence of Fe^{3+} ions.

A 5 Cobalt alloys

Various compounds exist in the two component Co-P system of which dicobalt monophosphide is particularly interesting because of its chemical resistance, abrasion resistance and catalytic activity. This compound is prepared by reaction with PCl_3 at high temperatures of up to 1273 K (1000 °C). It has a Vickers hardness of up to 900 kg/mm^2. Co_2P, obtained by complete reaction with phosphorus, has an enhanced oxidation and acid resistance. Figure 23 shows the mass loss depending on the time in 1 M HCl, 0.5 M H_2SO_4, 1 M HNO_3 and 1 M NaOH at 303 K (30 °C) compared to metallic cobalt [36].

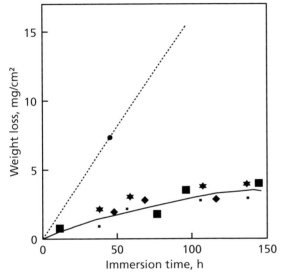

Figure 23: Mass loss of Co_2P in, ■ 1 M HCl; ★ 0.5 M H_2SO_4; ◆ 1 M HNO_3; ■ 1 M NaOH at 303 K (30 °C); ● pure cobalt sheet (in H_2SO_4) [36]

Antimony and bismuth compounds have a high inhibition efficiency for iron, nickel and zinc in hydrochloric acid. They form stable corrosion-protection coatings consisting of metallic antimony or bismuth and metal oxides. The work in [37] investigated the corrosion inhibition of cobalt in 1 M $HClO_4$, 0.5 M H_2SO_4 and 1 M HCl by bismuth(III)chloride. The current density/potential curves were recorded and the Co^{2+} concentration was determined depending on the time in the acidic solutions. Cobalt (99.9 %) was used for the tests. The test temperature was 303 K (30 °C).

The efficiencies were obtained from the polarisation measurements in the three acids and are summarised in Figure 24 in dependence on the inhibitor concentration. The highest efficiencies of almost 90 % were obtained at concentrations of 10^{-5} and 10^{-4} mol/l in 1 M $HClO_4$ and 0.5 M H_2SO_4. The curve for 1 M HCl is very different. Even at very low concentrations between 10^{-8} and 10^{-5} mol/l, relatively high efficiencies of up to 85 % were obtained. At higher concentrations of 10^{-4} and 10^{-3} mol/l, $BiCl_3$ acts increasingly as a corrosion catalyst.

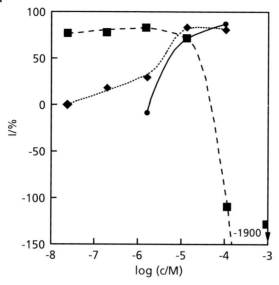

Figure 24: Inhibition efficiency (I) depending on the bismuth-(III)-chloride concentration for cobalt in 1 M HClO$_4$ (●), 0.5 M H$_2$SO$_4$ (♦) and 1 M HCl (■) [37]

In the 1 M HCl solution with 10^{-4} mol/l BiCl$_3$, a dark colloidal layer initially forms on the cobalt sample. Under this layer, islands of metallic bismuth are produced along with bare cobalt surfaces without the layer. Accelerated cathodic hydrogen evolution takes place on the bismuth islands that increases the dissolution of the metal.

The cobalt alloy Haynes® alloy 25 (CoCr20W15Ni, UNS R30605, DIN-Mat. No. 2.4964) with 20 % Cr, 15 % W, 10 % Ni, 3 % Fe, 2 % Mn, 1 % Si, 0.15 % C, exhibited a corrosion rate of 0.0025 mm/a (0.1 mpy) in 2 % hydrochloric acid at room temperature and 0.61 mm/a (24 mpy) in 5 % HCl at 339 K (66 °C). The following corrosion rates are reported [38] for the multiphase alloy MP 35N (DIN-Mat. No. 2.4999) with 35 % Ni, 35 %, Co, 20 % Cr, 10 % Mo at 323 K (50 °C):

10 % HCl	0.56 mm/a (22 mpy)
37 % HCl	0.43 mm/a (17 mpy)

A 6 Chromium and chromium alloys

Approximately 3 months before being commissioned, the above-ground pipelines together with the fittings and the gas preheater of a gasometer system were pickled in approx. 8 % hydrochloric acid with a commercially available inhibitor (Rodine 213, Collardin, Cologne) at 283-293 K (10-20 °C) to remove the scale. The parts of the system were exposed to the pickling agent for a total of approx. 30 h (14 to 26 % of the time with circulation) and also a further 14 h in dilute acid with a concentration of at least 1 %. After rinsing twice, the parts were treated for a period of 1 to

1.5 h with an alkaline NaOH solution of undefined concentration (pH values > 10.5 were measured).

During commissioning, leakages were found in numerous fittings. The investigation of the dismantled valves and ball valves from two different manufacturers showed that the hard chromium coatings, applied to the sealing surfaces during manufacturing, were damaged or were no longer present. It was suspected that the damage was caused during pickling of the parts; this was confirmed in a laboratory investigation [39].

The corrosion potential and the polarisation resistance in air saturated with 2 % hydrochloric acid (pH = 0.5), was measured at 293 K (20 °C) with and without inhibitor addition (0.2 % Rodine 213), on pure chromium samples (compacts made of chromium powder) as well as on remainders of the hard chromium coatings from ball valves whose surroundings had been covered with paint.

The compacts made of pure chromium exhibited an initial passive behavior in the uninhibited acid; however, after varying periods of time they became active and quickly dissolved. In the inhibited acid, these samples were passivated or they only exhibited partial activity.

The behavior of the hard chromium coating is shown in Figure 25 and Figure 26. In the non-inhibited acid, there was a sudden and complete activation after approx. 1 h. This was indicated by the steep decrease in the potential and the very low polarisation resistance in conjunction with rapid dissolution of the hard chromium coating. In the inhibited acid, the same fundamental process occurs except that the dissolution rate is lower so that the dissolution time of the chromium layer is longer. This is indeed the time range that the hard chromium coating of the fittings was exposed to the inhibited acid during the pickling process.

		2 % HCl / 293 K (20 °C)			2 % HCl + 0.2 % Rodine 213 / 293 K (20 °C)	
		Sample 1	Sample 2	Sample 4	Sample 3	Sample 5
Start of the rapid pickling attack (time of activation, indicated by a sudden drop in the potential as well as liberation of H_2) in hours		48	13	0.2	–	–
Resting potential U_R of the samples in mV_H	before activation	– 75...– 235	– 25...– 225	– 210	– 85...– 230	– 155...+ 325
	after activation	– 490	– 485	– 480	–	–
Polarisation resistance Rp in $\Omega\,cm^2$	before activation	100	90	360	580...80	450...9,000
	after activation	2	not measurable	4		
Average corrosion rate K [$g/m^2\,h$] (from mass loss measurements)		98	*	*	8	0.05

Table 22: Results of the tests of the electrochemical behavior of pure chromium in pickling solutions (* sample completely dissolved during the test) [39]

This case of practical damage and the laboratory tests proved that the hard chromium coatings had been damaged during pickling. Before a plant pickling is carried out, not only should the pickling behavior of all substrate materials be investigated, but any surface coatings should also be tested. This problem is also known for the pickling of power station systems.

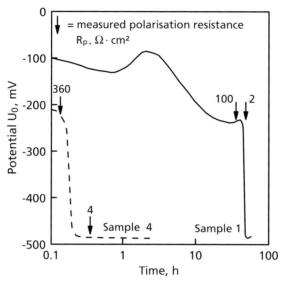

Figure 25.1: Electrochemical tests on the pickling resistance of pure chromium: chronological progression of the measured values in non-inhibited 2 % HCl at 293 K (20 °C) [39]

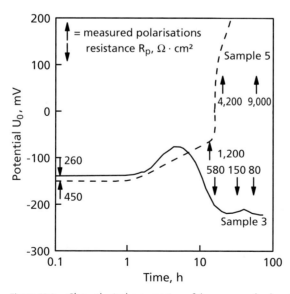

Figure 25.2: Chronological progression of the measured values in inhibited 2 % HCl [39]

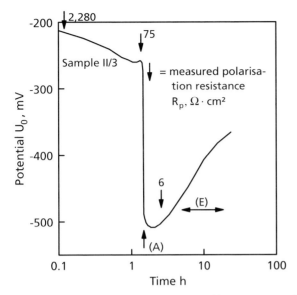

Figure 26.1: Electrochemical tests on the pickling resistance of hard chromium coatings: chronological progression of the measured values in non-inhibited 2 % HCl at 293 K (20 °C) [39]

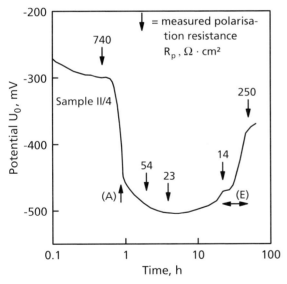

Figure 26.2: Chronological progression of the measured values in inhibited 2 % HCl. (A) = activation (start of the pickling attack. (E) = removal of most of the chromium coatings and attainment of the values measured for the substrate material [39]

A 7 Copper

Copper has a slightly more noble potential [E_o (Cu/Cu^{2+}) = + 0.33 V] than a standard hydrogen electrode and should therefore be resistant in non-oxidising acids. The attack by hydrochloric acid is thus not only dependent on the concentration and the temperature but, in particular, on the presence of oxidising agents. Copper is strongly attacked by aerated hydrochloric acid.

The behavior of copper in mixtures of oxidising and reducing acids is important with regard to pickling and electropolishing. In reference [40], the dissolution behavior of copper in acid mixtures HCl-HNO$_3$, HCl-H$_2$CrO$_4$ and H$_3$PO$_4$-HNO$_3$ was investigated with thermometrics, by measuring the rate of mass loss and the free mass loss at 298 K (25 °C). The total concentration of each acidic solution was 5 M.

Figure 27 and Figure 28 give the results for the three acid mixtures.
In HNO$_3$-H$_3$PO$_4$ mixtures, the removal rate constantly increases from zero in pure phosphoric acid to the corrosion rate in pure nitric acid (Figure 29).

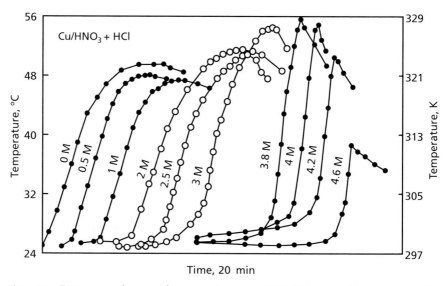

Figure 27: Temperature change in the copper / aqueous HNO$_3$-HCl system. The numbers on the curves denote the molarity of HCl [40]

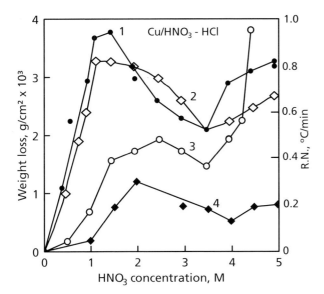

Figure 28: Changes:
1) of the reaction number (RN; see also A3),
2) of the thermometrically determined mass loss, and
3) of the free mass loss in a stirred solution,
4) in an unstirred solution as a function of the HNO_3 concentration [40]

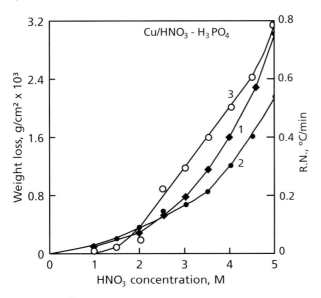

Figure 29: Changes:
1) of the reaction number (RN; see also A 3),
2) of the thermometrically determined mass loss, and
3) of the free mass loss as a function of the HNO_3 concentration (remainder H_3PO_4; total concentration 5 M) [40]

In HCl-H$_2$CrO$_4$ mixtures, even for a low addition of chromic acid, the corrosion rate is greatly increased, and already reaches a maximum below 1 M. It then drops slowly at first and then rapidly drops towards zero above 3 M H$_2$CrO$_4$. The strongly oxidising chromic acid passivates the copper in spite of the Cl$^-$ ions still present (Figure 30).

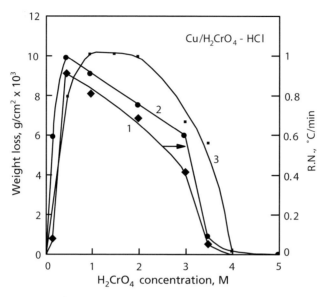

Figure 30: Changes:
1) of the reaction number (RN; see also A3),
2) of the thermometrically determined mass loss, and
3) of the free mass loss as a function of the H$_2$CrO$_4$ concentration (remainder HCl; total concentration 5 M) [40]

In HCl-HNO$_3$ mixtures, there is a similar tendency initially, namely a maximum at a relatively low addition of HNO$_3$ followed by a decrease. However, the values increase again between 3 to 4 M HNO$_3$ up to the value in pure nitric acid. Nitric acid in the concentration used here is, however, not able to passivate copper.

If the results from the current density/potential curves that were also recorded are included, it can be seen that the optimum acid mixture to pickle copper consists of 3 to 3.5 M HCl and 1.5 to 2 M H$_2$CrO$_4$. A mixture of the acids H$_3$PO$_4$ and HNO$_3$ is especially suitable for electropolishing of copper.

A 8 Copper-aluminium alloys

CuAl alloy

The addition of iron and manganese to aluminium bronze with approx. 7% aluminium improves the mechanical properties; however, it slightly decreases the corrosion resistance. The addition of tin can largely compensate this. In the study [41], comparative corrosion tests were carried out in HCl, H_2SO_4, NaCl and seawater on an alloy with the percentage composition 89.5 Cu, 7.0 Al, 2.0 Sn, 1.0 Fe, 1.0 Mn and on a corresponding alloy without tin. The corrosion rate was determined by the rates of mass loss and by means of the current density/potential curves. In 5 and 10 % solutions of the investigated media at 289 and 353 K (16 °C and 80 °C), the corrosion rate was approximately halved by the addition of 2 % Sn. The heat-treated state of the material, which affects the microstructure, also had a slight influence (Table 23).

	5 % HCl	10 % HCl	2 % NaCl	5 % NaCl	10 % NaCl	5 % H_2SO_4	10 % H_2SO_4	Seawater
Initial state	0.254 (10)	0.203 (8.0)	0.086 (3.4)	0.172 (6.8)	0.183 (7.2)	0.058 (2.3)	0.107 (4.2)	0.076 (3.0)
Annealed*	0.205 (8.1)	0.178 (7.0)	0.078 (3.1)	0.147 (5.8)	0.178 (7.0)	0.056 (2.2)	0.119 (4.7)	0.089 (3.5)
Alloy without added tin **	0.226 (8.9)	0.064 (2.5)	0.203 (8.0)	0.305 (12)	0.366 (14.4)	0.147 (5.8)	0.203 (8.0)	0.127 (5.0)

* slowly cooled from 1223 K (950 °C); ** initial state

Table 23: Corrosion rate of aluminium bronze in mm/a (mpy) at 289 K (16 °C) [41]

Aluminium bronze

The study [42] investigated how the corrosion behavior of an aluminium bronze (type B-150) with the base composition 91.2 Cu, 6.9 Al, 1.8 Fe changes when small amounts of Ta, La and Nd are added. The results of the mass losses in 1 M HCl at 303 K (30 °C) are summarised with respect to the test duration in Table 24. A low addition of 0.1 % Ta and La and of 0.05 % Nd improved the corrosion resistance by approx. 60 %. A higher addition of Ta and La decreased the favorable effect.

In H_2SO_4 and HNO_3, the addition of this element only gave a much smaller improvement in the corrosion resistance.

The same authors as in [41] investigated in [43] the effect of the addition of the rare earth metals La, Ce and Nd to an aluminium bronze of the base composition 90.1 Cu, 9.5 Al and 0.2 Mn in air-saturated 1 M HCl at 303 K (30 °C). The result is summarised in Table 25. As in [41], the lowest concentrations of additives gave the greatest improvement of the corrosion resistance. Compared to the results in [41], the small improvement of the corrosion resistance by the addition of 0.05 % Nd stands out. The addition of 0.1 % Nd even decreased the corrosion resistance.

Amount added Mass%	Testing time, h						
	24	48	72	96	120	144	168
	Mass loss rate, g/m^{2}d						
Basic material B-150	5.90	6.35	6.07	5.90	5.62	4.92	4.27
Ta 0.1	2.30	2.75	2.43	2.05	1.74	1.90	1.88
	(61)	(57)	(60)	(65)	(69)	(61)	(56)
0.2	4.80	4.40	3.64	3.83	3.09	2.90	2.60
	(19)	(31)	(40)	(35)	(45)	(41)	(39)
La 0.1	3.90	3.25	2.30	2.25	1.90	2.02	2.08
	(34)	(49)	(62)	(62)	(66)	(59)	(51)
0.25	4.40	3.50	2.80	2.87	2.58	2.42	2.33
	(25)	(45)	(54)	(51)	(54)	(51)	(46)
Nd 0.05	5.90	3.20	2.27	1.87	1.54	1.63	1.64
	(49)	(62)	(68)	(73)	(67)	(61)	

Numbers in brackets denote an improvement of the corrosion resistance, %

Table 24: Influence of the elements Ta, La and Nd on the mass loss rates of aluminium bronze of type B-150 in 1 M HCl at 303 K (30 °C) [42]

Amount added Mass%	Testing time, h				
	24	48	72	96	120
	Mass loss rate, g/m^{2}d				
Base alloy	3.30 ± 0.12	1.85 ± 0.07	1.30 ± 0.04	1.20 ± 0.02	1.66 ± 0.04
La 0.05	2.50 ± 0.10	1.05 ± 0.05	0.70 ± 0.03	0.85 ± 0.02	0.98 ± 0.02
	(24)	(43)	(46)	(29)	(41)
La 0.15	2.10 ± 0.08	1.45 ± 0.04	0.96 ± 0.04	1.12 ± 0.01	1.22 ± 0.02
	(36)	(22)	(26)	(6)	(26)
La 0.2	2.70 ± 0.10	1.50 ± 0.04	0.83 ± 0.02	0.75 ± 0.02	1.20 ± 0.04
	(18)	(19)	(36)	(37)	(28)
Ce 0.2	3.0 ± 0.07	1.50 ± 0.07	1.0 ± 0.05	0.75 ± 0.02	0.94 ± 0.04
	(9)	(19)	(23)	(37)	(43)
Ce 0.3	3.0 ± 0.06	1.75 ± 0.02	1.33 ± 0.04	1.10 ± 0.02	1.26 ± 0.02
	(9)	(5)	(–)	(8)	(24)
Nd 0.05	3.40 ± 0.10	1.70 ± 0.03	1.17 ± 0.02	0.97 ± 0.03	1.36 ± 0.03
	(–)	(8)	(10)	(19)	(18)
Nd 0.1	3.50 ± 0.11	2.15 ± 0.10	1.60 ± 0.04	1.57 ± 0.02	2.08 ± 0.05
	(–)	(–)	(–)	(–)	(–)

Numbers in brackets denote the improvement of the corrosion resistance, %

Table 25: Results of the investigations of the mass loss rates in air-saturated 1 M HCl, 303 K (30 °C) [43]

The change of the mass losses of propeller bronze of the base composition 82 Cu, 9 Al, 4 Fe, 4 Ni, 1 Mn by the addition of Ta and Nd in 4 % HCl at 303 K (30 °C) is given in Table 26. As the concentrations of the additives increased, the corrosion rate also increased compared to that of the base alloy [44]. However, in a later investigation by the same authors on the effect of addition of Ta and Nd to the base alloy 91.3 Cu, 6.9 Al, 1.8 Fe (type B-150) in 1M HCl at 303 K (30 °C), up to 60 % lower corrosion rates were obtained compared to the base alloy [42].

Amount added	Testing time, h						
	24	48	72	96	120	144	168
Mass%	Mass loss mg/cm^2						
Base material	0.36	0.68	1.05	1.57	2.40	3.43	4.61
Tantalum							
0.10	0.30	0.65	0.78	1.50	2.08	5.38	8.23
0.20	0.37	0.50	0.74	1.14	5.46	8.26	12.64
0.30	0.32	0.69	1.48	2.23	4.20	6.15	9.24
Neodymium							
0.05	0.19	0.70	0.90	1.40	2.34	3.84	6.05
0.10	0.30	0.80	1.22	2.06	3.42	9.32	15.72
0.15	0.47	0.95	1.55	2.70	5.12	10.8	16.51

Table 26: Influence of added tantalum and neodymium on the mass loss of propeller bronze in 4 % HCl at 303 K (30 °C) [44]

A 9 Copper-nickel alloys

The attack of CuNi alloys with 10 to 30 % Ni in hydrochloric acid is moderate to strong. The presence of oxidising substances noticeably enhances the attack.

In multistage evaporator systems to produce drinking water from seawater, one of the main problems is the formation of deposits in the condenser pipes, which are frequently made of CuNi 70/30. These deposits must be periodically removed to avoid a reduction of the heat transfer and flow rate. The best method is to subject the system to acid washing. For this, seawater acidified with HCl with a controlled pH is circulated through the system. The dissolution of $CaCO_3$ and $Mg(OH)$, the main components of the deposits, consumes HCl. This is replaced by adding more HCl. During acid washing of a seawater flash distillation plant in Abu-Dhabi in the United Arab Emirates, the metallic components of the plant were protected from corrosion by the acidic wash solution by the addition of dibutylthiourea as an inhibitor [45]. A number of different metallic materials are used in such a plant, such as titanium, stainless steels, CuNi alloys, bronze, aluminium bronze as well as large quantities of steel. Because no exact knowledge of the effectivity of the inhibitor was available, investigations were carried out with the main materials used in the plant, namely CuNi 70/30 and carbon steel. The following materials were available for the investigation:

CuNi 70/30: 66.47 % Cu, 29.54 % Ni, 1.92 % Fe

Steel: 0.21 % C, 0.10 % Si, 0.35 % Mn, 0.025 % P, 0.014 % S, 0.02 % Cr, 0.27 % Ni

The mass losses were determined in pure HCl solutions at pH 1.8 to 2.0 and in a correspondingly acidic HCl solution made from seawater from the Gulf of Arabia that had a total salt content of 55,000 ppm. The inhibitor was added in concentrations of 20 to 300 ppm.

Figure 31a gives the chronological progression of the mass loss depending on the temperature in pure hydrochloric acid. After a certain incubation time, which decreases with increasing temperature, the linear removal of material starts, the rate of which becomes increasingly faster as the temperature rises. Figure 31b gives the chronological progression of the mass loss rates at 333 K (60 °C), which is the usual temperature for acid washing, in dependence on the inhibitor concentration. For the curves marked with an "A", the same solution was used for the whole test period. However, for the curves marked with a "B", a new solution was used each time the mass was determined. The different shapes of the curves for the two methods show that the inhibitor used was consumed in a relatively short time and thus became ineffective (A curves). However, if the concentration of additive was maintained at a level of at least 200 ppm, then the inhibition efficiency was sufficient over a long period.

Figure 32 gives the corresponding inhibition curves in HCl solution with seawater from the Gulf. Under these conditions with an additional amount of 30,000 ppm Cl⁻ ions, that originate from the seawater, the inhibition efficiency is low and is effective at higher concentrations only after longer periods. Because acid washing is carried out within a period of 6 to 8 h, the suitability of dibutylthiourea as the inhibitor for CuNi 70/30 is questionable.

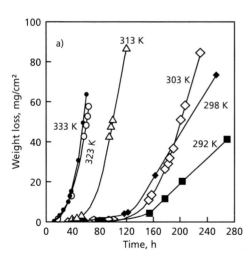 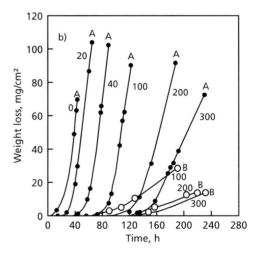

Figure 31: Time-dependent mass loss for CuNi 70/30 immersed in pure HCl at pH 1.8-2.0:
a) at temperatures between 292 and 333 K (19 and 60 °C);
b) at 333 K (60 °C) with a specified concentration of dibutylthiourea (numbers on the curves [ppm]) after mass determination in A: the same solution and B: immersed in fresh solution [45]

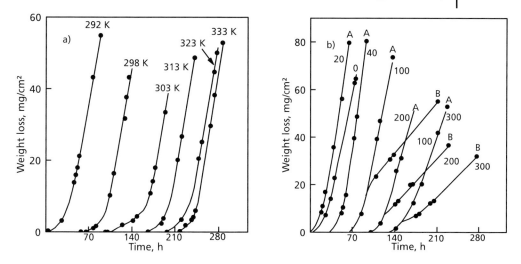

Figure 32: Time-dependent mass losses for CuNi 70/30 immersed in seawater acidified with HCl to a pH of 1.8-2.0
a) at temperatures between 292 and 333 K (19 and 60 °C);
b) at 333 K (60 °C) with a specified concentration of dibutylthiourea (numbers on the curves [ppm]) after mass determination in A: the same solution and B: fresh solution [45]

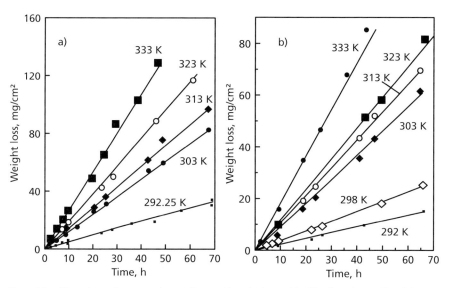

Figure 33: Time-dependent mass losses for steel in solutions with pH values from 1.8 to 2.0
a) pure HCl;
b) in seawater acidified with HCl
(numbers on the curves correspond to the individual temperature) [45]

The time-dependence of the mass losses for steel are shown in Figure 33 and Figure 34. From the start of immersion of the samples in the solution there is a linear time-dependence of the mass losses. The corrosion rates in the pure HCl solution are greater than those in the acid of the same strength but made up with seawater from the Gulf. The inhibition efficiency in the pure acid determined using method A at concentrations from 100 to 300 ppm is 85 %, on average, and in the acidic Gulf water solution it is 70 %. Method B shows almost complete inhibition in both solutions. For steel, dibutylthiourea is thus a serviceable inhibitor for acid washing of the seawater desalination plant.

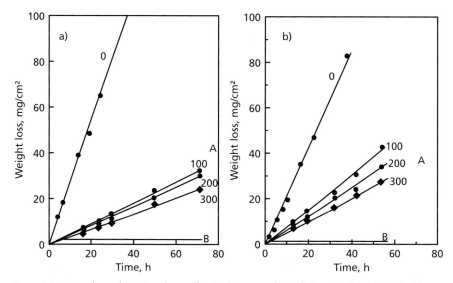

Figure 34: Time-dependent mass losses for steel immersed in solutions at 333 K (60 °C) with pre-specified concentrations of dibutylthiourea (numbers on the curves [ppm]) and pH values of 1.8 to 2.0 according to the mass determination in A: the same solution and B: in fresh solution [45]
a) pure HCl;
b) with seawater acidified with HCl

A 10 Copper-tin alloys (bronzes)
A 11 Copper-tin-zinc alloys (red brass)

The copper-tin alloys and copper-tin-zinc alloys have only low resistance in hydrochloric acid and are not suitable for continuous exposure.

A 12 Copper-zinc alloys (brass)

Hydrochloric acid and sulphamic acid are used to remove salt deposits in heat exchangers made of CuZn alloys (brass). Sulphamic acid is easy to handle and dissolves carbonate and phosphate deposits well. However, it has been shown that in the presence of oxidising Cu^{2+} ions, which are enriched during the pickling process, hydrogen and oxygen are absorbed by brass and have a detrimental effect on the behavior of the material.

The study in [46] investigated the corrosion of α–brass (CuZn37, CW508L, old DIN-Mat. No. 2.0321) in 5 % HCl and sulphamic acid with additions of $CuCl_2$ and $CuSO_4$ at ambient temperature. Before and after the corrosion tests, the oxygen and hydrogen contents in the material were determined by hot extraction. The results are given in Table 27. The mass loss rates, which are also given, are approx. 10 times greater in hydrochloric acid than in sulphamic acid. (Information on the duration of the tests is not given.) However, with regard to the uptake of the gases, the two acids behave conversely. Thus the oxygen content is 15 to 35 times greater in hydrochloric acid and 25 to 55 times greater in sulphamic acid compared to the initial content of the brass samples. The uptake of the gas destroys the intergranular bonding and thus impairs the mechanical behavior. In the initial state, the samples can be bent backwards and forwards by 180° for 14 to 15 times until they fracture. After the corrosions tests, the samples immersed in hydrochloric acid fracture after 4 to 5 bendings and in sulphamic acid after only 1 to 2 bendings.

	Mass loss rate $g/m^2\ h$	[O]	[H]
		Mass%	
Brass sample (initial value)	–	0.003 – 0.009	0.0002 – 0.0003
Brass sample after testing in HCl + 13 g/l $CuCl_2 \times 2\ H_2O$	34.4	0.082 – 0.117	0.0008 – 0.0012
Brass sample after testing in SA + 20 g/l $CuSO_4 \times 5\ H_2O$	3.2	0.140 – 0.236	0.0020 – 0.0052

Table 27: Oxygen and hydrogen content in the brass samples before and after the corrosion tests in 5 % HCl and sulphamic acid (SA) solutions in the presence of divalent copper ions [46]

The reason for the impairment of the mechanical properties is the formation of compounds that contain oxygen, hydrogen and sulphur. The formation of Cu_2O is known; it forms a galvanic element with copper, in which the Cu_2O acts as a cathode. This leads to destruction of the material. Cu_2O also reacts with hydrogen, which is formed by acid corrosion and penetrates into the material, according to the equation

Eq. 4 $\quad Cu_2O + 2\ H \rightarrow 2\ Cu + H_2O.$

This process also leads to destruction of the material. Since these processes are favored in sulphamic acid, hydrochloric acid should be preferred over sulphamic acid for acid washing of brass components in spite of the higher mass loss rates. Unfortunately, the report does not provide information on the duration of the tests.

Brass grades with high zinc contents are selectively corroded by hydrochloric acid with dezincing.

A 13 Other copper alloys

Silicon bronze

Silicon bronzes with 1% Si (e.g. UNS C64900) or 3% Si (e.g. UNS C65500, CW116C formerly referred to as CuSi3Mn, DIN-Mat. No. 2.1525) have a superior resistance in hydrochloric acid than the other copper-based alloys because they form a silicon-containing protective layer. At room temperature, these bronzes are regarded as sufficiently resistant to hydrochloric acid in concentrations of up to approx. 20%.

A 14 Unalloyed and low-alloyed steels/cast steel

Iron reacts with hydrochloric acid to produce hydrogen according to the reaction

Eq. 5 $Fe + 2\,HCl \rightarrow FeCl_2 + H_2$

Depending on the concentration and temperature of the acid, the removal rates lie between 2 and 20 mm/a (78 and 787 mpy). As the content of iron chloride in the solution increases the corrosion rate decreases; however, hydrochloric acid cannot be used without a suitable inhibitor to clean the iron and steel surfaces. Only after effective inhibitors had been developed could hydrochloric acid be used to pickle steel. With more advances in the development of inhibitors and pickling bath monitoring, hydrochloric acid pickling baths, which dissolve the rolling and annealing scale on steel much better than sulphuric acid, have increasingly displaced the previously used sulphuric acid pickling baths in recent years.

The following requirements are specified for the inhibitors:

- good solubility in hydrochloric acid
- simple dosability
- stability at the application temperature over longer operating periods
- insensitive to hydrogen
- reduction of hydrogen absorption into steel
- no reaction with the iron(II) and iron(III) ions produced during pickling
- no significant lengthening of the pickling time
- simple rinsing off from the surface
- no problems in the further processing of the pickled material.

In addition, they should, of course, be as inexpensive as possible.

Most of the inhibitors used in hydrochloric acid pickling baths for steel and which have been tested in practice are nitrogen-containing organic compounds [7, 8]. They are mostly based on the following substances:

- acetylene derivatives
- pyridine derivatives
- urotropine and its derivatives
- thiourea and its derivatives

- quaternary ammonium compounds
- alkylamines and arylamines
- aldehydes (e.g. formaldehyde)
- ketones (e.g. cyclohexanone).

Acid inhibition is a classical field of corrosion protection engineering and therefore has been extensively investigated in the older literature [47].

In recent literature, numerous reports have been published of investigations and developments of new compounds that are suitable as inhibitors for the reaction of unalloyed steels in hydrochloric acid [48, 49, 50, 51, 52]. Attempts are often made to correlate the relationship between the effect and the structure of the compound.

Indeed, it is possible that the substances have differing efficiencies in hydrochloric acid and sulphuric acid [53, 54, 55]. There are even cases in which they act as an effective inhibitor against metal corrosion and against hydrogen absorption by the steel in hydrochloric acid, but are less effective or even promote corrosion in sulphuric acid. An example of this is given in Table 28 [55].

Inhibitor concentration mmol/l	Inhibition efficiency %			
	1 M HCl		0.5 M H$_2$SO$_4$	
	Mass loss	Gas evolution	Mass loss	Gas evolution
ortho-Anisidine				
20	62.7	62.1	21.5	22.9
50	68.6	70.2	39.6	38.1
75	80.2	80.3	46.9	50.9
100	87.3	85.9	66.9	61.8
meta-Anisidine				
20	70.5	71.2	−21.9	−10.0
50	81.1	80.8	3.1	1.4
75	86.6	84.3	24.2	38.1
100	90.0	89.9	58.1	63.6
para-Anisidine				
20	84.5	81.8	−51.5	−39.0
50	91.4	90.4	−45.8	−33.1
75	95.0	93.9	−37.7	−16.1

Table 28: Inhibition efficiency of anisidine isomers on the corrosion of unalloyed steel in hydrochloric acid and in sulphuric acid [55]

For practical applications, tested products of well-known manufacturers should be used because they are based on many years of experience and competent customer services can be expected.

The study in [56] reports in detail on the removal of deposits and corrosion products on pipelines and plant components. In chemical cleaning, in addition to hydrofluoric acid, hydrochloric acid, in particular, plays a decisive role because it dissolves many deposits that originate from water-carrying systems, e.g. boiler scale, rust or magnetite, but not silicic acid or silicates. Figure 35 and Figure 36 show the dissolution rates of magnetite and natural rust on unalloyed steel in solutions of hydrochloric acid compared to other acid solutions. After hydrofluoric acid, hydrochloric acid is the most effective. However, it is not recommended to use hydrochloric acid to clean stainless steel components, even in combination with effective corrosion inhibitors, because there is a risk of local corrosion due to residual chloride ions.

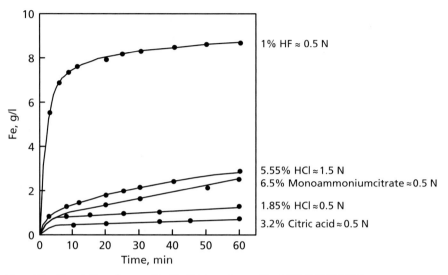

Figure 35: Dissolution of magnetite (Fe_2O_3) in various acids at 343 K (70 °C) [56]

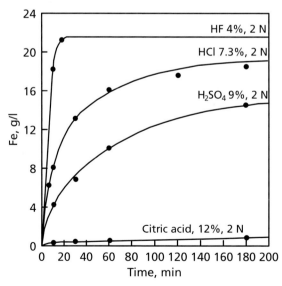

Figure 36: Dissolution of natural rust in various acids at 333 K (60 °C) [56]

The hydrolysis of chlorinated hydrocarbons is already catalysed at room temperature by rusting steel and by rust. The thus liberated hydrochloric acid is very corrosive [57]. If the action of external energy (heat, light) is excluded, a good hydrolysis resistance can be generally assumed. This was confirmed by tests with 50 ml H_2O and chlorinated hydrocarbons (CCl_4, $CHCl_3$, CH_2Cl_2, C_2HCl_3, C_2Cl_4) at 323 K (50 °C) stored in the dark in glass-stoppered flasks. After one week, the aqueous phase had the same pH and Cl^- ions could not be detected. When strips of sheet (35 × 70 × 1 mm from St 34) were added, then after 2 days there was strong rust formation with noticeable hydrolysis and formation of hydrochloric acid. Table 29 gives the corresponding corrosion rates.

Medium	g/m^2	mm/a (mpy)
CCl_4	190	4.4 (173)
$CHCl_3$	24.2	0.56 (22)
CH_2Cl_2	18.6	0.43 (17)
C_2HCl_3	35.0	0.82 (32)
C_2Cl_4	9.3	0.22 (8.7)

Table 29: Corrosion rates of unalloyed steel as a result of hydrolytically produced HCl in various chlorinated hydrocarbons [57]

These corrosion rates are much too high for normal rusting in H_2O.

The hydrolytic formation of hydrochloric acid in aqueous chlorinated hydrocarbons in the presence of unalloyed steel is also confirmed in [58]. In a two-component mixture of carbon tetrachloride and water in a 5 : 2 ratio, the corrosion rates of samples of unalloyed steel DIN-Mat. No. 1.0402 (UNS G10200, C 22) were measured at the boiling point in darkened rooms for test periods of 8-30 days in the three phases listed in Table 30.

Organic phase	0.7 mm/a (27.5 mpy)
Aqueous phase	9.0 mm/a (354 mpy)
Vapor phase	9.0 mm/a (354 mpy)

Table 30: Corrosion rates of steel C 22 in the two-component system CCl_4/H_2O at the boiling point [58]

These results were obtained by means of a radionuclide method in which the mass loss was calculated from the activity increase in the corrosion media caused by the corrosion products from a reactor-activated material sample.

The authors of [59] investigated the problem of stress corrosion cracking of high-strength carbon steels in hydrochloric acid solutions. The compositions and tensile strengths of the investigated steels are given in Table 31.

Hydrochloric Acid

Steel	C %	Mn %	Si %	P %	S %	Cr %	Mo %	V %	R_m MPa
J-55	0.42	1.07	0.37	0.029	0.019	0.13	0.01	–	593
C-75	0.37	1.48	0.29	0.014	0.016	0.14	0.07	–	753
N-80 DIN-Mat. No. 1.0564	0.33	1.34	0.29	0.011	0.013	0.16	0.05	–	746
P-105 DIN-Mat. No. 1.0670	0.40	1.59	0.33	0.012	0.012	0.57	0.20	0.09	936

Table 31: Chemical compositions and tensile strengths (R_m) of the investigated high-strength steels [59]

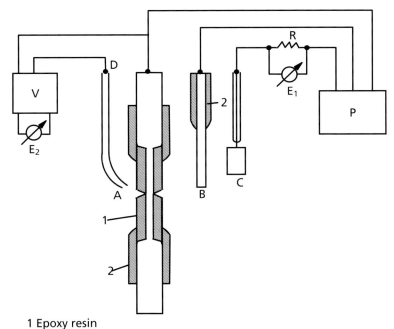

1 Epoxy resin
2 Teflon sealing tape

Figure 37: Diagram of the set-up of the equipment for electrochemically controlled SCC testing [59]
A – strained microtension specimen
B – unloaded macroelectrode
C – Pt counterelectrode
D – SCE reference electrode
R – resistance
E_1 and E_2 – recorders
V – voltmeter
P – potentiostat

The investigations were carried out on round tensile samples with a diameter of 3 mm in the test direction that could be polarised in an electrolysis cell and subjected to tensile loading (Figure 37). The incubation time, crack advancing speed and fracture time were derived from the course of the current density/potential curves. As expected, the fracture time decreased with increasing voltage, temperature and acid concentration. The fracture time also depended in a characteristic manner on the potential of the samples (Figure 38). This dependency results from the interaction of iron dissolution at the crack tip in the anodic potential range and on hydrogen-induced crack formation in the cathodic potential range. The influence of anodic dissolution was the major factor.

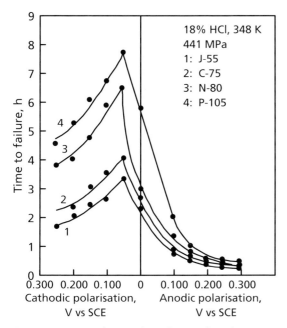

Figure 38: Time to fracture depending on the polarisation in 18% HCl at 348 K (75 °C) and 441 MPa [59]

In reference [60], the influence of molybdenum additions of up to 1% on the corrosion behavior of highly pure carbon steels with up to 0.4% C was investigated in 5, 10, 15 and 20% HCl at ambient temperature. The current density/potential curves of the samples were recorded. The Stern-Geary equation

Eq. 6 $\quad i_{cor} = \dfrac{\beta_a \beta_c}{\beta_a + \beta_c} \cdot \dfrac{1}{2,3 \cdot R_p}$

where β_a and β_c are the anodic and cathodic Tafel constants, R_p is the slope of the polarisation resistance lines and i_{cor} is the corrosion current density, was used to calculate the surface-related removal rate in mpy (1 mpy = 0.0254 mm/a).

The electrochemical principles are explained in [21] in an understandable manner.

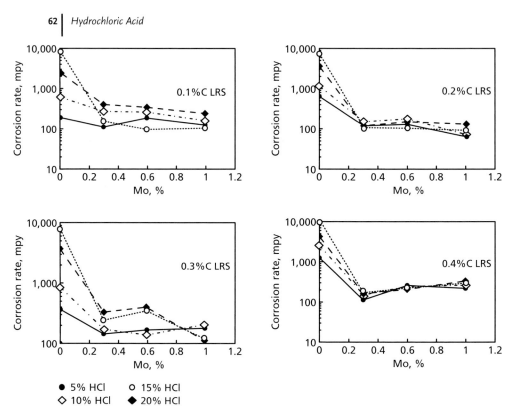

Figure 39: Corrosion rate of steels with differing C contents depending on the Mo content in various HCl solutions (1 mpy = 0.0254 mm/a) [60]

Figure 39 gives the corrosion rate of the steels with differing carbon contents depending on the molybdenum content in various HCl solutions. Independent of the C content, the corrosion rate deceased by up to approx. 0.3 % Mo as the acid concentration increased. It attained values of up to two powers of ten. No statement is made concerning the extent to which this favorable effect was influenced by Si, Mn, P, S etc. within the usual limits.

Corrosion protection in oil refineries

The corrosion of condensers, columns and connecting systems by overhead streams from the distillation of crude oil is a major problem in refineries.

Reference [61] discusses corrosion in the head region of fractionating columns used for oil distillation. The main cause is the generation of hydrochloric acid by hydrolysis of $MgCl_2$ and $CaCl_2$

Eq. 7 $MgCl_2 + H_2O \rightarrow 2\ HCl + MgO$

Eq. 8 $CaCl_2 + H_2O \rightarrow 2\ HCl + CaO$

and by absorption of HCl in the condensation water in the head region. In the region of the preheater, where the hydrolysis occurs, corrosion is low because water is not present as a liquid. The NaCl present in large amounts in crude oil is stable and only hydrolyses to a small extent in the preheater stage.

Because of the importance of this problem in oil refineries, the NACE (National Association of Corrosion Engineers) questioned many operators with regard to this problem with the aim of finding out how to deal with this type of corrosion. Possible measures are desalting of the crude oil and neutralisation with sodium hydroxide or sodium carbonate in the preheater circuit. Furthermore, ammonia and neutralising amines can be added to the heads of the columns to increase the pH value from less than 1 to 6 – 7. Because the neutralisation products are often solids that can also be corrosive and cause blockages, they are flushed out with water. Amine-based inhibitors are also added in the head region. These inhibitors are more effective in neutral environments than in acidic ones. Therefore, they are usually added together with neutralisers.

There are particularly aggressive corrosion conditions in the head region of the fractionating columns used to distil crude oil. The nitrogen and sulphur compounds present in crude oil decompose at high temperatures to form corrosive mercaptans and hydrogen sulphide. Any chlorides present are hydrolysed to hydrogen chloride. Hydrogen sulphide and hydrogen chloride tend to become enriched in the head region of the fractionating columns together with hydrocarbons and heated steam. Therefore, in the head region, corrosion occurs wherever water can condense. This preferentially absorbs the HCl arriving in the head region to form hydrochloric acid. The reaction is additionally complicated by the presence of H_2S in the head region where iron sulphide is deposited from the reaction between H_2S and soluble iron chloride, which is produced by the corrosion of the steel construction by HCl.

Nitrogen compounds, such as quaternary ammonium compounds, amines, amino salts and heterocyclic compounds, are used as corrosion inhibitors in the oil industry.

On the basis of several cases of damage in the head region of such crude oil processing plants, it is demonstrated in [62] that water-soluble inhibitors can be used to avoid these corrosion problems. These water-soluble inhibitors are more favorable than oil-soluble ones because if the temperature drops below the dew point of water, they are directly active in the critical HCl-containing condensates. Inhibitors based on modified fatty acids have also proved suitable in this field [63, 64].

The work in [65] aimed to investigate the behavior of 3-amino-1,2,4-triazole (ATR), 2-amino-thiazole (ATH) and 2,6-diamino-pyridine (DAP) as corrosion inhibitors for carbon steel exposed to a kerosine-water mixture (10 vol% water) with 3 ppm HCl and 800 ppm H_2S/per day at a pH value of 6 to 6.5 and a temperature of 328 to 333 K (55 to 60 °C) using ammonia as a neutralising agent. The three heterocyclic compounds were selected on the basis of their thermal stability up to approx. 573 K (300 °C), which facilitates the formation of a coating on the metal and the possibility of forming stable complexes with the surface of the carbon steel.

The three substances were used in the technically pure state (> 95 %) without purification. They are soluble in aromatic and aliphatic hydrocarbons and partly soluble in water. In the amounts used, there is no significant emulsification.

The investigated steel had the composition: 0.2 % C, 0.7 % Mn, 0.04 % P, 0.04 % S. Table 32 gives the results after a test period of 50 h. For an addition of 10 ppm of all three substances, the highest inhibition efficiency was 82.7-88.2 %. Higher additions lowered the inhibition efficiency.

The effectiveness of the organic compounds (ATR, ATH, DAP) was also investigated using air as the carrier gas because some commercially available inhibitors are also recommended for systems that contain oxygen in addition to H_2S, HCl and cyanides. For additions of 10 ppm and a test duration of 100 h, the inhibition efficiencies were 60.3 % for ATR, 55.8 % for ATH and 50.7 % for DAP.

Reference [65] gives the results for additions of 10 ppm after various test periods with nitrogen as the carrier gas along with ammonia as the neutralising agent and H_2S. Furthermore, octylamine was added in order to test the behavior of the three compounds in the presence of amines. Octylamine has a high boiling point compared to other aliphatic amines that are sometimes used to control the pH value.

The results show that the addition of octylamine alone gave a low inhibition efficiency, particularly after a test duration of 100 h. Without the addition of octylamine, the three investigated substances exhibited a high inhibition efficiency of approx. 90 % after 50 h, which dropped to approx. 70 % after 400 h. On adding 10 ppm octylamine, the values after an exposure time of 100 h were lower than these. The best inhibition efficiency was exhibited by ATH. The investigation showed that the three substances formed Fe(II) and Fe(III) complexes that form protective layers on the steel surface. They are suitable as inhibitors for fractionating columns in the oil industry.

It is known that the combustion of polyvinylchloride (PVC) produces hydrogen chloride which condenses together with water vapor as hydrochloric acid that can lead to strong corrosion of metallic objects. In reference [66], it is shown that the thermal decomposition of freones (hydrofluorocarbons, HFC), which can also contain chlorine, and also of halones (halogenated hydrocarbons), which in addition to fluorine also contain chlorine and particularly bromine, can lead to the formation of not only hydrochloric acid but also hydrofluoric and hydrobromic acids. Iodine-containing HFCs can also lead to the formation of hydroiodic acid.

These compounds are used as extinguishing agents for fires. A massive dosage of halons in the event of a fire produces a cloud of halogen halide gas with a relatively high density around the source of the fire. This blocks off any oxygen from reaching the fire so that it stagnates after a short period. The halogen halide gas liberated from the extinguishing agents are deposited on metal parts, such as iron, and can thus attack them to form the corresponding iron halogen salt. Any further processes depend on the relative humidity. Figure 40 gives the results of laboratory tests in which the surface of unalloyed steel was exposed for 1 hour to the vapor of concentrated halogen halide followed by 95 hours exposure to an acid-free atmosphere at various levels of relative humidity. Depending on the type of acid, there was a more or less strong corrosion for the different relative humidities. The corrosive attack

Test	Additive	3-Amino-1,2,4-triazole		2-Amino-thiazole		2,6-Diamino-pyridine	
		Corrosion rate mm/a (mpy)	Inhibitor efficiency %	Corrosion rate mm/a (mpy)	Inhibitor efficiency %	Corrosion rate mm/a (mpy)	Inhibitor efficiency %
1	None	0.295 (11.5)	–	0.295 (11.5)	–	0.295 (11.61)	–
2	Ammonia	0.113 (4.5)	–	0.113 (4.5)	–	0.113 (4.45)	–
3	Ammonia + 10 ppm inhibitor	0.017 (0.7)	84.7	0.013 (0.51)	88.2	0.020 (0.79)	82.7
4	Ammonia + 20 ppm inhibitor	0.018 (0.71)	84.0	0.025 (0.98)	77.6	0.030 (1.18)	73.1
5	Ammonia + 30 ppm inhibitor	0.033 (1.3)	70.9	0.024 (0.94)	79.1	0.051 (2.01)	55.1
6	Ammonia + 40 ppm inhibitor	0.048 (1.9)	57.1	0.029 (1.14)	74.7	0.058 (2.28)	48.7

The calculation of the inhibitor efficiency in tests 3-6 is based on the results from test 2

Table 32: Corrosion rates and inhibition efficiencies for steel in the presence of 3-amino-1,2,4-triazole, 2-amino-thiazole and 2,6-diamino-pyridine after an exposure time of 50 h [65]

was the strongest for hydrochloric acid and hydrobromic acid. In comparison, hydrofluoric acid gives a corrosion maximum only at 90 % relative humidity, which is significantly lower than the corrosion values for HCl and HBr.

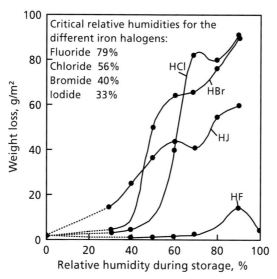

Figure 40: Corrosion experiment with unalloyed steel in an acidic atmosphere. Firstly infection for one hour in acidic vapors above solutions of the concentrated acid. Then storage for 95 hours in an acid-free atmosphere at various relative humidities [66]

The work in [67] describes chemical cleaning without interrupting operations (on-stream cleaning) in heat-exchangers and other plant sections in an oil refinery. This method, introduced at the start of the 1960s, injects inhibited concentrated hydrochloric acid into the flow of cooling water in sufficient quantities to obtain an acid concentration of 5 to 10 % at the outlet. The exact procedure used in nine cases and the corresponding improvements in the results obtained with this type of cleaning are described. Apart from savings of heating oil costs, advantages include lower operating downtimes and lower maintenance costs.

No information is given on the inhibitor used. However, it is mentioned that hydrochloric acid should not be used for equipment made of aluminium and stainless steels. Therefore, it can be concluded that the inhibitor used is only effective for carbon steel.

A 15 Unalloyed cast iron and low-alloy cast iron

The behavior of cast iron in hydrochloric acid can be compared to that of unalloyed steels, so that the information given in A 14 is also largely applicable to these materials.

A 16 High-alloy cast iron
A 16.1 Silicon cast iron

Grades of cast iron alloyed with silicon (approx. 14.5 % Si) with low contents of Mo and Cu have good resistance at room temperature to hydrochloric acid of all concentrations [68].

A 16.2 Austenitic cast iron (and others)

The austenitic, high-nickel cast-iron grades (Ni-Resist®, ASTM A 436) exhibit a certain resistance to very dilute non-aerated hydrochloric acids. However, aerated solutions or higher temperatures lead to strong attack [68].

A 17 Ferritic chromium steels with < 13 % Cr

Constructional steels with up to 10% chromium

Chromium often serves as an alloying additive in carbon steels that are used for heat-exchanger tubes in steam generators. To maintain optimum heat-exchange conditions, the pipes must be regularly cleaned with a dilute acid solution with or without inhibitors.

Although hydrofluoric acid is expensive and difficult to handle, it does have certain advantages in acid washing compared to other acids, particularly with regard to the dissolution of silicate compounds. Because little was known about the influence of chromium on the corrosion behavior of steel in contact with hydrofluoric acid, this was studied in [69] in a comparative investigation with other acids, particularly hydrochloric acid.

The mass losses were determined by solution analysis and from polarisation curves. Table 33 gives the compositions of the investigated steels with increasing chromium and molybdenum contents.

	Corresponding DIN-Mat. No.	C	Si	Mn	P	V	Cr	Mo
Armco iron		0.012	trace	0.017	0.025	0.005	0	0
ASTM A 106	1.0256, 1.0481	0.16	0.17	0.89	0.024	0.027	0	0
ASTM A 335 (P 11)		0.13	0.61	0.51	0.017	0.019	1.35	0.54
ASTM A 335 (P 22)	1.7375, 1.7380	0.13	0.34	0.46	0.021	0.017	2.28	0.93
ASTM A 335 (P 5)	1.7362	0.11	0.33	0.51	0.015	0.020	4.82	0.52
ASTM A 335 (P 9)	1.7386	0.15	0.25 – 1.0	0.3 – 0.6	0.03	0.03	10	0.9 – 1.1
AISI	1.4016	0.12	1.0*	1.0*	0.025	0.03	17	0

Table 33: Composition of the investigated steels, mass% [69]

The acids listed in Table 34 were used in the comparative investigations at 318 K (45 °C). The acid concentrations of 1.7 M HF (3 %) or 1.4 M HCl (5 %) correspond to the concentrations that are generally used for cleaning with these acids.

Table 34 compares the corrosion rates obtained by solution analysis and from the polarisation curves. From the results, which are all in good agreement, it can be concluded that dilute hydrochloric acid is advantageous for the cleaning of heat-exchangers made of chromium steels because of the considerably lower corrosion rate compared to that obtained in hydrofluoric acid. However, it is worthwhile reducing the corrosion rate further by adding a suitable inhibitor (see A 14); this is also what is done in practice.

Figure 41 and Figure 42 give the mass losses depending on the time for hydrofluoric acid and hydrochloric acid. In the case of hydrofluoric acid, the addition of chromium to the steel results in a considerable increase of the mass loss compared to the chromium-free steel. Hydrochloric acid exhibits the reverse tendency:
the mass loss decreases with increasing chromium content. The result from steel ASTM A 355 P22 with 2.28 % Cr is anomalous. The reason for this is probably the relatively high molybdenum content of 0.93 %.

	1.4 M HCl		1.7 M HF	
	■	◆	■	◆
Armco iron	1.2	2.0	1.3	–
ASTM A 106	2.1	1.8	1.8	3
ASTM A 335 (P 11)	0.4	0.3	11	10
ASTM A 335 (P 22)	0.9	1.2	58	40
ASTM A 335 (P 5)	0.3	0.3	> 26	25
ASTM A 335 (P 9)	0.5	0.5	30	30
AISI 430	0.4	–	37	–

■ From solution analysis; ◆ From polarisation curves

Table 34: Corrosion rates of alloys, mg/cm² h [69]

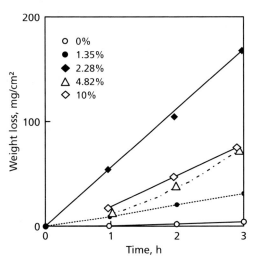

Figure 41: Mass losses of steels with differing chromium contents as a function of the exposure time in 1.7 M HF at 318 K (45 °C) [69]

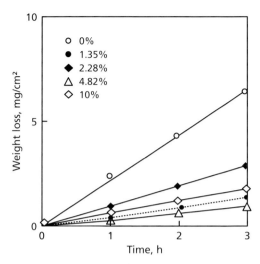

Figure 42: Mass losses of steels with differing chromium contents as a function of the exposure time in 1.4 M HCl at 318 K (45 °C) [69]

Constructional steels with up to 12% chromium

High-strength tempering steels contain not only an increased amount of carbon, but also chromium alone or in combination with nickel as a significant alloying element. Such tempering steels are used for high-strength screws, for example. Because of the material state (high material strength with limited toughness) and because of the operational loading (transfer of high tensioning forces and superimposed operating forces), these materials are susceptible to hydrogen-induced cracking if there is sufficient atomic hydrogen present in the environment. To prevent this, these screws are usually protected from corrosion by a galvanic coating. However, to ensure sufficient adhesion of these coatings, a bare, active metal surface is necessary and this can be obtained by acid pickling. It is known that, in the event of incorrect pickling and possibly during the subsequent galvanic coating, hydrogen is able to penetrate the high-strength screws in such amounts as to produce irreversible cracking.

This problem is studied in detail in [70]. Table 35 lists the investigated steels. Screws with different surface states were pickled in 15 % HCl at room temperature for different periods of time, and without or after increasing exposure times in creep tests for which the sample was pretensioned to 90 % of its 0.2 % yield strength.

Figure 43 shows the influence of the exposure time on the creep behavior of pickled M8 screws made of tempering steel 35B2 (1.5511) after 8 min picking time in 15 % HCl at room temperature. As the exposure time increased, there was a continuous increase in the critical tensile strength (tensile stress) R_{mcrit}, below which there was no longer any delayed hydrogen-induced cracking. After 138 h there was no longer a negative effect as a result of continuous hydrogen effusion.

Hydrochloric Acid

Material	DIN-Mat. No.		C	Si	Mn	P	V	Cr	Ni	Mo	Al	B
35B2	1.5511	standard analysis values	min.	0.32	0.15	0.50	–	–	–	–	–	–	0.005
			max.	0.40	0.40	0.80	0.035	0.035	–	–	–	–	0.008
		actual value	0.37	0.20	0.71	0.019	0.021	0.12	0.03	<0.01	0.05	0.0023
34Cr4	1.7033	standard analysis values	min.	0.30	0.15	0.60	–	–	0.90	–	–	–	–
			max.	0.37	0.40	0.90	0.035	0.035	1.20	–	–	–	–
		actual value	0.34	0.11	0.82	0.025	0.009	0.92	0.14	0.05	0.02	0.0014
34CrMo4 G34CrMo4	1.7220	standard analysis values	min.	0.30	0.15	0.50	–	–	0.90	–	0.15	–	–
			max.	0.37	0.40	0.80	0.035	0.035	1.20	–	0.30	–	–
		actual value	0.34	0.09	0.71	0.026	0.021	1.00	0.10	0.16	0.026	0.0016
30CrNiMo8	1.6580	standard analysis values	min.	0.26	0.15	0.30	–	–	1.80	1.80	0.30	–	–
			max.	0.33	0.40	0.60	0.035	0.035	2.20	2.20	0.50	–	–
		actual value	0.31	0.34	0.45	0.016	0.019	1.96	1.93	0.37	0.02	<0.001

Table 35: Chemical composition (mass%) of the investigated tempering steels [70]

A 17 Ferritic chromium steels with < 13% Cr

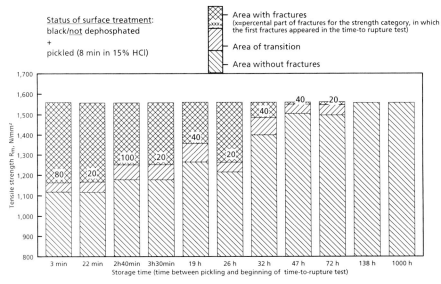

Figure 43: Influence of the exposure time on the creep behavior of pickled M8 screws made of tempering steel 35B2 [70]

The effect of pickling time on the creep behavior of M8 screws made of tempering steel 35B2 is shown in Figure 44. For a pickling time of 1 min, no negative effect of hydrogen can be seen; however, it can be seen above a pickling time of 4 min.

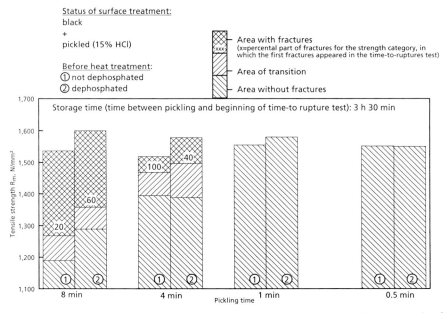

Figure 44: Influence of the pickling time on the creep behavior of pickled M8 screws made of tempering steel 35B2 [70]

To avoid delayed hydrogen-induced crack formation of high-strength screws it is important to keep the pickling time as short as possible and/or to store the screws for a long time before applying a galvanic zinc coating to protect against corrosion because the zinc layer acts as an effusion barrier. However, corrosion protection of high-strength screws is necessary because, depending on the aggressiveness of the surrounding conditions, hydrogen-induced cracking can also occur in rusting screws.

As can be seen from the analysis in Table 35, steel 34Cr3 with 0.009 % S has an extremely low sulphur content. Under otherwise identical test conditions, this steel proved to be the least susceptible to hydrogen-induced cracking. This is attributed to the fact that, for this low content of manganese sulphide, pickling with hydrochloric acid produces hardly any hydrogen sulphide, which is known to act as a promotor for hydrogen absorption into steel.

A 18 Ferritic chromium steels with ≥ 13 % Cr

The work in reference [71] reports on experience gained with 17 % ferritic, stainless chromium steels in an oil refinery. Advantages compared to 18/8-CrNi-steels are the lower thermal expansion, the higher thermal conductivity and the insensitivity to transgranular stress corrosion cracking. A disadvantage is the limited resistance to acidic, chloride-containing corrosive media that occur in some sections of the refinery.

Therefore, it appeared promising to use stainless ferritic steels with higher chromium contents that also contained molybdenum, known as superferrites, that were developed in the 1970s. The behavior of the steels, listed in Table 36, that were used as heat-exchanger tubes in various sections of the oil refining process was investigated.

The evaluation of the practical behavior proved this extended application possibility for the high chromium steels. Because the tendency of embrittlement at 748 K (475 °C) increases as the chromium content increases and at even higher temperatures embrittlement increases due to the precipitation of intermetallic phases, its use is limited to temperatures of up to 700 K (427 °C).

The following tables give the results of comparative laboratory tests with austenitic chromium-nickel steels and ferritic chromium steels. Table 37 shows the excellent resistance to pitting corrosion, and Table 38 shows the resistance to crevice corrosion compared to the austenitic chromium-nickel steels X5CrNi18-10 (AISI 304, DIN-Mat. No. 1.4301) and X5CrNiMo17-12-2 (AISI 316, DIN-Mat. No. 1.4401). As expected, the superferrites exhibited low corrosion rates of < 0.15 mm/a (< 6 mpy) in the Huey test (ASTM A 262 – Practice C).

UNS No.: Element	AISI 430 (S43000)	AISI 439 [1] (S43025)	AISI 444 [2] (S44400)	26-1S [3] (S44626)	E-BRITE® [4] (S44627)	29-4 (S44700)	29-4-2 (S44800)
Carbon	0.07	0.03	0.02	0.02	0.002	0.005	0.005
Nitrogen	0.025	0.020	0.025	0.025	0.010	0.013	0.013
Chromium	17	18	18	26	26	29	29
Molybdenum	–	–	2	1	1	4	4
Nickel	–	0.20	0.20	0.25	0.10	0.10	2
Manganese	0.45	0.35	0.35	0.30	0.05	0.05	0.05
Silicon	0.45	0.30	0.15	0.30	0.25	0.10	0.10
Phosphorus	0.040	0.020	0.020	0.030	0.010	0.015	0.015
Sulphur	0.010	0.010	0.010	0.010	0.010	0.010	0.010
Titanium	–	0.80	0.30	0.50	–	–	–
Niobium	–	–	0.35	–	0.10	–	–

[1] ASTM XM-8
[2] ASTM 18-2
[3] ASTM XM-33
[4] ASTM XM-27

Table 36: Composition of the stainless steels, mass% [71]

Alloy	U_{SCE}, mV		
	pH 10	pH 6	pH 2
AISI 304	+ 40	– 50	– 50
AISI 316	+ 120	+ 10	– 20
ASTM XM-27	+ 400	+ 420	+ 430
29-4	+ 1020	+ 940	+ 880
29-4-2	+ 990	+ 990	+ 860

Table 37: Critical pitting corrosion potentials of the steels in saturated NaCl solution at 38 °C (100 °F) [71]

Alloy	Critical crevice corrosion temperature	
	K	°C
AISI 304	< 270.5	< – 2.5
AISI 316	275.5	2.5
18-2	275.5	2.5
ASTM XM-27	293-298	20-25
Inconel® 625	< 323	< 50
29-4	323	50
29-4-2	323	50
Alloy C	338-347	65-74
Titanium	350	77

Table 38: Resistance of the steels to crevice corrosion [71]

Hydrochloric Acid

Alloy		1% HCl		10% H$_2$SO$_4$	
		mm/a	mpy	mm/a	mpy
AISI 304		81	3,189	400	15,748
AISI 316		71	2,795	22	866
AISI 430		1,500	59,055	6,400	251,969
18-2		850	33,465	2,400	94,488
ASTM XM-27	active range	2,000	78,740	3,400	133,858
	passive range	0.7	28		
29-4	active range	500	19,685	1,300	51,181
	passive range	0.2	8		
29-4-2		0.2	8	0.2	8

Table 39: Corrosion rates in a boiling 1% HCl and 10% H$_2$SO$_4$ solution for austenitic and ferritic steels [71]

Table 39 gives the corrosion rate in a boiling 1% HCl and 10% H$_2$SO$_4$ solution for austenitic and ferritic steels. With the exception of superferrite 29-4-2, high corrosion rates are exhibited by austenitic CrNi steels, and particularly by the ferritic Cr steels. Therefore, they are not suitable for use in reducing mineral acids, particularly at elevated temperatures. However, steel 29-4-2 was still resistant in 1.5% HCl and 2.5% H$_2$SO$_4$ from room temperature to the boiling point.

Furthermore, in crevice corrosion tests in acidic chloride-containing solutions, which occur in certain refinery processes, the superferrite 29-4-2 proved to be very resistant after four years in a condenser.

In summary, the behavior of superferrite 29-4-2 is evaluated as follows: on account of the combination of resistance to stress corrosion cracking, chloride corrosion and acid corrosion, this steel is particularly suitable for demanding conditions in heat-exchangers, such as those found in the main systems of petroleum distillation units.

A 19 Ferritic/pearlitic-martensitic steels

Reference [72] investigated the relationship between heat-treated state and corrosion or erosion-corrosion behavior of stainless tempering steel X17CrNi16-2 (AISI 431, DIN-Mat. No. 1.4057) in solutions with 0.1 M HCl, 0.05 M H$_2$SO$_4$ and 3% NaCl at 295 K (22 °C). The composition of the steel was 0.14% C, 0.36% Si, 0.41% Mn, 0.011% P, 0.01% S, 16.42% Cr, 2.24% Ni. After solution annealing for one hour at 1333 K (1060 °C) and quenching in oil, the samples were each subjected to two-hour tempering treatment between 473 and 1023 K (200 and 750 °C) with air quenching.

The measured Vickers microhardness depending on the heat-treatment for the microstructural components martensite, ferrite and residual austenite are shown in Figure 45. Above approximately 823 K (550 °C) there was a steep drop in hardness as a result of martensite decomposition due to the precipitation of chromium carbide. The conversion of residual austenite at 673 K (400 °C) produced a secondary hardening. Due to the chromium carbide precipitation, there was a depletion of chromium in the vicinity of the precipitation centres that greatly reduced the corrosion resistance. Figure 46 shows the mass loss rates in solutions with 0.1 M HCl and 0.05 M H_2SO_4, which indicates the large dependence of the corrosion behavior on the tempering temperature.

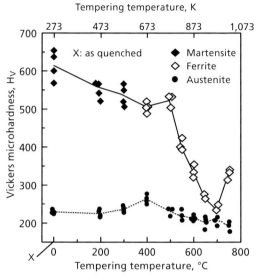

Figure 45: Vickers microhardness depending on the heat treatment for the microstructural components martensite (◆), ferrite (◇) and residual austenite (●) after cooling from 1333 K (1060 °C) (x = without further heat treatment) [72]

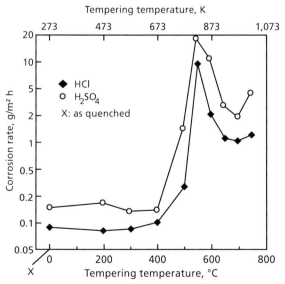

Figure 46: Results of the corrosion tests in 0.05 M H_2SO_4 (○) and 0.1 M HCl (◆) solutions at 295 K (22 °C) after 24 h (x = without further heat treatment) [72]

Figure 47 gives the result of the vibration cavitation test in a 0.1 M HCl solution at a frequency of 20 kHz and an amplitude of 50 μm depending on the test duration. As already observed for the corrosion rate, above approximately 823 K (550 °C) there was a considerable increase in the mass loss which was caused by the combination of a drop in hardness and lowering of the corrosion resistance. A quantitatively similar result was also found in the 0.05 M H_2SO_4 solution. The mass loss in the 3 % NaCl solution was lower by approximately one third.

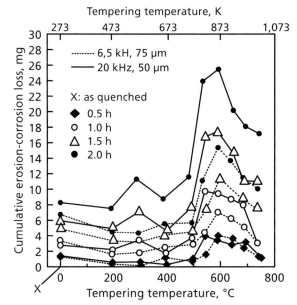

Figure 47: Result of the vibration cavitation test in a 0.1 M HCl solution at a frequency of 20 kHz and an amplitude of 50 μm depending on the test duration after 0.5 (◆), 1.0 (○), 1.5 (△) and 2.0 (●) h (x = without further heat treatment) [72]

Figure 48 shows the mass loss in the vibration cavitation test in the non-corrosive medium ethylene glycol. Because of the secondary hardening caused by the decomposition of residual austenite at 673 K (400 °C), there is a minimum in the mass loss. Above approx. 823 K (550 °C), an increase starts, although it is more than one power of ten less than in the 0.1 M HCl solution. Therefore, there is a strong synergetic effect for erosion corrosion depending on the hardness and corrosion resistance.

The investigated steel contained 0.14 % C, which is somewhat less than the carbon content of 0.15 – 0.23 % specified for X17CrNi16-2 (AISI 431, DIN-Mat. No. 1.4057). The dependence of the corrosion and erosion-corrosion behavior on the heat treatment would have been even more pronounced if the investigated steel had had a higher carbon content.

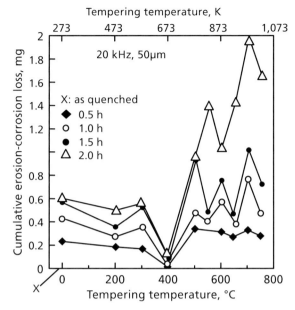

Figure 48: Mass loss in the vibration cavitation test in the non-corrosive medium ethylene glycol after 0.5 (◆), 1.0 (○), 1.5 (●) and 2.0 (△) h [72]

A 20 Ferritic-austenitic steels/duplex steels

Since the beginning of the 1970s, stainless duplex steels based on 25Cr-5Ni have been used as pipes in oil boreholes because of their approximately two-fold higher 0.2 % yield strength compared to stainless austenitic chromium-nickel steels and because of their good resistance to chloride corrosion (pitting, crevice and stress corrosion cracking). In those cases in which hot concentrated inorganic acids, especially hydrochloric acid, are used to increase the oil production, the corrosion resistance of the duplex steel is insufficient. The acid must be inhibited in order to protect the pipes made of this steel from premature failure as a result of corrosion. Since the proven inhibitors for low-alloyed steels are not sufficiently effective for duplex steels, new inhibitors had to be developed [73].

Reference [73] reports on investigations with a steel having the composition 0.03 % C, 0.33 % Si, 1.51 % Mn, 0.013 % P, 0.0045 % S, 23.3 % Cr, 6.0 % Ni, 3.0 % Mo, 0.16 % N (DIN-Mat. No. 1.4462, X2CrNiMoN22-5-3, 2205). The mass loss rates were determined along with the polarisation resistance, which was measured at certain intervals during these tests. The current density/potential curves were recorded in separated tests.

Condensation products of aromatic aldehydes and amines were tested as inhibitors. In order to increase their effectivity, surfactants were added that improved the distribution of the inhibitors and the wettability of the metal surface. Finally, potassium iodide was also tested as an inhibitor. The following pure substances were used:

Aldehydes

trans-cinnamaldehyde (TCA)
benzaldehyde (BAL)
salicylaldehyde (SAL)

Amines

aniline (PHA)
benzylamine (BAM)

Surfactants

N-dodecylpyridine chloride (DDPC)
N-dodecylquinoline bromide (DDQBr)
N-denzylquinoline chloride (BZQC)

Auxiliary inhibitor

potassium iodide (KI)

The listed aldehydes and amines were tested individually as well as in mixtures for their inhibition efficiency. Their reaction products were obtained after refluxing the mixtures for 3 hours.

	Corrosion rate mg/cm^2 h	Inhibition efficiency %
HCl 20 %	144.1	–
+ TCA-PHA	6.55	95.45
+ TCA-BAM	2.65	98.16
+ BAL-PHA	75.7	47.50
+ BAL-BAM	58.8	59.20
+ SAL-PHA	84.2	41.57
+ SAL-BAM	102.8	28.66

Table 40: Corrosion rates of the duplex steel and the inhibition efficiencies for the various combinations of substances with an inhibitor addition level of 0.2 % [73]

	Corrosion rate mg/cm^2 h	Inhibition efficiency %
TCA-PHA 0.2 %	6.55	95.45
TCA-PHA 0.2 % + DDPC 0.2 %	2.21	98.47
TCA-PHA 0.2 % + KI 0.2 %	1.28	99.11
TCA-PHA 0.2 % + DDPC 0.2 % + KI 0.2 %	0.21	99.85

Table 41: Corrosion rates of the duplex steel and the inhibition efficiencies in 20 % HCl at 363 K (90 °C) [73]

	Corrosion rate mg/cm² h	Inhibition efficiency %
TCA-PHA + DDPC + KI	0.21	99.85
TCA-BAM + DDPC + KI	0.20	99.86
BAL-PHA + DDPC + KI	0.71	99.50
BAL-BAM + DDPC + KI	1.30	99.10
SAL-PHA + DDPC + KI	0.66	99.54
SAL-BAM + DDPC + KI	0.70	99.51

Table 42: Corrosion rates of the duplex steel and the inhibition efficiencies in 20% HCl at 363 K (90 °C) [73]

	Corrosion rate mg/cm² h	Inhibition efficiency %
TCA-PHA 0.4%	1.98	98.63
TCA-PHA 0.4% + DDPC 0.2% + KI 0.2%	0.093	99.94

Table 43: Corrosion rates of the duplex steel and the inhibition efficiencies with an addition of 0.4% [73]

	Corrosion rate mg/cm² h	Inhibition efficiency %
BzQC 0.2%	0.88	99.39
BzQC 0.2% + TCA-PHA 0.4% + KI 0.2%	0.20	99.86
DDQBr 0.2%	59.62	58.60
DDQBr 0.2% + TCA-PHA 0.4% + KI 0.2%	0.10	99.93
DDPC 0.2%	13.10	91.00
DDPC 0.2% + TCA-PHA 0.4% + KI 0.2%	0.093	99.94

Table 44: Corrosion rates of the duplex steel and the inhibition efficiencies with different additives [73]

Table 40 to 44 give the mass loss rates and the inhibition efficiencies for the different material combinations as well as the additional effects due to the surfactants and the auxiliary inhibitor KI. Efficiencies of > 99% were obtained for a series of substance combinations with additions of the surfactant DDPC and of KI at an individual level of 0.2% (Table 42).

A 21 Austenitic CrNi steels

The austenitic standard steels (AISI series 300) are already attacked by dilute hydrochloric acid. The isocorrosion curve for a removal rate of 0.1 mm/a (4 mpy), e.g. for the steel AISI 316 L, lies below an acid concentration of 2.5% at room temperature.

These steels can undergo both pitting corrosion as well as stress corrosion cracking in acids whose concentration is not sufficient for general surface corrosion [68].

High-temperature corrosion in combustion gases at elevated temperatures

Vapors of vinyl chloride monomers are extremely hazardous to health and must therefore be made as ineffective as possible. Thermal oxidation is the most suitable method for this. However, this produces HCl and thus creates corrosion problems. If the temperature of the combustion gas is less than 393 to 423 K (120 to 150 °C), then equipment made of carbon steel can be protected against corrosion by means of linings made of rubber, plastics, graphite etc. At higher temperatures and in equipment used for heat transfer, metallic materials have to be used. In reference [74], a laboratory set-up and sample exposure tests in an industrial plant were used to find out whether stainless steels AISI 304 (X5CrNi18-10, DIN-Mat. No. 1.4301) and AISI 316 (X5CrNiMo17-12-2, DIN-Mat. No. 1.4401) are economically viable materials for heat exchangers in such combustion plants.

The samples were evaluated gravimetrically and visually after 50 h. At temperatures of up to 773 K (500 °C) and with 6000 ppm HCl, steel 1.4301 exhibited very low uniform surface corrosion rates of <0.1 mm/a (<4 mpy). The authors concluded that the standard stainless steels 1.4301 and 1.4401 can be used for heat exchangers if the process conditions are adjusted correspondingly. The testing equipment and operating units are described.

In mixtures of hydrochloric acid and nitric acid, the corrosion of materials and also the inhibiting effects are difficult to predict because reactions between the acids must be expected. Thus, at certain quantitative ratios of the acid mixture, it is possible that nitrosyl chloride is produced, which strongly attacks chromium-nickel steels with generation of a lot of heat. In reference [75], the temperature increase of mixtures of HCl-HNO_3 acids was used to characterise the corrosion of the steel AISI 304 (X5CrNi18-10, DIN-Mat. No. 1.4301).

For better reproducibility, samples of the same size were activated before the tests by contact with a zinc anode in 0.5 M H_2SO_4. They were then exposed to a constant quantity of an acid mixture in an open beaker. The temperature profile was measured with a thermometer over the course of one hour.

Figure 49 gives the temperature-time curve in 6 M mixtures of HCl-HNO_3 with differing mixing ratios. The rapid and large temperature increase for the mixture 40 HCl + 60 HNO_3 indicates a particularly high aggressiveness. In mixtures with a ratio of 20 HCl + 80 HNO_3 there was no temperature increase because the steel is passive. Figure 50 shows the result for a constant mixing ratio of 60 : 40 and variable mixture concentrations of 1 to 8 M. As expected, the aggressiveness increases as the concentration of the mixture increases.

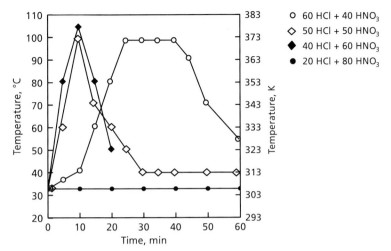

Figure 49: Temperature profile depending on the time for AISI 304 in mixtures of 6 M HCl with 6 M HNO$_3$ at various mixing ratios (○ 60 HCl : 40 HNO$_3$, ◇ 50 HCl : 50 HNO$_3$, ◆ 40 HCl : 60 HNO$_3$ and ● 20 HCl : 80 HNO$_3$) [75]

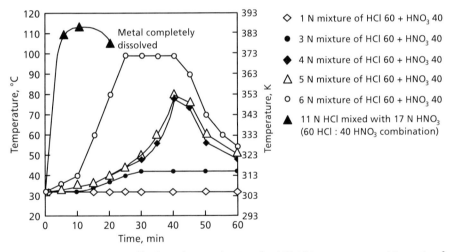

Figure 50: Temperature profile depending on the time for AISI 304 at a constant mixing ratio of 60 HCl : 40 HNO$_3$ and variable concentration of the mixture from 1 to 8 M [75]

A mixture of concentrated HCl and HNO$_3$ in a 60 : 40 ratio dissolved the sample with a large temperature increase of the solution after 20 min. At a 6 M mixture concentration and a mixing ratio of 60 HCl : 40 HNO$_3$, thiourea and hexamines gave a large inhibition (Figure 51 and Figure 52). At the critical ratio of 40 HCl : 60 HNO$_3$, the addition of hexamine up to 4 % did not produce any inhibition effect, instead it promoted the reaction (Figure 53). Therefore, there are complicated effects for acid mixtures (mixed acids) that can only be elucidated by means of tests.

Hydrochloric Acid

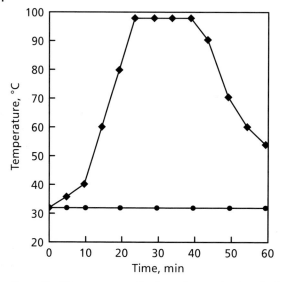

Figure 51: Temperature profile depending on the time for AISI 304 in 6 M mixtures of HCl-HNO$_3$ (60 HCl : 40 HNO$_3$) with (●) and without addition (◆) of 1 % thiourea [75]

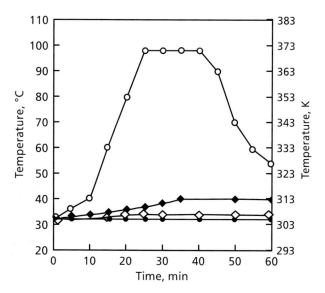

○ 60 HCl + 40 HNO$_3$
◆ 60 HCl + 40 HNO$_3$ + 0.1% Hexamine
◇ 60 HCl + 40 HNO$_3$ + 1% Hexamine
● 60 HCl + 40 HNO$_3$ + 2 and 4% Hexamine

Figure 52: Temperature profile depending on the time for AISI 304 in 6 M mixtures of HCl-HNO$_3$ (60 HCl : 40 HNO$_3$) with and without addition of various amounts of hexamine [75]

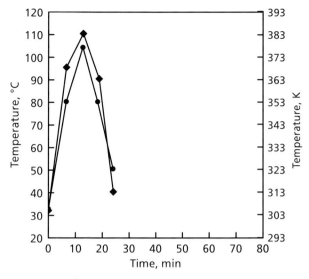

- ● 40 HCl + 60 HNO₃
- ◆ 40 HCl + 60 HNO₃ + 1%, 2% and 4% hexamine

Figure 53: Temperature profile depending on the time for AISI 304 in 6 M mixtures of HCl-HNO₃ (40 HCl : 60 HNO₃) with and without addition of various amounts of hexamine [75]

Atmosphere in a swimming pool hall

The collapse of a ceiling in a swimming pool hall in Ulster/Switzerland in 1985 resulted in twelve fatalities. The ceiling was suspended by wires made of stainless austenitic steel AISI 304 (X5CrNi18-10, DIN-Mat. No. 1.4301) with a diameter of 10 mm [76]. Because of the chlorine-containing atmosphere in the swimming pool hall, there was stress corrosion cracking and this caused the catastrophe. The occurrence of stress corrosion cracking in stainless austenitic steels at elevated temperatures >353 K (> 80 °C) is well known. The stress corrosion cracking occurs in the passive state without recognizable general corrosive attack. Because the hall temperature was well below this temperature, it was suspected that the stress corrosion cracking occurred in the active state because this is possible from the action of HCl-containing media at ambient temperature. As a result of the accident in Ulster, a group of researchers investigated active stress corrosion cracking of stainless austenitic steels [77].

Table 45 lists the investigated steels. Apart from the basic steel AISI 304 (1.4301), that was damaged in Ulster, more highly alloyed austenitic steels, a semi-austenic steel and a chromium steel (AISI 443, 1.4522) were investigated. Round rods, some of which had been cold-formed, were used as test objects. They had a neck that was electropolished to remove the surface hardening resulting from mechanical processing. The tests were carried out at constant load. The test solutions were mixtures of HCl-NaCl, based on reference [46].

Material	DIN-Mat. No.		C	Si	Mn	P	V	Cr	Mo	Ni	N	Ti	Nb	Cu
X2CrNiMnMoNbN23-17-6-3	1.3974		0.022	0.21	6.19	0.02	0.003	23.66	2.79	15.58	0.448	–	0.21	–
X5CrNi18-10	1.4301	AISI 304	0.035	0.71	1.65	0.025	0.019	18.31	0.45	8.79	0.06	–	–	–
X4CrNi18-12	1.4303	AISI 308	0.028	0.42	1.35	0.029	0.002	18.10	0.82	10.80	–	–	–	–
X2CrNiMo18-14-3	1.4435	cf. AISI 316 L	0.017	0.25	1.76	0.019	0.001	17.58	2.82	14.18	0.08	–	–	–
X2CrNiMoN17-13-5	1.4439	AISI 317 LMN	0.015	0.14	1.51	0.020	0.003	17.69	4.19	13.68	0.160	–	–	–
X2CrNiMoN22-5-3	1.4462	2205	0.024	0.44	1.71	0.023	0.003	22.14	3.11	5.43	1.11	–	–	–
X2CrMoNb18-2	1.4522	AISI 443	0.010	0.38	0.38	0.015	0.003	18.30	2.15	0.13	0.008	–	0.32	–
X1NiCrMoCu25-20-5	1.4539	AISI 904 L	0.018	0.10	1.72	0.019	0.003	20.68	4.59	24.44	0.07	–	–	1.35
X6CrNiMoTi17-12-2	1.4571	AISI 316 Ti	0.046	0.44	1.43	0.025	0.003	16.89	2.12	11.54	0.011	0.48	–	–

Table 45: Composition of the investigated steels, mass% [77]

Table 46 gives the result of steel AISI 304 (1.4301) depending on the electrolyte composition for a constant Cl⁻ ion concentration of 1.5 mol/l. It was found that the endurance of the samples to fracture was very dependent on the HCl content because general attack as well as the stress corrosion cracking are both dependent on it. This is also shown by the results in Table 47. Here, the HCl concentration was varied from 0.01 to 1.0 mol/l in a saturated NaCl solution at 298 K (25 °C).

Composition of the medium mol/l		Time to fracture h	Type of corrosion		
HCl	NaCl		Surface corrosion	SCC	Pitting
1.5	0	25	yes	yes	no
1.0	0.5	35	yes	yes	no
0.3	1.2	188	yes	yes	no
0.1	1.4	1,011	yes	no	no
0	1.5	> 4,000	no	no	yes

Table 46: Results from steel AISI 304 (1.4301) depending on the composition of the test solution for a constant concentration of Cl⁻ ions of 1.5 mol/l [77]

HCl concentration mol/l	Time to fracture h	Type of corrosion		
		Surface corrosion	SCC	Pitting
1.0	348		weak	
0.3	507		yes	
0.1	887	yes	yes	no
0.03	2,537		no	
0.01	> 4,000		no	

Table 47: Influence of the HCl concentration on the stress corrosion cracking behavior of steel AISI 304 (1.4301) in a saturated NaCl solution at 298 K (25 °C) for a loading level of 0.2 % of the yield strength [77]

In agreement with earlier publications [45], the appearance of stress corrosion cracking in the active state is most strongly pronounced in a solution with 1.0 mol/l HCl and 0.5 mol/l NaCl. Therefore, it was used as the standard solution in further tests. Figure 54 gives the endurance times in the standard solution at 323 K (50 °C) for the individual 0.2 % yield strength of the steels. Figure 55 gives these times for a constant load of 355 N/mm² for samples with 30 % cold deformation. The results show that stress corrosion cracking in the active state is always accompanied by a more or less pronounced general corrosion and that the endurance time is largely dependent on the composition of the steel, like passive stress corrosion cracking of these materials (the nickel content is decisive).

The authors conclude that the type of active stress corrosion cracking in HCl-NaCl solutions does not correspond to that observed in the damaged materials at Ulster where the stress corrosion cracks initiated at the corrosion pits; however, in these investigations, this type of corrosion only occurred in the pure NaCl solution without stress corrosion.

Hydrochloric Acid

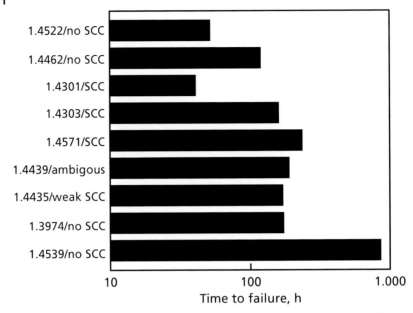

Figure 54: Endurance times in the standard solution at 323 K (50 °C) at the individual 0.2 % yield strength limit of the steels [77]

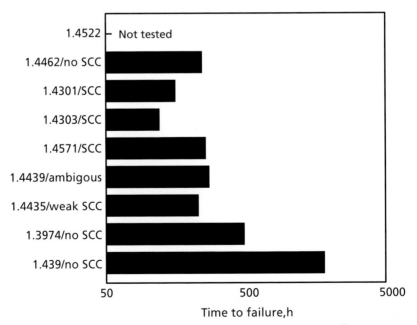

Figure 55: Endurance times of the steels for constant load of 355 N/mm^2, standard solution at 323 K (50 °C), 30 % cold deformation [77]

Measures to avoid the risk of stress corrosion cracking of stainless steels in swimming pool halls include the use of materials that have improved passivation properties or that have a more stable passive layer and increased resistance to pitting corrosion. These include austenitic steels with increased contents of molybdenum and nickel (e.g. DIN-Mat. No. 1.4439 (AISI 317 LMN , X2CrNiMoN17-13-5), 1.4539 (AISI 309 L, X1NiCrMoCu25-20-5), 1.4529 (UNS N08926, X1NiCrMoCuN25-20-7)), a ferritic-austenitic steel DIN-Mat. No. 1.4462 (2205, X2CrNiMoN22-5-3) with enhanced SCC resistance as well as nickel-based alloys [76]. According to Table 10 of the Building Inspectorate Approval Z-30.3-6, only the materials 1.4539, 1.4529, 1.4565 (X2CrNiMnMoNbN25-28-5-4) and 1.4547 (UNS S31254, X1CrNiMoCuN20-18-7) are qualified [78] for use as load-bearing components exposed to atmospheres in a swimming pool hall that are not cleaned regularly and upon which highly aggressive critical deposits can form.

In reference [79], the influence of the pH value on the stress corrosion cracking behavior of stainless austenitic steels 1.4300 (X 12 CrNi18 8) and 1.4301 (AISI 304, X5CrNi18-10) was investigated in the system NaCl-Na$_2$SO$_4$ + HCl with the aim of optimising the pH value. The standard solution had a boiling point of 381 K (108 °C), and contained 4.8 M NaCl, 0.15 M Na$_2$SO$_4$ and had a pH value of approx. 1 (3 ml conc. HCl/l). The solution was deaerated by continuous purging with nitrogen. The samples were tensile sheet specimens that were loaded to 70 % of their 0.2 % yield strength.

The potential-time profile during the stress corrosion cracking test of a sample of steel AISI 304 (corresp. to DIN-Mat. No. 1.4301, 18.32 % Cr, 8.48 % Ni, 0.19 % Mo, 0.07 % C, 1.68 % Mn, 0.020 % P, 0.004 % S, 0.51 % Si) is given in Figure 56. The

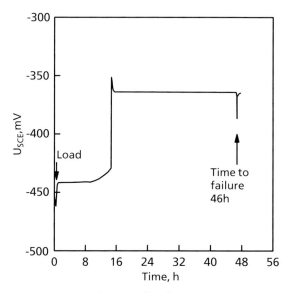

Figure 56: Potential-time profile of stress cracking samples of steel 304 in 4.9 M NaCl, 15 M Na$_2$SO$_4$, pH approx. 1 at 380 K (107 °C) [79]

sample was activated during use with a corrosion potential of approx. $U_H = -0.2$ V. When the load was applied there was a short-term drop in the potential, then the active starting potential was maintained over many hours with liberation of hydrogen on the sample. After 16 h and within a period of 3-5 min, the potential rose by approx. 0.2 V, with concomitant suppression of the hydrogen evolution. This indicates that the steel has become more or less passivated. After 46 h, the sample fractured due to stress corrosion cracking. The positive potential jump is a surprising result for a chloride-containing electrolyte; however, it does show that the stress corrosion cracking also occurs in this acid system with formation of an almost continuous protective layer, namely it is in a state that is almost passive and not active. The presence of Na_2SO_4 is probably significant.

Whereas stress corrosion cracking has been investigated in detail for highly concentrated chloride solutions, there are only a few investigations on stress corrosion cracking in relatively weak chloride-containing solutions at elevated temperatures, although it is exactly this type of corrosive conditions that frequently occurs in practice. This problem is dealt with in [80]. U-bending specimens of the materials 1.4301, 1.4306, 1.4401, 1.4406 (see Table 48), Carpenter® 20 Cb-3 (corresp to. DIN-Mat. No. 2.4660), Incoloy® 825 (corresp to. DIN-Mat. No. 2.4858), Haynes® 20 Mod. alloy; Hastelloy® alloy G (corresp to. DIN-Mat. No. 2.4618); Hastelloy® alloy C-276 (corresp to. DIN-Mat. No. 2.4819, 2.4887) were exposed for up to 30 days in an autoclave at 414 K (141 °C) to 4 % NaCl solutions that had been acidified with phosphoric acid, acetic acid and hydrochloric acid to a pH value of ~ 2. None of the investigated materials exhibited stress corrosion cracking in the pure NaCl solution. On the other hand, there was pronounced stress corrosion cracking of all four 18/8CrNi-based steels in all three acidified solutions. In contrast, steels containing higher amounts of nickel and nickel-based alloys Carpenter® 20 Cb-3 to Hastelloy® alloy C-276 were free of stress corrosion cracking. The results confirmed the finding that as the nickel content increases, the resistance to stress corrosion cracking in chloride-containing media is improved. This is also known from tests in highly concentrated, hot chloride solutions.

AISI	DIN-Mat. No.	Designation acc. to DIN EN 10028 T7	Designation acc. to EU 88
304	1.4301	X5CrNi18-10	X6 CrNi 18 10
304 L	1.4306	X2CrNi19-11	X3 CrNi 18 10
316	1.4401	X4CrNiMo17-12-2	X6 CrNiMo 17 12 2
316 LN	1.4406	X2CrNiMoN17-11-2	X3 CrNiMoN 17 12 2

Table 48: Designation of the materials according to various standards

As already known, crack formation in the U samples of the 18/8Cr/Ni steels can be stopped by contact with less noble materials (carbon steel, copper, nickel) and accelerated by contact with more noble materials with higher nickel contents.

The results agree with practical experience according to which pipes made of DIN-Mat. No. 1.4401 (AISI 316) failed in less than 3 months in a hot medium containing 1 to 2 % NaCl and approx. 7 % acetic acid as a result of stress corrosion cracking.

Sulphuric acid in all concentrations is widely used in the chemical industry. The corrosion behavior of stainless steels in pure sulphuric acid solutions has been investigated in detail. However, it is often contaminated with other acids. The effect on the corrosion behavior of materials is difficult to predict. Reference [81] reports an investigation of how the corrosion behavior of steel 1.4301 (AISI 304) changes if small amounts of hydrochloric acid and nitric acid are added to sulphuric acid.

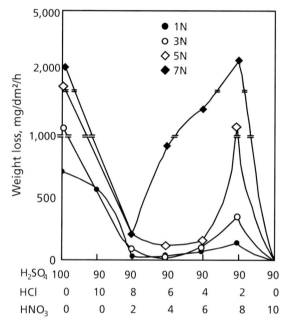

Figure 57: Influence of the acid concentration on the corrosion rate of steel AISI 304, 1 h test duration [81]

Starting from the mass loss rates in pure sulphuric acid solutions of samples that had been previously activated by contact with zinc in a 0.5 M H_2SO_4 solution, the mass loss rates were then determined in the acid mixtures given in Figure 57. The total acid concentration was kept constant between 1 M and 7 M. It can be seen that the corrosion rate decreases up to the mixing ratio $H_2SO_4 : HCl : HNO_3 = 90 : 2 : 8$, and strongly increases for a ratio 90 : 2 : 8, particularly for the more highly concentrated 5 N and 7 N solutions and drops to zero for all concentrations of the HCl-free composition $H_2SO_4 : HNO_3 = 90 : 10$, i.e. the steel is passivated in these solutions. The most critical composition was 90 : 2 : 8. The dependence of the corrosion rate on the composition is caused by the interaction between activating and passivating effects of the acids, the result of which cannot be predicted.

Further test results show that corrosion rates in the 1 M acid solution can considerably increase according to the mixing ratio with increasing test time and temperature. Under these test conditions, the highest corrosion rates were found for the mixing ratio $H_2SO_4 : HCl : HNO_3 = 90 : 2 : 8$.

The work described in [82] concerned stress corrosion cracking of stainless austenitic steels in the active state at ambient temperature. The materials given in Table 49 were tested both as sheets in electrochemical studies and as round tensile specimens in slow tensile tests (CERT test). The test solution contained HCl and NaCl with $c(Cl^-)$ = 1.5 mol/l and $c(H^+)$ = 1 mol/l. Tests were also carried out with steel 1.4301 in solutions with lower acid concentrations of 0.1 and 0.01 mol/l. The influence of martensite was also tested in separate experiments for steel 1.4301. For this, the specimens were cooled in liquid nitrogen.

No *	DIN-Mat. No. (abbreviation)	AISI	C	Si	Mn	P	V	Cr	Ni	Mo
1a, b	1.4301 (X5CrNi18-10)	304	0.07	1.00	2.00	0.045	0.030	20.0	10.5	–
2a, b	1.4571 (X6CrNiMoTi17-12-2)	316 Ti	0.10	1.00	2.00	0.045	0.030	18.0	13.5	2.5
3a	1.4439 (X2CrNiMoN17-13-5)	317 LMN	0.04	0.41	1.15	0.030	0.018	17.1	13.8	4.15
3b	1.4439	317 LMN	0.01	0.44	0.92	0.020	0.007	16.6	13.2	4.15
4a	1.4558 (X2NiCrAlTi32-20)		0.07	0.54	1.01	0.012	0.002	20.0	31.1	–
4b	1.4558		0.01	0.31	0.59	0.009	0.003	22.5	33.1	–

* a) Sheet specimens, b) tensile specimens
Table 49: Designation and chemical composition (mass%) of the investigated materials [82]

The measured current density/potential curves show that martensite has extended the active range somewhat, and has increased activation current density with a shift to more positive potentials.

The strain rate in the CERT test was 10^{-6} s^{-1}, the samples were elongated until they fractured. At $c(H^+)$ < 0.1 mol/l there were no transgranular cracks like those found for steel 1.4301. In the standard solution, steels AISI 304 and 316 (1.4301 and 1.4401) exhibited approximately the same types of corrosion and crack progression. In the martensitic variant of steel AISI 304, in addition to the transgranular cracks running perpendicularly from the surface, there was also corrosion in the longitudinal direction, which is due to attack of the martensite. There are only transgranular notches in steels AISI 317 LMN (1.4439) and 1.4558. The resistance of the materials correlates with that in hot $MgCl_2$ solutions. For these systems, the CERT test does not provide any advantages compared to tests with a constant load.

Comparative studies of stress corrosion cracking using the CERT and CCT tests in 1 M HCl at ambient temperature on steel AISI 304 (1.4301) were carried out in reference [83]. Under these corrosive conditions and for both type of test, when a relative elongation of 0.006 % of the test length was attained there was a permanent local destruction of the passive layer and cracking started immediately.

Phosphoric acid is the main component in the manufacture of phosphate fertiliser. During manufacture of phosphoric acid, the phosphate rock is dissolved by addition of hydrochloric acid. This leads to contamination with chloride ions, which initiate corrosion damage of stainless steels. Thus, e.g. steel AISI 304 (1.4301) exhibits pitting in a solution of 3.3 M H_3PO_4 + 0.1 M HCl. To avoid corrosion damage in phosphoric acid-chloride systems, guanidines and malonamides are added as inhibitors. In reference [84] some substituted dithiomalonamides were investigated for their inhibition efficiency towards steel 1.4301 in 3.3 M H_3PO_4 + 0.1 M HCl.

The structures of the investigated compounds were:

(I)   1,5-diphenyl-2,4-dithiomalonamide
(II)   1,5-di-p-methylphenyl-2,4-dithiomalonamide
(III) 1,5-dimethoxyphenyl-2,4-dithiomalonamide
(IV)   1,5-di-p-chlorophenyl-2,4-dithiomalonamide.

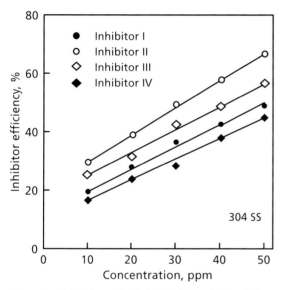

Figure 58: Inhibition action in H_3PO_4/HCl solutions [84]

Potentiodynamic current density/potential curves were recorded at 298 K (25 °C). The corrosion current densities, obtained by extrapolation of the anodic and cathodic polarisation curves, were used to calculate the inhibition efficiency. The results are summarised in Figure 58. All compounds exhibited a linear dependency for additions of 10 to 50 ppm. Compound II had an efficiency of 65 % at the highest addition of 50 ppm. The authors accounted for the differing efficiencies of the compounds on the basis of their structure using Auger electron spectra of the steel samples.

Chemical cleaning in the sugar industry

In former times in the sugar industry, unalloyed steels were used almost exclusively for pipes in evaporator units; however, from 1967 onwards stainless ferritic chromium steels and austenitic chromium-nickel steels were used increasingly as well.

Here too, it is necessary to regularly remove deposits. The most effective results were obtained with dilute hydrochloric acid. Since the steels are active in these solutions and are consumed at a high rate, they cannot be used without inhibitors [85].

Figure 59 demonstrates how the dissolution rate of materials that lie within the active potential range can be reduced in acids by inhibitors. Table 50 gives the results for the unalloyed steel St 35 (AISI 1010, DIN-Mat. No. 1.0308), the ferritic 17 % chromium steel AISI 430 (DIN-Mat. No. 1.4016, X6Cr17) and the austenitic 18/8 CrNi steel AISI 321 (DIN-Mat. No. 1.4541, X6CrNiTi18-10) in 5 % HCl solution with and without addition of 2.5 g/l of the commercially available inhibitors LITHSOLVENT® (Keller and Bohacek, Düsseldorf), that are used in the sugar industry. The mass loss rates were greatly reduced for all steels and temperatures by the three inhibitors used. Thus, at 373 K (100 °C) LITHSOLVENT® EB-II gave the following efficiencies for the three steels:

St 35 (AISI 1010)   99.1 %
1.4016 (AISI 430)   99.5 %
1.4541 (AISI 321)   98.4 %

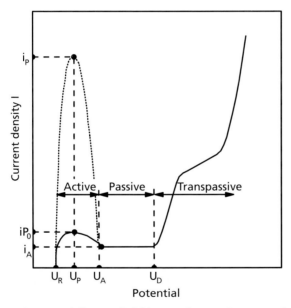

Figure 59: Influence of inhibitors on the corrosion current density in the active range (schematic) [85]
iP_0 = inhibition current density
i_A = passive dissolution current density
i_P = passivating current density
U_R = resting potential
U_P = passivating potential
U_A = activating potential
D_U = breakdown potential

The tests were carried out in static solutions. A decrease in the efficiency must be expected for higher flow rates of the solution. For the relatively low flow rates used in chemical cleaning, however, no great influence on the inhibition effect of the inhibitors can be expected from any movement. In addition, the report gives a detailed description of the execution of chemical cleaning in the sugar industry.

Inhibitor	Temp. K (°C)	Duration h	Area-related mass loss rate, $g/m^2\,h$		
			St 35 (AISI 1010)	1.4016 (AISI 430)	1.4541 (AISI 321)
without	313 (40)	1	65	117	1.5
	333 (60)	1	580	440	5.0
	353 (80)	1	800	760	27.0
	373 (100)	1	600	1000	63.0
LITHSOLVENT® EB-II	313 (40)	6	0.36	0.17	0.2
	333 (60)	6	0.80	0.42	0.2
	353 (80)	4	1.5	1.9	0.4
	373 (100)	2	5.0	5.1	1.0
LITHSOLVENT® EB	313 (40)	6	0.5	0.2	0.2
	333 (60)	6	1.3	1.4	0.25
	353 (80)	6	2.2	3.1	0.3
	376 (103)	3	7.5	9.5	3.2
LITHSOLVENT® HEN	293 (20)	6	0.15	0.15	0.04
	313 (40)	6	0.7	0.6	0.15
	333 (60)	6	2.0	1.0	0.32

$0.9\ g/m^2 \times h = 1\ mm/a = 39.4\ mpy$

Table 50: Area-related mass loss rates from chemical cleaning of boiler tubes with hydrochloric acid – with and without addition of inhibitors [85]

Mass loss measurements and current density/potential curves were used to investigate the inhibition efficiency of potassium iodide in hydrochloric acid on the four stainless steels listed in Table 51 [86]. Table 52 gives the results in 1 M HCl at 303 K (30 °C) in dependence on the inhibitor concentration. Except for steel AISI 304 (1.4301), efficiencies of > 93 % were obtained for high and intermediate inhibitor concentrations. Table 53 gives the mass loss rates in 1 M HCl + 0.1 M KI in dependence on the temperature. Even at elevated temperatures, there is a high efficiency. Thus from the mass loss rates for all four steels, an efficiency of > 98 % was calculated at the highest temperature of 353 K (80 °C).

	Corresponds to	Cr	Ni	Mo	C
AISI 304	DIN Mat. No. 1.4301	18.20	8.12	–	0.08
AISI 316	DIN Mat. No. 1.4401; 1.4436	16.18	10.14	2 – 3	0.08
AISI 430	DIN Mat. No. 1.4016	14.18	–	–	0.12
AISI 440		16.18	–	0.75	0.60

Table 51: Composition of the investigated stainless steels, mass% [86]

KI M	AISI 304		AISI 316			AISI 430			AISI 440		
	Corrosion rate mg/cm² h	Efficiency %	Corrosion rate mg/cm² h	Efficiency %		Corrosion rate mg/cm² h	Efficiency %		Corrosion rate mg/cm² h	Efficiency %	
–	0.036	–	0.064	–		0.580	–		0.260	–	
0.1	0.011	69.4	0.023	64.0		0.018	96.9		0.015	94.22	
10^{-2}	0.012	66.6	0.002	96.8		0.019	96.7		0.031	88.0	
5×10^{-3}	0.012	66.6	0.004	93.7		0.036	93.8		0.060	76.9	
10^{-3}	0.015	58.3	0.009	85.9		0.140	75.8		0.130	50.0	
10^{-4}	0.015	58.3	0.009	85.9		0.230	60.3		0.150	42.3	

Table 52: Influence of various potassium iodide (KI) concentrations on the corrosion rate of the stainless steels in 1 M HCl at 303 K (30 °C) [86]

Temperature, K (°C)	AISI 304 Corrosion rate, mg/cm² h		AISI 316 Corrosion rate, mg/cm² h		AISI 430 Corrosion rate, mg/cm² h		AISI 440 Corrosion rate, mg/cm² h	
	HCl	HCl + KI	HCl	HCl + KI	HCl	HCl + KI	HCl	HCl + KI
303 (30)	0.036	0.011	0.064	0.023	0.58	0.018	0.26	0.015
313 (40)	0.27	0.019	0.250	0.008	3.50	0.040	2.60	0.020
323 (50)	0.40	0.018	0.500	0.016	10.50	0.060	6.20	0.020
333 (60)	0.52	0.020	0.700	0.021	17.50	0.600	14.00	0.060
343 (70)	1.05	0.024	1.500	0.021	32.00	0.200	23.00	0.160
353 (80)	1.53	0.025	2.800	0.022	41.00	0.520	37.00	0.460

Table 53: Influence of the temperature on the corrosion rate of the stainless steels in 1 M HCl with 0.1 M potassium iodide [86]

The determination of the current density from the Tafel lines in the current density/potential curves gave comparable results to those given in Table 52.

Furthermore, the investigations show that the addition of very small quantities of 10^{-4} M KI in combination with other known inhibitors (such as: hexamine, quinoline and thiourea) in 1 M HCl at 303 K (30 °C) gave a considerable improvement in the efficiency for the steels AISI 304, AISI 316 and AISI 430. Therefore, there is a synergistic effect.

As a matter of principle, it is not recommended to use hydrochloric acid, whether with or without inhibitor, to clean components made of stainless austenitic steels.

Decomposition of chlorinated hydrocarbons

Halogenated hydrocarbons are frequently used to clean metallic objects. Anhydrous chlorinated hydrocarbons exhibit a completely inactive corrosion behavior at ambient temperature. In the presence of small quantities of moisture, however, they can be hydrolysed to form HCl and thus become aggressive [87, 88, 89].

Table 54 [90] shows how the corrosion rate of steel 1.4541 (AISI 321, X6CrNiTi18-10) is increased by HCl in dichloroethane. Thus the corrosion rate in the dichloroethane-HCl system (the mixture was made with 35 % HCl) with a content of 5 % HCl, is approx. 20 times greater than in an aqueous solution of the same concentration (5.16 : 0.26 mm/a (203 : 10 mpy)). Table 55 gives the temperature-dependent corrosion rates of steel 3 (unalloyed carbon steel) and the steel 1.4541 in dichloroethane with 0.11 % water. The corrosion rate increases 4- to 5-fold at 293 to 353 K (20 to 80 °C) for differing initial values. At even higher temperatures, the corrosion rate increases much more strongly.

Dichloroethane-HCl			Aqueous HCl solution			0.1 % HCl solution in dichloroethane		
Proportion of HCl %	Corrosion rate		Proportion of HCl %	Corrosion rate		Proportion of dichloroethane %	Corrosion rate	
	mm/a	mpy		mm/a	mpy		mm/a	mpy
0.00	0.004	0.16	–	–	–	11.0	0.066	2.6
0.01	0.015	0.6	0.01	–	–	55.0	0.093	3.7
0.05	0.122	4.8	0.06	–	–	75.0	0.108	4.3
0.10	0.980	38.6	0.11	0.002	0.1	99.7	0.980	38.6
1.00	1.860	73.2	0.33	0.002	0.1	–	–	–
5.00	5.160	203.1	5.00	0.26	10.2	–	–	–

Table 54: Change of the corrosion rate of steel 1.4541 (AISI 321, X6CrNiTi18-10) by HCl in dichloroethane (temperature 293 K (20 °C), test period 360 h) [90]

Temperature, K (°C)	Steel 3 *	1.4541
293 (20)	0.011 (0.43)	0.003 (0.12)
303 (30)	0.011 (0.43)	0.006 (0.24)
313 (40)	0.019 (0.75)	0.006 (0.24)
323 (50)	0.019 (0.75)	0.006 (0.24)
333 (60)	0.028 (1.10)	0.006 (0.24)
343 (70)	0.033 (1.30)	0.011 (0.43)
353 (80)	0.051 (2.01)	0.011 (0.43)

* unalloyed carbon steel

Table 55: Corrosion rate in mm/a (mpy) of the two steels in dichloroethane with 0.11 % water in dependence on the temperature (test duration 12 h) [90]

The cleavage of HCl at elevated temperatures can also initiate pitting and stress corrosion cracking in stainless steels.

δ–Ferrite attack by inhibited 7-10 % HCl at 393 K (120 °C)

In paper mills, white-liquor clarifiers are used to separate the solid lime sludge ($CaCO_3$) from the white liquor. White liquor is a strong base containing NaOH (approx. 100 g/l), Na_2S (30 g/l), Na_2CO_3 (25 g/l), Na_2SO_5 (10 g/l), $Na_2S_2O_3$ (40 g/l) and undissolved $CaCO_3$ (170 g/l) at 363 K (90 °C). At certain time intervals, it is necessary to clean the filters with dilute 7 to 10 % HCl (inhibited) at 393 K (120 °C) [91]. After cleaning 27 times in a period of seven months, a crack approx. 1 m long appeared in the longitudinal seam of a feed pipe made of austenitic steel 1.4306 (AISI 304 L, GX2CrNiN18-9, X2CrNi19-11) containing 0.024 % C, 18.5 % Cr, 9.3 % Ni. An inspection showed that a shut-off valve was leaking so that acid was able to enter the feed pipe during cleaning. Metallographic investigations showed that there was a pronounced selective corrosion of the δ-ferrite by the inhibited hydrochloric acid and that this had led to the damage. The longitudinal seam had been welded without a filler and thus had the same composition as the pipe material.

The corrosion of the welded material itself was low in circumferential assembly seams welded with a molybdenum-containing weld filler (1.7 and 2.6 % Mo). However, on the melt line to the substrate material there was enhanced attack as a result of increased δ-ferrite formation. The boiler itself, made of steel AISI 316 (1.4401), was not attacked except for slight pitting on the air/white liquor interface.

This damage shows that the inhibitor was able to provide sufficient corrosion protection for the austenitic microstructure, but not for the δ-ferrite.

A 22   Austenitic CrNiMo(N) and CrNiMoCu(N) steels

The behavior of molybdenum-alloyed austenitic standard steels in hydrochloric acid is essentially the same as that of the molybdenum-free grades. Some of the highly alloyed austenitic CrNiMoN steels (superaustenites), such as 254 SMO (UNS

S31254, corresp to. DIN-Mat. No. 1.4547) N08367, N08926, can be used in hydrochloric acids with concentrations below approximately 3%. Alloy 654 SMO (UNS S32654, corresp to. DIN-Mat. No. 2.4652) with nominal 7.3% Mo can also be used at room temperature at an acid concentration of approx. 8%. Figure 60 gives the isocorrosion curves for a corrosion rate of 0.1 mm/a (3.94 mpy) for a series of these materials [68].

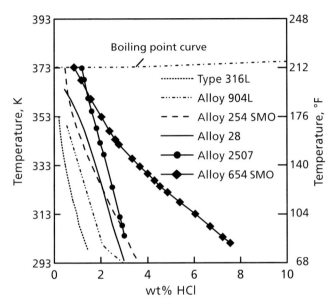

Figure 60: Isocorrosion diagram (0.1 mm/a (3.94 mpy)) for various austenitic steels in hydrochloric acid [68, p. 18]

A 23 Special iron-based alloys

In reference [92], the corrosion behavior of an austenitic cast alloy 18Cr-37Ni was investigated in alkaline, acidic and neutral solutions. Such austenitic steels and cast grades are suitable for nuclear power stations because of their enhanced resistance against stress corrosion cracking. The composition of the investigated cast alloy is given in Table 56. It was investigated in NaOH, H_2SO_4, HCl, NaCl and seawater in the cast state and after a homogenisation annealing for 100 and 400 h at 1173 K (900 °C). The mass loss rates at 289 K (16 °C) and at 373 K (100 °C) are summarised in Table 57 and Table 58. As expected, the corrosion rate in the acid solutions was high. The homogenisation treatment improved the corrosion resistance in these media.

C	Mn	Si	Cr	Ni	Mo	V	P	Fe
0.1	0.71	1.25	19.7	36.6	0.026	0.016	0.010	rest

Table 56: Chemical composition of the investigated alloy

Heat treatment	NaOH			H$_2$SO$_4$			HCl			NaCl		
	5%	10%	15%	5%	10%	15%	5%	10%	15%	5%	10%	15%
Initial state	0.025 (0.98)	0.04 (1.57)	0.07 (2.76)	1.075 (42.32)	1.25 (49.21)	2.9 (114.17)	0.85 (33.46)	1.35 (53.15)	1.9 (74.8)	0.05 (1.97)	0.02 (0.79)	0.02 (0.79)
100 h, 1173 K (900 °C)	0.0085 (0.33)	0.01 (0.39)	0.0225 (0.89)	0.85 (33.46)	1.125 (44.29)	2.9 (114.17)	0.83 (32.68)	1.125 (44.29)	1.2 (47.24)	0.055 (2.17)	0.04 (1.57)	0.04 (1.57)
400 h, 1173 K (900 °C)	0.0025 (0.1)	0.0075 (0.3)	0.0075 (0.3)	0.65 (25.59)	1.125 (44.29)	2.35 (92.52)	0.475 (18.7)	0.4 (15.75)	0.525 (20.67)	0.005 (0.20)	0.015 (0.59)	0.015 (0.59)

Table 57: Corrosion rate at 289 K (16 °C), mm/a (mpy) [92]

Heat treatment	H$_2$SO$_4$	HCl	NaCl	Seawater
untreated	1129.1 (44,453)	53.9 (2,122)	0.26 (10.24)	0.20 (7.87)
100 h, 1173 K (900 °C)	1152.8 (45,386)	119.2 (4,693)	0.06 (2.36)	0.0025 (0.10)
400 h, 1173 K (900 °C)	1283.4 (50,528)	121.2 (4,772)	0.21 (8.27)	0.005 (0.20)

Table 58: Corrosion rate at 373 K (100 °C), mm/a (mpy) [92]

A 24 Magnesium and magnesium alloys

Magnesium and magnesium alloys are strongly attacked by hydrochloric acid of practically every concentration even at room temperature. Therefore, they are unsuitable for applications in such solutions.

A 25 Molybdenum and molybdenum alloys

Molybdenum and its alloys have quite good resistance in dilute and moderately concentrated hydrochloric acid and are occasionally used for special purposes. More information on the resistance is given in reference [47].

A 26 Nickel

Depending on the metal composition, the corrosion resistance of nickel and nickel-based alloys covers a wide range of very aggressive media. While pure nickel is predominantly used in situations where there is exposure to very alkaline media, the NiCrMoFe and NiMo alloys are particularly suitable for acidic media.

Pure nickel (nickel 200, UNS N02200, Ni99.6 DIN-Mat. No. 2.4060, Ni99.2 DIN-Mat. No. 2.4066) exhibit a certain resistance in dilute hydrochloric acid at room temperature. The attack is enhanced by aeration and the presence of oxidising agents in the acid.

A 27 Nickel-chromium alloys
A 28 Nickel-chromium-iron alloys (without Mo)

The molybdenum-free NiCr or NiCrFe alloys, e.g. NiCr16Fe (UNS N06600, Inconel® 600, Alloy 600, DIN-Mat. No. 2.4816) or NiCr21Fe (Inconel® 800), are not used in hydrochloric acid because in dilute solutions they are susceptible to pitting corrosion and in concentrated solutions at elevated temperatures they are also susceptible to general surface corrosion. Their corrosion resistance is lower than that of pure nickel or alloy 400. More information on the resistance of NiCr or NiCrFe alloys in hydrochloric acid is given in reference [47].

A 29 Nickel-chromium-molybdenum alloys

The nickel-chromium alloys used in chemical process engineering generally contain molybdenum as well to increase the pitting and crevice corrosion resistance. The most commonly used materials are given in Table 59 with the frequently used and familiar manufacturer's designations.

Manufacturer's designation	UNS	DIN-Mat. No.	Abbreviation
Incoloy® 825 Nicrofer® 4221 Alloy 825 Uranus® 825	N08825	2.4858	NiCr21Mo
Inconel® G-3 Alloy G-3 Nicrofer® 4823 hMo	N06985	2.4619	NiCr22Mo7Cu
Inconel® 625 Alloy 625 Nicrofer® 6020 hMo Uranus® 625	N06625	2.4856	NiCr22Mo9Nb
Alloy C 276 Hastelloy® C-276 Nicrofer® 5716 hMoW	N10276	2.4819	NiMo16Cr15W
Hastelloy® C-4 Alloy C-4 Nicrofer® 6616 hMo	N06455	2.4610	NiMo16Cr16Ti
Nicrofer® 5923 hMo Alloy 59	N06059	2.4605	NiCr23Mo16Al

Table 59: Nickel-chromium-molybdenum alloys frequently used in chemical process engineering

As the molybdenum content increases, the resistance of the NiCrMo alloys decreases in hydrochloric acid so that alloy 625 has a better resistance than alloy G-3, which in turn is more resistant than alloy 825.

As the isocorrosion diagram in Figure 61 shows for alloy 825, it can be expected that this material has insufficient resistance at low acid concentrations and temperatures [68].

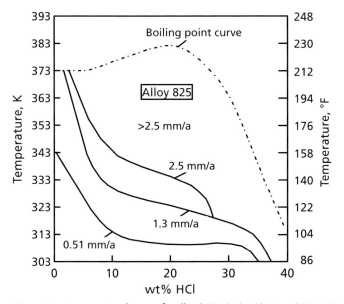

Figure 61: Isocorrosion diagram for alloy 825 in hydrochloric acid [68, p. 32]
> 2.5 mm/a (> 100 mpy)
2.5 mm/a (100 mpy)
1.3 mm/a (50 mpy)
0.51 mm/a (20 mpy)

The isocorrosion diagrams for alloy G-30 (UNS N06030, a modified G-3 variant) in Figure 62, alloy C-276 in Figure 63, alloy C-4 in Figure 64 and alloy 59 in Figure 65 demonstrate the considerably improved behavior of these materials [68].

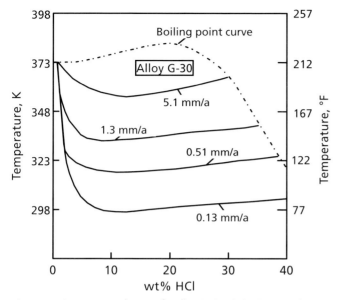

Figure 62: Isocorrosion diagram for alloy G-30 in hydrochloric acid [68, p. 33]
5.1 mm/a (200 mpy)
1.3 mm/a (50 mpy)
0.51 mm/a (20 mpy)
0.13 mm/a (5 mpy)

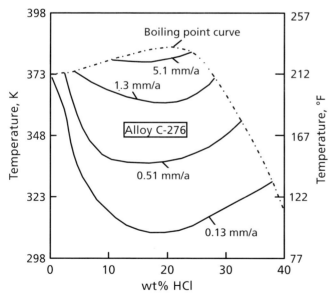

Figure 63: Isocorrosion diagram for alloy C-276 in hydrochloric acid [68, p. 34]
5.1 mm/a (200 mpy)
1.3 mm/a (50 mpy)
0.51 mm/a (20 mpy)
0.13 mm/a (5 mpy)

Hydrochloric Acid

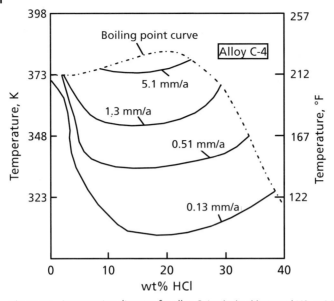

Figure 64: Isocorrosion diagram for alloy C-4 in hydrochloric acid [68, p. 36]
5.1 mm/a (200 mpy)
1.3 mm/a (50 mpy)
0.51 mm/a (20 mpy)
0.13 mm/a (5 mpy)

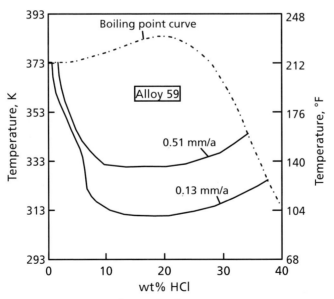

Figure 65: Isocorrosion diagram for alloy 59 in hydrochloric acid [68, p. 39]
0.51 mm/a (20 mpy)
0.13 mm/a (5 mpy)

Reference [93] discusses the factors that led to the development of the Ni-Cr-Mo-W alloy Hastelloy® C-22 (UNS N06022). Compared to other nickel-based alloys, it exhibits a very high corrosion resistance both in very aggressive reducing and oxidising acids.

Table 60 gives the composition of alloys in the development sequence of alloys developed prior to alloy C-22. During welding of alloy C with max. 0.08 % C and max. 1.0 % Si, chromium carbides were precipitated on the grain boundaries in the substrate material next to the weld seam, making this zone susceptible to intergranular corrosion. Therefore, the utilisation of this material was greatly limited because a subsequent solution annealing and quenching to eliminate the susceptible zones was only possible to a limited extent.

Alloy	Cr	Mo	W	Fe	Nb	Ti	V	Mn	C	Si	Ni
Alloy C	15	16	4	5	–	–	0.35 [1]	1.0 [1]	0.08 [1]	1.0 [1]	balance
Alloy C-276	16	16	4	5	–	–	0.35 [1]	1.0 [1]	0.01 [1]	0.08 [1]	balance
Alloy C-4	16	16	–	3 [1]	–	0.70 [1]	–	1.0 [1]	0.01 [1]	0.08 [1]	balance
Alloy C-22	22	13	3	3	–	–	0.35 [1]	0.50 [1]	0.015 [1]	0.08 [1]	balance
Alloy 625	21.5	9	–	5.0 [1]	3.5	0.40 [1]	–	0.5 [1]	0.10 [1]	0.50 [1]	balance
Alloy HC new	20	15	–	2	–	–	–	–	0.01	0.03	balance

[1] Maximum

Table 60: Composition of the Ni-Cr-Mo/W alloys, mass% [93]

By reducing the carbon content to max. 0.01 % and the silicon content to 0.08 %, intergranular susceptibility was eliminated in alloy C-276 [94]. In spite of its excellent general corrosion resistance, this alloy is still somewhat susceptible to preferred attack in the welding seam and in the heat-affected zone of the welded material because in the temperature range from 973 to 1373 K (700 to 1100 °C) a chromium and molybdenum-rich intermetallic phase (μ-phase) can be precipitated that is also susceptible to intergranular corrosion in aggressive acids.

Alloy C-4 was developed at a later date. This tungsten-free alloy with a reduced iron content is particularly thermally stable. An intermetallic phase is precipitated only after 100 h in the temperature range from 923 to 1373 K (650 to 1100 °C). The intergranular susceptibility due to precipitation of carbides is avoided by the low carbon content and by stabilisation with titanium. However, due to the lack of tungsten, the resistance to local corrosion (pitting and crevice corrosion) is reduced. Additionally, the resistance to oxidising acids is limited.

The increasing understanding of the effect of the individual alloying elements on the corrosion behavior allows the development of alloys with optimum corrosion properties. In nickel-based alloys, chromium enhances the resistance in oxidising acids, whereas molybdenum and tungsten enhances that in reducing acids. Based on this knowledge, the atomic percent factor (APF) was developed in order to compare the corrosion resistance of alloys with differing compositions. It expresses the opposing effect of chromium compared to that of molybdenum and tungsten. The ratios of the atomic weights of the relevant elements are

Cr (52) : Mo (96) : W (184) = 1 : 2 : 4

The APF is defined as the ratio of four times the weight% of chromium to the sum of twice the weight% of molybdenum + one times the weight% of tungsten, namely

APF = 4 Cr/(2 Mo + 1 W).

A high APF indicates a high resistance to oxidising acids, whereas a low APF indicates a high resistance to reducing acids. For an intermediate value of 2.5 to 3.3, the resistance to oxidising as well as reducing acids is high. The ratios are represented graphically in Figure 66. Alloy C-22 has an APF ≈ 3, which is the most favorable composition with respect to the corrosion behavior in oxidising and reducing acids.

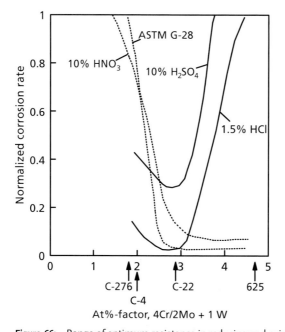

Figure 66: Range of optimum resistance in reducing and oxidising environments [93]

C-22 has a higher thermal stability than C-276. On tempering at 1200 K (927 °C), the temperature at which the fastest precipitation of the intermetallic phase occurs, intergranular susceptibility starts in less than 2 min for C-276, and in less than 15 min for C-22.

Table 61 gives comparative corrosion rates of the alloys listed in Table 60 in oxidising and reducing acidic aggressive media. Alloy C-22 exhibited the highest corrosion resistance in all media.

Alloy C-22 also exhibited the highest resistance in pitting and crevice corrosion tests in the FeCl$_3$ test with increasing temperature [95]. The lower molybdenum content compared to alloy C-276 is more than overcompensated by the higher chromium content.

Acid mixture	Temperature K (°C)	C-22	C-276	C-4	625
5 % HNO$_3$ + 6 % HF	333 (60)	1.68 (66.14)	5.18 (203.94)	5.1 (200.79)	1.825 (71.85)
5 % HNO$_3$ + 1 % HCl	boiling	0.0125 (0.49)	0.2 (7.87)	0.275 (10.83)	0.025 (0.98)
5 % HNO$_3$ + 25 % H$_2$SO$_4$ + 4 % NaCl	boiling	0.3 (11.81)	1.6 (62.99)	2.425 (95.47)	17.83 (701.97)
23 % H$_2$SO$_4$ + 1.2 % HCl + 1 % FeCl$_3$ + % CuCl$_2$	boiling	0.175 (6.89)	1.375 (54.13)	57.35 (2,257.87)	96.18 (3.786,61)

Table 61: Corrosion rates under oxidising conditions, mm/a (mpy) [93]

Even in the stress corrosion cracking tests, C-22 was generally better than alloys C-276 and 625 in aggressive media that trigger stress corrosion cracking.

However, as the values in Table 62 show for pure hydrochloric acid, the corrosion rates of alloy 59 and alloy C-276 were lower than those of alloy C-22 [68].

Acid concentration %	Temperature K (°C)	Corrosion rate mm/a (mpy)		
		C-22	C-276	59
1.5	Boiling	0.87 (34.2)	0.69 (27.2)	0.38 (14.9)
2	363 (90)	0.051 (2.01)	0.53 (20.8)	0.00 (0.0)
2	Boiling	2.18 (85.8)	1.14 (44.9)	0.99 (39)
3	363 (90)	1.41 (55.5)	0.96 (37.8)	0.053 (2.1)
3	boiling	4.55 (179)	2.06 (81.1)	2.05 (80.7)
5	366 (93)	5.07 (200)	1.99 (78.4)	2.81 (110.6)
10	328 (55)	0.77 (30.3)	0.44 (17.3)	0.55 (21.7)
10	355 (82)	3.03 (119)	1.56 (61.4)	2.47 (97.2)
10	Boiling	9.26 (365)	6.07 (239)	7.50 (295)

Table 62: Corrosion rates of the NiCrMo alloys C-22, C-276 and 59 in hydrochloric acid of differing concentrations and temperatures [68]

Acid mixtures

Acid mixtures are frequently used in chemical and metallurgical processes. For example, mixtures of nitric acid and hydrofluoric acid are used to pickle stainless steels. Although there have been many publications on the corrosion behavior of many alloys in pure acids, not enough is known about the corrosion behavior in mixed acids. Reference [96] reports on the corrosion behavior of nickel-based alloys in mixed acids. The results are based on the determination of the mass loss rates in static solutions that have not been deaerated.

Alloy	UNS No.	Fe	Cr	Ni	Mo	C	Others
AISI 316 L (corresp to. DIN-Mat. No. 1.4404)	S31603	balance	17	12	2.5	0.93	–
Ferrallium® Alloy 255 (corresp to. DIN-Mat. No. 1.4492)	S32550	balance	26	5.5	3.0	0.04 [1]	Cu = 1.7 N = 0.2
Alloy C-276	N10276	5.0 [1]	16	balance	16	0.01 [1]	W = 4.0
Alloy C 22 (corresp to. DIN-Mat. No. 2.4602)	N06022	3.0	22	balance	13	0.015 [1]	W = 3
Alloy G-3 (corresp. to DIN-Mat. No. 2.4619)	N06985	20	22	balance	7	0.015 [1]	Cu = 2.0 Nb = 0.5 [1]
Hastelloy® Alloy G-30 (corresp to. DIN-Mat. No. 2.4603)	N06030	15	29.5	balance	5	0.3 [1]	Nb = 0.7 W = 2.5 Cu = 1.7
Alloy B-2 (corresp to. DIN-Mat. No. 2.4617)	N10665	2.0 [1]	1.0 [1]	balance	28	0.02 [1]	–

[1] Minimum

Table 63: Composition of the investigated alloys, mass% [96]

Table 63 gives the maximum chemical composition of the investigated alloys in mass%. For comparison purposes, the stainless steel AISI 316 L (1.4404) and the duplex steel Ferallium® Alloy 255 were also included.

Nickel-based alloys exhibited high corrosion rates, particularly in strongly reducing acids. They are designed for such operating conditions. The corrosion rate is more or less strongly increased if oxidising media are also present. Therefore, it is particularly important to determine the corrosion behavior of these materials in acid mixtures.

Figure 67 gives the isocorrosion curves for alloy Hastelloy® C-276 at 353 K (80 °C) in the three-component representation H_2O – HCl – HNO_3. The data are limited to one corner of the triangular representation because the maximum concentrations of the analytical grade acids used was 71 % for HNO_3 and 38 % for HCl. This form of representation shows the corrosion behavior of an alloy in a wide range of acid combinations; however, it is not very suitable for a comparison of the corrosion behavior of different alloys. Further results are thus given as a function of the proportion of an acid in another acid of constant concentration.

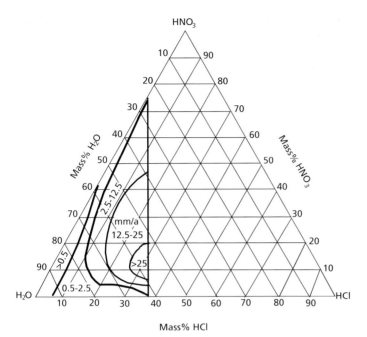

Figure 67: Isocorrosion curves for the alloy Hastelloy® C-276 at 353 K (80 °C) in the three-component representation $H_2O - HCl - HNO_3$ (175 °F) [96]
   < 0.5 mm/a   (< 20 mpy)
0.5-2.5 mm/a   (20-100 mpy)
2.5-12.5 mm/a   (100-500 mpy)
12.5-25 mm/a (500-1000 mpy)
   > 25 mm/a   (1000 mpy)

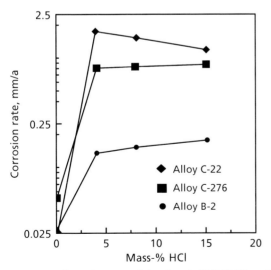

Figure 68: Corrosion rates of the alloys in 15 % H_2SO_4 at 80 °C (175 °F) [96] (1 mm/a ≙ 39.37 mpy)

Figure 68 gives the corrosion rates of alloys C-22, C-276 and B-2 in 15 % H_2SO_4 at 353 K (80 °C) with increasing HCl concentration. Due to the reducing effect of hydrochloric acid, the corrosion rate increases by differing amounts, and increases more strongly with decreasing Mo content in the alloy. The attack is in the form of pitting corrosion.

On the other hand, with respect to the favorable behavior of B-2 in this mixture of two reducing acids, it must be pointed out that the corrosion resistance of this alloy is reduced in the presence of oxidising media, e.g. Fe^{3+} ions which are frequently present.

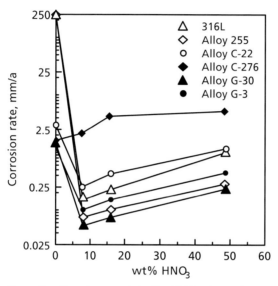

Figure 69: Corrosion rates in boiling 30 % H_2SO_4 with increasing concentration of HNO_3 [96] (1 mm/a ≙ 39.37 mpy)

With the exception of C-276, the addition of approx. 10 % of oxidising nitric acid to boiling 30 % H_2SO_4, a reducing acid, led to a large reduction in the corrosion rate, as shown in Figure 69. The effect is particularly large for the stainless steels AISI 316 L (1.4404) and Ferralium® Alloy 255, which are strongly attacked by boiling 30 % H_2SO_4. The corrosion rate is decreased by almost four powers of ten. If the HNO_3 concentration is increased further, then the corrosion rate increases again. In this range, the value of the corrosion rate is mainly determined by the amount of chromium and molybdenum contained in the alloy. The corrosion rate decreases with increasing chromium content and decreasing molybdenum content. Alloy C-276, with the lowest chromium (16 %) and highest molybdenum contents (16 %) has the highest corrosion rate and shows abnormal behavior. Alloy B-2 with 28 % Mo was not included in the investigation.

The addition of HNO_3 to 4 % HCl at 353 K (80 °C) leads to results, that are comparable to those in systems with boiling 30 % H_2SO_4 + HNO_3 (Figure 69 and Figure 70). There is a corrosion minimum for 10 % HNO$_3$. For further addition there is a stronger increase, whereby the level of the corrosion rate corresponds to the sequence in Figure 69, namely, the same alloying relationships are decisive for the corrosion rate. The attack is of the pitting type.

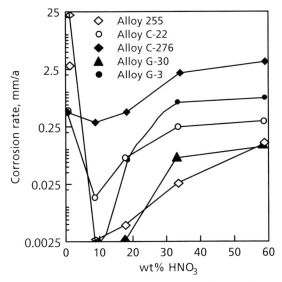

Figure 70: Corrosion rate in a 4 % HCl solution with increasing HNO$_3$ concentration at 80 °C (175 °F) [96]
(1 mm/a ≙ 39.37 mpy)

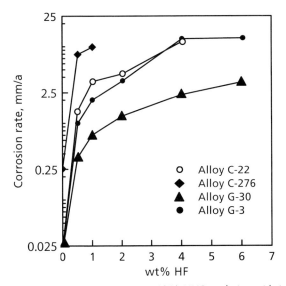

Figure 71: Corrosion rate in a 20 % HNO$_3$ solution with increasing HF concentration at 80 °C (175 °F) [96]
(1 mm/a ≙ 39.37 mpy)

To supplement the results, those for the nitric acid – hydrofluoric acid system are also given (Figure 71). This system is important for the pickling of stainless steels. The addition of HF to 20% HNO_3 at 353 K (80 °C) gives a large continual increase of the corrosion rate already commencing at 1% HF, whereby the sequence with respect to the high chromium and low molybdenum contents also applies.

High-performance alloys

Although extensive corrosion data is available for long-used stainless steels, there is still a lack of corresponding data for the "high-performance alloys". These include corrosion-resistant NiCr alloys with relatively high molybdenum content of up to approx. 30% for the chemical process industry and superalloys for the gas turbine industry. The work in reference [97] reports on the nickel-based alloys given in Table 64, the corrosion resistant alloys Ferralium® Alloy 255, Haynes® No. 20-Mod. and, for purposes of comparison, the stainless steels AISI 304 (X5CrNi18-10, DIN-Mat. No. 1.4301) and AISI 316 L (X2CrNiMo17-12-2, DIN-Mat. No. 1.4404).

	Ni	Fe	Cr	Mo	W	Co	Others
Hastelloy® Alloy G	bal.	19.5	22	6.5	< 1	< 2.5	Cu-2, Nb + Ta-2
Haynes® Alloy No. 625	bal.	5	21.5	9	–	< 1	Nb + Ta-3.5
Hastelloy® Alloy C-276	bal.	5	16	16	4	< 2.5	V-0.35
Hastelloy® Alloy B	bal.	5	< 1	28	–	< 2.5	–
Hastelloy® Alloy B-2	bal.	< 2	< 1	28	–	< 1	low carbon content
Ferralium® Alloy 255	5	bal.	25.5	3.5	–	–	Cu-1.7, N-0.17
Haynes® Alloy No. 20-Mod.	26	bal.	22	9	–	–	Ti-4×C Min.
AISI 304	9	bal.	19	–	–	–	–
AISI 316 L	12	bal.	17	2.5	–	–	low carbon content

Table 64: Composition of the alloys, mass% [97]

The corrosion rates in various acids, obtained from measurements of the mass losses, are summarised in Table 65. As expected, the highest corrosion rates were found in boiling 10% HCl, the most strongly reducing acid. By far the highest corrosion resistance was exhibited by B-2 in hydrochloric acid. In contrast, it had the lowest resistance by far in boiling 10% HNO_3. As already found in other work, the level of molybdenum is decisive for the corrosion behavior in hydrochloric acid, whereas the level of the Cr content is decisive in nitric acid.

Alloys	10% HCl	10% H$_2$SO$_4$	10% HNO$_3$	99% CH$_3$COOH	40% HCOOH	88% HCOOH	55% H$_3$PO$_4$	85% H$_3$PO$_4$
Hastelloy® Alloy G	28.85 (1,136)	0.635 (25)	0.02 (0.79)	0.041 (1.61)	0.13 (5.12)	0.109 (4.29)	0.099 (3.9)	0.686 (27.01)
Haynes® Alloy No. 625	12.9 (508)	0.168 (6.61)	0.025 (0.98)	0.01 (0.39)	0.185 (7.28)	0.236 (9.29)	0.114 (4.49)	1.803 (70.98)
Hastelloy® Alloy C-276	5.7 (224)	0.584 (22.99)	0.432 (17.01)	0.01 (0.39)	0.074 (2.91)	0.046 (1.81)	0.071 (2.8)	0.508 (20)
Hastelloy® Alloy B-2	0.175 (6.89)	0.046 (1.81)	490.5 (19,311)	0.03 (1.18)	0.011 (0.43)	0.015 (0.59)	0.091 (3.58)	0.102 (4.02)
Ferralium® Alloy 255	478.2 [1] (18,827)	7.04 (277.17)	0.048 (1.89)	0.03 (1.18)	0.196 (7.72)	0.965 (37.99)	0.130 (5.12)	44.35 (1.746)
Haynes® Alloy No. 20-Mod.	281.2 (11,071)	4.45 (175.2)	0.015 (0.59)	0.018 (0.71)	0.356 (14.02)	0.711 (27.99)	0.279 (10.98)	10.62 (418.11)
AISI 316 L	449.2 [1] (17,685)	60.88 (2,397)	0.053 (2.09)	0.097 (3.82)	1.067 (42.01)	0.483 (19.02)	0.201 (7.91)	16.38 (644.88)

[1] Sample dissolved

Table 65: Corrosion rates (in boiling media), mm/a (mpy) [97]

The degree of corrosion resistance of Hastelloy® Alloy B-2 is not negatively influenced by cold deformation and welding. Table 66 shows that for cold forming of up to 50 %, cold forming with subsequent welding as well as cold forming after welding, the corrosion rate remained < 0.25 mm/a (< 10 mpy).

State	mm/a (mpy)	State	mm/a (mpy)	State	mm/a (mpy)
0 % cold-formed	0.165 (6.5)	0 % cold-formed + welded	0.165 (6.5)	welded + 0 % cold-formed	0.155 (6.1)
10 % cold-formed	0.133 (5.24)	10 % cold-formed + welded	0.188 (7.4)	welded + 10 % cold-formed	0.138 (5.43)
30 % cold-formed	0.145 (5.71)	30 % cold-formed + welded	0.235 (9.25)	welded + 30 % cold-formed	0.14 (5.51)
50 % cold-formed	0.108 (4.25)	50 % cold-formed + welded	0.155 (6.1)	welded + 50 % cold-formed	0.15 (5.91)

Table 66: Corrosion rates for Hastelloy® Alloy B-2 in 10 % HCl [97]

Furthermore, this work also investigated the behavior of the alloys with regard to types of local corrosion, namely pitting, crevice and stress corrosion cracking. For passive materials, the level of the chromium and molybdenum contents is decisive for the resistance to pitting and crevice corrosion. These types of corrosion are caused by chloride ions, and the presence of any oxidising agents enhances the effect. A 10 % $FeCl_3$ solution is generally used to test for pitting and crevice corrosion of stainless and chemically resistant steels. However, this test solution is not aggressive enough for nickel-based alloys that often contain large amounts of molybdenum. To test the pitting corrosion of nickel-based alloys, a solution consisting of 7 vol% H_2SO_4 + 3 vol% HCl + 1 % $FeCl_3$ + 1 % $CuCl_2$ was used to simulate the conditions in scrubbers. Table 67 gives the results of samples with artificial crevices and of samples that are not in contact with another body. In both test solutions, and also at the higher temperatures of 323 and 375 K (50 and 102 °C), C-276 did not exhibit either pitting or crevice corrosion because of its high pitting index of almost 70 (calculated according to W = % Cr + 3.3 x % Mo). It must be mentioned at this point that in the presence of crevices, crevice corrosion always occurs before pitting corrosion. The samples become active in the crevice, and pitting corrosion on the surface is prevented because of cathodic protection by the anodic crevice region. The results in Table 67 confirm this fundamental corrosion mechanism for passive materials.

The author of [97] gives a very detailed report on the testing of local corrosion of nickel-based alloys in reference [98]. In addition to the materials given in Table 64 [97], the alloy Incoloy® 825 with the nominal percentage composition 42 % Ni, 39 % Fe, 21 % Cr, 3 % Mo, max. 1 % Mn, max. 0.5 % Si, max. 0.05 % C, 0.8 % Ti, 2 % Cu was included in the investigation. The main aim was to correctly determine the "critical pitting corrosion potential" using electrochemical methods because this potential (if correctly measured) would allow a reliable comparison of the pitting corrosion behavior of various materials in one medium. The difficulty concerned with the premature appearance of crevice corrosion on covers must be pointed out as it prevents the correct determination of the critical pitting corrosion potential. In addition, it is shown that the potentiodynamic quick test with a ramping rate of 1000 mV/min, which is often used, does not allow differentiation of the pitting corrosion behavior, particularly of the nickel-based alloy.

Alloy	10% FeCl$_3$; Sample with artificial crevice		Smooth samples 24 h 7 vol% H$_2$SO$_4$ + 3 vol% HCl + 1% FeCl$_3$ + 1% CuCl$_2$	
	25 °C (77 °F) 240 h	50 °C (122 °F) 100 h	25 °C (77 °F)	102 °C (215 °F)
Hastelloy® Alloy G	0.4 [15.8]	0.575 [22.6] (crevice corrosion)	0.01 [0.39]	43.95 [1,730] (pitting corrosion)
Haynes® Alloy No. 625	0.038 [1.50]	3.1 [122.1] (crevice corrosion)	0.008 [0.31]	47.3 [1,862] (pitting corrosion)
Hastelloy® Alloy C-276	0.005 [0.20]	0.005 [0.20] (crevice corrosion)	0.008 [0.31]	0.6 [23.6]
Ferralium® Alloy 255	0.015 [0.59]	19.95 [785.4] (crevice corrosion)	0.003 [0.12]	75.4 [2,968] (pitting corrosion)
Haynes® Alloy No. 20-Mod.	0.105 [4.13]	2.35 [92.5] (crevice corrosion)	0.058 [2.3]	57.95 [2,281] (pitting corrosion)
AISI 316 L	7.8 [307.1] (crevice corrosion)	11.5 [452.8] (crevice corrosion)	2.53 [100.8] (pitting corrosion)	97.73 [3,848] (pitting corrosion)

Table 4 [68], p. 35

Table 67: Corrosion rates, mm/a [mpy] [97]

The work focused on potentiostatic holding tests with holding times of 24 h for each potential step of 100 mV, from − 200 mV to + 800 mV, measured against a saturated calomel electrode. The sample was checked for local corrosion (pitting and/or crevice corrosion) after every holding test. For practical applications, this is a damage limit, regardless of whether one or the other type of corrosion occurs. Table 68 summarises the results of the 24 h holding tests in five different solutions at 343 K (70 °C) for Hastelloy® G. The shaded area represents the region of the free corrosion potential. The acidic and at the same time oxidising solutions of columns 1 and 3 are suitable as test solutions for nickel-based alloys because their corrosion potential, due to a high redox potential of the solution, is polarised to the damage potential, whereas the corrosion potentials of the three other solutions without oxidising agents are more or less negative and lie below the corresponding danger potential.

U_{SCE} mV	7 vol% H_2SO_4 + 3 vol% HCl + 1% $CuCl_2$ + 1% $FeCl_3$ (pH ≈ 0.2)	1 vol% HCl + 3.8% NaCl (pH ≈ 1.2)	3.8% $FeCl_3$ (pH ≈ 1.3)	4% NaCl (pH ≈ 5.5)	3% Na_2SO_3 + 3% $Na_2S_2O_5$ + 4% NaCl (pH ≈ 12)
+800	*	*	*	*	o
+700	*	*	*	*	o
+600	*	*	*	*	o
+500	*	*	*	*	o
+400	*	*	*	*	o
+300	o	*	*	*	o
+200	o	*	o	*	o
+100	o	o	o	*	o
0	o	o	o	o	o
−100	o	o	o	o	o
−200	o	o	o	o	o

* = local corrosion; o = no local corrosion

Table 68: Hastelloy® alloy G; Results of the 24 h holding tests in five different solutions at 70 °C (158 °F) [98]

Table 69 gives the potential ranges for damage or resistance of all investigated materials in a 3.8% $FeCl_3$ solution at 343 K (70 °C), determined with long-term potentiostatic tests. The resistance ranges are extended to more noble potentials as the molybdenum content in the alloys increases. This result agrees with theoretical and practical findings and confirms the serviceability of this test method.

U_{SCE} mV	2 to 3% Mo AISI 316	3% Mo Incoloy® Alloy 825	6.5% Mo Hastelloy® Alloy G	9% Mo Haynes® Alloy 625	16% Mo Hastelloy® Alloy C-276
+800	*	*	*	*	o
+700	*	*	*	*	o
+600	*	*	*	*	o
+500	*	*	*	o	o
+400	*	*	*	o	o
+300	*	*	*	o	o
+200	*	*	o	o	o
+100	*	*	o	o	o
0	*	*	o	o	o
−100	*	o	o	o	o
−200	o	o	o	o	o

* = local corrosion; o = no local corrosion

Table 69: 24 h exposure at constant voltage in 3.8% $FeCl_3$ solution at 70 °C (158 °F) [98]

In reference [99], the comparative corrosion resistance of alloys 625, C-276, Hastelloy® alloy C-22 and ALLCORR® was determined in HCl, H_2SO_4 and formic acid, the last two partly with low additions of HCl, HF, $FeCl_3$ and $CuCl_2$, at differing concentrations and temperatures over a test period of 48 h. The tolerance range for the composition of the alloys and the actual contents of the elements are summarised in Table 70. Welded samples were used that were prepared using the filling wires and welding conditions recommended by the manufacturers. The reason for the investigation was to select suitable materials for handling highly radioactive waste solutions, some of which are exceptionally corrosive.

	Alloy 625 (UNS N06625)	Alloy C-276 (UNS N10276)	Hastelloy® Alloy C-22 (UNS N06022)	ALLCORR® (UNS N06110)
Ni	60.20 (balance)	56.42 (balance)	57.52 (balance)	56.03 (balance)
Cr	22.34 (20.0-23.0)	16.11 (14.5-16.5)	21.31 (20.0-22.5)	30.93 (27.0-33.0)
Mo	8.48 (8.0-10.0)	15.99 (15.0-17.0)	13.55 (12.5-14.5)	9.97 (8.0-12.0)
W	– (–)	3.85 (3.0-4.5)	3.17 (2.5-3.5)	1.97 (4.0 max.)
Fe	4.53 (5.0 max.)	5.76 (4.0-7.0)	3.52 (2.0-6.0)	0.11 (–)
C	0.04 (0.10 max.)	0.002 (0.02 max.)	0.003 (0.015 max.)	0.030 (0.15 max.)
To	0.25 (0.40 max.)	– (–)	– (–)	0.24 (1.5 max.)
Nb (+ Te)	3.51 (3.15-4.15)	– (–)	– (–)	0.42 (2.0 max.)
P	0.010 (0.015 max.)	0.004 (0.030 max.)	0.007 (0.02 max.)	0.005 (–)
Al	0.14 (0.40 max.)	– (–)	– (–)	0.23 (1.50 max.)
Si	0.34 (0.50 max.)	0.03 (0.08 max.)	0.021 (0.08 max.)	0.02 (–)
V	0.001 (0.015 max.)	0.002 (0.030 max.)	0.002 (0.02 max.)	0.001 (–)
Mn	0.16 (0.50 max.)	0.47 (1.0 max.)	0.21 (0.50 max.)	0.01 (–)
Co	– (–)	1.22 (2.5 max.)	0.52 (2.5 max.)	0.01 (12.0 max.)
Cu	– (–)	– (–)	0.07 (–)	0.01 (–)
V	– (–)	0.14 (0.35 max.)	0.01 (0.35 max.)	0.01 (–)

Numbers in brackets give the maximum content or the possible fluctuation

Table 70: Composition, mass% [99]

Hydrochloric Acid

Solution [1]	%	Temperature °C (°F)	Alloys	Corrosion rate (μm/a)
HCl	35	84 (183)	625	28,000
			C-276	4,100 (2)
			C-22	6,800 (2)
			ALLCORR®	7,900 (2), (3)
HCl	15	108 (228)	625	25,000
			C-276	5,600 (2)
			C-22	7,100 (2), (3)
			ALLCORR®	15,000 (2), (3)
HCl	15	55 (131)	625	840
			C-276	280
			C-22	360 (3)
			ALLCORR®	710 (3)
HCl	10	104 (219)	625	36,000
			C-276	6,100 (3)
			C-22	11,000 (3)
			ALLCORR®	22,000 (2), (3)
HCl + HNO$_3$	28 17	109 (228)	625	> 99,000
			C-276	17,000 (3)
			C-22	2,000 (3)
			ALLCORR®	1,400 (2)

(1) Balance water (2) Intergranular corrosion (3) Welding seam corrosion

Table 71: Corrosion rates in HCl solutions [99]

Table 71 gives the results in HCl. The corrosion resistance is essentially determined by the amount of molybdenum contained in the alloy. The addition of 17% HNO$_3$ considerably lowers the corrosion rate, particularly that of alloy ALLCORR®, which has the highest Cr content of 31%. For the formic acid-sulphuric acid mixture, the low additions of HCl, HF, FeCl$_3$ and CuCl$_2$ increase the corrosion rate to different extents. It must be pointed out that, after the tests, many cases of intergranular corrosion and strong attack of the weld seam were found on the samples.

Investigations of various NiCrMo alloys for their possible use in supercritical water oxidation (SCWO) of organic waste is reported in [100]. This process is typically operated above the critical point of water (647 K (374 °C) and 221 bar for pure water). In the presence of compounds containing halogens and sulphur, the oxidation products are the corresponding acids, e.g. HCl, HF and H$_2$SO$_4$, in addition to CO$_2$ and H$_2$O. The materials listed in Table 72 were included in the investigation.

The tests were carried out in autoclaves in a mixture of water with 10 g/l dichloromethane (CH$_2$Cl$_2$) and 16 g/l H$_2$O$_2$ as the oxidising agent at a temperature of 693 K (420 °C) and a pressure of 400 bar. The test duration was 24 h. The mass losses of the samples for the 2 test runs are given in Table 73.

Material	UNS code	Cr	Ni %	Fe %	Al %	Mo %	W %	Others
G-30	N06030	29.4	40.1	14.5	–	5.1	2.9	Co, Cu
214	N07214	16.0	75.4	3.7	4.4	0.1	–	Zr, Cs*
602 CA		25.3	62.4	9.5	2.1	–	–	Ti
625	N06625	21.6	62.7	2.3	0.1	9.0	–	Ti
686	N06686	20.4	58.1	1.0	–	16.2	3.9	Ti

* in the commercially available variant, alloy 214 contains yttrium instead of caesium

Table 72: Composition of the investigated materials [100]

Material	Mass loss mg/cm^2	Mass loss mg/cm^2
G-30	2.8	2.9
214	26.6	–
214 preoxidised	50.8	39.9
602 CA	46.0-	–
602 CA preoxidised	15.5	–
625	–	33.5
686	73.3	23.8

Table 73: Mass losses of the samples after 24 h test period (2 test runs) [100]

According to this, only alloy G-30 can be expected to exhibit satisfactory corrosion behavior under SCWO conditions. The other alloys already show unacceptably high corrosion rates after 24 h.

A 30 Nickel-copper alloys

Nickel-copper alloys, e.g. Monel® Alloy 400 (UNS N04400, corresp to. DIN-Mat. No. 2.4360, 2.4366) exhibit lower resistance in hydrochloric acid than nickel itself. Some older data on the behavior of Monel® in different hydrochloric acid solutions at different temperatures is given in [47].

A 31 Nickel-molybdenum alloys

The NiMo alloy B-2 (UNS N10665, corresp. to. DIN-Mat. No. 2.4617), which has a low carbon content, was developed as a new version of the older grade alloy B (UNS N10001, corresp. to. DIN-Mat. No. 2.4800, 2.4810) especially for applications where there is exposure to hydrochloric acid. Newer variants are the grades alloy B-3 (UNS N10675) and alloy B-4 (UNS N10629) with improved thermal stability. Figure 72 shows an isocorrosion diagram for the grade B-2 [68]. Because the solubility of oxygen and thus the corrosion rate decreases at higher temperatures, there is a second line for the corrosion rate of 0.13 mm/a (5 mpy) at higher temperatures close to the boiling point.

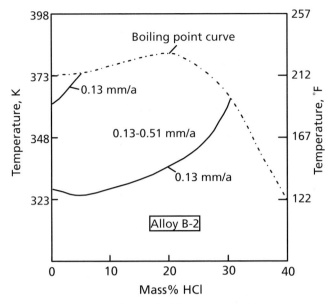

Figure 72: Isocorrosion diagram for alloy B-2 in hydrochloric acid [68, p. 28]
0.13 mm/a (5 mpy)
0.13-0.51 mm/a (5-20 mpy)

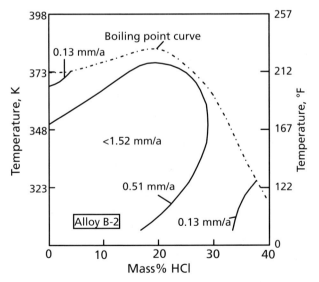

Figure 73: Isocorrosion diagram for alloy B-2 in hydrochloric acid purged with oxygen [68, p. 29]
0.13 mm/a (5 mpy)
0.51 mm/a (20 mpy)
< 1.52 mm/a (< 60 mpy)

Because of its high molybdenum content of 28 %, alloy B-2 and its variants exhibit the best resistance of all nickel alloys in hydrochloric acid. In non-aerated hydrochloric acid and in the absence of oxidising agents, alloy B-2 is one of the few metallic materials with a corrosion rate of < 0.5 mm/a (< 20 mpy) for all acid concentrations and temperatures.

Oxidising agents in the acid can have a noticeable detrimental effect on the resistance of the alloy, as shown in Figure 73 by the isocorrosion diagram for alloy B-2 in hydrochloric acid purged with oxygen [68].

The presence of a strong oxidising agent, such as e.g. Fe^{3+} or Cu^{2+} ions can lead to even higher corrosion rates in the acid, as shown in Figure 74 [68].

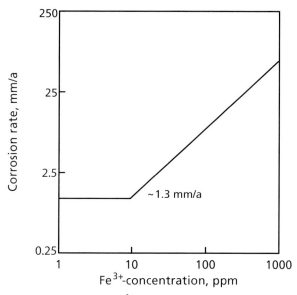

Figure 74: Influence of Fe^{3+} ions in boiling 20 % hydrochloric acid on the corrosion of alloy B-2 (1.3 mm/a ~ 50 mpy) [68]

Therefore, before NiMo alloys are used in hydrochloric acid, it is strongly recommended that welded samples are tested under near-service conditions.

Reference [101] mentions that melts of NiMo alloy B-2 (UNS N10665) that are within the permissible analysis limits can greatly differ in their mechanical properties and that some melts cause problems during production whereas others can be processed without problems. Table 74 gives the chemical composition of the NiMo alloys. The decisive difference between Hastelloy® B-2 and alloy B (UNS N10001) is the reduction of the carbon and silicon contents as well as the reduction of the iron content to max. 2 %. Because of the low carbon content of max. 0.01 %, it is not necessary to stabilise alloy B-2 with vanadium as it is for alloy B. Since the knife-line-corrosion and the attack of the heat-affected zone is prevented in B-2, in contrast to alloy B, alloy B-2 can be used in the welded state. Therefore, B-2 can be used for a much wider range of applications in modern process engineering. However, the occasional problems during processing (cracking) with individual melts necessitated a detailed investigation to elucidate this phenomenon.

Alloys (UNS No.)	Ni	Mo	Fe	Cr	Mn	Si	C
Alloy B (N10001)	balance	26 – 30	4 – 6	1 [A]	1 [A]	1 [A]	0.05 [A]
Alloy B-2 (N10665)	balance	26 – 30	2 [A]	1 [A]	1 [A]	0.08 [B]	0.01 [A]
Alloy B-3 (N10675)	65 [B]	27 – 32	1 – 3	1 – 3	3 [A]	0.10 [A]	0.01 [A]
Alloy B-4 (N10629)	balance	27 – 30	2 – 5	0.5 – 1.5	1.5	0.05 [A]	0.01 [A]
Alloy B-4 [C]	balance	28	3	1.3	0.6	0.03	0.006

[A] Maximum, [B] Minimum, [C] Typical composition

Table 74: Composition of the NiMo alloys, mass% [101]

As can be seen in Figure 75, the two brittle intermetallic phases Ni_4Mo and Ni_3Mo exist in the two-phase Ni-Mo system in addition to the face-centred cubic Ni-Mo matrix. Embrittlement and thus processing problems occur if intermetallic phases are formed, particularly Ni_4Mo. Earlier publications report that the formation of Ni_4Mo in Hastelloy® B depends on the iron content [102, 103]. The formation of this phase is suppressed at 4 % Fe. There is a significant reduction in the amount of Ni_4Mo for 2 % Fe in Hastelloy® B-2. Chromium has a similar effect on the formation of Ni_4Mo.

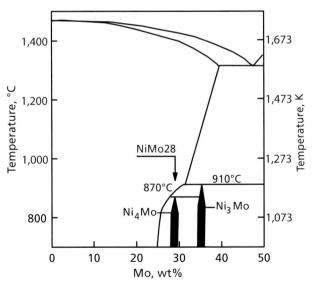

Figure 75: Phase diagram of the Ni-Mo system [101]

The influence of iron and chromium on the precipitation kinetics of Ni_4Mo was investigated with the melts summarised in Table 75. Melts 1 – 3 lie within the composition range of Hastelloy® B-2, whereas melts 4 – 6 lie outside it. The results show that embrittlement due to Ni_4Mo is prevented for a controlled content of 2 – 5 % Fe and 0.5 – 1.5 % Cr (see Table 74) and thus cracking during manufacture is avoided.

Melts	Mo	Fe	Cr	Mn	Si	Al	C
1	28.15	0.11	0.02	0.12	0.02	0.40	0.002
2	27.6	1.13	0.47	0.42	0.01	0.30	0.003
3	26.9	1.75	0.68	0.68	0.01	0.24	0.003
4	26.9	3.17	1.42	0.61	0.03	0.26	0.006
5	26.9	3.23	0.72	0.61	0.03	0.27	0.006
6	26.9	5.68	0.78	0.60	0.03	0.28	0.006

Table 75: Composition of the investigated melts of NiMo alloys, mass% [101]

Within these limits, the corrosion resistance is not negatively influenced as compared to that of Hastelloy® B-2 (Table 76). The resistance to stress corrosion cracking, which is influenced in NiMo alloys by the precipitation of intermetallic phases, was investigated by means of the ASTM G 30 test (boiling 10 % H_2SO_4) in the tempered state (1 h at 973 K and 6 h at 973 K (700 °C)). There was an improvement of a few hours endurance for melts 1 – 3, which correspond to B-2, whereas the endurance for melts 4 – 6, which correspond to B-4, increased to > 100 h (Table 78).

	Corrosion rate mm/a (mpy)	
Melts	Boiling 10 % HCl 24 h (A)	Boiling 20 % HCl 24 h (A)
1	0.29 (11.42)	0.58 (22.83)
2	0.31 (12.20)	not detectable
3	0.37 (14.57)	0.60 (23.62)
4	0.25 (9.84)	0.56 (22.05)
5	0.24 (9.45)	0.46 (81.11)
6	0.38 (14.96)	0.58 (22.83)

(A) Intergranular attack was less than 50 μm for all melts

Table 76: Corrosion rates in boiling HCl solutions [101]

	Content mass%			Time to fracture h	
Melts	Fe	Cr	Solution-annealed	Tempered 1 h, 973 K (700 °C)	Tempered 6 h, 973 K (700 °C)
1	0.08	≤ 0.01	> 100	3	2
2	1.13	0.47	> 100	3	3
3	1.75	0.68	> 100	45	3
4	3.17	1.42	> 100	> 100	> 100
5	3.23	0.72	> 100	> 100	> 100
6	5.86	0.78	> 100	> 100	> 100

(A) Investigations according to ASTM G 30 in 10 % boiling H_2SO_4

Table 77: Results of the stress corrosion cracking tests on the NiMo alloys[A] [101]

	Corrosion rate, mm/a (mpy)					
Tempering temperature	Melt 3 (1.75 % Fe, 0.68 % Cr)			Melt 4 (3.17 % Fe, 1.42 % Cr)		
K (°C)	0.5 h	1 h	8 h	0.5 h	1 h	8 h
873 (600)	0.32 (12.6)	0.29 (11.42)	0.39 (15.35)	0.18 (7.09)	0.33 (12.99)	0.31 (12.2)
923 (650)	0.33 (12.99)	0.38 (14.96)	0.33 (12.99)	0.27 (10.63)	0.37 (14.57)	0.31 (12.2)
973 (700)	0.35 (13.78)	0.36 (14.17)	0.87 [A] (34.25)	0.27 (10.63)	0.32 (12.6)	0.26 (10.24)
1023 (750)	0.35 (13.78)	0.35 (13.78)	0.39 (15.35)	0.24 (9.45)	0.35 (13.78)	0.29 (11.42)
1073 (800)	0.33 (12.99)	0.31 (12.2)	0.33 (12.99)	0.25 (9.84)	0.23 (9.06)	0.25 (9.84)
1123 (850)	0.32 (12.6)	0.32 (12.6)	0.31 (12.2)	0.34 (13.39)	0.20 (7.87)	0.25 (9.84)
1173 (900)	0.30 (11.81)	0.27 (10.63)	0.34 (13.39)	0.21 (8.27)	0.24 (9.45)	0.18 (7.09)

[A] Intergranular attack was less than 50 μm; and less than 25 μm for all other samples

Table 78: Corrosion rates of melt 3 and melt 4 in boiling 10 % HCl with respect to the tempered state [101]

The corrosion rates of melt 3 (corresponds to B-2) and melt 4 (corresponds to B-4) in boiling 10 % HCl are given in Table 78 depending on the tempered state. The results show that the higher iron and chromium contents in B-4 have a favorable effect on the resistance after tempering.

The alloy Hastelloy® B-4 (UNS N10629) represents a further improvement in the series of Ni-28 Mo alloys. It has advantages with regard to production and practical utilisation compared to Hastelloy® B-2.

In reference [104], comparative tests with non-heat-treated welding samples of Hastelloy® B and Hastelloy® B-2 proved the superiority of B-2 in boiling 20 % HCl for a test duration of 16 days. Mass loss measurements gave a corrosion rate of 1.5 mm/a (59 mpy) for Hastelloy® B and 0.6 mm/a (23.6 mpy) for Hastelloy® B-2. The corrosion of alloy B was mostly in the form of deeply penetrating knife-line corrosion at the transition between the welding seam and the parent material. Mass was lost uniformly from the surface of B-2 as it does not have any knife lines. Figure 76 gives the isocorrosion curve of B-2 in hydrochloric acid. From this it follows that the mass loss rate is < 0.5 mm/a (< 20 mpy) in the entire concentration and temperature range. It is pointed out that Hastelloy® B-2 also reacts sensitively to oxidising auxiliary substances in HCl. It is not recommended for use in direct contact with iron and copper in chloride-containing environments. The presence of these metals can lead to the formation of iron(III) and copper(II) chlorides which increase the corrosion rate.

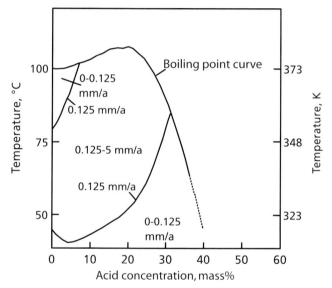

Figure 76: Isocorrosion curve of alloy B-2 in HCl [104]
0-0.125 mm/a   (0-5 mpy)
 0.125 mm/a   (5 mpy)
0.125-5 mm/a (5-200 mpy)

The susceptibility of NiMo alloys to intergranular corrosion in the strongly reducing acids HCl and H_2SO_4 is essentially due to the precipitation of molybdenum carbides on the grain boundaries and the associated molybdenum depletion in the region of the grain boundary. Thus the behavior is similar to that of CrNi steels that are also susceptible to intergranular corrosion due to chromium carbide precipitation and the associated chromium depletion.

In reference [105], the carbide precipitation and the associated intergranular susceptibility of NiMo alloys is discussed on a thermodynamic basis. Here, the solubility product of the carbide and the carbon activity are the decisive factors. The latter does not only depend on the main components, but also on other elements that are present in small amounts and which can be regarded as impurities. The tendency towards carbide precipitation increases as the solubility product of the carbide decreases and the carbon activity increases. Therefore, the carbon content must be kept as low as possible to counteract the tendency to precipitate carbides. Because impurities, particularly silicon, increase the carbon activity, they must also be reduced to a minimum. The theoretical results of this work are in agreement with the practical measures taken during the development of Hastelloy® B-2.

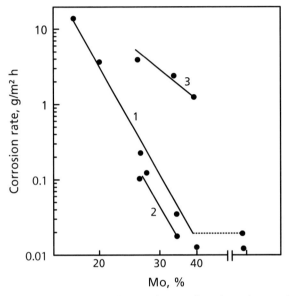

Figure 77: Corrosion behavior of NiMo alloys depending on the molybdenum content in the following aqueous acids:
1) 21 % boiling HCl;
2) 40 % boiling H_2SO_4;
3) 70 % boiling H_2SO_4;
Exposure time 200 h [105]

In [105], the corrosion behavior of NiMo alloys was investigated with respect to the molybdenum content of up to 40 % in boiling 21 % HCl (Figure 77). At 40 % Mo, the corrosion rate was approx. 20 times lower than that of the industrially used alloy with 28 % Mo. The alloy with 40 % Mo had a corrosion rate of 0.01 g/m² h, which is the same resistance as that of pure molybdenum. In the technical alloy, the level of molybdenum was limited to 28 % Mo because the intermetallic phases that occur above this value lead to precipitates, particularly during welding, that impair the corrosion resistance.

Recently, a Ni alloy with 25 % Mo and 8 % Cr (UNS N10242) was developed for use at higher temperatures. It has a low thermal expansion and high strengths over a wide temperature range. The behavior of this material in various acids of differing concentrations was investigated with respect to the heat-treatment and the results were compared to those from the NiMoCr alloy B-2 and the NiCrMo alloy C-2000 [106]. The compositions of the three materials are given in Table 79.

Alloy	UNS	Ni %	Cr %	Mo %	C max. %	Others max. %
242	N10242	65	8	25	0.03	2.5 Co; 2 Fe; 0.8 Mn; 0.8 Si; 0.5 Al; 0.5 Cu
B-3	N10675	65	1.5	28.5	0.01	3 Co; 1.5 Fe; 3 W; 3 Mn; 0.1 Si; 0.2 Cu
C-2000	N06200	59	23	16		1.6 Cu

Table 79: Nominal composition of the investigated alloys [106]

The strength values of alloy 242 can be noticeably increased by a corresponding heat treatment. They are then two to three times higher than those for the other two alloys, as shown in Figure 78.

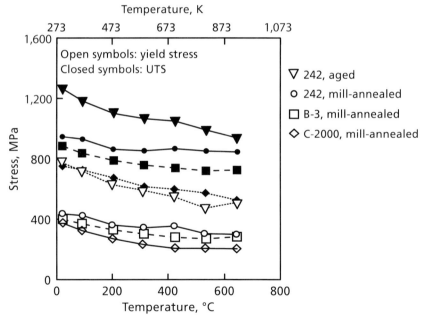

Figure 78: Values of the tensile strengths and yield strengths of the investigated alloys depending on the temperature [106]

All materials were tested in the as-delivered state. The strength of alloy 242 was further increased by tempering at 923 K (650 °C) for 24 h with subsequent cooling in air. Electrochemical investigations were carried out as well as immersion tests according to ASTM G 3 at 339 K (66 °C), 352 K (79 °C) and at the respective boiling point of the test solutions.

The following conclusions can be drawn from the results:

- Alloy 242 exhibits good corrosion resistance in reducing hydrochloric and sulphuric acids
- The corrosion resistance of alloy 242 in hydrofluoric acid is similar to that of alloy B-3
- The corrosion rates of alloy 242 in phosphoric acid, formic acid and acetic acid are very low and comparable to those of alloy B-3 under the same conditions
- Under moderately oxidising conditions, e.g. in aerated strong acids, alloy 242 exhibits a better corrosion resistance than alloy B-3
- The use of alloy 242 is not recommended in strongly oxidising acids (e.g. HNO_3)
- The heat-treated variant of alloy 242 exhibited higher corrosion rates than the alloy in the as-delivered state. This is particularly noticeable under strongly reducing conditions
- In both heat-treated states, alloy 242 is not susceptible to stress corrosion cracking in boiling acids and in boiling 45 % $MgCl_2$ solutions
- Alloy 242 appears to be susceptible to hydrogen embrittlement, particularly in the heat-treated state

Information on the concentration- and temperature-dependent corrosion behavior is given in Figures 84 to 87.

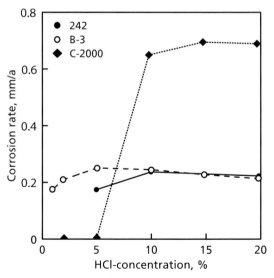

Figure 79: Corrosion rates depending on the HCl concentration at 339 K (66 °C) [106] (1 mm/a ≙ 39.37 mpy)

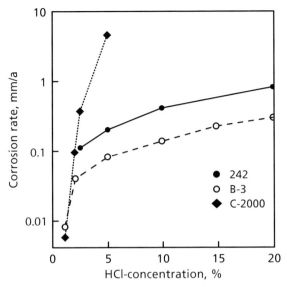

Figure 80: Corrosion rates of the alloys depending on the HCl concentration at the boiling point of the individual solution [106]
(1 mm/a ≙ 39.37 mpy)

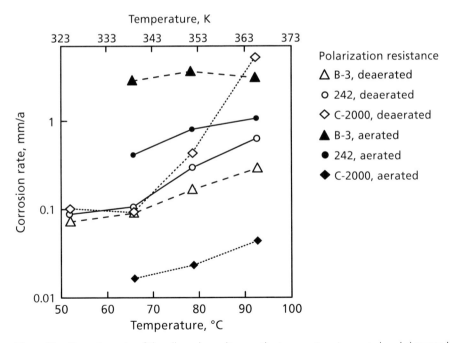

Figure 81: Corrosion rate of the alloys depending on the temperature in aerated and deaerated 1 M HCl [106]
(1 mm/a ≙ 39.37 mpy)

Hydrochloric Acid

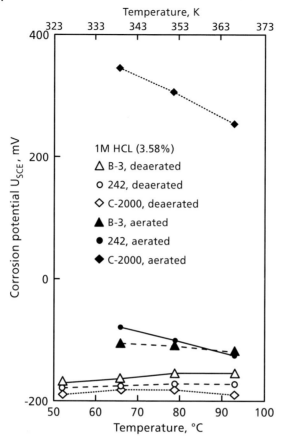

Figure 82: Corrosion potential of the alloys depending on the temperature in aerated and deaerated 1 M HCl [106]

Reference [107] reports on a case of damage to alloy B-2 in a Venturi scrubber of a catalytic reformer unit. The scrubber is used to neutralise the acidic process gases (CO_2, HCl, Cl_2) with an alkaline solution (pH 8 to 9). Leakages appeared in the upper region of the scrubber after an operating time of 9 months.

The damage was caused by the combined effect of corrosion and cavitation. A large temperature drop from 503 to 343 K (230 to 70 °C) in the affected area led to condensation of the very acidic gas components with simultaneous collapse of the gas bubbles. The cavitation effect of the bursting gas bubbles destroyed the protective surface coatings, enabling corrosive attack by the acidic condensates.

Because in addition to hydrochloric acid, to which alloy B-2 has a high resistance, high proportions of strong oxidising agents, such as oxygen, chlorine and Fe^{3+} ions were also present in the condensates, so that the corrosion resistance of the material would not have been sufficient under these conditions even without the cavitation effect. Subsequent laboratory investigations confirmed that a corrosion rate of

0.9 mm/a (35.43 mpy) must be expected for the material under these conditions. Investigations with the NiCr alloys C-276 (N10276; corresp. to: DIN-Mat. No. 2.4819), C-22 (N06022; corresp. to: DIN-Mat. No. 2.4602), C-2000 (N06200) Alloy 59 (N06059; corresp. to: DIN-Mat. No. 2.4605) and Alloy 686 (N06686; corresp. to: DIN-Mat. No. 2.4606); which usually exhibit better resistance in oxidising acids, also exhibited unsatisfactory behavior under these critical conditions. Thus, the same damage occurred in a replacement scrubber made of C-276 after an operating time of one year. Recommended countermeasures are either a strict control of temperature to prevent the temperature dropping below the dew point thus avoiding the formation of the aggressive condensates, or to test the use of acid-resistant ceramic coatings.

A 33 Lead and lead alloys

Lead is only moderately resistant to hydrochloric acid because the lead chloride produced by the reaction with the acid is relatively soluble so that it cannot form a protective layer. A limited resistance can only be expected at low acid concentrations and low temperatures (see Table 80) [68, p. 43 and 72]. The attack of lead by hydrochloric acid is noticeably increased in the presence of chlorine.

Acid concentration %	Temperature K (°C)	Corrosion rate mm/a (mpy)	
		HCl	HCl/Cl$_2$
0.5	359 (86)	0.18 (7.1)	0.94 (37.0)
0.5	395 (122)	0.33 (13)	1.8 (70.9)
1.0	359 (86)	0.15 (5.9)	1.1 (43.3)
1.0	395 (122)	0.30 (11.8)	5.59 (220)
2.5	359 (86)	0.18 (7.1)	2.2 (86.6)
2.5	395 (122)	0.41 (16.1)	6.25 (246)

Table 80: Corrosion rates of lead in dilute hydrochloric acid [68]

A 34 Platinum and platinum alloys
A 35 Platinum metals (Ir, Os, Pd, Rh, Ru) and their alloys

Platinum and the metals of the platinum group are resistant at room temperature to hydrochloric acid up to a concentration of 36 %. Corrosion rates for the metals of this group are given in [47].

A 36 Tin and tin alloys

Tin and its alloys are not resistant to hydrochloric acid

A 37 Tantalum, niobium and their alloys

Tantalum

Tantalum, along with tungsten and molybdenum, is a refractory metal. They are known for their high-temperature resistance and corrosion resistance in many aggressive media. The excellent corrosion resistance of tantalum is based on the formation of a passive layer of tantalum pentoxide, Ta_2O_5. Only media that are able to attack this oxide layer can corrode tantalum. These include strong alkalis, such as concentrated sodium and potassium hydroxide solutions, fuming sulphuric acid with free SO_2, fluorine, hydrofluoric acid and solutions that contain more than 5 ppm fluoride ions. Tantalum has excellent resistance to hydrochloric acid of all concentrations up to 463 K (190 °C). However, there may be strong attack at higher temperatures [108].

Tantalum exhibits the best resistance to hydrochloric acid of all special materials. With regard to this, Figure 83 shows an isocorrosion diagram (corrosion rate 0.13 mm/a (5 mpy)) for various materials depending on the concentration and temperature of the hydrochloric acid [68].

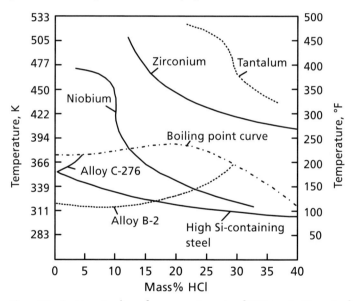

Figure 83: Isocorrosion lines for a corrosion rate of 0.13 mm/a (5 mpy) of various materials in hydrochloric acid depending on the concentration and temperature [68, p. 51]

Figure 84 shows the isocorrosion lines of tantalum for the corrosion rates 0 mm/a, 0.013 mm/a and 0.025 mm/a (0, 0.5 and 1.0 mpy) in hydrochloric acid depending on the concentration and temperature [68].

However, the great tendency of tantalum towards hydrogen embrittlement, even at room temperature, is a problem. Therefore, if tantalum is used it must be ensured that it does not undergo cathodic polarisation in solutions that contain hydrogen ions. Cathodic polarisation is quite possible in the case of electrically conducting contact with less noble metals, which includes nickel alloys, stainless steels, and even graphite in the case of tantalum. Because of these problems, components made of tantalum must be electrically insulated from other components [68, 108].

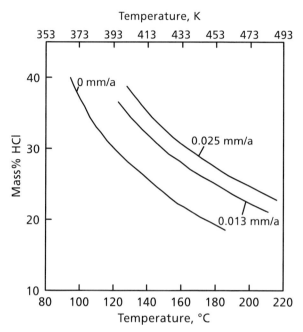

Figure 84: Isocorrosion lines of tantalum for corrosion rates of 0 mm/a, 0.013 mm/a and 0.025 mm/a (0, 0.5 and 1.0 mpy) in hydrochloric acid depending on the concentration and temperature [68, p. 52].

The possibility of hydrogen absorption must always be taken into consideration in hydrochloric acid at high concentrations and temperatures. Therefore, it is customary practice to check the suitability of tantalum for the planned application if the acid concentration and the wall temperature of the component lie above the isocorrosion line for 0 mm/a in Figure 84.

Hydrogen embrittlement of tantalum can be avoided by anodic polarisation. This can be carried out by means of an electrically conducting contact with a metal that has a lower hydrogen overpotential and which is electrochemically cathodic with respect to tantalum in an acidic environment. In this case, the hydrogen ions are discharged on the more noble metal and not on tantalum, so that H atoms cannot diffuse into the tantalum lattice. The noble metals platinum and palladium are especially effective for this.

As the data in Table 81 show, contact with platinum or palladium not only stops hydrogen embrittlement, but also the corrosion rates on platinum as well as on the two other noble metals are reduced [68].

Metal	Corrosion rate mm/a (mpy)		
	Contact-free	Contact with Pt	Contact with Pd
Tantalum	0.30 (11.81)	0.08 – 0.10 (3.15 – 3.94)	0.10 – 0.13 (3.94 – 5.12)
Platinum	2.0 (78.74)	< 0.025 (< 0.98)	
Palladium	1.6 (62.99)		< 0.025 (< 0.98)

Table 81: Corrosion rates of tantalum, platinum and palladium on their own and in contact with hydrochloric acid at 463 K (190 °C) [68]

Embrittlement must be expected in hydrochloric acid above an acid concentration of approx. 30 % and above a temperature of 413 K (140 °C) [68, 109]. If the surface is damaged and if a corresponding amount of hydrogen is available, then embrittlement can already occur at lower HCl concentrations and lower temperatures. In [110], hydrogen absorption of sintered and electron-beam melted tantalum wires was investigated by cathodic loading in 1 M HCl at 295 K (22 °C). The results showed that plastic deformation and mechanical surface damage (e.g. scratches) before the cathodic treatment did not have a significant effect on the hydrogen absorption. However, hydrogen absorption is greatly influenced if deformation and scratching take place during loading.

The most important conclusion for practical applications is that plastic deformation of the material, and in particular, direct damage to the surface must be avoided during the operation of equipment made of tantalum.

The electrical resistance of tantalum changes on absorption of hydrogen. This makes non-destructive indirect measurement of the hydrogen absorption possible.

Reference [111] reports on two cases of damage that were caused by hydrogen embrittlement of equipment made of tantalum. One case of damage occurred in a column that was used to concentrate acid. It was operated at a temperature of 453 to 493 K (180 to 220 °C) and a pressure of 6 bar.

The other case of damage occurred on a tantalum heat exchanger that was operating with 17.5 % hydrochloric acid and 5 bar pressure and was heated with steam at 460 K (187 °C) and 11 bar pressure. Leakages were found approx. 7 months after commissioning. Some of the tantalum tubes were so embrittled that they broke during transport. The investigation showed that the damage to the tantalum tubes had been caused by hydrogen. The hydrogen content of the damaged tubes was not homogeneous; it increased from low to high values in the direction towards the breakage point.

The authors assumed in both damage cases that there was a locally confined hydrogen source originating from leakage currents.

Because of these cases of damage, the behavior of tantalum in hydrochloric acid was tested in autoclave experiments. The test parameters were:

HCl concentration: 35 %
Temperature: 473 K (200 °C)
Material: tantalum, electron-beam melted
a) grain size according to ASTM 6
b) grain size according to ASTM 8
c) 90 % deformed
d) recrystallised.

The mass loss and the gas content were measured along with changes in the mechanical properties. After 15 days, there was a noticeable embrittlement (indicated by changes in the mechanical properties). There was no indication of a recognisable dependence on the grain size.

The tests showed that hydrogen absorption and embrittlement evidently take place at high hydrochloric acid concentrations and at high test temperatures.

Hydrogen loading tests in a direct stream of hydrogen at 873 K (600 °C) indicated that there was only a slight dependence on the Vickers hardness up to 600 ppm hydrogen (Figure 85). The yield point, tensile strength and elongation, determined in tensile tests, were independent of the hydrogen content for up to 600 ppm H_2. In contrast, a hydrogen content within these limits has a very pronounced effect on the electrical conductivity (Figure 86). The electrical conductivity can thus be used for non-destructive testing of the hydrogen absorption both in laboratory tests as well as in operating checks.

Reference [112] describes the special properties of tantalum compared to other materials. Tantalum is mainly produced by electron-beam melting. Furthermore, recycled tantalum metal is produced by arc melting. Cast arc-melted tantalum has an approx. 10 to 12 % higher strength than electron beam-melted grades.

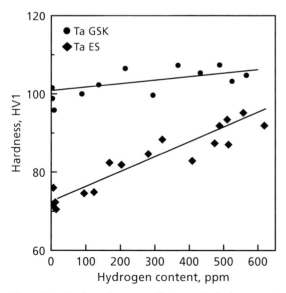

Figure 85: Hardness increase with increasing hydrogen content for Ta ES and Ta GSK [111]

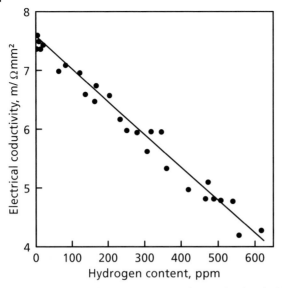

Figure 86: Electrical conductivity depending on the absorbed hydrogen [111]

In a comparison of procurement, repair and replacement costs over 15 years for tube bundles made of graphite and tantalum, the procurement costs of the tantalum bundle were three times higher, but the maintenance costs were only a third of those of the graphite tube bundle after the 15 years.

Tantalum-plated pump

According to [113], a centrifugal pump made of X5CrNiMo17-12-2 (AISI 316, DIN-Mat. No. 1.4401), lined with tantalum and tantalum carbide with a thickness of at least 0.25 mm, is particularly suitable for fluids that contain up to 10 % solids. It can pump to a height of 20 m and has a maximum capacity of 130 l/min. The pump is suitable for laboratory and pilot plants and can be used with glass or porcelain pipe systems.

Pump electroplated with tantalum

Reference [114] describes the high durability of ball valves made of cast iron. Their inner surfaces, which are in contact with the medium, have an electrolytic coating of tantalum approx. 15 µm thick produced from a fused salt bath using a Union Carbide process. The coating adheres firmly to the substrate and does not flake off, is free of inclusions and pores and has excellent ductility.

The ball valves at the base of the glass-lined reactor vessels are exposed to hydrochloric acid, organic chlorides, halogenides and alkaline substances at temperatures of up to 523 K (250 °C). Only the gaskets made of TFE have had to be replaced.

A tantalum alloy with 2.5 % W, produced by the arc-melting process, has an approx. 30 to 35 % higher strength. Its corrosion behavior is the same or even better than that of tantalum in most corrosive media. Because of the higher strength, costs can be reduced by reducing the weight.

Niobium

In the process that oxidises organic compounds with supercritical water (supercritical water oxidation SCWO), the organic compounds are dissolved in water above the critical point (647 K (374 °C) and 22.1 MPa) and oxidised (e.g. with oxygen) to carbon dioxide and water. Under these critical conditions, the solubility of oxygen in water is very high. The foreign atoms chlorine and sulphur react to give HCl and H_2SO_4. High-temperature water containing these acids and oxygen is extremely corrosive for NiCr alloys and stainless steels. Titanium, tantalum and niobium metals can be expected to have a better corrosion resistance than these materials under oxidising SCWO conditions.

The behavior of niobium samples was tested under simulated SCWO conditions in a pressurised reactor lined with an alumina ceramic [115]. The test conditions and results are given in Table 82.

Acid	O_2 mol/kg	T K (°C)	Testing time h	Mass loss mg/cm^2
0.05 mol/kg HCl	0.48	623 (350)	0.75	0.1
	0.48	623 (350)	5	0.6
	0.48	623 (350)	14.5	3.1
	0.48	623 (350)	20	5.7
	0.48	623 (350)	50	18.3
	2	623 (350)	45	40.2
	2	873 (500)	62	X
0.05 mol/kg H_2SO_4	2	623 (350)	49	49.6
	2	873 (500)	12	X
		873 (500)	29	X

X = complete conversion to the oxide

Table 82: Test conditions and results. The test pressure was always 24 MPa [115]

According to these results, niobium cannot be used as a reactor material for supercritical water under oxidising conditions, i.e. under SCWO conditions. In the subcritical temperature range, the corrosion rates were, however, quite low and noticeably lower than those of the nickel-based alloys, so that niobium can indeed be an alternative material in subcritical water.

Nb-Ta

Niobium has an excellent resistance to numerous mineral acids, except for hydrofluoric acid; however, in contrast to tantalum, it is also attacked by hot non-oxidising mineral acids, such as hydrochloric acid and sulphuric acid, or at higher temperatures. Thus in azeotropic boiling hydrochloric acid (20.2 % HCl, 328 K (55 °C)), the corrosion behavior of niobium and tantalum differs considerably [116]. However, under strongly oxidising conditions, the corrosion behavior of niobium is similar to that of tantalum.

The addition of niobium to tantalum noticeably reduces the acid resistance and thus limits the application range of TaNb alloys to low acid concentrations and temperatures. However, a series of niobium-tantalum alloys have been developed that are less expensive than tantalum but which have a better corrosion resistance than niobium. Since Nb and Ta form solid solutions in all ratios, these alloys are easy to produce and process.

Corrosion tests of NbTa alloys in hydrochloric and sulphuric acids show that the corrosion rate drops steeply as the tantalum content increases. This is probably due either to an increase of the tantalum concentration in the passive layer during the corrosion process as a result of preferential dissolution of niobium or to a concentration shift to higher tantalum contents in the surface of the metal from the outset.

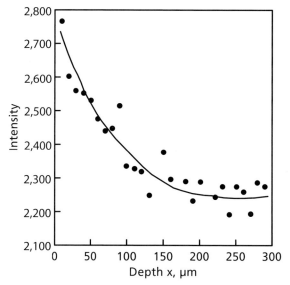

Figure 87: Intensity/depth profile of the tantalum signal of a 95 Nb-5 Ta alloy sample after depletion annealing [116]

Additional enrichment of tantalum at the surface also occurs during high-temperature annealing. Thus, Figure 87 shows the intensity/depth profile of the tantalum signal from a 95Nb-5Ta alloy sample (atom%) determined by an electron-beam microprobe after three-hour annealing at a temperature of 2573 K (2300 °C) and a pressure of 2×10^{-7} bar. A pronounced enrichment of Ta can be seen up to a depth of 100 µm. The corrosion rate of annealed samples is correspondingly lower (Figure 88 and Figure 89), because as the tantalum content increases, the protecting passive layer increasingly assumes the properties of a passive layer of Ta$_2$O$_5$ on pure tantalum. The practical utilisation of this measure to increase the corrosion behavior is probably limited.

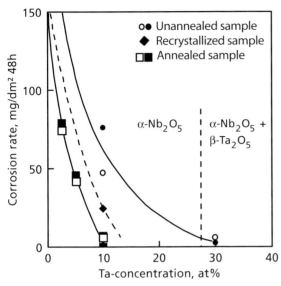

Figure 88: Corrosion rates of NbTa alloys in azeotropic boiling hydrochloric acid (20.2 vol%, T = 381.6 K (108.6 °C). The solid symbols correspond to gravimetrically determined corrosion rates [116]

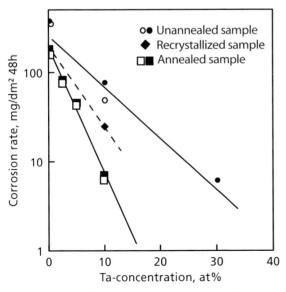

Figure 89: Logarithmic plot of the corrosion rate K against the tantalum concentration of NbTa alloys. The solid symbols correspond to gravimetrically determined corrosion rates [116]

Reference [117] reports on investigations of the corrosion behavior of tantalum, niobium and various mixtures of the two in hot hydrochloric acid and phosphoric acid. The composition of the investigated materials is given in Table 83.

Material	Nb %	Ta %	N ppm	O ppm
Nb	98.6	–	25 ± 2	156 ± 15
Nb-20Ta	77.5	20.4	42 ± 12	599 ± 41
Nb-40Ta	59.5	39.4	33 ± 6	643 ± 42
Nb-60Ta	38.8	60.9	29 ± 5	653 ± 24
Nb-80Ta	18.3	80.8	20 ± 4	966 ± 15
Ta	–	99	13 ± 2	204 ± 3

Table 83: Chemical composition of the investigated samples of Nb, Ta and NbTa alloys [117]

Corrosion tests in hydrochloric acid were carried out with acid concentrations of 5, 10, 15 and 20 % at the boiling point as well as in autoclave tests at 423 and 473 K (150 and 200 °C). For the tests at the boiling point, the test duration was varied between 3 and 14 days. The test duration was always 14 days at the higher temperatures. The results are presented as the corrosion rates (µm/a) calculated from the mass losses with respect to the sample surface, the material density and the test duration. The results of pure niobium and alloy Nb-20Ta in boiling hydrochloric acid are given in Figure 90 to Figure 92. For both materials, the corrosion rate increased with increasing acid concentration but decreased with increasing test

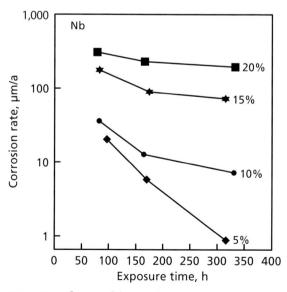

Figure 90: Influence of the test duration on the corrosion behavior of niobium in boiling hydrochloric acid [117]

duration. An addition of 20% Ta increased the corrosion resistance compared to tantalum-free niobium.

For the other alloys and pure tantalum, no mass losses were measurable after a test period of 14 days in boiling hydrochloric acid.

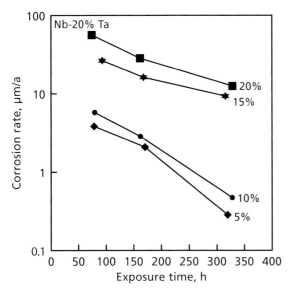

Figure 91: Influence of the test duration on the corrosion behavior of Nb-20Ta in boiling hydrochloric acid [117]

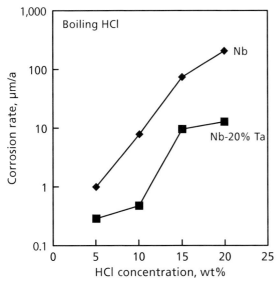

Figure 92: Influence of the acid concentration on the corrosion behavior of Nb and Nb-20Ta in boiling hydrochloric acid [117]

At the test temperatures of 423 K (150 °C), Figure 93 and 473 K (200 °C), Figure 94, the corrosion rate also increased with increasing acid concentration and temperature and decreased with increasing proportion of tantalum. In all cases, a sudden increase in the corrosion rate was observed on increasing the acid concentration from 15 to 20 %.

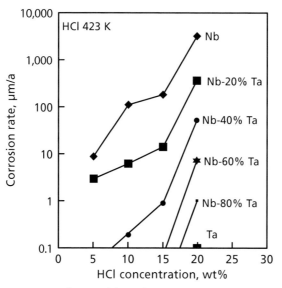

Figure 93: Influence of the acid concentration on the corrosion behavior of Nb, Ta and their alloys in boiling hydrochloric acid at 423 K (150 °C) [117]

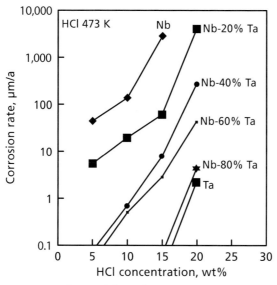

Figure 94: Influence of the acid concentration on the corrosion behavior of Nb, Ta and their alloys in boiling hydrochloric acid at 473 K (200 °C) [117]

The attack was always in the form of uniform surface corrosion; local attack was not observed.

The results of all the data obtained for the materials Nb, Nb-20Ta, Nb-40Ta and Nb-60Ta are given as isocorrosion curves for the corrosion rates 25, 125, 250 and 500 μm/a in Figure 95 to Figure 98. No corrosion rates above 25 μm/a were found for alloy Nb-80Ta and for tantalum.

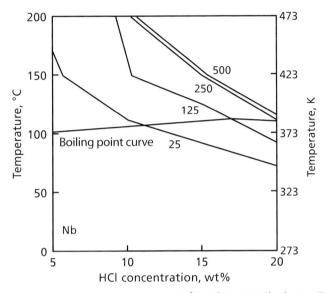

Figure 95: Isocorrosion curves (μm/a) for niobium in HCl solutions [117]

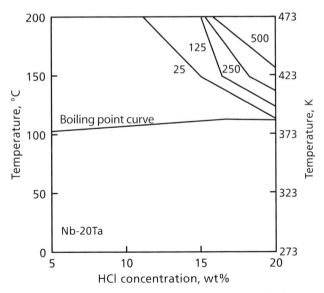

Figure 96: Isocorrosion curves (μm/a) for Nb-20Ta in HCl solutions [117]

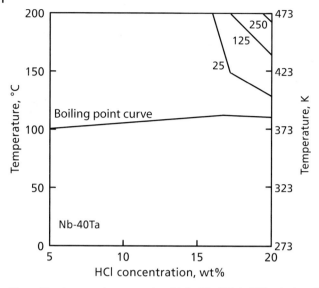

Figure 97: Isocorrosion curves (μm/a) for Nb-40Ta in HCl solutions [117]

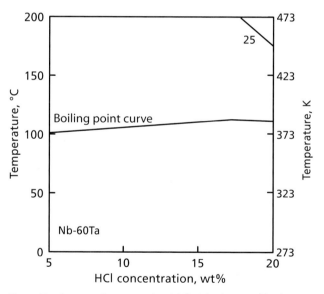

Figure 98: Isocorrosion curves (μm/a) for Nb-60Ta in HCl solutions [117]

The results can be summarised with the following conclusions:

- attack increases with increasing acid concentration for all materials and temperatures
- attack increases with increasing temperature for all materials and acid concentrations

- the corrosion rate decreases with increasing test duration due to the formation of protective covering layers
- the addition of tantalum increases the corrosion resistance of niobium
- NbTa alloys have a comparable corrosion rate to tantalum above 60 mass% Ta.

The question of whether hydrogen embrittlement of tantalum can be prevented by alloying additions was investigated by studying the corrosion behavior of tantalum and tantalum alloys in concentrated sulphuric acid at 473 and 523 K (200 and 250 °C) and in boiling hydrochloric acid of azeotropic composition of (20.2 % HCl, 382 K (109 °C) as well as in concentrated HCl (38 %) at 423 K (150 °C) in an autoclave with a Teflon lining [118]. The following alloys were investigated:

TaMo, TaW, TaNb, TaHf, TaZr, TaRe, TaNi, TaV, TaWMo, TaWNb, TaWHf, TaWRe

In contrast to concentrated sulphuric acid at 523 K (250 °C), the addition of 0.5 to 1.5 % Mo can completely suppress embrittlement. In 38 % HCl at 423 K (150 °C), none of the investigated alloys exhibited a significant reduction of the hydrogen absorption and thus no reduction of the embrittlement was achieved. In boiling azeotropic HCl, no attack and no embrittlement was detectable after a test period of 3 months.

A 38 Titanium and titanium alloys

Titanium is one of the very reactive metals and thus reacts spontaneously with oxygen in the air, even at room temperature, to form a thin but very compact and firmly adhering oxide layer on the surface. Its excellent resistance to a variety of aggressive media is solely due to the resistance of this protective passive layer. Therefore, it can be expected that the resistance is predominant in oxidising environments, whereas it is retained only to a limited extent in reducing media. Thus only limited resistance can be expected in strongly reducing hydrochloric acid.

As Table 84 shows, the corrosion rate of titanium in hydrochloric acid increases nearly steadily with increasing acid concentration [68].

Acid concentration %	Corrosion rate mm/a (mpy)
1	0.003 (0.12)
5	0.009 (0.35)
7.5	0.28 (11.0)
15	2.4 (94.5)
20	4.4 (173.2)

Table 84: Corrosion rate of titanium in hydrochloric acid at 308 K (35 °C) depending on the acid concentration [68]

Figure 99 shows that titanium can only be used in hydrochloric acid only up to a concentration of 7.5 % at room temperature, up to 3 % at 333 K (60 °C) and only up to 0.5 % at 373 K (100 °C).

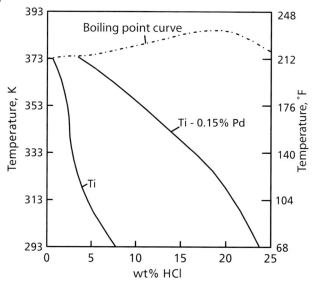

Figure 99: Isocorrosion diagram (0.1 mm/a (3.94 mpy)) of pure titanium and for titanium with 0.15 % Pd [68, p. 45]

Alloying with small amounts of palladium noticeably increases the corrosion resistance of titanium. Paired with titanium, palladium forms cathodic areas on the surface and the resulting galvanic current polarises them in the anodic direction. This anodic polarisation promotes the formation and conservation of the passivating oxide layer and also increases the resistance in hydrochloric acid. The corresponding isocorrosion line is plotted in Figure 99 as an example of the titanium grade with 0.15 % Pd, thus demonstrating the remarkably improved resistance of Pd-alloyed titanium grades.

Pitting corrosion of Ti and Ti alloys in chloride solutions

Reference [119] reports on investigations of the resistance of pure titanium and the Russian alloys 4200 (Ti-0.2 %Pd) and 4207 (Ti-2.5 %Ni-2 %Zr) to chloride solutions acidified with HCl at 373, 393, 413 and 433 K (100, 120, 140 and 160 °C). The following chloride solutions were used. Mass% 25 KCl, 25 NH_4Cl, 40 LiCl, 40 $CaCl_2$, 30 $BaCl_2$, 30 $MgCl_2$, 30 $NiCl_2$ and 50 $CdCl_2$ with additions of up to 0.3 % HCl. The tests were carried out in glass tubes that were sealed with a Teflon® tape and Teflon® stoppers. The temperature of the glass tubes was kept constant in drying cupboards. The duration of each test was 100 h.

Figure 100 and Figure 101 show the corrosion behavior depending on the temperature and the hydrochloric acid concentration of the solution. Above the curve, titanium is not resistant to pitting corrosion or to uniform and non-uniform surface corrosion. Below the curve, titanium is corrosion-resistant. As the hydrochloric acid concentration increases, the resistance is always more or less strongly limited depending on the type of salt solution.

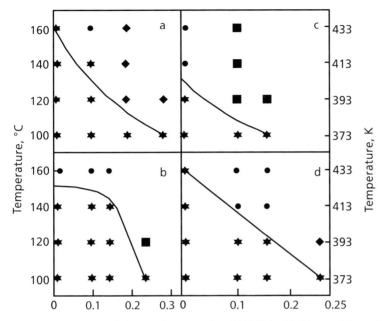

Figure 100: Corrosion behavior of pure titanium (VT 1-0) in chloride-containing solutions [119]
a) 25 % KCl; b) 25 % NH$_4$Cl; c) 30 % BaCl$_2$; d) 30 % NiCl$_2$
★) no corrosion; ●) pitting corrosion; ◆) uniform surface corrosion; ■) non-uniform surface corrosion

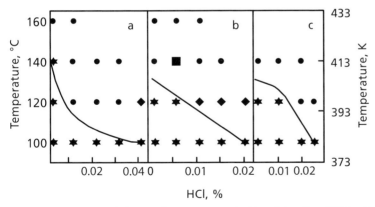

Figure 101: Corrosion behavior of pure titanium (a, b) and the alloy 4207 in chloride-containing solutions [119]
a) 40 % CaCl$_2$; b) 40 % LiCl; c) 30 % MgCl$_2$.
★) no corrosion; ●) pitting corrosion; ◆) uniform surface corrosion; ■) non-uniform surface corrosion

Intergranular hydrogen embrittlement

If titanium and its alloys are to be used in hydrochloric acid, then it must be taken into consideration that the corrosion reaction liberates hydrogen that can lead to embrittlement of the metal. The absorption of hydrogen can be enhanced by contact with less noble materials that polarise titanium cathodically, or by stray currents.

Reference [120] reports on investigations of the observation that if pure titanium is exposed in the unstressed state to a methanol-hydrochloric acid solution, it exhibits intergranular brittle fractures if it is subjected to slow elongation immediately afterwards. The investigations were carried out on pure titanium IMI 25 with the following composition in mass%:

0.13 % O; 0.0061 % N; 0.0022 % H; 0.001 % C; 0.02 % Fe; rest Ti

The test solution was analytical grade CH_3OH with 1 vol% analytical grade HCl at ambient temperature. After the samples had been taken out of the solution, they were broken apart in air by means of a slow tensile tester with a strain rate of 5.56×10^{-6} s^{-1}. Figure 102 shows the maximum intergranular penetration depth depending on the exposure time. When the samples were heat-treated in an argon atmosphere after exposure at 696 K (423 °C) for 20 h, the slow tensile tests showed that they no longer exhibited strain-rate dependent embrittlement. This proves that the substance which is decisive for the embrittlement and which is absorbed during exposure is very mobile, so there is little doubt that it is hydrogen.

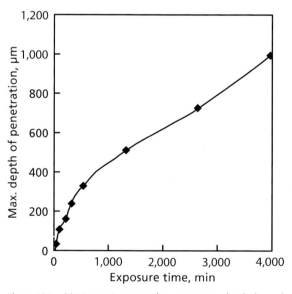

Figure 102: Maximum intergranular penetration depth depending on the exposure time [120]

At very slow strain rates, effusion of hydrogen takes place during the tensile test and a strain rate–dependent embrittlement is no longer observed.

The proposed mechanism for the strain-rate dependent embrittlement suggests that during exposure the adsorbed hydrogen mainly diffuses in along the grain boundaries. This report proposes that during subsequent elongation, the hydrogen absorbed from the environment is transported along the grain boundaries by the movement of the dislocations at low strain rates. The high hydrogen concentration in the grain boundary region together with the high stresses caused by the increase of dislocations leads to intergranular cracks.

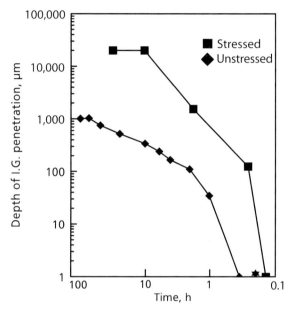

Figure 103: Maximum intergranular penetration depth for titanium samples exposed to a methanol-hydrochloric acid solution (1) with strain loading and (2) for unloaded samples. The difference between the two curves represents stress-induced intergranular fracture [120]

If the samples are exposed to the methanol-hydrochloric acid solution for the same length of time with simultaneous strain loading, then the depth of intergranular fracture is approx. 10 times greater than that measured for pre-exposed unstrained samples (Figure 103). The difference between the two curves represents intergranular fracture induced by stress/strain. This difference also results from hydrogen embrittlement. The same topic is treated in [121], and the results are essentially the same. Such strains can occur both by alternating internal pressures as well as by mechanical loading, so that this phenomenon is certainly of practical importance.

Inhibition by oxidising agents

The corrosion of titanium and zirconium as well as their alloys can be noticeably inhibited in acidic media, particularly in hydrochloric acid of various concentrations, by the addition of multivalent metal ions as well as inorganic and organic oxidising agents [122]. It is important that the inhibitor concentration is high enough to prevent damage. The inhibiting action is mainly due to oxidation which passivates the surface of the metal. Thus, in boiling 2 M HCl in the presence of metal ions of the platinum group or gold, titanium can be passivated by the addition of a concentration of 10^{-3} mol/l. The influence of various noble metal salts is shown in Table 85.

Additive	Concentration mol/l	Corrosion rate mm/a (mpy) 1% H_2SO_4	Corrosion rate mm/a (mpy) 3% HCl
None	–	9.04 (355.9)	5.38 (211.8)
$CuSO_4$	0.01	0.015 (0.59)	–
Ag_2SO_4	0.01	0.0	–
Hg_2SO_4	0.01	0.0	–
$PtCl_4 \cdot 2HCl$	0.01	0.008 (0.31)	0.02 (0.79)
$HAuCl_4$	0.01	0.008 (0.31)	0.02 (0.79)
$HgCl_2$	0.01	–	0.008 (0.31)
$CuCl_2$	0.01	–	0.008 (0.31)

Table 85: Influence of various noble metal salts on the corrosion rate of commercially available titanium in boiling 1% H_2SO_4 and 3% HCl [122]

Additive	Concentration mol/l 10^{-5}	Corrosion rate mm/a (mpy)
None	–	0.45 (17.7)
Cu^{2+}	0.05	0.49 (19.3)
	0.10	0.62 (24.4)
	0.20	0.76 (29.9)
	0.30	0.74 (29.1)
	0.40	0.00
Pt^{4+}	0.050	1.07 (42.1)
	0.075	1.29 (50.8)
	0.100	1.15 (45.3)
	0.200	0.00

Table 86: Corrosion rates of titanium in 15% HCl at ambient temperature depending on the concentration of Cu^{2+} and Pt^{4+} [122]

The concentration of the noble metal ions must exceed a critical value. Table 86 gives the corrosion rates of titanium in 15% HCl at ambient temperature depending on the concentration of Cu^{2+} and Pt^{4+}. Passivation starts above a certain concentration.

Oxidising inorganic acids are also effective inhibitors for titanium. Thus, the addition of 1% HNO_3 to 20% HCl at 308 K (35 °C) decreases the corrosion rate from

3.48 to less than 0.025 mm/a (1.0 mpy). The addition of 1 % CrO_3 to 5 % HCl at 363 K (90 °C) decreases the corrosion rate from 71 to 0.025 mm/a (2795 to 1 mpy). In 2.5 M H_2SO_4 and 5 M HCl at 353 K (80 °C), $SbCl_3$ is an effective inhibitor at a concentration of 10 mmol/l. The passivation mechanism of Sb^{3+} is similar to that of heavy metal cations because Sb is deposited on the corroding titanium surface.

Utilisation of Ti and alloy Ti-Code 12

Reference [123] describes the properties of the alloy Ti-Code 12 with 0.8 % Ni and 0.3 % Mo. This alloy has a higher strength than pure titanium (American designation ASTM Ti Grade 2 (UNS R50400)) and the 0.25 % Pd-containing titanium alloy ASTM Ti Grade 7 (UNS R52400, Ti2Pd). The corrosion behavior lies between that of the two alloys frequently used in the chemical process industry, namely ASTM Grade 2 and Grade 7. However, for some applications, e.g. for heat exchanger tubes in oxidising media, such as nitric acid, Ti-Code 12 is superior to the other two alloys. This is due to the higher strength of Ti-Code 12 both at ambient temperature as well as at higher temperatures. For example, the strength at 590 K (317 °C) is 65 % higher than that of Grades 2 and 7. This means that the tube walls can be thinner and this results in lower costs.

The crevice corrosion resistance of Ti-Code 12 in concentrated chloride-containing salt solutions is similar to that of Grade 7 (Grade 2 undergoes crevice corrosion under these conditions). However, when in contact with reducing acids, Ti-Code 12 is not better than unalloyed titanium if there are no passivating compounds present, e.g. nitrates and chromates. Therefore, if a higher strength is useful in certain cases, Ti-Code 12 has essentially the same range of applications as pure titanium with respect to its corrosion behavior.

Stress corrosion cracking of Ti-15 Mo

References [124] and [125] report on investigations of stress corrosion cracking of a Ti-15 %Mo alloy in $HCl-CH_3OH-CH_3OPSCl_2$ solution. At high temperatures, titanium and molybdenum have complete mutual solubility. At low temperatures, the mixed crystal becomes instable so that α-titanium precipitation occurs in alloys with less than 30 % Mo. The high-temperature-β-mixed crystal microstructure can be conserved, even at room temperature, by quenching.

Ti-15 Mo alloys have a good corrosion resistance and lower susceptibility to stress corrosion cracking in $HCl-CH_3OH$ solutions in contrast to pure titanium. However, in solutions which additionally contain methyloxythiophosphoryl dichloride, they are susceptible to stress corrosion cracking. The precipitation of α- or ω-phases not only decreases the corrosion resistance, but it also increases the susceptibility to stress corrosion cracking. This is in agreement with the general experience that homogeneous alloy systems generally exhibit the best possible corrosion resistance because structurally caused local elements cannot form on the surface which could lead to a reduction of the corrosion resistance.

The alloy was produced by mixing and pressing of pure titanium sponge (interstitial impurities: O < 0.05, C < 0.01, N < 0.01, H < 0.005 mass%) and pure molybdenum chips (99.9 mass% purity) to electrodes and melting in a vacuum arc furnace with a sacrificial electrode. The resulting cast samples were remelted in a vacuum into 140 mm round samples which were hot-forged into 16 mm ⌀ round samples and then hot-rolled into sheets.

Since the microstructure is dependent on the heat treatment, the stress corrosion cracking behavior was investigated with respect to the solution annealing temperature. The annealing solution contained 10% HCl, 40% CH_3OH and 50% CH_3OPSCl_2. Because the boiling point of CH_3OPSCl_2 is approx. 273 K (0 °C), the solution and the samples were kept at approximately this temperature in a cooling bath.

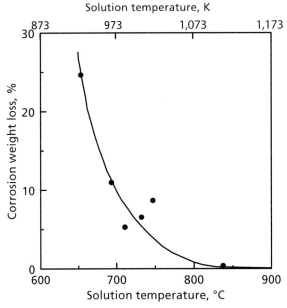

Figure 104: Influence of the solution annealing temperature on the corrosion resistance of Ti-15Mo [124]

The test, which lasted 500 h, investigated the influence of the solution annealing temperature on the corrosion resistance of Ti-15Mo in the aforementioned test solution. Figure 104 shows the percentage mass loss depending on the solution annealing temperature [124]. The homogeneous β-microstructural state, which is produced at a solution annealing temperature of 1123 K (850 °C) with subsequent quenching, exhibits a high corrosion resistance. Figure 105 shows the endurance in the stress corrosion test depending on the solution annealing temperature [124]. A large increase in the time-to-fracture is associated with the low mass loss above a solution annealing temperature of 1123 K (850 °C). Mass loss rate and resistance against stress corrosion cracking thus greatly depend on the microstructure which in turn depends on the heat treatment. As expected, the homogeneous β-microstructure produced by solution annealing above 1123 K (850 °C) and subsequent rapid quenching is the most favorable with respect to corrosion and resistance against stress corrosion cracking. This also applies to other media.

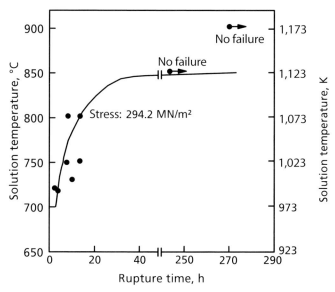

Figure 105: Influence of the solution annealing temperature on the time-to-fracture of Ti-15Mo in HCl-CH$_3$OH-CH$_3$OPSCl$_2$ [124]

Amorphous microstructure

In general, the microstructure and the composition are the main factors that decide the properties of an alloy, even the corrosion properties. Titanium has a high corrosion resistance in various media, but not in non-oxidising acids such as HCl. Very rapid cooling from the liquid state considerably improves the corrosion resistance of titanium alloys, as shown by the current density/potential curves in 1 M HCl at ambient temperature by alloys of the systems

| Ti-Al-Zr-W | and | Ti-Al-Zr-W-B, |
| Ti-Al-Zr-Mo | and | Ti-Al-Zr-Mo-B, |

that were cooled from the liquid state at a rate of 106 K/s [126].

The Ti-Al-Zr-W system has a finely dispersed microstructure after quenching, and has a corrosion rate in the passive potential range that is lower by approx. two powers of ten compared to that of the cast state. The boron-containing alloy solidifies in the amorphous state and exhibits the same improvement (Figure 106 and Figure 107).

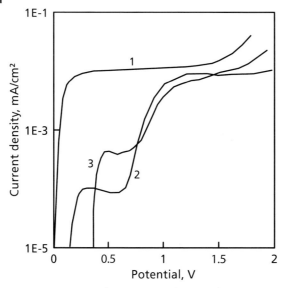

Figure 106: Current density/potential curves [126]
1 – Ti-Al-Zr-W cast alloy
2 – Ti-Al-Zr-W cast alloy (quenched)
3 – Ti-Al-Zr-W-B alloy (amorphous)

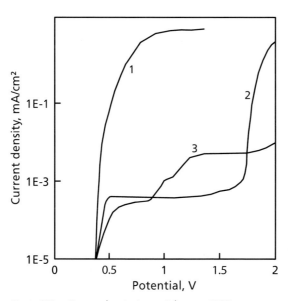

Figure 107: Current density/potential curves [126]
1 – Ti-Al-Zr-Mo cast alloy
2 – Ti-Al-Zr-Mo cast alloy (quenched)
3 – Ti-Al-Zr-Mo-B alloy (amorphous)

For the Ti-Al-Zr-Mo system with and without boron, the difference in the corrosion behavior between the cast state and the rapidly quenched state is even more pronounced. The microstructures that can be obtained by very rapid quenching from the liquid state are very homogeneous and free of precipitates. This inhibits corrosive attack that is greatly favored by the presence of microelements on the metal surface.

Utilisation of titanium under special loading conditions

Behavior of titanium towards chlorine and hydrogen chloride

In a literature evaluation on the corrosion behavior of titanium and titanium alloys in chlorine (gaseous), hydrochloric acid and chlorine-containing solutions, particular attention was paid to the extensive recent Russian literature [127]. The good corrosion resistance of titanium to chlorine and hydrogen chloride is due to the existence of a compact oxide layer with few lattice defects. However, titanium only has good corrosion resistance to chlorine if the chlorine contains at least 0.5 – 1 % water. Dry chlorine violently attacks titanium, even at room temperature.

Because of its good corrosion behavior with respect to chlorine and chlorine-containing compounds, titanium is used in electrochemical chlorine gas production plants and for heat exchangers with chlorinated seawater. Likewise, titanium can be used in seawater desalination plants in which chlorine is added in concentrations between 5 and 10 ppm to prevent plant growth.

As long as there is a sufficient amount of water present, the corrosion resistance of titanium to chlorine and chlorine-containing hydrochloric acid is excellent.

However, titanium is strongly attacked in dry chlorine, even at 300 K (27 °C, ambient temperature). But a water vapor content of 0.4 % extends the resistance to temperatures of up to 473 K (200 °C) and a content of 4 % extends it to 673 K (400 °C). If a corrosion rate of 0.1 mm/a (3.94 mpy) is specified as tolerable, then titanium can be used in dry hydrogen chloride up to approx. 450 K (177 °C). In very damp hydrogen chloride, the resistance is extended to 773 K (500 °C). However, in the lower temperature range at approx. 380 K (107 °C), there is an additional steep increase in the corrosion rate. This is based on the fact that if the temperature is too low, water can condense and hydrochloric acid can be formed which is very aggressive to titanium above 380 K (107 °C). The presence of water thus has a decisive influence on the existence of the protective TiO_2 layer on the surface of the metal. Titanium is not sufficiently erosion resistant and is thus unsuitable as a material for mixers and pumps.

The application possibilities of titanium and a titanium-aluminium alloy AT-6 (Russian grade: Ti 5-6.5Al-0.3-0.9Cr-0.1C) were investigated with regard to its use as a material for a shaft in stirred autoclaves in which organic products were chlorinated at approx. 363 K (90 °C). While the titanium shaft was destroyed at 80 to 100 rpm after 8 h, the shaft made of AT-6 did not show any signs of attack, even after an operating time of 1200 h. In the presence of chlorine under 3 bar, the stirring autoclave made of titanium was sufficiently corrosion-resistant in 12 % HCl at 363 K (90 °C).

Table 88 shows the interplay of hydrochloric acid and chlorine as well as the action of potassium bichromate additives in hydrochloric acid on the corrosion rate of titanium (VT 1-1) and the titanium alloys OT 4-1 and VT 5-1 (see Table 87).

Materials	Main components			Maximum content		
	Al	Mn	Sn	Fe	Si	C
VT 1-1	–	–	–	0.3	0.15	0.1
OT 4-1	2-3.5	0.8-2.0	–	0.4	0.15	0.1
VT 5-1	4-5.5	–	2-3	0.3	0.15	0.1

Table 87: Composition of the materials, mass% (rest titanium) [128]

As the test results show, the addition of 0.01 mol/l $K_2Cr_2O_7$ to hydrochloric acid of up to 30 % gave approximately the same inhibitive action as the saturation of the acid with chlorine. As the temperature is increased from 303 to 363 K (30 to 90 °C), however, the inhibitive effect of both the chlorine as well as that of potassium bichromate greatly decreased if the concentration of hydrochloric acid exceeded 15 %. The disappearance of the passivity, combined with a reappearance of considerable corrosion is indicated by a drop in the potential of approx. 1 V. A corresponding voltage measurement device is not complicated, and therefore, it can be recommended as a warning indicator.

The corrosion resistance of titanium is decidedly improved by the addition of 0.2 to 1 % Pd. Thus the addition of 1 % Pd decreases the corrosion rate of titanium in 20 % HCl at 393 K (120 °C) from 7.8 mm/a (307 mpy) for unalloyed titanium to 0.039 mm/a (1.5 mpy).

Material	200 mg/m^3 chlorine + 50 mg/m^3 HCl	aqueous solution of 1 mg/l HCl + 0.5 mg/l chlorine + 1.4 mg/l MgCl$_2$
99.5 % titanium VT 1-0	5×10^{-3} (0.19)	0.8×10^{-3} (0.03)
OT 4-1	6×10^{-3} (0.24)	below the limit of detection
VT 5-1	0.13 (5)	below the limit of detection
0Ch18N10T (corresp. to: AISI 321, 1.4541)	10.2 (402)	0.03 (1.2)

Table 89: Corrosion rate (in mm/a (mpy)) of titanium and titanium alloys as well as a chromium-nickel steel in gases and solutions containing chlorine and hydrochloric acid at 353 K (80 °C) [127]

With the aim of finding materials for the flue gas cleaning of gases and solutions containing chlorine and hydrogen chloride, the corrosion behavior of titanium and titanium alloys (OT 4-1 and VT 5-1, Russian grade, composition given above) was tested together with that of the austenitic chromium-nickel steel 0Ch18N10T (corresp. to: AISI 321, 1.4541) under operating conditions. The aqueous solution also contained magnesium chloride in addition to chlorine and hydrochloric acid. The results in Table 89 show the superiority of titanium and the two titanium alloys compared to austenitic CrNi steel. Titanium and titanium alloys are being used increas-

Temperature K (°C)	HCl %	Without additives			With chlorine			With potassium bichromate		
		VT 1-1	OT 4-1	VT 5-1	VT 1-1	OT 4-1	VT 5-1	VT 1-1	OT 4-1	VT 5-1
303 (30)	10	0 (0)	0.4 (15.7)	0.7 (27.5)	0 (0)	0 (0)	0 (0)	0.1 (4)	0 (0)	0 (0)
	15	0.9 (35.4)	1.2 (47.2)	1.1 (43.3)	0 (0)	0 (0)	0.02 (0.8)	0.1 (4)	0.1 (4)	0.1 (4)
	20	1.7 (66.9)	1.4 (55.1)	2.4 (94.5)	0 (0)	0 (0)	0 (0)	0.1 (4)	0.1 (4)	0.1 (4)
	30	6.8 (267.7)	13.8 (543.3)	25.1 (988.2)	0.1 (4)	0.2 (8)	0.1 (4)	0.2 (0.8)	0.1 (4)	0.1 (4)
333 (60)	10	3.3 (129.9)	4.7 (185.0)	7.0 (275.6)	0.1 (4)	0.1 (4)	0.1 (4)	0.1 (4)	0.1 (4)	0.1 (4)
	15	14.8 (582.7)	10.5 (413.4)	15.2 (598.4)	0 (0)	0.1 (4)	0 (0)	0.1 (4)	0 (0)	0.1 (4)
	20	28.9 (1,137)	27.3 (1,074)	41.6 (1,638)	0.1 (4)	0 (0)	0.6 (23.6)	0.1 (4)	0.1 (4)	0.1 (4)
	30	136.8 (5,385)	99.2 (3,905)	224.8 (8,850)	32.3 (1,271)	29.2 (1,150)	137 (5,394)	0.1 (4)	1.3 (51.2)	103.7 (4,082)
363 (90)	10	39.3 (1,547)	30.7 (1,209)	68.6 (2,701)	0.1 (4)	0 (0)	0 (0)	0 (0)	0 (0)	0 (0)
	15	88.7 (3,492)	80.7 (3,177)	–	0.3 (11.8)	0 (0)	0.4 (15.7)	0 (0)	0.1 (4)	0.1 (4)
	20	194 (7,637)	165 (6,496)	234 (9,213)	108 (4,252)	115 (4,527)	211 (8,307)	117 (4,606)	241 (9,488)	213 (8,385)
	30	331 (13,031)	308 (12,126)	299 (11,772)	310 (12,204)	238 (9,370)	314 (12,362)	309 (12,165)	307 (12,086)	309 (12,165)

Table 88: Corrosion rate of titanium VT 1-1 and the titanium alloys OT 4-1 and VT 5-1 (Russian grades) in mm/a (mpy) in hydrochloric acid of various concentrations either saturated with chlorine or with added 0.01 mol/l $K_2Cr_2O_7$; calculated from 24-h tests [127]

ingly as anode materials in electrochemical chlorine production. The electrodes are covered with platinum or ruthenium dioxide to reduce the overvoltage during chlorine production. With the same electrochemical behavior, platinum-iridium coatings should be even more corrosion resistant.

Stray currents can cancel the passivating action of chlorine in hydrochloric acid. The application of an alternating current to titanium immersed in a chlorine-saturated solution with 3 to 6 M HCl will initiate pitting. Table 90 summarises a corresponding result in chlorine-containing 6 M HCl at 333 K (60 °C) after a test duration of 24 h. Pitting corrosion starts above an alternating current density of 7.5 mA/cm^2. At 15 mA/cm^2, the maximum pit density is reached with an almost constant average pit diameter. This shows that titanium and titanium alloys must also be protected against stray currents just like other metals.

Alternating current mA/cm^2	Number of pits per cm^2	Maximum pit depth mm	Average pit diameter mm
7.5	0	–	–
11.3	120	0.36	0.1 – 1
15	200	0.87	~ 0.1
30	200	2.48	~ 0.1

Table 90: Pitting corrosion of titanium in chlorine-containing 6 M hydrochloric acid at 333 K (60 °C) after a test duration of 24 h [127]

Damage to titanium equipment

Reference [129] reports on three cases of damage to components made of titanium. The first concerns crevice corrosion in the sealing region of a plate heat exchanger made of pure titanium (ASTM Ti Grade 2, UNS R50400) with sealing material based on Viton® (Teflon®). In the plate heat exchanger, a 26 % NaCl solution was heated from 277 – 293 K to 348 K (4 – 20 °C to 75 °C) by water vapor condensate at 368 K (95 °C) that cooled to 288 to 303 K (15 to 30 °C). Within 30 days after the start of operations, crevice corrosion occurred in the region of the seal. This corrosion led to breakage of the 0.7 mm (27.5 mpy) thick titanium sheet in many places. Under the given conditions of the medium, titanium should be corrosion-resistant, even against crevice corrosion.

Scanning electron microscope investigations showed that the Viton® was treated with PbO (Litharge lead plate) to obtain an optimum resistance to water vapor and acid. Under the given conditions of attack, PbO is leached out and reduced to metallic Pb which greatly reduces the exchange current density of the reaction $2H^+ \rightarrow H_2$ resulting in acidification and thus activation of the titanium underneath the seal.

In laboratory investigations, this damage was simulated with sandwich samples made of titanium and Viton® in a chloride-containing solution. When the Viton® sealing rings contained MgO instead of PbO, there was no crevice corrosion underneath the seals. In this case, no alkalisation takes place.

Similar problems also occurred with PTFE seals; however, only at temperatures above 473 K (200 °C). In this case, the cause is hydrogen fluoride that is produced by pyrolysis of the plastic. This hydrogen fluoride leads to corrosion in the region of the seal. Fluoride-sensitive transition metals, titanium, zirconium and tantalum, are potentially at risk.

The second case of damage [129] concerns intergranular stress corrosion cracking of a shell-and-tube heat exchanger made of pure titanium (ATSM Ti Grade 2, UNS R50400). It is used to keep the temperature of an exothermic reaction mixture consisting of 62% ethanol, 14% NaCl, 128 ppm HCl and 24% other organic substances at 283 K (10 °C) or below. The reaction product entered the tubes with a temperature of 283 K (10 °C) and left the heat exchanger with 283 K (10 °C). The shell was in contact with cooled circulating methylene dichloride. Three weeks after commissioning, leakages occurred in the heat exchanger.

An inspection showed that cracks started in the region of the welded tube joints from the side in contact with the product. The investigations ruled out causes other than stress corrosion cracking. Up to that point, stress corrosion cracking was known for titanium in methanol-HCl, but not for ethanol-HCl. Laboratory investigations of titanium in the ethanol-NaCl-HCl system used in the process by means of slow tensile tests (constant extension rate tensile test, CERT test), verified the cracking. A certain minimum HCl concentration in ethanol is necessary for cracking; this is cited as 1.2%. HCl probably influences the start of cracking by the attack of the oxide layer. However, according to the results of the CERT tests, however, it is suspected that HCl also plays a role in crack propagation. The influence of NaCl is unknown.

Reference [129] describes a third case of damage to longitudinally welded tubes made of Ti Grade 12 (UNS R53400) in a multistage shell-and-tube heat exchanger. They are used to cool an alkaline salt solution on the tube side from 383 to 313 K (110 °C to 40 °C) using a neutral 20% $CaCl_2$ salt solution, which is thereby heated from 303 to 368 K (30 to 95 °C). Preceding laboratory tests in combination with results given in the literature lead to the conclusion that no corrosion is to be expected for the selected material under the given attack conditions. The plant was put into operation after cleaning and passivation of the heat exchanger with 2% nitric acid (20 h). After approx. 60 days, severe leakages occurred between the two circuits. The hottest of the multistage heat exchangers exhibited pitting holes in 86 of the 563 tubes initiated on the cooler side. The holes were arranged in a line along the axis of the tube, mostly in the longitudinal welding seam. Iron inclusions were found at the corrosion sites. These inclusions had been generated during manufacture of the tubes. This is possible if unsuitable grease is used during forming of the tubes as it leads to cold welds between the titanium material and the steel rollers and to splintering on the tube rollers being pressed into the titanium material. Subsequent cleaning with nitric acid should dissolve these iron particles. However, if they are deeply embedded, they cannot be completely removed during cleaning with nitric acid, but are exposed instead. These sites are then attacked during subsequent operation of the exchanger. They form active centres to give a distinct local element with the passive titanium material. Hydrolysis of the corrosion products, essentially

FeCl$_2$, causes acidification of the inclusion site thus activating the surrounding titanium material, which can perforate the tube material after a short time. This type of damage due to iron inclusions is also known for stainless steels.

Nitriding of titanium and titanium alloys

Titanium and most titanium alloys have only a limited wear resistance. Reference [130] reports on investigations for improving the wear and corrosion resistance by nitriding pure titanium (VT 1-0) and the two aluminium-alloyed grades AT 3 and AT 6 (Russian designation). The two alloys had the composition:

AT 3: 2.7 Al, 0.2 Fe, 0.28 Cr, 0.24 Si, 0.02 C, 0.16 O, 0.03 N, 0.005 H
AT 6: 5.0 Al, 0.28 Fe, 0.33 Cr, 0.29 Si, 0.03 C, 0.88 O, 0.02 N, 0.007 H

Nitriding was carried out by gas nitriding in pure nitrogen as well as by glow discharge in pure ammonia and in a mixture of ammonia and argon at temperatures of 1073, 1173 and 1273 K (800, 900 and 1000 °C) and for periods between 1 and 20 h. This produces titanium mononitride TiN. Aluminium does not form its own nitride, but it can substitute into titanium nitride as a solid solution.

In a hydroabrasion apparatus, the wear resistance was determined in a liquid medium with the following composition: tap water, 0.5 mass% quartz sand with a grain size of 0.2 to 0.4 mm, flow rate through the inflow nozzle 27.4 m/s, inflow angle 30°, test time 20 min. Nitriding improved the wear resistance by a factor of 2 to 3 for optimum nitride levels. The corrosion resistance in 20% HCl at 333 K (60 °C) and test times of 100 h was increased by a factor of up to 100 (Table 91) for optimum nitride levels.

On account of the synergistic effect of improved wear and corrosion resistance, the nitrided nozzles in a centrifugal separator used for vitamin manufacture exhibited longer endurance times without a significant wear on the exposed inner surface of the nozzles that were exposed to an acidic medium containing chloride and solids.

Nitriding conditions		Corrosion rate, g/m^2 h					
Temperature K (°C)	Time h	40% H$_2$SO$_4$	20% HCl	40% H$_2$SO$_4$	20% HCl	40% H$_2$SO$_4$	20% HCl
		Alloy VT 1-0		Alloy AT 3		Alloy AT 6	
Untreated		9.55	10.22	14.25	17.15	15.45	20.73
1123 (850)	20	0.0	0.027	0.213	0.227	0.683	0.965
1173 (900)	1	3.25	15.65	15.602	18.35	8.546	18.52
1173 (900)	5	0.009	11.97	0.011	19.31	0.186	19.75
1173 (900)	10	0.011	0.075	0.055	0.219	0.14	0.02
1273 (1000)	1	5.029	13.57	0.004	13.72	0.204	18.5
1273 (1000)	5	0.009	0.75	0.004	0.096	0.152	16.55

Table 91: Corrosion rates of pure titanium (VT 1-0) as well as alloys AT 3 and AT 6 after gas nitriding in pure nitrogen. Test duration 100 h at 333 K (60 °C) [130]

Nitriding is also recommended to increase the corrosion resistance of commercially available titanium to hydrochloric acid solutions. The influence of the nitriding temperature on the corrosion behavior of nitrided titanium was investigated in aqueous HCl solutions in [131]. For this, commercial titanium materials were nitrided at three temperatures 1223, 1123 and 1023 K (950 °C, 850 °C and 750 °C) for 10 hours in a nitrogen atmosphere at normal atmospheric pressure.

The samples were then exposed to 20 % and 30 % aqueous HCl solutions at room temperature. The total test duration was 60 days. The mass losses were determined every 10 days. The corrosion rates after 60 days are given in Table 92.

Nitriding	α-Titanium		Pseudo-α-titanium		(α + β)-Alloy		β-Alloy	
	20% HCl	30% HCl	20% HCl	30% HCl	20% HCl	30% HCl	20% HCl	30% HCl
Without	79.5	71.5	132	1200	129	1242	271	2647
1223 K (950 °C)	8.6	8.1	9.7	10.0	7.8	8.1	9.2	7.6
1123 K (850 °C)	5.2	6.3	7.6	8.0	6.7	9.5	5.2	6.4
1023 K (750 °C)	4.2	5.3	6.2	4.7	5.8	6.3	3.1	4.8

Table 92: Corrosion rates in mg/m^2 h of titanium and nitrided titanium samples after exposure (60 days) to HCl solutions [131]

For all investigated titanium materials, nitriding led to a pronounced improvement of the corrosion behavior in hydrochloric acid. Compared to the non-nitrided samples, the influence of the acid concentration on the nitrided samples is still very low. As the nitriding temperature decreases, the corrosion resistance increases so that low-temperature nitriding is an effective measure to protect titanium materials from corrosion in hydrochloric acid. Since nitriding at a low temperature of 1023 K (750 °C) only takes place very slowly, preheating in a vacuum is recommended.

Anodic protection

It is pointed out in reference [132], that in contrast to stainless steels, titanium can be passivated by anodic polarisation (anodic protection) in most non-oxidising acids and even in hydrochloric acid. Thus, the corrosion rate of titanium in concentrated 37 % HCl at 337 K (64 °C) can be kept at values < 0.1 mm/a (< 4 mpy) by anodic polarisation at $U_H = 1.7$ V. Under these conditions, the corrosion rate is approximately 140 mm/a (5,500 mpy) at the resting potential.

A 39 Zinc, cadmium and their alloys

Even at room temperature, hydrochloric acid of every concentration strongly attacks zinc with liberation of hydrogen. However, with regard to the use of zinc or zinc-coated components exposed to the atmosphere, the influence of HCl in the surrounding air and its influence on the behavior of such components is of interest.

Zn-Al alloys

Because of the increased aggressiveness of the atmosphere in the 1970s and 1980s, the service life of protective zinc coatings on components was found to be decreasing. Galvanised wires are of particular importance. Since the protection period of the zinc layers essentially depends on the thickness of the zinc layer, a double coating gave a corresponding lengthening of the protection time. A further lengthening was achieved with a Zn/Al eutectic coating containing 5 % Al and having a melting point of approx. 653 K (380 °C). The decisive factor for the aggressiveness of the atmosphere is primarily its acid content in the form of SO_2. Even if present in low concentrations, the content of HCl contributes to the accelerated corrosion of zinc coatings. In reference [133], it is shown that Zn-Al coatings in the presence of SO_2 and HCl are much more corrosion-resistant than wires with thick zinc coatings. The Kesternich test in a SO_2-containing medium showed that the mass loss of the Zn-Al coatings was only approx. 15 % of that lost by wires with thick zinc coatings after 20 cycles.

The immersion tests in 0.01 M H_2SO_4 and 0.01 M HCl at ambient temperature also showed the superior corrosion behavior of the Zn-Al protective coating. Figure 108 shows the mass loss in the 0.01 M HCl depending on the test duration. Up to a test duration of approx. 50 h, the Zn/Al coating is more strongly attacked than the pure zinc coating. Subsequently, the mass loss decreases, whereas it con-

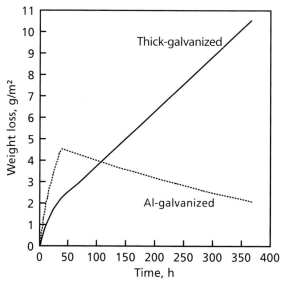

Figure 108: Area-related mass loss [g/m^2] of zinc-coated and aluminium/zinc-coated steel wires during long-term immersion in 0.01 M HCl depending on the test duration [133]

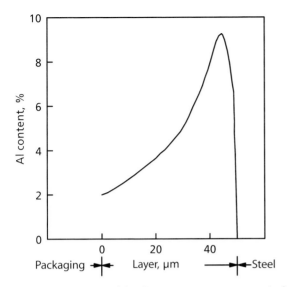

Figure 109: Change of the aluminium content in mass% of an aluminium/zinc-coated steel wire depending on the coating thickness [133]

tinuously increases for the thick zinc coating. This is explained on the basis of the aluminium distribution within the coating. It increases to almost 10% next to the steel (Figure 109). As the aluminium content increases, there is an increasing protective effect of the Zn/Al coating with regard to attack by HCl. A similar type of behavior was observed for the immersion test in 0.01 M H_2SO_4.

Although the influence of atmospheric SO_2 on zinc has been frequently investigated, there are few publications that deal with the influence of atmospheric HCl on the corrosion behavior of zinc. This is due to the fact that the concentration of HCl is lower than that of SO_2 in the atmosphere. The main source of HCl is flue gases produced by the combustion of chloride-containing coals. Coal may contain up to 1% chlorine, mostly in the form of chloride ions. Furthermore, the conversion of hydrochlorocarbons is important. They are produced by the incineration of waste, particularly PVC and chlorinated organic solutions.

In reference [134], laboratory tests on zinc samples were carried out in a sealed chamber with a parallel air stream to which gaseous HCl was added with various percentages of relative humidity and at various temperatures. The decisive test parameter was the flow rate in mg HCl per cm^2 sample surface per second in the testing chamber. On the basis of the HCl concentrations measured in a British city atmosphere of 1 to 10 µg/m^3 and an average wind speed of 2 m/s, this gives loading rates of 2×10^{-7} to 2×10^{-6} mg/cm^2 s. The lowest loading rate used in the experiments was 2.5×10^{-6} mg/cm^2 s. Although this is somewhat greater than the upper value in the city atmosphere, this value might occasionally occur. In industrial atmospheres, the HCl concentrations may be 3 to 30 times greater, particularly in the vicinity of HCl sources. This was taken into account in the experiments with loading rates of 8.3×10^{-6} and 83×10^{-6} mg/cm^2 s.

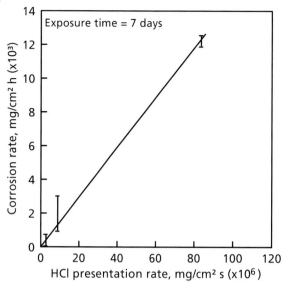

Figure 110: Corrosion rate of zinc for a test duration of 7 days at 80 % relative humidity depending on the HCl loading [134]

HCl loading rate mg/cm^2 s	Test duration d	Temperature K (°C)	Mass loss mg/cm^2	Corrosion rate mg/cm^2 h
0	7	298 (25)	0.03 ± 0.02	0.2 ± 0.2 × 10^{-3}
2.5 × 10^{-6}	7	298 (25)	0.04 ± 0.01	0.24 ± 0.6 × 10^{-3}
8.3 × 10^{-6}	7	298 (25)	0.34 ± 0.17	2.0 ± 1.0 × 10^{-3}
83.0 × 10^{-6}	7	298 (25)	2.10 ± 0.04	12.5 ± 0.2 × 10^{-3}
8.3 × 10^{-6}	2	298 (25)	0.11 ± 0.06	2.3 ± 1.3 × 10^{-3}
8.3 × 10^{-6}	7	298 (25)	0.34 ± 0.17	2.0 ± 1.0 × 10^{-3}
8.3 × 10^{-6}	30	298 (25)	2.67 ± 0.58	3.7 ± 0.8 × 10^{-3}
2.5 × 10^{-6}	7	298 (25)	0.04 ± 0.02	0.24 ± 0.01 × 10^{-3}
2.5 × 10^{-6}	7	318 (45)	0.03 ± 0.01	0.17 ± 0.06 × 10^{-3}

Table 93: Corrosion rate of zinc samples in HCl-containing atmospheres [134]

Figure 110 and Table 93 give the corrosion rates of zinc for a test duration of 7 days at 80 % relative humidity depending on the hydrochloric acid loading rate. For a loading rate of 2.5×10^{-6} mg/cm^2 s, which is slightly above the decisive range for a city atmosphere, the corrosion rate only differed slightly from that of the HCl-free atmosphere (Table 93). At this low loading rate, an increase in the temperature from 298 to 318 K (25 to 45 °C) did not increase the corrosion rate (Table 93). However, there was an almost linear increase when the loading rate was increased to 8.3×10^{-6} and 88×10^{-6} mg/cm^2 s.

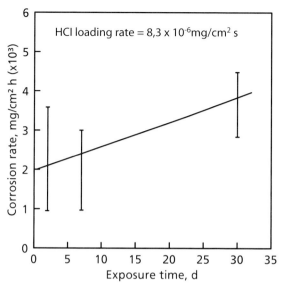

Figure 111: Time-dependence of the corrosion rate of zinc for a loading rate of 8.3×10^{-6} mg/cm^2 s and 80 % relative humidity [134]

The time-dependence of the corrosion rate of zinc for a loading rate of 8.3×10^{-6} mg/cm^2 s and 80 % relative humidity is given in Figure 111. After 2 days, the corrosion rate reached approximately 10 times that in the HCl-free atmosphere. The further increase up to 30 days was, however, slight.

The corrosion products on the samples with the 2.3×10^{-6} mg/cm^2 s loading rate were zinc hydroxide (Zn(OH)$_2$) and almost insoluble basic zinc chloride (ZnCl$_2 \times$ 4 Zn (OH)$_2$). These corrosion products essentially protect the substrate metal from further corrosion. Therefore, the corrosion rate of zinc is low in usual city atmospheres. At higher loading rates, the initially existing protective Zn(OH)$_2$ is increasingly converted to the readily soluble ZnCl$_2$. This leads to increased corrosive attack, as is the case in an industrial atmosphere in the vicinity of HCl sources. This starts at a loading rate of 8.3×10^{-6} mg/cm^2 s. Earlier investigations were carried out with unrealistically high HCl concentrations in the test atmospheres.

The same authors as in reference [134] used the same apparatus and the same loading method to investigate the effect of fly ash particles in air streams contaminated with HCl and SO$_2$ on the corrosion of the metal surface of iron and zinc-coated samples. It is well-known that fly ash particles can initiate local corrosion. This was confirmed in reference [135]. In addition to fly ash particles, which are produced by the combustion of coal and oil, inert glass spheres of a comparable size were also used in the investigations. The leachability of the fly ash particles is important with regard to their effect on the corrosion behavior. This is greater for oil ash particles than for coal ash particles. The investigations gave the following results:

- Generally, atmospheric corrosion of the metals, regardless of whether there is a contamination of the surface by particles or not, is determined by the conductivity of the superficial electrolyte layer.
- Under atmospheric conditions with low contamination levels, inert particles can increase the corrosion of zinc and steel due to differing aeration in the region of the particles. In accordance with the rate at which atmospheric pollution is increasing, the effect on the particle deposition is overlaid by a general increase in the corrosion.
- The increase in the atmospheric corrosion rate by deposited particles is mainly dependent on the amount of leachable substances from the particles. If leaching gives rise to a high pH value or a significant amount of inhibiting ions, such as Ca^{2+}, then the corrosion rate cannot increase much.
- Coal fly-ash particles with a leachable ion content of < 0.3 % are generally less corrosive than oil fly-ash particles with a corresponding leachable ion content of > 1.5 %.

Reference [136] reports on investigations of the efficiency of some N-arylpyroles as inhibitors for zinc in hydrochloric acid using electrochemical (current density/potential curves and impedance spectroscopy) and gravimetric methods. The following compounds were used in the investigations:

(1) 1-(2-Chlorophenyl)-2,5-dimethylpyrrol-3-carbaldehyde
(2) 1-(2-Methylphenyl)-2,5-dimethylpyrrol-3-carbaldehyde
(3) 1-Phenyl-2,5-dimethylpyrrol-3-carbaldehyde
(4) 1-(2-Chlorophenyl)-2,5-dimethylpyrrol-3,4-dicarbaldehyde
(5) 1-(2-Methylphenyl)-2,5-dimethylpyrrol-3,3-dicarbaldehyde
(6) 1-Phenyl-2,5-dimethylpyrrol-3,4-dicarbaldehyde

The purity of zinc was 99.98 %.

There was good agreement between the electrochemical and gravimetric results. Table 94 gives an example of the protective effect in 0.5 M hydrochloric acid with an addition of 5×10^{-3} M of compound (1) at 293, 313 and 333 K (20, 40 and 60 °C) using all three methods of investigation. The investigation on the effect with respect to the inhibitor concentration between 10^{-2} to 10^{-4} M gave an optimum concentration of 5×10^{-3} M that was then used in all further investigations.

The protective efficiency of all investigated inhibitors in 0.5 M hydrochloric acid for an addition of 5×10^{-3} M of the individual compound according to an evaluation by the Tafel method is given in Table 95. All compounds exhibited a high protective efficiency at 313 and 333 K (40 and 60 °C). However, the inhibition efficiency is low for compounds (4) and (6) at 293 K (20 °C). Above all, the authors point out that the substances are completely non-toxic and can thus be handled without health hazards. They are present as natural components in foodstuffs such as coffee, chocolate, eggs, rice, tea, tobacco and potatoes.

Solution	Protection efficiency %		
	393 K (20 °C)	313 K (40 °C)	333 K (60 °C)
0.5 M HCl	–	–	–
0.5 M HCl + inhibitor 1	94.8	97.1	97.8

(a) protective inhibitor efficiency according to the evaluation of the current density/potential curves

Solution	Protection efficiency %		
	393 K (20 °C)	313 K (40 °C)	333 K (60 °C)
0.5 M HCl	–	–	–
0.5 M HCl + inhibitor 1	95.0	96.9	96.8

b) on the basis of impedance spectroscopic measurements according to Epelboin et al. [137]

Solution	Protection efficiency %		
	393 K (20 °C)	313 K (40 °C)	333 K (60 °C)
0.5 M HCl	–	–	–
0.5 M HCl + inhibitor 1	80.1	94.8	97.8

(c) protective inhibitor efficiency of 1-(2-chlorophenyl)-2,5-dimethylpyrrol-3-carbaldehyde according to the gravimetric method

Table 94: Protective inhibitor efficiency for the corrosion of zinc in 0.5 M HCl, with and without addition of 1-(2-chlorophenyl)-2,5-dimethylpyrrol-3-carbaldehyde [136]

Solution	Protection efficiency %		
	393 K (20 °C)	313 K (40 °C)	333 K (60 °C)
HCl	–	–	–
1	95.3	99.3	98.9
2	94.8	97.0	97.8
3	86.0	97.1	98.4
4	38.6	83.6	98.3
5	91.5	97.8	98.7
6	56.7	94.5	99.1

(1) 1-Phenyl-5-dimethylpyrrol-3-carbaldehyde
(2) 1-(2-Chlorophenyl)-2,5-dimethylpyrrol-3-carbaldehyde
(3) 1-(2-Methylphenyl)-2,5-dimethylpyrrol-3-carbaldehyde
(4) 1-Phenyl-2,5-dimethylypyrrol-3,4 dicarbaldehyde
(5) 1-(2-Chlorophenyl)-2,5-dimethylpyrrol-3,4-dicarbaldehyde
(6) 1-(2-Methylphenyl)-2,5-dimethylpyrrol-3,4-dicarbaldehyde

Table 95: Protective inhibitor efficiency of the various inhibitors for the corrosion of zinc in aqueous HCl [136]

A 40 Zirconium and zirconium alloys

Zirconium has a high resistance to inorganic acids and has been used for approx. the last 40 years in chemical process engineering. In addition to tantalum, zirconium is one of the few metals that is corrosion-resistant in hydrochloric acid of all concentrations and up to high temperatures well above the respective boiling points, as can be seen from Figure 83 and Figure 112. At a concentration of 12 M, the corrosion rate below the boiling curve is < 0.127 mm/a (< 5 mpy) [68, 138].

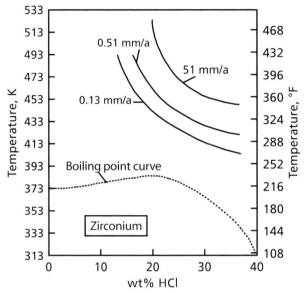

Figure 112: Isocorrosion diagram for zirconium in hydrochloric acid [68, p. 48]
0.13 mm/a (5 mpy)
0.51 mm/a (20 mpy)
5.1 mm/a (200 mpy)

Heat exchangers, stirrers and various other parts made of zirconium have been used for years in azo dye manufacturing involving reactions that alternate between hot hydrochloric acid and sulphuric acid and alkaline solutions [138].

The cost of zirconium corresponds to that of titanium and Hastelloy® alloys. On the other hand, tantalum is approx. 20 times more expensive. The processing properties are similar to those of pure titanium. It can be deformed, machined and welded using conventional machining equipment and well-known techniques.

Zirconium grade 705 is a zirconium alloy containing up to 3% niobium. In addition to the excellent corrosion resistance, which corresponds to that of pure zirconium, grade 705 has a higher strength and greater toughness. Welded tubes of this alloy are recommended as cost-effective products for heat exchangers and condensers [139].

Zirconium has a corrosion resistance to HCl similar to that of tantalum. Whereas tantalum is prone to hydrogen embrittlement in HCl of higher concentrations and high temperatures, this applies to zirconium to a much lesser extent. However, zirconium is susceptible to pitting and stress corrosion cracking in oxidising HCl, e.g. in the presence of oxidising impurities, such as iron(III) and copper(II) ions, or at positive electrode potentials. The work in [140, 141, 142] discusses these problems. In reference [140], tests were carried out with technically pure zirconium according to the ASTM R60702 standard in 10, 20 and 37 % HCl with 50, 100 and 500 ppm Fe^{3+} from 303 K (30 °C) up to the boiling point. The following results were obtained:

- In immersion tests on samples whose surfaces had been sanded with SiC (600 grit), the penetration speed was < 0.21 mm/a (< 8.3 mpy) in 10 and 20 % HCl containing up to 100 ppm Fe^{3+} and at temperatures from 303 K (30 °C) to the boiling point.
- Pickled surfaces are much less susceptible to pitting than non-pickled ones.
- Electrochemical protection at a potential 50 mV below the critical repassivating potential prevents pitting and stress corrosion cracking in 10 and 20 % HCl containing 500 ppm Fe^{3+}.
- In immersion tests coupled with slow tensile tests of U-bending samples, no embrittlement was found if electrochemical protection had been applied.

Table 96 summarises the results of the stress corrosion cracking behavior of zirconium in oxidising hydrochloric acid [141]. The test solution was HCl containing 500 ppm Fe^{3+} that had been added as $FeCl_3$.

HCl mass%	Temperature K (°C)	Corrosion rates and intergranular attack		Intergranular attack in the heat-affected zone (HAZ) after 32 days
		Initial state[1] (mm/a)	Pickled and ground (mm/a)	
10	333 (60)	0.086 (< 0.003)	0.005 (0.18)	no
	375 (102)	0.11 (0.1)	0.005 (1.3)	no
20	333 (60)	(−) (−)	0.008 (0.1)	no
	380 (107)	0.1 (0.1)	0.006 (1.5)	no
24	333 (60)	0.25 (−)	0.032 (0.17)	−
	377 (104)	0.135 (0.9) [2]	0.007 (1.25)	no
28	303 (30)	0.305 (−)	0.012 (0.01)	−
	333 (60)	0.18 (np)	0.017 (0.36)	no
	371 (98)	0.056 (2.8) [2]	0.01 (np)	yes
32	303 (30)	0.254 (np) [2, 3]	0.013 (np)	no
	350 (77)	0.056 (2.8) [3]	0.064 (np)	yes
37	303 (30)	0.124 (np)	0.048 (np)	yes
	326 (53)	0.114 (np) [3]	0.069 (np)	yes

(1) corrosion rate and (pitting penetration rate); np – no pitting corrosion
(2) pitting corrosion on the cut edge
(3) undermining of the edges

Table 96: Summary of the immersion tests of zirconium grade 702 in HCl with 500 ppm Fe^{3+} [141]

Immersion tests of welded samples (tungsten inert-gas butt-weld seam, without weld filler) exhibited a zone with intergranular attack directly adjacent to the weld seam. Thus zirconium exhibits sensitisation of the parent material due to welding heat that can lead to intergranular corrosion under certain conditions of attack. From the results summarised in Table 96 of the immersion tests after a test period of 32 days, it can be seen that the overall penetration rate of pitting corrosion is low and that the sensitisation in the HAZ only leads to intergranular attack at higher concentrations and temperatures.

The stress corrosion cracking behavior of Zircaloy® 2 with a composition of 1.40 % Sn, 0.15 % Fe, 0.08 % Cr, 0.051 % Ni, 0.115 % O; and 0.0075 % N was investigated in a methanol solution with 0.4 vol% HCl in the temperature range from 300 to 320 K (27 to 47 °C) on annealed and cold-rolled sheet samples [143]. Figure 113 gives the results of the time-to-fracture depending on the relative stress (as a percentage of the tensile strength). Above 45 %, annealed longitudinal and transverse samples exhibit the same very short time-to-fracture. At lower stresses, there are limiting stresses below which stress corrosion cracking no longer occurs. The limiting stress is noticeably higher for longitudinal samples than for the transverse ones. In contrast to this, cold-rolled longitudinal and transverse samples do not exhibit stress corrosion cracking, even for a loading of 80 % of the tensile strength. Cold-rolled samples are, however, sensitive to notches and notched samples exhibit brittle fracture.

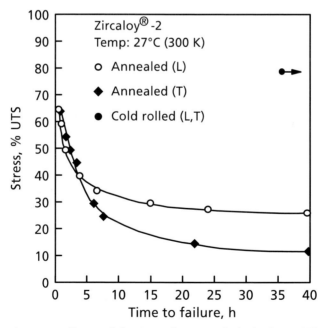

Figure 113: Change of the time-to-fracture with the loading (10 % of the tensile strength) in annealed and cold-rolled Zircaloy® 2. The arrow indicates that the samples are not yet fractured, and would probably hold even longer; L = longitudinal samples, Q = transverse samples [144]

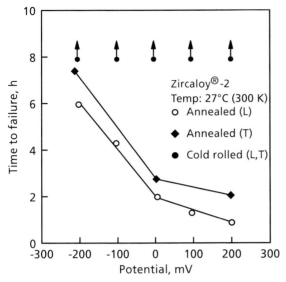

Figure 114: Influence of an applied (anodic or cathodic) potential on the time-to-fracture for various types of Zircaloy® 2 samples in a methanol solution with 0.4 vol% HCl; L = longitudinal samples, Q = transverse samples (50 % of the tensile strength) [144]

Figure 114 shows the dependence of the time-to-fracture on the potential and indicates an anodic crack initiation. The cracks are initially intergranular and then continue as brittle fractures. The service life of cold-rolled samples is independent of the potential at a stress of 50 % of the tensile strength that was chosen in these tests.

In [144], the influence of the tensile and torsional loading was investigated on Zircaloy® 2 under the same electrolyte conditions. Intergranular cracking occurred for both types of loading; in addition, there was transgranular cracking under tensile loading. The intergranular cracking is assumed to be caused by anodic dissolution at the crack tip, and transgranular cracking by hydrogen adsorption at the bottom of the crack.

In reference [145], it was found that the addition of osmium and iridium and a suitable heat treatment can produce a homogeneous microstructure with fine precipitates in these alloys that can improve the pitting resistance of zirconium in hydrochloric acid solutions. The pitting resistance was determined in 1 M HCl at room temperature by recording current density/potential curves. The results are given in Table 97. Highly pure zirconium exhibited a pitting potential of $U_H = 0.37$ V. For the highest addition of 7 atom% Os and 3 atom% Ir and the corresponding suitable heat treatment, the pitting potential increased to $U_H = 0.65$ and 0.68 V, respectively, i.e. by approx. 0.3 V, which indicates a distinct increase of the pitting resistance.

Test no.	Composition	Heat treatment	Phase composition	U_H V
1	Zr – 1 atom% Os	annealed at 1473 K (1200 °C), 5 h	α + β (traces)	0.41
		quenched from 1273 K (1000 °C)	α′	0.45
2	Zr – 5 atom% Os	annealed at 1473 K (1200 °C), 5 h	α + β	0.43
		quenched from 1273 K (1000 °C)	β + ω–quenched	0.33
		tempered at 673 K (400 °C) for 40 min after quenching	β + ω–quenched	0.46
3	Zr – 7 atom% Os	cooled in a furnace from 1273 K (1000 °C)	β	0.65
4	Zr – 1 atom% Ir	annealed at 1473 K (1200 °C), 5 h	α +Zr$_3$Ir	0.41
		quenched from 1273 K (1000 °C)	α′	0.48
5	Zr – 3 atom% Ir	annealed at 1473 K (1200 °C), 5 h	α +Zr$_3$Ir	0.46
		quenched from 1273 K (1000 °C), tempered	β + ω–quenched	0.39
		after quenching	β + ω–tempered	0.68
6	highly pure zirconium	annealed at 873 K (600 °C), 1 h	α	0.37

Table 97: Pitting potential for Zr-Os and Zr-Ir alloys [145]

A 41 Other metals and their alloys

The cracking tendency of a 13 mm wide and 45 μm thick ribbon of Metglas 2826 (Fe-40Ni-14P-6B), industrially manufactured by the melt-spin process was investigated in 1 M HCl, 0.39 M FeCl$_3$ and polythionic acid [146]. The ribbon was delivered elastically tensioned, but not plastically deformed, and rolled up on a spool. Such ribbons have slight longitudinal grooves on the side in contact with the drum during melting. The air side is comparatively smooth. Stress was applied by wrapping it around a glass cylinder with differing diameters. The ribbon was bent in both the transverse direction as well as the longitudinal direction.

The time until cracks appeared was very dependent on the potential (Figure 115). There was no measurable attack and there are no cracks at the resting potential in 1 M HCl at 296 K (23 °C). Cathodic polarisation gave increasingly faster cracking due to abrupt hydrogen embrittlement. Anodic polarisation from the resting potential caused cracking within short times that were triggered by the stress corrosion cracking mechanism.

In aqueous polythionic acid, the alloy is cathodically polarised with respect to the resting potential in 1 M HCl and there are spontaneous hydrogen embrittlement fractures. The resting potential in the 0.39 M FeCl₃ solution has a more noble potential than the resting potential in 1 M HCl and so stress corrosion cracking takes place. The cracking behavior of the amorphous alloy Fe-40Ni-14P-6B is comparable to that of high-strength steels.

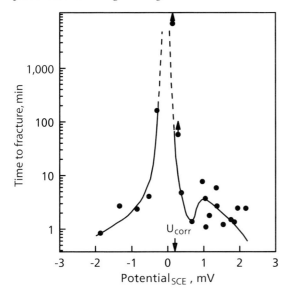

Figure 115: Time-to-fracture as a function of the potential U_{SCE} (V) for the Fe-40Ni-14 P-6B sample in the as-delivered state in 1 M HCl. U_R = resting potential; initial strain ε = 0.0063 [146]

In reference [147], which deals with the corrosion behavior of sheet material in the alloy system Fe-Cr-Al-P-C, a high corrosion resistance of the alloys was found in the amorphous state. This was proved by recording current density/potential curves in various corrosive media, particularly in 1 M HCl at 303 K (30 °C), with a ramping rate of 1 V in 4×10^2 s.

Figure 116 shows the result for the alloy system $Fe_{72}Cr_{8-x}Al_xP_{13}C_7$ with variable Cr and Al contents. Some of the originally amorphous melted alloys were subsequently recrystallised by annealing. The recrystallisation temperatures for the individual alloys were determined. They lie slightly above 700 K (427 °C). The dissolution currents in the passive potential range are approx. four powers of ten smaller for the amorphous alloys than for the crystallised alloys with the same composition. Furthermore, chromium can be largely replaced by Al in the amorphous alloys without lowering the high corrosion resistance (compare a-8-0, $Fe_{72}Cr_8P_{13}C_7$ and a-1-7, $Fe_{72}Cr_1Al_7P_{13}C_7$).

In [148], alloys based on Cr-Ni-20P with chromium contents between 40 to 80 atom% were produced by rapid quenching using the rotating drum method. The alloys are purely amorphous up to 60 atom% Cr. Above this value, they have a duplex structure with crystalline and amorphous regions. The mass losses and the current density/potential curves were determined in 6 M and 12 M HCl at 303 K (30 °C). Figure 117 shows the corrosion rate depending on the chromium content.

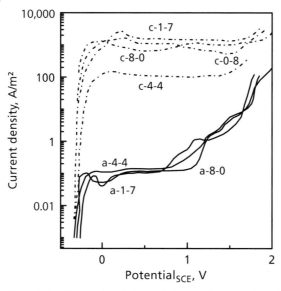

Figure 116: Current density/potential curves for amorphous (a) and crystalline (c) samples of the alloy system $Fe_{72}Cr_{8-x}Al_xP_{13}C_7$ in 1 M HCl (the numbers represent the variable Cr and Al contents) [147]

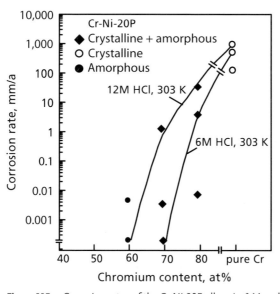

Figure 117: Corrosion rates of the Cr-Ni-20P alloys in 6 M and 12 M HCl at 303 K (30 °C) [148]

The removal rate lies below the detection limit for the amorphous alloys. The corrosion rate increases steeply as the crystalline proportion increases. The corrosion rate of crystalline chromium metal is also included for comparison. Its value of 600, respectively 900 mm/a (23,622, respectively 35,433 mpy), is very high. The progression of the current density/potential curves depending on the chromium content and thus on the microstructure of the alloys in 12 M HCl at 303 K (30 °C) (Figure 118) is revealing. The amorphous alloys exhibit no active potential region whatsoever. The current density/potential curves of the partially crystalline alloys 70Cr-10Ni-20P and 80Cr-20P increasingly approach that of the crystalline metallic chromium with extended active potential regions.

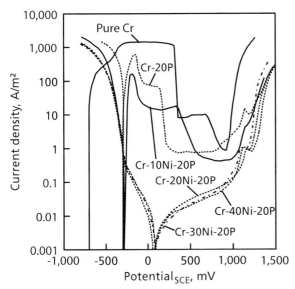

Figure 118: Progression of the current density/potential curves depending on the chromium content and thus on the microstructure of the alloys in 12 M HCl at 303 K (30 °C). The curve for pure chromium is also included for comparison [148]

In the initial quenched state, alloy 80Cr20P exhibits regions with the crystalline phase Cr_3P. The proportion of this phase increases on annealing at 823 and 1073 K (550 and 800 °C) for 3 h each. At 1073 K (800 °C), crystalline chromium precipitates additionally form in the amorphous matrix. The corrosion rate is increased by both precipitates, as proved by tests in 47 % HF at 303 K (30 °C).

In [149], the corrosion behavior of $Fe_{100-x}Si_x$ alloys rapidly quenched from the melt with silicon contents of 14 to 35 atom% (7.6 to 21.3 mass%) was investigated in 5 M HCl at 293 K (20 °C). As the cooling rate was increased, the microstructure became increasingly finely crystalline; however, an amorphous state was not attained. The rotation speed of the steel drum onto which the melt is poured and thus solidifies can be used as a measure of the cooling rate.

The course of the potentiodynamically recorded current density/potential curves (1 V/h) of the $Fe_{65}Si_{35}$ alloy depending on the rotation speed of the drum is shown in Figure 119. The results show that, as the cooling rate of the samples increases, the active dissolution current density, and particularly the passive dissolution current density, decrease by up to three powers of ten at an approximate potential of $U_H = 0.9$ V.

This shows that a microcrystalline microstructure gives an increase in the corrosion resistance of a magnitude similar to that obtained with an amorphous microstructure. The improvement of the corrosion resistance is additionally increased with increasing silicon content.

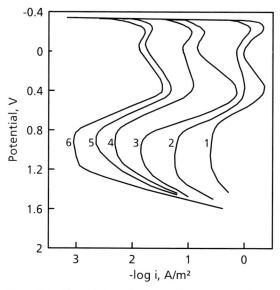

Figure 119: Potentiodynamic current density/potential curves of the $Fe_{65}Si_{35}$ alloy depending on the rotation speed (cooling rate; m/s) in 5 M HCl at 293 K (20 °C) [149]
1) not quenched; 2) 12.5; 3) 25; 4) 50; 5) 67; 6) 83

Tests in 10 % $FeCl_3 \times 6\ H_2O$ (pH 1) at 293 K (20 °C, corrosion potential U_H = 0.9 V) and at 333 K (60 °C) for 10 or 72 h showed that all investigated FeSi alloys were resistant to pitting under these stringent test conditions.

The high corrosion resistance of these microcrystalline alloys is attributed to the highly homogeneous distribution of the basic components in the matrix and the absence of segregations and of impurities, particularly on the grain boundaries of these alloys.

In [150], the corrosion behavior of sputter-deposited amorphous titanium-chromium alloys was investigated by recording current density/potential curves in 1 M and 6 M HCl at 303 K (30 °C). Similar to amorphous CrP-alloy and CrB-alloys produced by the PVD process in [152], the amorphous TiCr alloys were also much more corrosion-resistant than the crystalline starting materials. These alloys are amorphous from 37 to 73 atom% Cr. In 1 M and 6 M HCl, they exhibit spontaneous passivity and remain in this state, while Cr and Ti are activated in 6 M HCl. The passive layer that forms on amorphous TiCr alloys consists of Cr-Ti oxihydroxides.

The same authors investigated the two-component Cr-Zr system in the same way in [151]. These alloys are amorphous from 11 to 66 % Zr. Figure 120 gives the corrosion rate of the sputter-deposited CrZr alloys in 6 M HCl at 303 K (30 °C) depending on the zirconium content. The corrosion rates of chromium and zirconium are also given for comparison. Chromium actively dissolves with a high corrosion rate of 48 mm/a (1,890 mpy). The corrosion rates of the amorphous alloys are approximately 1×10^{-4} mm/a (0.004 mpy), while that of zirconium is 2.9×10^{-4} mm/a (0.01 mpy).

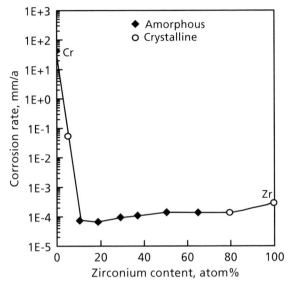

Figure 120: Corrosion rate of the sputter-deposited CrZr alloys in 6 M HCl at 303 K (30 °C) depending on the zirconium content. The corrosion rates of Cr and Zr are also given for comparison [151]

The passive layer formed on the amorphous CrZr alloys consists of double oxi-hydroxides of the elements chromium and zirconium. This type of protective layer is probably responsible for the fact that these alloys have, in particular, a greater resistance to pitting corrosion in chloride-containing media compared to zirconium. A difference between the two two-component systems Cr-Zr and Cr-Ti is that the amorphous Cr-Ti alloys are activated by cathodic polarisation in 6 M HCl, whereas the amorphous Cr-Zr alloys remain passive.

In [152], the corrosion behavior of amorphous (glass structure) CrP and CrB alloys in 12 M HCl, 0.39 M $FeCl_3$ and in the 20:1 mixture of both solutions was investigated and compared to that of crystalline, elemental chromium. These investigations were triggered after the work in [153] from 1974 reported for the first time that amorphous alloys of type $Fe_{(1-x)}Cr_xP_{13}C_7$ exhibited extraordinary corrosion behavior. Only 8 % Cr is necessary to generate spontaneous passivation compared to 12.5 % for stainless steels. These alloys also exhibit considerably greater resistance to pitting corrosion in chloride-containing solutions. The general composition of these amorphous alloys consists of

(Fe-Ni-Co) basic elements,
(Cr-Ti-Mo-W-Zr-V) passivating elements and
(B-Si-C-P) glass-forming elements.

They are single phase and contain 20 atom% of the glass-forming elements. As for the crystalline materials, the corrosion resistance of these amorphous materials is also due to the formation of protective passive layers. For the chromium-containing materials with an amorphous structure, the composition of the glass-forming metal oxides has a large influence on the corrosion behavior. Phosphorus has the highest influence and boron the lowest.

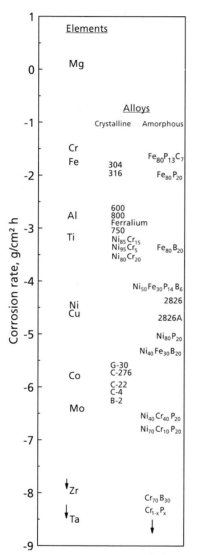

Figure 121: Corrosion rates of the various crystalline and amorphous alloys in 12 M HCl (see also Table 98). The arrows indicate the detection limit of the mass loss measurements [152]

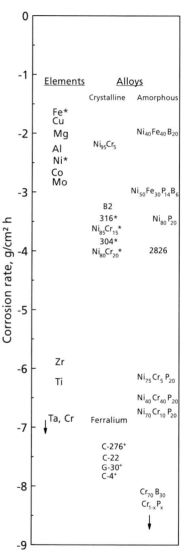

Figure 122: Corrosion rates of the various crystalline and amorphous alloys in 0.39 M $FeCl_3$ (see also Table 98). The arrows indicate the detection limit of the mass loss measurements; (*) pitting corrosion, (+) mass increase [152]

Crystalline alloys	Amorphous alloys
AISI 304 (X5CrNi18-10, DIN-Mat. No. 1.4301)	Ni-20P
AISI 316 (X5CrNiMo17-12-2, DIN-Mat. No. 1.4401)	Ni-10Cr-20P
Inconel® 600	Ni-40Cr-20P
Inconel® X 750	Ni-40Fe-20B
Alloy 800	Ni-30Fe-14P-6B
Ferralium®	Fe-40Ni-14P-6B (2826)
Hastelloy® G 30 alloy	Fe-20B
Alloy C 276	Fe-20P
Alloy C 22	Fe-13P-7C
Alloy C-4	Fe-36Ni-12Cr-14P-6B (2826A)
Alloy B-2	

Table 98: Investigated alloys [152]

The samples were produced by means of a PVD process (physical vapor deposition). Table 98 gives a comparison of the investigated crystalline and amorphous materials. Figure 121 and Figure 122 give the corrosion rates in g/cm² h of important alloying elements and the investigated crystalline and amorphous alloys in 12 M HCl and in 0.39 M FeCl$_3$. Whereas crystalline chromium is spontaneously activated in reducing 2 M HCl and rapidly dissolves, the corrosion rate of the amorphous alloys $Cr_{70}B_{30}$ and $Cr_{1-x}P_x$ lie below the detection limit of 10^{-8} g/cm² h. Their corrosion resistance is as good as that of crystalline zirconium and tantalum (Figure 121).

Crystalline chromium is also passive in an oxidising 0.39 M FeCl$_3$ solution, with a corrosion rate below the detection limit, similar to tantalum. This also applies to $Cr_{70}B_{30}$ and $Cr_{1-x}P_x$.

These results show that amorphous alloys have a surprisingly high corrosion resistance, even in highly concentrated HCl.

B
Non-metallic inorganic materials

B 3 Carbon and graphite

Structure and general corrosion properties

On account of their chemical and thermal stability, carbon and graphite are similar to refractory materials and because of their thermal and electrical conductivity they are also similar to metals.

Carbon and graphite products are mainly produced in a two-stage process that is described in detail in [154]. In the first stage, coke or other materials are produced as intermediate products by pyrolysis of carbon-containing raw materials (pitch and coal). In the second stage, these products are mixed with a binder, usually tar pitch, and are then fired at approximately 1273 K (1000 °C) or graphitised by further heating to 3273 K (3000 °C). In addition to tar pitch, phenolic and furan resins are also used as binders

Approximately 25 % of the volume of carbon and graphite products produced in this way consist of pores which can negatively influence the volumetric properties of the material. The pores are filled with carbon tar pitch or petroleum pitch and the raw products are pyrolysed again at 1073 K (800 °C). Graphite materials intended for utilisation at lower temperatures can be impregnated with curable phenol-formaldehyde or furan resins and can be used up to 473 K (200 °C). They are suitable for heat exchangers and other thermal equipment. If PTFE is used as an impregnating agent, application temperatures above 473 K (200 °C) can also be tolerated [154].

Carbon and graphite are stable in most corrosive media. They are attacked by strong oxidising agents, by chemicals that form intercalation compounds (for graphite e.g. sulphuric acid) and by materials that form carbides. Their reactivity mainly depends on the crystallinity of the materials. Their weak point is the binder phase, which has a less ordered structure. Corrosion is not limited to the outer surface, but also takes place inside the material as a result of pore diffusion.

Applications in processing plants

The high corrosion resistance and good thermal conductivity of graphite materials make them especially suitable for heat exchangers, evaporators, absorbers and columns. Further applications include pumps, fittings, pipes, accessories and packings for absorption towers [155]. Diabon®, a classical material in hydrochloric acid engineering, has been described in detail in older literature, and is still of importance today [47].

A summary of application examples is given in Table [154].

Process equipment	Maximum dimensions	Maximum operating data	Preferred applications
Shell-and tube heat exchangers	length 10 m diameter 2 m heat exchanger surface area 1000 m^2	6 (12) bar 473 K (200 °C)	equipment with large heat exchanging surface areas
Block heat exchangers	length 9 m diameter 1.5 m surface area 500 m^2	6 (15) bar 473 K (200 °C)	smaller surface areas with large heat-exchange rates
Plate heat exchangers	length 2.5 m diameter 0.7 m surface area 50 m^2	8 bar 473 K (200 °C)	large heat-exchanging rate, small pressure losses and temperature differences
Columns and reactors	diameter 1.5 m	1 bar 453 K (180 °C)	production of anhydrous HCl and pure hydrochloric acid
HCl production plants	115 t HCl gas per day	1 bar 623 K (350 °C)	production of anhydrous HCl and pure hydrochloric acid
Rotary pumps	400 m^3 / h	11 bar 453 K (180 °C)	process pumps for corrosive media
Burst disks	diameter 0.5 m	80 bar 623 K (350 °C)	safety equipment for process plants

Table 99: Processing equipment made of graphite and carbon [154]

The properties of graphite are described in [156] with particular regard to equipment manufacturing together with a description of graphite technology. The same reference also describes HCl synthesis plants in which graphitic materials play a central role.

In contrast to metals, graphite and carbon respond to mechanical stress with microcracks. Therefore, stresses must be avoided by means of a suitable design [157]. This applies, in particular, to connecting elements. In pressurised equipment, the mechanical loading tolerance can be improved by fiber reinforcement [158] and the erosion resistance with coatings [159].

Reference [160] reports on advances in pressure vessel construction. To obtain gas- and liquid-tight materials, the carbon-graphite bodies have to be impregnated with a synthetic resin, as described above.

An inevitable consequence of this impregnation is that the corrosion- and temperature-resistance is limited compared to that of pure carbon and graphite. The mechanical properties, such as bending, tensile, and compressive strengths as well as the permeability etc. are noticeably improved by impregnation with synthetic resin. Optimisation of the pore structure (as few small pores as possible) extended the application range of resin-impregnated graphite. Whereas the temperature range for resin-impregnated graphite (phenol-formaldehyde, furan resins) was formerly 438 K (165 °C) for continuous operation, the new grades impregnated with fluoropolymers (polytetrafluoroethylene, PTFE) are approved for continuous use at temperatures of up to 473 K (200 °C). Their corrosion resistance is also enhanced, particularly towards hydrochloric acid. The impregnated graphite grades are resistant to hydrochloric acid up to the highest permissible continuous temperature in the entire concentration range.

Chlorine-containing hydrochloric acid and aqueous chlorine solutions oxidatively attack graphite materials impregnated with phenolic resins. However, they are resistant to hydrochloric acid contaminated with chlorinated organic solvents.

The latest status of application and materials engineering for graphite heat exchangers operating with HCl is described in reference [4]. Both shell-and-tube heat exchangers as well as block heat exchangers (cross-flow heat exchangers) are used. Shell-and-tube heat exchangers are cheaper for large designs, whereas block heat exchangers are cheaper for small designs. Parts having phenol-formaldehyde resins as the standard impregnation material can be used up to 443 K (170 °C), those impregnated with fluoropolymers can be used up to 503 K (230 °C). However, fluoropolymers, in contrast to phenol-formaldehyde resins, exhibit a slight HCl gas permeability at higher temperatures, and this must be taken into consideration depending on the particular application.

The strength of graphite heat-exchanger tubes impregnated with fluoropolymers is three to four times greater than that of non-impregnated graphite.

Graphitic materials are not sensitive to thermal shocks by virtue of their very small coefficient of thermal expansion and elasticity modulus in combination with their simultaneously high thermal conductivity. However, because of their low ductility, mechanical problems can occur due to steam hammering [4].

New developments in sealing technology are based on so-called expanded graphite. A corresponding thermal treatment of metal mesh-reinforced graphite intercalation compounds produces a sealing material that can withstand purely thermal loads of up to 823 K (550 °C) at operating pressures of up to 250 bar and which is also resistant to hydrochloric acid (at correspondingly lower temperatures). In another development, graphite is impregnated with Kevlar® (polyphenylene terephthalamide) which allows a further increase in the possible operating temperature in acids. These two developments are discussed in [161].

The latest developments concerning further applications of graphite materials in equipment for HCl absorption, HCl production from hydrochloric acid, HCl stripping and HCl concentration are described in [162]. A typical usage as a cooler in a stripper plant to remove hydrochloric acid-containing hydrochlorocarbons from water is shown in Tables 116 and 117, Section C.

An example of the use of a carbon product is the utilisation of carbon bricks as a lining in a reaction vessel [163].

Graphite and carbon products are important in packings; suitable materials include graphite-filled PTFE yarns, carbon yarns and combination meshes made of carbon and PFTE yarns [164]. The behavior of fluoropolymers is particularly advantageous in case of exposure to hydrochloric acid.

In spite of the excellent hydrochloric acid resistance of PTFE–impregnated graphite, its low wear resistance must be taken into account. For example, the utilisation of impregnated graphite in centrifugal pumps used to transport highly viscous media (e.g. fat) for the impellers and spiral casing made these components subject to rapid wear. This disadvantage can be overcome with a wear- and corrosion-resistant surface coating.

One possibility is to expose prefabricated parts of a pump in the non-impregnated state to silicon monoxide vapor at high temperatures. The vapor diffuses into the graphite, including the pores, and forms a SiC layer that is just as corrosion-resistant in most media as graphite and which also exhibits greatly increased wear resistance. Further information on coated graphitic materials is given in Section B 13 "Metal-ceramic materials (carbides, nitrides)".

B 4 Binders for building materials (e.g. concrete, mortar)

The work in [165, 166] deals with the development of laboratory test equipment to investigate the influence of individual air contaminants on the destruction of stone materials under conditions as similar as possible to those in practice. Whereas the influence of the main component of air contamination (SO_2) was investigated in reference [165], the work described in [166] deals with the destructive action of both SO_2 and HCl. Although the proportion of HCl usually amounts to only a tenth of that of SO_2 in the air, the damaging effect is significant for the following reasons, as demonstrated by tests with lime mortar. When HCl gas comes into contact with damp lime mortar, it dissolves to give hydrochloric acid and reacts with the $CaCO_3$ to form $CaCl_2$ which is readily soluble in rain water. Gaseous SO_2 must first be oxidised to SO_3, which then reacts as sulphuric acid with $CaCO_3$ to yield $CaSO_4$. This reacts further to give sparingly soluble gypsum $CaSO_4 \times 2\ H_2O$ that mostly adheres to the surface and thus inhibits further reaction with SO_2 in the air.

Both reports give numerical data on air pollutants in England and in Italy. The action of air pollutants is of primary significance in Mediterranean countries with regard to the destruction of buildings and statues made of marble.

B 5 Acid-resistant building materials and binders (putties)

Portland concrete is attacked by hydrochloric acid-containing media. The silicate filler (sand, gravel) is resistant, but not the binder, i.e. cement. However, because concrete is a cost-effective building material, there have been plenty of attempts to improve its resistance to acidic media by the addition of plastics. For example, polymer concrete has been used successfully for many years e.g. as pipes for chemical wastewaters. Reference [167] reports on investigations of polymer concrete exposed to 0.36 % (0.1 M) HCl at 363 K (90 °C) for more than 1000 days. The polymer additive at a level of 12 % consisted of 55 mass% styrene (ST), 36 mass% acrylonitrile (ACN), and 9 mass% trimethylolpropane trimethacrylate (TMPTMA). The ratio of sand : cement was 9 : 1.

After the test, it was found that the water absorption of 1.5 to 2.0 % in the initial state had increased to 4.0 to 4.5 %, while the compressive strength had decreased by 10 to 15 %.

Sample no.	Base components 88 mass%		Polymer additives 12 mass%				Sample dimensions cm
	Sand %	Cement %	ST * %	ACN * %	AA * %	TMPTMA * %	
1	90	10	55	36		9	4.3 × 2.4 × 0.40
2	90	10	50	33		17	6.2 × 2.5 × 0.30
3	90	10	50	20		30	3.0 ⌀ × 2.16
4	80	20	55	35	5	5	3.0 × 2.3 × 0.35
5	70	30	55	35	5	5	3.1 × 2.8 × 0.55
6	60	40	55	35	5	5	3.7 × 3.0 × 0.55
7	50	50	55	35	5	5	3.0 ⌀ × 1.30
8	–	100	50	33		17	3.0 ⌀ × 1.04
9	90 glass	10	50	33		17	3.0 ⌀ × 1.00
10	90 quartz sand	10	50	33		17	3.0 ⌀ × 1.05
11	100 quartz sand	0	50	33		17	3.0 ⌀ × 1.00

* ST = styrene; ACN = acrylonitrile ; AA = acrylamide; TMPTMA = trimethylolpropane trimethacrylate

Table 100: Composition and dimensions of the polymer concrete samples [168]

Reference [168] reports on investigations of the samples that were made up on the basis of the data given in Table 100. The utilised sand had a sieve analysis of 1180, 600 and 150 µm in the ratio 50 : 25 : 25. The cement was type III Portland cement with the following composition: 19.1 % SiO_2, 7.1 % Al_2O_3, 61.8 % CaO, 2.8 % MgO and 3.7 % SO_3. A silane binder (A-174) with methacrylate functional groups was added to obtain a stronger chemical bonding between the organic and the inorganic phases. Before the start of the tests, the samples were subjected to controlled annealing in a vacuum at 343 K (70 °C) and then the initial weight was measured. The samples were then exposed to 30 % HCl at 333 K (60 °C) for various periods of up to 30 days. After the tests, the samples were annealed again at 343 K (70 °C) and their weight was measured. In the evaluation, it was assumed that only the cement component was attacked. Therefore, Table 101 gives the initial cement content in mass% compared to the mass loss in % after the 25-day test period. The samples were grey in the initial state and became yellowish during the tests. This indicates that cement was dissolved.

Sample no.	Initial cement content mass%	Mass loss after 25 days mass%
1	8.5	5.7
2	8.5	5.7
3	8.5	2.8
4	17.0	16.3
5	25.5	15.0
6	34.0	21.3
7	42.5	15.5
8	85.0	35.5
9	8.5	5.1
10	8.5	2.7
11	0.0	0.2

Table 101: Summary of the mass loss measurements of polymer concrete samples [168]

Taking account of the sample dimensions, it can be concluded that the properties of polymer concrete are very dependent on the sand : cement ratio. The hardest and most durable polymer concrete was obtained with 65 to 90 parts of sand and 35 to 10 parts of cement. The coherent contact between the sand and the polymer-cement mixture decreased when the proportion of sand dropped below 65 %. This destroyed the mechanical strength and the chemical stability of the polymer concrete.

It can also be seen from Table 101 that the corrosion rate depends on the shape of the samples. Thin samples corrode more quickly. Thus, the thinnest sample, 4, with a thickness of 0.35 cm, exhibited corrosion through the entire sample, as indicated by the yellowish appearance of the fracture surface.

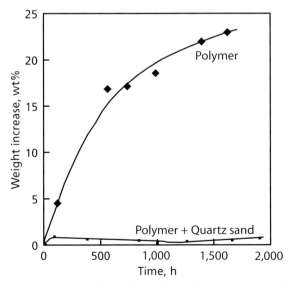

Figure 123: Mass change depending on the exposure time of polymer concrete and polymer samples in a HCl/Cl$_2$ environment at 363 K (90 °C) [168]

Furthermore, Table 101 shows that samples 10 and 11, which contained polymer and quartz sand, exhibited a particularly high resistance to HCl in this test. Damp chlorine gas is particularly aggressive because dilute hydrochloric acid is produced in addition to the strongly oxidising chlorine. Therefore, sample 11, containing pure quartz sand plus 12 % polymer, was tested in an environment with damp chlorine (HCl/Cl_2) at 363 K (90 °C) for 80 days (Figure 123). This sample exhibited an extremely high corrosion resistance. In contrast, the sample of pure polymer material was strongly attacked. The tests show that polymer concrete can be a cost-effective material in hydrochloric acid environments. Compare the information given on PMMA in Section C.

B 6 Glass

Structure and general corrosion properties

Silicate glass consists of tetrahedral silicon structural units that are linked by siloxane bridges ($\equiv$Si-O-Si$\equiv$). Although pure quartz glass has a high chemical resistance but disadvantageous technical properties (high melting point, low formability), the melting point is decreased by the addition of alkali ions. Alkali ions break the oxygen bonds and lead to ionic bonds.

$$\equiv\text{Si-O-Si}\equiv + \text{Na}_2\text{O} \rightarrow 2 \equiv\text{Si-O}^- \text{Na}^+$$

Further possible components are alkaline earth ions, which increase the number of linkages in the glass network ($\equiv$Si-O-Ca-O-Si$\equiv$). However, alkali and alkaline earth ions in the network reduce the chemical resistance. If silicon is partially replaced by aluminium, then anionic aluminium tetrahedra are built into the SiO_2 lattice. This gives rise to excess negative charges that are compensated by sodium ions. A detailed description of possible structures in glasses with reference to the alkali : alkaline earth : aluminium : boron ratio is given in [169].

The corrosion resistance depends on the composition, the structure and the surface state of the glass as well as on the composition of the medium (pH value, vapor pressure), temperature, hydrodynamic conditions and mechanical loading.

Glass corrosion mechanisms are complex and consist of several partial steps of which hydration of the glass network, ion exchange and dissolution of the network are the most important ones. A schematic representation of the various surface layers of a corroding sodium silicate glass is given in Figure 124.

The corrosion process in aqueous solutions starts with hydration of the surface by water molecules without affecting the glass structure. This is followed by hydrolysis (hydration) of network bonds and formation of a gel layer. This converted layer is porous and exhibits a high mobility for alkali ions and water molecules. A diffusion layer is thereby formed on the interface to the intact glass phase, within which the concentration changes of sodium and water take place. In acidic media, ion exchange between protons and sodium ions plays a decisive role and represents an important step in the corrosion mechanism. Coatings of insoluble corrosion products are formed outside the gel layer. In the present case it consists of silicic acid,

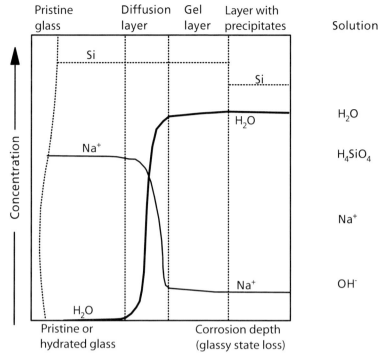

Figure 124: Schematic representation of the corrosion of a sodium silicate glass as a function of the water and sodium concentration [169]

but in alkaline earth, aluminium and heavy-metal containing glasses it consists of the corresponding oxides and hydrated oxides, all of which can influence the corrosion rate. This mechanism mainly applies to alkaline media. A detailed description of the relationships is given in [169].

Vitreous SiO_2 has a solubility of 100 ppm in acidic and neutral solutions at room temperature; this greatly increases above pH = 9. The lower solubility in acidic media means that the corrosion soon comes to a standstill, and the ratio of the surface to the volume has to be taken into consideration in the tests. The larger this ratio is, the earlier is saturation reached.

Amorphous Al_2O_3 is amphoteric, which means that it is soluble in both acidic and alkaline media and is practically insoluble only in the pH range from 6 to 8. The solubility of alkali aluminosilicate glasses can be qualitatively derived from the behavior of SiO_2 and Al_2O_3 [170].

The temperature coefficient of glass corrosion is approximately 1.5 for every 283 K (10 °C) increase in temperature. In addition to quartz glass, borosilicate and aluminosilicate glasses have the best resistance to acids [171]. It is essentially determined by the content of SiO_2. To increase the refractive index of optical glasses, SiO_2 is partially replaced by the heavy metal oxides BaO and PbO. However, this decreases the chemical resistance of the glasses.

Hydrochloric acid test to determine the chemical resistance of glasses

In the USA, the following test is used to determine the stability of optical glasses:

A sample is partially immersed for 30 s in 10 % HCl at 298 K (25 °C) to see whether any color changes take place. A second sample is completely immersed in 10 % HCl at 275 K (2 °C) for 10 min, and the mass loss is determined. The immersion time is extended to 2 h for particularly resistant glasses.

Reference [172] reports on the mass loss of the construction and crystal glasses listed in Table 102 compared to the optical glasses listed in Table 103 determined on basis of this test, but under more stringent conditions, namely in water and in 1 M HCl at 369 K (96 °C) and test durations of 3, 5 and 9 h. Figure 125 shows a graphic evaluation of the results. In water, the overall mass loss was small; however, there was a slight increase of the mass loss with increasing content of BaO and PbO. In hydrochloric acid, the mass loss of constructional and crystal glass was approximately the same as that in water. For the optical glasses containing BaO, the mass loss strongly increased with increasing BaO content. The sample with the

Oxide	SiO_2	R_2O_3 (Al_2O_3 + Fe_2O_3)	CaO	MgO	Na_2O	K_2O	PbO	Fe_2O_3	Total
Construction glass	71.13	1.42	8.80	3.15	15.50	–	–	(0.14)	100
Crystal glass	72.35	–	5.90	0.65	6.70	9.60	4.80	–	100

Table 102: Chemical composition of the investigated construction and crystal glasses [172]

Optical glasses (English designations)	Refractive index n_d	Oxide content mass%		
		SiO_2	BaO	PbO
Light Barium Crown L. B. C.	1.5415	57.1	26.9	–
Medium Barium Crown M. B. C.	1.5725	45.6	32.5	–
Dense Barium Crown D. B. C.	1.6575	36.2	45.9	–
Extra Light Flint E. L. F.	1.5484	66.8	–	7.1
Light Flint L. F.	1.5764	52.5	–	37.5
Dense Flint D. F.	1.6263	47.5	–	45.6

Table 103: Investigated optical glasses [172]

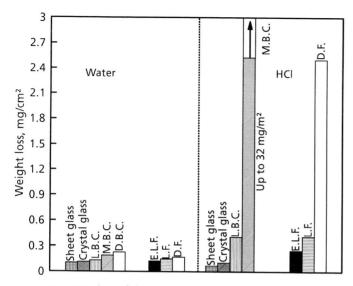

Figure 125: Mass loss of sheet, crystal and optical glasses in water and in HCl at 369 K (96 °C) after 9 hours [172]

highest content of BaO (45.9 %) was already strongly attacked after a test period of 3 h and the glass had a gelatinous consistency so that an accurate numerical statement of the mass loss is meaningless. Increasing the content of PbO also led to increasing mass losses; however, the effect is not as pronounced as in the case of BaO content.

Glass in chemical apparatus engineering

Glass is used for handling hydrochloric acid on its own and as a coating on unalloyed steel. Borosilicate glass plays an important role for glass equipment made entirely of glass because it is resistant to hydrochloric acid of all concentrations up to the boiling point [4].

The average composition of borosilicate glass in mass% is: 80.5 SiO_2, 13.0 B_2O_3, 4.0 Na_2O, 0.5 K_2O, 2.0 Al_2O_3. This composition is stipulated in the US Federal Specification DD-G-541B and in the ASTM standards C-600-70 and C-599-70. A glass of this composition is used exclusively for the construction of process plants and pipelines.

The hydrochloric acid resistance of a coating glass is characterized by the isocorrosion lines in Figure 126 [173]. It can be seen that the 0.1 mm/a (4 mpy) line is exceeded only above 393 K (120 °C) and 0.5 mm/a (20 mpy) in the temperature range above 433 K (160 °C). Furthermore, the maximum corrosive attack is in 20 % hydrochloric acid.

In addition to hydrochloric acid, glasses also have a good resistance in sulphuric acid, nitric acid and acetic acid, which is documented by corresponding diagrams of the isocorrosion lines. In phosphoric acid, the glass exhibits a somewhat increased

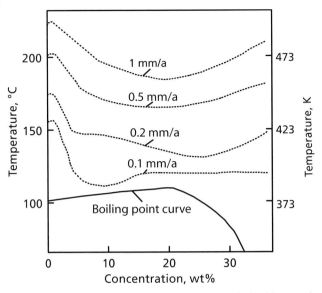

Figure 126: Isocorrosion lines of a coating glass in hydrochloric acid [173]
0.1 mm/a (4 mpy)
0.2 mm/a (8 mpy)
0.5 mm/a (20 mpy)
1.0 mm/a (40 mpy)

sensitivity compared to hydrochloric acid. This is indicated by the isocorrosion line of 0.1 mm/a (4 mpy), which lies at approx. 393 K (120 °C) in hydrochloric acid and which drops to 363 K (90 °C) in phosphoric acid. In contrast, this glass is strongly attacked by hydrofluoric acid or acidic fluoride solutions, even in concentrations in the ppm range. The same applies to attack by alkalis above pH = 9, which greatly increases with increasing pH value and increasing temperature.

The toughness of glass is so low as to be practically immeasurable. Its large modulus of elasticity (69×10^9 Pa) compared to other non-metallic materials means that even small strains are associated with high tensile stresses and the material is very susceptible to fracturing.

Chemical apparatus made of individual pieces of glass is assembled with PTFE (Teflon®) seals and taking account of the brittle behavior of glass. The size and capacity of such equipment is limited. However, this is not a significant disadvantage for the production of high-quality substances in small amounts, e.g. as is common in the pharmaceutical industry. In contrast, advantages include the high corrosion resistance, the smooth surface and the possibility of visual observation of the process course. The smooth surface means that hardly any deposits can adhere so that the glass apparatus can almost always be completely cleaned without too much effort.

Glass has a high thermal conductivity, similar to that of tantalum and graphite, so that a combination of these materials in heat-exchanger equipment can be worthwhile because of their similar corrosion resistance.

References [174 – 178] describe the properties of borosilicate glass, apparatus components made of this material and procedures to assemble complete processing units. In addition, application examples are given, including those with exposure to hydrochloric acid, also at elevated temperatures, with the comment that borosilicate glass has proven suitable for these units.

An important application field of glass is its use as a fiber in glass fiber-reinforced plastics. Glasses generally used for this are C glass, an alkali-lime glass with an increased boron content, and ECR glass, an aluminium silicate glass with added titanium. For further information on fiber reinforcements in GRP, see Section D 3, Composites.

Glass coatings on steel

Glass coatings on steel are used for the corrosion protection of chemical reactors, vessels, pipes and fittings, heat exchangers, columns, evaporators, pumps and accessories. Alkali silicate glasses are generally used for this purpose. The composition, manufacture and range of applications has much in common with enamel and is described in detail in [173].

Glass-coated equipment seldom exhibits damage due to strong hydrochloric acid attack during operation. Thermal and mechanical shock, attack by hydrofluoric acid and alkalis as well as abrasion are the primary causes of failure. Another damage process, which is frequently overlooked and which is attributed to external corrosion, is hydrogen attack. If atomic hydrogen is present on the outside of steel apparatus due to corrosion as a result of leakages, it can diffuse through the steel wall and recombine at the steel/glass interface. The resulting pressure can lead to flaking of the glass coating [4].

B 8   Enamel

The work in [179 – 184] deals exhaustively with the corrosion behavior of chemical-resistant enamel in acids, particularly in hydrochloric acid. Chemical-resistant enamel consists of layers of highly corrosion-resistant borosilicate glass on steel and thus has essentially similar corrosion properties as borosilicate structural glass that is mainly utilised in aggressive acidic media at elevated temperatures.

The resistance of the glass to attack by acids is due to the fact that the main component of most glasses is silicon dioxide, which has an extremely low solubility in acids. Investigations have shown that the entire glass network, including SiO_2, is dissolved in acids [182]:

$$[\equiv\text{Si-O-M}]_{glass} + H^+ \times aq \leftrightarrow SiO_2 \times aq + M^+ \, aq$$

However, because the solubility of SiO_2 in acids is usually in the ppm range [181], even relatively small amounts of glass corrosion products are sufficient to have an inhibiting effect on the continuation of the corrosion reaction. The consequence of this is that the slower ion-exchange phase follows after the primary exposure phase:

$$[\equiv\text{Si-O-M}]_{\text{glass}} + \text{H}^+ \times \text{aq} \leftrightarrow [\equiv\text{Si-O-H} \times \text{aq}]_{\text{glass}} + \text{M}^+ \times \text{aq}$$

The same as for glasses, on account of the silicic acid saturation effect, the corrosion rate is greatly dependent on the exposure time as well as the ratio between the volume and the surface area V/A. If this is not taken into account in laboratory tests, then the significance of the results is considerably limited for practical applications.

The acid resistance of enamel is tested according to DIN ISO 2743 [185] using the apparatus shown in Figure 127. The cylindrical vessel made of borosilicate glass releases SiO_2. Since the ratio between the acid volume and the surface area of the sample is low, rapid saturation of the acid with SiO_2 can be expected. This inhibits corrosive attack on the base of the sample which is totally immersed in the acid. Because of the very low saturation vapor pressure of SiO_2, the condensate phase is free of SiO_2 and so there is no inhibition of the upper sample exposed to the vapor phase. The corrosive attack is thus much greater there compared to that of the base of the sample. Since the volume/surface ratio of enamelled containers is generally large in practice, the larger corrosion rate measured for the vapor phase sample is relevant in practical applications.

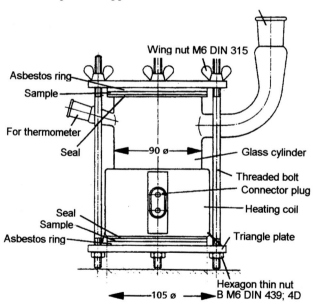

Figure 127: Test apparatus according to DIN ISO 2743 to determine the acid resistance of enamel [182]

The investigations in references [179 – 184] were carried out in tantalum–lined autoclaves to prevent this effect being caused by glass apparatus and also to be able to carry out tests at higher temperatures under pressure. Figure 128 shows a schematic representation of the set-up of the autoclaves, whereby the enamel sample can be exposed to both the vapor phase and the liquid phase. Another variation of this test is in autoclaves that can be additionally rotated so that the influence of moving liquids on the corrosion rate of enamel can be determined [182]. In these tests, the influence of the test period, the volume/surface area ratios (samples enamelled on all sides), acid concentration, acid temperature, pressure and motion of the acid were investigated.

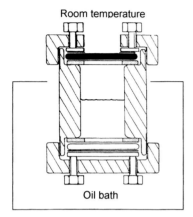

Figure 128: Enamel corrosion by condensing acids; Set-up of the test apparatus; ▬ = enamel sample [182]

Figure 129 shows the influence of the test time on the corrosion rate of chemical-resistant enamel I in 20% HCl at 413 K (140°C) with various V/A ratios. For the small V/A = 5.5 ml/cm², there is already considerable inhibition within 8 days, whereas the inhibition effect for V/A = 45 ml/cm² was still small, even after 20 days. Extrapolation gives a corrosion rate of 0.23 mm/a (9 mpy). This is the value which must be used in the calculations for batch operation of a vessel in which the acid product is changed on a daily basis. The relationship between V/A and the nominal volume V of enamelled vessels is illustrated in Figure 130.

The quality of chemical-resistant enamel is expressed by the corrosion rate according to DIN ISO 2743 in condensing 20% HCl at the boiling point 381.6 K (108.6°C). The five enamel grades, listed in Table 104 in the order of increasing corrosion rates according to DIN ISO 2743, were used in further autoclave tests to determine the temperature dependence of the corrosion rate:

	V_1, mm/a	(mpy)
Chemical-resistant enamel I	0.042	(1.65)
Chemical-resistant enamel II	0.058	(2.28)
Chemical-resistant enamel III	0.070	(2.76)
Chemical-resistant enamel IV	0.100	(3.94)
Chemical-resistant enamel V	0.165	(6.5)

Table 104: Corrosion rates of the investigated chemical-resistant enamel grades [182]

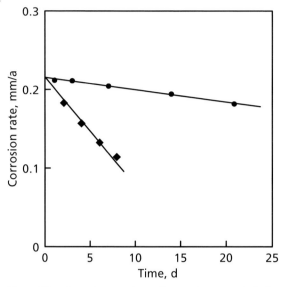

Figure 129: Corrosion rate depending on the test period; chemical-resistant enamel I, 20% hydrochloric acid; 413 K (140 °C), ●: V/A = 45 ml/cm^2, ◆: V/A = 55 ml/cm^2 [182]

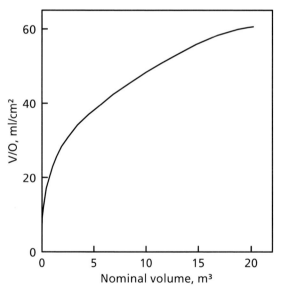

Figure 130: Volume/surface area ratio V/A of an enamelled vessel filled to its nominal volume V [182]

The temperature dependence of the corrosion rate on the enamel follows the Arrhenius equation, as shown in Figure 131. The lines are almost parallel, i.e. the activation energy is independent of the enamel grade. This gives a useful consequence for practical applications. Any differences in the quality found between two enamels by means of a DIN ISO 2743 test also applies in a corresponding manner to higher temperatures. The behavior of an enamel can be forecast with relative accuracy from the temperature dependence of its behavior without having to test it at higher temperatures. The factors F given in Table 105 from DIN ISO 2743 at 381.6 K (108.6 °C) can be used to calculate the increased attack rates at temperature T with regard to exposure to 20 % HCl:

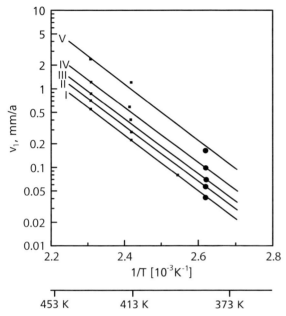

Figure 131: Corrosion rate v_1 as a function of $1/T$ (T = corrosion temperature); chemical-resistant enamels I – V, 20 % hydrochloric acid; ■: autoclave values, V/A = 45 ml/cm², t = 1 d; ●: value according to DIN ISO 2743 (condensate attack, t = 2 d [182]

T, K (°C)	Factor, F
393 (120)	1.9
413 (140)	5.5
433 (160)	14.3

Table 105: Factors for estimating the effect of temperature on the corrosion of enamel [182]

For example, the standard value for chemical-resistant enamel I of 0.042 at 413 K (140 °C) multiplied by 5.5 gives 0.231 mm/a (9 mpy), as determined in autoclave tests (see Figure 129).

The corrosion rate plotted against the acid concentration has a maximum at approx. 20 % (Figure 132). This runs parallel to the maximum of the hydrogen ion activity, which can be represented by the electrical conductivity.

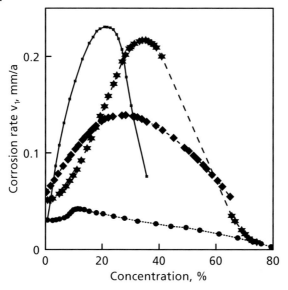

Figure 132: Corrosion rate v_1 as a function of the concentration of various acids; chemical-resistant enamel I, V/A = 45 ml/cm^2; t = 1 d, T = 413 K (140 °C), ■: hydrochloric acid, ★: sulphuric acid, ◆: nitric acid, ●: acetic acid [182]
(1 mm/a = 3.9 mpy)

The temperature-dependence of the system pressure for 20 and 36 % HCl is shown in Figure 133. Tests with additional pressures of up to approx. 15 bar did not lead to a significant change in the corrosion rate. For example:

Chemical-resistant enamel I
20 % HCl – 433 K (160 °C), 5.8 bar system vapor pressure, V = 0.369 mm/a (14.5 mpy)
20 % HCl – 433 K (160 °C), additional pressure with a total pressure of 20 bar;
V = 0.352 mm/a (13.8 mpy)

The addition of colloidal silicic acid to the acid products can greatly reduce the corrosion rate of chemical-resistant enamel on account of the inhibiting effect of SiO$_2$. Only small quantities of silicic acid are necessary for this because the solubility, particularly in highly concentrated acid solutions, is low (Figure 134) [183]. The strong inhibition by SiO$_2$ can be seen in Figure 135. In 20 % HCl, the corrosion rate at 433 K (160 °C) is reduced from 0.6 mm/a (23.6 mpy) for an enamel without additional SiO$_2$ to approx. 0.03 mm/a (1.2 mpy) [183]. This does not apply to the condensate region. Inhibition in the condensate region is only possible by the addition of silicone oil, which has a higher vapor pressure than silicic acid and thus enters the condensate in sufficient quantities. Silicone oil itself does not have an inhibiting effect; however, it releases SiO$_2$ in aqueous acid. The results given in Table 106 demonstrate the favorable effect of silicone oil.

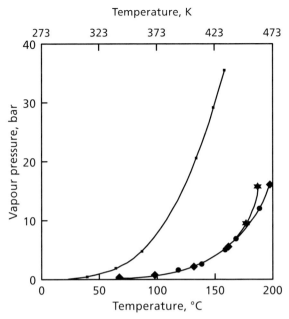

Figure 133: System vapor pressure of various acids in test autoclaves as a function of the temperature T; ■: 36% hydrochloric acid, ●: 20% hydrochloric acid, ◆: 20% nitric acid, ★: 65% nitric acid [182]

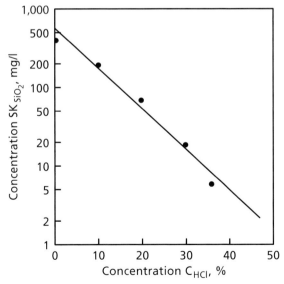

Figure 134: Saturation concentration of silicon dioxide (SK_{SiO2}) in hydrochloric acid at 413 K (140 °C) as a function of the acid concentration (C_{HCl}) [183]

Hydrochloric Acid

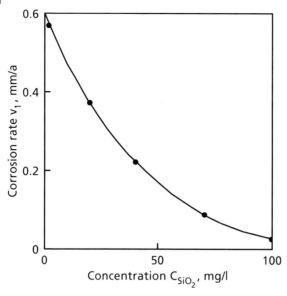

Figure 135: Corrosion rate (v_1) of chemical-resistant enamel I in 20 % hydrochloric acid at 433 K (160 °C) as a function of the silicon dioxide concentration (C_{SiO2}) of hydrochloric acid [183] (1 mm/a = 3.9 mpy)

Enamel grade	Liquid phase	Additive	Condensate phase	Condensate temperature (K) °C	v_1 mm/a (mpy)
Chemical-resistant enamel I	20 % HCl	– 3 ml/paraffin oil 0.3 ml/l silicone oil	20 % HCl	381 (108)	0.040 (1.57) 0.038 (1.50) 0.01 (0.39)
Chemical-resistant enamel III	20 % HCl	– 1.5 ml/l silicone oil	20 % HCl	381 (108)	0.10 (3.94) 0.01 (0.39)

Table 106: Influence of oil additives on the corrosion of enamel [183]

Although paraffin oil enters the condensate, it does not have an inhibiting effect because no SiO_2 is formed. However, for all enamels and for different temperatures, the addition of silicone oil in relatively small quantities can reduce the corrosion rate to a tenth of that of the acid without additives.

In spite of the generally high corrosion resistance to acids, inhibition by silicic acid is of considerable importance in practical applications because the enamel layers are only 1.5 mm thick.

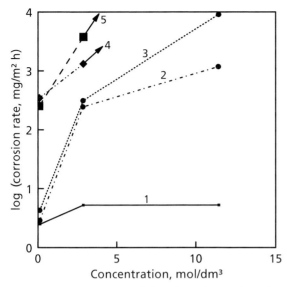

Figure 136: Corrosion rates of alloys depending on the HCl concentration at 353 K (80 °C)
1) α-Al_2O_3; 2) SUS 316 (AISI 316); 3) SUS 304 (AISI 304); 4) S 45 C (AISI 1042, 1045); 5) SUS 430 (AISI 430) [189]

B 12 Oxide ceramic materials

Because of their temperature, corrosion and erosion resistance, oxide ceramic materials are used under particularly aggressive conditions, including, in particular, hydrochloric acid media at elevated temperatures.

Flowmeters made of aluminium oxide ceramics for liquids up to 473 K (200 °C) and pressures up to 1000 psi (68.95 bar) are able to withstand concentrated HCl [186].

Gas filter elements made of ceramic (trade name CERAFIL-S®) are also able to withstand high temperatures and extremely corrosive conditions. They have a low pressure loss and no tendency to clog, even at high flow rates. The product has excellent filtration properties [187]. As part of the treatment of a waste pickling bath from a steelworks, 74 % iron chloride solution is circulated at 403 K (130 °C) during which the solution reacts with hot combustion gases at 1523 K (1250 °C). This decomposes the iron chloride into hydrochloric acid and iron oxide so that there are enhanced corrosion and erosion conditions, especially for the pump. The iron oxide content can reach values of up to 260 g/l. The impeller and casing consist of Cer-Vit®, a mixture of lithium, silica and alumina which is melted to a glass. In this form it can be cast and processed like a glass. It is then crystallised at high temperatures after casting. In this state, the thermal expansion is almost zero and the resistance to thermal shocks, corrosion and abrasion is very high. The fracture strength and the strength are also increased. No visible signs of erosion or corrosion were found during an inspection after the filter had been in operation for more than one year [188].

In the Thermite process, long tubes of unalloyed steel (S 45 C, corresp. to: AISI 1042, 1045, DIN-Mat. No. 1.0503, 1.1191) are lined under vigourous rotation according to the reaction

$$3 \text{ MeO} + 2 \text{ Al} \rightarrow \text{Al}_2\text{O}_3 + 3 \text{ Me}$$

with a dense, homogeneous, compression-stressed α–Al_2O_3 coating (corundum) containing approx. 20 % $FeO\text{-}Al_2O_3$ spinel inclusions [189].

The lining layer has a high corrosion resistance and wear resistance. Figure 136 shows the corrosion rates depending on the hydrochloric acid concentration at 353 K (80 °C) compared to S 45 C (corresp. to: AISI 1042, DIN-Mat. No. 1.0503, 1.1191), SUS 430 (AISI 430, corresp. to: DIN-Mat. No. 1.4016), SUS 304 (AISI 304, corresp. to: DIN-Mat. No. 1.4301) and SUS 316 (corresp. to: AISI 316, DIN-Mat. No. 1.4401, 1.4436). The corrosion rate of the coating at higher concentrations is approx. 100 times less than that of the most resistant molybdenum-containing austenitic steel AISI 316. The actual attack of the coating is limited to the FeO phase.

B 13   Metal-ceramic materials (carbides, nitrides)

Reference [190] reports on investigations of the corrosion behavior of single-phase compacts of chromium carbides $Cr_{23}C_6$, Cr_7C_3 and Cr_3C_2. The samples were prepared by hot pressing of synthetic powders. The current density/potential curves in 11 M HCl (conc. hydrochloric acid) at 295 K (22 °C) and at 345 K (72 °C) show that the strong passivation tendency of the chromium carbides increases in the sequence $Cr_{23}C_6$ – Cr_7C_3 – Cr_3C_2. In contrast, pure chromium cannot be passivated in 11 M HCl either at 22 or at 345 K (72 °C).

The high passivation tendency of chromium carbides suggests that this is also the case in anhydrous media containing halogen hydracid, for which the choice of sufficiently corrosion-resistant materials is very limited. Tests in anhydrous methanolic 1 M HCl showed that the carbides are passive. This is indicated by the more noble corrosion potentials and by the corrosion rates of the carbides

$C_{23}C_6$	0.0014 mm/a (0.055 mpy)
C_7C_3	0.0013 mm/a (0.051 mpy)
Cr_3C_2	0.0016 mm/a (0.063 mpy)

Stainless steel 1.4541 (AISI 321) and pure chromium metal cannot be passivated under these conditions of attack. Like SiC coatings, the corrosion and wear behavior of chromium carbide coatings can be greatly improved if the structure consists of a dense coherent layer [190].

The results of the investigations on the chromium carbides also suggest that the intergranular susceptibility is not caused by the precipitation of $M_{23}C_6$ on the grain boundaries of austenitic steels, but can be attributed to the associated chromium depletion.

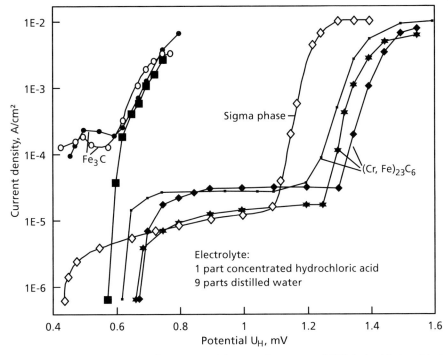

Figure 137: Stationary current density/potential curves in aqueous 3.7% hydrochloric acid at room temperature on compacts of the carbides Fe_2C, $M_{23}C_6$ and the sigma phase [191]

The results in [190] agree with those of earlier investigations in which current density/potential curves were used to determine the transpassive dissolution potentials of precipitates in stainless austenitic steels, and particularly of the carbide $M_{23}C_6$ in 3.7% HCl at room temperature [191]. Here too, dissolution potentials of $U_H = 1.3$ V were found for $(Cr, Fe)_{23}C_6$ (Figure 137). Figure 138 gives an overview of the dissolution potentials in 3.7% HCl at room temperature of various carbides, nitrides and intermetallic phases that occur in conventional steel materials [192]. The more noble the dissolution potential, the wider is the corrosion resistance range. Note the low dissolution potential of WC, which is a main component of known Co-Cr-W cladding materials, of $U_H \approx 0.4$ V. Even in weakly oxidising corrosive media, WC is selectively dissolved from the microstructure. For example, the rapid failure of valve seatings and valve balls in power station equipment on changing to the combined mode of operation is attributed to this. Oxygen (150 to 300 ppb) is added to the boiler water, which has a pH of 8 to 8.5 adjusted with ammonia, to give oxidising corrosive conditions [193].

Figure 139 gives the results of peening wear tests on impregnated graphite, graphite with a SiC layer, glass fiber-reinforced plastic, titanium and the stainless steel 1.4401 (AISI 316) [194]. Impregnated graphite already exhibited a mass loss of approximately 20% after a test period of 1 min. This documents the low wear

200 | Hydrochloric Acid

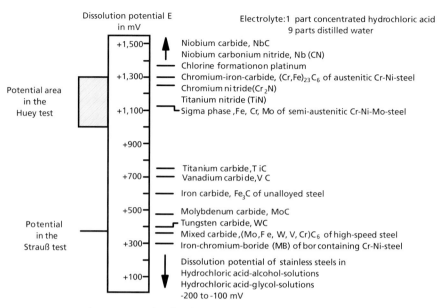

Figure 138: Dissolution potentials of carbides, nitrides and intermetallic phases in hydrochloric acid solutions [192]

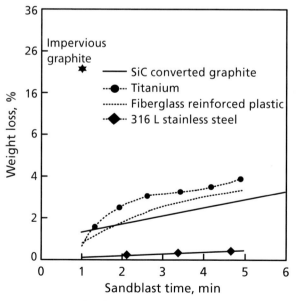

Figure 139: Results of the peening wear tests [194]

resistance of graphite. The wear resistance of the coated graphite is more than 10 times greater and exceeds that of titanium and glass fiber-reinforced plastic. The particularly high wear resistance of the austenitic steel 1.4401 in the peening test is based on the cold-hardening of the austenitic microstructure as the peening material impinges upon it. These comparative tests show that the SiC layer on graphite can considerably improve the wear resistance with a more or less constant high corrosion resistance.

Flow tests in water with 3 to 5% casting sand as the solid and with high (1100 l/min) and low (200 l/min) flow rates, and thus differing particle speeds, for a period of 120 h demonstrated the clear superiority of the impellers coated with silicon. Also after a test duration of 1200 h and a flow rate of 200 l/min, hardly any material had been removed from the impeller.

SiC-coated graphite pumps have proven suitable for HCl recovery in waste pickling solutions from the iron and steel industry using the Keramchemie-Lurgi process [195]. The media conditions were: 2 to 3% HCl, 20 to 22% $FeCl_3$, iron oxide particles, 378 K (105 °C). Pumps made of impregnated graphite, which were used because of their high corrosion resistance, failed because of their low wear resistance after 2 to 4 weeks, or even within a few days at particularly critical points. On the other hand, SiC-coated pumps did not fail within a period of 6 months. The inspection showed that there was no significant corrosion of the most highly loaded places on the impeller. The pumps remained in use, and the impregnated graphite pumps were gradually replaced by SiC–coated pumps.

The corrosion resistance of ceramic materials is very high in most corrosive media. Determination of the corrosion rate by means of the mass loss in immersion tests is therefore time-consuming. Test periods of 200 h and more are necessary to determine the mass loss gravimetrically with reasonable accuracy. In [196], it is shown that in such cases the corrosion rate can be determined more easily and more rapidly by electrochemical methods.

Pressureless-sintered single-phase α–SiC and reaction-sintered SiC (RSSiC) with 8 to 13 weight% of free silicon (two-phase microstructure) were used for the tests.

The current density/potential curves were recorded in 5, 14, 18.5, 27.8 and 37% HCl, 70% HNO_3, 85% H_3PO_4 and in aqua regia with a ramping rate of 0.2 mV/s at an average ambient temperature of 296 K (23 °C). The position of the intersection of the anodic and cathodic Tafel lines gives the corrosion current density, from which the technical corrosion rate (unit mm/a (mpy)) can be calculated.

The corrosion rate is very low for α-SiC and is essentially independent of the HCl concentration (Figure 140). The corrosion rates for RSSiC are much larger and are greatly dependent on the concentration. A maximum is attained in the middle concentration range at 18.5% (Figure 141). The corrosion behavior of RSSiC is determined by the corrosion resistance of the metallic binder Si. The dependency on the acid concentration is parallel to the dependency of the hydrogen ion activity on the acid concentration. Titanium nitride has a high hardness, a high wear resistance and sufficient corrosion resistance in various corrosive media. However, the use of pure TiN is limited due to its high brittleness. The addition of metallic binders,

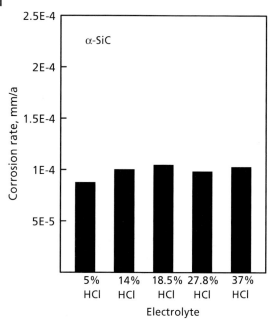

Figure 140: Influence of the HCl concentration on the corrosion rate of α-SiC at 296 K (23 °C) [196]

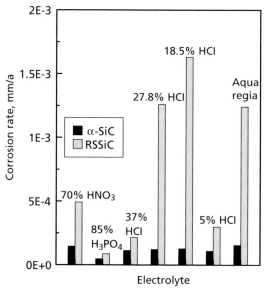

Figure 141: Comparison of the corrosion resistance of sintered α-SiC and RSSiC in acidic solutions [196]

particularly nickel and its alloys, reduces the brittleness. There is then a two-phase microstructure whose corrosion resistance is decisively dependent on the type of alloy used for the binder.

The work in [197] investigated the effect of the alloy composition of the binder in sintered TiN samples in 5% HCl at ambient temperature on the course of the current density/potential curves and thus on the corrosion behavior. These are potentiodynamic curves with a ramping rate of 0.5 mV/s recorded with an Ag/AgCl$_2$ standard electrode (Figure 142).

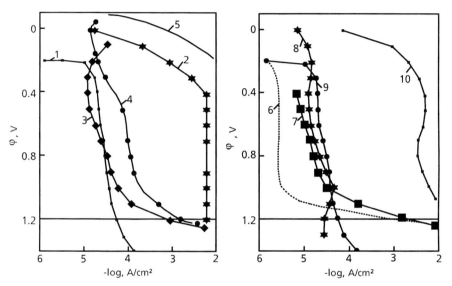

Figure 142: Current density/potential curves for the anodic dissolution of TiN (1), TiN-15 Ni (2), TiN-30 Cr (3), TiN-30 (Ni, Cr) (4), Ni (5), Cr (6), TiN-16 (Ni, Mo)-30 Cr (7), TiN-16 (Ni, Mo)-10 Cr (8), TiN (9), TiN-16 (Ni, Mo) (10) in HCl against a standard Ag/AgCl$_2$ electrode [197]

Pure TiN (curve 1) exhibits passive behavior with a very low dissolution rate in the investigated potential range of + 0.2 to + 1.4 V. The addition of 15 % nickel binder increases the dissolution current by a factor of 100 because the binder is strongly attacked (curve 2). Curve 5 for nickel is very active, and this explains the strong attack of TiN-15Ni because there is only a slight element exchange between the binder and TiN. TiN-30Cr is a passive binder and exhibits dissolution current densities that are somewhat less than those of pure TiN (curve 3) in some ranges. Compared to curve 1 for pure TiN, curve 3 exhibits a steep increase in the current density, i.e. a characteristic transpassive breakthrough occurs for chromium, as clearly demonstrated by curve 6 for pure chromium. However, TiN-30Cr is extremely brittle and hard so that it appears to be unsuitable for use as an alloy binder in practical applications.

The other curves in Figure 142 show that TiN-16(Ni, Mo)-10Cr (curve 8) has a similarly low dissolution rate as the basic material TiN over the entire potential range, thus it also does not have a transpassive breakthrough. Therefore, a 25 % Ni-Mo-Cr binder containing 10 to 15 % Cr gives an optimum corrosion resistance and an acceptable level of mechanical properties for a metal-ceramic material based on TiN.

In an oxidising atmosphere, silicon carbide materials are protected by the formation of dense SiO_2 surface layers not only from further oxidation, but also additionally from corrosion by other aggressive gases. The stability and integrity of such oxide layers is of decisive importance for the effectiveness of their corrosion–protecting function. Corrosion tests of non-preoxidised SiSiC bending specimens in a reducing atmosphere (40 mbar H_2, 1 mbar H_2O) with differing proportions of hydrogen chloride (1 – 40 mbar HCl) at temperatures between 1373 to 1573 K (1100 to 1300 °C) exhibited volatilisation of the Si phase as $SiCl_4$ depending on the HCl partial pressure and the temperature (Figure 143) [198]. The bending strength is given for the individual measuring points. In every case, the formation of SiO_2 was also observed, whereby the samples that were not attacked exhibited a coherent and dense SiO_2 layer. The corroded samples had a defective oxide layer or it was not present in a continuous layer because $SiCl_4$ volatilisation is faster than the formation of the SiO_2 layer. The selective dissolution of the silicon from the material led to pore formation that resulted in a large decrease in the bending strength compared to the initial state of the samples. Pre-oxidised samples were not attacked by HCl under the same conditions.

The work in [199] also deals with the dissolution of silicon from the SiSiC structure in an HCl-containing atmosphere. Whether silicon is dissolved depends on

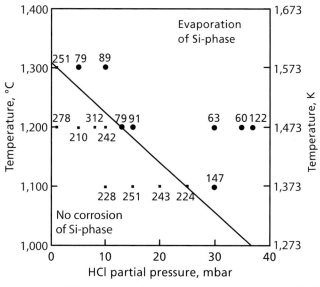

Figure 143: SiSiC corrosion by volatilisation of the silicon phase ($SiCl_4$) by HCl-containing gases. The numbers on the points correspond to the individual bending strength in MPa (p_{H2} = 40 mbar, p_{H2O} = 1 mbar) [198]

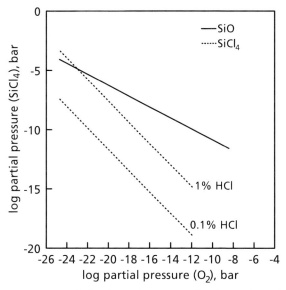

Figure 144: Calculated partial pressures of SiCl$_4$ over the oxidised silicon phase of SiSiC at 1373 K (1100 °C) as a function of the oxygen partial pressure [199]

whether the partial pressure of SiCl$_4$, the volatilisation product, is greater than that of the silicon phase in SiSiC which is being oxidised. This depends on the oxygen partial pressure and the concentration of HCl. Figure 144 shows these relationships at 1373 K (1100 °C). There is selective dissolution of silicon only at very low pO$_2$, i.e. in a reducing atmosphere and for a higher HCl concentration of 1 % HCl. This was proved by the following three experiments:

- No corrosive attack of the SiSiC samples was found after 168 h at 1373 K (1100 °C) in air (high pO$_2$ partial pressure) with 1 % HCl.
- Under the same conditions, but in a reducing atmosphere consisting of 4 % H$_2$, 96 % N$_2$ and 110 vppm H$_2$O, there was strong corrosive attack. The fracture strength dropped from an average value of 322 MPa to 99 MPa with increased scattering of the values.
- If the concentration of HCl was reduced from 1 % to 0.1 % under the same conditions as for b), then there was no corrosive attack.

In agreement with thermodynamic calculations, there was corrosion of SiSiC by 1 % HCl at 1373 K (1100 °C) only for very low oxygen contents (pO$_2$ pressures) in dry H$_2$/N$_2$ atmospheres. However, there was no corrosion of SiSiC if the HCl concentration was reduced to 91 % or if 1 % HCl was mixed with air.

C
Organic materials / plastics

General

Organic materials, just like metallic and non-metallic inorganic materials, exhibit an extremely wide spectrum of differing resistance properties with respect to hydrochloric acid solutions. The reason for this is not only the differing molecular structure of organic materials and the resulting differing resistances, but also the large concentration and temperature dependencies of the corrosive aggressiveness of hydrochloric acid.

To facilitate the understanding of the basic principles of the effects of chemicals, in particular hydrochloric acid, some knowledge of the types of material, the molecular structure and damage mechanisms of plastics are necessary. They form the basis for the explanation of the material behavior under the individual conditions. The material behavior essentially depends on the methods used for investigation and testing, and this must always be taken into account when evaluating resistance data.

In addition to a rather indiscriminate evaluation of the material behavior in resistance tables, the behavior of various material groups and individual materials are discussed in detail in the corresponding sections.

Types of material

Organic materials consist of macromolecular molecular chains of the elements carbon, hydrogen, oxygen, nitrogen and sulphur. Their physical, mechanical and chemical properties depend on the molecular structure, particularly on the length and linkages in the macromolecules.

A useful classification of plastics is based on their mechanical behavior and their temperature dependence, e.g. on the basis of the torsional oscillation test [200, 201]. Plastics are divided into thermoplasts, duroplasts, elastomers and thermoelastomers.

In contrast to metallic and ceramic materials, it is typical for plastics that their temperature sensitivity has a pronounced structural dependence, and their physical and mechanical properties have a pronounced time-dependence.

These changes in properties arise without the action of specific media and are designated as **ageing**.

The temperature-dependence of the mechanical properties of plastics is reflected by the permissible continuous temperature which can be tolerated by a plastic component [202] (Table 107). The individual types of plastic and plastic composites exhibit great differences in their permissible temperature loading. It must be taken into account that this data only applies if there is no chemical loading. In the discussions of the effect of hydrochloric acid given below, it will be shown that the permissible upper temperature limits can be significantly lower.

Material type	Designation	Temperature, K (°C)
Thermoplast	PVC-U	333 (60)
	PVC-C	353 (80)
	PE-HD	333 (60)
	PP	353 (80)
	PVDF	393 (120)
Duroplast, GRP	Polyester resin	353 (80)
	Vinyl ester resin	373 (100)
	Epoxy resin	393 (120)
	Furan resin	403 (130)
	Phenol resin	383 (110)
Composites GRP / Thermoplast	GRP / PVC-U	343 (70)
	GRP / PVC-C	363 (90)
	GRP / PE-HD	343 (70)
	GRP / PP	368 (95)
	GRP / PVDF	408 (135)

Table 107: Permissible continuous temperature for plastic components without exposure to chemicals [202]

Furthermore, it must be taken into consideration that plastics are not usually used as pure substances. Stabilisers, plasticizers, pigments, fillers and other additives are added to improve the application properties of the plastics. These materials not only influence the chemical stability, but they must also be sufficiently stable themselves.

Thermoplasts

Thermoplasts, such as polyolefins, polyvinyl chloride, fluoropolymers, polystyrene, polyamides etc. are hard to rubbery-soft materials at room temperature. They can be plastically deformed, even at moderate temperatures, and exhibit viscous flow above the crystallite melting temperature. They are amorphous to partially crystalline. In amorphous thermoplasts, the linear or branched macromolecules are only cross-linked by physical bonds. They form random tangled bundles which align into a certain order in the partially crystalline state. The crystallinity is decisive for the mechanical properties, particularly for the brittleness.

Duroplasts

Duroplasts include, for example, phenol-formaldehyde resins, unsaturated polyester resins, epoxy resins etc. Duroplasts are high-strength polymers with a great number of chemical cross-linkages. Above the glass temperature they can only be deformed to a limited extent. On heating they do not become viscous but decompose instead. Their structure arises from chemical cross-linking reactions of reactive components and formation of saturated valancies. This causes a high thermal and chemical resistance. They are classified further on the basis of the most important synthesis reactions, namely polycondensation and polyaddition, to produce polycondensates (e.g. phenol-formaldehyde resins, urea-formaldehyde resins, melamine-formaldehyde resin and unsaturated polyester resins) and polyadducts (e.g. epoxy resins). Silicones belong to the former group. Their polymer backbone contains silicon instead of carbon.

Elastomers

Elastomers, such as natural rubber, synthetic rubber, silicone rubber, have stretchable macromolecules with only a few cross-linkages. Their glass temperature or crystallite melting temperature lies below 273 K (0 °C). Therefore, their rubber-elastic domain covers a large temperature range up to the decomposition temperature and there is no viscous state.

Thermoelastomers (Thermoplastic elastomers)

Thermoelastics (also known as thermoplastic rubbers) are an intermediate between thermoplastics and elastomers. They have widely-spaced cross-linkages and exhibit a distinct rubber-elastic plateau between the glass temperature or the crystallite melting temperature and the decomposition temperature. In contrast to elastomers, the glass or crystallite melting temperature lies above 273 K (0 °C). Their great advantage compared to elastomers is that they can be processed like thermoplasts above certain temperatures. Thermoelastomers consist of a combination of soft and hard polymer segments. Examples of this are polymer blends of crystalline polypropylene and amorphous elastomers, such as ethylene-propylene-diene rubber and acrylonitrile-butadiene rubber. Thermoelastomers are described under elastomers in the discussions of the chemical resistance.

Abbreviations / Chemical structure

To simplify dealings with plastics, it is customary to use abbreviations. These are summarised in Table 108. They are partly based on standards; however, the abbreviations are also commonly used [203, 204, 205]. Reference [205] also contains a detailed list of trade names.

Abbreviation	Chemical structure
ABS	acrylonitrile/butadiene/styrene copolymer
BS	butadiene/styrene copolymer
CR	chloroprene rubber
CSM	chlorosulfonylpolyethylene
CTFE	chlorotrifluoroethylene
ECTFE	ethylene/chlorotrifluoroethylene copolymer
EP	ethylene/propylene copolymer
ETFE	ethylene/tetrafluoroethylene copolymer
EP	epoxy resin
EP-GF	epoxy resin, glass fiber-reinforced
EPDM	ethylene-propylene-diene rubber
FPM	fluorinated rubber
FEP	tetrafluoroethylene/hexafluoropropylene copolymer
Fu	furan resin
GRP	glass fiber-reinforced plastic
IR	polyisoprene
IIR	isobutylene-isoprene rubber
MBS	methylmethacrylate/butadiene/styrene copolymer
MF	melamine-formaldehyde plastics
NBR	acrylonitrile-butadiene rubber
NR	natural rubber
PA 6	polyamide, homopolycondensate of ε-caprolactam
PA 66	polyamide, homopolycondensate of hexamethylenediamine and adipic acid
PAN	polyacrylonitrile
PAS	polyarylsulfone
PB	polybutene – I
PBT	polybutylenterephthalat
PC	polycarbonate
PE	polyethylene
PE-C	chlorinated polyethylene
PE-HD	polyethylene, high density (> 0.940 g/cm^3)
PE-LD	polyethylene, low density (< 0.930 g/cm^3)
PE-LLD	polyethylene, low density (0.918 – 0.935 g/cm^3) and linear structure
PE-VLD	polyethylene very low density
PEI	polyetherimide
PEK	polyetherketone
PES	polyarylethersulfone

Table 108: Abbreviations and chemical structure of important plastics [203 – 206]

Table 108: Continued

Abbreviation	Chemical structure
PET	polyethylene terephthalate
PFA	PTFE-perfluoroalkoxy copolymer
PF	phenol-formaldehyde
PF-GF	phenol-formaldehyde resin, glass fiber-reinforced
PI	polyimide
PIB	polyisobutylene
PMMA	polymethylmethacrylate
POM	polyacetal, polyoxymethylene
PP	polypropylene
PPS	polyphenylene sulphide
PPSU	polyphenylene sulphone
PS	polystyrene
PSU	polysulphone
PTFE	polytetrafluoroethylene
PUR	polyurethane
PVAC	polyvinyl acetate
PVAL	polyvinyl alcohol
PVC	polyvinyl chloride
PVC-C	chlorinated polyvinyl chloride
PVC-P	soft (plasticised) polyvinyl chloride
PVC-U	hard (unplasticised) polyvinyl chloride
PVDF	polyvinylidene fluoride
S/B	styrene-butadiene copolymer
SAN	styrene/acryl copolymer
SBR	styrene/butadiene rubber
SBS	styrene/butadiene/styrene block copolymer
SI	silicones
TPE-A	thermoplastic elastomers (polyetheramide grades)
TPE-O/V	thermoplastic elastomers (polyolefin grades)
TPE-U (TPU)	thermoplastic elastomers (polyurethane grades)
UP	unsaturated polyester resins
UP-GF	unsaturated polyester resins, glass fiber-reinforced
VE	vinyl ester

Table 108: Abbreviations and chemical structure of important plastics [203 – 206]

Basic processes with respect to the action of chemicals

The following mainly deals with the effects of chemical loading, namely swelling, stress cracking and chemical degradation. The effects of thermal and thermal-oxidative degradation in the presence of oxygen must always be taken into account. In contrast, in the present considerations, the resistance to photooxidation and ionising radiation only plays a subordinate role. Corresponding data is given in [203 and 204]. Attack by microorganisms, which is important in many types of waters, is described in [207].

In the evaluation of the chemical resistance, it is useful to divide processes into two basic types [203]:

- physically acting media that give rise to changes in the properties of the materials that are partially reversible, e.g. from swelling up to irreversible damage by cracking
- chemically acting media with irreversible changes as a result of the cleavage of chemical bonds.

Physical damage processes depend on the viscoelasticity of the polymer material [208]. Whereas metals, ceramic materials and glass react elastically to mechanical loading and only show slight long-term changes at low temperatures, the elasticity modulus of polymer materials decreases on loading, as indicated by plastic deformation of the material. This is especially pronounced when the temperature is increased. The glass temperature is decisive for this as it must be exceeded. It lies between 353 and 473 K (80° and 200 °C) for the most common polymer materials. The effect of chemicals can also cause changes in the viscoelasticity and thus the mechanical resistance.

Swellability and stress cracking

A widespread type of damage to plastics is the formation of stress cracks which are particularly associated with swelling processes if the material is in contact with solvents.

During **swelling** the solvent molecules enter the material by diffusion into the interstitial spaces between the tangled molecules. This pushes the molecular chains apart, which reduces the cohesion and thus the mechanical strength. In general, these processes are reversible since the solvent can be removed by effusion from the polymer to give the original structures. Diffusion processes in plastics are active processes, i.e. they are not due to concentration differences in the infiltrating medium but depend on the possibility of forming free volumes between the molecule chains. Such free volumes are created more easily in amorphous polymers than in crystalline polymers.

Of course, it must be taken into consideration that a high degree of swelling can lead to loss of shape or even to dissolution of the material in extreme cases. On the other hand, partial dissolution of the surface of the material is desired for adhesive bonding.

The old chemist's rule applies to the swellability of a polymer in solvents "A material will dissolve in a solvent of similar structure". This means that the ability of a polymer to absorb solvent molecules is greatest when the chemical structures of the two materials are most similar. This rule applies to organic solvents and is only a coarse approximation. Further evaluation factors, such as polarity of the solvent, solubility parameters and cohesion energies of the material and the media must also be taken into account [203].

The swellability of plastics in water does not follow any particular rules. The values vary greatly depending on the manufacturing process and additives. Data on the water absorption capacity are given in the corresponding literature for the individual materials.

If the swelling process takes place under tensile stress or elongation which lies under the yield strength in a short-time test, **stress cracks** can occur as an additional complication. The interaction between increased mobility of the macromolecules due to the solvent or diminished energy of the fracture surface on the one hand and residual or external stresses on the other hand, leads to cracking. Time also plays a decisive role here: cracks may occur immediately due to the action of certain chemicals or they may appear only after a long time.

A certain group of cracking processes cannot be explained by swelling processes alone. Thus, polyolefins exhibit cracks if they come into contact with aqueous solutions containing detergents. The solution wets the surface and decreases the surface energy so that the energy barrier for the formation of crazes (networks of microscopic cracks) or single cracks can be overcome more easily.

According to the present state of knowledge, it is assumed that stress cracking is caused by a combination of both mechanisms (swelling + lowering of the surface energy) [208].

A further undesirable effect is **permeation**. This plays a role in all media with a high vapor pressure and small molecular size. The gases are able to diffuse through the material to different extents and exit on the other side. If the material is e.g. a coating on unalloyed steel within which there is a temperature gradient so that the lower temperature is on the steel side, then there may be condensation of liquid and formation of blisters on the boundary surface between the two materials which destroys the adhesion. This permeation effect is particularly marked for gases with a small molecule size, e.g. HCl and HF. Concentrated hydrochloric acid and hydrofluoric acid with high partial pressures of these gases exhibit this permeation effect.

There are only a few systematic investigations on the permeation of HCl through organic corrosion protection materials [209]; however, more recent investigations on the permeation of water provide good clues to their fundamental behavior [210]. The tests were carried out at temperatures up to 363 K (90 °C) on various materials and the permeation coefficients and activation energies were determined. Table 109 gives the data for a few selected materials. The results show that there are considerable differences in the permeation coefficients.

However, no systematic relationship can be derived on the basis of the structure of the materials. The activation energies, which represent a measure of the temperature dependence of the permeation, also vary very strongly. Some of the values lie well over 5 kJ/mol and thus indicate the existence of so-called activated diffusion, as mentioned above.

Material	Temperature K (°C)	Permeation coefficient ng/cm h Torr	Activation energy of permeation kJ/mol
PE-HD	363 (90)	4.1	42
PP	363 (90)	10.0	18 to 26
PVDF	363 (90)	20.0	23
FEP	363 (90)	1.8	38
VE-GF (based on bisphenol A)	363 (90)	8.0	18
PE-GF	363 (90)	23.4	5.1
EP-GF	363 (90)	6.2	32.3
Soft rubber CR	343 (70)	37.6	–
Hard rubber NR	343 (70)	5.2	–

Table 109: Permeation coefficients and activation energies for water permeation through some corrosion protection materials [210]

Swelling, stress cracking and permeation are temperature-dependent: with increasing temperature an amplification of these effects is generally observed.

Chemical degradation processes

Most technical applications concern the handling of hydrochloric acid in aqueous solutions. However, non-aqueous hydrochloric acid solutions must also be considered, particularly alcoholic acid solutions that can be highly corrosive. As a consequence, in the evaluation of the chemical resistance of organic materials in hydrochloric acid solutions, the type of solvent must be considered in addition to the concentration and the conditions, e.g. temperature and mechanical loading.

The chemical reactions of macromolecular substances follow the same rules as low molecular-weight organic compounds. The rate of reaction and the resulting changes in properties are, however, very different. This depends on the fact that the attacking medium attacks only the surface (duroplasts) or must first diffuse into the material in order to attack the macromolecules (thermoplasts and elastomers). The resulting processes inside the material are also very different. The decisive factor is the type and number of reactive groups on the macromolecule that can react with the attacking medium.

For example, it is known from general organic chemistry that hydrocarbons are not very reactive. Similarly, high-molecular weight compounds, whose macromolecules are formed of hydrocarbon chains, are not very reactive. Thus, polyethylene and polypropylene are very resistant to attack by acids, bases, salts and weak oxidising agents in aqueous solutions. The resistance is considerably increased if the hydrogen in the macromolecule is replaced partially or completely by fluorine which shields the carbon atoms from chemical attack. It is this effect that gives rise to the high resistance of fluorinated plastics and which allows their increasing use in acid technology of chemical apparatus engineering.

If the hydrocarbon chain contains OH groups, as in the case of polyvinyl alcohol (PVAL), then these provide places that can be attacked in aqueous solutions and this leads to corrosion. If the carbon atoms are partly replaced by oxygen or nitrogen, then attack also occurs here and the chains are degraded. Correspondingly, such compounds are readily attacked by hydrolytically active media (aqueous acids and bases) with the result that polyamides, polyurethanes and polyesters have little chemical resistance to these media [204]. Further information on hydrolytic degradation are given at the end of this section.

Similar to swelling processes, mechanical tensile stresses or oscillation loads with simultaneous action of chemicals lead to stress cracks whose generation mechanisms have already been described above.

Because this has an unfavorable effect on the viscoelastic behavior (e.g. creep strength) of thermoplastic polymers, the designer must use so-called reduction factors that take account of the reduced mechanical resistance when dimensioning equipment and components with regard to strength. Details of the stress crack are given in the discussions of the resistance of the individual plastics and plastic groups.

The detailed physical-chemical attack mechanisms are very complex and some have not been elucidated. Even purely **thermal degradation** of a polymer without the presence of oxygen in the air or chemicals is based on chemical processes. There are three basic chain degradation processes [208]:

- Chain cleavage with chain degradation

If the temperature exceeds a certain limit, the macromolecules break apart and form radicals. The molecules become smaller starting from the ends and monomers are formed that react further following different pathways to give various products.

- Statistical chain cleavage

The molecular chain is not cleaved from the ends but breaks apart at weak points. Such points can be functional groups, for example those found in polyesters, polyamides and polyurethanes. Statistical chain cleavage is frequently observed if, in addition to thermal loading, there is also chemical attack in the form of hydrolysis, e.g. by the action of hydrochloric acid, by chemical oxidation processes or if there is biological attack.

- Thermal reactions without chain cleavage

If substituents on the macromolecule react before the chain cleaves, then new structures are formed that have new polymer properties. A typical example of this is dehydrochlorination (release of hydrochloric acid) from PVC to form double bonds. This reaction mechanism dominates in thermolysis of PVC, e.g. during fires. An interesting feature of this is the autocatalytic effect due to the hydrogen chloride formed which leads to enhancement of the polymer degradation. Although chain cleavage can hardly be prevented, it is possible to successfully control thermal substituent cleavage reactions by means of suitable additives (fatty acid salts of heavy metals).

At room temperature, the purely thermal degradation of polymers is more or less insignificant, whereas **oxidative degradation** is important and determines the long-term behavior in the presence of air. Because oxygen is present in many technical applications, a few comments on the chemistry of polymer oxidation are necessary. Oxidation takes place via a radical chain reaction to form peroxides at easily oxidisible groups (e.g. tertiary H atoms of polypropylene). These intermediate products are converted into alkoxy radicals and finally into alkyl radicals, and chain cleavage occurs at this point. Chain cleavage means a decrease in the molecular weight and a reduction of the glass temperature. The consequences are changes in the mechanical properties, such as a decrease in toughness, fracture length, bending strength and thus an overall increasing brittleness.

The details of these processes are very complex and differ, for example, in semicrystalline polyolefins and amorphous PVC. The understanding of these processes provides the basis for the development of so-called polymer blends and block copolymers.

Likewise, the addition of antioxidants to improve the oxidation resistance is based on this knowledge. A good overview of the present state of knowledge regarding the mechanisms is given in [208].

In oxidising chemicals, such as oxidising acids e.g. nitric acid, concentrated sulphuric acid and peroxy compounds, oxidative degradation is not a radical chain reaction, but instead a one-step process [211]. Since hydrochloric acid is not an oxidising acid, oxidation processes are only important in the presence of additional oxidising impurities. The effect of such impurities is discussed together with the chemical resistance of the individual polymers.

Hydrolytic degradation is of particular importance for polymer materials in contact with hydrochloric acid. Hydrolysis is the cleavage of C-O, C-N, C-Cl and C-S bonds by water which thereby degrades the polymer chain. Depending on the type of reaction, they can be catalysed by acids or bases. The great sensitivity of certain plastics to highly concentrated hydrochloric acid and sodium hydroxide solutions is based on this. Hydrolysis is a special type of solvolysis, which is the cleavage of polymers by proton donors, such as alcohols. In the last case, this degradation reaction in alcholic hydrochloric acid is known as alcoholysis.

The prerequisite for hydrolytic attack is the presence of water and the existence of the above-mentioned carbon-heteroatom bonds in the macromolecule. Hydrocarbons do not contain such bonds so that polyolefins are essentially resistant to hydrolysis. Polyesters, polyamides, polycarbonates and polyurethanes all contain hydrolysible groups in the main chain and are thus correspondingly sensitive to acids and alkalis. In addition to the cleavage of the main chain, hydrolysable substituents, such as the C-Cl group in PVC, can be attacked; however, only under correspondingly enhanced conditions on account of their stability.

An important phenomenon is the differing hydrolysis resistance of crystalline and amorphous regions in polymers. Amorphous regions can undergo hydrolytic attack whereas crystalline regions remain resistant. This effect is used in analysis, e.g. in the determination of the degree of crystallinity in a polymer material [211].

The differing susceptibility of crystalline and amorphous regions to hydrolysis means that as the temperature increases the transition from crystalline into amorphous regions becomes important and additionally contributes to a reduction in the chemical resistance. In addition to the chemical/catalytic temperature effect, this phenomenon also contributes to the high temperature sensitivity of the chemical resistance of polymers.

Table 110 gives a rough summary of the hydrolysis resistance of polymers, whereby common materials are divided into three groups on the basis of their resistance [211]. The above-mentioned principles are clearly visible in this list.

Resistant in acidic and alkaline media	Not resistant in acidic and alkaline media	Not resistant only in alkaline media
Polyolefins	PVC soft	unsaturated polyester resins (phthalate and bisphenol resins)
PVC hard	polyamides, polycarbonates and polyurethanes	phenol-formaldehyde resins
Fluoropolymers	polysulfones	polyacrylonitrile

Table 110: Hydrolysis resistance of common plastics [203, 208]

Testing of the chemical resistance

In contrast to metallic and non-metallic inorganic materials, the chemical resistance of polymeric materials is not evaluated by means of quantitative corrosion rates or penetration rates (as for local corrosion of metals), but only by the changes in a number of properties. The tests deal primarily with changes in the mechanical properties.

Storage tests are used to determine the mass changes of shaped parts on exposure to chemicals [204]. The mass change is measured after differing periods, whereby the mass increase due to diffusion and swelling is of particular importance. The resistance criteria given in Table 111 apply here.

Characteristic mechanical values from short-term tests are compared after contact with the attacking media to the corresponding data of the material in the as-delivered state. These include ultimate elongation (Table 111), tensile strength, impact toughness, notch impact toughness, bending strength and indentation hardness. For example, the criteria for the evaluation "resistant", "limited resistance" and "not resistant" for the determination of the ultimate elongation are given in the corresponding DIN and ISO standards and are summarised in Table 111 [212]. A brief description of the individual tests as well as their designations are given in the DIN, ISO, and ASTM regulations [204].

The most reliable values are obtained from **endurance tests** (creep tests). They are used to determine the long-term behavior of plastics in dependence on the temperature and the environment [205]. This long-term behavior cannot be determined on the basis of the short-term tests described above; however, it is very important for

Evaluation	Relative mass change (increase = swelling)	Range of the ultimate elongation ratios after exposure to the medium for 112 d
Resistant	≤ + 3.6 %	50 % to 125 %
	≥ – 0.8 %	
Fairly resistant	from + 3.6 to + 10 %	30 % to 50 % and
	from – 0.8 to – 2 %	125 to 150 %
Not resistant	above + 10 %	< 30 % and > 150 %
	below – 2 %	

Table 111: Resistance criteria for chemical exposure of PVC-U pipes in an immersion test according to DIN 16888, Part 2 [212]

apparatus engineering. Because the fundamental viscoelasticity can be decisively dependent on the effect of chemicals, the creep rupture test was developed for pipes and is described in detail in the literature [212, 213] and is also specified in the rules and regulations [214, 215]. An example of the creep behavior of PE-HD in water and hydrochloric acid is shown in Figure 145. A time-related resistance factor of $f_{CRt} = 0.35$ was found, i.e. the endurance is decreased by this factor. A stress-related resistance factor of $f_{CRt} = 0.73$ was calculated from the same diagram along with the inverse value of the reduction factor $A_2 = 1.37$.

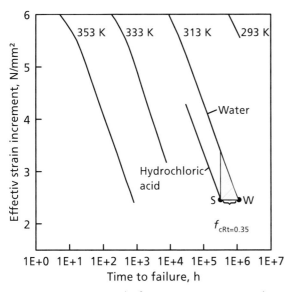

Figure 145: Creep strength of PE-HD pipes in contact with water and 33 % hydrochloric acid with resistance factor f_{CRt} [229]

Resistance tables

Table 112 to Table 115 give an approximate overview of the resistance of polymer materials in various concentrations of aqueous hydrochloric acid at various temperatures. The HCl concentrations of saturated hydrochloric acid vary between 35 and 37 mass% at room temperature. This is probably the result of small variations in the temperature and total pressure during the experiments in the various laboratories. The statement made in Section V 1 is important: temperatures above the boiling point of concentrated hydrochloric acid 338 K (65 °C) can only be attained at corresponding pressures.

Because of its qualitative nature, this type of resistance table can only provide approximate information on certain application cases. More detailed information on the individual materials is given in the following section under the individual descriptions of the materials and in the section on materials with special properties.

The tables are positive lists, i.e. examples for possible applications are given, and complete failure of the materials is not mentioned. Failure must always be assumed to be the case if the given upper limits for the temperature and concentration are exceeded. The numerical values of the upper temperature limits for safe application differ depending on how conservative the evaluations for the individual sources are. Since these are mainly corporate publications, they always recommend further consultations. In some cases, the resistance potential is not fully utilised because the tests were not carried out at the upper limit (Table 112).

Most data applies to chemical loading without an additional mechanical loading [216], whereby, e.g. for highly concentrated hydrochloric acid, the permissible temperatures can be higher compared to values with simultaneous mechanical loading (Table 115).

Table 115 contains so-called reduction factors A_2 of thermoplasts for vessels and pipes. These represent the reduction of the permissible stress exposed to a medium that has to be taken into account in the design of equipment and plants. The reduction factor A_2 is thus a reciprocal resistance factor and is included in the creep strength.

The listed reduction factors are based on extensive compilations of a renowned institute [217]. Recently, additional and more extensive data have been published under the European standard DIN EN 1778, which can be used instead of the national design regulations [218]. Further details on the definition and utilisation of reduction factors, joining factors (welding factors) and safety coefficients are given in [219].

Thermoplast / aqueous hydrochloric acid				
	Concentration limit mass%	Temperature limit K (°C)	Resistance	Reference
PE-HD	37	313 (40)	+	217
	36	333 (60)	+	206, 216, 219, 220
PP	10	313 (40)	+	206
	10	333 to 353 (60 to 80)	⊕	206
	20	60 (333)	+	216, 219
	20	333 to 353 (60 to 80)	⊕	216, 219
	30	353 (80)	+	217
	37	338 (65)	+	220, 221
PVC-U	5	313 (40)	+	206
	5	333 (60)	⊕	206
	10	313 (40)	+	206
	10	333 (60)	⊕	206
	20	293 (20)	+	219
	20	333 (60)	⊕	219
	36	313 (40)	+	206
	36	333 (60)	⊕	206
PVC-C	37	333 (60)	+	217
	36	353 (80)	+	206, 216, 219, 221
PTFE	35	358* (85)	+	209
ETFE	35	343* (70)	+	209
FEP	35	85* or 93 (358 or 366)	+	209, 221
PFA	35	353* (80)	+	209
PVDF	36	100 (373)	+	206, 216, 217, 219, 221

Table 112: Resistance of thermoplastic polymers to hydrochloric acid; swelling and stress cracking: see particular section. + ≙ resistant, ⊕ ≙ fairly resistant,
* ≙ values lie below the maximum loading capacity in the immersion test

Duroplast / aqueous hydrochloric acid				
	Concentration limit mass%	Temperature limit K (°C)	Resistance	Reference
UP	20	393 (20)	+	219
	36	339 (66)	+	221
UP-GF	37	333 (60)	+	217
EP	10	353 (80)	+	221
	20	373 (100)	+	219
EP-VE	20	383 (110)	+	222
	37	353 (80)	+	222
EP (Novolac)	37	353 (80)	+	223
PF-GF	37	403 (130)	+	220

Table 113: Resistance of duroplasts to hydrochloric acid; swelling and stress cracking: see particular section. + ≙ resistant, ⊕ ≙ fairly resistant

Elastomer / aqueous hydrochloric acid				
	Concentration limit mass%	Temperature limit K (°C)	Resistance	Reference
NR	37	293 to 353 (20 to 80)	+	220, 224
CR	10	293 (20)	⊕	206
NBR	10	293 (20)	⊕	206
Bromobutyl rubber	25	333 (60)	+	224
	37	343 (70)	⊕	220
EPDM	10	353 (80)	+	206
	37	353 (80)	⊕	220
FPM	30	333 (60)	⊕	206
	36	373 to 393 (100 to 120)	+	220, 225
CSM	30	293 (20)	+	206
	30	313 (40)	⊕	206
	37	328 (55)	+	220
IIR-PVC	25	233 to 353 (- 40 to 80)	+	226
	37	233 to 313 (- 40 to 40)	+	226

Table 114: Resistance of elastomers and sealing materials to hydrochloric acid; swelling and stress cracking: see particular section. + ≙ resistant, ⊕ ≙ fairly resistant

Material / aqueous hydrochloric acid			
	Concentration limit mass%	Temperature limit K (°C)	Reduction factor A_2
PE-HD	37	303 (30)	1.2
	37	313 (40)	1.2
PP	20	353 (80)	1.1
	30	353 (80)	2.2
PVC-U	37	333 (60)	1.0
PVDF	37	373 (100)	1.6
GRP laminate of unsaturated UP/PHA resins	37	313 (40)	1.2
GRP laminate with PP lining	37	313 (40)	1.1
GRP laminate with PVC-U lining	37	313 (40)	1.1
GRP laminate with PVDF lining	37	373 (100)	1.1

Table 115: Maximum application temperatures and reduction factors A_2 for various polymers in aqueous hydrochloric acid [217, 222]

The resistance data listed in Tables 112 to 115 apply to aqueous hydrochloric acid. If HCl-containing organic or inorganic media, e.g. solvents, are to be used, then the resistance of the polymer in such media must also be taken into account. Because there are many material / medium combinations, they cannot be discussed here individually. It is important that a hydrochloric acid-containing organic medium e.g.

alcoholic hydrochloric acid, is also characterised by its swellability. If this is pronounced, then HCl can be expected to have a large effect.

There are no resistance data available on polymer materials from investigations in non-aqueous media containing hydrochloric acid. Therefore, it is recommended that the fundamental resistance in the corresponding organic or inorganic media is determined initially, e.g. using the tables given in [207]. If the material is not resistant, then the presence of HCl will amplify the effect and the material cannot be used. If the material is resistant in the organic medium, then the existence of HCl can drastically reduce the resistance. In most practical cases it is necessary to determine the resistance by corresponding experimental investigations under conditions similar to those in practice.

Resistance of individual material groups and materials

Thermoplastics

The following selection of thermoplastic polymers (see also Table 112) is primarily based on their resistance to hydrochloric acid. Therefore, the discussions mainly concentrate on polyolefins, vinyl polymers and fluoropolymers that have a good-to-excellent resistance within given limits.

The discussion is based on the primary literature as well as corporate publications that have a high standard of information, such as those found at the ACHEMA 2003 exhibition in Frankfurt (Germany).

The selection of a material must also be based on cost considerations. It is interesting to do this using the example of acid-resistant plastic pumps [227]. If the costs of the pump made of PP are set to 100%, then the relative costs for PVC-U are 110%, for PVC-C they are 125% and for PVDF they are approx. 150%.

Polyolefins and their copolymers

Polyolefins are made of saturated hydrocarbon chains with a high molecular weight. They have a low tendency to undergo chemical reactions. For this reason they are suitable for use in contact with hydrochloric acid if the temperature does not exceed intermediate values from 313 to 333 K (40 to 60 °C). As expected, they exhibit a concentration-dependent susceptibility to attack by oxygen.

Process-related impurities in the hydrochloric acid, mainly halogenated hydrocarbons (HHC) and free chlorine, may make a considerable contribution to the aggressiveness of the medium. Polyolefins cannot be used if there is a separate HHC phase present. At room temperature, there is a certain resistance to HCC levels below 100 ppm dissolved in the medium [220].

Polyethylene (PE)

Of the various polyethylene grades of differing density, PE-HD has proven suitable for the handling of hydrochloric acid. Copolymers with a few percent butene or hexene improve the creep strength and the stress cracking resistance [220].

Hydrogen chloride can easily diffuse through polyethylene. This is indicated by an increase in weight, e.g. 0.2 % after exposure to 35 % hydrochloric acid at 333 K (60 °C) after 96 days. PE-HD turns yellow after a short time in concentrated hydrochloric acid; PE-LD does not exhibit this effect [203].

The technically permissible upper temperature limits for PE-HD is 333 K (60 °C) for hydrochloric acid concentrations of up to 30 % and 313 K (40 °C) for concentrated acid (37 %) [220].

The media lists from the German Institute for Building Technology give quantitative statements on the chemical resistance of polymer materials as they are needed to obtain building inspectorate approvals for vessels and pipes. The following boundary conditions apply to PE-HD [217]:

Hydrochloric acid concentrations up to 37 %; temperatures up to 313 K (40 °C); reduction factor 1.2

This means that the material can be used up to these permitted maximum conditions if the reduction factor is taken into account. The permissible stress for the individual load case is then calculated according to the known equation derived from the strength of the materials and which is explained in detail in [219].

PE-HD is the classical polymer material whose creep strength was determined qualitatively by a creep rupture test for the first time in [228, 229, 230]. Figure 145 (see the section "Testing of the chemical resistance") shows the creep behavior in water and in 33 % hydrochloric acid measured in the internal pressure test of a pipe at 5 bar and at 313 K (40 °C) [229]. The influence of hydrochloric acid can be seen from the shift of the curve to the left. The most important result is that the service life in water (point W, Figure 145) is decreased from 50 years to 35 years in hydrochloric acid (point S, Figure 145). A time-related resistance factor of $f_{CRt} = 0.35$ is calculated from the ratio of the service life in hydrochloric acid to the service life in water. This resistance factor is included in the reduction factor A_2.

Detailed information on the current level of knowledge of determining the chemical resistance of plastics is given in [212].

Copolymers of ethylene and vinyl acetate exhibit improved application properties. However, as expected, they are less resistant to acids because of the C-O bonds [205]. Information on improved application engineering on account of the use of block copolymers and random copolymers with PE is given in [231, 232].

Chlorination of PE can greatly improve the properties with regard to wear and elasiticity. PE-C has been used successfully for diaphragms in sand-slurry pumps that handle up to 15 % hydrochloric acid. Diaphragms of this type should be resistant up to 423 K (150 °C) [233].

Polypropylene (PP)

Polypropylene is preferred over polyethylene because of its somewhat better resistance in the handling of hydrochloric acid. It is available as a homopolymer and as a copolymer with ethylene. The latter exhibits better low-temperature properties. Differing values for the upper application temperatures in concentrated hydrochloric acid are reported; they vary between 313 and 338 K (40° and 65 °C), Table 112. The media lists of the German Institute for Building Technology [217] give an upper temperature limit of 353 K (80 °C) with a reduction factor of 2.2 for hydrochloric acid with a concentration of up to 30 %. Polypropylene can also be used with glass-fiber reinforcement. A relatively high upper temperature limit of 373 K (100 °C) is given for this material [220].

More recent investigations have shown that the diffusion behavior and the resistance in hydrochloric acid depends on the pigments in the PP. Zinc sulphide pigment dissolves and this process reduces the resistance of the PP. This can be remedied by the use of titanium dioxide as a pigment [234].

Copolymers of propylene and ethylene (EPDM) have rubbery properties and are discussed in the section dealing with elastomers.

A good application example is the use of PP in a plant to treat wastewater containing chlorinated hydrocarbons. Undesired HCl is produced in this process by hydrolysis. This HCl in combination with the organic solvent exerts high demands on the resistance of the materials. In the plant, the pure gas pipelines, stripper pumps and droplet separators, that were operated at temperatures between 293 and 358 K (20 °C and 85 °C), were made of PP, whereas a phenol-formaldehyde resin coating or graphite was necessary for the extreme demands in the stripper column (403 K (130 °C)) [235]. An overview of the materials used in the various sections of the plant is given in Table 116. Practically all hydrochloric acid-resistant materials are used in this plant depending on the chemical/mechanical/thermal demands.

Plant component	Loading	Material
Stripper reservoir	293 K (20 °C), HCC in water	steel, coated with furan resin
Stripper	293 K (20 °C), HCC in water	phenol resin duroplast
Stripper pumps	293 K (20 °C), HCC in water, gaseous HCC	polypropylene
Droplet separator after the stripper, pipeline after the stripper, packing in the stripper	293 K (20 °C), HCC in water, gaseous HCC	polypropylene
Heater for stripping air	293 K (20 °C), HCC in water, gaseous HCC	stainless steel
Clean gas pipeline	293-358 K (20-85 °C), traces of HCl, damp	polypropylene
Adsorption vessel	403 K (130 °C), condensing steam, condensing HCC, HCl	phenol resin duroplast
Cooler/condenser	403 K (130 °C), condensing steam, condensing HCC, HCl	graphite
Regenerate pipeline	403 K (130 °C), condensing steam, condensing HCC, HCl	PVDF (solid)
Safety valve	403 K (130 °C), condensing steam, condensing HCC, HCl	PTFE/glass

Table 116: Materials for a stripper unit to remove hydrochlorocarbons [235]

Plant component	Loading	Material
Process air dampers, dry blower dampers	403 K (130 °C), condensing steam, condensing HCC, HCl	phenol resin duroplast
Regenerate dampers	403 K (130 °C), condensing steam, condensing HCC, HCl	steel/PTFE–sheathed
Steam dampers	steam, 403 K (130 °C)	steel/PTFE–sheathed
Solvent separator	303 K (30 °C), condensed HCC, HCl	phenol resin duroplast
Solvent pump	condensed HCC	stainless steel

Table 116: Materials for a stripper unit to remove hydrochlorocarbons [235]

The mechanical properties of PP change on long-term temperature loading in dilute aqueous hydrochloric acid (up to 5 %) in the same way as in pure water. For example, the decrease in the strain in the tensile test after contact with dilute acid is only slight. In 10 % acid and at 373 K (100 °C), the strain decreases by 30 % over a period of three months. Stress cracks are initiated at higher temperatures in hydrochloric acid concentrations above 30 % [203]. This behavior is covered in the design by higher reduction factors (Table 114).

Vinyl polymers

Polyvinyl chloride (PVC)

PVC-U (hard PVC) is resistant in all hydrochloric acid concentrations up to 333 K (60 °C). However, the material turns white in concentrated acid at 313 and 333 K (40 and 60 °C) and exhibits blistering on the surface at 333 K (60 °C). No changes of the mechanical properties were found up to 313 K (40 °C). PVC-P (soft PVC) has only limited resistance at 293 K (20 °C) in concentrated hydrochloric acid [203].

In the media lists of the German Institute for Building technology, PVC-U is permitted in concentrated hydrochloric acid up to 333 K (60 °C) with a reduction factor of 1.0 [217]. This documents its superiority regarding its chemical resistance compared to polyolefins.

Experience gained in practice is interesting: the resistance of PVC increases in dilute aqueous solutions of acids, bases and salts up to intermediate concentrations [222]. This is attributed to the fact that only water, but not inorganic compounds, can diffuse into PVC. Because the water activity decreases with increasing concentration, the chemical aggressiveness of the aqueous solution also decreases [203].

The upper operating temperature of PVC can be increased by glass-fiber reinforcement. If used as a liner in vessels, temperatures of up to 368 K (95 °C) are permissible if the coatings are free of pores [220]. PVC-U has only slight susceptibility to stress cracking.

In contrast to polyolefins, PVC is resistant in chlorine-containing hydrochloric acid. However, it cannot be used in hydrochloric acid containing hydrochlorocarbons with a separate solvent phase on account of diffusion and swelling [220].

The dominating role of PVC-U for components, equipment and machine parts in many industrial branches is well-documented in the literature. With regard to its resistance to hydrochloric acid, which is the main focus of this report, this dominance is based on its relatively good resistance, its high compressive and tensile strengths and its simple processability by adhesive bonding, welding, thermoforming and cutting.

With regard to adhesive joints in aqueous hydrochloric acid, it must be taken into consideration that ordinary PVC joints are not sufficiently resistant in hydrochloric acid concentrations above 25 %. An adhesive based on PVC-C in methylene chloride is used instead of one based on PVC-U in tetrahydrofuran [236]. Further important application engineering information is also given in [237].

PVC has been used as a classical coating and lining material for many applications from an early stage. A useful listing with older literature references is given in [47].

Successful examples of a PVC-based material engineering are reported for their use in many different types of pumps in the processing industry and electroplating industry [227], in a unit used for hydrochloric acid recovery in surfactant production plants [238], in filling equipment for corrosive products [239] as well as in chloralkali and HCl electrolysis plants that have to handle media containing hydrochloric acid and chlorine under extremely different conditions [240]. PVC-U and PVC-C have proved to be suitable for this at temperatures below 313 and 323 K (40 and 50 °C), respectively.

The favorable properties with regard to the corrosion resistance are contrary to the behavior of PVC in the event of a fire. Although it does not burn, it releases HCl as a result of thermolysis. Plasticizers support these processes as they are combustible. Decomposition starts at 413 to 523 K (140° to 250 °C), depending on the type of PVC. At 573 K (300 °C), 30 to 85 % HCl is liberated and at 773 to 873 K (500 to 600 °C) the entire chlorine is liberated as HCl [241]. The HCl combines with the steam, which is also released by the fire, to form hydrochloric acid and can thus lead to enormous corrosion damage, particularly of electrical and electronic equipment, whereby the main damage is to metallic and other inorganic materials. HCl contamination in the event of PVC burning requires a great effort in damage evaluation and in remediation measures, which are summarised in [242].

The cleavage of HCl from chlorinated polymers that are used as binders for coatings which are subjected to high temperatures has also caused problems [243].

The formation of hydrochloric acid is also a problem in waste incineration because HCl has to be eliminated from the flue gases.

Fluoropolymers

Fluoropolymers are one of the most chemically resistant groups of organic materials. Their thermal, oxidative and photooxidative resistances are excellent. The introduction of thermoplastically processable copolymers considerably increased the range of available fluoropolymers. A summarising overview of the physical properties, the resistance to chemicals (tabulated) and comments on application engineering of these materials are given in [244, 245, 246] as well as in a series of more recent corporate publications [164, 206, 209, 222, 225, 247 – 251]. Further information on fluoropolymers is given in the following specialised sections on elastomers, as well as in those dealing with coatings and linings (D1) and seals (D2).

Fluoropolymers are first-choice materials for hydrochloric acid equipment at elevated temperatures and concentrations. They are particularly resistant in hydrochloric acid in the presence of chlorinated hydrocarbons (HCC), as found in solvent recovery units. As already described in the section on PP, the activated carbon used for absorption in solvent recovery processes is regenerated with steam, whereby hydrochloric acid is produced due to hydrolysis of the hydrochlorocarbons. The following fluoroplastics are recommended for plant components in contact with the solvent in accordance with Table 117 [252]:

Plant component	Material
Vessels, piping for solvents, water separator	PVDF
Condensers and coolers	glass
Condensate valves and air valves	PTFE sheaths
Seals and hoses	vinylidene fluoride-hexafluoropropylene copolymer (Viton®)

Table 117: Materials for a solvent recovery plant [252]

A phenol-formaldehyde resin coating or graphite can be used instead of glass for the condensers and coolers [235].

The upper temperature limits for the use of fluoropolymers in the presence of hydrochloric acid and HCC are given in Table 118 [220]. The type of HCC is not given in more detail; however, the effect of individual representatives of this group of solvents and the aggressiveness are comparable.

Material	Abbreviation	Upper temperature limit K (°C)
Polytetrafluoroethylene	PTFE	400 (127)
Perfluoroalkoxy copolymer	PFA	400 (127)
Tetrafluoroethylene / hexafluoropropylene copolymer	FEP	369 (96)
Polyvinylidene fluoride	PVDF	338 (65)
Ethylene / chlorotrifluoroethylene copolymer	ECTFE	boiling point of the corresponding HCC

Table 118: Upper temperature limits for the use of fluoropolymers in the presence of hydrochloric acid and hydrochlorocarbons (HCC) [220]

The stress cracking of thermoplasts decreases in the order PP, PE, PVC, PVDF and ECTFE [222], so that fluoroplastics also have excellent properties in this field.

Polytetrafluoroethylene (PTFE)

PTFE is the most chemically resistant fluoropolymer. It has no thermoplastic properties and must therefore be processed by means of pressure sintering, powder extrusion and paste extrusion. Its decomposition temperature lies above 673 K (400 °C), and the permissible operating temperature lies between 73 and 523 K (− 200 °C and + 250 °C). However, exact upper operating temperatures in concentrated hydrochloric acid that go up to the limit of loading are not given in the literature. This is probably due to the high pressures needed for the determination which makes this too complicated. However, the information supplied by the chlor-alkali industry that PTFE is resistant in damp chlorine (thus containing hydrochloric acid) up to 473 K (200 °C) can be assumed to be reliable and can be directly applied to aqueous hydrochloric acid [240].

The upper operating temperature of 523 K (250 °C) applies to pure thermal loading without mechanical loading and exposure to chemicals and thus only applies to conditions without these additional loads. It is important that the creep strength of PTFE is low ("cold flow"), and this must be taken into consideration in apparatus engineering. Therefore, its main application fields are linings of vessels and pipelines in which metallic materials act as the mechanical support [253, 254] or in special components in which the mechanical strength can be ensured in other ways, e.g. valves [255]. The successful use of PTFE in shell-and-tube exchangers for hydrochloric acid separation in chlorinating plants at temperatures up to 423 K (150 °C) is described in [256].

PTFE linings and coatings are described in detail in Section D 1.

PTFE has no chemical resistance only in a few very special media, such as fused alkali metals, fluorine, chlorine trifluoride. Similar to all fluoropolymers, they have a high resistance to stress cracking [204].

PTFE is susceptible to permeation by gaseous HCl and HF, and this has to be taken into account in concentrated acids. Reference [209] reports a permeation coefficient for concentrated hydrochloric acid at 343 K (70 °C). It lies approximately 20 % above the values for inert gases, oxygen and nitrogen.

Pyrolysis starts at temperatures above 473 K (200 °C) and can lead to the formation of traces of hydrogen fluoride and thus to problems for fluoride-sensitive materials if they are in contact with PTFE. For example, PTFE seals in titanium equipment can have a corrosive effect on the contact surfaces of this material at high temperatures because fluorides are a problem for transition metals such as titanium, zirconium and tantalum.

Polyvinylidene fluoride (PVDF)

In contrast to PTFE, this material can be processed thermoplastically and can be welded and adhesively bonded. Therefore, it is one of the most frequently used fluoropolymers. Its resistance properties lie between those of PTFE and polyolefins. The operating temperature range is reported as extending from 213 to 413 K (−60 °C to +140 °C). However, an upper limiting temperature of 373 K (100 °C) should be reliable for concentrated hydrochloric acid (Table 112). This also agrees with the information given in the media lists published by the German Institute for Building Technology which report a maximum operating temperature of 373 K (100 °C) with a reduction factor of 1.6 for PDVF in concentrated hydrochloric acid [217]. A value of 393 K (120 °C) is reported for its use as a lining material in pumps [257].

The pressure/temperature diagram (Figure 146) takes account of the mechanical loading capacity of PVDF. Since this diagram only applies to water, the pressure must be reduced in the presence of hydrochloric acid; however, no qualitative information is given for this [250]. Analogous diagrams (for other materials) are also published elsewhere [206, 258].

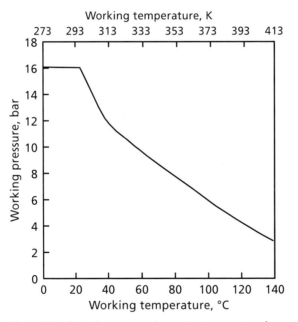

Figure 146: Operating pressure / operating temperature diagram for PVDF in water [250]

PVDF reacts sensitively to hot alkaline solutions by stress cracking. This must be taken into account for the neutralisation of hydrochloric acid with sodium hydroxide solutions. The permissible loading limit in alkalis is only pH = 12 for 313 K (40 °C) [222].

A number of practical applications of PVDF in the paper industry and chemical industry are listed in [259]. The use of this material in the pump industry is especially advantageous [260].

Tetrafluoroethylene / hexafluoropropylene copolymer (FEP)

FEP has a similar chemical resistance to PTFE; however, it can be processed like common thermoplastics by injection moulding and extrusion. It has a high resistance to stress cracking [205, 209].

Perfluoroalkoxy copolymer (PFA)

PFA consists of a PTFE backbone linked to fluorinated side-chains via oxygen atoms so that this polymer can be thermoplastically processed like FEP. The thermal and chemical resistances are very good [205, 209]. PFA is a standard material used in plastic-lined chemical pumps [257].

Tetrafluoroethylene / fluorinated cyclic ether copolymer

This amorphous, highly transparent material is a combination of tetrafluoroethylene with a fluorinated cyclic ether and was launched onto the market in 1989 as Teflon AF®. The structure contains covalent C-F and C-O bonds which give rise to a high deflection temperature and a high resistance to attack by chemicals. The possibility of dissolving Teflon AF® in fluorinated solvents establishes new application fields for coatings [205].

Ethylene / tetrafluoroethylene copolymer (ETFE)

In addition to PTFE, the copolymer contains approx. 25 % PE that has to be protected by stabilising additives against thermal and photochemical degradation. ETFE was the first fluorinated plastic that could be reinforced with glass fibers. It exhibits a very good chemical resistance [205, 209].

Ethylene / chlorotrifluoroethylene copolymer (ECTFE)

Because of its easy processability, this polymer is used in pore-free, electrostatic powder coatings and as woven filters. Its good mechanical behavior is maintained even in cryotechnical applications, 233 K (– 40 °C) [205, 209]. ECTFE has a chemical resistance superior to PVDF [222].

The resistance of a series of fluoropolymers, such as FEP, PFA, ETFE and ECTFE, to concentrated hydrochloric acid at 368 K (95 °C) was proved by absorption tests over a period of 50 days [209]. Since saturation was already reached at low concentrations of HCl after 20 days, it was possible to extrapolate for longer times so that the behavior in concentrated hydrochloric acid can be classified as being very good.

Styrene polymers (PS), polyacrylates

Polystyrene (PS)

In addition to polyolefins and vinyl polymers, styrene polymers (PS) and polyacrylates, e.g. polymethylmethacrylate (PMMA) are frequently used plastics.

Polystyrene (PS) is resistant in weak hydrochloric acid solutions; however, in many media it is susceptible to stress cracking [204, 220] and therefore it cannot be used as a material in hydrochloric acid applications.

Acrylonitrile-butadiene-styrene copolymers (ABS)

Acrylonitrile-butadiene-styrene copolymers have good mechanical properties; however, they are also resistant only in dilute acid and are sensitive to HCl in the atmosphere.

Polymethylmethacrylate (PMMA)

Because of the ester group in the macromolecule of polymethylmethacrylate (PMMA), this material is susceptible to corrosion, even in dilute hydrochloric acid, and has a greater tendency than styrene to stress cracking [203]. However, good experience has been obtained with PMMA as a binder in polymer cement. It was used successfully as a two-component system for the remediation of a concrete foundation for hydrochloric acid pumps [261].

For this group of polymer materials, it is advisable to avoid contact with hydrochloric acid solutions because noticeable attack already starts above concentrations of 10 % at 293 K (20 °C) [47].

High-temperature thermoplasts

The presence of sulphur as a sulphide (S) or a sulfone group (SO_2) in a carbon backbone that contains aromatic rings gives rise to very heat-resistant polymers that are marketed as high-performance plastics. Their main advantages are high stiffness, strength and hardness over a large temperature range (173 K to maximum 473 K (– 100 °C to maximum + 200 °C) for long-term applications) as well as flame-resistance and a low level of waste gas. On account of their chemical structure they are relatively resistant to chemicals, but are attacked by strongly oxidising acids and oxidising agents. There is not enough data available regarding their use in hydrochloric acid at higher concentrations and temperatures.

Polysulfone (PSU)

PSU has been reported to have a good chemical resistance, whereby this includes aqueous inorganic acids. In some solvents, including water at higher temperatures, this material has a tendency to form stress cracks [205].

Polyphenylene sulphide (PPS)

PPS is becoming increasingly important, particularly in combination with reinforcing glass and carbon fibers. The maximum operating temperatures are relatively high: a value of 533 K (260 °C) is given for short-term loadings, 411 K (138 °C) for long-term applications and 493 K (220 °C) for fiber-reinforced grades. The material is resistant to chemicals, such as dilute mineral acids, and stress cracking is not observed [205].

Its successful use in a heat recovery unit of an incineration plant for hydrochlorocarbons is reported in [262].

Polyarylethersulfone (PES)

Polyethersulfone is reported as being resistant in hot water up to 413 K (140 °C) for continuous exposure, as having a low water absorption capacity and a high heat deflection temperature. It is used in rotary pumps [263] and can be used in hydrochloric acid media, although the concentration limits are not stated.

Polyarylsulfone (PAS), polyetherimide (PEI)

These materials have very interesting technical properties, but differing chemical resistances. Their behavior in hydrochloric acid of higher concentrations and temperatures still needs to be investigated.

Further thermoplastic polycondensates

Polyetherketone (PEK, PEEK)

PEEK is used in chemical plants and as a composite with carbon fibers in the aircraft industry. For thermal long-term loading without mechanical loading, an upper temperature limit is given as 433 K (160 °C), or 523 K (250 °C) when reinforced with C fibers [204]. Extensive tests have demonstrated its resistance in most chemicals as well as its hydrolysis resistance in water at 523 K (250 °C) and 18 bar [205].

Polyamide (PA)

The properties of polyamides are determined by the carboxyamide group, CO-NH, which is the weak point with regard to hydrolytic attack. The classic materials PA 66 and PA 6 (Nylon® and Perlon®) differ only slightly in their structures but exhibit differences in their melting points and in their water absorption capacities (lower for PA 66). As already described in the general information on chemical degradation of polymers, C-N bonds in the polymer backbone are susceptible to hydrolysis and are therefore not resistant in dilute hydrochloric acid [203, 205].

Aromatic polyamides of the Aramide family have become important as fibers to replace asbestos in fiber-reinforced materials. These and other types of polyamides or modified polyamides do not exhibit different behavior with regard to their resistance.

Polyester

Polyesters all contain the typical ester group, which can be readily hydrolysed. Representatives of this group of polymers include polycarbonate (PC) and polyethylene terephthalate (PET). Thermoplastic polyesters with a linear chain must not be mistaken for unsaturated polyester resins (UP), which are duroplasts.

Polycarbonate (PC)

PC is resistant to dilute acids, but it is not resistant in acids of higher concentration or in hot water. Residual stresses resulting from forming can lead to stress cracking in media that dissolve or swell the polymer [205]. The material can withstand 20% hydrochloric acid at room temperature, but is already susceptible to stress cracking in 9% acid [203].

Polyethylene terephthalate (PET)

This material, mainly used as a transparent packaging plastic, is sensitive to hydrolysis in water above 343 K (70 °C) and does not have sufficient resistance to hydrochloric acid.

Duroplasts

Within the group of duroplasts (see also Table 113) there is a series of materials which exhibit a high resistance to hydrochloric acid because of their three-dimensional cross-linked structure.

Epoxy resins (EP)

Epoxy resins are basically polyethers that have been cross-linked with a curing agent (aliphatic or aromatic amines, acid anhydrides) to produce temperature- and chemical-resistant duroplasts. Carbon C-C and ether C-O-C bonds confer a high chemical stability so that the three-dimensional cross-linked network does not offer many points of attack for an acid. As for the unsaturated polyester resins, there are many formulations for different technical applications. These are described in detail in [205].

In spite of their general chemical stability, their resistance in hot water is limited. No changes in the appearance or significant changes in the mechanical properties of EP were found in water up to 313 K (40 °C). However, on exposure to water at 343 K (70 °C), the bending strength dropped by 25 % and the impact toughness by 35 %. In contrast, 20 % NaCl solutions at 373 K (100 °C) did not produce any changes [204], probably because of the lower water activity in salt solutions (this effect was also observed with PVC). As for UP materials, their resistance depends on the type of curing agent, on the quantity and type of fillers and on other additives.

Acids diffuse into the polymer and lead to reactions, particularly if amine and polyamide curing agents are used. For this reason, epoxy resin coatings are not suitable for the protection of metals in acidic media [203]. The material fails even in 10 % hydrochloric acid at room temperature [264]. Further information on its use as a liner is given in Section D 1 as well as [246].

Phenol-formaldehyde resins (phenoplasts / PF)

The technical application grades as curable PF moulding materials allows the production of a variety of components and serviceable objects that are resistant to dilute acids, but not to strong acids and boiling water (depending on the type). On the other hand, the so-called technical phenol resins are used as cost-effective binders with a high proportion of acid-resistant fillers, such as graphite, carbon, feldspar or glass fibers, for manufacturing acid-resistant equipment, pumps, fittings, columns and acid-resistant coatings [205]. In this type of application, the material is reported as being resistant to concentrated hydrochloric acid up to 403 or 373 K (130 °C or 100 °C) [220, 265], although only limited resistance is reported in another reference [246]. In contrast, the material is not resistant to hydrochloric acid containing oxidising agents and in oxidising acids.

There are many operating examples of this classic coating material on unalloyed steel to be found in the older literature [47].

Furan resins (FU)

Furan resins are duroplasts synthesised from furfuryl alcohol with formaldehyde, phenol, furfural and other reaction partners. In contrast to phenol-formaldehyde resins, furan resins are stable in acids at high temperatures. The material is resistant up to 353 K (80 °C) in concentrated hydrochloric acid, and up to 373 K (100 °C) in 15 % acid [47]. Operating temperatures of up to 433 K (160 °C) are reported for furan resins reinforced with glass or carbon fibers [220]. In contrast to phenol resins, furan resins have lower fracture stresses, which must be taken into account in the construction, manufacture and operation of equipment. Its use as a coating is described in Section D 1.

Unsaturated polyester resins (UP)

As the name implies, unsaturated polyfunctional starting materials are polymerised to linear esters and then cross-linked, e.g. with styrene, using a peroxide catalyst to produce three-dimensionally networked final products (curing).

The variety of chemical components that can be used for the polycondensation and curing reactions leads to a wide range of properties, particularly with regard to the chemical resistance. The various types of UP resins are described in [246]. As mentioned above, UP resins must not be mistaken for linear saturated polyesters (e.g. PET, PC).

Reinforcement with glass fibers leads to a considerable improvement of the properties. The wide variety of these chemically resistant, high-temperature resistant, elastic and high-strength products are classified in standards and guidelines [205]. These mouldable materials are also known as alkyd resins.

Glass-fiber reinforced, mouldable UP materials (UP-GF) have only a slight susceptibility to stress cracking, but they can be chemically degraded. The following processes are characteristic for their degradation in chemicals: hydrolysis of the ester bonds by acids and hot water, swelling by organic solvents and oxidative degradation by oxidising agents, such as oxidising acids.

The general damage sequence in UP-GF can be described as follows: the aqueous hydrochloric acid diffuses as a hydrophilic medium along the glass fibers in the composite. This decreases the bonding between the resin and the glass and may also cause hydrolytic degradation of the resin. This is detrimental to the mechanical properties. Since the weak point is the interface between the resin and the glass, the glass fibers must not protrude out of the composite but have to be covered with a chemically resistant layer of resin, a so-called chemical-resistant coating that acts as a barrier.

Of the various damage mechanisms, detailed investigations have been made of cracking and blistering in hydrochloric acid solutions as well as investigations concerning types of damage in aggressive media of the chlor-alkali industry. The results of these investigations are given in Section D 3 under glass-fiber-reinforced plastics.

Depending on the type of UP, it can also have different resistances in hot water, as indicated, e.g. by a decrease in the bending strength after a test period of one year (see Table 119). Of the various types of resin, those based on a vinyl ester exhibit the most favorable behavior [203]. Their terminal double bonds are joined to an epoxy resin chain and are cross-linked with styrene. Their properties and application engineering are described in [223, 266].

Resin type	UP, unfilled Bending strength at 373 K (100 °C) in % of the initial value	UP-GF Bending strength at 353 K (80 °C) in % of the initial value
Phthalate resin	7	20
Isophthalate resin	10	32
Bisphenol resin	60	50
Vinyl ester resin	70	70

Table 119: Decrease in the bending strength of unfilled and glass fiber-reinforced polyester resins after one year of exposure to hot water at 373 or 353 K (100 or 80 °C) [203]

Vinyl esters also exhibit the best behavior according to the results of exposure tests in hydrochloric acid. The permissible upper operating temperatures of various UP-GF are summarised in Table 120 [203].

Hydrochloric acid concentration	Upper operating temperatures in K (°C) depending on the type of resin			
Mass%	Phthalate	Isophthalate	Bisphenol	Vinyl ester
20	298 (25)	313 (40)	343 (70)	363 (90)
36	298 (25)	298 (25)	313 (40)	353 (80*)

* Value is probably too high

Table 120: Permissible upper operating temperatures of various glass fiber-reinforced polyester resin types in hydrochloric acid of differing concentrations [203]

According to the media lists of the Institute for Building Technology, i.e. according to the latest state of the art, GRP laminates of UP/PHA resins can be used in concentrated hydrochloric acid up to 313 K (40 °C) if resins types 5/6 according to DIN 18 820-1 are used [217].

Examples of successful applications of UP-GF, e.g. as vessels, pickling towers, chimneys, ventilator casings, fume hoods and of UP as binders in polymer concrete are given in [47]. Further reported application fields are: handling of concentrated hydrochloric acid, wet chlorine, chlorine dioxide and hypochlorite in chlorine production [220, 266 – 268], as a pipe material and apparatus material for large-sized units used to handle acid-containing flue gases up to a maximum temperature of 343 K (70 °C) [269] and for waste acid tanks for a very wide range of media compositions [270]. Its use as a coating with fiber-reinforcement is treated separately in Section D 1 and in [246].

Urea-formaldehyde resins (aminoplasts / UF) and melamine-formaldehyde plastics (MF)

These groups of materials are not resistant in strong acids and in hot water. Therefore, they cannot be used to handle hydrochloric acid.

Polyurethane (PUR)

Because of the -O-C-N- group in the PUR structure, they are not resistant to hydrolysis or oxidation and are thus not resistant in hydrochloric acid solutions.

Silicones (SI)

Silicones contain silicon atoms instead of carbon atoms in the polymer backbone (often in combination with oxygen) along with methyl, ethyl and phenyl groups as organic components. Depending on the starting materials, they are either linear or three-dimensionally cross-linked polymers with a property spectrum that is favorable for application engineering. Because of their structure, however, their resistance to acids is strongly limited. Even dilute hydrochloric acid at room temperature attacks these materials [271].

Elastomers

Both natural and synthetic rubber grades are often used as linings and coatings of chemical equipment, pipes and components that handle hydrochloric acid. An important application area is their use as seals. Linings and coatings are treated separately in Section D 1 and in [246]. Section D 2 contains further information on seals (see also Table 114).

All rubber grades exhibit good resistance in aqueous acids. In contrast, they are chemically attacked in media containing oxidising agents. They are widely used as rubber coatings on steel. For this, the hard rubber grades that have a closely cross-linked structure and tough vulcanisates based on natural rubber and styrene-butadiene rubber have better chemical and thermal resistance compared to the more widely cross-linked and more flexible soft rubber grades [254, 272]. Of the soft rubber grades, mixtures of butyl rubber and PVC are chiefly used for hydrochloric acid-containing media. References [254, 272] also provide extensive information on application engineering.

Soft rubber coatings based on isobutylene-isoprene rubber (butyl rubber) and PVC as well as those based on chlorosulphonated polyethylene are used for oxidising media.

Because of their high level of cross-linking, the hard rubber grades have a greater brittle hardness than tough vulcanisates with low ultimate elongation values of approx. < 5 %. The main component of hard rubber grades is usually natural rubber in combination with styrene-butadiene rubber or isoprene rubber [273].

Vulcanisation plays an important role as a cross-linking reaction. The macromolecules of the plastically deformable rubber are cross-linked via sulphur bridges to form an elastically deformable rubber. Hard rubber grades are vulcanised with a high level of sulphur to obtain hard vulcanisates that have a duroplastic character. Low levels of sulphur are used for the soft rubber grades, frequently in combination with vulcanisation catalysts.

Hot vulcanisation is carried out in autoclaves at temperatures ranging from 393 K (120 °C) to approx. 418 K (145 °C, hard and soft rubber coatings). Cold vulcanisation is possible only with soft rubbers and only with certain types of rubbers, particularly chloroprene rubber as well as bromo- and chlorobutyl rubber. Sulphur and certain vulcanisation catalysts are used for cold vulcanisation. The corrosion resistance depends on the level of vulcanisation and the additives (fillers, plasticizers, anti-ageing agents, bonding agents).

The hard rubbers described below that are based on natural rubber, isoprene rubber and styrene-butadiene rubber, also belong, in principle, to the duroplasts because they have similar physical properties.

Preliminary information is given in Table 114 regarding the upper temperature limits of the resistance of elastomers to hydrochloric acid at concentrations above 10 %.

Natural rubber

Data on the maximum operating temperatures of hard rubbers based on natural rubber as a lining vary, as shown in Table 114. A soft rubber lining can withstand aqueous hydrochloric acid of all concentrations up to 333 K (60 °C), whereas hard rubber can withstand it up to 358 K (85 °C) [220, 271]. A conservative user reports good resistance of hard rubber in concentrated hydrochloric acid at only 293 K (20 °C) [224], at the same time, the same manufacturer gives the upper limit for products based on natural rubber as being 353 K (80 °C) in 25 % acid.

The reason for these differing values is probably due to the vapor pressure of 37 % hydrochloric acid being several times greater than that of 25 % hydrochloric acid so that a large increase in the HCl permeation can be expected.

The large range of data values support the well-known fact that if rubber coatings are used then further operating factors and particularly the know-how of the manufacturer must be taken into account because resistance tables can only provide initial approximate values.

The action of hydrochloric acid on soft rubber results in a hard surface layer that is impenetrable to hydrochloric acid; however, it only acts as an effective barrier in the absence of flexural loading, which is usually the case [220].

There are, of course, many application examples for the use of classical rubber materials that are still valid today. Corresponding references are given in [47]. The current level of experience and knowledge is available in corporate publications, e.g. in [224, 226, 265, 274].

Butyl, chlorobutyl and bromobutyl rubber

These synthetic elastomers are reported as having good resistance in all concentrations of hydrochloric acid up to 343 K (70 °C); however, they tend to lose elasticity and to blister [220]. This is also known from an older publication [47]: butyl rubber looses 15 % of its original ultimate elongation in 37 % hydrochloric acid at 293 K (20 °C) after approximately 4 weeks [275], it is resistant in 33 % hydrochloric acid up to 339 K (66 °C) [276]. Butyl rubber / PVC products are marketed especially for use in concentrated hydrochloric acid [226, 274].

Acrylonitrile-butadiene rubber (NBR)

This is not resistant to concentrated hydrochloric acid, and has only limited use in dilute acids. Therefore, this elastomer is not suitable for handling hydrochloric acid [276].

Styrene-butadiene rubber (SBR)

The chemical resistance of styrene-butadiene rubber is similar to that of grades based on natural rubber.

Ethylene-propylene-diene rubber (EPDM)

This elastomer, which is synthesised by copolymerisation of ethylene, propylene and diene monomers, is used in many fields in preference to natural rubber. The numerous variants (amorphous and partly crystalline EPDM, blends of PP + EPDM) allow the manufacture of customised elastomer materials. The temperature limit of EPDM in hydrochloric acid lies at 353 K (80 °C). It has better mechanical properties and acid resistance than butyl rubber [220].

Fluoroelastomers (FPM)

As for the thermoplasts, fluoropolymers also have a special status amongst elastomers on account of their hydrochloric acid resistance. The recently introduced, thermoplastically processable, physically cross-linked fluoroelastomers are an example of this [205, 209]. Vulcanisable dipolymers of vinylidene fluoride and hexafluoropropylene as well as polymers based on PVDF, hexafluoropropylene and perfluoromethyl vinyl ether (Viton®) guarantee operating temperatures of up to 393 K (120 °C) in concentrated hydrochloric acid. Upper temperature limits of up to 503 K (230 °C) are reported for the copolymer of PTFE and perfluoromethyl vinyl ether (Kalrez®) [220]. Thus, fluoropolymers are the best elastomers for handling hydrochloric acid under extreme conditions.

Chloroprene rubber (CR)

Chloroprene rubber is not resistant in concentrated hydrochloric acid, even at room temperature, and is resistant in 20 % acid only up to 323 K (50 °C) [47]. For cost reasons, the elastomer is not interesting compared to the application possibilities of natural rubber [220].

Chlorosulfonylpolyethylene (CSM)

This thermoplastic elastomer CSM is synthesised by modifying polyethylene. It is reported as being resistant up to 328 K (55 °C) in all hydrochloric acid concentrations; however, above this temperature it tends to swell and to blister [220]. CSM has a resistance comparable to that of EPDM. Older publications report that it is not resistant in concentrated hydrochloric acid, even at room temperature, and at medium concentrations the limit lies at 323 K (50 °C). Special grades can withstand concentrated acid up to 50° C (323 K) [47].

Polysulphide rubber

This elastomer has only limited application possibilities. The upper limit lies at a temperature of 303 K (30 °C) in 15 % hydrochloric acid.

Silicone rubber

The same statements apply to silicone rubber as for the silicones: this group of elastomers is not suitable for use with hydrochloric acid.

D
Materials with special properties

D 1 Coatings and linings

Metallic coatings

Because most of the metallic materials that are sufficiently resistant to hydrochloric acid are usually very expensive, e.g. tantalum, zirconium and NiMo alloys, they are frequently used as platings or linings instead of in the solid form. If correctly processed, particularly during welding of plated parts, the same resistance as for the pure plating material can be expected.

Titanium is frequently used as a medical implant to replace bones and is also regarded as being a promising replacement for teeth. However, at pH values below 2, which do indeed occur in dental applications, the corrosion resistance of titanium is limited. The deposition of niobium films on titanium is one possibility to increase the corrosion resistance in acidic media, as investigations of such films in 5 M HCl have shown [277]. In these investigations, niobium films with a thickness of some 10-20 nm were applied to samples of titanium and the titanium alloy TiAl6V4 by sputtering. In all cases, tests in a 5 M HCl solution showed a reduction of the corrosion current by two powers of ten for the coated samples compared to the uncoated samples.

Inorganic non-metallic coatings and films

Please refer to the Chapters "B 6 Glass" and "B 8 Enamel".

Organic coatings and linings

The information given in this report is based on fundamental literature publications, standards and directives listed below:

- Monograph [219] with extensive information on standards and guidelines.
- The American monograph [246] with the corresponding standards.
- Relevant DIN standards concerning coatings and linings.
- The reviews in references [254, 272, 278] and the corporate publication [279].

- Further special publications and corporate publications, which were largely cited under the individual polymers in Section C Organic materials / plastics.

The concept of this report concentrates, of course, on the chemical resistance of coatings and linings, particularly in hydrochloric acid solutions. Not included are the similarly important directives, codes and standards for construction and design as well as testing and quality assurance. In this respect, please refer to the monographs [219, 278] and [246], in which both German/European as well as the American state of the art are described in detail.

Organic coatings

According to DIN EN ISO 12944-1 [280], an organic coating is a collective term for one or more coherent layers on a substrate material that are not made of prefabricated materials and whose binder is usually of an organic nature [200]. The DIN standards [281, 282] provide extensive information on the requirements and tests of coatings made of organic materials. These are classified into

- cold-curable or catalytically curable liquid coating materials based on unsaturated polyester resins (EP), vinyl esters (VE), epoxy resins (EP) as well as (not resistant to hydrochloric acid) polyurethane resins (UP). They are mainly applied by spreading [283] and are then cured by cross-linking at room temperature.
- Stove enamels made of PF, EP and PF/EP resins, which are cured at 433 to 483 K (160 to 210 °C) on the surface to be protected.
- Powder coatings made of duroplasts EP, UP, the fluoropolymers PVDF, ECTFE, PFA and the thermoplasts PE, PVC, that are melted on to form protective layers with closed pores [284].
- Laminate coatings made of glass fiber-reinforced duroplasts, such as unsaturated polyester resins, as well as vinyl ester, furan, phenol-formaldehyde and epoxy resins [285].

When selecting a corrosion protection coating to be used in the presence of hydrochloric acid, the chemical resistance must be guaranteed above all else. Information on this is given in Section C under the individual polymers as well as in the resistance tables at the end of every section in the monograph [246].

The resistance of coatings is divided into the resistance to swelling, diffusion as well as chemicals. If there is no swelling, diffusion and hydrolytic degradation, which often occur in dilute acids at low temperatures, then the properties can be expected to remain unchanged, even after many years of exposure. Hydrolytic degradation in highly concentrated acid is critical for its use along with HCl diffusion along a decreasing temperature gradient in the direction of the substrate at elevated temperatures. This leads to blistering and delamination if the adhesion of the coating is poor. Materials for coatings on unalloyed steel for condensers and heat exchangers must therefore be carefully selected, whereby phenol-formaldehyde resins have proven suitable.

Systematic investigations on the rusting underneath organic coatings and on their permeation and adhesion behavior have been carried out with epoxy resin, alkyd resin and PE-C products [286]. The results showed that the water vapor permeability of free clear-varnish films at 96 % humidity and at 313 K (40 °C) paralleled the loss due to corrosion underneath the coating, and decreased in the sequence alkyd resin, EP, PE-C. Although there are no corresponding investigations for hydrogen chloride vapors, the material can be assumed to exhibit a similar behavior, whereby the effects are certainly enhanced. Further data on the permeation of water in corrosion protection materials are summarised in Table 109 and discussed in detail in [210]. At this point, it must be mentioned that the tests were carried out at temperatures of up to 363 K (90 °C), which is a particularly interesting temperature with regard to technical applications.

In addition to new information on the resistance behavior of polymer materials, advances in application engineering have been made and are discussed in detail in [254, 279]. Thus, cold curable and catalytically curable coatings that contain solvents or are free of solvents and have a thickness of 150 to 800 µm, or even up to 1000 µm in special cases, have been processed. Several ground coats and top coats can be applied on-site to components of any size.

A particular advantage of thermally curable coating materials is their lack of pores. Coatings applied in several layers are individually dried and heat-treated to obtain coating thicknesses between 120 and 300 µm. PF-based stove enamels have proven successful for internal pipe coatings in heat exchangers because they are resistant to diffusion and do not allow delamination [279].

The best-known processes for powder coating are powder sintering, fluidised bed sintering and flame spraying. The basic principle is to apply the powder to the hot substrate to which it adheres. It is then melted to a homogeneous film in the second step. The advantage of electrostatic spraying is that, as a result of the electrical field between the spray gun and the component, the entire surface, including the rear side, is covered by a uniform layer of powder which is then melted homogeneously in an enamelling furnace.

Laminate coatings made of glass fiber-reinforced unsaturated polyester resins, vinyl ester resins, furan resins, phenol-formaldehyde resins and epoxy resins are applied by hand-laid lamination. The structure of a so-called mat laminate is shown in Figure 147. The approx. 3-mm thick coating consists of several layers, whereby the reactive resin can be cured catalytically so that an oven is not needed. The layer in contact with the medium (sealing layer) must be particularly resistant to chemicals.

An essential criterion for the final selection of materials is the cost. A cost comparison of various coatings and linings is given in Table 121 [287]. It is based on the current level of experience gained in the workshop of a large chemical company. The calculations were based on a stirred tank reactor, 2 m high, 2 m diameter, on three supports. The costs are referenced to a hard rubber lining, which is set to 1. As expected, there are large differences in the costs, particularly for the fluoropolymers.

Hydrochloric Acid

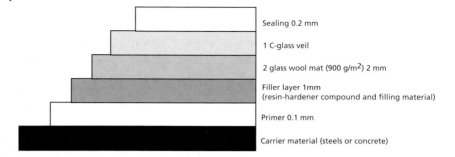

Figure 147: Diagram of the structure of a mat laminate on a layer to be protected

Material	Cost factor
0.2 to 0.3 mm stove enamel	0.8
3 mm hard rubber	1.0
3 to 4 mm hand-laid laminate	1.3
2.3 mm PVC soft	2.4
3 mm hard rubber with 2 mm polypropylene	5.5
3 mm hard rubber with 2.3 mm ECTFE	6.5
3 mm PVDF	5.0
2.3 mm ECTFE	5.0
3 mm PTFE	6.0

Table 121: Cost overview of various surface protection processes using organic materials [287]

The requirements regarding the metal surfaces are given in DIN 28053 [282]. A good pretreatment of the surface is the basic prerequisite for the adhesion of the coating. This consists of derusting, degreasing and application of adhesion-promoting coatings, e.g. the classical phosphating process according to the specifications in DIN EN ISO 12944-1 [280]. Advances in the field of surface pretreatment have been made by the introduction of the Silicoater process in which a very thin (< 0.1 µm) silicate layer is produced by flame-pyrolysis decomposition of an organosilicon compound. Waterborne paints on steel exhibit very good wet strength after this pretreatment [288].

Linings of organic materials

A corrosion-protection lining (liner) is a chemically resistant polymer bonded to a supporting substrate, e.g. unalloyed steel or GRP [289]. The number of materials that can be used for linings is large and includes representatives of all groups of materials. If there is exposure to hydrochloric acid, then the range of lining materials is, of course, limited to those with a suitable chemical resistance that are given in Table 122. According to the information given in Section C Organic materials / plastics and on the basis of the resistance tables given in [246], this list is reduced further depending on the degree of exposure. Thus, for example, only the following thermoplasts can be used in concentrated hydrochloric acid: PVC, polyolefins, fluoropolymers, duroplasts UP, EP, FU, PF, VE and the elastomers CR, CSM, and FPM. There are further limitations if the upper operational temperature limits are relatively high, so that ultimately only fluoropolymers can fulfil the requirements. This is, for example, the case for dew point corrosion in electrostatic separators for flue gases in coal-fired power stations [5].

Abbreviation	Chemical name	Trade name (not binding)
Thermoplastics		
PVC	polyvinyl chloride	Hostalit®, Vestolit®, Trovidur®
PE	polyethylene	Hostalen®, Vestolen®, Lupolen®
PP	polypropylene	Hostalen PP®, Trovidur PP®
PVDF	polyvinylidene fluoride	Kynar®, Solef®, Foraflon®
FEP	tetrafluoroethylene / hexafluoropropylene	Teflon FEP®, Neoflon®, Hostaflon FEP®
PTFE	polytetrafluoroethylene	Hostaflon®, Teflon®, Halon®
E-CTFE	ethylene / chlorotrifluoroethylene	Halar®
PFA	perfluoroalkoxy polymers	Teflon PFA®, Hostaflon PFA®
ETFE	ethylene / tetrafluoroethylene	Hostaflon ET®, Tefzel®, Aflon®
PCTFE	polychlorotrifluoroethylene	Halon®, Kel-F®, Voltalef®
Duroplasts (reinforced or filled)		
UP	unsaturated polyester resins	Alpolit UP®, Palatal®
EP	epoxy resins	Beckopox®, Araldit®
FN	furan resins	QuaCorr®
PF	phenol-formaldehyde resins	Genakor®, Bornumharz®, Vetrodur-N®, Asplit®
Elastomers		
IR	polyisoprene	Natsyn®, Cariflex IR®
SBR	styrene-butadiene rubber	Buna Hüls®, Cariflex®
NBR	acrylonitrile-butadiene rubber	Perbunan N®, Baypren®, Buna®
IIR	isobutylene-isoprene rubber	Genakor®, Butylkautschuk®
CR	chloroprene rubber	Baypren®, Neopren®
CSM	chlorosulphonated polyethylene	Hypalon®, Baypren®
FPM	fluorinated rubber	Viton®, Fluorel®, Kalrez®
NR	(natural rubber)	Smoked sheets, First Latex Crepe

Table 122: Polymer materials for linings and coatings that can be used where there is exposure to hydrochloric acid (based on [202])

Linings are usually 1.5 to 6 mm thick. They are applied as prefabricated sheets, plates, pipes and shaped parts onto the supporting construction. The quality of the composite components is decisively determined by the processing possibilities as well as the experience of the craftsmen.

It must be expected that the adhesion of firmly adhering thermoplast linings on steel is impaired if there is a temperature gradient with the lower temperature on the steel side because hydrochloric acid can diffuse to the surface of the steel. According to a conservative evaluation, this type of lining should not be used above 323 K (50 °C) [219]. This can be remedied by multilayer linings with a diffusion-resistant rubber coating as the bottom layer.

Rubber coatings

The oldest, classical form of lining is a rubber coating, which therefore has a special status. Hard and soft rubber coatings are still the most commonly used organic coating materials for steel because of their corrosion resistance, their simple handling and their favorable price [254, 272]. Experience has shown that hard rubber grades of the types NR, SBR and NBR are more chemically and thermally resistant than the soft rubber grades IIR, CR and CSM.

The non-crosslinked and thus easily processable sheets are glued to the surface to be protected and are subsequently vulcanised. In the workshop, the rubber coatings are vulcanised at 393 to 418 K (120 to 145 °C) and a pressure of 3 to 6 bar in an autoclave. On building sites, the coatings are vulcanised with hot water or by flowing steam. Another possibility for vulcanisation at the building site is the addition of vulcanisation accelerators to the catalytically cross-linkable rubber grades, namely chloroprene, chlorobutyl, bromobutyl and nitrile rubbers. These accelerators already vulcanise the rubber at room temperature. Cross-linking by vulcanisation not only increases the corrosion resistance, but also produces good bonding between the lining and the substrate. Experience has shown that such an intensive cross-linking can produce adhesive forces that are far superior to conventional adhesive systems [290].

Prevulcanised rubber sheets are widely used for coating with rubber at the building site. These sheets are vulcanised in autoclaves in the factory and are then adhesively bonded to the components at the building site by means of suitable adhesives. Further information is given in [272].

Rubberised steel has proved to be excellent in practice for hydrochloric acid storage tanks. A rubberised storage tank with a capacity of 8000 m^3 for 32 % hydrochloric acid is reported. It has a three-layer rubber coating based on a natural rubber, whereby soft rubber was used as an elastic adhesive bridge between the steel and the hard rubber that acts as a diffusion barrier to hydrogen chloride vapor [254, 279]. The tank is still in operation along with further objects of the same type.

If the hydrochloric acid contains even traces of organic solvents, then the tank is given a welded foil lining of PP on hard rubber [279].

In addition to many technical applications, rubber coatings are also used as linings in pumps. For this NR/SBR with hard rubber dust as a filler has good properties, particularly its wear resistance [291].

Thermoplast linings

These linings are classified into loose and adhesive types. Loose linings are simply clamped to the flanges. The sealing surfaces must be designed to prevent flow of the material under operating pressures and temperatures. Typical materials for this are PE, PP and particularly PTFE, which cannot be adhesively bonded. The design of a loose PTFE lining on unalloyed steel in the region of a flange is shown in Figure 148 [202].

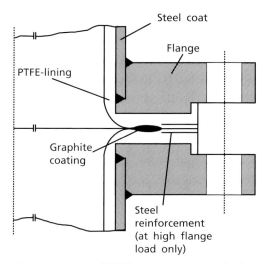

Figure 148: Loose PTFE lining in the region of a flange of the pipeline made of unalloyed steel according to [202].

The correct design of a PVC-P lining on alloyed steel is described in detail in [292].

Thermoplast linings have become very important due to the development of multilayer linings and the introduction of new fluorinated plastics.

Adhesively bonded linings made of PVC, PP, PVDF, ECTFE and FEP are bonded to the steel surface with epoxy resin-based adhesives [219]. To improve the adhesive bond, a textile sheet made of glass fibers or synthetic fibers can be applied to the surface and acts as a mechanical adhesive bridge. The hydrochloric acid resistance of the adhesive must be taken into account for bonded linings [236].

In addition to adhesive bonding, welding of thermoplasts is also used extensively. In the hot gas welding process, butt joints of semi-finished products are welded together. Detailed information on the welding technology is given in [219]. In each case, trained specialists should carry out the welding to ensure that the joints are functional and durable.

Incorrectly executed welding joints are often the cause of damage. This was the case, e.g. for damage to a PVDF lining of transport tanks made of stainless steel for 31% hydrogen peroxide. The tanks were used for road transport at temperatures between 313 K (+ 40 °C) and 253 K (− 20 °C). After a short time, stress cracking occurred in the region of the welded seams of the PVDF sheets. This was caused by incorrect welding (overheating) during fabrication of the lining.

Fluoropolymers are the first choice of materials for hydrochloric acid equipment at elevated temperatures and concentrations. They are particularly resistant in hydrochloric acid in the presence of chlorinated hydrocarbons (HCC), as required in solvent recovery units.

Problems with fluoropolymer linings can be caused by mechanical stresses, such as those occurring on cooling after operating at elevated temperatures [220]. Since the expansion coefficients of the polymers are approximately 10 times greater than those of the unalloyed steel, an increase in temperature can give rise to plastic deformation of both loose as well as adhesively bonded linings. On cooling, the associated contraction gives rise to tensile stresses which are so pronounced that an operating temperature above 363 – 373 K (90 –100 °C) is not recommended. Thus if fluoropolymers are used, it is not the chemical resistance but the mechanical stress characteristics of the equipment which are the limiting factor for the maximum permissible temperature loading.

Multilayer linings with a rubber coating on steel and an adhesively bonded PP or PVDF layer are used for particularly safety-relevant transport of chemicals in road tankers.

The testing of linings for thickness, hardness, adhesion, pores etc. is specified in DIN 28055, Part 2 [293]. Glass fiber-reinforced plastics bonded to thermoplastic linings are described in Section D 3 Composites.

D 2 Seals and packings

Sealing materials are elastomers whose general hydrochloric acid resistance has already been discussed in Section C "Organic materials/Plastics (Table 114)" and in Section D 1 under rubber coatings.

In addition to being chemically resistant, a seal must be elastic in the largest temperature range possible. Additionally, it should be plastic within certain limits to ensure microsealing. Depending on the loading scenario, these properties are met by various elastomers, such as natural rubber and synthetic rubber, fluoropolymers and by combinations of graphite or carbon with organic materials.

A property of seals which requires particular attention is their tendency to creep, i.e. to relaxation. This is a function of the stress, temperature and time, and is more pronounced for thermoplasts than for duroplasts. As described in Section C, the creep behavior is influenced by the action of chemicals. The relaxation of seals is determined by special tests, e.g. flat seals are placed between two flanges that are screwed together with a specified tightening torque of the screws. Such tests can be used to determine the decrease of the tightening torque on relaxation of sealing materials, which is important in practice, and this is then specified in pT-diagrams [219].

There are particular requirements for applications in high-pressure and high-temperature engineering for simultaneous loading with chemicals. The above-described pT-diagram is the basis for this. A further loading to be taken into account is wear caused by frictionally loaded sealing elements, e.g. machine parts that move backwards and forwards or rotate.

Even the product overview of a single manufacturer [164] gives a large variety of seals, which are divided into:

- seals for static surfaces, e.g. O-rings, flat seals
- seals for reciprocating machine parts, e.g. compact seals, lip seals, O-rings, bellows, diaphragms
- seals on rotating machinery, e.g. radial shaft seals, Simmerings®, packings
- shaped parts and constructional elements for static and dynamic applications, e.g. constructional elements for vibrational damping, accumulators.

Seals on static surfaces, e.g. flanged joints, are divided further into flat seals, cambered flat seals with a steel insert, profile packings, shaft seals, round seals and profile packings with an inserted steel supporting ring [219].

The materials that can fulfil the requirements of sealing technology are, for example [164]: NBR, FPM, PTFE, PTFE compound, the material pairs NBR/FKM and FKM/FKM. PTFE seals are used extensively in the construction of glass apparatus.

Packings can be made of graphite-filled PTFE yarn, carbon yarns and combination braids of carbon and PTFE yarns. If the material comes into contact with hydrochloric acid, preference should be given to fluoropolymers, which have particularly favorable temperature properties, see Table 125.

Elastomers are the basic materials used in hose engineering [294]. Thus, for chemical hoses, the elastomers UP, EPDM, NBR, FEP as well as the thermoplasts PP and PTFE are used for the core. The cover of the hose can be made of CR and EPDM. Inserts (supporting elements) can be made of special foils, textiles, copper braids and stainless steel. If the hose is exposed to hydrochloric acid, then fluoropolymers should again be given preference.

Finally, it must be pointed out that the range of loading scenarios (chemical/thermal/mechanical) arising from the individual function of the component is indeed extensive. A detailed discussion goes well beyond the scope of this publication. Therefore, for answers to specific questions, please refer to the information supplied by the relevant manufacturer, e.g. in corporate publications [164, 206, 225, 258, 294, 295]. Since sealing technology in case of exposure to hydrochloric acid requires a lot of know-how, it is recommended that any problems are discussed with experienced manufacturers.

D 3 Composite materials

Composites are combinations of at least two materials in which a variation of form, size and spatial distribution gives rise to material properties that are superior to those of the individual materials. Composites can be subdivided into

- fiber-reinforced composites
- laminate composites
- particle-reinforced composites
- penetration composites.

From the corrosion protection point of view, fiber and laminate composites made of combinations of metallic and organic materials are particularly interesting and were briefly discussed in Section C Organic materials/plastics under unsaturated polyester resins as the resin component.

According to the individual applications, commercially available material combinations include [296]:

- glass fibers, carbon fibers, Aramid® and natural fibers as reinforcing materials
- and polyester, vinyl ester, and epoxy resins as well as high-temperature resistant phenol and polyimide resins as matrix materials.

In the following, glass fiber-reinforced plastics are discussed again because of their importance in corrosion protection. Possible damage mechanisms are also described in detail.

Glass fiber-reinforced plastics (GRP)

An organic matrix material, e.g. a resin, is reinforced by the inclusion of fibers. For certain fiber and matrix materials, the properties can be optimised by varying the fiber orientation and the fiber content. The fibers must be uniformly distributed and uniformly aligned within the matrix. In addition, the fiber must be sufficiently bonded to the matrix. Finally, the fiber and the resin must be resistant to corrosion. There are many different manufacturing processes [200] and they are of great importance for the resistance.

For plastics reinforced with long fibers, these are:

- winding of impregnated fibers and fiber bundles (rovings)
- hot-pressing of pre-impregnated mats (prepregs)
- pultrusion of previously impregnated rovings, mats and tapes.

For plastics reinforced with short fibers, these are:

- fiber spraying in a jet of resin
- spinning processes in a rotating die
- hot-pressing processes of pre-impregnated mats.

The resin mainly influences the following properties: heat deflection temperature, chemical and ageing resistance as well as susceptibility to cracking.

The glass fibers determine the strength, elastic modulus, ultimate elongation, coefficient of thermal expansion and directional dependence of the properties.

The appropriate manufacturing process will produce a composite with the desired properties. For the construction of apparatus and pipes, these are usually so-called laminates.

The failure analysis of fiber-reinforced composites is complex [219]. Damage is classified into failure of the individual layers and failure of the multilayered composite. An analysis of the individual layers shows the following damage phenomena: fiber fractures, matrix fractures and fractures at the fiber/matrix interface.

It is obvious that a purely mechanical consideration of the damage sequence in GRPs is insufficient if there is exposure to chemicals (hot water, dilute and concentrated hydrochloric acid), so that the following processes must be taken into consideration:

- swelling of the matrix by absorption of water, aqueous hydrochloric acid and any organic solvents (as impurities)
- hydrolytic degradation of the matrix by aqueous hydrochloric acid
- chemical attack of the glass fibers by hydrochloric acid diffusing into the material
- loss of adhesion between the glass fiber and the matrix as a result of adsorption processes and chemical processes on the interface.

The failure of a multilayer composite cannot be explained by simple superposition of the strength characteristics of the individual layers [219]. Therefore, failure is empirically divided into first ply failure (FPF) and last ply failure (LPF). The criterion of first ply failure is used to design the laminate for components with high safety requirements, for example in apparatus engineering if the glass fibers have to be protected from corrosive media (use of a chemical-resistant coating).

Glass fiber-reinforced plastics with a thermoplast lining combine the advantageous GRP properties (high strength with a low weight, freedom in designing the shape) with the chemical resistance of the lining. The upper operating temperature limits of thermoplasts, duroplasts and GRP/thermoplast composites have already been given in Table 107, whereby it must be mentioned that a chemical loading was not taken into account. This can be seen from the relatively high values of the temperature compared to the values given in the resistance tables 112 to 115 as well as in [203]. If there is exposure to chemicals, then the particular information given in Section C Organic materials / plastics must be taken into account. The permissible upper operating temperatures may then be considerably lower and may also vary, as the data in concentrated hydrochloric acid sometimes show.

A typical GRP material, as used successfully in chlor-alkali electrolysis plants as pipelines for hydrochloric acid up to 35 %, chlorine-containing aqueous condensate and NaCl solutions, had the following characteristics [268]: a vinyl ester resin was applied to a prefabricated glass-fiber layer by a spin-coating process and a further, especially corrosion- and abrasion-resistant resin layer (chemical-resistant coating)

2 mm thick was spun-coated in the same production step. The pipes had ultimate elongations of up to 25 % and were flexible enough to prevent wetting during operation. The upper temperature limits for dilute hydrochloric acid and the NaCl solution was 363 K (90 °C), that for the chlorine-containing condensate was 323 K (50 °C).

A GRP with a chemical-resistant coating is a supporting laminate with a duroplastic core at least 2.5 mm thick [219]. It is applied as a resin layer reinforced with a non-woven containing < 30 mass% of glass.

The following types of glass can be used in apparatus engineering for reinforcements of GRPs [219]:

- E glass: aluminium-boron-silicate glass with < 1 % alkali oxides for general reinforcement of plastics
- C glass: alkali-lime glass with an increased amount of boron and which is particularly resistant to inorganic acids
- ECR glass: aluminium-silicate glass with added titanium that has a good chemical resistance, suitable for the chemical-resistant coating in GRP equipment and pipelines
- R glass: special glass for high mechanical requirements, even at elevated temperatures.

The glass fibers are manufactured by various melting / nozzle processes followed by a sizing process. An important part of this process is the generation of good adhesion between the glass and the resin using a bonding agent. Silane or chromium compounds are used for this [288]. The hydrochloric acid resistance of glass is discussed in detail in Section B 6.

Various cases of stress corrosion cracking have been observed in glass fiber-reinforced polyester resins exposed to mineral acids [297]. The damage occurred in conjunction with mechanical stress that was well below the permissible limits. Two processes were decisive for this: 1) Ion exchange between the metal cations in the glass fiber and the hydrogen ions of the acid and 2) direct leaching of the glass surface. Systematic investigations with articulated lever specimens in 2 % hydrochloric acid at 293 K (20 °C) were evaluated by fracture mechanics. The results showed that as the toughness of the matrix increased, the crack propagation rate in the specimens greatly decreased. At the same time, as the toughness increased, the effective stress in the crack tip appeared to decrease for a given stress intensity factor. Crack propagation is influenced by HCl diffusion in the region of the crack tip. The diffusion of the medium into the matrix is, however, insufficient to explain the crack propagation rate.

Another damage phenomenon is blistering. This represents the first stage in the attack of hydrochloric acid on certain grades of glass fiber-reinforced, unsaturated polyester resins. This phenomenon occurs particularly in products made of glass fibers with a relatively low SiO_2 content (type E glass) and has been investigated in detail [298]. The results obtained with UP-GF samples with a polyester barrier layer (chemical-resistant coating) can be summarised as follows:

- Attack starts with diffusion of the hydrochloric acid through the inner barrier layer to the glass fibers. Here it reacts with the glass by ion exchange to form chlorides. The volume required by the corrosion products leads to blistering. Influencing parameters are the temperature, the concentration and the thickness of the barrier layer.
- Laminates made of Novolak® epoxy vinyl ester withstand blistering the longest compared to bisphenol A-polyester, vinyl ester and other modified vinyl esters.
- For an optimum resin and a thickness of the barrier layer of at least 800 µm, there is no blistering in hydrochloric acid of up to 10 % and temperatures of 338 K (65 °C) for at least 10 years. These products always exhibited blistering in concentrated hydrochloric acid.

Damage investigations in the chlor-alkali industry, where there are often problems regarding the handling of hydrochloric acid, showed that diffusion of the acid into the polymer matrix was always the primary step at higher hydrochloric acid concentrations [267]. This is followed by attack of the glass fibers. Finally, acid penetrates the cracks that have formed and enters the core of the material. All these partial steps in the mechanism depend on the type of glass and resin, on the laminate structure and on the level of curing, so that plants and components made of UP-GF can have very different service lives.

A solution to this problem is the use of UP-GF constructions with internal liners made of the thermoplasts PVC or PP or even PVDF in extreme cases. Further information is given in [217].

E
Materials recommendations

Warranty disclaimer

This chapter of the DECHEMA Corrosion Handbook has been compiled from literature data with the greatest possible care and attention. The statements made in this chapter only provide general descriptions and information.

Even for the correct selection of materials and correct processing, corrosive attack cannot be excluded in a corrosion system as it may be caused by previously unknown critical conditions and influencing factors or subsequently modified operating conditions.

No guarantee can be given for the chemical stability of the plant or equipment. Therefore, the given information and recommendations do not include any statements, from which warranty claims can be derived with respect to DECHEMA e.V. or its employees or the authors.

DECHEMA is liable – regardless of the actual reason – for all damage caused by their official representatives and their employees on purpose or by gross negligence.

If there is only light negligence DECHEMA is only liable for infringing essential duties that arise from the use of the DECHEMA Corrosion Handbook ("cardinal duties"). DECHEMA can not be made liable for any indirect or subsequent damage and for any damage arising from production stops or delays, which are caused by using the DECHEMA Corrosion Handbook.

Materials recommendations are an integral part of the technology of chemical plants and affect industrial and occupational safety. Therefore, they are important with regard to approval procedures for chemical plants. When handling hydrochloric acid in concentrations above 10 mass%, the statutory directive regulating protective measures against hazardous materials (Hazardous Materials Ordinance) must be observed [3].

A
Metallic materials

The resistance of metallic materials to hydrochloric acid depends not only on the concentration of the acid and the temperature, but also greatly on the purity of the acid, even for materials that are regarded as being resistant. Aeration and the presence of oxidising auxiliary substances or impurities play a decisive role here.

The silicon bronzes with 1 % Si (e.g. UNS C64900) or 3 % Si (e.g. UNS C65500, CW116C formerly known as CuSi3Mn, DIN-Mat. No. 2.1525) are regarded as being sufficiently resistant at room temperature to hydrochloric acid concentrations of up to approx. 20 % (see A 13). Cast iron with high silicon contents of 14 – 15 % is also regarded as being resistant at room temperature to hydrochloric acid of all concentrations (see A 16.1).

Of the multitude of highly alloyed stainless steels, only a few of the superduplex steels and superaustenites can be used at room temperature in relatively low concentrations of hydrochloric acid (Table 123). These materials can be considered if only small amounts of hydrochloric acid are produced in the chemical processes. However, it must be taken into consideration that oxidising impurities can considerably increase the corrosive attack.

Of the nickel-based alloys (A 27 – A 31), NiMo alloys exhibit the best resistance to hydrochloric acid.

The NiMo alloy alloy B-2 (UNS N10665, NiMo28, DIN-Mat. No. 2.4617) was specially developed for use in hydrochloric acid, even at high temperatures up to the boiling point. This alloy is one of the few metallic materials that can be expected to have corrosion rates of less than 0.5 mm/a (20 mpy) in all concentrations of non-aerated hydrochloric acid that is free of oxidising agents, even at the boiling point. Alloy B-2 and the more developed variants B-3 (UNS N10675, DIN-Mat. No. 2.4600) and B-4 (UNS N10629, DIN-Mat. No. 2.4600) with increased thermal stability, are used in numerous processes to produce and handle hydrochloric acid in which high levels of heat input may occur during processing e.g. during welding. But here too, there is a pronounced dependence of the resistance on the level of aeration of the acid or on the presence of oxidising agents. Therefore, these materials should only be used in practice after the suitability of samples, including welded samples, has been tested under operating conditions.

Alloy	Type	Coresp. to: DIN-Mat. No.	UNS	Limiting HCl concentration %
Zeron 100®	FeCr25Ni7Mo3.5	1.4501	S32760	6 – 7
Alloy 2507		1.4410	S32750	
254 SMO	Fe55Cr20Ni18Mo6.1	1.4547	S31254	max. 3
Al-6XN	Fe49Cr20Ni24Mo6.2		N08367	
Al-6XN PLUS	Fe46Cr22Ni25Mo6.7		N08367	
A 269	Fe47Cr20Ni25Mo6.2	1.4529	N08926	
Alloy 904 L	Fe47Cr21Ni25Mo4.5	1.4539	N08904	
Alloy 28	Fe37Cr27Ni25Mo4.5	1.4563	N08028	
Alloy 31 (USA)	Fe31Cr27Ni31Mo6.5	1.4562	N08031	8
654 SMO	Fe42Cr24Ni22Mo7.3	2.4652	S32654	

Table 123: Austenitic and austenitic-ferritic steels that can be used in hydrochloric acid at room temperature and their operating limits

The resistance of the NiCrMo alloys in hydrochloric acid is distinctly less than that of NiMo alloys. Table 1245 lists the limiting concentrations and temperatures for the most important NiCrMo alloys. Even for the NiCrMo alloys, the presence of oxygen and other oxidising agents increases the corrosion rate; however, these affect them less than for the NiMo alloys.

Alloy	Type	DIN-Mat. No.	UNS	% HCl	HCl K (°C)
Alloy C 276	Ni59Cr16Mo16Fe6W4	2.4819	N10276	8	323 (50)
				8	308 (35)
Alloy C-4	Ni67Cr16Mo16Fe1	2.4610	N06455	8	323 (50)
				8	293 (20)
Alloy C 22	Ni57Cr21Mo13Fe4W3	2.4602	N06022	8	323 (50)
				8 – 15	293 (20)
Alloy 59	Ni59Cr23Mo16Fe1	2.4605	N06059	30	313 (40)
Alloy C-2000	Ni59Cr23Mo16Fe1Cu1.6	2.4675	N06200	10	313 (40)
				10 – 20	
Alloy 625	Ni61Cr21Mo9Fe4	2.4856	N06625	25	303 (30)

Table 124: NiCrMo alloys that can be used in hydrochloric acid and their operating limits

Titanium and titanium alloys (see A 38) are known to have good corrosion resistance in essentially oxidising media. Therefore, their resistance in hydrochloric acid is limited. Pure titanium should therefore only be used in hydrochloric acid up to a concentration of approximately 5 % and temperatures up to approximately 303 K (30 °C). Palladium-alloyed titanium grades have sufficient resistance at the same temperature up to acid concentrations of approx. 20 %. At higher temperatures, not only does the corrosive attack distinctly increase, but also the material properties must be taken into consideration on account of the hydrogen produced by the corrosion reaction. In contrast to the nickel-based alloys, oxidising auxiliary substances in the acid have a positive effect on the corrosion behavior of titanium materials.

Zirconium (see A 40), in addition to tantalum, is one of the few metals that is corrosion resistant in hydrochloric acid of all concentrations and up to high temperatures well above the respective boiling point. At a concentration of 12 M and below the boiling curve, the corrosion rate is < 0.127 mm/a (< 5 mpy). Whereas tantalum is susceptible to hydrogen embrittlement in higher concentrations of HCl and at higher temperatures, this applies to zirconium to a much lesser extent. However, zirconium and its alloys are susceptible to pitting and stress corrosion cracking in oxidising HCl, e.g. in the presence of oxidising impurities or at positive electrode potentials. Therefore, components made of zirconium should be insulated electrically from parts made of other metals to avoid an unfavorable shift in the potential.

Tantalum (see A 37) exhibits the best resistance of all special materials to hydrochloric acid of all concentrations up to 463 K (190 °C). However, there may be strong attack at higher temperatures. Unfortunately, tantalum is extremely susceptible to hydrogen embrittlement, even at room temperature. Therefore, if tantalum is used, it must be ensured that it does not undergo cathodic polarisation in solutions that contain hydrogen ions. Components made of tantalum must be electrically insulated from other components on account of these problems.

B
Non-metallic inorganic materials

Graphite is the first choice for materials used for high chemical/thermal loading scenarios of components used in units for HCl absorption, HCl recovery from hydrochloric acid, HCl stripping and HCl concentration. In addition to the good thermal conductivity, the excellent hydrochloric acid resistance is important and this is mainly determined by the type of impregnation. Parts having phenol-formaldehyde resins as the standard impregnation material can be used up to 443 K (170 °C), those impregnated with fluoropolymers can be used up to 503 K (230 °C). However, fluoropolymers, in contrast to phenol-formaldehyde resins, exhibit a slight HCl gas permeability at higher temperatures, and this must be taken into consideration depending on the particular application.

Chlorine-containing hydrochloric acid and aqueous chlorine solutions oxidatively attack graphite materials impregnated with phenolic resins. In contrast, they are resistant to hydrochloric acid contaminated with chlorinated organic solvents.

Graphite and **carbon** respond to mechanical stresses with microcracks, but these can be avoided with a suitable design. This particularly applies to connecting elements. The mechanical loading tolerance of pressurised equipment and seals can be improved by fiber reinforcement, and the erosion resistance with coatings.

By virtue of the very small coefficient of thermal expansion and elasticity modulus in combination with a high thermal conductivity, graphitic materials are not sensitive to thermal shocks. However, because of their low ductility, mechanical problems can occur due to steam hammering.

Glass is used for handling hydrochloric acid on its own and as a coating on unalloyed steel. Borosilicate glass plays an important role for equipment made entirely of glass because it is resistant to hydrochloric acid of all concentrations up to the boiling point.

Alkali silicate glasses are especially suitable for glass coatings because they have a corrosion rate of more than 0.1 mm/a (4 mpy) only above 393 K (120 °C) and a rate of more than 0.5 mm/a (20 mpy) only above 433 K (160 °C). The maximum corrosive attack is in 20 % hydrochloric acid.

In all applications, it must be taken into consideration that there may be strong attack if hydrofluoric acid or fluorides are present as impurities in the hydrochloric acid, even at concentrations in the ppm range. Corrosion by alkalis first starts above pH = 9, but greatly increases with increasing temperature.

Chemical apparatus made of individual pieces of glass is assembled with PTFE seals and taking into account the brittle behavior of glass. The size and the capacity of such units is limited; however, they are often used in the synthesis of valuable substances e.g. in the pharmaceutical industry.

Glass coatings on steel are used for corrosion protection of chemical reactors, vessels, pipes and fittings, heat exchangers, columns, evaporators, pumps and accessories.

An important application of glass is its use as a fiber in fiberglass-reinforced plastics. Glasses generally used for this are C glass, an alkali-lime glass with an increased boron content, and ECR glass, an aluminium silicate glass with added titanium.

If a **chemical-resistant enamel** is to be exposed to hydrochloric acid of various concentrations in the application, then the resistance should be tested under conditions similar to those in practice according to DIN ISO 2743 before it is used.

C
Organic materials / plastics

Organic materials play an important role in hydrochloric acid engineering on account of their favorable resistance properties. The following materials recommendations depend on the loading scenario, which has to be defined with respect to the hydrochloric acid concentration, temperature and mechanical loading. For reasons of clarity, the list is divided into the most important plant components, equipment and machines that have a central importance in the manufacture and utilisation of aqueous hydrochloric acid in the chemical industry and in other industrial fields. In addition to a brief discussion of the individual uses of the materials, the results are tabulated as a positive list with particular attention to the upper temperature limits during use, which is one of the most important parameters for the utilisation of plastics. For applications under pressure, it must be taken into account that the pressure/temperature behavior in dilute acid is similar to that in pure water, but that in concentrated hydrochloric acid over the boiling point of approx. 338 K (65 °C) it can lead to high HCl vapor pressures, as shown in Figure 2 in the introduction.

The recommendations are based on the information given in the previous Sections C and D with particular regard to the reviews [217, 219, 220, 254, 272]. Further sources of information, such as corporate publications, are mostly not listed again, but can be found under the individual materials in the special sections.

Vessels

Vessels and tanks are used to store mostly concentrated hydrochloric acid at ambient temperature. If they are outdoors they can experience large differences in the temperature, so that the temperature can vary between 253 and 313 K (- 20 °C and + 40 °C) depending on the size of the vessel. This particularly applies to tankers. The individual maximum concentration and temperature as well as the presence of impurities (e.g. organic solvents, heavy metals, oxidising agents, solids) must be taken into account in the operation of vessels because they often influence the resistance behavior.

Smaller objects can be made of thermoplasts, glass fiber-reinforced plastics (GRP), rubber-coated steel or GRP with a thermoplastic inner liner. Rubber-coated steel is generally used for large storage tanks. The information given in reference [219] should be taken into account for the construction and design of the equipment. This reference also provides calculation examples for the design.

The constructional pre-requisites must be fulfilled for linings used for steel vessels: there should be no unevenness, welding seams should be ground, there should be no riveted connections, supports, bases, edges and corners must be correctly executed [292].

It is recommended that **rubber coatings** [220, 254, 272] of soft and/or hard rubber are applied in several layers with overlapping edges followed by spark testing to check that they are free of pores. In the case of hot vulcanisation, attention must be paid to delayed cracking of the lining due to the very different coefficients of thermal expansion of rubber and steel and because of the tensile stresses occurring on cooling. Cracking can occur due to over-vulcanisation of hard rubber, which must always be avoided.

Of the synthetic rubber grades (butyl, chlorobutyl and bromobutyl rubbers), which are also used for liners, bromobutyl rubber has the best lining and curing characteristics. Both ethylene-propylene as well as ethylene-propylene-diene rubber (EPDM) are superior to the brominated rubber grades by virtue of their better mechanical properties and higher acid resistance.

Instead of a rubber coating, smaller vessels can be made of thermoplasts or GRPs as well as multilayers and composites of GRP with thermoplastic inner liners. **Polypropylene** can be used without mechanical reinforcement up to 338 K (65 °C) in all hydrochloric acid concentrations and is superior to polyethylene, which has an upper limiting temperature of 313 K (40 °C). With reinforcement (as a loose liner or as a GRP composite), temperatures up to 373 K (100 °C) should be possible [220]. As a GRP composite, the operating temperature must lie 293 K (20 °C) below the softening temperature of the GRP resin.

Plasticiser-free **polyvinyl chloride** (PVC-U) is resistant up to 333 K (60 °C) in concentrated hydrochloric acid. In contrast to polyolefins, PVC is resistant in chlorine-containing hydrochloric acid. However, it cannot be used in hydrochloric acid that contains hydrochlorocarbons with a separate solvent phase because of diffusion and swelling.

The advantages of PVC compared to polyolefins lie in its high compressive and tensile strengths as well as its simple processability by adhesive bonding, welding, thermoforming and machining.

The favorable properties with regard to the corrosion resistance are in opposition to the behavior of PVC in the event of a fire. Although it does not burn, it releases HCl by thermolysis. In combination with water, this HCl forms hydrochloric acid and can thus cause a great deal of damage. The generation of hydrochloric acid from PVC waste is also a problem in waste incineration.

Fluoropolymers are the first choice of materials for hydrochloric acid equipment at elevated temperatures and concentrations. They are particularly resistant in hydrochloric acid in the presence of chlorinated hydrocarbons (HCC) as found in plants for solvent recovery.

PTFE is frequently used as a butt-welded loose liner for highly stressed vessels and columns. Fluoropolymers PVDF, FEP, PFA and ECTFE can be used for adhesively bonded liners. To ensure the bonding of the fluoropolymers, which cannot be adhesively bonded to the steel surface, a glass-fiber mat is embedded in the polymer surface and this mat is glued to the steel surface with a polyester resin. PVDF, which has a lower melting point, can be bonded to the steel surface with a hot melt adhesive made of polyester resin. In the last step, the edges of the liner are welded with hot gas and then a spark test is carried out to check that there are no pores.

The problem regarding fluoropolymer linings are the mechanical stresses resulting from shrinkage that occurs due to cyclic temperature changes [220]. This problem is discussed again under thermal equipment.

Glass fiber-reinforced polyester resins are the most frequently used construction materials for vessels used to handle concentrated hydrochloric acid at ambient temperature. A typical design consists of a supporting structure made of R or ECR glass with 45 % to 70 % proportion of resin and a protective coating with 80 % to 90 % resin, preferably an epoxy-vinyl ester resin. Possible problems with the resistance of a chemical-resistant coating made of unsaturated polyester resins can be avoided by using glass fiber-reinforced PVC or PP instead. Another solution to this problem is the use of UP-GF constructions with internal liners made of the thermoplasts PVC or PP or even PVDF in extreme cases.

According to the media lists of the German Institute for Building Technology, GRP laminates made of UP/PHA resins can be used in concentrated hydrochloric acid up to 313 K (40 °C) if resin grades 5/6 according to DIN 18 820-1 are used [217].

Because of their excellent chemical resistance, glass fiber- and carbon fiber-reinforced phenol resins and furan resins have replaced the traditional asbestos fiber-reinforced products (health risk). When deciding between phenol or furan resins for the construction, manufacture and operation of vessels, it should be taken into account that although furan resins are more resistant to chemicals, they have lower fracture strengths than phenol resins.

In addition to self-supporting GRP constructions, good experience has been made with GRP liners on unalloyed steel because, in this case, the problem of mechanical load tolerance is not relevant.

In the design of GRP vessel constructions, the corresponding guidelines in the AD Information sheet N1 [219] and in the ASME standard [299] must be observed.

Pipelines and fittings

Thermoplasts and elastomer materials have become established in pipeline construction either as a solid material or as coatings and linings. Their classification according to their chemical resistance is part of an ISO standard for pipes and fittings [216]. The standard contains information on the chemical resistance of the materials that are relevant for hydrochloric acid, namely PE-HD, PE-LD, PP, PB, PVC-U, PVC-C, PVDF without additional mechanical loads. It is important to note that if there is simultaneous mechanical loading in more concentrated hydrochloric acid, the values can be less than those without loading. pT-diagrams for pipes and fittings show that initial pressures of approx. 10 bar are permissible above a temperature of 293 K (20 °C). However, even at 323 K (50 °C), they drop to approximately half the value [206]. These permissible application limits only apply to water and thus also to dilute hydrochloric acid as the flowing medium. Increased safety factors are necessary if there is exposure to more concentrated acids. Under these conditions, PVDF exhibits the best behavior.

In hydrochloric acid engineering, the resistance data for vessels also applies to pipelines and fittings. Therefore, the information in preceding sections applies here. In principle, the conservative limiting data of the German Institute for Building Technology, for example, can be used with regard to concentrated hydrochloric acid both for vessels and for pipes [217]. This data is compiled together with other values in Table 125. Further resistance data is given in Section C under the individual materials. Information from the appropriate manufacturers of pipes, hoses and fittings was included. e.g. [206, 222, 274, 294].

If the material is exposed to very corrosive media, loose linings made of PTFE, PVDF and FEP can be used. Similar to the situation for thermal equipment, their maximum operating temperature is limited by their thermal expansion, but it is higher than for the adhesively bonded linings [220]. Maximum operating temperatures of linings are also given in Table 125.

In elastic pipe connections, e.g. hoses and bellows, the core can be made of the elastomers UP, EPDM, NBR, FEP as well as the thermoplasts PP and PTFE. The cover of the hose can be made of CR and EPDM [294]. It is obvious that fluoropolymers should be used in preference to other materials if there is exposure to hydrochloric acid.

Thermal equipment

In conventional hydrochloric acid engineering, non-metallic inorganic and metallic materials are preferred over self-supporting plastics for heat exchangers, distillation units, reactors, absorption columns and stripper towers (as the most important representatives of this type of apparatus). The reason for this is the limited mechanical stability of the plastics at higher temperatures. Recently, however, composite constructions of steel with a polymer lining, preferentially fluoropolymers, are becoming increasingly important. However, the products from the various manufacturers differ with regard to their microporosity, tensile strength and creep properties, and this has to be taken into consideration depending on the application case.

The advantage of a column made of steel lined with fluoropolymers, mainly PTFE, instead of graphite is that the units can be operated at a higher pressure. On the other hand, their failure leads to a rapid destruction of the steel sheath. To avoid this, several small units are used that can be rapidly exchanged in the event of damage.

Thermal equipment is generally a special construction with special requirements for the materials, e.g. the poor thermal conductivity of plastics must be taken into consideration for heat exchangers which makes larger heat-exchanging surfaces necessary. For reactors, safety factors must be used with regard to unstable behavior (sudden increase in the temperature and pressure). Because of the inherent higher temperatures of thermal equipment, HCl permeation through the polymer materials must be taken into consideration in hydrochloric acid engineering.

There are various examples of the utilisation of fluoropolymers in thermal units. The use of PTFE was reported at an early stage for hydrochloric acid separators in shell-and-tube exchangers in chlorination units. Plate heat exchangers made of solid PVDF for flue gas desulphurisation plants are available that can, in principle, be used in media containing hydrochloric acid. In addition to these materials, the thermoplastically processable types FEP, PFA and ECTFE are becoming increasingly important. The particular advantage of fluoropolymers lies in the fact that they are more or less insensitive to chlorinated solvents and oxidising agents (e.g. chlorine) that are often present in contaminated hydrochloric acid. An increasingly important and wide application field for fluoropolymers are techniques using highly pure chemicals, e.g. those used to manufacture electronic components, especially chips.

The problem regarding fluoropolymer linings is the mechanical stresses resulting from shrinkage that occur on cooling after operating at relatively high temperatures [220]. Since the expansion coefficients of the polymers are approximately 10 times as high as those of the unalloyed steel, an increase in temperature can give rise to plastic deformation of loose as well as adhesively bonded linings. On cooling, the associated contraction gives rise to tensile stresses which are so pronounced that an operating temperature above 363-373 K (90-100 °C) is not recommended. Thus, if fluoropolymers are used where there is cyclic temperature loading, it is not the chemical resistance but the mechanical stress characteristics of the equipment which are the limiting factors for the maximum permissible temperature loading. However, this is a conservative limitation. Because of the know-how necessary for the manufacture and operation of thermal equipment, it is absolutely necessary to take advantage of the experience described in the corporate publications listed under fluoropolymers in Sections C and D.

In addition to fluoropolymers, some other materials can be used that have retained their importance for reasons of cost. For example, coatings of phenol-formaldehyde resin on steel have been used successfully in stripper columns for the recovery of halogenated hydrocarbons. This material was formerly used frequently as a classical coating material in hydrochloric acid technology, as indicated by a series of older references [47].

If the process is carried out under atmospheric pressure or slightly increased pressure, then absorption columns made of phenol-formaldehyde resin with a fiberglass-reinforced polyester resin sheath are used. Stripper columns to clean chlorinated hydrocarbons are designed in the same way.

The advantage of GRP sheath constructions as reinforcement for reaction equipment and columns made of corrosion- and temperature-resistant thermoplasts compared to steel as a supporting material is that they have better corrosion resistance if a leakage occurs.

Pumps and fittings

Pumps and fittings e.g. valves, belong to the central and more sensitive components in a chemical plant. As the plant productivity increases, with the corresponding necessity for reduced downtimes and maintenance times as well as with continuously changing environmental obligations, the requirements on the pump construction and selection of materials also increase.

Pumps in the process industry are generally rotary pumps whose construction materials must fulfil the usual requirements regarding chemical and temperature-resistance and must also be resistant to abrasion and erosion-corrosion (due to the high flow rates). Furthermore, the problem of sealing between rotating and stationary parts, using rotating mechanical seals for example, is of great importance.

Materials must be selected for the following pump components: impellers, casing, casing covers, flanges and pipes for installation, instrumentation as well as supporting constructions for free-standing pumps and submersible pumps. In addition, elastic materials for seals and the liner must also be selected.

Common organic materials for rotary pumps are rubber coatings, vinyl polymers (PVC and PVC-C), polyolefins (PP and PE) and fluoropolymers (PTFE, PVDF, ECTFE, PFA). Their individual application possibilities together with further aspects for the use of plastic pumps are discussed in general in [227] and in detail in [257, 301, 301, 302].

Materials for peristaltic pumps, which have very elastic hoses as the central component, are the elastomers NBR, NR, EPDM, NBR as well as PP, PVDF, PTFE, PE and PVC for the solid parts.

A multilayer composite for a stirrer shaft with stirring blade (similar loading to that for a pump) consisting of 3 mm PVDF on 3 mm hard rubber has been reported as being used successfully in sulphuric acid and should also be suitable for hydrochloric acid [279].

Rubberised steel is a classic pump material for the handling of hydrochloric acid. Single-layer rubber coatings with hard rubber are used for chemical loading, and multilayer rubber coatings with soft rubber are used for wear protection. NR/SBR with hard rubber dust as a filler has exhibited good properties for this, particularly with regard to wear resistance [291].

However, the general trend is to use rubberised pumps rather than solid plastic pumps, whereas those made of laminates are being used less and less.

Table 125 contains pump-relevant characteristics of the individual materials for their use in hydrochloric acid. These are for:

- **rubber coatings** the chemical resistance in hydrochloric acid according to the resistance data in Section C.
- **PVC** the chemical resistance in hydrochloric acid according to the resistance data; they are more resistant than rubber coatings in some solvents.
- **PVC-C** has a comparable resistance to PVC combined with an increased mechanical stability and higher operating temperatures.
- **PE** has a somewhat higher temperature resistance in concentrated hydrochloric acid compared to PVC. It is sensitive to oxidising media and some solvents.
- **PTFE** has the maximum chemical resistance of all polymers. However, its tendency to creep means that it can only be used in small units or sufficiently dimensioned pump casings, instrument casings and seals.
- **PVDF** is the most frequently used fluoropolymer. It is chemically very stable and can be used without pigments for handling very pure chemicals (chip industry).
- **PFA** is a thermoplastically processable fluoropolymer that is resistant to hydrochloric acid.
- **ECTFE** has similar chemical resistance to PVDF, but there are still limitations regarding certain shapes (pipes and castings) [227].

Plastic pumps are lighter than metal pumps: PE and PP are 1/6 of the weight of high-alloy steels, fluoropolymers are approx. 1/3 of the weight. Chemical resistance and temperature are decisive for application possibilities. The abrasion resistance in process streams containing solids can be considerably better for plastics than for high-alloy steels. Thus, PE, PVDF, PP and PVC have a 3- to 10-fold abrasion resistance compared to high-alloy steel. PTFE and FRP are less favorable by a factor of 7 to 20 (results with the so-called Taber abraser [227]). However, this only applies to low energy inputs. For flow rates above 20 m/s and impact wear, the wear resistance, particularly of PE, can be considerably reduced [303]. GRP-reinforced pumps are also sensitive to abrasion because glass fibers are exposed if the chemical-resistant coating has eroded away so that the medium can penetrate the material.

An important advantage of plastic pumps made of solid material compared to metal pumps is the possibility of perfectly adjusting the construction to the hydrodynamic requirements and thus to correspondingly increase the efficiency.

Fittings include valves, gate valves, taps, flaps and accessories (already partially described under pipes in this section). The technical material requirements that apply to rotary pumps also apply to valves and other fittings. A typical example of this is the brief discussion on selecting materials for shut-off gates in concentrated acid [304]. A 17% chromium steel with a PTFE coating was selected for the gate for which temperatures up to 403 K (130 °C) are permissible. The ring bellows were made of Teflon PFA®, and have the same upper temperature limits.

The following maximum permissible temperatures apply to fittings in contact with hydrochloric acid:

- valves lined with hard rubber 353 K (80 °C)
- PE-HD values of solid material 323 K (50 °C), for PP 338 K (65 °C)
- fluoropolymer valves of solid material 363-373 K (90-100 °C), conservative value from [220]).

Steel constructions lined with PVDF, FEP, PFA, ECFE can be used for larger units.

Seals

Seals were described in Section D 2. Basic resistance properties that are relevant for hydrochloric acid engineering are given under elastomers in Section C and in Table 114.

The most important types of seals are O-rings and flat seals. The following temperature limits are given for O-ring seals in concentrated hydrochloric acid (conservative values):

- natural rubber, soft   313 K (40 °C)
- chloroprene and butyl rubber   333 K (60 °C)
- chlorosulfonylpolyethylene CSM   328 K (55 °C)
- ethylene-propylene rubber, EPDM   353 K (80 °C)
- perfluoroelastomers   393 K (120 °C)

Flat seals consist of the above-mentioned O-ring materials on their own or as a multilayer seal together with PTFE. In the latter case, PTFE determines the chemical resistance. Seals made of flexible PTFE felt and graphite-filled PTFE have great importance.

Rotating mechanical seals in pumps need special attention. In concentrated hydrochloric acid, double-action rotating mechanical seals made of the combinations SiSiC / Kalrez® / Hastelloy C® / carbon graphite / Viton® can be used up to 353 K (80 °C) [257]. The complex material design of such a sealing system illustrates the amount of know-how required for their use in concentrated hydrochloric acid. The upper temperature limit of magnetically driven pumps without a sealed shaft [301] is 353 K (80 °C) on account of the limited temperature resistance of the housing seals [303].

Tabulated summary

Materials recommendations are given in Table 125 for dilute, concentrated and contaminated hydrochloric acid. The differentiation between dilute / concentrated hydrochloric acid is not exact. Dilute hydrochloric acid is regarded as having a concentration of approx. 7 mass% and lower. Concentrated hydrochloric acid usually has a concentration of 30% and 37% (commercially available). Contaminated hydrochloric acid can contain:

- halogenated hydrocarbons that can lead to swelling of the plastics
- oxidising agents, such as chlorine, oxidising acids and heavy metal ions that oxidatively degrade the plastic
- solids that cause abrasion in pumps and fittings.

Cost comparisons for the use of polymer materials for corrosion protection are given in [278]. Coatings and linings of the vessels as well as the pipeline systems made of organic and metallic materials are compared to unalloyed steel.

Table 121 gives the cost factors for coatings and linings of organic materials for a stirred tank reactor [287]

Similarly interesting relative costs of acid-resistant plastic pumps are given in [227]. If the costs of a pump made of PP are set to 100%, then the relative costs for PVC-U are 110%, for PVC-C they are 125% and for PVDF they are approx. 150%.

Component, apparatus, machine	Medium	Upper operating temperature, K (°C)	Material
Vessels, tanks	dilute HCl	333 (60)	soft rubber lining, PE-HD
		353 (80)	PVC-C, EP, PP, VE-GF
	concentrated HCl	313 (40)	PE-HD
		333 (60)	PVC-U
		338 (65)	UP
		353 (80)	PP
		358 (85)	hard rubber lining
		373 (100)	PVDF
	contaminated HCl		only fluoropolymers
Pipes and fittings (hoses, see text)	dilute HCl	up to 353 (80)	same materials as for vessels
		353 to 373 (80 to 100)	fluoropolymers
	concentrated HCl	40 (313)	PE-HD
		60 (333)	PVC-U
		80 (353)	PP
		100 (373)	PVDF
		100-150 (373-423)	FEP lining, loose
		150 (423)	PTFE lining, loose
Thermal equipment	dilute HCl	80 (353)	VE-GF, PP, PVC-C
		100 (373)	PVDF
	concentrated HCl	80 (353)	PVDF
		100 (373)	GRP / PVDF lining
		130 (403)	PTE, PFA, FU linings, PF-GF
	contaminated HCl	130 (403)	only fluoropolymers
Pumps and fittings	dilute HCl	60 (333)	PE-HD
		80 (353)	PP, PVC-C
		85 (358)	hard rubber
	concentrated HCl	60 (333)	PVC-U
		80 (353)	PP, PFA
		120 (393)	PVDF
		130 (403)	PTFE
	contaminated HCl	100 (373)	PVDF

Table 125: Selection of plastic materials used in equipment and machines for handling hydrochloric acid (for material abbreviations, see Table 108)

Bibliography

[1] Ullmann's encyclopedia of industrial chemistry
5th ed., vol. A13, "Hydrochloric Acid" (1989) p. 283 – 296
VCH Verlag, Weinheim

[2] Römpps Chemie Lexikon
9. ed., vol. 5, "Salzsäure" (1992) p. 3978–3980
Franckh'sche Verlagshandlung, Stuttgart

[3] Verordnung zum Schutz vor gefährlichen Stoffen (Ordinance on hazardous substances-GefStoffV) of November 15, 1999
(Ordinance on the protection of hazardous substances) (in German)
(BGBL 1, p.2059)
Jedermann-Verlag, Heidelberg, 2000

[4] Materials Selector for Hazardous Chemicals, vol. 3, Hydrochloric Acid, Hydrogen Chloride and Chlorine
MIT Publication MS-3, Elsevier Science, 1999

[5] Crowe, D. C.
Corrosion of electrostatic precipitators
Tappi Journal 70 (1987) No. 6, p. 87–90

[6] Heitz, E.
Stimulation subpermeabler Gesteine durch Korrosionsprozesse
(Stimulation of subpermeable stones by corrosion processes) (in German)
DGMK-research report 491
DGMK e.V., Hamburg, 1996

[7] Schmitt, G.
Application of inhibitors for acid media
Br. Corros. J. 19 (1984) 4, p. 165–176

[8] Schmitt, G.
Wasserstoff und Korrosion, 2nd. ed.
(Hydrogen and Corrosion) (in German)
Irene Kuron, Bonn (2000) p. 341–378

[9] Pound, B. G.; Macdonald, D. D.; Tomlinson, J. W.
The electrochemistry of silver in KOH solutions at elevated temperatures
I. Thermodynamics
Electrochim. Acta 24 (1979) p. 929

[10] Product information
Silber und Silberlegierungen – Fortschritte durch Edelmetalle
(Silver and silver alloys, progress by noble metals) (in German)
DEGUSSA, Frankfurt am Main, edition 1971, EH 1-0-7–271

[11] Product information
Schatz, J.
DEGUSSA Festschrift "Aus Forschung und Produktion", offprint, p. 10

[12] Product information
Edelmetalle im Apparatebau, p. 6, 26
(Noble metals in apparatus engineering) (in German)
Fr. Kammerer AG, Pforzheim

[13] Dobbs, B.; Slenski, G.
Investigation of Corrosion Related Failures in Electronic Systems
Mater. Performance 35 (1984) 3

[14] Fukuda, Y.; Fukushima, T.; Sulaiman, A.; Musalam, I.; Yap, L. C. et. al.
Indoor Corrosion of Copper and Silver Exposed in Japan and ASEAN Countries
Journal of the Electrochemical Society 138 (1991) 5, p. 1238–1243

[15] Graedel, T. E.
Corrosion Mechanisms for Silver Exposed to the Atmosphere
Journal of the Electrochemical Society 139 (1992) 7, p. 1963–1970

[16] DIN EN ISO 8044 (11/1999)
Korrosion von Metallen und Legierungen: Grundbegriffe und Definitionen

(Corrosion of metals and alloys: Basic idea and definitions) (in German)
Beuth Verlag GmbH, Berlin

[17] Haagen, H.; Gaszner, K.; Heinrich, M.
Filiformkorrosion von Aluminium
(Filiform corrosion of aluminium)
(in German)
Farben + Lack 100 (1994) 3, p. 177–180

[18] Heinrich, M.; Haagen, H.; Schuler, T.
Filiformkorrosion von Aluminium
(Filiform corrosion of aluminium)
(in German)
Farben + Lack 100 (1994) 4, p. 249–252

[19] Aluminium Taschenbuch, 15. edition, vol. 1
(Aluminium pocket book) (in German)
Aluminium Verlag, Düsseldorf, 1998, p. 330

[20] Bhattamishra, A. K.; Mishra, G. S.; Banerjee, M. K.
On the Effect of Mixed Metal Addition on Corrosion Behavior of Aluminium in Inorganic and Organic Acids
Zeitschrift für Metallkunde 83 (1992) 10, p. 766–768

[21] Heitz, E.; Henkhaus, R.; Rahmel, A.
Korrosionskunde im Experiment: Untersuchungsverfahren-Meßtechnik-Aussagen, 2. edition
(Corrosion science in the experiment: Investigation methods – measurement-technique – conclusions) (in German)
VCH Verlagsgesellschaft, Weinheim (1990)

[22] Koch, U.; Hack, T.
Korrosionsverhalten von technischen AlLi-Legierungen
(Corrosion behavior of technical AlLi-alloys) (in German)
in: Korrosion verstehen – Korrosionsschäden vermeiden, ed.: Gräfen, A.; Rahmel, A.
Verlag Irene Kuron, 1994, p. 112–119

[23] Aluminium-Taschenbuch, 14. edition
(Aluminium pocket book) (in German)
Aluminium-Zentrale, Düsseldorf (1988)
p. 712–730

[24] Murthy, K. S. N.; Ambat, R.; Dwarakadasa, E. S.
The role of metal cations on the corrosion behavior of 8090 – T 851 alloy in a pH 2.0 solution
Corros. Sci. 36 (1994) 10, p. 1765–1775

[25] Talati, J. D.; Patel, G. A.; Gandhi, D. K.
Maximum Utilization Current Density in Cathodic Protection
Part I: Aluminium in Acid Media
Corrosion-Nace 40 (1984) 2, p. 88–92

[26] Al-Suhybani, A. A.
Corrosion of aluminium in hydrochloric acid solutions
Part 1 – The inhibitive action of monoamines
Corros. Prev. Control 36 (1989) 3, p. 71–77

[27] El-Nader H. M. A.; Moussa, M. N. H.; Shalaboratory, A. M.; Fakhry, E. M.
Heterocyclic hydrazone derivatives as inhibitors for aluminium corrosion in hydrochloric acid solutions
Indian Journal of Technology 31 (1993) 10, p. 730–733

[28] Moussa, M. N.; El-Tagoury, M. M.; Radi, A. A.; Hassan, S. M.
Carboxylic acids as corrosion inhibitors for aluminium in acidic and alkaline solutions
Anti-Corrosion 37 (1990) 3, p. 4–8

[29] Ahmed, A. I.; Soliman, M. S.
Inhibition of corrosion of aluminium in 2 M hydrochloric acid by pyridine derivatives
Indian J. Technol. 26 (1988) 11, p. 541–545

[30] Ahmed, A. I.; Basahel, S. N.; Khalil Cemistry, R. M.
Inhibition of the acid corrosion of aluminium with some morpholine and thiosemicarbazide derivatives
Anti-Corrosion 35 (1988) 8, p. 4–8

[31] Mourad, M. Y.; Seliman, S. A.; Ibrahim, E. H.
The Inhibitive Action of Dimethyltin Dichloride Towards the Corrosion of Aluminium in Hydrochloric Acid and Sodium Hydroxide Solutions
J. Chem. Technol. Biotechnol 46 (1989) 1, p. 27–40

[32] El-Tagouri, M. M.; Mostafa, M. R.; Abu El-Nader, H. M.; Abu El-Reash, G. M.
Efficiency of some 2-heterocarboxaldehyde-2′-pyridyl-hydrazones as corrosion inhibitors for Al dissolution in HCl solution
Anti-Corrosion 36 (1989) 9, p. 10–14

[33] Elewady, Y. A.; El-Tagoury, M. M.; Bekheit, G. E.; Hassan, S. M.
Effect of 4-benzylidene, 3-methyl, 5-oxo-1-phenyl-pyrazoline derivatives on the corrosion of aluminium in HCl solutions
Anti-Corrosion 39 (1992) 3, p. 4–7

[34] Aksüt, A. A.; Bayramoglu, G.
The effect of propargyl alcohol on the corrosion of pure aluminium and aluminium alloys in aqueous solution
Corros. Sci. 36 (1994) 3, p. 415–422

[35] White, B.
Corrosion in acid chloride
Corros. Prev. Control 37 (1990) 3, p. 78–79

[36] Motojima, S.; Nakayama, Y.
Phosphidation of cobalt plate and some of its properties
Journal of the Less-Common Metals 118 (1986) 1, p. 109–115

[37] Itoh, M.; Ihara, M.; Nishihara, H.; Aramaki, K.
Corrosion Inhibition of Cobalt in Some Acid Solutions by Bismuth (III) Chloride
Journal of the Electrochemical Society 141 (1994) 2, p. 352–358

[38] Haynes Alloy no. 25
Kobalt (1972) 3, p. 3–10

[39] Hickling, J.; Gabor, G.
Untersuchungen zur Beizbeständigkeit von Hartchrom-Schichten
(Investigations on the etching resistance of hard chromium layers) (in German)
Der Maschinenschaden 56 (1983) 124, p. 24–28

[40] Khedr, M. G. A.; Gaamoune, B.
Corrosion behavior of Cu in oxidizing and non oxidizing acid mixtures
Metaux – Corrosion Industrie 60 (1985) 713, p. 1–8

[41] Ahmad, Z.
Effect of tin addition on the corrosion resistance of aluminium bronze
Anti-Corrosion 24 (1977) 1, p. 8–12

[42] Singh, R. N.; Tiwari, S. K.; Singh, W. R.
Effects of Ta, La and Nd additions on the corrosion behavior of aluminium bronze in mineral acids
Journal of Applied Electrochemistry 22 (1992) 12, p. 1175–1179

[43] Singh, R. N.; Verma, N.; Tiwari, S. K.; Singh, W. R.
Effect of some rare earth additives on corrosion of aluminium bronze in hydrochloric acid solution
Indian Journal of Chemical Technology 1 (1994) 2, p. 103–107

[44] Singh, R. N.; Tiwari, S. K.; Verma, N.
The influence of minor additions of Ta and Nd on the corrosion behavior of propeller bronze in mineral acids
Corros. Prev. Control 35 (1988) 2, p. 43–48

[45] Saber, T. M. H.; Tag El Din, A. M. K.
Dibutyl thiourea as corrosion inhibitor for acid washing of multistage flash distillation plant
Br. Corros. J. 27 (1992) 2, p. 139–143

[46] Yakovleva, L. A.; Vakulenko, L. I.; Vdovenko, I. D.; Lisogor, A. I.; Kalinyuk, N. N.; Novitskaya, G. N.
Corrosion disintegration of α-brass in hydrochloric and sulfamic acid solutions
Ukrainskii Khimicheskii Zhurnal 53 (1987) 7, p. 709–711

[47] DECHEMA-WERKSTOFF-TABELLE
Salzsäure
(Hydrochloric Acid) (in German)
DECHEMA e.V., Frankfurt a.M. (1976)

[48] Hanna, F.; Sherbini, G. M.; Barakat, Y.
Commercial fatty acid ethoxylates as corrosion inhibitors for steel in pickling acids
Br. Corros. J. 24 (1989) 4, p. 269–272

[49] Trabanelli, G.; Zucchi, F.; Brunoro, G.; Rochini, G.
Corrosion inhibition of carbon and low alloy steels in hot hydrochloric acid solutions
Br. Corros. J. 27 (1992) 3, p. 213–217

[50] Zuchi, F.; Trabanelli, G.; Brunoro, G.
Iron corrosion inhibition in hot 4 m HCl solution by t-cinnamaldehyde and its structure-related compounds
Corros. Sci. 36 (1994) 10, p. 1683–1690

[51] Horvath, T.; Kalman, E.; Kutsan, G.; Rauscher, A.
Corrosion of mild steel in hydrochloric acid solutions containing organophosphonic acids
Br. Corros. J. 29 (1994) 3, p. 215–218

[52] Ajmal, M.; Miden, A. S.; Quraishi, M. A.
2-Hydrazino-6-methyl-benzothiazole as an effective inhibitor for the corrosion of mild steel in acidic solutions
Corros. Sci. 36 (1994) 1, p. 79–84

[53] Rengamani, S.; Vasudevan, T.; Venkatakrishna Iyer, S.
Beta-phenylamine as an inhibitor for corrosion of mild steel in acidic solutions
Indian J. Technol. 31 (1993) 7, p. 519–524

[54] Rengamani, S.; Muralidharan, S. et al.
Inhibiting and accelerating effects of aminophenols on the corrosion and permeation of hydrogen through mild steel in acid solutions
J. Appl. Phys. (London) 24 (1994) 4, p. 355–360

[55] Rengamani, S.; Muralidharan, S.; et al.
Influence of anions on the performance of anisidines as inhibitors for the corrosion of mild steel in acidic solutions

Indian J. Chem. Technol. 1 (1994) 3, p. 168–174

[56] Reimann-Dubbers, V.
Entfernung von Ablagerungen und Korrosionsprodukten aus Rohrleitungen und Anlagenteilen
(Elimination of deposits of corrosion products from pipes and plant components)
(in German)
3 R International 24 (1985) 3, p. 133–144

[57] Schwenk, W.
Untersuchungen über die Hydrolyse von C-Cl- und C-S-Bindungen durch rostenden Stahl
(Investigations on hydrolysis of C-Cl- and C-S-bonds by rusting steel) (in German)
Werkst. Korros. 33 (1982) 10, p. 551–553

[58] Vehlow, J.
Korrosion von Metallen in halogenierten Kohlenwasserstoffen
(Corrosion of metals in halogenated hydrocarbons) (in German)
Ergebnisse des Forschungs- und Entwicklungsprogramms "Korrosion und Korrosionsschutz" 1 (1980), p. 175–177
(Results of the research and development program "Corrosion and corrosion protection")
Korrosion verstehen – Korrosionsschäden vermeiden
(ed. Gräfen, H.; Rahmel, A.) vol. 1, p. 267–276
Verlag Irene Kuron, Bonn (1994)

[59] Radovici, O.; Popa, M. V.
The electrochemical study of stress corrosion cracking of some high strength carbon steels
CORROSION-NACE 37 (1981) 8, p. 443–449

[60] Al-Hajji, J. N.; Nawwar, A. M.
Corrosion behavior of molybdenum alloyed low residual carbon steels in HCl solutions
Br. Corros. J. 26 (1991) 2, p. 127–132

[61] Merrick, R. D.; Auerbach, T.
Crude Unit Overhead Corrosion Control
Mater. Performance 22 (1983) 9, p. 15–21

[62] French, E. C.; Fahey, W. F.
Water soluble filming inhibitor system for corrosion control in crude unit overheads
Mater. Performance 22 (1983) 9, p. 9–14

[63] Badran, B. M.; Abdel Fattah, A. A.; Abdul Azim, A. A.
New corrosion inhibitors based on fatty materials – I. Epoxidized fatty materials modified with aliphatic amines
Corros. Sci. 22 (1982) 6, p. 513–523

[64] Badran, B. M.; Abdel Fattah, A. A.; Abdul Azim, A. A.
New corrosion inhibitors based on fatty materials – II. Epoxidized fatty materials modified with aromatic amines
Corros. Sci. 22 (1982) 6, p. 525–536

[65] Al-Kharafi, F. M.; Al-Hajjar, F. H.
Control of the corrosion of carbon steel in crude oil distillation units
Br. Corros. J. 25 (1990) 3, p. 209–212

[66] Bresel, Ake
Korrosionsschäden an Metall infolge von Halonen und Fluorkohlenwasserstoffen
(Corrosion damages on metal as a consequence of halones and fluorocarbons)
(in German)
Maschinenmarkt, Würzburg 87 (1981) 43, p. 888–890

[67] Axsom, J. F.
On-stream cleaning of refineries pays off
Oil and Gas Journal 71 (1973), p. 106

[68] Materials Selector for Hazardous Chemicals, vol. 3
Hydrochloric Acid, Hydrogen Chloride and Chlorine
MIT Publication MS-3, Elsevier Science, 1999, p. 15

[69] Frignani, A.; Monticelli, C.; Brunoro, G.
Influence of chromium content on corrosion resistance of low alloyed chromium steels in hydrofluoric acid and in other acid environments
Br. Corros. J. 22 (1987) 3, p. 190–194

[70] Kloos, K. H.; Landgrebe, R.; Speckhardt, H.
Untersuchungen zur wasserstoffinduzierten Rißbildung bei hochfesten Schrauben aus Vergütungsstählen
(Investigations on the hydrogen-induced crack formation in high-strength screws of tempering steels) (in German)
VDI-Z 127 (1985) 19, p. S92-S102

[71] Moller, G. E.; Franson, I. A.; Nichol, T. J.
Experience With Ferritic Stainless Steel In Petroleum Refinery Heat Exchangers
Mater. Performance 20 (1981) 4, p. 41–50

[72] Ogino, K.; Hida, A.; Kishima, S.; Kumanomido, S.
Susceptibility of type 431 Stainless Steel to erosion-corrosion by vibration cavitation in corrosive media
Wear 116 (1987) 3, p. 299–307

[73] Zucchi, F.; Trabanelli, G.; Brunoro, G.
Corrosion inhibition of duplex steel in hot HCl solutions
Eur. Congr. Corros. 9 (1989), p. 1–6

[74] Machida, M.; Nakajima, S.
Corrosion prevention for incineration treatment plant of vinyl chloride number
AIChE Meeting Porg., Pachec (1977), p. 1306–1313

[75] Gupta, S.; Kumar, Y.; Sanyal, D. B.; Pandey, G. N.
The exothermic reaction of stainless steel (AISI 304) in mixtures of HCl and HNO_3, and the inhibition of its corrosion
Corros. Prev. Control 30 (1983) 1, p. 11–14

[76] Nürnberger, Ulf
Korrosion und Korrosionsschutz im Bauwesen, 1. ed., vol. 2
(Corrosion and corrosion protection in civil engineering) (in German)
Bauverlag, Wiesbaden und Berlin, 1995, p. 997

[77] Hirschfeld, D.; Busch, H.; Stellfild, I.; Arlt, N.; Michel, E.; Grimme, D.; Steinbeck, G.
Stress corrosion cracking behavior of stainless steels with respect to their use in architecture, part 1: corrosion in the active state
Steel Research 64 (1993) 8/9, p. 461–465

[78] Allgemeine bauaufsichtliche Zulassung
(General permission of the building inspection) Z-30.3–6 of August 3, 1999
Bauteile und Verbindungselemete aus nichtrostenden Stählen
(Components and joining elements of stainless steels) (in German)
Deutsches Institut für Bautechnik, Berlin
Offprint 862 of the Informationsstelle Edelstahl Rostfrei, Düsseldorf 2000

[79] Vaccaro, F. P.; Hehemann, R. F.; Troiano, A. R.
Stress Corrosion Cracking of Austenitic Stainless Steel in an Acidified Chloride Solution
Corrosion-Nace 36 (1980) 10, p. 530–537

[80] Asphahani, A. I.
Effect of Acids on the Stress Corrosion Cracking of Stainless Materials in Dilute Chloride Solutions
Mater. Performance 19 (1980) 11, p. 9–14

[81] Vajpeyi, M.; Gupta, S.; Pandey, G. N.
Corrosion of stainless steel (AISI 304) in H_2SO_4 contaminated with HCl and HNO_3
Corros. Prev. Control 32 (1985) 5, p. 102–104

[82] Liu, X.; Wu, Y.; Dahl, W.; Schwenk, W.
Untersuchungen der Spannungsrißkorrosion austenitischer Stähle durch saure Chloridlösungen bei niedrigen Temperaturen
(Investigation of stress corrosion cracking of austenitic steels in acid chloride solutions at low temperatures) (in German)
Werkst. Korros. 44 (1993) 5, p. 179–186

[83] Davies, D. E.; Dennison, J. P.; Odeh, A. A.
The assessment of stress corrosion damage in austenitic stainless steel by measurements of cracks formed during constant strain rate and constant load tests in 1 M HCl
Corros. Sci. 24 (1984) 11/12, p. 953–964

[84] Kumar, A.; Singh, M. M.
Substituted dithiomalonamides as inhibitor for the corrosion of AISI 304 SS in phosphoric acid-hydrochloric acid mixture
Anti-Corrosion 40 (1993) 12, p. 4–7

[85] Heppner, H.-J.
Korrosionsverhalten chemisch beständiger Stähle in der Zuckerindustrie unter besonderer Berücksichtigung der chemischen Reinigung
(Corrosion behavior of chemically resistant steels in the sugar industry under special consideration of the chemical cleaning) (in German)
Z. Zuckerind. 25 (1975) 11, p. 622–626

[86] Sanad, S. H.; Ismail, A. A.; Mahmoud, N. A.
Efecto inhibidor del ioduro potasico sobre la corrosion del acero inoxidable en contacto con soluciones de acido clorhidrico
Revista de Metalurgia (Madrid) 28 (1992) 2, p. 89–97

[87] Schober, J.; Sandmann, H.; Kaufmann, W.
Korrosionsverhalten rostfreier Stähle gegenüber halogenhaltigen Lösungsmitteln
(Corrosion behavior of stainless steels in halogen containing solutions) (in German)
Werkst. Korros. 33 (1982) 7, p. 404–410

[88] DECHEMA-WERKSTOFF-TABELLE
32. Ergänzungslieferung
Chlorkohlenwasserstoffe- Chlorethane
(Chlorinated Hydrocarbons – chloroethanes) (in German)
DECHEMA e.V., Frankfurt (1993)

[89] DECHEMA-WERKSTOFF-TABELLE
33. Ergänzungslieferung
Chlorkohlenwasserstoffe- Chlormethane
(Chlorinated Hydrocarbons – chloromethanes) (in German)
DECHEMA e.V., Frankfurt (1993)

[90] Ovchiyan, V. N.; Abalyan, N. P.; Buldukyan, A. M.; Solunkina, G. P.; Dzhanumov, A. N.
Corrosion of steels in acid mixtures based on dichloroethane
The Soviet Chemical Industry 47 (1971) 10, p. 748–749

[91] Karjalainen, L. P.; Suutala, N.
Corrosion Damage in White Liquor Filter Piping – Korrosion in Filterrohren für Weißlauge
Praktische Metallographie 20 (1983) 7, p. 360–364

[92] Ahmad, Z.; Davami, P.
The Corrosion Resistance of 18 Cr-37 Ni Austenitic Steel in Alkaline, Acid and Neutral Solutions
Boshoku Gijutsu 29 (1980) 12, p. 595–601

[93] Manning, P. E.; Smith, J. D.; Nickerson, J. L.
New Versatile Alloys for the Chemical Process Industrie
Mater. Performance 27 (1988) 6, p. 67–73

[94] Class, I.; Graefen, H.; Schell, E.
Highly Corrosion-Resistant Ni-Cr-Mo Alloy with Improved Resistance to Intergranular Corrosion
U. S. Patent No. 3, 203, 792 (1965)

[95] Manning, P. E.; Schoebel, J. D.
Hastelloy alloy C-22 – A New and Versatile Material for the Chemical Process Industries
Werkst. Korros. 37 (1986), p. 137

[96] Sridhar, N.; Wu, J. B. C.; Corey, S. M.
The effect of acid mixtures on corrosion of nickel-base alloys
Mater. Performance 26 (1987) 10, p. 17–23

[97] Asphahani, A. I.
Corrosion Resistance of High Performance Alloys
Mater. Performance 19 (1980) 12, p. 33–43

[98] Asphahani, A. I.
Localized Corrosion of High Performance Alloys
Mater. Performance 19 (1980) 8, p. 9–21

[99] Corbett, R. A.; Morrison, W. S.
Comparative Corrosion Resistance of Some High-Nickel, Chromium-Molybdenum Alloys
Mater. Performance 28 (1989) 2, p. 56–59

[100] Konys, J.; Fodi, S.; Hausselt, J.; Schmidt, H.; Casal, V.
Corrosion of high-temperature alloys in chloride-containing supercritical water oxidation systems
Corrosion 55 (1999) 1, p. 45–51

[101] Agarwal, D. C.
UNS N10629: A New Ni-28 % Mo Alloy
Mater. Performance 33 (1994) 10, p. 64–68

[102] Brooks, C. R.; Spruiell, J. W.; Stansbury, E. E.
Int. Metals Rev. 29 (1984) 3, p. 210

[103] Lei, T. S.
The Effect of 0 – 8 Wt % Fe on Transformation of Phases in Ni-Mo-Fe Alloys
University of Tennessee, Knoxville, Tennessee, 1979, PhD diss.

[104] Anonymous
Improved version of Hastelloy Alloy B
Anticorros. Methods Mater. 22 (1975) 11, p. 19–20

[105] Golovanenko, S. A.; Svistunova, T. V.
Relaxation between intercrystalline corrosion resistance of Nickel-Molybdenum alloys and thermodynamic activity of carbon
Zashch. Met. 20 (1984) 5, p. 690–697

[106] Rebak, R. B.; Srivastava, S. K.
Corrosion performance of a nickel-molybdenum-chromium alloy: effects of aging, alloying elements and electrolyte composition
Corrosion 55 (1999) 4, p. 412–421

[107] Groysman, A.; Kaufman, A.; Feldman, B.; Man, Y.
Cavitation corrosion in a continuous catalytic reformer unit
Mater. Performance 39 (2000) 6, p. 62–65

[108] Rowe, D.
Temptated by tantalum
Process Industry Journal 9 (1994) 3, p. 37–39

[109] Hörmann, M.; Lupton, D.; Heinke, H.; Horn, E.-M.
Sondermetalle und ihre Anwendung im Chemieapparatebau
(Special steels and their use in chemical construction) (in German)
Z. Werkstofftechnik 18 (1987) 6, p. 186–194

[110] Fromm, E.; Hörz, G.
Wasserstoffversprödung von Tantal in Säuren unter mechanischer Beanspruchung
(Hydrogen embrittlement of tantalum in acids under mechanical stress) (in German)
Ergebnisse des Forschungs- und Entwicklungsprogramms "Korrosion und Korrosionsschutz", Project no. 1.5/5 (Results of the research and development program "Corrosion and corrosion protection")
Max-Planck-Institut für Metallforschung, Stuttgart, 1984, p. 47–51
see also: Gräfen, H.; Rahmel, A. (ed.)

Korrosion verstehen – Korrosionsschäden vermeiden
Verlag Irene Kuron, Bonn (1994) p. 38–51

[111] Sperner, F.; Liesner, Chr.
Das Korrosionsverhalten von Tantal in heißer Salzsäure
(Corrosion behavior of tantalum in hot hydrochloric acid) (in German)
Werkst. Korros. 32 (1981) 2, p. 57–65

[112] Flanders, R. B.
Try tantalum for corrosion resistance
Chemical Engineering (New York 86 (1979) 27, p. 109–110

[113] Anonymous
Pump handles hot acids
Chemical Processing 18 (1972) Nov. p. 55

[114] Anonymous
Tantalum-plated ball valves resist attack of 1000 – 1500 organic chemicals
Chemical Processing 35 (1972) June, p. 19–20

[115] Kritzer, P.; Boukis, N.; Franz, G.; Dinjus, E.
The corrosion of niobium in oxidizing sub- and supercritical aqueous solutions of HCl and H_2SO_4
Journal of Materials Science Letters 1818 (1999) 1, p. 25–27

[116] Krehl, M.; Schulze, K.; Olzi, E.; Petzow, G.
Korrosionsverhalten von geglühten Niob-Tantal-Legierungen
(Corrosion behavior of niobium tantalum alloys) (in German)
Zeitschrift für Metallkunde 74 (1983) 6, p. 358–363

[117] Robin, A.; Rosa, J. L.
Corrosion behavior of niobium, tantalum and their alloys in hot hydrochloric and phosphoric acid solutions
International Journal of Refractory Metals & Hard Materials 18 (2000) 1, p. 13–21

[118] Gypen, L. A.; Brabers, M.; Deruyttere, A.
Corrosion resistance of tantalum base alloys. Elimination of hydrogen embrittlement in tantalum by substitutional alloying
Werkst. Korros. 35 (1984) 2, p. 37–46

[119] Ruskol, Yu. S.; Viter, L. I.; Balakin, A. I.; Fokin, M. N.
Corrosion of titanium alloys in concentrated chloride solutions at temperatures up to 160°C
Zashch. Met. 18 (1982) 4, p. 516–619

[120] Hollies, A. C.; Scully, J. C.
The stress corrosion cracking and hydrogen embrittlement of titanium in methanol-hydrochloric acid solutions
Corros. Sci. 34 (1993) 5, p. 821–835

[121] Ebtehaj, K.; Hardie, D.; Parkins, R. N.
The stress corrosion and pre-exposure embrittlement of titanium in methanolic solutions of hydrochloric acid
Corros. Sci. 25 (1985) 6, p. 415–429

[122] Petit, J. A.; Chatainier, G.; Dabosi, F.
Inhibitors for the corrosion of reactive metals: Titanium and Zirconium and their alloys in acid media
Corros. Sci. 21 (1981) 4, p. 279–299

[123] Cavaseno, V.
Outlook for titanium brightens with copper grains
Chemical Engineering 84 (27) 1977, p. 40–42

[124] Chuang, Y. C.; Liu, H. A.; Li, P. S.; Wang, S. L.
Metallographic Study of Stress Corrosion Cracking of Ti-15 Mo Alloy in HCl-CH_3OH-CH_3OPSCl_2 Solution
Praktische Metallographie 21 (1984) 3, p. 122–132

[125] Chuang, Y.-C.; Liu, H.-A.; Chang, T.; Chu, Z.-S.
Stress Corrosion Cracking of Ti-15 Mo Alloy in HCl-CH_3OH-CH_3OPSCl_2 Solution
Z. Metallkd. 75 (1984) 10, p. 796–800

[126] Kovneristyj, Y. K.; Bolotina, N. P.; Kazarin, V. I.; Nagorbina, L. A.
Effect of quenching from the liquid state on the structure and corrosion properties of titanium alloys
Izv. Akad. Nauk SSSR, Met. (1982) 4, p. 117–118 (UDC 669.295:539.213)

[127] Hauffe, K.
Über das Korrosionsverhalten von Titan und Titanlegierungen gegen Chlor, Salzsäure und chlorionenhaltige Lösungen
(On the corrosion behavior of titanium and titanium alloys in chlorine, hydrochloric acid and solutions containing chloride ions) (in German)
Metalloberfläche 36 (1982) 12, p. 594–598

[128] Strunkin, V. A.; Poret, E. N.; Tsejtlin K. L.
Einfluß von Oxidationsmitteln auf die Beständigkeit von Titan und dessen Legierungen in Salzsäure
(Influence of oxidizing agents on the resistance of titanium and its alloys in hydrochloric acid) (in German)
Metallschutz Moskau 5 (1969) p. 265

[129] Liening, E.
Unusual Corrosion Failures of Titanium Chemical Processing Equipment
Mater. Performance 22 (1983) 11, p. 37–44

[130] Boriskina, N.; Ksenina, E.; Tumanova, T.; Shashkov, D. P.; Mikhalin, V.
Influence of nitriding on the corrosion and wear resistance of alloys AT 3 and AT 6
Zashch. Met. 19 (1983) 1, p. 61–64

[131] Fedirko, V. M.; Pohrelyuk, I. M.; Yas'kiv, O. I.
Corrosion resistance of nitrided titanium alloys in aqueous solutions of hydrochloric acid
Materials Science (Russia) 34 (1998) 1, p. 119–121

[132] Cotton, J. B.
Practical Use of Anodic Passivation for the Protection of Chemical Plant
Brit. Corrosion 10 (1975) 2, p. 66–68

[133] Nünninghoff, R.; Fischer, H.
Korrosionsverhalten von Zink-Aluminium-Legierungen auf Stahldrähten
Stahl Eisen 105 (1985) 9, p. 517–522

[134] Askey, A.; Lyon, S. B.; Thompson, G. E.; Johnson, J. B.; Wood, G. C.; Cooke, M.; Sage, P.
The corrosion of iron and zinc by atmospheric hydrogen chloride
Corros. Sci. 34 (1993) 2, p. 233–247

[135] Askey, A.; Lyon, S. B.; Thompson, G. E.; Johnson, S. B.; Wood, G. C.; Sage, P. W.; Cooke, M. J.
The effect of fly-ash particulates on the atmospheric corrosion of zinc and mild steel
Corros. Sci. 34 (1993) 7, p. 1055–1081

[136] Stupnisek-Lisac, E.; Podbrscek, S.
Non-toxic organic zinc corrosion inhibitors in hydrochloric acid
J. Appl. Electrochem. 24 (1994) 8, p. 779–784

[137] Epelboin, I.; Keddam, M.; Takenouti, H.
J. Appl. Electrochem. 2 (1972) p. 71

[138] Knittel, D. R.
Zirconium: A corrosion-resistant material for industrial applications
Chem. Eng. 87 (1980) 11, p. 95–96, 98

[139] Anonymous
Zirconium Grade 705 resists corrosives
Chemical Processing 42 (1979) 13, p. 208

[140] Yau, T.-L.; Maguire, M.
Electrochemical Protection of Zirconium in Oxidizing Hydrochloric Acid Solutions
Corrosion-Nace 40 (1984) 6, p. 289–296

[141] Yau, T.-L.; Maguire, M.
Electrochemical Protection of Zr Against SCC by Oxidizing HCl Solutions
Corrosion-Nace 41 (1985) 7, p. 397–405

[142] Barkov, A. A.
Mechanochemistry and pitting on zirconium in hydrochloric acid solution
Protection of materials (Russia) 35 (1999) 5, p. 420–424
see also: Proceedings of the Symposium on Passivity and its Breakdown
Electrochemical Society, Inc., NJ (USA) 1998

[143] Hebbar, K. R.; Sudhaker Nayak, H. V.; Ramchandar, T.
Stress corrosion failure of Zircaloy-2 sheets in methanolic HCl solution: Role of crystallographic texture
Werkst. Korros. 33 (1982) 10, p. 554–562

[144] Golozar, M. A.; Scully, J. C.
The effect of loading mode on the stress corrosion cracking of zircaloy-2 in a CH_3OH/HCl Solution
Corros. Sci. 22 (1982) 11, p. 1015–1024

[145] Kolomets, L. L.; Skorokhod, V. V.; Solonin, Yu. M.; Shcherbakova, L. G.; Yagupol'skaya, L. N.
Effect of fine structure on the pitting of Zr-Os and Zr-Ir alloys
Zashch. Met. 24 (1988) 1, p. 115–118

[146] Archer, M. D.; McKim, R. J.
Unusual stress-corrosion cracks observed in glassy Fe-40Ni-14P-6B alloy
J. Mater. Sci. 18 (1983) 4, p. 1125–1135

[147] Cho, K.; Hwang, C.-H.; Ryeom, Y.-J.; Pak, C.-S.
Corrosion Behavior of Amorphous Fe-Cr-Al-P-C-Ribbon Alloys
Metallurgical Translations A 13A (1982) 5, p. 901–905

[148] Zhang, B.-P.; Habazaki, H.; Kawashima, A.; Asami, K.; Hashimoto, K.
The corrosion behavior of melt-spun Cr-Ni-20P alloys in concentrated hydrochloric and hydrofluoric acids
Corros. Sci. 34 (1993) 2, p. 201–215

[149] Kolotyrkin, V. I.; Sololov, S. A.; Novokhatskii, I. A.; Knyazheva, V. M.; Lad'yanova, V. I.; Usatyuk, I. I.
Corrosive and electrochemical behavior of rapidly quenched Fe-Si alloys with high silicon content
Zashch. Met. 23 (1987) 1, p. 75–81

[150] Kim, J. H.; Akiyama, E.; Yoshioka, H.; Habazaki, H.; Kawashima, A.; Asami, K.; Hashimoto, K.
The corrosion behavior of sputter-deposited amorphous titanium-chromium alloys in 1 M and 6 M HCl solution
Corros. Sci. 34 (1993) 6, p. 975–987

[151] Kim, J. H.; Akiyama, E.; Habazaki, H.; Kawashima, A.; Asami, K.; Hashimoto, K.
The corrosion behavior of sputter-deposited amorphous chromium-zirconium alloys in 6 M HCl solution
Corros. Sci. 34 (1993) 11, p. 1817–1827

[152] Moffat, T. P.; Latanision, R. M.
Production and Characterization of Extremely Corrosion Resistant Chromium-Metalloid Alloys
J. Electrochem. Soc. 138 (1991) 11, p. 3280–3288

[153] Naka, M.; Hashimoto, K.; Masumoto, T.
J. Jpn. Inst. Met. 38 (1974), p. 835

[154] Ullmann's encyclopedia of industrial chemistry
6th ed., "Carbon", (2001) Electronic Release
Wiley-VCH, Weinheim

[155] Schley, J. R.
Impervious Graphite for Process Equipment Part 1 and Part 2
Chem. Eng. 81 (1974) 4, p. 144; 81 (1974) 6, pp. 102

[156] Product information
Chemie baut auf Graphit, HCl-Synthese
(Chemistry counts on graphite, HCl synthesis) (in German)
SGL Carbon Group, SGL Technik GmbH, Meitingen (2000)

[157] Würmseher, H.; Swozil, A.
Graphit als Werkstoff im Apparatebau
(Graphite as material in apparatus construction) (in German)
Z. Werkstofftech. 9 (1978), p. 19–30

[158] Patent
Swozil, A.
Sigri Electrographit
Graphitrohr, DE 3 116 309 (1981)

[159] Patent
Swozil, A.; Krätschmer, W.; Ullmann, G.
Sigri Electrographit (Sigri electro graphite)
Blockwärmetauscher (Block heat exchanger)
DE 3 117 187 (1982)

[160] Würmseher, H.; Swozil, A.; Künzel, J.
Kohlenstoff und Graphit als Werkstoffe für hohe Korrosionsbeanspruchung im Druckbehälter- und Apparatebau
(Carbon and graphite as materials for high corrosion demands in vapor vessel and apparatus construction) (in German)
Swiss Chem 5 (1983) 10a, p. 17–22, 24–25

[161] Product information
Novaphit®, Novatec®, Medienliste (list of media)
Frenzelit-Werke, Bad Berneck (1999)

[162] Product information
Systeme für HCl-Gas, Salzsäure, Schwefelsäure, Abgas, Rauchgas
(Systems for HCl-gas, hydrochloric acid, sulphuric acid, waste gas)
SGL Carbon Group, SGL Technik, Meitingen (2000)

[163] Product information
Säureschutz (Acid protection)
Didier Säureschutz GmbH, Königswinter

[164] Product information
Simrit CD-Rom Katalog, Version 3.1
(Simrit catalogue on CD-Rom, version 3.1)
Freudenberg Simrit KG, Weinheim (2000)

[165] Johnson, J. B.; Haneef, S. J.; Hepburn, B. J.; Hutchinson, A. J.; Thompson, G. E.; Wood, G. C.
Laboratory exposure systems to simulate atmospheric degradation of building stone under dry and wet deposition conditions
Atmospheric Environment 24A (1990) 10, p. 2585–2592

[166] Hutchinson, A. J.; Johnson, J. B.; Thompson, G. E.; Wood, G. C.
Stone degradation due to dry deposition of HCl and SO_2 in a laboratory-based exposure chamber
Atmospheric Environment 24A (1992) 15, p. 2785–2793

[167] Muchamedbaeva, Z. A.; Atakuziev, T. A.
Some problems of producing thick and chemically resistant compounds on the basis of water glass
Sbornik Naucn. Trudov. Taskentsk. Politechnic. Inst. Im Aburajchana Beruni 178 (1977), p. 29–31

[168] Yeo, R. S.; Zeldin, A. N.; Kukacka, L. E.
Corrosion of Polymer-Concrete Composites in Hydrochloric Acid at Elevated Temperature
Journal of Applied Polymer Science 26 (1981) 4, p. 1159–1165

[169] Grambow, B.
Corrosion of Glas
in: Uhlig's Corrosion Handbook, 2nd ed.
p. 411–437
John Wiley, New York (2000)

[170] Scholze, H.
Glass – water interactions
J. Non-Cryst. Solids, 102 (1988) p. 1–11

[171] Ullmann's encyclopedia of industrial chemistry
5th ed., vol. A12 "Glass", p. 365–432
VCH-Verlag, Weinheim (1989)

[172] Salman, S. M.; El-Batal, H. A.; Salama, S. N.
Chemical Durability of Some Commercial Glasses with Special Reference of Optical Glasses
Sprechsaal 113 (1980) 5, p. 359–360, 365, 368

[173] Glass Linings and Porcelain Enamel Coatings
in: Coatings and Linings for Immersion Service
TCP publications 2, rev. edition (1998)
NACE International, Houston, Texas

[174] Wittenberger, W.
Glasapparaturen im Chemiebetrieb
(Glass apparatus in chemical plants) (in German)
Chemie für Labor und Betrieb 31 (1980) 4, p. 138–147

[175] Zimmermann, H.
Verfahrenstechnik in Glas
(Process engineering in glass) (in German)
Umschau 85 (1985) 6, p. 340–345

[176] Bucsko, R. T.
Glass as a Material of Construction
Chemical Engineering Progress 79 (1983) 2, p. 82–85

[177] Sayers, J. A.
Brittle Materials
Chemical Engineering 79 (1972) 4, p. 51–56

[178] Window, J. G.
Glass shell-and-tube heat exchangers take on coiled-coil units
Process Engineering October (1973), p. 81–82

[179] Lorentz, R.; DeClerck, D. H.
Emailkorrosionsprüfung mit Autoklaven
(Enamel corrosion test in the autoclave) (in German)
Chem.-Ing.-Tech. 51 (1979) 6, p. 671

[180] Lorentz, R.
Glass enamel – Efficient protection against corrosion
Trib. Cebedeau 35 (1982) 460, p. 111–115

[181] Lorentz, R.
Untersuchungen zum Einfluß des gelösten Siliziumdioxids auf die Korrosion von Chemieemail in Salzsäure
(Investigations on the influence of dissolved silica on the corrosion of chemical service glass enamel in hydrochloric acid) (in German)
Werkst. Korros. 33 (1982) 5, p. 247–253

[182] Lorentz, R.
Angriff wäßriger Säure auf Chemieemail
(Attack of aqueous acids on chemical service glass enamel) (in German)
Werkst. Korros. 34 (1983) 5, p. 219–230

[183] Lorentz, R.
Inhibition des Säureangriffs auf Chemieemail
(Inhibition of the acid attack on chemical service glass-enamel) (in German)
Werkst. Korros. 34 (1983) 9, p. 437–445

[184] Lorentz, R.
Emailkorrosion durch Säure bei höheren Temperaturen und Druck; Aufklärung des Korrosionsmechanismus und Entwicklung von praxisgerechten Prüfmethoden
(Enamel corrosion by acid at higher temperatures and pressure, investigation of the corrosion mechanism and development of practical investigation methods) (in German)
Ergebnisse des Forschungs- und Entwicklungsprogramms "Korrosion und Korrosionsschutz" 3 (1984), p. 439–441
(Results of the research and development program "Corrosion and corrosion protection")

[185] DIN ISO 2743 (03/1987)
Bestimmung der Beständigkeit von Email gegen kondensierenden Salzsäuredampf
(Vitreous and porcelain enamels; determination of resistance to condensing hydrochloric acid vapor) (in German)
Beuth Verlag GmbH, Berlin

[186] Anonymous
Rugged construction keynoted in flowmeter
Chemical Processing 46 (1983) 8, p. 136

[187] Anonymous
Gas Filtration Elements Capable of Withstanding extreme Temperatures and Corrosion
Filtration & Separation 28 (1991) 3, p. 162

[188] Phillips, A. D.; Weyermuller, G. H.
Ferric chloride and abrasion at 305 °F resisted by glass-ceramic pump
Chemical Processing 36 (1973) Sept., p. 40

[189] Odawara, O.
Long ceramic-lined pipes with high resistance against corrosion, abrasion and thermal shock
Materials & Manufacturing Processes 8 (1993) 2, p. 203–218

[190] Babich, S. G.; Knyazheva, V. M.; Kozhevnikov, V. B.; Yurchenko, O. S.; Kolosvetov, Yu. P.
Corrosion resistance of chromium carbides in chloride-containing nonaqueous media
Poroskovaja Metallurgija 305 (1988) 5, p. 45–48

[191] Bäumel, A.
Vergleichende Untersuchung nichtrostender Chrom- und Chrom-Nickel-Stähle auf interkristalline Korrosion in siedender Salpetersäure und Kupfersulfat-Schwefelsäure Lösung
(Comparative study of stainless chromium and chromium-nickel-steels on the intercrystalline corrosion in boiling nitric acid and copper sulfate-hydrochloric acid solution) (in German)
Eisen und Stahl 84 (1964) 13, p. 798–807

[192] Bäumel, A.
Korrosion – Definition, Gliederung, grundsätzliche Einflüsse, Auftragsschweißen zur Abwehr von Verschleiß und Korrosion
(Corrosion – definition, structuring, basic influence, build-up weld for the defence of abrasion and corrosion) (in German)
DVS-Berichte, vol. 105 (1986), p. 206–210

[193] Effertz, P.-H., Fichte, W.; Szenker, B.; Resch, G.; Burgmann, F.; Grünschläger, E.; Beetz, E.
Kombinierte Konditionierung von Wasser-Dampfkreisläufen in Blockanlagen mit Durchlaufkesseln durch Sauerstoff und Ammoniak
(Combined conditioning of water-vapor cycles in block plants with boilers by oxygen and ammonia) (in German)
Der Maschinenschaden 51 (1978) 1, p. 3–15

[194] Smerek, J. M.; Graves, T. P.
Silicon Carbide Surfacing of Graphite Pump Parts for Corrosive/Abrasive Applications
Mater. Performance 20 (1981) 5, p. 28–31

[195] Gaines, A.
Pumps of silicon carbide reduce maintenance costs in acid regenerating plant
Chemical Processing 45 (1982) 3, p. 36

[196] Divakar, R.; Seshadri, S. G.; Srinivasan, M.
Electrochemical Techniques for Corrosion Rate Determination in Ceramics
Journal of the American Ceramic Society 72 (1989) 5, p. 780–784

[197] Kaidash, O. N.; Marinich, M. A.; Kuzenkova, M. A.; Manzheleev, I. V.
Corrosion resistance of cermets based on titanium nitride
Poroshkovaya Metallurg. 337 (1991) 1, p. 77–81

[198] Ullmann-Papst, S.; Naoumidis, A.; Förthmann, R.; Nickel, H.
Korrosion von SiSiC durch HCl-haltige Gase
(Corrosion of SiSiC by HCl containing gases) (in German)
Forschungszentrum Jülich, Jülich, 1994, p. 59–65

[199] Förthmann, R.; Naoumidis, A.
Influence of the high temperature corrosion by different gaseous environments on the bending strength of silicon carbide materials
Werkst. Korros. 41 (1990) 12, p. 728–733

[200] Gräfen, H. (ed.)
Lexikon Werkstofftechnik, 2. ed.
VDI-Verlag, Düsseldorf (1993)

[201] DIN 7724 (04/1993)
Polymere Werkstoffe; Gruppierungen polymerer Werkstoffe aufgrund ihres mechanischen Verhaltens
(Polymeric materials; grouping of polymeric materials based on their mechanical behavior) (in German)
Beuth Verlag, Berlin

[202] Renneberg, H.; Schneider, W.
Kunststoffe im Anlagenbau: Werkstoffe, Konstruktion, Schweißprozesse, Qualitätssicherung Fachbuchreihe Schweißtechnik, vol. 135
(Plastics in plant construction: materials, construction, welding processes, quality assurance) (in German)
DVS-Verlag, Düsseldorf (1998) p. 4, 316, 317

[203] Doležel, B.
Die Beständigkeit von Kunststoffen und Gummi
(The resistance of plastics and rubber)
Carl Hanser Verlag, München (1978)

[204] Krebs, Ch.; Avondet, M.; Leu, K.
Langzeitverhalten von Thermoplasten
(Long-time behavior of thermoplastics) (in German)
Carl Hanser Verlag, München (1999)

[205] Dominighaus, H.
Die Kunststoffe und ihre Eigenschaften, 5. ed.

(Plastics and their properties) (in German)
Springer-Verlag, VDI-Buch, Berlin (1998)

[206] Product information
Kunststoff-Rohrleitungssysteme
Chemische Beständigkeit von Kunststoffen und Elastomeren
(Plastic pipeline systems – chemical resistance of plastics and elastomers) (in German)
Georg Fischer, Schaffhausen, Schweiz (1999)

[207] DECHEMA-WERKSTOFF-TABELLE
41. Ergänzungslieferung
Abwasser (industrielles)
(Waste water (industrial)) (in German)
DECHEMA e.V., Frankfurt a.M. (1999) p. 52

[208] Billingham, N. C.
Degradation and Stabilization of Polymers
in: Corrosion and Environmental Degradation, vol. 2
Ed.: Schütze, M.
Wiley-VCH, Weinheim (2000) p. 470–475

[209] Product information
Daikin Industries Fluoroplastics
Daikin Chemical Europe GmbH, Düsseldorf (2000)

[210] Gibbesch, B.; Schedlitzki, D.
Water Vapor Permeability of Organic Materials for Coatings, Rubber Linings and Equipment Components
PCE (Protective Coatings Europe) November 1996, p. 10–41

[211] Schnabel, W.
Polymer Degradation
Principles and practical Applications
Carl Hanser Verlag, München (1981)

[212] Barth, P.; Hessel, J.; Kempe, B.
Die Bestimmung der chemischen Widerstandsfähigkeit von Rohren aus thermoplastischen Kunststoffen
(Determination of the chemical resistance of thermoplastics pipes) (in German)
Werkst. Korros. 48 (1997) p. 273–288

[213] Kempe, B.
Prüfmethoden zum Verhalten von Polyolefinen bei der Einwirkung von Chemikalien
(Test methods on the behavior of polyolefines during the development of chemicals) (in German)
Z. Werkstofftech. 15 (1984) p. 157–172

[214] ISO 8584–1 (03/1990)
Druckrohre aus Thermoplasten für industrielle Anwendungen; Bestimmung der chemischen Beständigkeitsfaktoren und Basisspannungen
Teil 1 : Rohre aus Polyolefinen
(Pressure pipe of thermoplastics for industrial application; designation of chemical resistance factors and base voltage – Part 1: Pipes of polyolefines) (in German)
Beuth Verlag GmbH, Berlin

[215] ISO/TR 8584–2 (12/1993)
Thermoplastische Druckrohre für industrielle Zwecke; Bestimmung des chemischen Widerstandsfaktors und der Grundspannung
Teil 2: Rohre aus halogenierten Polymeren
(Thermoplastic pressure pipes for industrial purposes; designation of the chemical resistance factor and of the base voltage – Part 2: Pipes of halogenated polymers) (in German)
Beuth Verlag GmbH, Berlin

[216] ISO/TR 10358 (06/1993)
Plastic pipes and fittings – Combined chemical-resistance classification table
Beuth Verlag GmbH, Berlin

[217] Deutsches Institut für Bautechnik
Medienliste 40 (Media list 40) (in German)
Berlin (1998)

[218] DIN EN 1778 (12/1999)
Charakteristische Kennwerte für geschweißte Thermoplast-Konstruktionen – Bestimmung der zulässigen Spannungen und Moduli für die Berechnung von Thermoplast-Bauteilen)
(Characteristic values for welded thermoplastic constructions – Determination of allowable stresses and moduli for design of thermoplastic equipment) (in German)
German version of EN 1778, December 1999
Beuth Verlag GmbH, Berlin

[219] Renneberg, H.; Schneider, W.
Kunststoffe im Anlagenbau: Werkstoffe, Konstruktion, Schweißprozesse, Qualitätssicherung
Fachbuchreihe Schweißtechnik, vol. 135
(Plastics in plant contruction: Materials, construction, welding processes, quality assurance) (in German)
DSV-Verlag, Düsseldorf (1998)

[220] Materials Selector for Hazardous Chemicals, Hydrochloric Acid
Hydrogen Chloride and Chlorine
MIT Publication MS-3, Elsevier Science (1999)

[221] Rolston, J. A.
When and how to select plastics
Chem. Eng. 29 (1984) p. 70–75

[222] Product information
Chemische Widerstandsfähigkeit von Thermoplasten
(Chemical resistance of thermoplastics)
(in German)
SIMONA AG, Kirn (2000)

[223] Product information
Epoxy Vinyl Ester Resins
Chemical Resistance Guide
DOW Plastics, N-Amsterdam (1996)

[224] Product information
Gummierungen für den industriellen Korrosionsschutz
Tabellen Chemische Beständigkeit
(Rubber coating for the industrial corrosion protection – Tables chemical resistance)
(in German)
Tip Top Oberflächenschutz Elbe GmbH, Wittenberg (1997)

[225] Product information
Beständigkeitstabelle
(Resistance list) (in German)
Klinger, Idstein (1997)

[226] Product information
Beständigkeitsliste Werksgummierungen
(Resistance list factory gumming)
(in German)
KCH KERAMCHEMIE, Siershahn (1998)

[227] Besic, D.
Give Thermoplastic Pumps a Spin
Chem. Engin. Progress (May 2000) p. 61–68

[228] Gaube, E.; Müller, W.; Diederich, D.
Zeitstandsfestigkeit und chemische Beständigkeit von Rohren aus Hartpolyethylen und Polypropylen
(Creep strength and chemical resistance of pipes of hard polyethylene and polypropylene) (in German)
Werkst. Korros. 19 (1968) p. 22–25

[229] Gaube, E.; Diederich, G:
Kunststoffe im Rohrleitungs- und Apparatebau
(Plastics in pipeline and apparatus construction) (in German)
Chemie-Ingenieur-Technik 46 (1976) p. 273–318

[230] Ehrbar, J.; v. Meysenbug, C.-M.
Untersuchungen zur Zeitstandsfestigkeit von Thermoplaste-Rohren unter Chemikalieneinwirkung
(Investigations on the creep strength of thermoplastic pipes under the influence of chemicals) (in German)
Z. Werkstofftech. 7 (1976) p. 429–437

[231] Jansen, N.
Polypropylen – ein bewährter Rohrwerkstoff
(Polypropylene – a well proven pipe material)
(in German)
3R International 37 (1998) p. 113–116

[232] Bosche, L.
Kanalrohrsystem aus Polypropylen
(Sewer pipe system of polypropylene)
(in German)
3 R International 37 (1998) p. 597–600

[233] Drake, J.; Germain, A.
Hypalon-lined diaphragm pumps combat aggressive sand slurry and save $ 1000–1500 a day
Chemical Processing 51 (1988) p. 64

[234] Derxheimer, G.
Private Mitteilung
(Private note)
Interner Bericht, Bayer AG, Leverkusen

[235] Lichtenstein, R.; Schippert, E.
Entfernung chlorierter Kohlenwasserstoffe aus Abwasser
(Elimination of chlorinated hydrocarbons from waste water) (in German)
KCH, Keramchemie GmbH, Kunststofftechnik, Siershahn (1983)
Druckschrift Nr. K83–114

[236] Product information
Technische Unterlagen zu Rohrklebstoffen
(Technical data on pipe adhesives)
(in German)
Henkel KGaA, Düsseldorf (1999)

[237] Product information
Elastische Klebe- und Dichtsysteme
(Elastic adhesive and density systems)
(in German)
Sika Chemie GmbH, Stuttgart (1999)

[238] Beijk, R.; Klinke, G.
Salzsäuregewinnung mit Flüssigkeitsstrahl-Vakuumpumpen
(Extraction of hydrochloric acid with liquid jet vacuum pumps) (in German)
Chemie-Technik 11 (1982) p. 483–484

[239] Anonymous
Abfüllen korrosiver Produkte in der Chemie
(Filling of corrosive products in Chemistry)
(in German)
Chemie-Technik 19 (1990) 6, p. 60–62

[240] Ullmann's encyclopedia of industrial chemistry
5th ed., vol. A6 "Chlorine" p. 459 and 471
VCH-Verlag, Weinheim (1986)

[241] Atterby, P.
Plastics on Fire – Corrosion
FOA Reports 6 (1972) 5

[242] Pohl, M.; Burchard, W.-G.; Schiffers, H.
Elektronenmikroskopische Schadensanalyse bei Werkstoffbeeinflussung durch Brandschäden
(Electron-microscopical damage analysis at the influencing of materials by damages by fire) (in German)
Praktische Metallographie 23 (1986) p. 187–203

[243] Hare, C. H.
Trouble with paint: dehydrochlorination
Journal of protective Coatings & Linings 16 (1999) p. 22–25

[244] Homann, J.
Fluorkunststoffe in der chemischen Verfahrenstechnik, insbesondere als Auskleidungs-/Beschichtungswerkstoffe
(Fluoroplastics in chemical process engineering, especially as lining/coating materials) (in German)
Werkst. Korros. 37 (1986) p. 532–543

[245] Werthmüller, E.
Auswahlkriterien für Auskleidungen mit Fluorkunststoffen
(Selection criteria for linings with fluoroplastics) (in German)
Chemie-Technik 20 (1991) 4, p. 21–29

[246] Coatings and Linings for Immersion Service
TPC publications 2, rev. Edition (1998)
NACE International, Houston, Texas

[247] Product information
Thermoplastische Kunststoffe, Beständigkeitstabellen
(Thermoplastics, resistance lists) (in German)
Röchling Haren KG, Haren (1995)

[248] Product information
Dyneon Fluorelastomere
(Dyneon fluoroelastomers) (in German)
Dyneon GmbH, Neuss (2000)

[249] Product information
Fluorkunststoff-Produkte
(Products made of fluoroplastics) (in German)
HEUTE + Comp., Radevormwald

[250] Product information
Fittings aus PVDF
(Fittings of PVDF) (in German)
GLYNWED, Vertrieb: FRIATEC, Mannheim

[251] Product information
Fluor-Parts
Edgar Kluth, Remscheid (2000)

[252] Anonymous
Lösemittel-Rückgewinnungsanlage aus Kunststoff
(Solvent recovery plant of plastic) (in German)
Technika, Basel, vol. 32 (1983) p. 2110–2112

[253] Busse, H.; Schindler, H.
Polymerwerkstoffe im Chemie-Apparatebau
(Polymer materials in chemical apparatus construction) (in German)
Chemie-Ingenieur-Technik 62 (1990) p. 271–277

[254] Gräfen, H.
Fortschritte beim Einsatz organischer Werkstoffe im Korrosionsschutz
(Progress in the use of organic materials in corrosion protection) (in German)
VDI-Berichte 670 (1988) p. 451–468

[255] Anonymous
Valve controls Process flow
The Oil and Gas Journal 77 (1979) p. 294

[256] Pierson, T.; Wickersham, C. P.
Heat exchanger equipped with tubes of Teflon resists attack by HCl
Chem. Processing 42 (1976) p. 133–134

[257] Product information
Tyachem / Kunststoffausgekleidete Chemie-Normpumpen mit Beständigkeitsliste
(Tyachem / plastic lined chemical standard pumps with resistance list) (in German)
KSB AG, Pegnitz (2000)

[258] Product information
KLINGERtop-chem, KLINGERsil, KLINGERgraphit-Produkte (products)
KLINGER GmbH, Idstein (2000)

[259] Dukert, A. A.
Fight corrosion with polyvinylidene fluoride
Chem. Processing UK 22 (1976) p. 21–22

[260] Margus, E.
Polyvinyliden fluoride for corrosion resistant pumps
Chem. Eng. 82 (1975) p. 133–134

[261] Gaines, A.
Fast setting polymer cement protects concrete structures in acid environments
Chem. Processing 45 (1982) p. 145–147

[262] Roderick, W. O.
Graphite/PPS ball valves withstand harsh

media for flue gas scrubber
Chem. Processing 43 (1980) 3, p. 24–25

[263] KSB AG
Kreiselpumpenlexikon, 3. ed.,
Werkstoffe, p. 351
(Dictionary of centrifugal pumps, materials) (in German)
KSB Eigenverlag, Frankenthal (1989)

[264] Carlowitz, B.
Kunststofftabellen, 3. ed.
(Plastic tables) (in German)
Carl Hauser Verlag, München (1986)

[265] Product information
Resistenzliste
(Resistance list) (in German)
Harzer Apparatewerke, Bornum (1992)

[266] Anonymous
Vinylesterharze lösen Korrosionsprobleme
(Vinylester resins solve corrosion problems) (in German)
Ind.-Anz. 102 (1980) 17, p. 22–25

[267] Alt, B.
Zum Einsatz von glasfaserverstärkten ungesättigten Polyesterharzen in der Chlorelektrolyse
(On the use of glass-fiber-reinforced unsaturated polyester resins in chlorine electrolysis)
Chemie-Technik 4 (1975) p. 237–240

[268] Pringle, J. D.; Kramer, K. L.
FRP vinylester piping undergoes rigors of chloralkali plant corrosives
Chem. Processing 46 (1983) 4, p. 24–25

[269] Bültjer, U.
Materiaux composite pour tuyauterie de grandes dimensions resistantes à la Corrosion
Techniques Modernes 78 (1987) 1/2, p. 49–51

[270] Cannon, D.; Lawlor, L.
Vinyl ester resin storage tanks in top condition after 7 years of corrosive attack
Chem. Processing 47 (1984) 7, p. 92–93

[271] Schweitzer, P. A.
Corrosion resistance tables
Marcel Dekker, Inc., New York (1986)

[272] Hopp, A.; Schedlitzki, D.
Korrosionsschutz in verfahrenstechnischen Anlagen
(Corrosion protection in chemical engineering plants) (in German)
Protective Coatings Europe (2001)

[273] Schedlitzki, D.
Private Mitteilung
(Private note) (in German)
Limburg (2000)

[274] Product information
KCH-Werksgummierungen, KCH-Baustellengummierungen
(KCH factory rubber coatings, KCH building ground rubber coatings) (in German)
KCH KERAMCHEMIE, Siershahn (1999)

[275] Bourgois, P.
Revetement et Protection 47 (1956) p. 21–28

[276] NACE-Ber., Gruppe t-6A-1
Corrosion (Houston) 17 (1961) 9, p. 453t–459t

[277] Gunzel, R.; Mandl, S.; Richter, E.; Liu, A.; Tang, B. Y.; Chu, P. K.
Corrosion protection of titanium by deposition of niobium thin films
Surface and Coatings Technology (1999), p. 1107–1110

[278] Khaladkar, P. R.
Using Plastics, Elastomers and Composites for Corrosion Control
in UHLIG'S Corrosion Handbook
(Ed. Revie, R.W.) 2nd Edition
J. Wiley, New York (2000) p. 965–1002

[279] Product information
Kunststoffe in Chemieanlagen
(Plastics in chemical plants) (in German)
Bayer AG, Zentrales Ingenieurwesen, Leverkusen (1985)

[280] DIN EN ISO 12944–1 (07/1998)
Beschichtungsstoffe – Korrosionsschutz von Stahlbauten durch Beschichtungssysteme –
Teil 1: Allgemeine Einleitung
(Paints and varnishes – Corrosion protection of steel structures by protective paint systems – Part 1: General introduction) (in German)
Beuth Verlag GmbH, Berlin

[281] DIN 28054–1 (09/2000)
Chemischer Apparatebau – Beschichtungen mit organischen Werkstoffen für Bauteile aus metallischem Werkstoff – Teil 1: Anforderungen und Prüfungen
(Chemical apparatus – Coating with organic materials for metallic components – Part 1: Requirements and testing) (in German)
Beuth Verlag GmbH, Berlin

[282] DIN 28053 (04/1997)
Chemischer Apparatebau – Beschichtungen und Auskleidungen aus organischen Werkstoffen für Bauteile aus metallischem Werkstoff – Teil 1: Anforderungen an Metalloberflächen

(Chemical apparatus – Organic coatings and linings on metal components – Requirements for metal surfaces) (in German)
Beuth Verlag GmbH, Berlin

[283] DIN 28054–3 (06/1994)
Chemischer Apparatebau – Beschichtungen mit organischen Werkstoffen für Bauteile aus metallischem Werkstoff – Teil 3: Spachtelbeschichtungen
(Chemical apparatus; organic coatings on metal components; trowel applied coatings) (in German)
Beuth Verlag GmbH, Berlin

[284] DIN 28054–5 (12/1996)
Chemischer Apparatebau – Beschichtungen mit organischen Werkstoffen für Bauteile aus metallischem Werkstoff – Teil 5: Pulverbeschichtungen
(Chemical apparatus – Organic coatings on metal components – Part 5: Powder coating) (in German)
Beuth Verlag GmbH, Berlin

[285] DIN 28054–2 (01/1992)
Chemischer Apparatebau – Beschichtungen mit organischen Werkstoffen für Bauteile aus metallischem Werkstoff – Teil 2: Laminatbeschichtungen
(Chemical apparatus; organic coatings on metal components; laminate coatings) (in German)
Beuth Verlag GmbH, Berlin

[286] Kaiser, W.-D.; Pietsch, S.; Rudolf, A.
Zusammenhänge zwischen Permeations- und Adhäsionsverhalten sowie der Unterrostung organischer Beschichtungen
(Connection between permeation and adhesion behavior and the under-rusting of organic coatings) (in German)
Farbe + Lack 98 (1992) p. 182–187

[287] Kressin, V. J.
Private Mitteilung
(Private note) (in German)
Bayer AG, Leverkusen (2000)

[288] Kaiser, W.-D.; Rudolf, A.; Walther, B.
Untersuchungen zu Adhäsions- und Korrosionsvorgängen beschichteter Metalle
(Investigations on the adhesion and corrosion of coated metals) (in German)
Farbe + Lack 101 (1995) p. 285–289

[289] DIN 28055–1 (09/1990)
Chemischer Apparatebau – Auskleidungen aus organischen Werkstoffen für Bauteile aus metallischem Werkstoff – Teil 1: Anforderungen
(Chemical apparatus – Surface protection with organic linings for application to metallic components – Part 1: Requirements) (in German)
Beuth Verlag GmbH, Berlin

[290] Product information
Oberflächenschutz – Resistent gegen aggressive Angriffe
(Surface protection – resistant against aggressive attacks) (in German)
Keramchemie GmbH, Siershahn (2000)

[291] Product information
Genakor
SGL-ACOTEC, Siershahn

[292] Product information
Trovidur® W 2000, Auskleidungsfolie aus PVC-P, Beständigkeitsliste
(Trovidur® W 2000, lining foil of PVC-P, resistance list) (in German)
Röchling Trovidur, Troisdorf

[293] DIN 28055–2 (04/2002)
Chemischer Apparatebau – Oberflächenschutz mit Auskleidungen aus organischen Werkstoffen für Bauteile aus metallischem Werkstoff – Teil 2: Eignungsnachweis und Prüfung
(Chemical apparatus – Surface protection with organic linings for application to metallic components – Part 2: Proof of suitability and testing) (in German)
Beuth Verlag GmbH, Berlin

[294] Product information
Chemieschläuche
(Chemical flexible tubes) (in German)
Gummi-Roller GmbH, Eschborn (2000)

[295] Product information
Medienliste Frenzelit-Produkte; NOVATEC; novaphit
(Media list Frenzelit products; NOVATEC, novaphit) (in German)
Frenzelit-Werke GmbH, Bad Berneck (1998,1999)

[296] Product information
CTS-Produktinformationen, CTS Composite Technologie Systeme
(CTS product information, CTS composite technology systems) (in German)
Hamburg-Barsbüttel (2000)

[297] Price, J. N.; Hull, D.
Effect of Matrix Toughness on Crack Propagation during Stress Corrosion of Glass Reinforced Composites
Composites Science and Technology 28 (1987) p. 193–210

[298] Niesse, J. E.
Fiber reinforced Plastic in Hydrochloric Acid Service
Mater. Performance 21 (1982) 1, p. 25–32

[299] Conslik, P. J.; Niesse, J. E.
User's Guide to ASME Standards for Fiberglass Tanks and Vessels
MTI Publication No. 50 (1996)

[300] Product information
Chemie-Normpumpen copper K-U
(Chemical standard pumps copper K-U) (in German)
KSB AG, Pegnitz (1991)

[301] Product information
Tyamagno-Kunststoffausgekleidete Chemie-Normpumpen
(Tyamagno plastic lined chemical standard pumps) (in German)
KSB AG, Pegnitz (2000)

[302] Margus, E. A.
Plastic pumps for corrosive services
Pumps, Pompes, Pumpen May (1980) p. 202–207

[303] KSB Werkstofftechnik
Persönliche Mitteilung
(Personal note) (in German)
KSB, Pegnitz, 2000

[304] Product information
Absperrklappen, Werkstofffibel
(Shutting clack, material primer) (in German)
KSB AG, Pegnitz (1992)

Nitric Acid

Author: K. Hauffe †/Editor: R. Bender

		Page
Survey Table		285
Introduction		288
A	**Metallic materials**	291
A 1	Silver and silver alloys	291
A 2	Aluminium	293
A 3	Aluminium alloys	303
A 4	Gold and gold alloys	311
A 5	Cobalt and cobalt alloys	312
A 6	Chromium and chromium alloys	314
A 7	Copper	316
A 8	Copper-aluminium alloys	320
A 9	Copper-nickel alloys	320
A 10	Copper-tin alloys (bronzes)	321
A 12	Copper-zinc alloys (brass)	321
A 13	Other copper alloys	328
A 14	Unalloyed steels and cast steel	329
A 15	Unalloyed cast iron	335
A 16	High-alloy cast iron, high-silicon cast iron	335
A 17	Structural steels with up to 12 % chromium	336
A 18	Ferritic chromium steels with more than 12 % chromium	337
A 19	Ferritic-austenitic steels with more than 12 % chromium	345
A 20	Austenitic chromium-nickel steels	347
A 21	Austenitic chromium-nickel-molybdenum steels	383
A 22	Austenitic chromium-nickel steels with special alloying additions	406
A 23	Special iron-based alloys	410
A 24	Magnesium and magnesium alloys	411
A 25	Molybdenum and molybdenum alloys	412
A 26	Nickel	412
A 27	Nickel-chromium alloys	414
A 28	Nickel-chromium-iron alloys	418
A 29	Nickel-chromium-molybdenum alloys	423
A 30	Nickel-copper alloys	425
A 31	Nickel-molybdenum alloys	425
A 32	Other nickel alloys	426
A 33	Lead and lead alloys	428
A 34	Platinum and platinum alloys	430
A 35	Platinum metals (Ir, Os, Pd, Rh, Ru) and their alloys	431
A 36	Tin and tin alloys	431
A 37	Tantalum, niobium and their alloys	432
A 38	Titanium and titanium alloys	435
A 39	Zinc, cadmium and their alloys	450
A 40	Zirconium and zirconium alloys	450
A 41	Other metals and alloys	456
B	**Non-Metallic inorganic Materials**	460
B 2	Natural stones	460
B 3	Carbon and graphite	460
B 4	Binders for building materials (e. g. mortar and concrete)	462
B 5	Acid-resistant building materials and binders (putties)	463
B 6	Glass	463
B 7	Quartz ware and quartz glass	464
B 8	Enamel	464
B 9	Porcelain	466
B 10	Stoneware	467
B 11	Refractory materials	467
B 12	Oxide ceramic materials	467
B 13	Metallo-ceramic materials (carbides, nitrides)	468
B 14	Other inorganic materials	473
C	**Organic Materials**	475
C 4	Wood	475
C 6	Furan resins	475
C 7	Polyolefins and their copolymers	475
C 8	Polyvinyl chloride and its copolymers	480
C 9	Polyvinyl esters and their copolymers	482
C 10	Phenolic resins	482
C 11	Acrylic resins	483
C 12	Polyamides	483
C 13	Polyacetals	484
C 14	Polyesters	484
C 15	Polycarbonates	486
C 16	Polyurethanes	487
C 18	Epoxy resins	489
C 19	Fluorocarbon resins	493
C 22	Silicones	496
C 23	Other plastics	497

Corrosion Handbook: Hydrochloric Acid, Nitric Acid
Edited by: G. Kreysa, M. Schütze
Copyright © 2005 DECHEMA e.V.
ISBN: 3-527-31118-1

		Page		Page
D	**Materials with Special Properties**	500	**Bibliography**	510
D 1	Coatings and linings	500	**Key to materials compositions**	569
D 2	Seals and packings	503	**Index of materials**	585
D 3	Composite materials	504	**Subject index**	605
D 4	Heat-resistant and scaling-resistant alloys	506		
D 5	Natural and synthetic elastomers	507		
D 6	Powder metallurgical materials	508		

Survey Table

The table below contains general data on the corrosion behavior. For additional comments, refer to the material numbers in the text.

Material No.*	Type	Behavior**	Material No.*	Type	Behavior**
A	**Metallic Materials**		A 35	Platinum metals (Ir, Os, Pd, Ru) and their alloys	+ to –
A 1	Silver and silver alloys	–	A 36	Tin and tin alloys	–
A 2	Aluminium	+ to –	A 37	Tantalum, niobium and their alloys	+
A 3	Aluminium alloys	+ to –	A 38	Titanium and titanium alloys	+ to ⊕
A 4	Gold and gold alloys	+	A 39	Zinc, cadmium and their alloys	–
A 5	Cobalt and cobalt alloys	+ to –	A 40	Zirconium and zirconium alloys	+ to ⊕
A 6	Chromium and chromium alloys	+ to –	A 41	Other metals and alloys	+ to –
A 7	Copper	–	**B**	**Non-Metallic Inorganic Materials**	
A 8	Copper-aluminium alloys	–	B 1	Asbestos	
A 9	Copper-nickel alloys	–	B 2	Natural stones	+ to –
A 10	Copper-tin alloys (bronzes)	–	B 3	Carbon and graphite	+ to ⊕
A 11	Copper-tin-zinc alloys (red brass)	–	B 4	Binders for building materials (e.g. mortar and concrete)	+ to –
A 12	Copper-zinc alloys (brass)	–	B 5	Acid-resistant building materials and binders (putties)	+ to ⊕
A 13	Other copper alloys	⊕ to –	B 6	Glass	+
A 14	Unalloyed steels and cast steel	+ to –	B 7	Quartz ware and quartz glass	+ to ⊕
A 15	Unalloyed cast iron	⊕ to –	B 8	Enamel	+ to ⊕
A 16	High-alloy cast iron, high-silicon cast iron	+ to –	B 9	Porcelain	+
A 17	Structural steels with up to 12 % chromium	+ to –	B 10	Stoneware	+ to ⊕
A 18	Ferritic chromium steels with more than 12 % chromium	+ to ⊕	B 11	Refractory materials	+
A 19	Ferritic-austenitic steels with more than 12 % chromium	+ to ⊕	B 12	Oxide ceramic materials	+ to –
A 20	Austenitic chromium-nickel steels	+ to –	B 13	Metallo-ceramic materials (carbides, nitrides)	+ to –
A 21	Austenitic chromium-nickel-molybdenum steels	⊕ to –	B 14	Other inorganic materials	+ to –
A 22	Austenitic chromium-nickel steels with special alloying additions	⊕ to –	**C**	**Organic Materials**	
A 23	Special iron-based alloys	+ to –	C 1	Natural fibers (wool, cotton, silk etc.)	
A 24	Magnesium and magnesium alloys	–	C 2	Bituminous compositions (bitumen, asphalt, pitch)	
A 25	Molybdenum and molybdenum alloys	⊕ to –	C 3	Fats, oils, waxes	
A 26	Nickel	–	C 4	Wood	–
A 27	Nickel-chromium alloys	⊕ to –	C 5	Modified cellulosics	
A 28	Nickel-chromium-iron alloys	⊕ to –	C 6	Furan resins	+ to ⊕
A 29	Nickel-chromium-molybdenum alloys	+ to –	C 7	Polyolefins and their copolymers	+ to ⊕
A 30	Nickel-copper alloys	–	C 8	Polyvinyl chloride and its copolymers	+ to ⊕
A 31	Nickel-molybdenum alloys	–	C 9	Polyvinyl esters and their copolymers	+ to –
A 32	Other nickel alloys	+ to –	C 10	Phenolic resins	+ to ⊕
A 33	Lead and lead alloys	–			
A 34	Platinum and platinum alloys	+			

Material No.*	Type	Behavior**	Material No.*	Type	Behavior**
C 11	Acrylic resins	⊕	D	**Materials with Special Properties**	
C 12	Polyamides	⊕ to −	D 1	Coatings and linings	+ to −
C 13	Polyacetals	−	D 2	Seals and packings	+ to ⊕
C 14	Polyesters	+ to −	D 3	Composite materials	+ to −
C 15	Polycarbonates	+ to −	D 4	Heat-resistant and scaling-resistant alloys	+ to −
C 16	Polyurethanes	+ to −			
C 17	Alkyd resins		D 5	Natural and synthetic elastomers	⊕ to −
C 18	Epoxy resins	⊕ to −			
C 19	Fluorocarbon resins	+ to ⊕	D 6	Powder metallurgical materials	+ to −
C 20	Polyvinylidene chloride				
C 21	Amino resins				
C 22	Silicones	⊕ to −			
C 23	Other synthetic materials	+ to −			

* Any notes in the text are entered under the same number as the materials, (for example A1, B5, C7)
** + resistant / ⊕ fairly resistant / − unsuitable. Where no indication of corrosion resistance is made, experimental data are not available.

Warranty disclaimer

This chapter of the DECHEMA Corrosion Handbook has been compiled from literature data with the greatest possible care and attention. The statements made in this chapter only provide general descriptions and information.

Even for the correct selection of materials and correct processing, corrosive attack cannot be excluded in a corrosion system as it may be caused by previously unknown critical conditions and influencing factors or subsequently modified operating conditions.

No guarantee can be given for the chemical stability of the plant or equipment. Therefore, the given information and recommendations do not include any statements, from which warranty claims can be derived with respect to DECHEMA e.V. or its employees or the authors.

The DECHEMA e.V. is liable to the customer, irrespective of the legal grounds, for intentional or grossly negligent damage caused by their legal representatives or vicarious agents.

For a case of slight negligence, liability is limited to the infringement of essential contractual obligations (cardinal obligations). DECHEMA e.V. is not liable in the case of slight negligence for collateral damage or consequential damage as well as for damage that results from interruptions in the operations or delays which may arise from the deployment of the DECHEMA Corrosion Handbook.

Introduction

Not only is nitric acid one of the strongest mineral acids, it is also a liquid with vigorous oxidizing properties – especially in the anhydrous form. It can corrode certain metals and alloys which, as such, are resistant in concentrated nitric acid, resulting in explosions.

The behavior of nitric acid towards metals varies according to its concentration. Dilute nitric acid reacts with all metals to the left of hydrogen in the electrochemical series causing the evolution of the latter, as can be seen from the example of Zn:

$$Zn + 2\ HNO_3 \rightarrow Zn(NO_3)_2 + H_2 \uparrow$$
$$Zn + 2\ H^+ \rightarrow Zn^{2+} + H_2 \uparrow$$

The hydrogen ions have an oxidizing action in this reaction.

Copper, silver and mercury react with concentrated nitric acid. The non-dissociated nitric acid molecules have an oxidizing action in these reactions. Gold and platinum are not attacked even by concentrated nitric acid. Some non-noble metals, especially iron and chromium, are resistant to concentrated HNO_3 as a result of passivation.

Titanium, for example, has an excellent resistance in almost the entire concentration range of nitric acid up to 450 K (177 °C). However, caution is required when using highly concentrated, fuming nitric acid, since a pyrophoric reaction may occur, leading to explosive phenomena. Combination of titanium with steel in a plant can also lead to a pyrophoric reaction in the presence of concentrated nitric acid, flames resulting. Corrosion accompanied by explosion is also observed on uranium-niobium alloys.

Similar phenomena occur with plastics. Explosions occurred, for example, during processing of nuclear fuels with cation exchangers consisting of basic polystyrene resins, which are in themselves quite resistant to nitric acid, in 7 to 9 mol/l nitric acid in the presence of catalytic ions and radiation.

Combinations of fuming nitric acid with metals and plastic should be tested initially in laboratory experiments to establish the absence of danger, such as, for example, in rocket technology, where fuming nitric acid is used as a carrier for oxygen, or in the concentration of nitric acid as well as production and storage of the fuming acid.

Introduction

While aluminium is corroded to a considerable degree in dilute nitric acid of between 20 and 40 %, it has an excellent resistance to the concentrated acid (> 85 %), so that it can be used for tank wagons and drums containing highly concentrated nitric acid. Thus aluminium and some of its alloys are used in the nitric acid industry and the nitric acid-based explosives industry.

As a result of their cost, tantalum, niobium and their alloys, which are also very resistant to nitric acid, are used only where cheaper materials such as, for example, aluminium and steel fail.

The following chapter deals with the corrosion behavior of metallic and non-metallic, inorganic and organic materials in liquid and vaporous nitric acid and in nitric acid process solutions, e.g. in nitration reactions. Since electrolytic etching with solutions containing nitric acid can be regarded as a special case of electrochemical dissolution of metals, alloys and other materials, corrosion data from such studies are also reported.

Boiling, concentrated nitric acid is used for testing stainless, austenitic steels for their susceptibility to local corrosion. Local corrosive attack due to structure inhomogeneities can occur as well as intercrystalline corrosion. The test method, also known as the **Huey Test**, is specified in ASTM A 262, Practice C. Test criterion is the material consumption rate per unit area. The specimens are exposed to 65 to 67 % boiling nitric acid over 5 test periods of 48 h respectively. After each test period the weight loss is determined. For testing Mo-bearing austenitic stainless steels for their susceptibility to intercrystalline attack, the test according to ASTM A 262, Practice D, can be performed. A solution consisting of 10 % HNO_3/3 % HF is used as the test medium at 343 K (70 °C) (4 h). For exact details see the above-mentioned specifications.

Since low-alloy steels are attacked considerably in dilute nitric acid without an inhibitor and the resistant high-alloy steels and alloys, such as, for example, chromium-nickel-molybdenum steels and nickel-chromium-iron alloys, are often too expensive, protective coatings are applied to low-alloy steels. Metallic coatings, such as, for example, titanium, or enamel coatings, are also suitable. If temperatures are not too high, polyolefins and fluorocarbon resins can also be used.

The majority of the results from tests on numerous steels and alloys after various mechanical and thermal treatments in nitric acid solutions of varying composition, as well as pure acid, can be arranged to only a limited degree because the aims of the tests often overlap.

Users of the DECHEMA Corrosion Handbook should therefore examine each case individually in order to establish which material is most suitable for the particular application.

If strength requirements are not too high, glass, which has an excellent resistance to nitric acid in all concentrations up to the boiling point, can be combined with resistant alloys without problems. The same applies to porcelain and ceramic materials based on aluminium oxide.

When choosing materials, plastics should be given particular attention because of their low specific gravity, such as, for example, polypropylene for nitric acid scrubbers and polytetrafluoroethylene (Hostaflon®, Teflon®) for distillation columns in nitric acid/hydrogen fluoride solutions. Other materials which are quite resistant to nitric acid are polyvinyl esters, polyvinyl chloride, phenol-formaldehyde resins, fluorocarbon resins, polyesters, polyamides and epoxy resins.

To evaluate the protective action of resin coatings towards nitric acid it is not sufficient merely to determine the resistance of the resin layer. It is also necessary to determine its permeability to nitric acid. For this reason, the latter is also stated, where available.

The occasionally contradictory data on resistance of resins to nitric acid result from the type of manufacture, the degree of polymerisation, the density and also the additives or impurities.

The equation given in General Remarks and Instructions for Use is used for converting the material consumption rate or weight increase per unit area (g/m^2 h and g/m^2 d) to corrosion rates in mm/a (mpy). Unless stated otherwise, the symbol %, used to designate the concentration, means percent by weight.

A
Metallic Materials

A 1 Silver and silver alloys

Silver is attacked and dissolved even by dilute nitric acid. By alloying with relatively large quantities of a metal resistant to nitric acid, such as, for example, 50 atomic percent gold, virtually no attack occurs. All the silver can be dissolved out of thin AgAu sheets (40 atomic percent Au) by repeated boiling in nitric acid, the gold which remains approximately retains the shape of the initial sheet [9].

A silver-gold alloy (50:50) changes its structure and composition in 50% nitric acid due to selective dissolution of silver and the formation of islands of gold from the gold atoms which remain on the surface. These islands grow as the corrosion increases [6]. Further silver atoms are exposed by this process, so that the silver is dissolved out down to a depth of > 1 μm. The growth of islands from the gold remaining can be used to estimate the extent of selective corrosion and surface passivation with the aid of model calculations and micromorphological studies [7, 8, 85]. The gold deposits formed on the silver-gold alloy Ag-25Au in 50% nitric acid and influencing of these deposits by plastic deformation has been examined by electron microscopy (TEM) [86].

Figure 1 shows the dependence of the life of stress-corroded silver-gold alloys before cracking on the gold content under the action of nitric acid and stress (75% of the tensile strength). After passing through a minimum at a gold content of 25 atomic percent, the life increases with higher amounts of gold and becomes infinite when the resistance limit is reached at about 40 atomic percent, so that no general corrosion occurs [83].

Silver is not noticeably attacked by aqua regia ($HNO_3 : HCl = 1:3$ parts by volume) due to the formation of a thin passive layer of silver chloride (AgCl) [2]. Layers of gold can accordingly be detached quantitatively from silver by aqua regia, without the silver being attacked.

Brass waveguides silvered on the inside, in which stresses occur during soldering, are often attacked by nitric acid which forms as a result of electrical discharges in damp air, stress corrosion cracking occurring [3]. The silver layer can be protected by rhodium.

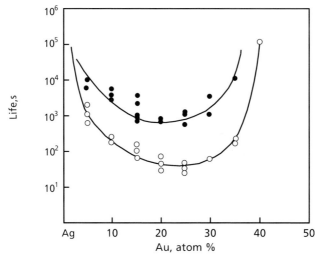

Figure 1: Influence of the gold content of AgAu alloys in ○ 1.4% and ● 1.52% HNO_3 on the service life of the alloy before cracking under stress (75% of the tensile strength) [83]

Copper and copper alloys are desilverized at 353 K (80 °C) in a solution consisting of 95% by volume H_2SO_4 (d = 1.84 g/ml) + 5% by volume HNO_3 (d = 1.40 g/ml). To prevent etching of the surface of the carrier alloy, the solution should not be diluted. Simple immersion in nitric acid (d = 1.33 g/ml) is sufficient for desilverizing aluminium. The dissolution of silver becomes noticeably slower when the solution has reached a silver content of about 15 g/l [115].

According to Figure 2, the rate of dissolution of silver in a solution of 0.95 mol/l KNO_3 + 0.05 mol/l HNO_3 at 298 K (25 °C) and 0.6 V_{SCE} decreases by more than 3 powers of ten if 20% palladium is alloyed with the silver. The potentiodynamic curves of the specimens 0.5 cm^2 in size were recorded at a rate of 10 m V/s [4]. At about 0.5 V_{SCE}, the alloy Ag-20Pd is passive in the above solution.

From the polarization curve of the anode with a matt and polished silver coating in an electrolyte consisting of 0.1 mol/l nitric acid + 0.9 mol/l potassium nitrate, the corrosion current density of the matt coating of 3.4×10^{-7} A/cm^2 was three times as high as that of the gloss coating of 1.01×10^{-7} A/cm^2, prepared in a cyanide electrolyte with nicotinic acid as the polishing agent [129]. The polished silver coating has the highest corrosion resistance, this being obtained at 0.005 A/cm^2. As the current density increases, the corrosion resistance decreases and becomes lower than that of the matt silver coating. The corrosion current density decreases further (6.6×10^{-8} A/cm^2) if other polishing agents are chosen (nicotinic acid/mercaptobenzothiazole in a ratio of 10:1).

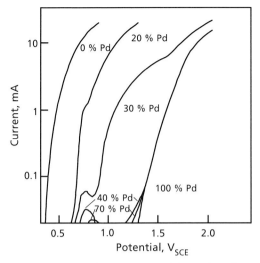

Figure 2: Potentiodynamic curves of the anodic polarization of Ag, AgPd alloys and Pd in 0.95 mol/l KNO_3 + 0.05 mol/l HNO_3 at 298 K (25 °C) and a rate of change of the potential of 10 mV/s [4]

The silver solder Ag19Cu21Zn20Cd used for welding chromium-nickel steel is etched by a rectified alternating current in methanol/nitric acid for metallographic evaluation [5].

As potentiostatic anodic polarization measurements on silver in 0.1 mol/l nitric acid show, the electrode process is characterized by the transition of silver into Ag^+ ions without side reactions. The dissolution is accelerated as the NO_3^- concentration increases and the rate of rotation of the electrode increases [144].

A 2 Aluminium

The corrosion rates of Al 99.5 (AA 1050A) decrease again from a corrosion maximum in 40 % HNO_3 as the concentration increases. This has been demonstrated for the temperature range 273 to 303 K (0 to 30 °C). The corrosion rates likewise fall as the temperature decreases. They reach a value of 0.1 mm/a (3.94 mpy) at the temperatures and concentrations listed in Table 1.

Temperature K (°C)	HNO_3 concentration %	Corrosion rate mm/a (mpy)
273 (0)	≥ 66	≤ 0.1 (3.94)
283 (10)	≥ 76	≤ 0.1 (3.94)
293 (20)	≥ 85	≤ 0.1 (3.94)
303 (30)	≥ 95	≤ 0.1 (3.94)

Table 1: Limiting concentrations of nitric acid at various temperatures at which the corrosion rate of Al 99.5 reaches 0.1 mm/a (3.94 mpy) [70]

The various strength levels of Al 99.5 (AA 1050A, W 7, F 11, F 15) had no influence on the corrosion rate. Like the temperature and concentration, the purity of aluminium had an effect on the corrosion behavior. In comparison with pure aluminium Al 99.5 (2.3×10^{-3} mm/a (0.09 mpy)), the corrosion rate of refined aluminium Al 99.99 (3 ppm Cu) in 99.8 % HNO_3 at 273 K (0 °C) was reduced to 0.7×10^{-3} mm/a (0.003 mpy) (test duration 100 d). As the Cu content rose to 50 ppm, an increase in the corrosion rate was again found (1.7×10^{-3} mm/a (0.07 mpy)).

In the case of pure and refined aluminium, no noticeable difference was found in the corrosion rates of the wrought and continuous-cast material.

A number of aluminium alloys (AlMg1 (AA 5005A), AlMgSi1 (AA 6082), AlMn, AlMg2Mn0.8 (AA 5049), AlMn1Mg0.5 (AA 3005) and AlMgSi0.5 (AA 6060)) showed corrosion rates comparable to those of pure aluminium in an exposure test lasting one year in 98.5 % HNO_3 at 303 K (30 °C) (acid changed every 8 to 10 d). The alloys AlMg2.5 (AA 5052) and AlMg2.7Mn (AA 5454) showed higher corrosion rates [70].

Since aluminium and its alloys show passive behavior in concentrated nitric acid (> 85 %) and therefore a good corrosion resistance, they are used in the nitric acid industry and in the production of explosives based on nitric acid. In dilute nitric acid of between 20 and 40 %, however, considerable corrosion occurs, occasionally reaching values of 3 to 5 mm/a (118 to 197 mpy) at room temperature [14].

According to Figure 3, the material consumption rate of aluminium (99.99 %) reaches its maximum in 30 % nitric acid between 313 and 353 K (40 and 80 °C) and decreases sharply as the concentration increases [25].

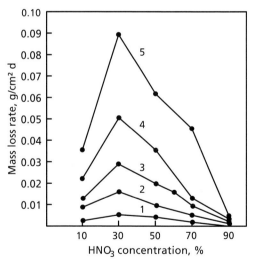

Figure 3: Material consumption rate of aluminium in nitric acid as a function of the concentration at 1) 313 (40), 2) 323 (50), 3) 333 (60), 4) 343 (70) and 5) 353 (80) K (°C) [25] (0.01 g/cm² d = 4.17 g/m² h ≙ 13.3 mm/a (524 mpy))

The corrosion rate of industrial-grade aluminium (AA 1060; Al-0.12Si-0.02Fe-0.04Mn) in various concentrations of nitric acid after tests lasting 48 h is shown for comparison in Figure 4 [11].

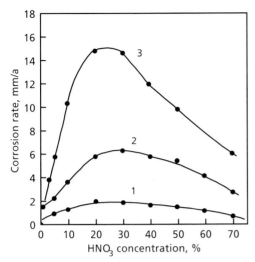

Figure 4: Corrosion rate of AA 1060 in nitric acid as a function of the HNO_3 concentration at 1) 298 (25), 2) 308 (35) and 3) 318 K (45 °C) [11]

According to the plot of the corrosion rate of AA 1060 in nitric acid at room temperature as a function of time shown in Figure 5, it decreases continuously in 70% nitric acid and approaches a stationary value of 0.2 mm/a (7.9 mpy) only after about 1000 h. In contrast, the corrosion rate in 20% nitric acid increases slowly and reaches 1 mm/a (39.4 mpy) after about 100 h and 1.5 mm/a (59.05 mpy) after 1000 h [11]. On the basis of these results, the experimental values in Figure 4 are set too low for < 40% HNO_3 and too high for > 50% HNO_3 [28].

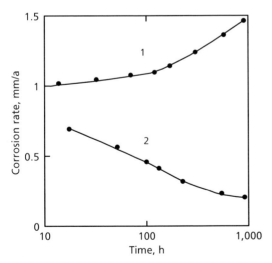

Figure 5: The corrosion rate of AA 1060 in nitric acid at 1) 20% and 2) 70% HNO_3 at room temperature as a function of time [11]

According to Table 2, the material consumption rate of aluminium and in particular that of a welded joint in 98 % nitric acid in both the liquid and the gaseous phase is less than that of the aluminium-magnesium alloys AlMg3 (AA 5754) and AlMg5 (AA 5056A) at 298 and 323 K (25 and 50 °C) [133].

Alloy	Medium phase	Temperature K (°C)	Metal	Welded joint
			Material consumption rate, g/m² h	
Al ADO	liquid	298 (25)	0.004	0.005
		323 (50)	0.042	0.081
	gaseous	298 (25)	0.017	0.028
	gaseous + liquid	298 (25)	0.022	0.050
AlMg3 (AA 5754)	liquid	298 (25)	0.006	0.019
		323 (50)	0.139	0.82
	gaseous	298 (25)	0.040	0.116
	gaseous + liquid	298 (25)	0.075	0.80
AlMg5 (AA 5056A)	liquid	298 (25)	0.026	0.048
		323 (50)	1.083	1.042
	gaseous	298 (25)	0.196	0.214
	gaseous + liquid	298 (25)	0.496	0.616

Table 2: Corrosion behavior of Al and AlMg alloys in 98 % HNO_3 at 298 and 323 K (25 and 50 °C) [133]

If a higher strength is not required, aluminium (e.g. Al-0.30Si-0.14Fe-0.06Cu) is suitable as a tank material for concentrated nitric acid (98 – 99 %). According to Table 3, attack in the vapor phase is considerably greater than that in the liquid phase and further increases under alternate wetting [69].

Aluminium of the Russian grade AV2 corrodes in boiling 98 % nitric acid at a material consumption rate which is independent of time (25 – 100 h) of 0.0091 to 0.0094 g/m² h (0.029 to 0.03 mm/a (1.14 to 1.18 mpy). The same applies to welded specimens (arc welding) in which the welding material is alloyed with 0.04 to 0.05 % titanium. The titanium content caused no change in the corrosion behavior. According to Table 4, pretreatment of aluminium specimens in a solution of 10 % HNO_3 + 7 g/l NaF reduces the attack by nitric acid to about 25 % [62].

Because of the periodic nature of dissolution of aluminium (active/passive) in nitric acid, the passivation period can be prevented, before it starts, by dipping the specimen in fresh acid [137].

The material consumption rate on aluminium (Al-0.1Si-0.11Fe) in a crevice 0.06 mm wide between aluminium/aluminium or aluminium/glass and aluminium/fluoro-plastic in boiling 98.9 % nitric acid was 0.52 g/m² h, regardless of the material combination. The corrosion rates of aluminium in boiling 98.9 % acid at various crevice widths after a test duration of 200 h are shown in Table 5 [26]. They are significantly higher than those in the absence of a crevice. Corrosion of aluminium in 98.9% HNO₃ at the boiling point as a function of the crevice width is reported in [26].

Crevice width mm	Material consumption rate g/m² h	Corrosion rate mm/a (mpy)
0.02	0.30	0.97 (38)
0.05	0.40	1.30 (51)
0.25	1.20	3.89 (153)
0.50	2.00	6.49 (255)
1.00	1.88	6.10 (240)
2.00	2.84	9.21 (363)

Table 5: Corrosion of aluminium in 98.9% HNO₃ at the boiling point as a function of the crevice width [26]

Plastic deformation and residual stress have no noticeable influence on the corrosion of aluminium in 98% nitric acid. The corrosion rate of the specimen in the delivery state after 500 h was thus 0.16 mm/a (6.3 mpy), and that in the deformed state was about 0.13 mm/a (5.12 mpy) [96]. No intercrystalline corrosion occurred.

Aluminium, which corrodes by itself in boiling 98% nitric acid at a rate of 0.11 g/m² h, suffers an increase in corrosion to 0.55 g/m² h in contact with the steel AISI 304 L. In contrast, the corrosion of the steel is reduced from 0.26 to < 0.01 g/m² h [578].

Corrosion of aluminium in white or red fuming nitric acid at 350 K (77 °C) can also be inhibited effectively by addition of 0.25 or 0.50% fluoride ions respectively, so that, for example, no corrosion is found after 26 d. By extrapolation of the test results, fuming nitric acid containing 1% fluoride ions could be kept in aluminium tanks at 350 K (77 °C) for 2 years without problems [124].

Medium	Aluminium				Welded joint	
	Corrosion rate mm/a (mpy)	Yield strength 0.2% offset N/mm²	Tensile strength N/mm²	Elongation at break %	Corrosion rate mm/a (mpy)	Tensile strength N/mm²
–	–	50	90	47.5	–	90
Acid	0.0016 (0.0254)	50	86	45.5	0.0015 (0.06)	100
Gas phase	0.031 (1.22)	–	170[*]	9.0[*]	0.028 (1.10)	70
Alternate wetting	0.050 (1.97)	–	165[*]	10.0[*]	0.052 (2.05)	70

[*] sample cut in the direction of rolling

Table 3: Corrosion behavior of Al as material and welded joint in concentrated HNO₃, in the gas phase and under alternate wetting with data on the mechanical values [69]

Metal	Welded specimen	Metal	Welded specimen	Test duration h
Without pretreatment		With pretreatment		
Material consumption rate, g/m² h (corrosion rate, mm/a)				
0.0098 (0.037)	0.0094 (0.03)	0.0023 (0.007)	0.0016 (0.005)	100
–	–	0.0013 (0.004)	0.0009 (0.003)	500

Table 4: Corrosion of Al-AV2 in boiling 98 % HNO_3 with and without pretreatment in 10 % HNO_3 + 7 g/l NaF [62]

Practically no change in the material consumption rate is found on addition of 0.015 to 1.5 mol/l ammonium nitrite to 40 % nitric acid (5.4 g/m² h). Only when the acid is saturated with NH_4NO_2 does the material consumption rate drop to 0.35 g/m² h. The corrosion rate is increased even by only small amounts of fluoride ions (0.003 mol/l fluoride). The material consumption rate reaches, for example, 37.5 g/m² h with 0.0266 mol/l fluoride in 40 % nitric acid [37]. In contrast, additions of 0.03 to 0.27 mol/l hydrogen chloride have no noticeable influence on the corrosion behavior.

An aqueous solution of 5 to 10 % HNO_3 + 0.4 to 0.7 % NaF is suitable for determining the corrosion of aluminium and its welded joint [47].

Low-alloy aluminium (Al-(1-1.5)Mn-0.15Cu-0.75Fe-0.6Si) corrodes at a material consumption rate of 1.15 g/m² h in 1 mol/l nitric acid at 305 K (32 °C). It was possible to reduce the material consumption rate to 0.26 g/m² h by addition of 10 ppm dithioglycolic acid, which corresponds to an inhibiting action of 78 % [76]. Higher inhibitor contents caused no further inhibition of corrosion. Diphenylthiourea achieved an inhibition of only 72 % even at 200 ppm.

The corrosion of AA 1060 in 20 % nitric acid at 308 K (35 °C) is inhibited to the extent of 70 % by 0.5 % 3,4,5-tri-hydroxybenzoic acid, to about 55 % by 2,4-dihydroxybenzoic acid, to 43 % by p-nitrobenzoic acid and only to about 20 % by benzoic acid itself [136].

Reducing the corrosion rates on aluminium in nitric acid (< 20 %) by inhibitors has so far been unsatisfactory. For example, the highest inhibition efficiency of some substituted thioureas with AA 1060 in 20 % HNO_3 at 298 K (25 °C) of only 70 % was achieved by 250 ppm p-chlorophenylthiourea [16]. The inhibition determined from weight measurements and by the galvanic method was in agreement.

The inhibition efficiency of phenylthiourea, o-, p- and m-tolylthiourea and diphenyl- and ditolylthiourea in concentrations of 250 ppm with AA 1060 in 20 % nitric acid was still unsatisfactory at 40 – 60 % [10]. At higher temperatures (318 K (45 °C)), the inhibition rose to about 70 % [89].

Aluminium AA 1100 (Al-0.068Mn-0.021Mg-0.13Si-0.52Fe-0.01Cu) corroded in 20 % nitric acid at 308 K (35 °C) with and without 200 ppm phenylthiourea at material consumption rates of 2.03 and 0.73 g/m² h respectively, which correspond to an inhibition efficiency of 75 %. A comparable figure of about 71 % was also obtained with 250 ppm p-tolylthiourea [29]. Ethyl-, methyl- and i-propylthiourea are considerably poorer inhibitors.

The inhibition efficiency of 1-aryl-3-formamidine-thiocarbamides with AA 1060 in the same medium at twice the concentration is somewhat better. The values achieved with 500 ppm 1-p-methoxyphenyl-3-formamidine-thiocarbamide were thus 71 % at 298 K (25 °C) and 80 % at 318 K (45 °C). None of the other thiocarbamides tested was as effective [22]. The current density which prevails in the passive state of aluminium decreases as the inhibition increases.

Hydroxyl and nitro derivatives of benzoic acid did not even achieve an inhibition efficiency of 50 % even at considerably higher concentrations (up to 1 %), with one exception (about 65 %) [23].

The inhibition efficiency of haloacetic acids with aluminium in 20 % nitric acid is also modest. The material consumption rate of Al-0.1Cu-0.1Mg-0.5Si-0.7Fe-0.1Mn-0.1Zn (Jindal aluminium, Bangalore, India) of 25.9 g/m^2 h at 303 K (30 °C) could thus be reduced to 12.5 g/m^2 h by 0.001 mol/l of the best inhibitor iodoacetic acid, which corresponds to an inhibition efficiency of 60 %. As the temperature increases, the latter also increases here (e.g. to about 81 % at 323 K (50 °C)) [59].

The inhibition efficiency of aromatic amines (p-toluidine, p-anisidine, p-chloroaniline, p-aminobenzoic acid and p-phenetidine) with AA 1060 in 20–40 % nitric acid at 308 K (35 °C) of 0.1 % is of no industrial importance [93].

Only in 20 % nitric acid do 400 ppm o-tolylthiourea achieve an inhibition efficiency with aluminium of about 80 %. This drops drastically at lower and higher concentrations of HNO_3 [95].

It has recently become possible to reduce the corrosion of the alloy AA 1060 in 20 % nitric acid at 308 K (35 °C) by 98 % by addition of 500 ppm 1-m-tolyl-3-tolyl-formamidine-thiocarbamide (Table 6) [116].

A corrosion-resistant coating is obtained on aluminium and its alloys if nitric acid of > 30 % is allowed to act on the degreased and chemically polished metal and the metal is then treated with an acrylic resin solution under direct current [48].

In a solution of 56 % HNO_3 + 44 % N_2O_4 (HDA = high-density acid), aluminium acquires a dense layer which prevents pitting corrosion. More prolonged exposure to nitric acid containing fluoride ions generates a thick film which, in addition to oxide, also contains α-AlF_3 [49]. For rapid formation of a colored film on aluminium, this is polarized anodically in oxalic acid containing 0.2 g/l nitric acid. The HNO_3 added to the oxalic acid stabilizes the oxide film formed, in spite of its corrosive action on aluminium. Comparable results are also obtained by replacing the oxalic acid with sulfoxalic acid [45]. Nitric acid is added to the chemical polishing and brightening baths for aluminium for the purpose of uniform oxide film formation and to dissolve away foreign oxides [50, 17].

Acid pickling solutions, such as, for example, HNO_3/H_2O (1:1, 2:1, 3:1), HNO_3/HF, HNO_3/NaF, HNO_3/CaF_2, H_3PO_4/HNO_3/$KClO_3$ and H_2CrO_4/HNO_3, fulfill a particular function in the electroplating process sequence of aluminium. They have, inter alia, relatively powerful oxidizing and activating properties in comparison with alkaline pickling solutions. In particular, the heavy-metal oxides formed by the heat treatment are removed [51, 61].

Inhibitor	Concentration ppm	Material consumption rate, g/m² h	Corrosion rate mm/a (mpy) 308 K (35 °C)	Inhibition efficiency	Material consumption rate, g/m² h	Corrosion rate mm/a (mpy) 318 K (45 °C)	Inhibition efficiency %
Without	–	2.18	7.1 (280)	–	3.39	11.0 (433)	–
1,3-di-p-tolylformadine thiocarbamide	250	0.69	2.2 (87)	68	0.68	2.2 (87)	80
	500	0.56	1.8 (71)	74	0.48	1.5 (59)	86
1-m-tolyl-3-tolylformadine thiocarbamide	250	0.64	2.1 (83)	71	0.63	2.0 (79)	81
	500	0.43	1.4 (55)	98	0.35	1.1 (43)	90

Table 6: Corrosion of AA 1060 in 20% HNO_3 and the inhibition efficiency of two substances at 308 and 318 K (35 and 45 °C) [116]

Treatment with nitric acid and, where appropriate, addition of HF provide two distinct advantages. Firstly, the alkali residues formed with water after degreasing, which can be removed only with difficulty, are neutralized. Secondly, the deposits of metal oxides and hydroxides (e.g. of Cu and Fe) and silicon, which form due to strong attack by the pickling solution, are detached [52].

Particularly highly concentrated HNO_3 containing solutions (70% HNO_3) are used at room temperature for decopperizing aluminium. Immersion in nitric acid (d = 1.33 g/ml) at 298 K (25 °C) is also adequate for removal of lead coatings on aluminium [130]. As a result of the good resistance of aluminium to concentrated nitric acid, this can be used quite generally for removal of metal contamination on the Al surface [17].

The following solution is recommended as a polishing bath for brightening pure aluminium (99.9%) between 333 and 343 K (60 and 70 °C): 870 ml H_3PO_4 (85%) + 65 ml HNO_3 (54%) + 65 ml acetic acid (99%) + 1 g/l $Cu(NO_3)_2$ [106].

Boehmite layers were generated by dipping aluminium specimens of 99.99, 99.8 and 99.3% Al into a solution of 6% HNO_3 + 2% H_2SO_4 and their corrosion potentials were determined, these becoming less noble as the contamination of the aluminium decreased. The same tendency is also observed by dipping specimens into water of increasing temperature [138].

Figure 6 shows the temperature dependence of the limiting current density of the dissolution of aluminium in 7% by volume HNO_3 (100%) at -0.6 V_{SCE} [12]. As well as the dissolution rate of aluminium (Curve 1), the corresponding rate of reduction of nitric acid is also shown.

The rate of dissolution of aluminium can be calculated from anodic and cathodic current density/potential curves of aluminium in an electrolyte consisting of 78 ml 87% H_3PO_4 + 15 ml 94% H_2SO_4 with increasing additions of HNO_3 (from 0 to 20 mol/l) [53].

The importance of aluminium in semiconductor technology is based on its formation of a defined oxide layer with a pore diameter of 120 to 180 Å and a barrier or cell wall of 10 to 13 Å/V. The following solution is mentioned for photographic etching of variously prepared oxide layers at 298 K (25 °C): 50 ml/l 40 % HF + 350 ml/l 65 % HNO_3 [20].

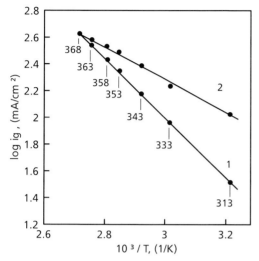

Figure 6: Temperature dependence of the limiting current density (i_g) of the dissolution of Al in 7 % by volume HNO_3 (100 %) at -0.6 V_{SCE}. Curve 1) corresponds to the dissolution of Al and Curve 2) corresponds to the reduction of HNO_3 [12]
log i_g = 2.0 ≙ 100 mA/cm^2 ≙ 670 g/m^2 h

The following solution has also been tested for polishing aluminium between 373 and 378 K (100 and 105 °C): 1100 – 1500 g/l H_3PO_4 + 25 – 60 g/l HNO_3 + 15 – 25 g/l Al (in salt form) with a density at 298 K (25 °C) of 1.66 – 1.70 g/cm^3. A number of other solutions on this basis have also been described [56].

The following HNO_3 containing solutions, amongst others, are recommended for removing the oxide layer on aluminium: 10 – 50 % by volume HNO_3 with and without 1 – 2 % HF or 1 – 10 % by volume HNO_3 + (0.1 – 1.0) % HF + (0.1 – 3.0) % iron(III) sulphate. The last-mentioned solution is particularly suitable for removing etching residues on aluminium. Appropriate immersion times are 2 – 3 min [92].

Pure aluminium (Al-0.0055Si-0.006Fe-0.002Cu-0.0005Mg-0.0015Mn) acquires a bright surface as a result of immersion in a solution of 6.0 % by volume HNO_3 + copper salt (1 g/l Cu) for 1 to 3 min. The presence of copper increases the dissolution of aluminium and largely prevents the settling of aluminium salt [90]. Comparable results were obtained with technical-grade aluminium (Al-0.06Mg-0.015Cu-0.15Si-0.19Fe-0.025Ti). After immersion of aluminium in 50 % nitric acid at room temperature for 1 min, a reflection of 96 % was achieved, which dropped to 90 % after longer action (3 min) and 69 % after 10 min [91].

An aluminium specimen (99.994 % Al + 0.004 % Si + 0.001 % Fe + 0.001 % Cu) was polished chemically in an HNO_3/H_3PO_4 solution (1:6) for the purpose of analysing the composition of the film generated by etching. ESCA measurements showed a layer of aluminium oxide containing small amounts of OH^- ions, in addition to NO_2-ions [94].

An addition of 0.02 – 3% 2,4-diamino-6-phenyl-1,3,5-triazine with 0.005 – 10% Cu^{2+} ions to nitric acid ($\leq$ 10%) is recommended for chemical polishing of aluminium [98]. A solution of 0.25% HNO_3 + 0.1% HF + 0.25% CrO_3 + 0.25% $H_2Cr_2O_7$ at 323 K (50 °C) is recommended elsewhere as being suitable for the surface treatment of aluminium and its alloys [99].

For cleaning the surface of aluminium and its alloys after electrolytic deposition of thicker coatings, treatment in 60% nitric acid for 3 hours and in 55% hydrofluoric acid for 1 hour at room temperature is suggested, e.g. for removal of silicon carbide from an aluminium cylinder in an engine [100].

A solution of 5% HNO_3 + 10% sulfamic acid is used at 313 K (40 °C) for removal of an aluminide layer 20 – 50 µm thick on aluminium. Solutions of HNO_3 + H_3PO_4 are recommended for thicker layers (> 80 µm) [101].

A solution of 75% H_3PO_4 and 66% HNO_3 in a ratio of 13:1, to which 5 mg copper nitrate are added per kg solution, is used as an etching agent for structuring aluminium layers for decorative effects. An etching layer about 1 µm thick is obtained by dipping for 150 s [103].

Aluminium components are also pretreated in 10 to 20% nitric acid at room temperature for chromation and phosphate chromation [104].

Before aluminium workpieces are anodized for application of firmly adhering plastic coatings, they are degreased in a hot solution, at 358 K (°C), of 9 g/l Na_2SiO_3 + 1 g/l NaOH for about 30 minutes, subsequently deoxidized by dipping in a solution of 400 g/l HNO_3 + 5 g/l HF for 5 min and then etched in 250 g/l HNO_3 for 1 min [58].

In order to clean the surface of cast aluminium, it is dipped into a sodium hydroxide melt at 773 K (500 °C) doped with 0.5% zinc for 20 min and, after being cooled in air for 2 min, subsequently washed in 3% potassium bichromate solution for 15 to 20 s. Treatment in 16% nitric acid for 5 min is then carried out, followed by dipping in 5% potassium bichromate solution for 20 min. The aluminium is then washed in cold water and dried in air [84].

A test solution of 5 to 10% HNO_3 + 0.4 to 0.7% NaF between 353 and 375 K (80 and 102 °C) is used to investigate intercrystalline corrosion of aluminium and its alloys [102].

Aluminium as a packaging material in the food industry is classified as not adequately resistant to nitric acid for normal use [134].

Pure aluminium (99.99%) which was polarized anodically at 2 V and 723 K (450 °C) in a potassium nitrate melt for 45 min (0.001 A/cm^2, charge flow 30 coulombs) corroded in 1 mol/l nitric acid after a test duration of 24 h with a material consumption rate of only 0.033 g/m^2 h (0.11 mm/a). After anodic treatment, the specimen had a shiny black appearance as a result of very imperfect aluminium oxide [24].

A 3 Aluminium alloys

Of 30 aluminium casting alloys analyzed (sand casting, permanent die casting and pressure die casting alloys), those containing magnesium and silicon had the highest corrosion resistance. The sand casting grade S-5A (Al-5.25Si) and the permanent die casting grades GS-42A (Al-1.8Si-4.0Mg) and SG-70A (Al-7.0Si-0.3Mg), for example, are attacked insignificantly by concentrated nitric acid at room temperature, but more severely by dilute nitric acid [97].

Table 2 shows the good resistance of the aluminium-magnesium alloys AlMg3 (AA 5754) and AlMg5 (AA 5056A), even as weld joints, in 98 % nitric acid at 298 K (25 °C). Welded AlMg alloys can no longer be used at 322 K (49 °C), with material consumption rates of 0.82 and 1.0 g/m^2 h. The increase in the corrosion rates of the alloys in the gas phase and especially in the liquid-gas phase is remarkable [133].

In addition to aluminium (see Table 2), the aluminium-magnesium alloys and the aluminium-manganese alloys AMts are suitable as a tank material for concentrated nitric acid (98 – 99 %) because of their strength. Some alloys are listed in Table 7 with their compositions and mechanical properties.

The corrosion behavior of the aluminium alloys listed in Table 7, both as compact alloys and weld joints, in concentrated acid (98 – 99 %) in the gas phase and under alternate wetting is summarized accordingly in Table 8 [69]. This table shows that the attack is lowest in the solution and highest with alternate wetting.

The corrosion behavior of heat-treated specimens of the Al alloys G-AlMn, G-AlMg2.7Mn and G-AlMg2Mn0.8 taken from continuous cast bars was affected by the heat treatment temperature after exposure to 98.5 % HNO$_3$ at 303 K (30 °C) for one year. Higher heat treatment temperatures at 823 K (550 °C) and a holding time of 4 h proved to be considerably more favorable than lower temperatures (673 K (400 °C)). At 673 K (400 °C), the corrosion rate decreased with the holding time (4 to 24 h), but the values of 823 K (550 °C) were not reached.

The corrosion rate of G-AlMn was thus reduced from 0.88 to 0.14 mm/a (34.65 to 5.51 mpy) by heat treatment, and that of the alloy G-AlMg2.7Mn was reduced from 0.22 to 0.1 mm/a (8.66 to 3.94 mpy). The corrosion rate of the heat-treated specimen of G-AlMg2Mn0.8 was 0.22 mm/a (8.66 mpy) [70].

The corrosion rates of some wrought alloys in industrial 98.5 % nitric acid at 303 K (30 °C) for a test duration of 1 year are listed in Table 9 [70].

The alloy AA 5052 (Al-0.17Si-0.04Fe-2.30Mg) shows a corrosion rate of 2.5 mm/a (98 mpy) in 30 % nitric acid at 298 K (25 °C) and 0.75 mm/a (30 mpy) in 70 % acid, comparable to the corrosion rates of the alloy AlMg3 (AA 5754) [11]. According to Figure 7, the corrosion rates of the aluminium-manganese alloy AA 3003 (Al-0.16Si-0.58Fe-0.78Mn) in 30 % nitric acid at 298 K (25 °C) and 318 K (45 °C) of 3.1 and 21 mm/a (122 and 827 mpy) respectively are too high for it to be used [11].

Alloy	Chemical composition, %							Mechanical properties		
	Mg	Mn	V, Ti	Si	Fe	Cu	Zn	Yield strength 0.2 off-set N/mm^2	Tensile strength N/mm^2	Elongation at break %
Amts	0.01	1.56	–	0.48	0.28	0.13	0.026	135	160	20.5
AlMg3 (AA 5754)	3.65	0.47	–	0.77	0.25	0.04	0.066	115	230	28.0
AlMg5 (AA 5056A)	5.06	0.52	0.02 – 0.20 V	0.20	0.0	0.03	–	166	325	22.0
AlMg6	6.35	0.64	0.06 Ti	0.37	0.2	0.10	0.03	175	335	25.0
AlMg6	5.88	0.54	0.05 Ti	0.17	–	–	0.05	225	365	26.0

Table 7: Chemical composition and mechanical properties of some Al alloys [69]

Alloy	Medium	Alloy				Weld joint	
		Corrosion rate mm/a (mpy)	Yield strength 0.2 offset N/mm^2	Tensile strength N/mm^2	Elongation at break %	Corrosion rate mm/a (mpy)	Tensile strength N/mm^2
Amts	acid	0.00065 (0.03)	135	160	24.5	0.0009 (0.04)	130
	gas phase	–	–	–	–	0.055 (2.17)	125
	w.a.w.*	–	–	–	–	0.053 (2.09)	120
AlMg3	acid	0.0011 (0.04)	110	215	26.0	0.0010 (0.04)	215
	gas phase	0.032 (1.26)	110	230	26.0	0.035 (1.38)	245
	w.a.w.*	0.051 (2.09)	110	220	26.0	0.053 (2.09)	215
AlMg5	acid	0.0036 (0.14)	160	300	22.5	0.00075 (0.03)	120
	gas phase	0.028 (1.10)	155	320	22.0	0.026 (1.02)	–
	w.a.w.*	0.053 (2.09)	165	305	23.5	0.064 (2.52)	–

*with alternate wetting

Table 8: Corrosion behavior and mechanical properties of the Al alloys in Table 7 in 98 to 99 % HNO_3 at room temperature [69]

Table 8: Continued

Alloy	Medium	Alloy				Weld joint	
		Corrosion rate mm/a (mpy)	Yield strength 0.2 offset N/mm²	Tensile strength N/mm²	Elongation at break %	Corrosion rate mm/a (mpy)	Tensile strength N/mm²
AlMg6	acid	0.0013 (0.05)	180	350	28.0	0.002 (0.08)	280
	gas phase	0.043 (1.69)	170	355	26.0	0.043 (1.69)	295
	w.a.w.*	0.060 (2.36)	170	335	25.0	0.068 (2.68)	260

*with alternate wetting

Table 8: Corrosion behavior and mechanical properties of the Al alloys in Table 7 in 98 to 99% HNO_3 at room temperature [69]

Al Alloy	Registration Aluminium Association		Old DIN-Mat.-No.	Corrosion rate mm/a (mpy)
AlMg1	AA 5005	EN AW-5005A	3.3315	0.07 (2.76)
AlMgSi1	AA 6082	EN AW-6082	3.2315	0.08 (3.15)
AlMn				0.085 (3.35)
AlMg2Mn0.8	AA 5049	EN AW-5049	3.3527	0.09 (3.35)
AlMn1Mg0.5	AA 3005	EN AW-3005	3.0525	0.1 (3.94)
AlMgSi0.5	AA 6060	EN AW-6060	3.3207	0.11 (4.33)
AlMg2.7Mn	AA 5454	EN AW-5454	3.3537	0.115 (4.53)
AlMg2.5	AA 5052	EN AW-5052	3.3523	0.15 (5.91)

Table 9: Corrosion of the AlMn alloys in acid mixtures at 305 K (32 °C), test duration 48 h [70]

Figure 8 shows the course of the material consumption per unit area of the alloy AlMg2Mn0.8 (AA 5049, Al-1.94Mg-0.99Mn-0.27Si-0.48Fe-0.09Cu-<0.01Zn-0.01Ti) in highly concentrated nitric acid (97.5 to 98.5% + 1.3 – 2.25% NO_2) at 278 and 303 K (5 and 30 °C) as a function of time [109]. As a result of changes in the starting concentration of 98.5% HNO_3 + 1.3% NO_2 to 95.9% HNO_3 + 3.2% NO_2, the material consumption rate at 303 K (30 °C) after 36 d increased from 0.012 (0.038 mm/a) to 0.016 g/m² h (0.051 mm/a). The corresponding material consumption rate at 278 K (5 °C) after 36 d was 0.00278 g/m² h (0.0089 mm/a). The ratio of the material consumption of Al/Mg or Al/Mn depends on the test duration [109].

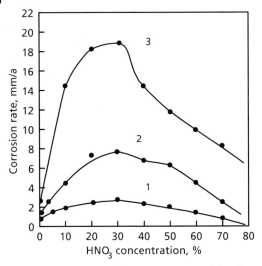

Figure 7: Dependence of the corrosion rate of the AlMn alloy AA 3003 on the HNO$_3$ concentration at 1) 298 (25), 2) 308 (35) and 3) 318 (45) K (°C) [11]

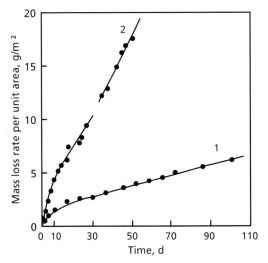

Figure 8: Course of the material consumption per unit area of AlMg2Mn0.8 in 97.5 to 98.5 % HNO$_3$ + 1.3 to 2.25 % NO$_2$ at 1) 278 K (5 °C) and 2) 303 K (30 °C) as a function of time [109]

The aluminium-magnesium-manganese alloy AlMg2Mn0.8 (AA 5049, Al-1.94Mg-0.99Mn-0.48Fe-0.27Si-0.09Cu-0.01Ti-<0.01-Zn) corroded in highly concentrated nitric acid (96.4 % HNO$_3$ + 2.5 % NO$_2$) at 303 K (30 °C) in two linear sections with different gradients, i.e. up to 36 d at 0.012 g/m^2 h and above this up to 50 d at 0.016 g/m^2 h. At 278 K (5 °C), the material consumption rate remained constant at 0.0023 g/m^2 h for up to 110 h with no change in the course with respect to time [561].

Aluminium alloys of the type Al-4.5Zn-(1.3-2.5)Mg are attacked in Keller's solution (10 ml 46% HF + 15 ml 100% HNO_3 + 10 ml 38% HCl + 25 ml H_2O) and suffer pitting corrosion. The pits tend to form at places where three grain boundaries meet. The pitting sites become shallower on alloying with 0.3% chromium or 0.13% copper. The susceptibility to pitting corresponds to that to stress corrosion cracking [32].

Table 10 shows the corrosion behavior of the AlMn alloy Al-3s (Al-0.6Si-1.0Mn-0.75Fe-0.15Cu) in acid mixtures of nitric acid and hydrochloric acid, sulfuric acid and phosphoric acid at 305 K (32 °C). The severe attack by 0.9 mol/l hydrochloric acid is reduced drastically by addition of 0.1 mol/l HNO_3. The attack by HNO_3/H_2SO_4 mixtures corresponds to that of pure acids and is approximately the same [15].

Acid mixture mol/l	HCl + HNO_3		H_2SO_4 + HNO_3		H_3PO_4 + HNO_3	
	Material consumption rate, g/m² h (corrosion rate, mm/a)					
1.0 + 0.0	13.4	(43)	0.104	(0.33)	0.567	(1.84)
0.9 + 0.1	0.53	(1.7)	0.109	(0.35)	0.501	(1.68)
0.8 + 0.2	0.15	(0.49)	0.112	(0.36)	0.431	(1.40)
0.5 + 0.5	0.119	(0.38)	0.111	(0.36)	0.307	(0.98)
0.2 + 0.8	0.113	(0.36)	0.109	(0.35)	0.155	(0.50)
0.0 + 1.0	0.107	(0.34)	0.107	(0.34)	0.107	(0.34)

Table 10: Corrosion of the AlMn alloy in acid mixtures at 305 K (32 °C), test duration 48 h [15] nitric and phosphoric acid

Structure studies were carried out in the heat-affected zone on an AlZnMg1 (cf. AA 7020) alloy welded with S-AlMg5 by the TIG method with 20% HNO_3. This zone ("white zone") is at risk from intercrystalline corrosion [77].

The aluminium alloy Al-7075–T73, which is otherwise insensitive to stress corrosion cracking in nitric acid at room temperature, suffers from the latter in certain rocket fuels, such as, for example, nitrogen tetroxide (N_2O_4) and fuming nitric acid between 344 and 347 K (71 and 74 °C) [73].

The influence of internal stress on the corrosion of an AlZnMg alloy specimen (Zn/Mg ≈ 0.5 and 2.0) was determined in a solution of 0.5 mol/l NaCl + H_2O_2 + HNO_3 by acoustic emission. It is possible in this way to detect the additional effect of internal stress during corrosion and to prevent its effect by suitable heat treatment [68].

Both intercrystalline corrosion and stress corrosion cracking were detected by electron microscopy studies on the following AlZnMg alloys in Keller's solution (10 ml HF (46%) + 25 ml HNO_3 (100%) + 15 ml HCl (38%) + 45 ml water) at room temperature: (Al-4.57Zn-1.35Mg), (Al-4.63Zn-2.46Mg), (Al-4.86Zn-1.44Mg-0.3Cr) and (Al-4.43Zn-1.34Mg-0.13Cu). The sensitivity to stress corrosion cracking runs parallel with the occurrence of triple-point pitting corrosion [105].

Stress corrosion cracking on Al4Cu crystals in a solution of 5.7% NaCl + 0.3% HNO_3 at room temperature occurs preferentially on the (100) surface. The number of cracks is independent of the tensile stress and temperature. The fracture length increases linearly with the load [71]. The dissolution along the (100) surface takes place at any anodic potential, the time to fracture decreasing exponentially as the potential increases. Different crack orientations result above and below the yield point [72].

A solution of 5 to 10 % HNO_3 + 6 g/l NaF at 353 K (80 °C) has proved suitable for testing the susceptibility of aluminium alloys (e.g. A-5, Russian grade) to intercrystalline corrosion on site in machine factories. A 50 and 98 % nitric acid solution is used only in exceptional cases [139].

The intercrystalline corrosion observed at weld joints of rolled, low-alloy aluminium A-5, A-7, A-8 and A-85 (Al-(0.066-0.220)Fe-(0.005-0.020)Cu-(0.014-0.050)Zn-(0.005-0.030)Ti-(0.060-0.18)Si) in 98 % nitric acid and 5 % HNO_3 + 6 g/l sodium fluoride at 353 K (80 °C) is based on $FeCl_3$ precipitates at the grain boundaries. The iron content should accordingly be kept as low as possible and the $FeAl_3$ should be uniformly distributed in the welding material [135].

Table 11 shows the corrosion of the alloy (Al-(1-1.5)Mn-0.15Cu-0.75Fe-0.6Si) in 1 mol/l nitric acid at 303 K (30 °C) with the inhibitors diphenylthiourea and dithioglycolic acid. The maximum inhibition efficiency of the second inhibitor of 78 % is achieved at only 10 ppm [76].

Inhibitor	Concentration ppm	Material consumption rate $g/m^2 h$	Corrosion rate mm/a (mpy)	Inhibition efficiency %
Without	–	1.15	3.7 (165)	–
Diphenylthiourea	10	0.419	1.3 (51.2)	64
	40	0.361	1.2 (47.2)	69
	100	0.337	1.1 (43.3)	71
	200	0.326	1.0 (39.4)	72
Dithioglycolic acid	10	0.262	0.84 (33.1)	78
	40	0.255	0.81 (31.9)	79
	100	0.279	0.89 (35.0)	76
	200	0.299	0.96 (37.8)	75

Table 11: Corrosion and inhibition of the Al alloy (Al-(1-1.5)Mn-0.15Cu-0.75Fe-0.6Si) in 1 mol/l HNO_3 at 303 K (30 °C) from 24 h tests [76]

Inhibition of corrosion on the aluminium alloy AA 3003 (Al-0.16Si-0.58Fe-0.78Mn) in 20 % nitric acid at 318 K (45 °C) by additions of 0.01 to 1.0 % benzoic acid and its derivatives (trihydroxy-, dihydroxy-, p-hydroxy-, o-nitrohydroxy-, m-nitrohydroxy- and p-nitrohydroxy-benzoic acid) was unsatisfactory, with inhibition values between 10 and 60 % resulting [23].

A comparably unsatisfactory inhibition on AA 3003 in 20 % nitric acid with a maximum of 70 % was obtained with 0.1 % isatin, thiosemicarbamide and isatin-3-(3-thiosemi-carbazone) at 303, 313 and 323 K (30, 40 and 50 °C) [75].

Through attack of dilute nitric acid on aluminium, an etching agent was developed for the Al12Si casting alloy for visualizing the silicon regions, which consists of a mixture of 400 ml ethanol + 20 ml H_2O + 50 ml HNO_3 + 7.5 g ammonium molybdate. After a short etching time, an MoO_2 layer is applied selectively to the silicon phase and the structure which segregates is thereby visualized [78].

The following solution was used at 293 K (20 °C) for dislocation etching of an A120Zn alloy (with Si < 0.003 + 0.0021 Cu + 0.0014 Fe) after high-temperature creep: 1 ml HNO_3 + 1 ml HF + 1 ml HCl + 100 ml water [107].

The development of the lamellar structure of eutectic aluminium-silicon alloys (Silumin®, Al-14Si) caused by addition of a mixture of NaF + NaCl at 1123 K (850 °C) is characterized by needle-shaped precipitation of silicon crystals. A pickling solution of, for example, (percent by volume) 1 % HF + 4 % HNO_3 + 1 % CH_3COOH is used to visualize this structure [49].

The alloys AA 5083 (Al-0.02Cu-0.16Si-0.30Fe-4.18Mg-0.75Mn-0.02Zn-0.024Ti-0.06Cr) and AA 7004 (Al-0.03Cu-0.17Si-0.23Fe-1.55Mg-0.49Mn-4.45Zn-0.02Ti-0.13Zr) were treated with a solution of 26 % HNO_3 + 6 % HF + 68 % HCl to investigate the grain boundaries. A solution of 475 ml methanol + 8 ml nitric acid + 21 ml perchloric acid and a cathode of chromium-nickel steel was used for electropolishing at 14 to 21 V and 0.8 to 1.2 A/cm^2 for detection of the microstructure [82].

Dipping for 30 minutes in concentrated nitric acid containing chromium trioxide (HNO_3 + 5 % CrO_3) has proved to be most suitable for removing the oxide layers on aluminium alloys, since with rapid removal of the oxide layer the alloy is attacked least by this procedure. The material consumption of a surface-oxidized specimen of LM4 (18 cm^2) (Al-5.15Si-3.0Cu-0.68Fe-0.16Ni-0.46Sn-0.36Mn-0.09Pb-0.06Sn-0.13Ti-0.01Mg) as a result of detachment of the oxide layer is therefore 14 mg after one

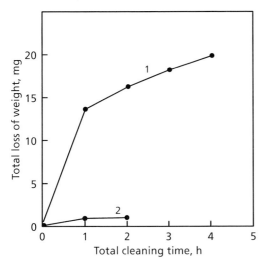

Figure 9: Influence of the treatment time (cleaning of the surface) of the LM4 specimen (18 cm^2) in HNO_3-5 % CrO_3 solution at room temperature in 1) the surface-oxidized state and 2) after removal of the oxide layer [65]

hour, while the corrosion of the worked (practically oxide-free) specimen results in a weight loss of only about 0.1 mg after dipping in the HNO_3-CrO_3 solution (Figure 9) [65].

The following bath is recommended for polishing aluminium and aluminium-magnesium alloys at 333 (60) to 343 (70) K (°C): 870 ml H_3PO_4 (85 %) + 65 ml HNO_3 (54 %) + 65 ml CH_3COOH (99 %) + 1 g/l $Cu(NO_3)_2$ [106].

The adhesion of a paint film based on polyurethane or acrylic resin to aluminium alloys is independent of the cleaning, whether H_2SO_4 + HNO_3 + HF or other acid mixtures have been used. To avoid over-etching, the alloy is sprayed and not dipped [108].

A nitric acid/hydrofluoric acid mixture is suitable for removal of hydrated oxide layers on aluminium and aluminium alloys, with simultaneous polishing, in the case of alloys containing copper and silicon. HNO_3 prevents copper deposits which impair the polish and HF prevents the formation of SiO_2, which is possible in oxidizing acids. A mixture of 3 parts 100 % HNO_3 with 1 part 20 % HF is recommended at room temperature, and 3 parts 70 % HNO_3 with 1 part 20 % HF or 3 parts 80 % HNO_3 with 1 part 10 % HF at somewhat elevated temperatures [142]. Zinc and magnesium in the alloys may make polishing difficult due to local cell formation.

For polishing aluminium-copper-magnesium alloys for the purpose of structure analysis under an electron microscope, the alloy is polarized anodically, after pretreatment in 60 % sodium hydroxide solution, in a solution of 40 % glacial acetic acid + 30 % conc. H_3PO_4 + 20 % conc. HNO_3 + 10 % distilled water at 263 K (–10 °C) and 6 V with a current density of 0.25 A/cm^2 for about 5 to 10 min [130].

A solution of 47 % HNO_3 + 50 % HCl + 3 % HF is recommended for electrolytic polishing of the three-layer system AlMg6–AD1-AlMg6 (Al-5.86Mg-0.45Mn/Al-Al-5.86Mg-0.45Mn) [141].

Treatment for 3 min at room temperature in a solution of 41 % HNO_3 + 25 % HCl is proposed for sealing surface-oxidized aluminium alloys [112]. A solution of 2 mol/l HCl + 1 mol/l HNO_3 + 5 g/l $CuCl_2$ is described as a suitable etching agent for the alloy Al-2.4Cu-1.5Mg-1.2Ni-1.0Fe [113].

Steel	Protective layer	Material consumption rate, g/m² h	Optimized coating according to Box-Wilson
C steel	–	72	–
	CrAlTi	15	0.57
	CrAl	2	0.8
U8	–	108	–
	CrAlTi	3	0.8
	CrAl	12	< 0.5

Table 12: Corrosion of C steel and U8 with and without protective Al diffusion layers in 10 % HNO_3 at room temperature, test duration up to 50 h [140]

Diffusion layers of CrAlTi and CrAl (produced in 4 h at 1373 K (1100 °C) in 40 % Al_2O_3 + 60 % (45 % Al + 55 % (70 % Cr_2O_3 + 30 % TiO_2)) and 30 % Al_2O_3 + 65 % (70 % Cr_2O_3 + 30 % Al) + 5 % AlF_3) are used for the purpose of reducing corrosion of carbon steel (C steel) and the low-alloy steel U8 (Russian grade) in 10 % nitric acid (see Table 12) [140].

A 4 Gold and gold alloys

Gold is resistant to nitric acid over the entire concentration range, even at higher temperatures. Silver and palladium with gold contents of about 40 % are dissolved only slowly by nitric acid. Some 14 carat gold alloys, such as, for example, Au-20Ag-21.5Cu, are attacked by aqua regia, but not dissolved, since the alloy surface is covered with a dense protective layer of AgCl which blocks access to the alloy by the acid [9].

Selective attack on the silver in gold-silver alloys can be expected.

A micromorphological study of gold-silver alloys (50:50) in 50 % nitric acid showed an epitactic relationship with the alloy substrate during growth of the gold islands. Heat treatment for a period of minutes at 723 K (450 °C) caused the formation of gold islands to disappear in favor of the formation of a thin homogeneous gold film. Treatment in 100 % nitric acid produced a homogeneous gold film without heat treatment [9].

A hard gold layer (Au-0.2Co) is resistant to nitric acid vapors [143]. In Figure 10, the gold content dependence of the life of gold-copper alloys, before cracking occurs under loading with 80 % of the tensile strength in nitric acid (d = 1.4 g/ml), is remarkable [83].

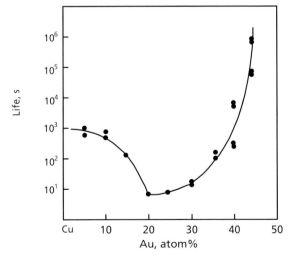

Figure 10: Dependence of the life of stress-corroded copper-gold alloys on the Au content under load (80 % of the tensile strength) in 1.4 g/ml HNO_3 at room temperature [83]

Layers of gold on chromium-nickel steel with an intermediate layer of nickel flake off after detachment of the nickel layer in tributyl-containing nitric acid with and without additions of carboxylic acid. The gold is not attacked here [114].

A 5 Cobalt and cobalt alloys

Cobalt corrodes in nitric acid in a similar manner to nickel. On addition of 11 to 12% chromium, the corrosion of the alloy in 1 mol/l nitric acid at 298 K (25 °C) becomes negligibly low. Table 13 summarizes the corrosion rates of some familiar cobalt alloys, the composition of which is given in Table 14, in boiling 10% and 65 to 70% nitric acid [117].

HNO_3 concentration %	Duratherm® 600	Havar®	Elgiloy®	Haynes® alloy 188	Haynes® alloy 6B	Stellite® alloy 25	MP35N
			Corrosion rate, mm/a (mpy)				
10	1.7 (66.93)	0.038 (1.5)	–	0.015 (0.59)	0.025 (0.98)	0.020 (0.79)	0.030 (1.18)
65 – 70	3.6 (141.73)	0.43 (17)	0.60 (24)	0.55 (22)	> 25 (984)	1.3 (51)	0.93 (37)

Table 13: Corrosion rates of some cobalt alloys in boiling HNO_3 [117]

Alloy	Chemical composition, %						
	Mo	W	Ni	Co	Fe	C	Cr
Duratherm® 600	4	4	26	40	14	–	12
Havar®	2	2.8	13	42.5	19.5	0.2	20
Elgiloy®	7	–	15	40	17.9	0.15	20
Haynes® alloy 25	–	15	10	51.9	3	0.1	20
Haynes® alloy 188	–	14	22	41.9	3	0.1	22
Stellite® alloy 6 B	1.5	4.5	3	56.8	3	1.2	30
MP35N	10	–	35	35	–	–	20

Table 14: Composition of the alloys tested in Table 13 [117]

In contrast to iron and nickel, cobalt is difficult to passivate. In an open circuit for example, the current density and corrosion rate of cobalt in 24 mol/l nitric acid (fuming HNO_3) are about 1 A/cm^2 and 1.35 mm/h respectively [122].

The behavior of cobalt alloys as regards stress corrosion cracking in boiling 70% nitric acid can be seen from Table 15.

Alloy	Pretreatment, state	Results of study
Havar®	mill annealed	general dissolution, no IC
Havar®	quenched, 30% cold reduction, aged at 750 K (477 °C)	intergranular corrosion (IC)
MP35N	annealed or 30% cold reduction	intergranular corrosion
Alloy C 276*	annealed or 30 – 50% cold reduction	general dissolution, no IC
Alloy 25	annealed or 30% cold reduction	neither dissolution nor IC detectable
Alloy 21**	cast or wrought, annealed or 30% cold reduction	general dissolution, no IC

* Ni-16Cr-16Mo-4W-5Fe
** Vitallium®, Co-27Cr-6Mo-2.5Ni

Table 15: Behavior of cobalt alloys and a nickel alloy towards stress corrosion cracking in boiling 70% HNO_3 [117]

Stress corrosion cracking in concentrated nitric acid can be largely prevented by the correct heat treatment.

The cobalt alloy Duratherm® (Co-26Ni-12Cr), a hardenable, resilient material, corrodes in the hardened state at 293 K (20 °C) with a rate of 0.1 mm/a (3.94 mpy) in 10% HNO_3 and 0.1 to 1.0 mm/a (4 to 40 mpy) in 65% HNO_3. The corrosion rates for 10 and 65% HNO_3 at 373 K (100 °C) are 1.0 to 10.0 mm/a (40 to 400 mpy) [143]. After preliminary pickling in 20 to 50% by volume sulphuric acid at 343 K (70 °C) and thorough rinsing with water, the main pickling is carried out in 25% HNO_3 + 10% HF + 65% water.

According to Figure 11, the current density/potential measurements on amorphous cobalt-zirconium-chromium alloys (Co-xCr-10Zr or $Co_{90-x}Cr_xZr_{10}$) in 1 mol/l

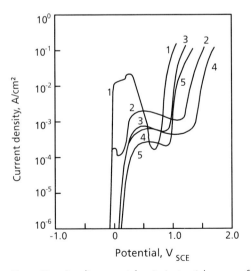

Figure 11: Anodic current density/potential curves of $Co_{90-x}Cr_xZr_{10}$ at room temperature as a function of x and the chromium content in 1 mol/l HNO_3 [145] x ≙ 1) 0. 2) 5, 3) 15, 4) 20, 5) 25

nitric acid show a decrease in the current density and corrosion as the chromium content increases. A current density of $< 10^{-6}$ A/cm^2, i.e. a material consumption rate of < 0.02 g/m^2 h, is observed at a potential around 0 V$_{SCE}$ [145].

The cobalt-chromium-molybdenum alloy Vitallium® (Co-28Cr-6Mo-Mn-Ni-Fe-<2W) used in orthopaedics, like chromium-nickel-molybdenum steels, preferentially forms (before cobalt oxide and molybdenum oxide) compact thin Cr$_2$O$_3$ layers which protect the alloy from corrosion [149].

The corrosion behavior of the complex Russian cobalt casting alloy 40KChNM (Co-(0.1-20)Cr-(14.3-15.3)Ni-(6.9-7.4)Mo-(1.9-2.0)Mn-(0.3-04)Si), quenched from 1370 K (1097 °C), deformed by 70 % and tempered at 823 K (550 °C), in 67 % nitric acid is independent of the smelting process (induction, vacuum melt or electroslag) [121].

The cobalt-platinum alloy Co-50Pt is resistant to nitric acid. Its wear resistance is also good [125].

A solution of 8 % HNO$_3$ + (1.5 to 2) % HF or Na-, K- or NH$_4$HF$_2$ containing a wetting agent (e.g. ethylene oxide condensate or an ethoxylate of a primary, linear alcohol) is recommended for detaching aluminide coatings from cobalt-chromium alloys. After an action time of this solution of 1 h at 343 K (70 °C) on a Co-(19-21)Cr alloy, the material consumption per unit area was < 0.06 mg/cm^2, which corresponds to a material consumption rate of < 0.6 g/m^2 h [120].

A 6 Chromium and chromium alloys

The passive state of chromium in 15 % nitric acid at 368 or 423 K (95 or 150 °C) lies between -0.05 and $+1.25$ V or $+0.25$ and $+1.1$ V$_{SHE}$ respectively. The temperature dependence of the material consumption rate of chromium in nitric acid can be seen from Table 16 [151]. An addition of 5 or 20 g/l potassium bichromate (K$_2$Cr$_2$O$_7$) increases the material consumption rate in 60 % nitric acid at 423 K (150 °C) to 3.5 and 30 g/m^2 h respectively.

Temperature K (°C)	HNO$_3$ concentration %	Material consumption rate g/m^2 h
373 (100)	15	approx. 0.015
408 (135)		approx. 0.018
423 (150)		0.10
373 (100)	60	approx. 0.07
408 (135)		0.35
423 (150)		1.3

Table 16: Material consumption rate of chromium in HNO$_3$ between 373 K (100 °C) and 423 K (150 °C), 4 h tests [151]

The material consumption rate of a chromium-nickel alloy at 353 K (80 °C) in a solution of 12 mol/l HNO_3 + 0.12 mol/l HF after 200 h was about 0.9 g/m² h. This value was achieved only after a minimum duration of about 50 h. Before that, it was significantly higher, for example approximately 6 g/m² h after 5 h [152].

The material consumption rate of chromium and an alloy with 1% lanthanum in dilute nitric acid (1 and 3%) in the presence of 0.7 or 2.2 g/l sodium sulfate under polarization with – 0.3 V_{SHE} was < 0.1 g/m² h. Once the passive state had been reached, the chromium alloyed with lanthanum corroded at a material consumption rate of ≤ 0.01 g/m² h [150].

Among the acid etching solutions used for thin layers of chromium, the following is recommended: (55 to 100) g/l $Ce(SO_4)_2$ 4 H_2O + 160 g/l concentrated HNO_3 + (0 to 7) g/l concentrated H_2SO_4. It operates between 298 (25) and 323 K (50 °C) with an etching rate of 80 to 250 nm/min [153]. The following etching solution is used between 298 (25) and 303 K (30 °C): 30 g/l $(NH_4)_4Ce(SO_4)_4$ 2 H_2O + 220 g/l concentrated HNO_3.

Diffusion layers of chromium and chromium-titanium on carbon steel and the low-alloy steel U8 greatly reduce the material consumption rate in 10% nitric acid (see Table 17) [140].

Steel	Protective layer	Material consumption rate, g/m² h	
		Before optimization < 50 h action	After optimization > 50 h action
C steel	–	72	–
	Cr	9	3
	CrTi	11	0.6
U8	–	108	–
	Cr	4	2
	CrTi	8	0.8

Table 17: Corrosion of C steel and U8 in 10% HNO_3 with and without a diffusion layer of Cr and CrTi [140]

Particles of halogenated chromium dispersed in gas can be applied to objects produced by powder metallurgy and consisting essentially of iron for the purpose of generating chromium layers which provide protection from corrosion in 32% nitric acid [154]. The diffusion-chromed and heat-treated steels 40 K (-233 °C) (Fe-(0.37-0.45)C-(0.8-1.1)Cr), ChVG (Fe-(0.9-1.05)C-(0.8-1.1)Mn-(0.9-1.2)Cr-(1.2-1.6) W) and St 35 (Fe-(0.32-0.40)C) have an exceptional resistance to 5% nitric acid at room temperature (≤ 0.001 mm/a (0.04 mpy)) [153].

The good protective action of chromium diffusion layers generated on the carbon steels 08KP and 09G2 (Russian grades, see A 14) in vacuum can be seen from Table 18. The resistance of the diffused chromium layer is comparable to that of the chromium-nickel steel Ch18N10T (cf. 1.4541, AISI 321 ; Fe-<0.2C-18Cr-10Ni, Ti-stabilized) [156].

Nitric Acid

Corrosive medium %	Test duration h	Temperature K (°C)	Cr-content of the protective layer, %				Ch18N10T
			20–22	30–32	38–41	60	
			Corrosion rate, mm/a (mpy)				
2–5 HNO$_3$	2100	303–313 (30–40)	0.008 (0.31)	0.00025 (0.01)	0.00029 (0.01)	–	0.0004 (0.02)
30–40 HNO$_3$	2100		0.0017 (0.07)	0.0009 (0.004)	0.00065 (0.03)	–	0.0008 (0.03)
Gases: HNO$_3$ + H$_2$SO$_4$ + fluorine compounds*	1032	368–373 (95–100)	–	0.064** (2.52)	0.045 (1.77)	0.023 (0.91)	0.116 (4.57)

*gases from fertilizer reactor
** 35 % Cr content

Table 18: Corrosion of welded samples of the chromized C steels 08KP and 09G2 in HNO$_3$ and St-3 in HNO$_3$ containing gases [156]

The amorphous chromium-boron alloy containing 20 atomic percent boron is resistant to nitric acid-hydrochloric acid solutions (1:1) at room temperature [458].

A 7 Copper

Copper is not resistant in nitric acid. It is attacked by 15 mg/l HNO$_3$ at 288 K (15 °C) with a material consumption rate of 20 g/m^2 h after 1 h and 160 g/m^2 h after 5 h. This rate could be reduced to about 27 g/m^2 h after 4 h by 2 mg/l maize extract (PH inhibitor, Russian grade, consisting of (40 – 52) % protein + (22 – 27) % soluble hydrocarbons + (1 – 3) % fat + < 0.5 % starch + (0.7 – 1.1) % lactic acid), which corresponds to an inhibition efficiency of 74 % [159]. At a concentration of 10 mg/l nitric acid, the inhibition efficiency at 2 g/m^2 h was 82 %.

The severe corrosion on copper in nitric acid can also be seen from current density/potential measurements. At a polarization voltage of + 0.23 V$_{SHE}$, the current density at room temperature could be reduced to < 10^{-3} mA/cm^2 [158], which corresponds to a corrosion rate of < 0.01 mm/a (< 0.4 mpy), if the current density was stationary.

To clarify the influence of the crystal structure of copper on its corrosion in nitric acid, dissolution tests were carried out on copper monocrystal beads in nitric acid (0.037 to 0.21 mol/mol) between 298 K (25 °C) and the boiling point [161]. The corrosion rate on the (100), (111) and (311) planes was between 1 and 10 μm/min.

The corrosion rate of copper in nitric acid was reduced in a magnetic field by 10 % at pH 2 and by 40 to 50 % at a pH 5 [168].

The inhibition efficiency of dimethylol thiourea on the corrosion of copper in 0.182 mol/l nitric acid between 293 (20) and 333 K (60 °C) at a concentration of 1 g/l was 90 and 78 %. The corrosion rate is thereby reduced from 0.093 to 0.011 mm/a (3.66 to 0.43 mpy) at 293 K (20 °C) and correspondingly from 0.30 to 0.065 mm/a (11.81 to 2.56 mpy) at 333 K (60 °C) [160].

The inhibition of copper corrosion in 1 to 5 mol/l nitric acid by amine-containing inhibitors after 0.5 and 2 h is summarized in Table 19 [166]. The inhibition efficiency is not sufficient for industrial use.

The inhibition efficiency of monosaccharides (glucose, fructose, mannose and galactose) regarding the corrosion of electrolytic copper in 0.01 to 1.0 mol/l nitric acid between 288 (15) and 308 K (35 °C) was completely inadequate even at 1 mol/l of the saccharide [167, 164].

The inhibition efficiency of 0.12, 0.5 and 1.0 g/l benzotriazole in 0.1 mol/l nitric acid at room temperature after 5 days was 61, 84 and 97 % [165]. Benzotriazole is unsuitable in corrosive solutions with a complexing or oxidizing action.

The inhibition efficiency of 0.001 mol/l tolyltriazole at 293 K (20 °C) in 0.1 mol/l nitric acid of 66 %, i.e. 0.0336 g/m² h, is greater than that of 0.001 mol/l benzotriazole (0.0494 g/m² h), based on the value without an inhibitor of 0.0991 g/m² h. Further information on the action and structure of the inhibitors can be found in a review [162].

HNO_3 concentration mol/l	Inhibitor	Inhibitor concentration %	Material consumption rate g/m^2 h	Inhibition efficiency %
1	–	–	2.41	–
	hexamine	2.0	0.45	81.0
2	–	–	65.9	–
	hexamine	2.0	8.2	87.5
	α-naphthylamine	0.20	1.87	97.2
	benzidine	0.10	0.42	99.4
4	–	–	1604	–
	hexamine	2.0	150	90.6
	α-naphthylamine	0.5	13.6	99.2
	benzidine	0.2	843	46.7
5	–	–	2335	–
	α-naphthylamine	1.0	6.6	99.7

Table 19: Corrosion of copper in HNO_3 at 308 K (35 °C) in the presence of amine-containing inhibitors [166]

Aromatic amines, such as aniline, o-toluidine, o-chloroaniline and o-phenetidine, show an inhibition efficiency of up to 99 % in 2 to 5 mol/l nitric acid, but do not reduce the material consumption rate of copper to a value which is of industrial interest. The lowest material consumption rate in 2 mol/l nitric acid of 2.21 g/m² h was obtained with 17.4 ml/l o-phenetidine [170].

Thiourea is unable to reduce the corrosion of copper in 3.5 mol/l nitric acid to values of industrial interest. Stimulation even occurs at higher concentrations, probably due to hydrolysis to diaminocarbonyl disulphide, which is reduced to thiocarbamic acid in the presence of the dissolving metal at cathodic sites [169].

According to Table 20, the inhibition efficiency of gallic acid regarding the corrosion of copper (Cu-0.0015P-0.003Cd-0.0005Zn-0.001Pb-0.0015S) in 1 mol/l nitric acid at room temperature was superior to that of tannic acid [164]. In contrast, in an acid mixture of 1 mol/l (5 ml HCl + 2 ml HNO$_3$ + 95 ml H$_2$O) the inhibition efficiency of 0.5 % tannic acid of 66 % was somewhat better than that of gallic acid of 62 %.

Inhibitor	Concentration %	Material consumption rate g/m^2 h	Inhibition efficiency %
Gallic acid	–	2.1	–
	0.1	0.45	79
	0.2	0.38	81
	0.5	0.35	82
Tannic acid	0.25	1.0	47
	0.5	0.83	60
	1.0	0.55	73

Table 20: Inhibition of the corrosion of copper in 1 mol/l HNO$_3$ by gallic and tannic acid at room temperature after 400 h [164]

The material consumption rates of copper in 0.5 and 2 mol/l nitric acid at 298 K (25 °C) of 0.468 and 46.1 g/m^2 h respectively can be reduced to 0.025 and 0.169 g/m^2 h by 2 g/l inhibitor KMA, which is based on waste products of coking plants [342].

To protect copper wires from corrosion in an HNO$_3$ containing atmosphere (0.01 %) at room temperature, they were provided with a gold layer 5 µm thick [171]. To bypass additions of HNO$_3$ as an oxidizing agent to sulfuric acid for pickling copper pipes, hydrogen peroxide is used, which reduces the loss of metal [172].

Oxide films generated by an alternating current of 50 Hz on copper in 15 g/l sodium nitrate and in 40 g/l sodium carbonate at about 330 K (57 °C) and a current density of 15, 6 or 0.5 A/dm^2 over a period of 10 min improve the corrosion behavior in solutions of 5 % acetic acid + nitric acid [173].

Before copper is coated with aluminium in an ester- and hydride-containing bath, the surface must be etched. The following mixture is recommended: (percent by volume) 58 to 60 % conc. HNO$_3$ + 2 to 3 % conc. HCl + 37 to 40 % 12 mol/l H$_2$SO$_4$ [174].

The following solution is recommended for cleaning and polishing copper at room temperature: (parts by volume) 1 HNO$_3$ (63 %) + 2H$_2$SO$_4$ (98 %) + 2 water + 1 ml/l HCl (36 %) [228].

To improve the uniform removal of copper during electro-polishing in a viscous electrolyte consisting of an aqueous HNO$_3$ containing glycerol solution at 293 K (20 °C), the following solution is recommended for increasing the solubility of the Cu(NO$_3$)$_2$ × 3H$_2$O formed for the purpose of optimum polishing conditions: (percent by volume) 75 % glycerol (d = 1.225 g/ml) + 25 HNO$_3$ (d = 1.342 g/ml) [206].

The dissolution rate of copper (99.99 %) in 0.1 and 1 mol/l nitric acid was drastically increased by additions of mercury nitrate in short-term tests of 15 min duration as shown in Table 21. This solution is used as a quick test for the susceptibility of copper alloys to stress corrosion cracking. No cracking occurs on unalloyed copper [175].

The wear resistance and corrosion resistance of copper and bronze in nitric acid were improved by diffusion layers of aluminium at 1220 (947 °C) and chromium at 1173 – 1203 K (900 – 930 °C) [775].

The action of resorcinol, o- and p-aminophenol, pyrocatechol, o-cresol and salicylaldehyde as inhibitors of copper corrosion is described in [42].

$Hg(NO_3)_2$, mol/l	0	0.001	0.01	0.1	0.4
Dissolution rate, mm/h	< 0.00001	0.00049	0.00059	0.018	0.048

Table 21: Dissolution rate of copper in 1 mol/l HNO_3 as a function of the $Hg(NO_3)_2$ content at room temperature [175]

Addition of polyvinyl alcohol, which improves chemical polishing of copper in dilute nitric acid and dilute mixtures of nitric and phosphoric acids, reduces the solubility of copper and increases the uniformity of material removal as a result of an increase in the solution viscosity. Some material consumption rates are summarized in Table 22 [144].

Electrolyte, %*	Polyvinyl alcohol, g/l				
	0	0.1	1.0	10	50
	Material consumption rate, g/m² h				
5 HNO_3 + 95 H_2O	2	0.1	0.1	0.1	0
10 HNO_3 + 90 H_2O	4	1.0	0.8	0.3	0.1
20 HNO_3 + 80 H_2O	16	32	40	52	12
40 HNO_3 + 60 H_2O	1072	1570	1683	1651	643
5 HNO_3 + 5 H_2O + 90 H_3PO_4	94	134	224	182	0.1
5 HNO_3 + 10 H_2O + 85 H_3PO_4	128	170	269	147	0.1
5 HNO_3 + 20 H_2O + 75 H_3PO_4	263	365	402	279	0.37
10 HNO_3 + 5 H_2O + 85 H_3PO_4	132	222	383	303	2.4
10 HNO_3 + 20 H_2O + 70 H_3PO_4	414	497	625	402	1.1
15 HNO_3 + 5 H_2O + 80 H_3PO_4	353	312	480	393	7.4

* percent by volume

Table 22: Corrosion of copper in HNO_3 and HNO_3/H_3PO_4 mixtures at room temperature with and without polyvinyl alcohol [144]

A 8 Copper-aluminium alloys

Some corrosion rates of copper-aluminium alloys with additions of manganese, iron, tin and nickel in 5 % nitric acid in the presence of air are summarized in Table 23. These alloys are not resistant to dilute nitric acid [176].

Alloy	Temperature K (°C)	Corrosion rate mm/a (mpy)
Cu-6.0Al-2.0Ni-1.95Sn-7.98Mn-2.5Fe	293 (20)	5.4 (213)
	343 (70)	44.3 (1744)
Cu-8.0Al-1.95Sn-8.0Mn-2.5Fe	293 (20)	18.6 (732)
	343 (70)	30.0 (1181)
Cu-10.6Al-3.1Fe-0.8Mn	293 (20)	79.9 (3146)

Table 23: Corrosion rates of some CuAl alloys in 5 % HNO_3 containing air [176]

A 9 Copper-nickel alloys

Additions of mercury nitrate increase the corrosion rate of the copper-nickel alloy (Cu-29.57Ni-0.88Mn-0.60Fe-<0.10Zn-0.005Pb-0.029P-0.036C) in 0.1 and 1 mol/l nitric acid. Some test results from short-term tests of 15 min duration are summarized in Table 24 [175].

$Hg(NO_3)_2$ concentration mol/l	0.1 mol/l HNO_3	1.0 mol/l HNO_3
	Corrosion rate, mm/h	
0	0.00037	0.00037
0.001	0.00044	0.00050
0.01	0.0025	0.0012
0.1	0.0044	0.0044
0.25	0.0051	–
0.4	–	0.013

Table 24: Dissolution rate of the CuNi alloy in HNO_3 as a function of the $Hg(NO_3)_2$ content at room temperature [175]

The corrosion resistance of copper-nickel-chromium layers at room temperature in a solution of 5 ml/l HNO_3 + 10 g/l $NaNO_3$ + 1.3 g/l NaCl was demonstrated by coulometric measurements [177].

A 10 Copper-tin alloys (bronzes)

According to Table 25, the high corrosion rate of the copper-tin alloys BrCh08 and SP19 (no composition data) in concentrated nitric acid at room temperature is drastically reduced by aluchromizing of the alloys at 1253 K (980 °C) for 6 and 9 h respectively [178].

Alloy	Duration of aluchromizing h	Test duration h	Material consumption rate g/m^2 h
BrCh08	–	0.67	3130
	6	3	67
	9	50	0.36
SP19	–	1	2330
	6	4	7.9
	9	100	0.12

Table 25: Material consumption rate of the CuSn alloys BrCh08 and SP19 in concentrated HNO_3 with and without aluchromizing to a thickness of about 4 – 7 µm [178]

A 12 Copper-zinc alloys (brass)

The Cu-37Zn alloy (Cu-36.6Zn-0.07Sn-0.01Pb) corrodes in 0.5 mol/l nitric acid at 308 K (35 °C) with a material consumption rate of 0.22 g/m^2 h. The inhibitors 2-mercaptobenzimidazole and 2-mercaptobenzothiazole, suitable for reducing the corrosion of brass in 0.5 mol/l hydrochloric acid, are unusable for corrosion in nitric acid (stimulation) [179, 315]. Brass casting alloys are not resistant to nitric acid even at room temperature [237].

The corrosion behavior of different CuZn alloys in varying concentrations of HNO_3 at various temperatures can be seen from Table 26. The inhibition efficiency of most substances is not sufficient to reduce the corrosion rates, in particular in more highly concentrated nitric acid, to values of industrial interest.

Meta-substituted aromatic amines are highly effective inhibitors of corrosion of Cu-30Zn in 2 mol/l nitric acid on the basis of current density/potential measurements (inhibition efficiency 98 to 99 %). m-Chloroaniline is also worth mentioning [192]. Of the heterocyclic amines 2-aminothiazole, 2-aminopyridine and 4-aminophenazone, only the first-mentioned is suitable (but only from 1 %) as an efficient inhibitor with Cu-30Zn. The inhibition efficiency at a material consumption rate of 0.1 g/m^2 h is > 99 % [193].

HNO$_3$ mol/l	Material	Temperature K (°C)	Inhibitor	Inhibitor Concentration mol/l	Inhibitor Efficiency %	Material consumption rate g/m^2 h	Literature
0.1[1]	Cu-30Zn	RT	none	–	–	0.0864	[185]
			triazole	0.001	–30[2]	0.112	
			benzotriazole	0.001	60	0.0351	
			naphthotriazole	0.001	30	0.0609	
0.1	Cu-40Zn	303 (30)	none	–	–	114	[186]
			chloroacetic acid	0.001	97	3.42	
			dichloroacetic acid	0.001	99.5	0.57	
			trichloroacetic acid	0.001	> 99.9	0.114	
			bromoacetic acid	0.00001	89	12.54	
0.1	Cu-40Zn	303 (30)	none	–	–	105	[190]
			diphenyl thiourea	0.0001	> 99.9	0.105	
			allyl thiourea	0.0001	< 95.2	< 1	
			phenyl thiourea	0.00001	94.5	5.6	
			dimethyl thiourea	0.00001	84	16.7	
0.5	Cu-37Zn	303 (30)	none	–	–	0.22	[189]
			p-thiocresol[3]	0.016	73	0.059	
0.5	Cu-40Zn	303 (30)	none	–	–	311	[186]
			chloroacetic acid	0.001	76	74.64	
			dichloroacetic acid	0.001	85	46.65	
			trichloroacetic acid	0.001	91	27.99	
			bromoacetic acid	0.00001	76	74.64	
0.5	Cu-36Zn	298 (25)	none	–	–	4	[342]
			KMA[4]	2 g/l	> 99.2	0.029	
1.0	Cu-40Zn	RT	none	–	–	18	[76]
			dithioglycolic acid	20 ppm[5]	99	0.18	

1) pH 1.4 to 1.7
2) stimulation
3) stimulating action in 1 mol/l HNO$_3$
4) inhibitor based on waste products from coking plants
5) higher inhibitor additions produce no improvement (≤ 200 ppm)
6) 30 min tests
7) low-copper brass

Table 26: Corrosion and inhibition of corrosion using different CuZn alloys in nitric acid

Table 26: Continued

HNO$_3$ mol/l	Material	Temperature K (°C)	Inhibitor	Inhibitor Concentration mol/l	Efficiency %	Material consumption rate g/m^2 h	Literature
1.0	Cu-40Zn	303 (30)	none	–	–	437	[186]
			chloroacetic acid	0.001	62	166.06	
			dichloroacetic acid	0.001	76	104.88	
			trichloroacetic acid	0.001	80	87.4	
			bromoacetic acid	0.00001	67	144.21	
1.5	Cu-40Zn	303 (30)	none	–	–	602	[186]
			chloroacetic acid	0.001	51	294.98	
			dichloroacetic acid	0.001	65	210.7	
			trichloroacetic acid	0.001	69	186.62	
			bromoacetic acid	0.00001	59	246.82	
2.0	Cu-30Zn	RT	none	–	–	72	[180]
			o-phenylenediamine	0.05 %	95	3.4	
				0.2 %	96	2.8	
			m-phenylenediamine	0.2 %	95	3.4	
			p-nitroaniline	0.2 %	96	2.6	
			m-toluidine	4.35 ml/l	94	4.6	
				43.5 ml/l	95	3.6	
			p-toluidine	1.0 %	96	2.6	
			m-aminobenzoic acid	0.5 %	99	0.4	
			p-aminobenzoic acid	0.05 %	99	0.4	
			o-chloroaniline	8.7 ml/l	96	2.6	
			p-chloroaniline	0.05 %	99	0.64	
				0.1 %	99.5	0.34	

1) pH 1.4 to 1.7
2) stimulation
3) stimulating action in 1 mol/l HNO$_3$
4) inhibitor based on waste products from coking plants
5) higher inhibitor additions produce no improvement (≤ 200 ppm)
6) 30 min tests
7) low-copper brass

Table 26: Corrosion and inhibition of corrosion using different CuZn alloys in nitric acid

Table 26: Continued

HNO$_3$ mol/l	Material	Temperature K (°C)	Inhibitor	Inhibitor Concentration mol/l	Efficiency %	Material consumption rate g/m^2 h	Literature
2.0[6]	Cu-30Zn	RT	none	–	–	71.6	[182]
			p-nitroaniline	0.2 %	96	2.56	
	Cu-37Zn	RT	none	–	–	63.4	[182]
			phenylhydrazine	4.35 ml/l	96.5	2.22	
			thiourea	0.2 %	98.7	0.82	
	Cu-40Zn		none	–	–	52.0	[182]
			p-anisidine	2.0 %	92	4.2	
2.0	brass[7]	RT	none	–	–	30.95	[211]
			hydrazine sulfate	0.05 %	95.8	1.3	
2.0	Cu-37Zn	RT	none	–	–	24.4	[183]
			o-toluidine	0.024	94	1.56	
			acridine	0.02	99	0.22	
2.0	Cu-37Zn	300 (27)	none	–	–	78.9	[191]
			p-chloroaniline	0.01 %	98	1.51	
				0.1	98.4	1.25	
			p-aminobenzoic acid	0.025	99.4	0.47	
				0.1	99.6	0.36	
2.0	Cu-36Zn		none	–	–	101	[342]
			KMA[4]	2 g/l	99.8	0.169	
2.0	Cu-40Zn	303 (30)	none	–	–	710	[186]
			chloroacetic acid	0.001	45	390.5	
			dichloroacetic acid	0.001	50	355	
			trichloroacetic acid	0.001	60	284	
			bromoacetic acid	0.00001	50	355	

1) pH 1.4 to 1.7
2) stimulation
3) stimulating action in 1 mol/l HNO$_3$
4) inhibitor based on waste products from coking plants
5) higher inhibitor additions produce no improvement (≤ 200 ppm)
6) 30 min tests
7) low-copper brass

Table 26: Corrosion and inhibition of corrosion using different CuZn alloys in nitric acid

Table 26: Continued

HNO₃ mol/l	Material	Temperature K (°C)	Inhibitor	Inhibitor Concentration mol/l	Efficiency %	Material consumption rate g/m² h	Literature
3.0	Cu-37Zn	300 (27)	none	–	–	1067	[191]
			p-chloroaniline	0.01 %	86.5	143	
				0.1	97.6	25.2	
			p-aminobenzoic acid	0.025	99.8	0.4	
				0.1	99.9	0.24	
4.0[6)]	Cu-30Zn	RT	none	–	–	1090	[182]
			p-nitroaniline	0.2 %	99.9	1.28	
	Cu-37Zn	RT	none	–	–	1359	[182]
			phenylhydrazine	4.35 ml/l	> 99.9	0.38	
			thiourea	0.2 %	> 99.9	0.44	
	Cu-40Zn		none	–	–	1972	[182]
			p-anisidine	2.0 %	99.9	2.78	
4.0	Cu-37Zn	308 (35)	none	–	–	2740	[189]
			N-methylaniline	17.4 ml/l	98.6	38.2	
				43.5 ml/l	99.4	15.4	

1) pH 1.4 to 1.7
2) stimulation
3) stimulating action in 1 mol/l HNO₃
4) inhibitor based on waste products from coking plants
5) higher inhibitor additions produce no improvement (≤ 200 ppm)
6) 30 min tests
7) low-copper brass

Table 26: Corrosion and inhibition of corrosion using different CuZn alloys in nitric acid

10^{-4} mol/l trichloroacetic acid can be recommended as an effective inhibitor of the corrosion of Cu-40Zn at 300 K (27 °C) in dilute nitric acid (< 0.1 mol/l) [186].

o-Phenetidine, α-naphthylamine, o-toluidine and o-anisidine are unusable as inhibitors of the corrosion of brass in nitric acid [181].

Phenylthiourea is also not effective enough as an inhibitor in HNO_3 [187].

High-alloy types of brass, such as, for example, cast brass Cu-28.3Zn-3.8Pb-0.9Sn-0.2Fe, suffer from high corrosion rates in nitric acid, e.g. about 119 mm/a (4685 mpy) in 5 % HNO_3 at 293 K (20 °C) [176].

Pipes made of the brasses Cu-28Zn-1Sn and Cu-20Zn-2Al are resistant to solutions of 0.5 mol/l NaCl + 0.02 HNO_3 at room temperature [195].

According to 1300 h tests at 408 K (135 °C), the copper alloy no. 445 (phosphorus-containing Admiralty brass (Cu-28Zn-1Sn-P)) can be used in a mixture of organic acids with nitric acid (80% propionic acid + 15.4% higher organic acids + 2.5% butyric acid + 0.1% acetic acid + 2% nitric acid). Its corrosion rate was 0.19 mm/a (7.48 mpy). Arsenic-containing Admiralty brass is destroyed [196].

The rate of dissolution of the brass types 220 (Cu-10.05Zn-0.004Pb-0.006Fe) and 260 (Cu-30.01Zn-0.02Pb-0.01Fe) in 0.1 and 1 mol/l nitric acid is greatly increased by additions of mercury nitrate of 0.1 mol/l or greater (Table 27) [175].

HNO_3 concentration mol/l	$Hg(NO_3)_2$ content, mol/l		
	None	0.01	0.1
	Corrosion rate, mm/h		
0.1	0.0011	0.0023	0.054
	(0.00039)*	(0.00059)*	(0.127)*
1.0	–	0.00089	0.021
	(0.0006)*	(0.00097)*	(0.031)*

*alloy 260

Table 27: Dissolution rate of the CuZn alloys 220 and 260 in HNO_3 at room temperature as a function of the $Hg(NO_3)_2$ content [175]

Figure 12 shows the embrittlement coefficient as a measure of the stress corrosion cracking. The latter is defined as the ratio of the fracture resistances of specimens after pre-exposure in the corrosive environment to that without pre-exposure. According to this figure, the decrease in the embrittlement coefficient with the duration of exposure on α-brass specimens along the direction of rolling is greater in comparison with specimens perpendicular to the direction of rolling [197].

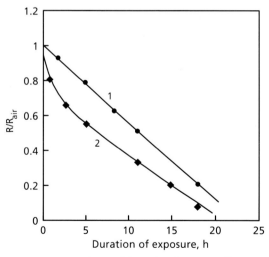

Figure 12: Dependence of the embrittlement coefficient R/R_{air} of α-brass specimens as a function of the exposure time in 4.5 mol/l HNO_3 at room temperature [197]
1) R_{air} (perpendicular to the rolling direction) 20 kJ/m^2
2) R_{air} (along the direction of rolling) 30 kJ/m^2

The dependence of the dissolution rate of 0.30 mm specimens of Cu-15Zn in nitric acid at room temperature on the degree of rolling is also remarkable. While a maximum material consumption rate of 43 g/m² h is observed at a degree of rolling of about 30%, a minimum of about 24 g/m² h occurs between 50 and 53%, followed by another maximum of about 40 g/m² h at 65% [198].

These results are understandable bearing in mind previous studies on the time to fracture carried out on brass (containing 15 atomic percent Zn) in HNO_3 containing and copper tetramine solutions under defined rolling conditions [199]. In the soft state, cracking is transcrystalline and the sensitivity to stress corrosion cracking is low.

According to Figure 13, an optimum resistance to stress corrosion cracking of loop samples of the brasses Ms-72 and Ms-80 in 0.1% nitric acid without alloying measures is achieved solely by choosing specific annealing temperatures or degrees of cold working. Cold-worked sheet metal usually shows increasing resistance to stress corrosion cracking with increasing deformation. Tempering up to about 570 K (297 °C) drastically increases the life; but only a perfectly recrystallized specimen (sheet 1 mm thick) shows the maximum values [200]. Higher copper contents in principle increase the life of the brass as regards stress corrosion cracking in 0.1 mol/l nitric acid. However, regardless of the copper content of the brass, the resistance to stress corrosion cracking can be increased by a factor of 100 or more (i.e. from a life of 50 min to one of > 2600 min) by suitable heat treatment and production measures (degree of cold rolling), that is to say without complex alloying measures [201].

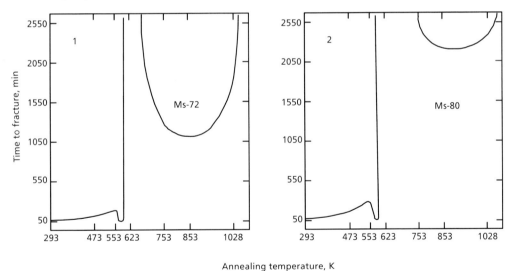

Figure 13: Time to fracture of loop samples in 0.1 mol/l HNO_3 at room temperature as a function of the annealing temperature of the brass sheets [200]

The start and course of stress corrosion cracking of a cold-drawn pipe made of α-brass cleaned by dipping in 40% nitric acid, washing and drying could be monitored by acoustic emission analysis as a function of time by means of a piezoelectric sound source [316].

Formed parts of copper and copper alloys, e.g. of brass, are polished chemically by mixed solutions of, inter alia, nitric and sulfuric acid. The treatment time can be between a few seconds and a few minutes, depending on the nature of the surface and the rate of polishing in the solution [202].

For surface treatment of aluminium brass (Cu-22Zn-2Al-0.03As), electropolishing is carried out in agitated 33 % nitric acid in methanol at room temperature and at 6 to 7 V using platinum as the cathode [236].

A 13 Other copper alloys

According to Table 28, of the phenylenediamines only o-phenylenediamine is of any interest as an inhibitor for copper-cadmium alloys in 0.1, 0.2 and 0.3 mol/l nitric acid at 303 K (30 °C) [203]. However, this inhibitor is of only limited use for industrial applications.

Alloy	Inhibitor concentration, %	Inhibition efficiency, %		
		0.1 mol/l HNO_3	0.2 mol/l HNO_3	0.3 mol/l HNO_3
Cu-6Cd	0.005	70	29	82
	0.010	85	55	89
Cu-20Cd	0.005	95	91	73
	0.010	92	94	87

Table 28: Inhibition of corrosion on CuCd alloys in HNO_3 at 303 K (30 °C) by o-phenylenediamine, test duration 18 h [203]

As has already been mentioned for the brass types, urea and thiourea cannot be used as inhibitors for reducing the corrosion rates of copper-cadmium alloys in 0.1 and 0.2 mol/l nitric acid at 303 K (30 °C) because their action is too slight [204].

The corrosion rate of amorphous and crystalline samples of copper containing 50 atomic percent Ti or Zr in 1 mol/l nitric acid at 303 K (30 °C) can be seen from Table 29. According to this table, the amorphous CuTi alloy has a higher resistance than the crystalline alloy [205].

Alloy	Structure	Corrosion rate, mm/a (mpy)
Cu-50Ti	amorphous	0.0022 (0.09)
	crystalline	0.017 (0.67)
Cu-50Zr	amorphous	0.0004 (0.02)
	crystalline	0.0004 (0.02)

Table 29: Corrosion of Cu containing 50 atomic percent Ti or Zr in 1 mol/l HNO_3 at 303 K (30 °C) [205]

According to Figure 14, the sensitivity of mono- and polycrystalline copper-gold alloys to stress corrosion cracking in 1:1 dilute aqua regia at room temperature reaches a maximum with 27 atomic percent gold. The time to fracture was determined on mono- and polycrystalline circular specimens (Ø 3 mm) [199].

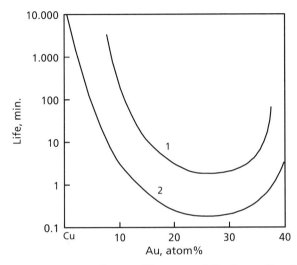

Figure 14: Time to fracture of 1) mono- and 2) polycrystalline circular specimens of CuAu alloys in 1:1 dilute aqua regia under load (50 % of the tensile strength) at room temperature [199]

According to potentiodynamic current density/potential curves, anodic dissolution of the alloys Cu-3Pd, Cu-3Y, Cu-1Pd-1Y and Cu-2Pd-2Y (additions in atomic percent) in 10 % nitric acid above + 1 V_{SHE} with 0.9 A/cm² is the same as that of unalloyed copper [815].

A 14 Unalloyed steels and cast steel

While carbon steel or low-alloy steel is severely attacked by dilute nitric acid, it is resistant in > 40 % concentrated nitric acid due to the formation of a protective passive layer [207].

If the concentration of nitric acid is increased to 7.9 mol/l, after a sharp increase in corrosion, the state of passivity is reached where the corrosion drops practically to zero. In the active range, e.g. at 0.5 mol/l HNO_3, the material consumption rate of iron can be reduced to 0.67 g/m² h by addition of about 100 g/l sulfamic acid. In current density/potential measurements, a pulsed current occurs in the range 0 to + 0.2 V_{SCE} [210].

Figure 15 shows the material consumption rate of a carbon steel in nitric acid at room temperature as a function of the HNO_3 concentration. Above 35 % HNO_3 there is a sharp drop in the material consumption rate due to the formation of a passive layer [146].

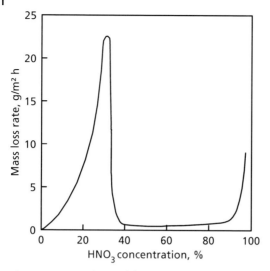

Figure 15: Dependence of the material consumption rate of C steel on the HNO_3 concentration at room temperature [146]

The corrosion of steel-10 and steel-45 in 50 % nitric acid at room temperature is increased by a factor of about 4 by 60 % deformation [858].

Dissolution of iron in 1 mol/l flowing nitric acid leads to the formation of ammonia, nitric oxide (NO) and nitrite ions (NO_2^-) in approximately equal amounts. Dinitrogen monoxide (N_2O) and nitrogen are additionally found in stationary acid [212].

Steel-20 corrodes in 1 mol/l nitric acid at room temperature at a material consumption rate of 193 g/m² h [546].

Hydrogen present both in soft iron and ferritic steel (Fe-0.15C-0.36Si-1.49Mn-0.017P-0.018S-0.02Ni-0.076Al) caused a further increase in corrosion in 0.5 mol/l nitric acid [227].

Normally killed steel ((%) Fe-0.013C-0.1Mn-0.005Al-0.002Si-0.065O) is resistant after deoxidation with aluminium only after careful heat treatment. Carbide-forming elements in the steel, such as, for example, titanium (Ti-stabilized steel containing 0.01 % C + 0.30 % Si + 0.50 % Mn) prevent cracking in boiling nitric acid. The resistance of low-alloy steels in boiling nitric acid generally depends largely on the structure [532].

Corrosion of iron cannot be prevented in 0.7 to 13.7 mol/l nitric acid at 300 and 333 K (27 and 60 °C) even by cathodic polarization between 30 and 40 A/cm². The material consumption rate of iron in 8.03 and 11.2 mol/l nitric acid at 300 K (27 °C) for example, thus increases from about 0.5 at 0 V to about 4 and 30 g/m² h respectively at $-2\ V_{SHE}$ [15].

The chromized steels 3931 (Fe-0.035C-0.54Mn-0.35Ti-1.7Ni-0.1Si) and 4238 (Fe-0.06C-0.13Mn-0.44Ti-0.23Si) corrode in 65 % nitric acid at 333 K (60 °C) with a material consumption rate of 0.0026 and 0.0052 g/m² h [218]. The material consumption rate of the steel 4240 (Fe-0.06C-0.25Mn-0.6Nb-0.23Si) is the lowest, at < 0.00002 g/m² h.

Low-alloy steels cannot be inhibited effectively against corrosion in 0.3 to 2 mol/l nitric acid at room temperature by amines (monoethyl-, monomethyl- and dimethylamine) and hydrazine [220]. Benzenesulfonic acid derivatives cannot reduce the corrosion of iron in 2 mol/l nitric acid to an industrially usable degree [221].

Additions of 0.05 to 0.5 g/l potassium salts of benzoic, salicylic, oxalic, succinic, adipic and p-nitrobenzoic acid have practically no effect on corrosion of the steel St 3 (1.0333) in 1 mol/l nitric acid at room temperature [385].

While the iodides of N-decyl- and N-dodecylquinoline greatly inhibit corrosion of St 3 in sulfuric acid at 293 K (20 °C) and 353 K (80 °C), no effect was found in 0.5 and 2 mol/l nitric acid [512].

The inhibition efficiency of some substances and the material consumption rates of unalloyed steels in various concentrations of nitric acid can be seen from Table 30.

HNO_3 concentration mol/l	Temperature K (°C)	Inhibitor	Inhibitor Concentration	Efficiency %	Material consumption rate $g/m^2\ h$	Literature
0.001[1]	303 (30)	none	–	–	0.081	[217]
		sodium petroleum sulfonate	200 ppm	36	0.052	
0.01[1]		none	–	–	0.229	[217]
		sodium petroleum sulfonate	50 ppm	27	0.167	
			200 ppm	49	0.117	
0.1[1]		none	–	–	6.25	[217]
		sodium petroleum sulfonate	50 ppm	33	4.17	
			200 ppm	83	1.04	
0.1[2]	RT	none	–	–	59	[774]
		benzylpyridine thiocyanate	0.0044 mol/l	98.8	0.7	
		allylpyridine thiocyanate	0.0056 mol/l	98.8	0.7	
		benzyl-2-methyl pyridine rhodanide	0.0041 mol/l	99.1	0.5	

1) Fe-0.35Mn-0.2Cu-0.025P-0.035S
2) Armco® iron
3) steel St 3
4) distillation residues obtained during ethanolamine purification

Table 30: Inhibition efficiency of various substances and material consumption rates of unalloyed steels in nitric acid

Table 30: Continued

HNO₃ concentration mol/l	Temperature K (°C)	Inhibitor	Inhibitor		Material consumption rate g/m² h	Literature
			Concentration	Efficiency %		
1.0	RT	none	–	–	573	[215]
		indole + thiocyanate	2.1 + 9.8 mmol/l	99.8	1.1	
			4.27 + 6.56 mmol/l	99.85	0.82	
1.0[3]	RT	none	–	–	885	[223]
		indole + sodium sulfide	0.2 + 0.8 g/l	99.7	2.5	
2.0[3]	298 (25)	none	–	–	1820	[1115]
		[4] + benzoquinoline thiocyanate (8:1)	2.0 + 0.25 mol/l	99.9	1.8	
3.0	RT	none	–	–	1820	[215]
		indole + thiocyanate	2.1 + 9.8 mmol/l	> 99.9	0.96	
			4.27 + 6.56 mmol/l	> 99.9	0.68	

1) Fe-0.35Mn-0.2Cu-0.025P-0.035S
2) Armco® iron
3) steel St 3
4) distillation residues obtained during ethanolamine purification

Table 30: Inhibition efficiency of various substances and material consumption rates of unalloyed steels in nitric acid

Benzotriazole and thiourea (0.010 mol/l) inhibit the corrosion of low-alloy steel (Fe-0.22C-0.79Mn-0.66Si-0.05S-0.03P) in 0.1 mol/l nitric acid at room temperature to the extent of 91 and 96 % [219]. Benzotriazole is considerably more effective than normal triazole or aminotriazole.

According to Table 31, the material consumption rate of Armco® iron (Fe-0.03C-0.10Mn-0.18Si-0.01P-0.015S), steel-45 (Fe-0.44C-0.23Cr-0.26Ni-0.69Mn-0.34Si-0.013P-0.02S) and the steels U7A (cf. 1.1520, C70U; Fe-0.70C-0.11Cr-0.09Ni-0.26Mn-0.28Si-0.013P-0.008S) and U10A (cf. 1.1545, AISI W1; Fe-0.98C-0.12Cr-0.20Ni-0.20Cu-0.26Mn-0.25Si-0.03P-0.02S) at 295 K (22 °C) in 1 mol/l nitric acid is reduced by hydrochloric acid. The greatest reduction occurs on Armco® iron [384, 513].

Material	HNO₃	HNO₃ + HCl		
	1 mol/l	100 : 1	10 : 1	1 : 1
	Material consumption rate, g/m² h			
Armco® iron	1050	980	770	4
Steel 45	1030	960	450	8
Steel U7A	1060	950	550	19
Steel U10A	1070	1010	95	23

Table 31: Corrosion of Armco® iron and three carbon steels in 1 mol/l HNO₃ with and without additions of 1 mol/l HCl at 295 K (22 °C) [384]

The material consumption rates of Armco® iron in nitric acid/hydrochloric acid mixtures listed in Table 32 are considerably lower. Corrosion is increased by the high additions of hydrochloric acid. A halving of the material consumption rate of iron in 0.5 mol/l nitric acid is found with 1.5 mol/l hydrochloric acid [208].

HNO₃ concentration mol/l	HCl addition mol/l	Material consumption rate g/m² h
0.5	–	18.85
0.5	1.5	9.74
0.5	3.0	26.3
0.5	6.0	97.8
1.0	–	40.35
1.0	6.0	285
2.0	–	147
2.0	6.0	641

Table 32: Corrosion of Armco® iron in HNO₃ with and without HCl at room temperature [208]

Corrosion of steel in pure hydrochloric acid (5.8 mol/l) at a rate of 0.8 g/m² h is increased to 2.1 g/m² h by additions of nitric acid (e.g. 40 percent by volume 5.8 mol/l HNO₃), and reduced again when the additions are increased further (e.g. to 0.63 g/m² h by 60 percent by volume HNO₃) [533].

According to Table 33, the corrosion of carbon steel at 323 K (50 °C) in concentrated sulfuric acid is intensified by small additions of nitric acid. At a content of 1128 ppm HNO₃, however, passivation of the steel occurs, associated with a sharp decrease in corrosion rates [297].

The corrosion rate of carbon steel at 294 K (21 °C) in 65% sulfuric acid of 0.89 mm/a (35 mpy) is increased to 10.5 and 14 mm/a (413 and 551 mpy) by additions of 2 and 2.6% HNO₃. After passing through a maximum of > 30 mm/a (1180 mpy), the corrosion rate drops to about 0.2 mm/a (7.9 mpy) on addition of 8% HNO₃. Regardless of the concentration of sulfuric acid (62 – 65), corrosion at 311 K

HNO₃ addition ppm	–	5.6	14	25	56	141	423	1128
Corrosion rate, mm/a (mpy)	0.11 (4.33)	0.23 (9.06)	0.36 (14.17)	0.47 (18.5)	0.66 (25.98)	0.78 (30.71)	1.55 (61.02)	0.034 (1.34)

Table 33: Influence of HNO₃ additions on the corrosion of C steel at 323 K (50 °C) in concentrated H₂SO₄ after 320 h [297]

(38 °C) reached a rate of < 0.03 mm/a (< 1.2 mpy) after addition of 0.5 % nitric acid after 6 d tests. The corresponding value for 70 % sulfuric acid + 0.5 % HNO₃ was 0.05 mm/a (1.97 mpy) [519].

Figure 16 shows the course of the corrosion rate of carbon steel in sulfuric acid solutions of various concentrations at 310 K (37 °C) as a function of small amounts of nitric acid. The corrosion rate is already < 0.02 mm/a (< 0.8 mpy) with 0.5 % HNO₃ [830].

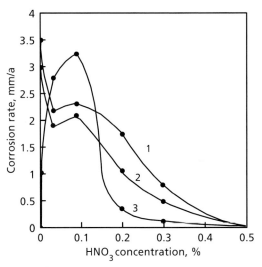

Figure 16: Corrosion rate of C steel in H₂SO₄ with small additions of HNO₃ at 310 K (37 °C) [830]
1) 62.2; 2) 64.9 and 3) 70.7 % H₂SO₄

At concentrations which are not too high (< 100 g/l HNO₃), silver nitrate causes chemical passivation of iron in nitric acid, as shown in Figure 17, whereby the material consumption rate does not exceed 0.1 g/m² h. In contrast, the corrosion of iron in 90 g/l AgNO₃ solution greatly increases with a growing concentration of nitric acid [222]. About 5, 15 and 40 g/l silver nitrate are needed in a solution containing 30, 75 and 150 g/l HNO₃ to inhibit the corrosion of iron to < 0.1 g/m² h.

An inhibitor solution of 5 – 10 % nitric acid + 15 g/l urea + 2 g/l thiourea is used to clean plants made of carbon steel and allowed to act at 293 or 313 K (20 or 40 °C) for 5 h. While the rate of dissolution is 460 or 895 g/m² h respectively without an inhibitor, it is reduced to 1.5 and 7.1 g/m² h in the presence of the inhibitor mixture (inhibition efficiency > 99 %) [516].

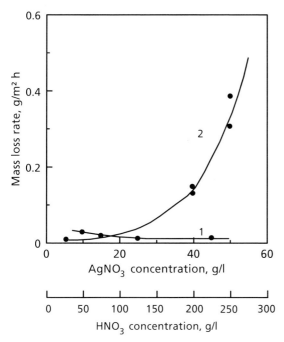

Figure 17: Dependence of the material consumption rate of passivated iron on the concentration of HNO_3 and $AgNO_3$ [222]
1) chemical passivation in 75 g/l HNO_3 at varying $AgNO_3$ concentrations
2) chemical passivation in 90 g/l $AgNO_3$ at varying HNO_3 concentrations

A 15 Unalloyed cast iron

Mixtures of 45 % H_2SO_4 + 50 % HNO_3 + 5 % H_2O attack cast iron at higher temperatures [230]. Silicon-containing cast iron, high-silicon cast iron, is superior to unalloyed cast iron in its corrosion behavior in nitric acid [229].

Cast iron tanks made of pearlitic gray cast iron with a lamellar graphite structure are attacked during regeneration of waste sulfuric acid (about 8 to 10 mm/a (314 to 394 mpy)). On the other hand, if 5 % nitric acid is added as a passivator at the start of the test, corrosion of the cast iron is reduced to about 1.5 mm/a (60 mpy) [234].

A 16 High-alloy cast iron, high-silicon cast iron

The material consumption rate of high-silicon cast iron containing 14.5 % Si in 5 to 65 % nitric acid between 293 and 373 K (20 and 100 °C) is < 0.1 g/m² h [231]. As the silicon content increases, the corrosion behavior of cast iron in nitric acid is increasingly improved [233].

According to Table 34, cast iron with a low silicon content (Fe-(3.4-3.8)C-(1.9-3.1)Si-(0.6-0.9)Mn) is severely attacked in nitric acid at 298 K (25 °C) [319].

Cast iron	HNO_3 concentration, %		
	1	5	10
	Material consumption rate, $g/m^2 \, h$		
Nodular graphite	42	285	490
Lamellar graphite	70	365	671

Table 34: Corrosion of nodular graphite cast iron containing 2.5 – 3.1 % Si and cast iron with lamellar graphite containing 1.9 – 2.4 % Si in HNO_3 at 298 K (25 °C) [319]

The corrosion rate of the gray cast iron ChS-13 with a carbon and silicon content of 0.9 and < 9 % respectively in 40 % nitric acid was > 1 mm/a (> 40 mpy), and that with an Si content of 12.3 % was 0.025 mm/a (0.98 mpy) (tempered at 1773 K (1500 °C) for 30 min before the test). Corresponding tests in a solution of 69 to 71 % sulfuric acid + 25 to 27 % nitric acid at 363 K (90 °C) resulted in a corrosion rate of 0.01 to 0.02 mm/a (0.39 to 0.79 mpy). At the same time, the improved flexural strength was 250 – 280 N/mm^2 and the hardness 100 N/mm^2 [343]. A high-silicon cast iron containing not less than 14 % silicon should be chosen for mixtures of nitric acid and sulfuric acid at the boiling point [233].

Iron-silicon alloys (Tantiron® N) containing 14.25 to 15.25 % Si are absolutely resistant to nitric acid. However, where particularly severe corrosive attack is feared, e.g. in mixtures of HNO_3 and H_2SO_4 at higher temperatures, Tantiron® E with a silicon content of 16 to 18 % is recommended [232].

Cast iron containing 15 % silicon has an adequate resistance to nitric acid containing potassium chloride at room temperature, with a corrosion rate of about 0.15 mm/a (5.9 mpy) [540].

Alloyed cast iron (Fe-3.3C-3.2Si-0.3Mn-0.3P-0.06S) is attacked by a solution of 0.3 % $FeCl_3$ + 0.1 % HCl + 0.1 % HNO_3 at room temperature with a material consumption rate of about 6.0 and 0.59 g/m^2 h after 10 and 90 d respectively. The rate decreases further after more prolonged exposure giving, for example, about 0.23 g/m^2 h after 800 d [319].

A 17 Structural steels with up to 12 % chromium

The ferritic chromium steel ((%) Fe-0.1C-9.0Cr-1Mo-0.47Mn-0.44Si-0.017P-0.017S) shows only mild general corrosion without preferential local dissolution after exposure to boiling 65 % nitric acid for 24 h. A test carried out on a weld point of the steel showed no intercrystalline corrosion [235].

The austenitic chromium-nickel steels predominantly used in the nitric acid industry can be replaced by the steel 1.4512 (AISI 409 (%) Fe-<0.08C-(10.5-12.5)Cr-<1Mn-<1.0Ti-<1.0Si-0.2Al-<0.045P-<0.030S) for moderately hot, (sub)-azeotropic nitric acid. In azeotropic nitric acid free from oxidizing agents, the steel corrodes at a material consumption rate of < 0.01 g/m² h due to the formation of protective surface layers of aluminium and chromium oxides [638].

Addition of 20 to 400 ppm chloride to 5, 23 and 56 % nitric acid has no influence on the corrosion rate of chromium steel ((%) Fe-(0.16-0.24)C-12Cr-0.6Mn-0.6Ni-0.6Si-<0.035P-<0.030S), which is about 0.79, 0.26 and 0.13 mm/a (31.1, 10.24 and 5.12 mpy) respectively at room temperature [317].

While hardened chromium steels ((%) Fe-(0.22-0.29)C-(4.1-14)Cr-(0.12-0.36)Ni-(0.25-0.5)Mn-(0.07-0.48)Si) with a chromium content of 4 to 11 % corrode in boiling 65 % nitric acid at a rate of about 9 g/m² h, this figure drops to < 0.015 g/m² h at 14 % chromium [344].

A 18 Ferritic chromium steels with more than 12 % chromium

All chromium steels with a chromium content of ≥ 14 % are resistant to all concentrations of nitric acid at room temperature [258]. Chromium steels of the types Fe-0.015(C+N)-26Cr-1Mo and Fe-0.025(C+N)-29Cr-4Mo (Ti- or Nb-stabilized) as well as Fe-0.025(C+N)-29Cr-4Mo-2Ni (Ti- or Nb-stabilized) are absolutely resistant in boiling 65 % nitric acid [256].

A chromium steel with 30 % chromium shows a material consumption rate of < 0.1 g/m² h in 5 to 65 % nitric acid in the temperature range 293 to 373 K (20 to 100 °C). At 373 K (100 °C), the corrosion rate is > 0.5 g/m² h in > 30 % nitric acid [231]. Addition of molybdenum leads to no increase in the resistance.

Table 35 lists some test results on the corrosion behavior of E-Brite® 26-1 (26Cr-1Mo) in nitric acid as the temperature rises [356].

HNO$_3$ concentration %	Temperature K (°C)	Corrosion rate mm/a (mpy)
10	394 (121)	0.012 (0.47)
	422 (149)	0.093 (3.7)
	436 (163)	0.059 (2.32)
	477 (204)	21.1 (831)
30	379 (106)	0.011 (0.43)
	422 (149)	0.151 (5.94)
	452 (179)	0.152 (5.98)
60	366 (93)	0.024 (0.94)
	393 (120)	0.097 (3.8)
	422 (149)	3.52 (139)
65	394 (121)	1.61 (63.4)
	422 (149)	18.9 (744)
70	366 (93)	0.12 (4.7)
	394 (121)	0.15 (5.9)

| | 90 | 364 (91) | 1.09 (42.9) |

Table 35: Corrosion rate of E-Brite® 26-1 in HNO₃ at various temperatures [356]

Corrosion rates of the chromium-molybdenum steel E-Brite® 26-1 ((%) Fe-0.001C-26.0Cr-1.0Mo-0.25Si-0.12(Ni+Cu)-0.01Mn-0.01P-0.01S-0.01N) in nitric acid can also be seen from Table 36. If hydrogen fluoride is present at the same time, the corrosion rate of the steel is increased considerably, rising, for example, to about 2.5 mm/a (98 mpy) in 2 mol/l HNO₃ + 0.5 mol/l HF [239]. E-Brite® 26-1 is suitable for coolers and condensers for nitric acid. The resistance of E-Brite® 26-1 to nitric acid is confirmed again and again [244].

HNO₃ concentration %	State steam/liquid	Temperature K (°C)	Corrosion rate mm/a (mpy)
10	steam	433 (160)	0.081 (3.19)
20			0.122 (4.8)
30			0.325 (12.8)
65	liquid	boiling	0.10 – 0.13 (3.94 – 5.12)
75			0.356 (14.02)
85			0.737 (29.02)
70	liquid	299 (26)	< 0.025 (0.98)
90	liquid		< 0.025 (0.98)
95	liquid, fuming		< 0.025 (0.98)
90	liquid	333 (60)	0.25 (9.84)

Table 36: Corrosion of E-Brite® 26-1 in various concentrations of HNO₃ [239]

The corrosion rate of the steel 07Ch13AG20 ((%) Fe-0.038C-14.3Cr-19.5Mn-0.7Ni-0.02Ti-0.35Si-0.12N) in 50 % nitric acid at 333 K (60 °C) is about 0.1 mm/a (4 mpy) after 300 h, regardless of the exposure time [251].

The corrosion behavior of the nickel-free or low-nickel chromium-manganese steel AS-43 ((%) Fe-0.05C-18.2Cr-16.3Mn-0.36Si-0.4Nb-0.005B-0.05Ce-0.008S-0.014P-0.83-Ni) and, for comparison, that of the austenitic chromium-nickel steel Ch18N10T (AISI 321, 1.4541) in 300 h tests in hot nitric acid is shown in Table 37. While the material consumption rates of the steels AS-43 and Ch18N10T in boiling 45 % nitric acid are 0.60 and 0.22 g/m² h after 100 h, the rates are 0.50 and 0.40 g/m² h after 500 h. After heat treatment between 873 and 1173 K (600 and 900 °C), chromium-manganese steel is not susceptible to intercrystalline corrosion [321].

The test developed by Huey with boiling 65 % nitric acid indicates intercrystalline corrosion damage due to chromium carbide precipitates and sigma phase formation in Mo-alloyed and Ti- or Ta/Nb-stabilized steels [411].

The Huey test is generally used to evaluate the resistance of stainless austenitic steels to exposure to nitric acid (65 to 67%). The specimens are exposed for 5 periods of 48 h each, the nitric acid being renewed and the weight loss being determined after each period. The steels on which the weight loss does not rise noticeably over the entire test duration are described as resistant.

Steel	HNO$_3$ concentration %	Temperature, K (°C)			
		333 (60)	363 (90)	373 (100)	Boiling point
		Material consumption rate, g/m^2 h			
AS-43	45	0.012	0.070	0.20	0.55
	65	0.028	0.180	0.55	7.0
Ch18N10T	45	0.045	0.014	0.062	0.33
	65	0.008	0.050	0.30	1.4

Table 37: Corrosion behavior of the steels AS-43 and Ch18N10T (AISI 321, 1.4541) in hot HNO$_3$ [321]

The "Superferrite" X1CrMoNb 30 2 (Japanese steel) with the composition (%) Fe-<0.004C-0.14Si-<0.02Mn-0.014P-0.012S-30.52Cr-1.75Mo-0.17Ni-0.009N-<0.02Cu-0.12Nb was investigated, inter alia, in [1] on this basis, but with an extended test duration of a total of 50 periods and with measurement of the depth of intergranular attack. The material showed increasing material consumption rates and an increase in the depth of intercrystalline corrosion both in the delivery state and after additional annealing (1123 K (850°C)/20 min/air). The material consumption rates reached 0.12 to 0.15 g/m^2 h after a test duration of 100 d. The material is not suitable for long-term exposure in hot nitric acid [1]. The "Superferrite" X1CrNiMoNb28-4-2 (25-4-4, UNS S32803, 1.4575) is also unsuitable.

The chromium steels Fe-35Cr and CrMo26-1 corrode in boiling 65% nitric acid with a corrosion rate of 0.15 and 0.20 mm/a (5.91 and 7.87 mpy) respectively [249]. No intercrystalline corrosion was found.

Vacuum-smelted, highly pure ferritic 25Cr steel ((%) Fe-(0.005-0.007)C-(24-25)Cr-(0.42-0.63)Mn-(0.04-0.35)Si-(0.014-0.015)S-(0.024-0.028)P-(0.005-0.010)N) with a low carbon content is superior to normal ferritic steels in its corrosion resistance. To keep the susceptibility to intercrystalline corrosion in boiling 65% nitric acid as low as possible, as well as the general corrosion, stabilization annealing at 1053 K (780°C) for about 1 h with subsequent quenching in water is necessary. The sensitivity to corrosion increases above 1073 K (800°C) and as a result of longer annealing times [504].

The high-chromium ferritic steel with a low carbon and nitrogen content 005Ch25B ((%) Fe-0.005C-0.007N-25Cr; Nb-stabilized), heat treated at 1573 K (1300°C) for 30 min, corroded in boiling 65% nitric acid with a rate of 0.21 mm/a (8.27 mpy) (5 cycles of 48 h each). The corrosion resistance, especially towards intercrystalline corrosion, is improved at a ratio of Nb/(C + N) 20 [421]. Addition of 0.6 to 4.7% molybdenum alters the corrosion behavior in 65% nitric acid giving corrosion rates of about 8 mm/a (315 mpy) for 0.6 % Mo, 0.4 mm/a (15.75 mpy) for 2.3 % Mo and 3.6 mm/a (142 mpy) for 4.7 % Mo.

Nitrogen-alloyed chromium-manganese steel ((%) Fe-0.04C-0.455N-19.2Cr-10.6Mn-0.6Si) corrodes in boiling 65 % nitric acid the slowest in the quenched state at 0.49 g/m^2 h, compared with steel tempered at 873 and 1173 K (600 and 900 °C). The steel tempered at 873 K (600 °C) for 20 or 2 h corrodes at a rate of 3.8 and 0.66 g/m^2 h respectively [424]. The susceptibility to intercrystalline corrosion caused in particular during longer tempering times at 873 K (600 °C) is reduced by higher tempering temperatures.

The chromium-manganese steel Ch18AG14 with 18 % chromium, 14 % manganese and 0.5 % nitrogen behaves similarly in boiling 65 % nitric acid. The lowest material consumption rate here is 1.2 g/m^2 h [359, 423].

The resistance of chromium-manganese steels of the type ((%) Fe-<0.1C-(13-21)Cr-(7-17)Mn-(0.03-0.88)N) to intercrystalline corrosion in boiling 65 % nitric acid is increased by nickel. Although the steel containing (%) Fe-18.4Cr-7.4Mn-0.3N is resistant to intercrystalline corrosion in the absence of nickel, it is attacked at a rate of about 0.7 mm/a (27.6 mpy) [254]. Replacement of the chromium by manganese makes the steel highly susceptible to intercrystalline corrosion.

The corrosion behavior of nitrogen-chromium-manganese steels ((%) Fe-0.03C-0.57N-11.0Mn-21.3Cr) with a ferrite content of 5 % in boiling 65 % nitric acid depends greatly on the temperature and duration of annealing. After an annealing time of 20 to 60 min at 873 K (600 °C), a material consumption rate of about 0.7 g/m^2 h results after 24 h, but after annealing for 120 and 480 min, values of about 3.2 and 10.7 g/m^2 h respectively occur. Annealing for 2 h at 1273 K (1000 °C) led to a material consumption rate of 0.6 g/m^2 h [831].

20CrMo steels with 3 or 5 % molybdenum and 1 to 5 % nickel show average material consumption rates of about 0.2 g/m^2 h in the Huey test. At the same time, however, mild to moderate intercrystalline attack is found, this occurring preferentially in the region of weld seams. In contrast, all 28CrMo steels with 2 % molybdenum and 2 to 4 % nickel are completely resistant to intercrystalline attack with an average material consumption rate of 0.1 g/m^2 h [507].

The titanium-stabilized chromium-molybdenum steel ELI 18-2CrMo(Ti) ((%) Fe-0.02C-17.5Cr-2.1Mo-0.23Mn-0.26Ti-<0.02(P+S)-<0.8(Ni+Cu)-0.60Si) gives a corrosion rate between 2.6 and 5.5 mm/a (102 and 217 mpy) in the welded state after exposure to boiling 65 % nitric acid for 3 periods of 48 h each. Heat treatment at 1113 K (840 °C) for 6 h reduces the figure to about 1.8 mm/a (71 mpy) [240].

The material consumption rate of pipe specimens of 18Cr2Mo steel ((%) Fe-0.018C-17.5Cr-2.52Mo-0.67Ti-0.4Sr-0.04Mn-0.009P-0.006S), produced from powder and cast ingots, in boiling 65 % nitric acid is 0.67 and 0.41 g/m^2 h respectively [577].

Molybdenum-free chromium steel ((%) Fe-(0.003-0.006)C-17.3Cr-0.003N) corrodes in boiling 65 % nitric acid, after quenching from 1323 K (1050 °C), with a material consumption rate of 0.657 g/m^2 h after 250 h. The same steel containing 3 % molybdenum gave a figure of 1.05 g/m^2 h under the same test conditions. If the two steels are tempered at 973 K (700 °C) for 15 h, the corrosion of the Mo-free steel is reduced to 0.42 g/m^2 h and that of the Mo-containing steel increased to 9.4 g/m^2 h [414].

The chromium steel Fe-<0.006C-17.3Cr, solution-annealed at 1323 K (1050 °C) for 0.5 h and then quenched in water, corroded in boiling 65 % nitric acid at a rate of 0.65 g/m^2 h in the Huey test. After additional tempering at 973 K (700 °C) for 1 h, the material consumption rate dropped to 0.44 g/m^2 h. An alloying addition of 3.08 % molybdenum to the Cr steel led to an increase in the first case to 1.2 and in the second case to about 6.0 g/m^2 h [238].

The corrosion of the chromium steels AISI 430 ((%) Fe-0.07C-17Cr-0.45Mn-0.45Si-0.04P-0.01S-0.025N) and ASTM XM-27 (E-Brite® 26-1: (%) Fe-0.002C-26Cr-1Mo-0.10Ni-0.05Mn-0.25Si-0.01P-0.01S-0.10Nb-0.01N) and of the CrMoNi steel 29-4-2 ((%) Fe-0.005C-29Cr-4Mo-2Ni-0.05Mn-0.10Si-0.015P-0.010S-0.013N) in boiling 65 % nitric acid (Huey test, ASTM A 262-C) is shown in Table 38 [241].

Steel	Cr-Content %	Corrosion rate mm/a (mpy)
AISI 430	17	0.5 – 0.9 (19.7 – 35.4)
ASTM XM-27	26	0.11 (4.3)
Steel 29-4-2	29	0.06 (2.4)

Table 38: Corrosion rates of some Cr steels in boiling 65 % HNO$_3$ [241]

Some data on the influence of heat treatment on the corrosion of the steel AISI 430 (1.4016 (%) Fe-0.079C-15.85Cr-0.63Ni-0.42Mn-0.05Cu-0.39Si-0.025N-0.015S-0.017P) and AISI 446 (cf. 1.4762 (%) Fe-0.098C-24.96Cr-0.38Ni-0.94Mn-0.05Cu-0.24Si-0.21N-0.013S-0.10P) in boiling 65 % nitric acid are shown in Table 39 [299].

Steel	Heat treatment temperature, K (°C)				
	920 (647)	1140 (867)	1255 (982)	1475 (1202)	1530 (1257)
	Material consumption rate, g/m^2 h				
AISI 430	0.54	1.50	1.10	4.58	3.75
AISI 446	0.13	0.12	0.12	0.63	0.60

Table 39: Influence of heat treatment for 1 h on the corrosion of AISI 430 and 446 in boiling 65 % HNO$_3$ after 240 h [299]

The chromium steels W 4027 ((%) Fe-0.14C-14.2Cr-1.45Mn-0.21Ni-0.64Si) and W 4059 ((%) Fe-0.19C-16.9Cr-1.19Mn-1.43Ni-0.67Si) are resistant in 1 to 67 % nitric acid after 12 h at room temperature; the corrosion rates being 0.035 and 0.017 in 1 % HNO$_3$, 0.0017 and < 0.001 in 10 % HNO$_3$, < 0.001 for both in 25 % HNO$_3$ and 0.0015 and 0.00011 mm/a (0.06 and 0.004 mpy) in 67 % HNO$_3$ [407].

A non-welded specimen of the steel 0Ch25T (Russian grade; (%) Fe-<0.01C-25Cr, Ti-stabilized) showed a material consumption rate of 2.7 g/m^2 h after 220 h at 453 K (180 °C) in 70 % nitric acid [242]. The niobium-stabilized CrMo steel 18 2 is used in preference in the processing of nitric acid because of its resistance. The same steel, but stabilized with titanium, is cheaper and has almost the same resistance [248].

The chromium-manganese steel 08Ch18G8N2T ((%) Fe-0.08C-18Cr-8Mn-2Ni; Ti-stabilized), which is resistant to sulfuric acid, is also said to be suitable for tank wagons used for transportation of dilute nitric acid and ammonium nitrate [322].

According to Table 40, the influence of the crystal structure on the corrosion of the alloy Fe-47Cr in nitric acid solutions at 338 K (65 °C) is considerable [298]. The sigma phase of the alloy thus corrodes faster than the ferrite phase by a factor of 42.

Medium	Temperature K (°C)	Ferrite phase	Sigma phase
		Material consumption rate, $g/m^2 h$	
65 % HNO_3	boiling	0.085	3.59
10 % HNO_3 + 3 % HF	338 (65)	0	0.25

Table 40: Influence of the crystal structure on the corrosion of Fe-47Cr in HNO_3 solutions [298]

To prevent intercrystalline corrosion of chromium-molybdenum steel ((%) Fe-0.012C-26.13Cr-1.1Mo-0.014N) in hot 10 % nitric acid + 3 % hydrofluoric acid at 343 K (70 °C), 0.26 to 0.30 % titanium was alloyed to the steel. A similar effect was also achieved by an addition of 0.33 – 0.44 % niobium [506].

Steels of the type (%) Fe-(0.023-0.032)C-(0.012-0.023)N-(0.23-0.48)Ti-18Cr-2Mo were exposed, in accordance with ASTM-262, Practice D, to a solution of 10 % HNO_3 + 3 % HF at 343 K (70 °C) for (2 + 2) h and the influence of the sheet thickness and the degree of stabilization Ti/(C + N) on the extent of corrosion was investigated (Table 41). While there is no relationship between general corrosion and intercrystalline corrosion, one exists between the degree of stabilization and the tendency to undergo intercrystalline corrosion [246].

Content, %				Sheet thickness, mm					
				4		2		1	
C	N	Ti	Ti/(C + N)	mm/d	μm	mm/d	μm	mm/d	μm
0.026	0.015	0.23	5.6	0.237	550	0.127	530	0.127	400
0.032	0.019	0.33	6.5	0.247	320	0.060	100	0.050	50
0.026	0.018	0.41	9.3	0.120	80	0.127	50	0.137	70
0.027	0.018	0.45	10.0	0.100	35	0.187	< 20	0.160	< 20
0.023	0.012	0.48	13.7	0.100	< 20	0.143	30	0.170	40

Table 41: Corrosion rate (mm/d) of sheets of the CrMoTi steel 18 2 with various thicknesses and different Ti, N and C contents, and the maximum depth of intercrystalline attack in μm [246]

The corrosion rate of the chromium-manganese steel AS-43 ((%) Fe-0.05C-18.4Cr-16.5Mn-1.60Ni-0.31Nb-0.26N-0.01Ce-0.004B) in hot nitric acid-hydrogen fluoride solutions passes through a minimum of 0.057 and 0.26 mm/a (2.24 and 10.24 mpy) at about 5 mol/l HNO_3 + 0.01 or 0.1 HF respectively. The corresponding values for 1 and 12 mol/l nitric acid are higher, 0.18 or 1.8 respectively and 0.14 or 1.23 mm/a (5.5 or 48.4 mpy) respectively [413].

The chromium-manganese steels 07Ch17G17DAMB ((%) Fe-0.06C-17.6Cr-15.2Mn-0.43Mo-0.3Nb-0.005B-0.38N) and 07Ch17G15NAB ((%) Fe-0.05C-18.4Cr-16.5Mn-1.6Ni-0.01Ce-0.005B-0.32N) corrode in 0.5 mol/l nitric acid at 293 K (20 °C) at a rate of 0.009 mm/a (0.35 mpy), and at rates of 0.35 and 0.17 mm/a (13.78 and 6.69 mpy) respectively with addition of 0.01 mol/l hydrogen fluoride. These values increase at 373 K (100 °C) to 0.034 and 0.029 mm/a (1.34 and 1.14 mpy) respectively without hydrogen fluoride, and to 0.46 and 0.33 mm/a (18.11 and 12.99 mpy) respectively with 0.01 mol/l hydrogen fluoride [285]. In the case of chromium-manganese steels, good resistance in solutions containing nitric acid can be achieved with additions of HF or HF + HCl only after addition of molybdenum.

Table 42 shows the corrosion behavior of the two chromium-manganese steels 07Ch17G17DAMB and 06Ch17G15NAB ((%) Fe-0.05C-18.36Cr-16.5Mn-1.6Ni-0.31Nb-0.12Si-0.01Ce-0.017P-0.014S) in nitric acid with and without hydrofluoric acid [252]. According to this table, both steels are unsuitable in boiling nitric acid.

Steel	Medium mol/l	Test duration h	Material consumption rate, g/m^2 h	
			Alloy	Weld sample
07Ch17G17DAMB	10 HNO_3	25	0.37	0.36
		200	–	0.52
	10 HNO_3 + 0.01 HF	25	0.68	2.01
		200	–	1.46
06Ch17G15NAB	10 HNO_3	25	0.36	0.43
		200	–	0.54
	10 HNO_3 + 0.01 HF	25	1.37	2.06
		200	–	1.53

Table 42: Average material consumption rates of two CrMn steels as alloy and weld specimens in nitric acid solutions at 373 K (100 °C) [252]

The influence of small additions of hydrogen fluoride (0.01 and 0.1 mol/l) and 5 g/l chloride on the corrosion of three chromium-manganese steels in 10 mol/l nitric acid at 293 and 323 K (20 and 50 °C) can be seen from Table 43. According to this, small additions of chloride improve the resistance of the steels in nitric acid containing HF [408].

The corrosion behavior of the chromium steels 10Ch13 (Fe-0.10C-13Cr, cf. 1.4006, AISI 410) and 10Ch17 (Fe-0.10C-17Cr, cf. 1.4571, AISI 316 Ti) from powdered and sintered material and as rolled steel specimens in solutions containing HNO_3 is summarized in Table 44 [250]. The corrosion is increased drastically if sulfuric acid is present at the same time.

According to Table 45, nickel-free chromium-manganese steel ((%) Fe-0.08C-17.7Cr-17.55Mn-0.015P-0.019S-0.53N) is not as corrosion-resistant in HNO_3 containing solutions at 382 K (109 °C) as the CrNi steel 18 8 (AISI 302 SS, cf. 1.4310).

Steel	Additions, mol/l	Corrosion rate, mm/a (mpy)	
		293 K (20 °C)	323 K (50 °C)
06Ch17G15NAB	none	0.009 (0.35)	0.025 (0.98)
	0.01 HF	0.074 (2.91)	0.40 (15.75)
	0.1 HF	0.29 (11.42)	1.52 (59.84)
	0.01 HF + 5 g/l Cl$^-$	0.037 (1.46)	0.14 (5.51)
	5 g/l Cl$^-$	0.019 (0.75)	0.079 (3.11)
07Ch17G17DAMB	none	0.009 (0.35)	0.014 (0.55)
	0.01 HF	0.029 (1.14)	0.22 (8.66)
	0.1 HF	0.30 (11.81)	1.49 (58.66)
	0.01 HF + 5 g/l Cl$^-$	0.027 (1.06)	0.14 (5.51)
	5 g/l Cl$^-$	0.016 (0.63)	0.024 (0.94)
08Ch18G8N2T	none	0.020 (0.79)	0.034 (1.34)
	0.01 HF	0.18 (7.09)	0.058 (2.28)
	0.1 HF	0.16 (6.3)	7.15 (281.5)
	0.01 HF + 5 g/l Cl$^-$	0.050 (1.97)	0.19 (7.48)
	5 g/l Cl$^-$	0.020 (0.79)	0.058 (2.28)

Table 43: Corrosion rate of CrMn steels in 10 mol/l HNO$_3$ with additions of HF and Cl$^-$ at 293 and 323 K (20 and 50 °C) [408]

Medium %	Temperature K (°C)	Test duration h	Sintered		Rolled	
			10Ch13	10Ch17	10Ch13	10Ch17
			Corrosion rate, mm/a (mpy)			
20 HNO$_3$	293 (20)	–	0.005 (0.2)	–	–	–
	323 (50)	72	0.005 (0.2)	0.002 (0.1)	0.1 (3.9)	0.1 (3.9)
20 HNO$_3$ + 60 H$_2$SO$_4$	323 (50)	72	0.40 (15.8)	0.37 (14.6)	–	0.21 (8.3)
4.7 HNO$_3$ + 87.7 H$_2$SO$_4$	323 (50)	47	0.18 (7.1)	0.04 (1.6)	0.13 (5.1)	0.13 (5.1)

Table 44: Corrosion of powdered and sintered material and rolled specimens (7.5 % pores) of two chromium steels in solutions containing HNO$_3$ [250]
10Ch13 (cf. 1.4006, AISI 410); 10Ch17 (cf. 1.4571, AISI 316 Ti)

Medium, mol/l	CrMn steel	AISI 302 SS
	Corrosion rate, mm/a (mpy)	
6 HNO_3	0.15 (5.9)	0.032 (1.3)
6 HNO_3 + 0.1 $HSO_3 NH_2$	0.60 (23.6)	0.050 (2.0)
6 HNO_3 + 0.3 $HSO_3 NH_2$	0.84 (33.1)	0.077 (3.0)
6 HNO_3 + 0.1 H_2SO_4	0.20 (7.9)	0.049 (1.9)
6 HNO_3 + 0.3 H_2SO_4	0.53 (20.9)	0.070 (2.8)

Table 45: Corrosion of the Ni-free CrMn steel and the CrNi steel 18 8 (AISI 302 SS) in solutions containing HNO_3 at 382 K (109 °C) after 240 h [253]

The chromium-manganese steels 08Ch18G8N2T ((%) Fe-0.08C-18Cr-8Mn-2Ni; Ti-stabilized), 10Ch14G14N4T ((%) Fe-0.10C-14Cr-14Mn-4Ni; Ti-stabilized) and 12Ch13G18D ((%) Fe-0.12C-13Cr-18Mn-Cu) corrode in a hot solution of 40.4 g/l HNO_3 + 18.2 g/l oxalic acid at 373 K (100 °C) with corrosion rates of 0.027, 0.021 and 0.023 mm/a (1.06, 0.83 and 0.91 mpy) respectively [412].

The material consumption rate of the Cr steel 1Ch13 (Fe-0.1C-13Cr, cf. 1.4006, AISI 410), which is about 7000 g/m² h in a boiling solution of 8 mol/l nitric acid containing 10 g/l NaF + 20 g/l $Al(NO_3)_3$ + 0.01 g/l K_2CrO_4, can be reduced again from about 6000 to about 5 g/m² h by addition of 4 g/l chromate [255].

A solution of 20 percent by volume nitric acid + 22 g/l sodium bichromate is used at about 320 K (47 °C) for passivation of chromium steel containing 12 to 14 % chromium [245].

To detect carbides finely distributed in the structure of the ultra high-strength steel X41CrMoV5-1 (1.7783, AISI 610, UNS T20811), the steel is etched with methanolic 2 % nitric acid at room temperature for about 5 s. A solution of 90 ml methanol + 10 ml HNO_3 is used at room temperature for 12 s for deep etching for scanning electron microscopy photographs [444].

A 19 Ferritic-austenitic steels with more than 12% chromium

The ferritic-austenitic chromium steel VLX 562® (1.4462, UNS S31803(%) Fe-0.03C-(21-23)Cr-(2.5-3.5)Mo-(4.5-6.5)Ni-2Mn-1Si-0.03P-0.02S-(0.08-0.20)N) with twice the tensile strength of the chromium-nickel steel AISI 316 L and a very good resistance to stress corrosion cracking corrodes in boiling 65 % nitric acid at a rate of < 0.6 mm/a (< 24 mpy) [257].

The ferrite phase present to the extent of 31 to 40 % in the ferritic-austenitic chromium-nickel steel Ch22N5 (Russian grade; (%) Fe-0.07C-21.54Cr-5.73Ni) is more severely attacked in 0.1 mol/l nitric acid at 303 K (30 °C) than the austenite with 19.5 % chromium, in spite of its higher chromium content. Heat treatment at 1323 or 1523 K (1050 or 1250 °C) produced no improvement in the corrosion behavior [403].

The ferritic-austenitic steel X2CrMnNiMoN26-5-4 (cf. 1.4467, A 905, (%) Fe-<0.04C-25.5Cr-5.8Mn-3.7Ni-2.3Mo-0.37N) has, in the quenched state, a structure with in each case 50% ferrite and austenite and a 0.2% offset yield strength of at least 590 N/mm^2. It is therefore superior in strength to the previous steels and has an equally good corrosion resistance (about 0.1 g/m^2 h in the Huey test). Its weldability and toughness are also good [409].

The corrosion behavior of three martensitic chromium-nickel steels 13-4 (C) (cf. 1.4313, UNS S41500, (%) Fe-0.082C-11.6Cr-4.2Ni-0.54Mn-0.31Mo-0.26Si-0.009P-0.012S-0.022N), 13-4 (B) ((%) Fe-0.032C-13.2Cr-3.9Ni-0.66Mn-0.36Mo-0.66Si-0.008P-0.002S-0.026N) and 13-4-1 ((%) Fe-0.043C-12.7Cr-3.9Ni-1.5Mo-0.68Mn-0.39Si-0.009P-0.013S-0.030N) in 5% nitric acid at room temperature is shown in Figure 18. The higher strength of the cast steels achieved by suitable heat treatment also resulted in a lower material consumption rate [323]. In addition to uniform attack, intercrystalline corrosion was found along segregations in more highly tempered states.

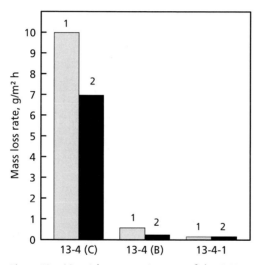

Figure 18: Material consumption rate of the CrNi steels 13-4 (C), 13-4 (B) and 13-4-1 in 5% HNO$_3$ at room temperature [323]
1) low strength; 2) higher strength

The corrosion resistance of the heat-affected zone can be increased considerably by increasing the rate of cooling of the weld seam on the ferritic-austenitic steel 0Ch21N5T to 450 K/s. The resulting corrosion rate in boiling 65% nitric acid corresponds to that of the unaffected steel in Table 46 [327]. Titanium-free, low-carbon steel (≤ 0.04% C) is more resistant to corrosion.

The CrNi steel 18 3 is resistant at 373 K (100 °C) in up to 40% HNO$_3$ with a corrosion rate of < 0.1 mm/a (< 4 mpy) [231].

	Composition of the steel, %*						Corrosion rate mm/a (mpy)
C	Cr	Ni	Si	Mn	Ti	Ti/C	
0.04	20.0	5.56	0.25	0.49	–	0	0.24 (9.5)
0.08	20.95	5.8	0.56	0.7	0.2	2.5	0.58 (22.8)
0.05	20.23	5.83	0.33	0.57	0.25	5	0.32 (12.6)
0.06	22.02	6.51	0.42	0.54	0.58	9.2	0.26 (10.2)

* balance Fe

Table 46: Corrosion rates of some steels of the type 0Ch21N5T in boiling 65 % HNO_3, test duration 40 h [327]

A 20 Austenitic chromium-nickel steels

The steel AISI 304 L ((%) Fe-0.03C-18.5Cr-10.5Ni-<0.5Mo-0.6Si-0.030P-0.030S) is generally used in nitric acid plants. By reducing the contents of Si, P and S ((%) Fe-0.20C-<0.1Si-18.5Cr-11Ni-<0.1Mo-0.015P-0.010S), the corrosion behavior of the steel AISI 304 L (HNO_3 grade) at higher concentrations of nitric acid can be improved, as shown in Figure 19. With increasing demands of corrosion resistance in nitric acid, either a modified steel AISI 310 ((%) Fe-0.020C-24.5Cr-20.5Ni-<0.3Mo-<0.3Si-0.020P-0.015S) or the steel 2RE 69 (1.4466, AISI 310 mod.,(%) Fe-0.20C-25Cr-22Ni-2.1Mo-<0.4Si-0.020P-0.015S-0.12N) is suitable [274]. The paper on corrosion problems in stainless steel heat exchangers in a nitric acid plant should also be mentioned [390].

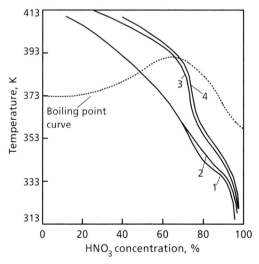

Figure 19: Isocorrosion curves (0.1 mm/a (3.94 mpy)) of the steels (1) AISI 304 L, (2) 304 L (HNO_3 grade), (3) AISI 310 (HNO_3 grade) and (4) Sandvik® 2RE-69 in nitric acid [274]

The steel AISI 304 is unsuitable for the housing and worm wheel shafts of nitric acid pumps, since the pump corrodes and the shaft seizes. The corrosion products consist mainly of complex salts and aluminium compounds [426].

In the production of highly concentrated nitric acid by the HOKO process by reaction of liquid nitric oxide with oxygen and water under an increased pressure of 5.0 MPa, the CrNiSi steel 18 15 2, which is resistant to nitric acid up to the boiling point, is used for the storage tank (5000 m^3 capacity) and the heat exchanger jackets and pumps [373].

The chromium-nickel steel 12Ch18N10T (1.4541, AISI 321) is used for the production of nitric acid and ammonium nitrate. In the presence of 56 % nitric acid, it can be replaced by the steel 10Ch14G14N4T ((%) Fe-0.10C-(13-15)Cr-(3.5-4.5)Ni-(13-15)Mn-0.65Ti), but only below 320 K (47 °C) [387].

The use of CrNi 18 8 steels is recommended for a nitric acid plant for oxidation of ammonia and further processing to 60 % acid [369].

The steel Fe-<0.01C-18Cr-15Ni-4Si was developed to keep intercrystalline and general corrosion low when using steels for factories processing nitric acid. However, since completely austenitic steels with a high silicon and low carbon content tend to crack in the weld region, the additional material must be alloyed so that 4 to 10 % δ-ferrite is present in the welding material [352, 361].

Doping a CrNi steel 18 9 with boron not only achieves an increase in the deformation resistance and tensile strength, but also increases the resistance to nitric acid [440].

Even high-alloy chromium-nickel steels, such as, for example, FeCr21Ni32TiAl, can be sensitive to intercrystalline corrosion in boiling concentrated nitric acid and corrode at a material consumption rate of > > 1 g/m^2 h. Cold-rolled specimens of these steels with 15 – 46 % deformation and subsequently heat-treated at 1173 K (900 °C) showed a reduction in the material consumption rate to values of < 1 g/m^2 h [456].

Chromium-nickel-titanium steel (Fe-0.1C-18Cr-8Ni-1Ti) is recommended as a substitute for the steels AISI 304 and 316 for the absorption column in a nitric acid plant [553].

The titanium-stabilized steel Fe18Cr10NiTi used for the absorption column (47 – 55 % HNO$_3$, 323 K (50 °C) and 70 N/mm^2) in a nitric acid plant showed intercrystalline corrosion after 5 years of operation, especially in the region of the weld joints, where TiC had formed. During repair, steel wires made of 01Ch19N9 were used for the welding. Nowadays, the more resistant steel 03Ch18N11 (1.4306, AISI 304 L) would be preferred [457].

A 5000 m^3 capacity, fixed-cover tank for storage of highly concentrated nitric acid was produced, inter alia, from the special alloy X1CrNiSi18-15-4 (cf. 1.4361, A 336) with silicon and showed no noticeable corrosion phenomena after 4 years of use [679].

The shaft cover on a chemical pump made of the steel X1CrNiSi18-15-4 for the delivery of concentrated nitric acid showed signs of local corrosion after 2 years. This was primarily due to contact with glass fiber-reinforced bellows made of polytetrafluoroethylene (PTFE), since the protective layer which formed was constantly damaged [455].

Figure 20 shows the dependence of the corrosion rate of some chromium-nickel steels in boiling nitric acid on their concentration. While the material consumption rate of the steel X1CrNiSi18-15-4 with about 4 % silicon passes through a maximum at about 1 g/m² h in 50 % HNO₃, decreases again and reaches about 0.05 g/m² h in 98 % HNO₃, those of the steels 1.4306 (X2CrNi19-11, AISI 304 L) and 1.4465 (X1CrNiMoN25-25-2, AISI 310 MoLN) increase [349].

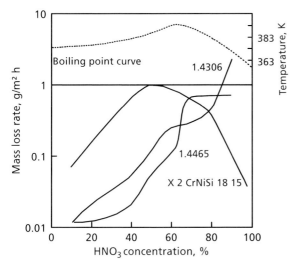

Figure 20: Dependence of the material consumption rate of the steels X2CrNiSi 18 15, 1.4306 and 1.4465 on the concentration of boiling HNO₃, including the boiling point curve [349]

The austenitic chromium-nickel cast steel Noricid® 9.4306 (GX3CrNiSiN20-13-5), which is resistant to nitric acid even at higher temperatures, is a suitable material for centrifugal pumps for solutions containing nitric acid [963].

If it is not possible to eliminate the internal mechanical stresses generated by weld joints in the steels 12Ch18N10T (cf. 1.4541, AISI 321) and 03Ch18N11 (AISI 304 L) during construction of the plant, intercrystalline corrosion occurs during production of dilute nitric acid. The steel 03Ch18N11 is still said to be better in its behavior [360].

Nitrogen tetroxide can be used as a heat transfer agent in nuclear energy plants, since it hardly attacks stainless steels, such as Ch18N10T, at 5.0 MPa and operating temperatures of about 870 K (597 °C). The material consumption rate in N_2O_4 under 5.0 MPa at 773 K (500 °C) for example, is thus 0.0002 g/m² h. Nitric acid formed due to moisture leads to a noticeable increase in the corrosive attack [394].

Contact with lead causes an even greater increase in the material consumption rate of the steel 1Ch18N10T (cf. 1.4541, AISI 321) in 4 % nitric acid at 368 K (95 °C). The rate was 0.005 without contact and 55.6 g/m² h with contact. Contact with nickel, which is readily polarizable, in contrast to lead, caused an increase to only 0.14 g/m² h [381].

Stainless steels, such as AISI 304 L, are preferentially attacked by hot nitric acid in the crevices which form for example during welding or lining, or between the steel and the plastic sealing discs. While the steel AISI 304 L corrodes at a rate of 0.05 and 0.07 mm/a (1.97 and 2.76 mpy) after 100 and 500 h respectively in 15 % nitric acid at 363 K (90 °C), rates of 0.27 and 0.54 mm/a (10.63 and 21.26 mpy) respectively occur after 500 h in the crevice at the weld and between the steel and the sealing disc. Corrosion tests were performed on the steels AISI 304 L, 310 L and a CrNi steel 18 10 in nitric acid to investigate the corrosion mechanism in the crevice. For more details see [604].

Regardless of the nickel content of between 2 and 5 %, the corrosion rate of chromium-manganese-nickel steels of low nickel content (e.g. (%) Fe-0.036C-13.5Cr-20.4Mn-(2-5)Ni-0.085N) at room temperature in 20 and 60 % nitric acid is about 0.002 mm/a (0.08 mpy). At 363 K (90 °C), the corresponding values are about 0.03 (20 % HNO_3) and 0.68 mm/a (1.18 and 26.77 mpy) (60 % HNO_3) [435].

The steel of the type AISI 310 ((%) Fe-0.020C-25.6Cr-19.41Ni-0.40Mo-0.060N) shows a corrosion rate in 20 and 40 % nitric acid at 422 K (149 °C) of about 0.075 and 0.20 mm/a (2.95 and 7.87 mpy) respectively. Corresponding values of 0.089 and 0.21 mm/a (3.5 and 8.27 mpy) were determined at 394 K (121 °C) in 60 and 80 % nitric acid respectively [356].

By reducing the dew point in a hydrogen-containing atmosphere to 233 to 223 K (–40 to –50 °C), passivating treatment of the steels 12Ch18N10T (1.4541, AISI 321) and 12Ch17G9AN4 ((%) Fe-0.12C-16.8Cr-4.1Ni-9.5Mn-0.27Si-0.02P-0.01S-0.22N) is not necessary, as comparative corrosion tests in boiling 30 % nitric acid have shown. The material consumption rates were 0.070 and 0.098 g/m^2 h respectively [454].

The oxide film which forms on the steel SUS 304 (cf. 1.4301, AISI 304, Fe-0.06C-18.25Cr-8.42Ni-0.93Mn-0.54Si) within 30 min in hot 30 % nitric acid at 333 K (60 °C) consists of a mixture of CrO_3 (or $Cr_2O_7^{2-}$), Cr_2O_3, Fe_2O_3 and a small amount of CrOOH and γ-FeOOH [461].

The relationships between cold working, precipitation properties and corrosion resistance in boiling 5 mol/l nitric acid were investigated on specimens taken from a pipe made of the steel X1CrNiSi18-15-4 (cf. 1.4361, A 336, (%) Fe-0.009C-4.24 Si-1.33Mn-0.008P-0.001S-17.32Cr-14Ni-0.01Mo-0.13-Nb-0.03Ti-0.012N). Cold-worked (4 to 20 %) steel subsequently heat-treated at 973 K (700 °C) showed higher material consumption rates than the unworked material. This is based on precipitates ($M_{23}C_6$, Cr-rich carbide and $Cr_5Ni_3FeSi_2$) produced at the grain boundaries by deformation and subsequent annealing, the amount of which increases with the degree of deformation. The specimens were annealed beforehand in air at 1393 K (1120 °C)/10 min/water to establish a recrystallized, precipitation-free starting structure. The material consumption rates per unit area are shown as a function of the treatment state and test duration in Table 47 [30].

As a result of galvanic contact of the unworked and deformed specimens with the same treatment state, the material consumption rate of the deformed specimen was 2 to 3 times higher than that of the unworked specimen. Deformation of the steel and stress-induced precipitates impair the corrosion resistance in nitric acid [30].

Treatment state	Surface cm²	Test duration h	Material consumption rate g/m² h
Non-deformed, non-sensitized	16.1 (16.1)*	48 (72)*	0.08 (0.10)*
Deformed (20 %), non-sensitized	12.1 (12.3)*	72 (72)*	0.12 (0.28)*
Non-deformed, sensitized	16.2 (16.4)*	48 (72)*	0.08 (0.09)*
Deformed (20 %), sensitized	11.6 (11.6)*	72 (72)*	0.08 (0.17)*

* values of specimens in galvanic contact

Table 47: Material consumption rates of specimens of the steel X2CrNiSi 18 15 with different treatment states in 5 mol/l boiling HNO_3 [30]

Table 48 shows that as the polarization resistance decreases, the material consumption rate of steel with a low carbon content ((%) Fe-0.02C-23.8Cr-20.67Ni-1.10Mn-0.02Mo-0.60Si-0.015P-0.014S) in boiling nitric acid increases [310]. After calibration, the polarization resistance can be used for rapid determination of the corrosion.

HNO_3 %	Temperature K (°C)	Material consumption rate g/m² h	Polarization resistance Ω/cm^2
35	382 (109)	0.028	12,000
45	386 (113)	0.049	7,390
55	390 (117)	0.100	5,210
65	393 (120)	0.270	2,578

Table 48: Corrosion rate and polarization resistance (R_p) of steel in boiling nitric acid [310]

The steels 03Ch18N11 (1.4306, AISI 304 L) and 12Ch18N10T (1.4541, AISI 321) can be used in up to 40 % boiling nitric acid, since the corrosion rates are about 0.12 and 0.10 mm/a (4.7 and 3.9 mpy), but 0.35 and 0.78 mm/a (13.8 and 30.7 mpy) respectively at 65 % [725].

On the other hand, in contrast to the steel 12Ch18N10T, 03Ch18N11 shows no crevice corrosion between room temperature and the boiling point in nitric acid (> 50 %). The corrosive attack on the steel and also on its welded joint is uniform [438].

Alternating current impedance with radioindicator measurements were used for short-term determination (over a few days) of the corrosion rate of the steel Fe-18Cr-13Ni-1Nb giving values of 0.00005 to 0.001 mm/a (0.002 to 0.039 mpy) in 7.8 mol/l nitric acid at room temperature [345].

The resistance of the chromium-nickel steel NAR-SN-5 (Jap. type) with 27 % chromium and 7 to 10 % nickel to stress corrosion cracking in nitric acid of moderate concentration can be improved by nitrogen contents of 0.09 to 0.13 %, which cause an increase in the ferritic phase of 40 to 65 %. Small additions of molybdenum, copper or niobium led to a deterioration in the corrosion behavior [451].

According to Table 49, tests under evaporation conditions with heat transfer in the vaporizer in a titanium tank with heating steam coils made of the steels

Ch18N10T and El-943 (0Ch23N28M3D3T: (%) Fe-<0.01C-23Cr-28Ni-3Mo-3Cu; Ti-stabilized) with 55 % nitric acid resulted in material consumption rates on the steels due to the acid which were up to 3 times higher in comparison with normal boiling tests when the temperature of the heating steam is between 403 and 418 K (130 and 145 °C) [261]. According to these tests, CrNiMo steel and not CrNi steel should be used for heating coils in the vaporizer.

Steel	Pressure kP	Temperature K (°C)	Wall temperature* K	Material consumption rate, g/m² h		
				On the pipe	On the wall	In the solution at 388 K (115 °C)
Ch18N10T	274.8	403 (130)	400	0.45	0.33	0.16
	421.7	418 (145)	409	0.59	0.58	0.16
El-943	274.8	403 (130)	400	0.10	0.13	0.08
	421.7	418 (145)	409	0.21	0.25	0.08

* calculated

Table 49: Corrosion rate of the steels Ch18N10T and El-943 after 25 h in 55 % HNO_3 under various conditions [261]

After 60 d tests in hot 56 % nitric acid at 333 and 391 K (60 and 118 °C) (boiling point), the nitrogen-alloyed chromium-nickel steel X5CrNiN19-9 (1.4315) corrodes at a rate of 0.005 and 0.42 mm/a (0.0254 and 16.54 mpy) respectively, which is comparable to that of the steels X5CrNi18-10 (1.4301, AISI 304) and X8CrNiTi18-10 (1.4878, SUS 321) of higher nickel content of 0.007, 0.01, 0.36 and 0.64 mm/a (0.28, 0.39, 14.17 and 25.20 mpy) respectively [466].

Welded joints of the steels 12Ch18N10T (cf. 1.4541, AISI 321), 08Ch18N10T (cf. 1.4541, AISI 321), 08Ch22N6T and 08Ch18G8N2T as well as 03Ch18N11 (cf. 1.4306, AISI 304 L) are attacked only insignificantly in nitric acid solutions containing dicarboxylic acids in the manufacture of adipic acid. After 11,000 operating hours the material consumption rate, for example, of the 03Ch18N11 steel still-head from the concentration column containing 57 % nitric acid at 343 K (70 °C) was 0.0024 and that of the absorption column containing 60 % nitric acid at 313 K (40 °C) was 0.0007 g/m² h [364]. The corrosion of the steel 03Ch18N11 also remained uniform, with material consumption rates between 0.0050 and 0.0071 g/m² h, in the column where the nitric oxides are driven off and the other steels mentioned showed intercrystalline and knife-line corrosion.

The ferritic phase occurring to the extent of about 10 % has no influence on the corrosion resistance of the stainless steels Ch18N10T, 000Ch18N11 (cf. 1.4306, AISI 304 L) and 02Ch19N9 in boiling 65 % nitric acid. According to potentiostatically recorded polarization curves, this behavior can be expected at potentials both in the passive region and in the passive/transpassive transition region [397].

Steels without a particular heat treatment but good strength, toughness and weldability as well as corrosion resistance usually contain 4 – 6 % Mn, 6 – 9.5 % Ni, 20 –

21.5 % Cr, 0.25 – 0.35 % N and possibly up to 2.5 % Mo and 1 % Nb. According to the Huey test, a steel of this type (completely austenitic and non-magnetic) corrodes at a rate of 0.15 mm/a (5.91 mpy) (tempered) and 0.17 mm/a (6.69 mpy) (sensitized) [452].

The steel X2CrNiSi18-15 (cf. 1.4361, A 336) with 2 % silicon has a material consumption rate in boiling concentrated nitric acid of 2 g/m^2 h and with 6 % silicon of 0.03 g/m^2 h [678].

Intercrystalline corrosion on weld seams caused by nitric acid can be prevented if 6 to 8 % silicon is added to steels of Fe ≤ 0.02C-(5.8-8.5)Cr-(21-25)Ni [374].

The resistance of the austenitic CrNiSi steel 17 14 4 to boiling 65 % nitric acid can be further improved by additions of niobium, zirconium or tantalum. However, nitrogen and titanium additions cause a deterioration. With its good mechanical properties and weldability, the steel (%) Fe-≥0.05C-17Cr-14Ni-4Si-0.8Nb exhibits an excellent resistance to concentrated nitric acid [375].

The corrosion rate of the CrNiSi steel 17 14 4 in boiling 65 % nitric acid only becomes independent of the heat treatment temperature between 870 and 1220 K (597 and 947 °C) if about 1 % niobium, zirconium or tantalum is alloyed with the steel [355]. The material consumption rate is between 0.4 and 0.6 g/m^2 h.

The Nb- and Ta-containing steel (%) Fe-0.05C-18.18Cr-9.06Ni-1.51Mn-0.7(Nb+Ta)-0.55Si-0.027P-0.006S-0.018-N-0.0031B is said to corrode in boiling 65 % nitric acid with a material consumption rate of only 0.09 g/m^2 h [445].

Titanium-stabilized CrNi steel ((%) Fe-0.09C-18.1Cr-13.4Ni-1.17Mn-0.37Mo-0.51Ti-0.017N), which precipitates TiC at the grain boundaries after sensitization at 1070 – 1170 K (797 – 897 °C), is resistant to intercrystalline corrosion, especially at low temperatures, but attacked at the $M_{23}C_6$ and TiC precipitates in the Huey test (boiling 65 % HNO$_3$) [1006].

The austenitic chromium-nickel steels Fe-0.08(C+N)-18Cr-8Ni (Ti-stabilized) and Fe-0.08(C+N)-18Cr-10Ni-2Mo (Ti-stabilized) are resistant in boiling 65 % nitric acid according to [256]. This conclusion is limited, however, by the material consumption rates, reported elsewhere, of the steels X8CrNiTi18-10 (1.4878, AISI 321 H) and X5CrNi18-10 (1.4301, AISI 304) in boiling 56 % nitric acid of 0.58 and 0.32 g/m^2 h [379]. When alloying elements dissolved in the acid become concentrated, the corrosion can be increased further by a factor of 2 to 3 [294, 301].

The corrosion resistance of the steel 03Ch18N11 (1.4306, AISI 304 L) depends greatly on the Si and C content, especially in the sensitized state. According to Table 50, the presence of more than 0.3 % silicon has an adverse effect on the resistance at carbon contents of about 0.03 %. A reduction in the carbon content leads to a decrease in the corrosion rate of the steel [291].

The steel 03Ch18N10 with 0.020 to 0.028 % C is improved in its austenitic stability by addition of 0.017 to 0.172 % N, the deformability remaining constant. At the same time, the susceptibility to intercrystalline corrosion in boiling 65 % nitric acid is reduced [432].

Alloy content, %		Corrosion rate, mm/a (mpy)		
Si	C	1348 K (1075 °C)	1393 K (1120 °C)	1423 K (1150 °C)
0.17	0.030	0.27 (10.6)	0.24 (9.5)	0.24 (9.6)
0.49	0.030	6.52 (256)	0.96 (37.8)	0.63 (24.8)
0.78	0.030	16.34 (643)	4.65 (183)	1.44 (57.0)
0.28	0.020	0.25 (9.8)	0.29 (11.4)	0.24 (9.6)
0.48	0.023	0.23 (9.1)	0.20 (7.9)	0.21 (8.3)
0.75	0.012	0.18 (7.1)	0.17 (6.7)	0.17 (6.7)

Table 50: Corrosion rates on test specimens from industrial smelting of the steel 03Ch18N11 (AISI 304 L) with various Si and C contents in boiling 65 % HNO_3 after prior annealing at 1348 to 1423 K (1075 to 1150 °C)/water quenching and subsequent tempering at 923 K (650 °C) [291]

In corrosion tests with solution-annealed and sensitized CrNi 18 8 steels, the steel with the lower carbon content ((%) Fe-0.035C-0.37Si-0.79Mn-16.65Cr-9.80Ni-0.005P-0.024S) also had the lower material consumption rates of 0.50 and 1.22 g/m^2 h, in comparison with the corresponding values of 0.80 and 5.5 g/m^2 h for the steel with the higher carbon content ((%) Fe-0.08C-0.60Si-1.40Mn-18.10Cr-9.20Ni-1.43Mo-0.015P-0.028S), after exposure for 3 times 47 h to boiling 65 % nitric acid. Steels of comparable composition but additionally doped with 0.66 or 0.78 % niobium gave higher corrosion rates [560].

The material consumption rate of the steel 12Ch18N10T (AISI 321) in boiling 65 % nitric acid could be reduced from 1.5 to about 1 g/m^2 h by addition of 0.01 % cerium or vanadium [417].

Chemical stabilization of the steel AISI 321 SS ((%) Fe-0.066C-18.2Cr-9.0Ni-1.1Mn-0.48Ti-0.47Si-0.001S-0.012P) can be lost by incorrect heat treatment. Undesirable sensitization both during processing and use of the steel in question, and therefore intercrystalline corrosion in boiling nitric acid, could be prevented by heat treatment at 1173 K (900 °C) for 2 h [300]. These circumstances also apply to other steels.

The steel AISI 308 (cf. 1.4303; (%) Fe-0.040C-20.95Cr-9.82Ni-1.76Mn-0.41Si-0.008S-0.016P) has an austenitic form after heat treatment up to 1473 K (1200 °C) for 1 h and a duplex form (ferritic + austenitic structure) above 1473 K (1200 °C). Table 51 shows the corrosion behavior after quenching in water and various forms of aging [308]. The lower the aging temperature, the more the two steels corroded. The material consumption rate of the steel AISI 308 SS (completely austenitic) in boiling 65 % nitric acid of 0.121 g/m^2 h was accelerated to 81.6 g/m^2 h by addition of 4 g/l sodium bichromate ($Na_2Cr_2O_7$) (test duration 144 h) [308].

Some material consumption rates for the steel (%) Fe-0.019C-18.60Cr-1.05Ni-1.01Mn-0.03Mo-0.49Si-0.017P-0.007S-0.022N in boiling 65 % nitric acid (Huey test) are shown in Table 52 as a further example of the strong influence of the tempering temperature over a period of 300 h on the corrosion of chromium-nickel steel. As the tempering time increases, the sensitivity to corrosion increases significantly, especially at temperatures below 800 K (527 °C). Above 1250 K (977 °C), the influence of the tempering time is no longer significant [331].

1473 K (1200 °C)/ Water aging	Material consumption rate g/m² h	1573 K (1300 °C)/ Water aging	Material consumption rate g/m² h
–	0.177	–	0.341
1 h at 973 K (700 °C)	0.802	1 h at 973 K (700 °C)	0.465
100 h at 973 K (700 °C)	2.88	100 h at 973 K (700 °C)	2.25
5 h at 873 K (600 °C)	0.358	2 h at 873 K (600 °C)	1.87
96 h at 873 K (600 °C)	> 10	96 h at 873 K (600 °C)	3.14
–	–	5 h at 873 K (600 °C)	4.92
–	–	300 h at 823 K (550 °C)	1.54

Table 51: Influence of aging on the corrosion of AISI 308 in boiling 65 % HNO_3 after heat treatment at 1473 and 1573 K (1200 and 1300 °C) with subsequent quenching in water [308]

Tempering temperature, K (°C)	673 (400)	773 (500)	823 (550)	873 (600)	923 (650)	973 (700)	1073 (800)	1123 (850)
Material consumption rate, g/m² h	0.1	1.05	12	20	1.5	0.5	0.15	0.13

Table 52: Corrosion rate of steel in the Huey test after solution annealing at 1323 K (1050 °C) for 15 min and after 300 h of tempering at various temperatures [330]

The corrosion behavior of stabilized steels moreover depends on the precipitation of carbonitride, since this is severely attacked by boiling nitric acid in the Huey test.

In agreement with this, after correct heat treatment (i.e. a sufficiently high tempering temperature), the steel Remanit® 4306 (X2CrNi19-11, 1.4306, AISI 304 L; (%) Fe-0.019C-19.09Cr-10.06Ni-1.59Mn-0.03Mo-0.003Al-0.16Si-0.018P-0.004S-0.0005B-0.042N with 5.03 % ferrite) showed, after annealing at 1393 K (1120 °C) for 10 min and quenching in water, material consumption rates of 0.10 and 0.11 g/m² h after 12 and 15 periods respectively of 48 h each in the Huey test. Subsequent annealing at 973 K (700 °C) for 30 min caused an increase in the material consumption rate to 0.26 g/m² h [730].

Remanit® 4335-So ((%) Fe-0.012C-25.46Cr-19.93Ni-<0.01Si-1.41Mn-<0.05P-<0.006S-0.140N) heat-treated at 1323 K (1050 °C) for 30 min and quenched in water corrodes in boiling 65 % nitric acid with a material consumption rate of about 0.05 g/m² h after 11 and 15 boiling periods of 48 h each. After a final heat treatment in the plant, the rate was 0.50 g/m² h [731].

According to the Huey test, AISI 304 can be used for waste tanks in the nuclear reactor industry if the steel is heat-treated at 813 K (540 °C) for 1 to 10 h (no longer), since the corrosion rate in boiling 65 % nitric acid does not exceed 0.24 mm/a (9.45 mpy). The steel AISI 304 L with a corrosion rate of 0.15 mm/a (5.9 mpy) can also be used if it is tempered either at 513 or 703 K (240 or 430 °C) for 100 h [306].

The corrosion behavior of the heat-treated steels AISI 304 L ((%) Fe-0.029C-18.37Cr-8.84Ni-1.56Mn-0.52Si-0.032N), AISI 304 ((%) Fe-0.071C-17.52Cr-8.57Ni-

1.80Mn-0.68Si-0.075N), LC-20 ((%) Fe-0.022C-21.62Cr-8.67Ni-1.79Mn-0.51Si-0.035N) and MC-20 ((%) Fe-0.06C-22.57Cr-8.55Ni-1.71Mn-0.55Si-0.036N) as cold-rolled specimens in boiling 65 % nitric acid can be seen from Table 53 [295]. The heat treatment temperature has a particular influence on the corrosion rate of the steel AISI 304 L.

Steel	Ferrite content %	Corrosion rate, mm/a (mpy)	
		1200 K (927 °C)	1339 K (1066 °C)
AISI 304 L	–	0.39 (15.35)	0.66 (25.98)
AISI 304	–	3.33 (131.1)	0.31 (12.2)
LC-20	20	0.17 (6.69)	0.17 (6.69)
MC-20	18 – 20	0.23 (9.06)	0.28 (11.02)

Table 53: Corrosion of some CrNi steels containing ferrite in boiling 65 % HNO_3 after heat treatment at 1200 and 1339 K (927 and 1066 °C) for 1 h [295]

According to Figure 21, cold working (here, for example, a decrease in the cylinder height of the steel specimen after deformation) has a varying influence on the corrosion in boiling 65 % nitric acid of the following non-sensitized steels solution-annealed at 1350 K (1077 °C) for 2 h and rapidly quenched: AISI 304 and AISI 304 L as well as AISI 316 ((%) Fe-0.05C-17.27Cr-13.09Ni-2.56Mo-1.76Mn-0.24Cu-0.13Co-0.04V-0.45Si-0.033P-0.015S) [302].

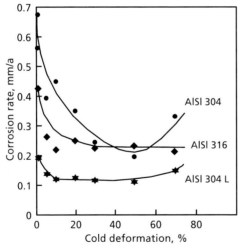

Figure 21: Corrosion of non-sensitized stainless steel in boiling 65 % HNO_3 as a function of the degree of cold working [302]

Compared with a corrosion rate of AISI 304 in boiling 65 % nitric acid of 1.1 mm/a (43.31 mpy) after a cooling rate of 1 K/s (1 °C/s), this drops to about 0.4 mm/a (15.75 mpy) after a cooling rate of 55 K/s (55 °C/s) [303].

According to Table 54, corrosion of the steel AISI 304 in boiling 65 % nitric acid (Huey test) was increased considerably by heat treatment at 923 K (650 °C) of increasing duration, but decreased at 1023 K (750 °C) [311].

Heat treatment		Corrosion rate, mm/a (mpy)
Temperature, K (°C)	Duration, h	
–	–	0.26 (10.3)
923 (650)	1	0.88 (34.6)
	2	4.2 (165)
	5	8.8 (346)
	50	21.4 (843)
1023 (750)	0.5	1.5 (59.1)
	1	1.3 (51.2)
	2	1.1 (43.3)
	5	1.1 (43.3)

Table 54: Corrosion of AISI 304 in boiling 65 % HNO_3 after various heat treatments [311]

The CrNi steels AISI 304 and AISI 304 L of comparable composition gave corrosion rates which approximately coincide after 240 h in the Huey test – 0.46 mm/a (18.11 mpy) for AISI 304 (in the delivery state) and for AISI 304 L 0.31 mm/a (12.2 mpy) (20 min at 950 K (677 °C)) and 0.61 mm/a (24.02 mpy) (1 h at 950 K (677 °C)) [1015].

AISI 304 L (as sheet or pipe), heat treated for 20 min or 1 h at 950 K (677 °C), corroded in boiling 65 % nitric acid at a rate between 0.20 and 0.22 or 0.26 mm/a (7.87 mpy) [354].

According to Figure 22, even small amounts of boron reduce the corrosion rate of AISI 304 in boiling 65 % nitric acid. An increase caused by sensitizing heat treatment between 922 and 1033 K (640 and 760 °C) is largely cancelled out by the presence of boron [309]. A content of 4 ppm boron had no influence on the corrosion behavior of a solution-annealed steel. The corrosion rate of this steel after 240 h was 0.065 mm/a (2.56 mpy) with and without boron [309].

Both sulfur (0.03 %) and phosphorus (0.06 %) have no influence on the corrosion behavior of the steel AISI 304 ((%) Fe-0.069C-18.6Cr-9.4Ni-0.01Si-0.003P-0.009S) in boiling 65 % nitric acid (Huey test). The material consumption rates were about 0.27 and 0.10 g/m^2 h, independent of the heat treatment of 100 to 1000 h at 923 and 973 K (650 and 700 °C) [304].

Contact between the steel AISI 304 and platinum, gold or graphite in 65 % nitric acid does not noticeably intensify the corrosion rate either at 293 K (20 °C) (from 0.008 with and without Pt contact to 0.007 with graphite and 0.002 mm/a (0.08 mpy) with Au contact) or 373 K (100 °C), but merely leads to a shift in the potential to more positive values, although these still remain within the passivity range [467].

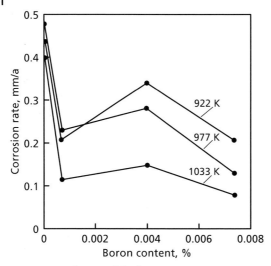

Figure 22: Influence of the boron content in AISI 304 on the corrosion rate of the steel in boiling 65 % HNO_3 as a function of the sensitization temperature [309]

To prevent the formation of carbides in the steel SUS 304 ((%) Fe-0.04C-18.4Cr-9.0Ni-1.62Mn-0.08Mo-0.02Cu-0.64Si-0.02P-0.007S) by quenching in oil after heat treatment at 1423 K (1150 °C), which causes increased corrosion, cooling in two steps – in air down to about 1200 K (927 °C) and then in oil – is recommended. It is possible to reduce corrosion in boiling 65 % nitric acid by a factor of about 3 by this procedure [276].

The stresses caused by welding have the lowest influence on intercrystalline corrosion on the steel 08Ch18N10T (1.4541, AISI 321) in boiling 65 % nitric acid after hardening at 1473 K (1200 °C). Tempering both at 923 and 1023 K (650 and 750 °C) increases the tendency to undergo intercrystalline corrosion. This also has an effect on the material consumption rate (from 8 to 9 up to 65 g/m² h). Corresponding tests with the steel 03Ch18N11 (1.4301, AISI 304 L) showed a rise in the material consumption rate from 0.51 to 29 g/m² h [328].

In another study, the influence of the dislocation structure on intercrystalline corrosion of the steels 08Ch18N10T (0.07 % C) and 03Ch18N11 (0.03 % C) in the elastoplastic state was investigated. This state was produced by welding the steel specimens, which had first been quenched from 1473 K (1200 °C) and sensitized at 923 and 1023 K (650 and 750 °C) for 6 h. According to the Huey test (3 × 48 h in boiling 65 % nitric acid), the most severe intercrystalline corrosion occurred on the specimen sensitized at 923 K (650 °C). The site of attack lay in the heat-affected zone about 10 – 15 mm away from the weld joint [431].

The welded joints of niobium-stabilized chromium-nickel steel ((%) Fe-0.06C-16.8Cr-11.2Ni-1.02Mn-0.02Ti-0.32Si-0.015P-0.008S-0.83Nb) showed a material consumption rate of 0.223 g/m² h in boiling 65 % nitric acid without sensitization, while after sensitization at 863 K (590 °C) for 1, 10 and 500 h, the steel corroded at rates of 0.304, 0.593 and 2.45 g/m² h respectively. In contrast to the Ti-stabilized steels, the Nb-stabilized steels are resistant to knife-line corrosion immediately after welding. Only after heating at critical temperatures at which chromium carbides precipitate does the Nb-stabilized steel also become sensitive. However, this can be eliminated by two heat treatments [391].

Spot welding of unstabilized steels ((%) Fe-0.10C-17.5Cr-9.5Ni-1.45Mn-0.5Si-0.05Ti) does not cause susceptibility to intercrystalline corrosion in boiling 65 % nitric acid even with very long welding times. The corrosion rate after 3 × 48 h periods in the Huey test is 0.2 to 0.3 mm/a (7.87 to 11.81 mpy). A potential of about + 1.0 V_{SHE} is established, with chemical resistance [362].

At a very low concentration of the corrosion products (metal nitrates), the corrosion rate of the steel K-299 ((%) Fe-0.04C-17.50Cr-10.12Ni-1.43Mn-0.73Nb-0.69Si) in boiling 65 % nitric acid is 0.20 mm/a (7.87 mpy). As the concentration of the corrosion products increases, so does the corrosion rate, for example, to about 0.78 mm/a (30.71 mpy) at 0.1 g/l [961].

When the steel Sandvik® 2R12 ((%) Fe-<0.020C-<0.10Si-19.5Cr-11Ni-<0.015P-<0.010S) is used, temperatures of 390, 360 and 350 K (117, 87 and 77 °C) should not be exceeded for an acceptable corrosion rate of 0.1 mm/a (3.94 mpy) in 65, 80 and 90 % nitric acid respectively [602].

According to Table 55, the corrosion rate of chromium-nickel steel in boiling 65 % nitric acid is practically unchanged at 0.5 mm/a (19.69 mpy) as the silicon content increases (from 3.8 to 5.9 %), but is reduced in boiling 98 % acid (at the hyperazeotropic point) by a factor of 6 at an Si content of 5.9 % [293].

Chemical composition, %									Corrosion rate
Cr	Ni	Mn	Mo	Si	N	C	S	P	mm/a (mpy)
21.71	15.34	0.88	–	3.78	0.098	0.010	0.011	0.002	0.12 (4.72)
20.70	15.65	0.87	0.21	4.29	0.077	0.009	0.013	0.004	0.08 (3.15)
20.95	15.87	0.92	0.05	4.80	0.132	0.007	0.012	0.003	0.06 (2.36)
20.57	16.71	0.96	–	5.62	0.117	0.006	0.011	0.002	0.04 (1.57)
21.80	17.54	1.07	–	5.87	0.112	0.017	0.013	0.005	0.02 (0.79)

Table 55: Corrosion rate of the CrNiSi steel 20 15 in boiling 98 % HNO_3 as a function of the Si content [293]

Figure 23 shows the corrosion behavior of the steels X 3 CrNi 18 10 (cf. 1.4306, AISI 304 L), X 2 CrNi 25 20 (X1CrNi25-21, 1.4335) and X 2 CrNiSi 18 15 (X1CrNiSi18-15-4, 1.4361, A 336) in boiling nitric acid solutions in the concentration range 67 – 98 % in three 48 h tests. An excellent resistance (about 0.04 g/m² h), even in 98 % acid, is achieved by addition of 4 % silicon to the steel [259]. Welding Si-containing steels requires particular measures.

After 220 h at 453 K (180 °C) in 70 % nitric acid, the steels 08Ch18N10T (1.4541, AISI 321) and 08Ch22N6T are unusable for industrial purposes either as compact alloy or welding material, with corrosion rates of 4.2 and 6.1, 3.9 and 4.5 g/m² h respectively. On the other hand, the low-carbon steel 000Ch18N11 (cf. 1.4306,

AISI 304 L; Fe-<0.001C-18Cr-11Ni) is suitable for industrial use, as compact alloy or in the welded state, with material consumption rates of 0.4 and 0.53 g/m² h respectively [242].

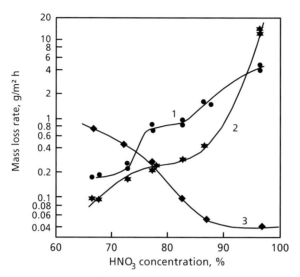

Figure 23: Material consumption rate of the steels 1) X3CrNi 18 10, 2) X2CrNi 25 20 and 3) X2CrNiSi 18 15 4 in 67 to 98 % boiling HNO_3 [259]

The CrNi cast steel 18 8 with a normal Si content of 0.5 to 1 % and a C content slightly above 0.15 % is resistant in up to 80 % nitric acid at 298 (25 °C). Above 80 % HNO_3, an Si content of at least 2.5 to 3 % is necessary. With more than 4 % Si and a C content of between 0.12 and 0.13 % the steel is practically resistant in the entire concentration range of nitric acid. Corrosion rates as a function of the Si and C content can be seen in Figures 24 an 25 [35].

The silicon-containing steel X 2 CrNiSi 18 15 has found use, above all, for tanks, pipelines, pumps and fittings for hot, highly concentrated nitric acid, since conventional austenitic steels are not suitable in above 70 % HNO_3 [405].

According to Table 56, the corrosion resistance of the steel 000Ch20N20 (Fe-0.03C-18.57Cr-19.40Ni-0.71Mn-0.26Si; austenitic) increases with an increasing silicon content in 24 mol/l nitric acid at 373 K (100 °C), but is reduced in boiling 12 mol/l nitric acid. Above 6 % silicon, an austenitic-ferritic structure is present. If the heat-treated steel is tempered at 1023 to 1173 K (750 to 900 °C), a sigma phase forms at silicon contents above 3 %, leading to susceptibility to intercrystalline corrosion and a reduction in the notched impact strength [324].

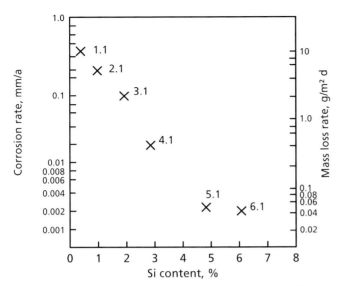

Figure 24: Influence of the Si-content on the corrosion rates of CrNi cast steel 18 8 with about 0.03 % C in 98 % HNO_3 at 298 K (25 °C), test duration 720 h [35]

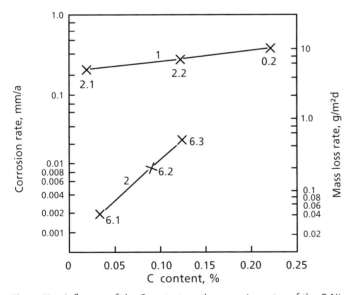

Figure 25: Influence of the C-content on the corrosion rates of the CrNi cast steel 18 8 with Si contents of about 1 and > 6 % in 98 % HNO_3 at 298 K (25 °C), test duration 720 h [35]
1) 1 % Si; 2) 6 % Si

HNO₃ mol/l	Temperature K (°C)	Si-content %	Material consumption rate, g/m² h
24	373 (100)	0.26	3.5
		2.0	2.3
		4.0	1.65
		5.0	0.3
		6.0	0.2
12	boiling point	0.26	0.15
		2.0	0.4
		4.0	0.5
		6.0	0.5

Table 56: Dependence of the corrosion rate of the steel 000Ch20N20S on the Si-content in 24 mol/l HNO₃ at 373 K (100 °C) and 12 mol/l HNO₃ at the boiling point after 100 h [324]

The material consumption rate of stainless steel in concentrated nitric acid (> 95 %) can be greatly reduced (e.g. from 0.26 to < 0.01 g/m² h) if ≥ 20 % of the surface of the steel is brought into contact with aluminium. The potential of the partly covered steel is reduced to a value close to that of aluminium in this way [501, 578].

The contributions to the material consumption rate of the metals in the steel X5CrNiSi 18 15 in 98 % nitric acid at 323 K (50 °C) correspond quite closely to the proportions present at an anodic polarization of 1.85 and 1.95 V_{SHE} [247].

The corrosion behavior of the special steel X 2 CrNiSi 18 15 (cf. 1.4361, A 336; (%) Fe-(0.014-0.015)C-(17.65-17.92)Cr-(15.0-15.32)Ni-(0.07-0.09)Mo-(0.72-0.79)Mn-(3.90-4.26)Si-(0.007-0.021)S-(0.017-0.021)P) in 98 % industrial nitric acid at 293 K (20 °C) is shown in Table 57 [348]. Iron makes the greatest contribution to the weight loss here.

During manual arc welding on the steel X 2 CrNiSi 18 15 with about 4 % silicon, careful control of the heating of the weld joints is necessary, largely in order to exclude intercrystalline corrosion in the presence of nitric acid [377]. The corrosion rate of the steel in 98 % nitric acid at 323 K (50 °C) rose slightly with the duration of exposure at the start, and then remained virtually constant as the test progressed, after the transpassive region had been reached [378].

Si-content %	Fe	Cr	Ni
	Material consumption rate, g/m² h		
3.90	0.0063	0.0015	0.0013
4.26	0.0114	0.0059	0.0028

Table 57: Corrosion of the steel X 2 CrNiSi 18 15 (A 336) with 3.9 and 4.26 % Si in 98 % HNO₃ at 293 K (20 °C) showing the contribution to the weight loss of Fe, Cr and Ni, from 800 h tests [348]

Corrosion studies with the Russian steels 02Ch8N22S6 ((%) Fe-0.02C-8Cr-22Ni-6Si), 02Ch8N22T (Ti-stabilized), 02Ch8N22S6B ((%) Fe-0.02C-8Cr-22Ni-6Si; Nb-stabilized), 02Ch12N10S5, 02Ch12N10S5T, 02Ch12N10S5B ((%) Fe-0.02C-12Cr-10Ni-5Si; Nb-stabilized) and 02Ch12N10S5T, which were rolled at 1370 K (1097 °C) to sheets 12 mm thick and then quenched from 1323 K (1050 °C) in water and finally tempered at 773 – 1123 K (500 – 850 °C) for about 5 – 120 min, also demonstrated the beneficial effect of niobium additions on intercrystalline corrosion in boiling 72 and 98 % nitric acid. The occurrence of a ferritic phase caused a further improvement in the resistance to intercrystalline corrosion [376].

By addition of 2.3 % aluminium to chromium-nickel steel ((%) Fe-0.018C-24.86Cr-19.96Ni-0.12Si), which had been solution-annealed at 1373 K (1100 °C) for about 15 min, the material consumption rate in boiling 98 % nitric acid could be reduced from 5.55 to 0.47 g/m² h [446].

According to Figure 26, corrosion of the steel ZI-52 (000Ch20N20S5, Russ. grade: (%) Fe-0.017C-19.2Cr-20.4Ni-5.4Si-0.33Mn) in 23 mol/l nitric acid reacts very sensitively to the temperature and duration of tempering treatment. At 1023 K (750 °C), the corrosion-sensitive sigma phase precipitates. Only after tempering at 1233 K (960 °C) is the susceptibility largely eliminated. This steel should therefore be used only in non-welded form for nitric acid plants [296].

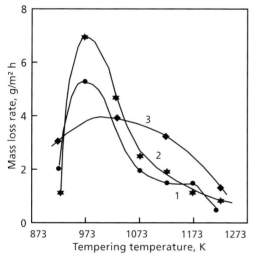

Figure 26: Dependence of the material consumption rate of the steel ZI-52 in 23 mol/l HNO_3 on the temperature and duration of tempering [296]
1) 10 min, 2) 1 h, 3) 100 h

Studies have shown that the sensitivity of chromium-nickel steels to intercrystalline corrosion in nitric acid-chromate solutions increases with increasing grain size, decreasing solution annealing temperature, decreasing carbon content, increasing test temperature, increasing nitric acid and chromate concentration and increasing mechanical stress [350].

The penetration depth of intercrystalline corrosion on the steel SUS 304 L (cf. 1.4306; (%) Fe-0.02C-18.74Cr-11.33Ni-0.98Mn-0.54Si-0.030P-0.008S) in boiling 21 % nitric acid + 4 g/l Cr^{6+} ions is about 80 µm after 27.8 h; it follows the duration of exposure linearly in the time interval 1 to 100 h [277].

The high-chromium carbides of the type $M_{23}C_6$ which occur in the steels Ch18N10 ((%) Fe-0.08C-18.4Cr-10.2Ni-1.08Mn-0.3Si-0.005P-0.014S-0.005N) and Ch18N14 ((%) Fe-0.035C-18.8Cr-14.6Ni-0.35Mn-0.75Si-0.005P-0.03S-0.004N) after quenching from 1323 K (1050 °C) and subsequent heat treatment of 973 K (700 °C) for 500 and 1000 h are attacked selectively by boiling 27 % nitric acid + 40 g/l Cr^{6+} (as $K_2Cr_2O_7$) [392].

The material consumption rate of the steel 03Ch18N14 (Russ. grade; Fe-0.03C-18Cr-14Ni) in a solution of 27 % nitric acid with 40 g/l Cr^{6+}, with renewal of the solution every 2 hours, was determined in 14 h tests at room temperature as a function of the phosphorus and silicon content. It was 19.1 g/m^2 h with 0.048 % P + 0.16 % Si and 11.6 g/m^2 h with 0.005 % P + 0.75 % Si [402]. Intercrystalline corrosion occurred in the first 6 hours.

The effect of the silicon content in the steel Ch20N20 ((%) Fe-(0.004-0.015)C-(19.4-21.8)Cr-(19.3-20.8)Ni-(0.05-5.40)Si-(0.002-0.1)P) on corrosion in boiling 27 % nitric acid + 40 g/l Cr^{6+} (as chromate) depends on the phosphorus content of the steel. In 48 h tests, curve (1) in Figure 27 is obtained at a content of 0.002 % P and curve (2) at a content of 0.1 % P [313]. Only above 4 % silicon is the material consumption rate of the steel, especially with a high phosphorus content (0.1 % P), significantly reduced. A relationship exists here between the corrosion rate and the cathodic polarization current density at 1.29 V_{SHE}. Similar corrosion rate values (as a function of the silicon content) to those in Figure 27, curve ①, are also found with the steel Ch18N11 in boiling 27 % nitric acid + 4 g/l Cr^{6+} [270].

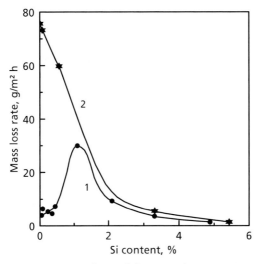

Figure 27: Dependence of the material consumption rate of the steel Ch20N20 in boiling 27 % HNO_3 on the silicon content at ① 0.002 % and ② 0.1 % phosphorus [313]

The dependence of corrosion behavior on the silicon and phosphorus content of the steel 12Ch18N10T (1.4541, AISI 321) in Cr^{6+}-containing nitric acid is also confirmed elsewhere [280].

After 96 h, the material consumption rate of the chromium-nickel steel (%) Fe-(0.006-0.011)C-20Cr-20Ni-0.002P-Si in boiling 27 % nitric acid + 40 g/l Cr^{6+} (as chromate) falls as the silicon content of the steel increases, from 7.6 g/m^2 h at 0.01 % Si to 1.4 g/m^2 h at 4.8 % Si. The corrosion-improving action of silicon recedes as the carbon content increases [505].

Additionally introduced phosphorus (0.002 – 0.097 % P) causes a marked increase in the intercrystalline corrosion of the steel Ch20N20 ((%) Fe-<0.02C-21Cr-19.5Ni-0.005Mo-0.002Mn-<0.01Cu) in boiling 27 % nitric acid + 40 g/l Cr^{6+} by a factor of about 20. A marked increase was already found with 0.025 % P. In contrast, sulfur has no noticeable influence [383].

The electrochemical corrosion behavior of the steel Ch20N20 in 27 % nitric acid at 313 K (40 °C) and 1.36 V_{SHE} with and without addition of 40 g/l Cr^{6+} is not substantially influenced by the carbon content of the steel (0.004 – 0.096 % C), regardless of whether the phosphorus content is low (0.002 % P) or high (0.1 % P). In contrast, the phosphorus content increases the susceptibility to intercrystalline corrosion in the transpassivity region. The susceptibility of the phosphorus-containing steel to corrosion also increase as the size of the austenite grain decreases [284]. The material consumption rate was 5.7 – 6.7 g/m^2 h at a phosphorus content of 0.002 % P and 59 g/m^2 h at 0.09 % P.

To be able to use the steel AISI 304 L (Fe-25Cr-20Ni-Nb) in boiling 8 – 12 mol/l nitric acid with radioactive material, the concentration of highly oxidizing metal ions, such as, for example, Cr^{6+} and Fe^{3+} of 100 – 1000 ppm, must be greatly reduced. The occurrence of vapor-liquid phase boundaries should also be prevented in heat exchanger pipes [439].

As a representative of other chromium-nickel steels, Remanit® 4306 (AISI 304 L, X2CrNi19-11: (%) Fe-0.025C-18.35Cr-10.24Ni-0.26Mo-1.54Mn-0.36Si-0.027P-0.003S-0.033B-0.043N; annealed at 1393 K (1120 °C) for 15 min and quenched in water) corrodes in 65 % nitric acid containing 0.5 % Cr^{6+} ions at 313 K (40 °C) with uninterrupted exposure of the specimen at a material consumption rate of 2.75 g/m^2 h after 72 h. If the test solution is renewed periodically every 4 h, the consumption rate is significantly higher at 4.2 g/m^2 h. This difference in corrosion is based on exhaustion of the test solution, which can largely be avoided by a sufficiently large solution volume [729].

To test the usability of the steel UHB 25 L (1.4845, AISI 310 S; (%) Fe-0.016C-24.9Cr-20.8Ni-1.63Mn-0.08Mo-0.03Cu-0.16V-0.45Si-0.012P-0.009S) for reprocessing nuclear fuels, it was investigated in a stringent Huey test (boiling 65 % (13.4 mol/l) nitric acid with and without addition of Cr^{6+} in the form of dissolved CrO_3). Some results are summarized in Table 58 [425].

The material consumption rate of the CrNi steel 14 14 also increases in Cr^{6+}-containing boiling 65 % nitric acid as the phosphorus content increases, from about 3 g/m^2 h at 10 ppm P to 83 g/m^2 h at 1000 ppm P. While general corrosion with a material consumption rate of about 3 and < 2 g/m^2 h is found at a silicon content of

Material state	CrO$_3$ addition g/l	Number of test periods			
		5	15	5	15
		Corrosion rate, mm/a (mpy)		Depth of local attack, mm	
Delivery state	–	0.06 – 0.07 (2.36 – 2.76)	0.06 – 0.07 (2.36 – 2.76)	–	0.005
Solution-annealed	–	0.05 – 0.07 (1.97 – 2.76)	0.065 (2.56)	0.002	0.005 – 0.010
Sensitized for 30 min at 973 K (700 °C)	–	0.119 (4.69)	0.267 (10.51)	0.018	0.064
		0.068 (2.68)	0.120 (4.72)	–	0.048
Solution-annealed	0.050 (1.97)	0.697 (27.44)	–	0.036	–
	0.100 (3.94)	1.616 (63.62)	–	0.075	–

Table 58: Corrosion of the steel UHB 25 L (AISI 310 S) in boiling 65 % HNO$_3$ with and without a CrO$_3$ addition in 5 to 15 periods of 48 h each [425]

< 10^3 and > 2 × 10^4 ppm, intercrystalline corrosion occurs between these silicon contents, with a maximum of about 63 g/m^2 h at 6 × 10^3 ppm Si [462].

While additions of 0.005 to 3.4 % molybdenum at a constant silicon and phosphorus content of 0.5 and 0.003 % hardly affect the corrosion rate of the steel (%) Fe-0.05C-17Cr-12.5Ni-(1.60-1.65)Mn-0.012N at all in boiling 65 % HNO$_3$ + 0.02 g/l Cr^{6+}, increasing phosphorus and silicon contents reduce the corrosion of the steel. The material consumption rate is thus reduced from 0.44 to 0.22 g/m^2 h when the Si content rises from 0.5 to 2.08 %, and increased from 0.41 to 0.56 g/m^2 h when the phosphorus content rises from 0.002 to 0.098 % [460].

In nitric acid solutions of ≥ 80 % or lower concentration, but containing Cr^{6+} ions, it is necessary to use a CrNi cast steel 18 9 or 18 13 with the lowest possible C content (≤ 0.03 % C) and an Si content of about 4 %. If it is not possible to produce a cast steel with such a low C content, steel with C contents of up to 0.12 % C, but only stabilized by Nb can also be used. The specimens of the materials investigated were solution-annealed (1353 K (1080 °C) ± 20 K (20 °C)/2 h/water). In 60 to 70 % HNO$_3$, the CrNi cast steels 18 9 and 18 13 with a normal Si content and C contents ≤ 0.035 C is adequate. The steel with a higher Ni content showed the better corrosion resistance in the HNO$_3$ solutions. Valves made of CrNi cast steel 18 9 and 18 13 with Si contents about 4.5 % have proved suitable in concentrated HNO$_3$ (about 96 %) at 318 K (45 °C) and were found to be several times more resistant than those of cast steel with normal Si contents of around 1 % [43].

An electrochemical method for the determination of the susceptibility of stainless steels to intercrystalline corrosion is based on the appearance of an activation branch in the potentiodynamic curve in the potential range – 0.15 to + 0.55 V (against AgCl) at a measurement rate of 12 V/h. The electrolyte comprises 20 g/l FeCl$_3$ 6 H$_2$O with a high redox potential, as well as 5 % nitric acid and 80 mg/l hydrogen chloride [264, 329].

A method for determining the tendency to undergo intercrystalline corrosion on the basis of potential measurements at room temperature under a drop of liquid consisting of 5 % HNO_3 + 20 g/l $FeCl_3 \times 6\ H_2O$ + 70 ml/l HCl (for CrNi steels 18 8) or 240 ml/l HCl (for El-943) at 0.74 – 0.80 V was tested on the steels Ch18N9T (1.4541, AISI 321) and 06ChN28MDT (Russ. grade El-943; (%) Fe-0.06C-22.8Cr-24.7Ni-0.50Ti-2.77Cu-3.5Mo). The conclusiveness of the results is increased by precise demarcation of the drop, electrical insulation, enlargement of the drop diameter and brief pickling of the steel surface before-hand at room temperature in hydrochloric acid. The method in question is also suitable for in-situ testing of workpieces [367].

In chloride-containing HNO_3 such as occurs, for example, in the fertilizer industry, CrNi steels also undergo intercrystalline corrosion and pitting corrosion, as well as general corrosion. The passive or transpassive state of the steel is eliminated and the surface activated by Cl^- ions.

The occurrence of intercrystalline corrosion and pitting corrosion is determined by the nitric acid and Cl^- ion concentration, the temperature, the flow conditions and the operating time, apart from the material parameters. Figure 28 shows the dependence of the material consumption rates of X6CrNiTi18-10 (1.4541, AISI 321) on the NaCl content and HNO_3 concentration. This figure clearly shows the critical chloride ion concentration at the particular HNO_3 concentration [365].

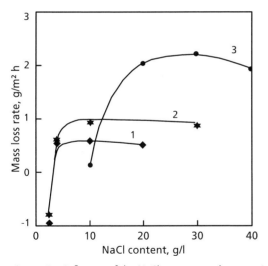

Figure 28: Influence of the NaCl content on the corrosion behavior of X6CrNiTi18-10 in various concentrations of HNO_3 at 300 K (27 °C), test duration 1 h [365]
(1) 1 mol/l; (2) 2 mol/l; (3) 4 mol/l

No pitting corrosion was found on the steel X6CrNiTi18-10 (AISI 321) at nitric acid concentrations between 1.3 and 54.7 % with 14.2 g/l chloride up to 333 K (60 °C), and between 3.1 to 4.9 % HNO_3 with 28.4 g/l chloride [371]. In 45 % HNO_3 containing HCl or Fe(III)chloride, the steels CrTi25, CrNiMoCu28-28, CrMnNiN17-19-4 and CrNiTi18-10 are attacked both by intercrystalline corrosion and pitting [406]. Intercrystalline corrosion on X6CrNiTi18-10 occurred above 323 K (50 °C) in 48.4 % HNO_3 with 3.65 g/l HCl [268].

While the corrosion rate of X5CrNi18-10 (1.4301, AISI 304) in boiling 30 % HNO_3 with 6.8 g/l chloride is 0.03 to 0.33 mm/a (1.18 to 13 mpy), in the vapor space it is considerably higher at 1.1 to 3.3 mm/a (43.3 to 130 mpy). This is due to the formation of nitrosyl chloride. At a chloride content of 13.6 g/l, the corrosion rate was > 3.3 mm/a (> 130 mpy) (see Table 59) [529, 859].

Steel	HNO_3 %	HCl g/l	Cl^- g/l	Temperature K (°C)	Corrosion rate mm/a (mpy)
X5CrNi18-10 (AISI 304)	30	–	6.8	boiling	1.1 – 3.3 (43.3 – 130)
	30		13.6	boiling	> 3.3 (130)
	3.7	14.6	–	305 (32)	4.2 (165)
	48.6	167.6	–	305 (32)	84.0 (3307)
CrTi steel, ferritic	38	1.9	–	313 (40)	0.3 – 1.1 (11.8 – 43.3)
	46	3.8	–	313 (40)	0.3 – 1.1 (11.8 – 43.3)
CrNiTi steel, austenitic	38	1.9	–	313 (40)	> 3.3 (130)
	46	3.8	–	313 (40)	> 3.3 (130)

Table 59: Corrosion behavior of steels in the vapor space above chloride-containing nitric acid solutions [529, 859]

The steel 1.4306 (AISI 304 L, nitric acid quality) was investigated as a cooling finger and aftercondenser in a nitric acid condensate with a low chloride content after a test duration of 12 weeks. To record irregularities in the sealing region, the cooling finger was made from pipe rings sealed with PTFE rings. The aftercondenser was produced from bar material. The material consumption rates per unit area of the starting material a) of the pipe rings and b) of the bar material in the delivery state or solution-annealed at 1333 K (1060 °C)/15 min/ water in the Huey test were a) 0.08 – 0.1 g/m^2 h and b) 0.12 or 0.09 g/m^2 h respectively. The condensates consisted of 6 mol/l HNO_3 + 45 mg/l Cl^-, 8.5 mol/l HNO_3 + 20 mg/l Cl^- and 10 mol/l HNO_3 + 15 mg/l Cl^- and all contained 1 mg/l F^-. The corrosion medium was renewed weekly. Although the studies showed very low material consumption rates, local roughening, shallow pits and pitting corrosion occurred. The steel 1.4466 (AISI 310 MoLN, X1CrNiMoN25-22-2) is recommended as it is resistant to pitting corrosion and at the same time has a good resistance to sulfuric acid [67].

According to Figure 29, chloride ions have a disastrous influence on the material consumption rate of the steels 04Ch18N10 ((%) Fe-0.013C-17.8Cr-9.9Ni-1.15Mn) and 12Ch18N10T (1.4541, AISI 321(%) Fe-0.09C-17.4Cr-9.4Ni-1.74Mn-0.41Ti; Ti-stabilized) in dilute nitric acid [279]. In order to keep the corrosion rate of the steels below 0.1 mm/a (4 mpy) in 1 mol/l nitric acid, the content of chloride ions in the acid should be < 0.1 g/l.

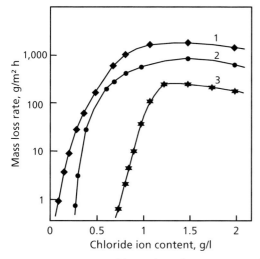

Figure 29: Corrosion of the steels 04Ch18N10 (1, 3) and 12Ch18N10T (2) in boiling 1 mol/l (1, 2) and 0.1 mol/l (3) HNO_3 as a function of the chloride ion content [279]

To reduce the corrosive action of chloride-containing 2.3 – 3.0 mol/l nitric acid, it is treated with potassium permanganate, the chloride largely being removed as chlorine. Subsequent distillation results in 13 % nitric acid with 0.15 g/l chloride ions. The corrosion which occurs during further concentration in a distillation column consisting of chromium-nickel steel was eliminated by further reduction of the chloride content to about 0.02 g/l [542].

The presence of chloride ions in nitric acid reveals complex processes in the low-carbon and Ti-stabilized steels. The cathodic process in chloride-containing nitric acid is thus inhibited only at the carbide inclusions. The kinetics of the reduction of nitric acid are not influenced on the passive surface of the steel 12Ch18N10T (AISI 321) or 04Ch18N10 [265].

Stainless steels are not suitable for plants in which mineral fertilizers are produced with reaction mixtures consisting of nitric acid and potassium chloride (sylvinite). Vitreous carbon, smelting slags and mullite are suitable here (see Sections B 3 and B 12) [390].

Table 60 shows the influence of hydrogen fluoride and chloride ions on the corrosion of the steels 12Ch18N10T (AISI 321, 1.4541; (%) Fe-0.09C-17.1Cr-10.4Ni-1.28Mn-0.78Ti), 08Ch22N6T ((%) Fe-0.08C-21.7Cr-5.4Ni-1.28Mn-0.78Ti) and 03Ch23N6 ((%) Fe-0.02C-22.7Cr-5.7Ni-1.42Mn) in 10 mol/l nitric acid. The steel 03Ch23N6 is the most resistant for such solutions.

The corrosion rates of the steels 12Ch18N10T, 08Ch22N6T and 03Ch23N6 in 0.5 mol/l nitric acid with and without additions of hydrogen fluoride are summarized in Table 61 [285]. With an addition of 0.1 mol/l HF to the nitric acid, the corrosion rate for the steel 12Ch18N10T of 0.69 mm/a (27.17 mpy) at room temperature is already too high.

Steel	Addition, mol/l	293 K (20 °C)	323 K (50 °C)
		Corrosion rate, mm/a (mpy)	
12Ch18N10T	–	0.018 (0.71)	0.020 (0.79)
	0.01 HF	0.12 (4.72)	0.48 (18.9)
	0.1 HF	0.77 (30.31)	5.08 (200)
	0.01 HF + 5 g/l Cl^-	0.052 (2.05)	0.16 (6.3)
	5 g/l Cl^-	0.014 (0.55)	0.036 (1.42)
08Ch22N6T	–	0.008 (0.31)	0.022 (0.87)
	0.01 HF	0.067 (2.64)	0.26 (10.24)
	0.1 HF	0.47 (18.5)	1.29 (51)
	0.01 HF + 5 g/l Cl^-	0.027 (1.06)	0.11 (4.33)
	5 g/l Cl^-	0.013 (0.51)	0.037 (1.46)
03Ch23N6	–	0.010 (0.39)	0.016 (0.63)
	0.01 HF	0.023 (0.91)	0.16 (6.3)
	0.1 HF	0.093 (3.66)	0.68 (26.77)
	0.01 HF + 5 g/l Cl^-	0.033 (1.3)	0.087 (3.43)
	5 g/l Cl^-	0.011 (0.43)	0.018 (0.71)

Table 60: Influence of HF and Cl^- on the corrosion rate of some CrNi steels in 10 mol/l HNO_3 at 293 and 323 K (20 and 50 °C) [408]

Steel	293 K (20 °C)			323 K (50 °C)			373 K (100 °C)		
	0	0.01 HF	0.1 HF	0	0.01 HF	0.1 HF	0	0.01 HF	0.1 HF
	Corrosion rate, mm/a (mpy)								
12Ch18N10T	0.014 (0.55)	0.040 (1.57)	0.61 (24.02)	0.020 (0.79)	0.043 (1.69)	0.82 (32.28)	0.031 (1.22)	0.087 (3.43)	1.95 (76.77)
08Ch22N6T	0.014 (0.55)	0.016 (0.63)	0.17 (6.69)	0.014 (0.55)	0.019 (0.75)	0.28 (11.02)	0.034 (1.34)	0.045 (1.77)	0.64 (25.2)
03Ch23N6	0.012 (0.47)	0.016 (0.63)	0.05 (1.97)	0.017 (0.67)	0.026 (1.02)	0.29 (11.42)	0.033 (1.3)	0.037 (1.46)	0.87 (34.25)

Table 61: Corrosion of CrNi steels in 0.5 mol/l HNO_3 with and without additions (mol/l) of HF between 293 and 373 K (20 and 373 °C) [285]

Figure 30 shows the corrosion of the steel 12Ch18N10T (AISI 321) in boiling HNO_3/HF solutions [281]. The complex relationship between the acid mixtures and corrosion of the steel can be seen from the isocorrosion curves. According to Figure 31, 0.2 mol/l iron nitrate in the HNO_3/HF solution has an inhibiting influence on corrosion of the steel AISI 304 SS ((%) Fe-0.059C-18.29Cr-9.36Ni-1.0Mn-0.48Si-0.029P-0.002S) at 223 K (-50 °C) [346]. According to ESCA studies, the oxide layer formed on the chromium-nickel steel in nitric acid consists of an accumulation of chromium oxide alongside iron oxides on the outside with a layer of SiO_2 of varying thickness [347].

Attack on the steel 12Ch18N10T (AISI 321) at 353 K (80 °C) in 12 mol/l nitric acid + 0.12 mol/l hydrogen fluoride, which initially proceeds at a material consumption rate of about 30 g/m² h, drops to about 3.5 g/m² h after 300 h [335].

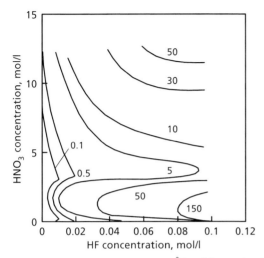

Figure 30: Isocorrosion curves (g/m² h) of the steel 12Ch18N10T (AISI 321) in boiling HNO_3/HF solutions [281]

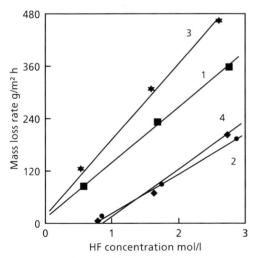

Figure 31: Influence of the HF concentration on the rate of dissolution of the steel AISI 304 SS in HNO_3/HF solutions with and without iron nitrate at 223 K (−50 °C) [346]
1) 3.5 mol/l HNO_3
2) 3.5 mol/l HNO_3 + 0.2 mol/l $Fe(NO_3)_3$
3) 0.8 mol/l HNO_3
4) 0.8 mol/l HNO_3 + 0.2 mol/l $Fe(NO_3)_3$

In a solution of (14 – 16) % HNO_3 + 3.8 % HF, the steel 12Ch18N10T (AISI 321) corrodes between 328 and 333 K (55 and 60 °C) with a material consumption rate of 3.83 g/m^2 h (4.3 mm/a) which is too high [416].

Thorium-containing fuel elements require nitric acid containing hydrogen fluoride for processing. While Inconel® 690 (2.4642, see Section A 28) is recommended for the reaction vessel at about 400 K (127 °C), AISI 304 L can also be used up to 368 K (95 °C) [449].

Solutions of 10 % HNO_3 + 39 % HF at 343 K (70 °C) and 5 % HNO_3 + 1 % $FeCl_3$ are used to investigate the susceptibility of weld seams of CrNiTi 18 9 0.5 steels to intercrystalline corrosion [307].

According to a previous report, stainless steel proved suitable for evaporation of the primary solution of HNO_3 + HF + $Al(NO_3)_3$ + plutonium nitrate, as well as for the heating pipes [368].

The corrosion problems which arise during processing of nuclear fuels (plutonium) with 12 mol/l nitric acid + 0.44 mol/l hydrogen fluoride + 0.48 mol/l aluminium nitrate (for the purpose of complexing with fluorides), especially during evaporation and recovery of the nitric acid, cannot be solved by CrNi steels, but more favorably by special nickel-chromium-alloys, such as, for example, Ni-34Cr-1.0Ti-3.7Si-0.6Mn [340].

In addition to this, the material consumption rates of the steels 12Ch18N10T (AISI 321, 1.4541) and 03Ch23N6 in boiling 1 mol/l nitric acid with 1 g/l hydrogen fluoride + 2.7 g/l aluminium nitrate are 3.2 and 0.55 g/m^2 h [289].

In nitric acid solutions containing additions, such as, for example, sodium nitrate, secondary reactions have an influence on the corrosion process and the redox potential of steels [272].

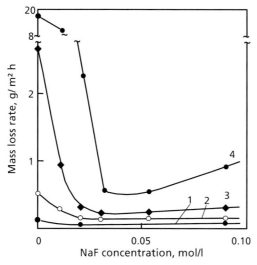

Figure 32: Material consumption rate of the steel 12Ch18N10T (AISI 321) in 4 mol/l HNO_3 + 10 g/l $K_2Cr_2O_7$ + NaF as a function of the NaF concentration at (1) 323 K (50 °C), (2) 348 K (75 °C), (3) 363 K (90 °C) and (4) the boiling point [267]

Isocorrosion curves (1 and 0.5 mm/a (39.4 and 19.7 mpy)) for various CrNi and CrNiMo steels in nitric acid solutions containing sodium fluoride show that the steels of lower nickel content are more resistant than the chromium-nickel steel 18 10 [334].

According to Table 62, corrosion of the three chromium-nickel steels, which differ only in their carbon content, in hydrazine-containing nitric acid at 388 K (115 °C) is increased drastically by sodium fluoride. In its absence the influence of 2 g/l hydrazine can be ignored [283].

Figure 32 shows the dependence of the corrosion of the steel 12Ch18N10T (AISI 321) in 4 mol/l nitric acid + 10 g/l $K_2Cr_2O_7$ + NaF on the temperature and the sodium fluoride concentration. The material consumption rates of the steel at the boiling point are given in Table 63 [267]. The optimum passivating action of the fluoride ions lies between 0.03 and 0.05 mol/l.

Concentration, g/l			Material consumption rate, g/m² h		
HNO_3	NaF	N_2H_4	12Ch18N10T *	08Ch18N10T *	04Ch18N10T *
75	0	2	0.11 ± 0.04	0.16 ± 0.04	0.2 ± 0.1
75	0.05	2	0.7 ± 0.1	2.4 ± 0.1	5.1 ± 0.6
75	0.5	2	11 ± 1	16 ± 1	15 ± 1
75	5	2	96 ± 2	183 ± 8	131 ± 11
75	2	0	1.0 ± 0.1	1.2 ± 0.4	2.2 ± 0.7
75	2	2	47 ± 1	46 ± 5	68 ± 2
75	2	10	53 ± 2	66 ± 7	56 ± 3
100	2	2	51 ± 1	95 ± 7	81 ± 5
500	2	2	5.8 ± 1	3.8 ± 0.8	3.6 ± 0.8

Table 62: Corrosion rate of some chromium-nickel steels in HNO_3 with NaF and hydrazine [283]
* cf. AISI 321 with different carbon content

$K_2Cr_2O_7$ mol/l	NaF, mol/l					
	0	0.01	0.02	0.03	0.05	0.10
	Material consumption rate, g/m² h					
0.017	10.5	2.0	0.4	0.4	0.5	0.7
0.034	16.5	8.0	2.7	0.7	0.8	1.2
0.068	33.0	18.2	9.7	2.0	1.2	1.7

Table 63: Dependence of the corrosion of the steel 12Ch18N10T in boiling 6 mol/l HNO_3 + $K_2Cr_2O_7$ on the concentration of NaF [267]

Pitting and knife-line corrosion occurs on welded joints of chromium-nickel steels in a solution of 500 g/l nitric acid + 600 g/l sodium nitrate + 50 g/l copper nitrate + 5 g/l sodium fluoride (Table 64) [341].

Steel	Chemical composition, %						Corrosion rate mm/a (mpy)	Type of corrosion
	C	Mn	Si	Cr	Ni	Others		
00Ch18N10	0.015	1.7	0.7	17.3	10.4	–	1.2 (47.2)	pitting corrosion
0Ch18N10T	0.06	1.5	0.8	18.0	9.8	0.6 Ti	0.98 (38.6)	pitting corrosion
0Ch18N12B	0.05	0.9	0.3	17.7	12.1	0.7 Nb	0.79 (31.1)	pitting corrosion
1Ch18N9T	0.11	1.3	0.8	17.2	7.9	0.7 Ti	1.13 (44.5)	knife-line corrosion and pitting corrosion
Ch28N18	0.16	1.7	1.1	22.6	18.0	0.4 Ti	2.90 (114)	intercrystalline and pitting corrosion

Table 64: Corrosion rate of welded joints of austenitic steels in a solution containing 500 g/l HNO_3 + 600 g/l $NaNO_3$ + 50 g/l $Cu(NO_3)_2$ + 5 g/l NaF at room temperature [341]

Corrosion of the steel 12Ch18N10T (AISI 321) in boiling 6 mol/l nitric acid is inhibited by fluorides if oxidizing agents, such as, for example, 0.2 mol/l CrO_3 or ammonium vanadate (NH_4VO_3), are present at the same time. At a constant acid concentration, the corrosion rate reaches a minimum at a fluoride ion concentration of between 0.05 and 0.15 %, the position of the minimum depending on the composition and nature of the oxidizing agent and the structure of the particular alloy [282].

Nevertheless, the material consumption rate of the steel in 0.5 mol/l nitric acid can be increased considerably (130 to 160 g/m^2 h) by between 0.1 and about 0.35 mol/l fluoride ions (KF, NaF, ammonium fluoride and hydrogen fluoride). As the fluoride ion concentration increases further, the material consumption rate then decreases to 30 to 40 g/m^2 h. Corrosion in the presence of fluoride ions proceeds differently in dilute nitric acid to that in concentrated acid, in which the material consumption rate is < 0.1 g/m^2 h [263].

According to the temperature/concentration graph of corrosion of the steel Sandvik® 2RE10 ((%) Fe-<0.02C-24.5Cr-20.5Ni; corresponds to AISI 310 L) in hot nitric acid, its resistance can be described as good. In hot 65 % nitric acid at 393 K (120 °C), in 80 % HNO_3 at 353 K (80 °C) and in 90 % HNO_3 at 325 K (52 °C), the corrosion rate is about 0.1 mm/a (4 mpy). This steel is also used for reaction tanks in the explosives industry for the production of nitroglycerine in a solution of 55 % nitric acid and 45 % sulfuric acid, and in the production of TNT (trinitrotoluene) in a solution of 25 % HNO_3 + 75 % H_2SO_4, where temperatures between 350 and 365 K (77 and 92 °C) occur during the course of the reaction [508].

Grinding grooves caused intercrystalline corrosion on a storage tank made of AISI 321 (1.4541) in an anhydrous mixed acid solution containing 90 % HNO_3 + 10 % H_2SO_4. The same result is also observed on coarse grinding grooves on the steel Fe-18Cr-15Ni-4Si, although if polished adequately, this is resistant to highly concentrated nitric acid [550].

According to Figure 33, the corrosion rate in 10 % sulfuric acid is reduced from 1.2 or 10 mm/a (47.2 or 394 mpy), depending on the type of steel, to about 0.1 mm/a

(4 mpy) by addition of 1 to 3.5% HNO_3. For comparison, two chromium-nickel-molybdenum steels are shown alongside the two chromium-nickel steels 08Ch22N6T and 12Ch18N10T (AISI 321). Above an addition of 8% HNO_3, the corrosion rate is the same for all four steels at about 0.001 mm/a (0.04 mpy) [370].

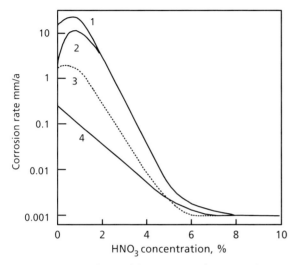

Figure 33: Dependence of the corrosion of some steels in HNO_3 containing 10% H_2SO_4 on the HNO_3 content at room temperature [370]
(1) 08Ch22N6T
(2) 12Ch18N10T (AISI 321)
(3) 08Ch21N6M2T
(4) 06ChN28MDT

If nitrous gases are washed out with a mixture of sulfuric acid with 1 to 5% nitric acid instead of with 10 to 45% sulfuric acid in the production of sulfuric acid, the corrosion of the steels used can be greatly reduced. The concentration of the nitric acid required is lower, the higher the sulfuric acid concentration. The increased resistance occurs in particular at a somewhat elevated temperature (328 K (55 °C)) [370]. Some test values are summarized in Table 65 (06ChN28MDT: (%) Fe-0.06C-22.4Cr-27.9Ni-2.6Mo-2.9Cu-0.76Ti-0.30Mn-0.69Si and 08Ch21N6M2T (%) Fe-0.08C-21Cr-6Ni-2Mo, Ti-stabilized).

In agreement with the results in Table 65, the steels 08Ch22N6T and 12Ch18N10T (AISI 321, 1.4541) corroded in 50% sulfuric acid at 333 K (60 °C) with an addition of 0.1% HNO_3 at an expected rate of 0.03 and 0.01 mm/a (1.18 and 0.39 mpy). This rate is also approximately maintained at 1% HNO_3 [441].

The steel AISI 304, which corrodes in pure sulfuric acid at room temperature at a material consumption rate of about 70 g/m^2 h, gives values of < 0.1 g/m^2 h in a mixture of 90% H_2SO_4 + 10% nitric acid [459]. It also behaves similarly in a mixture of 90% 1.5 mol/l H_2SO_4 + 6% HCl + 4% HNO_3.

Steel	Temperature K (°C)	10% H$_2$SO$_4$ +1% HNO$_3$	10% H$_2$SO$_4$ +5% HNO$_3$	30% H$_2$SO$_4$ +1% HNO$_3$	30% H$_2$SO$_4$ +5% HNO$_3$	50% H$_2$SO$_4$ +1% HNO$_3$	50% H$_2$SO$_4$ +5% HNO$_3$
		\multicolumn{6}{c}{Corrosion rate, mm/a (mpy)}					
08Ch22N6T	295 (22)	32 (1260)	0.003 (0.12)	0.004 (0.16)	0.003 (0.12)	0.004 (0.16)	0.001 (0.04)
	328 (55)	0.3 (11.81)	0.009 (0.35)	0.01 (0.39)	0.02 (0.79)	0.009 (0.35)	0.02 (0.79)
12Ch18N10T	295 (22)	2.1 (82.68)	0.004 (0.16)	0.002 (0.08)	0.001 (0.04)	0.002 (0.08)	0.001 (0.04)
	328 (55)	0.3 (11.81)	0.009 (0.35)	0.01 (0.39)	0.01 (0.39)	0.008 (0.31)	0.02 (0.79)
06ChN28MDT	295 (22)	0.2 (7.87)	0.003 (0.12)	0.001 (0.04)	0.001 (0.04)	0.004 (0.16)	0.001 (0.04)
	328 (55)	0.01 (0.39)	0.005 (0.20)	0.006 (0.24)	0.006 (0.24)	0.003 (0.12)	0.002 (0.08)
08Ch21N6M2T	295 (22)	3.0 (118.11)	0.002 (0.08)	0.002 (0.08)	0.001 (0.04)	0.004 (0.16)	0.001 (0.04)
	328 (55)	0.03 (1.18)	0.009 (0.35)	0.01 (0.39)	0.008 (0.31)	0.002 (0.08)	0.008 (0.31)

Table 65: Corrosion rate of some steels in H$_2$SO$_4$/HNO$_3$ mixtures at 295 and 328 K (22 and 55 °C) [370]

The corrosion behavior of the steel 12Ch18N10T in 10 % sulfuric acid with nitric acid additions at 333 K (60 °C) after 25 and 100 h tests is summarized in Table 66 [372]. Here also, corrosion of the steel in sulfuric acid is reduced by additions of nitric acid.

HNO$_3$ content mol/l	With mixing	Without mixing
	\multicolumn{2}{c}{Material consumption rate, g/m^2 h}	
–	2.83	8.51
		2.39*
0.003	3.62	9.6
0.100	0.16	0.22
0.150	0.11	0.19
0.300	0.02	0.02
		0.001*

Table 66: Influence of HNO$_3$ on the corrosion of 12Ch18N10T (AISI 321, 1.4541) in 10 % H$_2$SO$_4$ at 333 K (60 °C) after 25 and 100 h* [372]

According to Table 67, contact between the steel Ch18N10T and graphite or platinum in HNO_3 or sulfuric acid/nitric acid mixtures at 363 K (90 °C) causes an increase in the material consumption rate [380]. Additions of potassium bichromate (1 – 10 g/l) in turn increase the corrosion of the steel in the H_2SO_4/HNO_3 mixtures.

The material consumption rate could be reduced from 18.6 to 1.07 g/m^2 h by an addition of 0.3 mol/l nitric acid to a solution of 10 % sulfuric acid + 1.2 % fluoride ions at 333 K (60 °C) [372].

In 5 % sulfuric acid with additions of 8 % NaCl + 8 % Na_2SO_3 + 2 % HNO_3, the steels 12Ch18N10T and 08Ch22N6T corrode with corrosion rates of 37 and 41 mm/a (1456.69 and 1614.17 mpy) at 313 K (40 °C) [290].

The corrosion rate of the steel 12Ch18N10T (AISI 321, 1.4541) in a solution of 36.6 % H_2SO_4 + 23.3 % HNO_3 + 39.1 % SO_3 after 200 h tests is about 0.1 mm/a (4 mpy) [386]. In a solution of 28 % H_2SO_4 + 12 % HNO_3 + 60 % SO_3, the corrosion rate was only 0.06 mm/a (2.36 mpy). At 403 K (130 °C), the corrosion of the steel in the first and second solution achieves rates of 0.26 and 2.1 mm/a (10.24 and 82.68 mpy) after 250 h tests [386].

According to Table 68, the steel Ch18N10T, which is resistant in 2.5 mol/l nitric acid at room temperature, corrodes with a high material consumption rate after addition of 0.22 mol/l H_2SO_3, but this decreases again after addition of 0.82 mol/l H_2SO_3 at potentials of + 0.34 and + 0.03 V_{SHE} respectively [382].

Acid mixture, % $H_2SO_4/HNO_3/H_2O$	Contact material	S_1/S_2*	Potential V_{SHE}	Material consumption rate g/m^2 h
0/57/43	–	–	1.160	0.05
	Pt	5	1.290	1.07
	graphite	1	1.280	1.05
10/51/39	–	–	1.100	0.28
	Pt	0.17	1.285	1.4
	graphite	1.0	1.300	2.1
40/33/27	Pt	0.17	1.355	3.0
	graphite	0.38	1.395	5.0
70/16/14	Pt	0.17	1.405	3.1
	graphite	1.0	1.546	3.8

* S_1/S_2 = contact area/total area

Table 67: Corrosion rate and potential of the steel Ch18N10T in H_2SO_4/HNO_3 mixtures at 363 K (90 °C) with and without graphite or Pt contact [380]

Concentration, mol/l		Potential V_{SHE}	Current Density mA/cm^2	Material Consumption Rate $g/m^2\,h$
HNO_3	H_2SO_3			
2.5	0.22	−0.11	310	275
	0.38	+0.07	1.0	5
	0.82	+0.34	0.2	0.02
5.0	0.22	−0.11	300	270
	0.38	−0.09	80	90
	0.82	+0.03	2.0	10

Table 68: Corrosion of the steel Ch18N10T at room temperature in HNO_3 with additions of H_2SO_3 and different potentials [382]

The steels 03Ch18N11 (AISI 304 L, 1.4306), 03ChN28MDT, 03Ch21N21M4GB and 08Ch22N6T can be recommended in the fertilizer industry for solutions of superphosphoric acid + nitric acid up to 333 K (60 °C) (the test solution consisted of 100 g $H_4P_2O_7$ + 246 g 47 % HNO_3) [437].

HNO_3 concentration %	Addition mol/l	Corrosion rate mm/a (mpy)	
		353 K (80 °C)	383 K (110 °C)
15	–	0.002 (0.08)	0.017 (0.67)
30	–	0.004 (0.16)	0.038 (1.5)
15	0.001 oxalic acid	0.003 (0.12)	0.022 (0.87)
15	0.1 oxalic acid	0.007 (0.28)	0.023 (0.91)
30	0.001 oxalic acid	0.006 (0.24)	0.044 (1.73)
30	0.1 oxalic acid	0.009 (0.35)	0.052 (2.05)
15	0.1 glutaric acid	0.001 (0.04)	0.013 (0.51)
15	0.1 glutaric acid	0.001 (0.04)	0.010 (0.39)
30	0.01 glutaric acid	0.003 (0.12)	0.032 (1.26)
30	0.1 glutaric acid	0.002 (0.08)	0.024 (0.94)
15	0.01 adipic acid	0.003 (0.12)	0.007 (0.28)
15	0.1 adipic acid	0.002 (0.08)	0.005 (0.2)
30	0.01 adipic acid	0.006 (0.24)	0.033 (1.3)
30	0.1 adipic acid	0.003 (0.12)	0.026 (1.02)
15	0.001 azelaic acid	0.001 (0.04)	0.011 (0.43)
15	0.015 azelaic acid	0.001 (0.04)	0.015 (0.59)
30	0.001 azelaic acid	0.003 (0.12)	0.024 (0.94)
30	0.015 azelaic acid	0.004 (0.16)	0.026 (1.02)

Table 69: Corrosion of the steel Ch18N10T in a mixture of HNO_3 and dicarboxylic acid after 200 h [286]

The steels 12Ch18N10T and 08Ch22N6T corrode in a hot solution of 40.4 g/l HNO_3 + 18.2 g/l oxalic acid at 373 K (100 °C) with a corrosion rate of 0.007 mm/a (0.28 mpy) [412].

According to the corrosion rates in Table 69, the steel Ch18N10T can be used as a material for the reactor for the production of saturated dicarboxylic acids by oxidation of kerogen by means of nitric acid [286].

The steel AISI 304 was treated with 25.7 g/l U (as $UO_2(NO_3)_2 \times 6H_2O$) + 81 g/l Cd ($Cd(NO_3)_2 \times 4H_2O$) + 2.5 g/l Te ($K_2TeO_4 \times 2H_2O$) + 4.1 g/l K ($KNO_3$) + 3.8 g/l Sr ($Sr(NO_3)_2$) + 16.0 g/l Zr ($ZrO(NO_3)_2 \times 2H_2O$) + 19.4 g/l Mo ($K_2MoO_4 \times H_2O$) + 6.1 g/l Fe ($Fe(NO_3)_3 \times 9H_2O$) + 47.2 g/l lanthanides ($Ce(OH)_4$) + 6.9 g/l Ba ($Ba(NO_3)_2$) in boiling 0.8 mol/l nitric acid, activated with neutrons and left for a maximum of 90 d for the purpose of simulating a solution containing radioactive waste. The corrosion rate on this steel was between 0.13 and 0.7 mm/a (5.1 and 27.6 mpy) [386]. These results are approximately the same as those from the tests carried out on AISI 304 L in boiling 0.9 mol/l nitric acid under comparable conditions without radioactive substances [337, 338].

Figure 34 provides further information on the influence of uranium (UO_2^{2+}) in nitric acid/fluoride solutions from 333 K (60 °C) up to the boiling point on the material consumption rate of 12Ch18N10T (AISI 321, 1.4541). According to Figure 34 b), the corrosion drops sharply as the ratio of metal and fluoride ions increases [292].

The susceptibility of the steels 08Ch18N10 (AISI 304, 1.4301) and 06ChN40B ((%) Fe-0.055C-17.01Cr-39.04Ni-1.99Mn-0.50Nb-0.60Si-0.013S-0.022P) to intercrystalline corrosion is intensified by irradiation with neutrons and subsequent heat treatment. Damage caused by the Huey test has less effect, the higher the resistance of the steel in the non-irradiated state [339].

If the boiling nitric acid contains 400 g/l uranyl nitrate, corrosion rates on the steel in the delivery state of 0.044 mm/a (1.73 mpy) and on the sensitized steel of

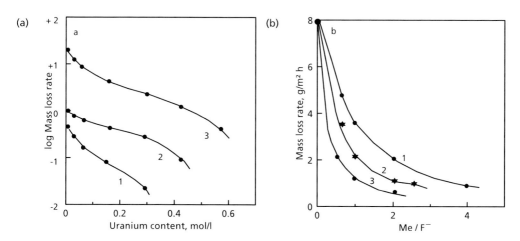

Figure 34: Influence of the uranium content on the corrosion behavior of the steel 12Ch18N10T in HNO_3 + 0.05 mol/l F^- at 353 K (80 °C) (Figure a: (1) 0.5, (2) 5 and (3) 10 mol/l HNO_3) and influence of the Me/F^--ratio (Figure b: (1) U, (2) Zr and (3) Al) on the corrosion in 5 mol/l HNO_3 + 0.05 mol/l F^- [292]

0.06 mm/a (2.36 mpy) result after 240 h tests; i.e. the presence of uranium ions causes only a slight increase in the corrosion rates. When boiling 5.5 mol/l nitric acid was used, the corresponding corrosion rates were only 0.006 and 0.008 mm/a (0.31 mpy). Mild intercrystalline attack was found on the sensitized specimens [425].

For the preparation of pure silver nitrate for the photographic industry, the chromium-nickel steel 08Ch22N6T ((%) Fe-0.08C-22Cr-6Ni; Ti-stabilized) is recommended for the reaction chamber where the salting-out of $AgNO_3$ (2.2 kg/l in 70% HNO_3) takes place [962].

According to Table 70, corrosion of the steel Ch18N10T in 5 mol/l nitric acid is greatly increased by addition of cerium nitrate and ozone. When 0.02 mol/l silver nitrate was added instead of cerium nitrate, the material consumption rate was 10 to 13 g/m^2 h [332].

Temperature K (°C)	$Ce(NO_3)_2$ mol/l	Ozone %*	Material consumption rate g/m^2 h
293 (20)	–	–	0.004
	0.02	2 – 6	5.05
368 (95)	–	–	0.015
	0.02	2 – 6	11.5

* percent by volume

Table 70: Influence of cerium ions on corrosion in ozonized 5 mol/l HNO_3 [332]

According to Table 71, laboratory tests in nitric acid with and without additions of ammonium nitrate show that the steel 12Ch18N10T is a suitable material for the production of ammonium nitrate by the one-stage process [396]. The steel 12Ch18N9T behaves similarly.

To avoid additions of hydrofluoric acid to nitric acid for cleaning pipelines made of the steel 0Ch18N10T (1.4541, AISI 321), a solution of 75 ml/l nitric acid (d = 1.32) + 25 ml/l hydrogen superoxide (30%) is proposed. This produces good results after a period of 20 min at 333 K (60 °C) [404].

Addition of thiourea ($CS(NH_2)_2$) accelerated corrosion of the steel 12Ch18N10T in 0.5 mol/l nitric acid between 293 and 333 K (20 and 60 °C) [288].

An increase in the phosphorus content from 0.001 to 0.11 or 0.12% in the CrNi steels 04Ch18N10, 04Ch18N27 and 04Ch18N40 increases the material consumption rate in boiling 65% nitric acid only insignificantly, i.e. from 0.20 to 0.25, from 0.12 to 0.17 and from 0.10 to 0.14 g/m^2 h respectively [363]. The phosphorus content has a considerably greater influence on the material consumption rate of the three steels in 5 mol/l HNO_3 + 22 g/l $K_2Cr_2O_7$ (4 to 24 g/m^2 h, 4.5 to 37 g/m^2 h and 5.5 to 50 g/m^2 h respectively).

If fluoride-containing nitric acid is used for etching CrNi steels (e.g. in the case of the steel 12Ch18N10T in 1.5 mol/l nitric acid containing 0.1 mol/l sodium fluoride at various polarization potentials), the corrosion rate above 370 K (97 °C) with heat transfer remains practically constant, while without heat transfer it increases sharply as the temperature rises. The corrosion rate of the steel in the passive range (+ 0.7 V_{SHE}) at temperatures up to 373 K (100 °C) is only slightly below 1.2 mm/a (47 mpy), regardless of the heat transfer [269].

State of steel	HNO_3 %	Temperature K (°C)	NH_4NO_3 %	Material consumption rate $g/m^2\ h$
Not welded	47	boiling	0	0.00088
			10	0.00092
			20	0.00278
Welded	57	boiling	0	0.00128
			10	0.00148
			20	0.00248
Not welded	50	373 (100)	0	0.00045*
			5	0.00036*
			20	0.00021*
Welded	58	373 (100)	0	0.00040
			20	0.00030

* test duration 100 h

Table 71: Corrosion of the steel 12Ch18N10T (AISI 321, 1.4541) in HNO_3 with and without NH_4NO_3 in laboratory tests at various temperatures, test duration 200 h [396]

Nitric acid/hydrofluoric acid solutions (10:1) at 303 K (30 °C) and suitable potentials are used for rapid removal of the oxide layer on the steel AISI 304. Maximum etching with a reflection of 85 % was achieved after only 3 min between − 0.3 and − 0.5 V_{SMSE}, while a reflection value of only about 5 % was obtained without a polarization potential under otherwise identical conditions, even after 5 min [393].

To avoid crevice corrosion on chromium-nickel and chromium-nickel-molybdenum steels, combined treatment of 10 min in 10 % sulfuric acid at 333 K (60 °C) + 30 min in 25 % nitric acid at 313 K (40 °C) is recommended, instead of surface treatment in the usual nitric acid/hydrogen fluoride solution (65 % HNO_3 + 1 percent by volume HF), for workplace-friendly reasons (avoidance of hydrofluoric acid) [351].

A hot solution of 10 % nitric acid + 1 % hydrogen fluoride at 368 K (95 °C) and a treatment time of about 6 min has also proved appropriate for removal of the oxide layer on rolled and heat-treated AISI 304. Replacement of HF by hydrogen chloride caused pitting corrosion during the pickling operation [549].

Treatment for 1 − 3 h in a solution of 20 % HNO_3 + 2 % H_2O_2 at 343 K (70 °C) is also proposed for removal of oxide layers on the steels 08Ch17T and 12Ch18N10T [419].

It was demonstrated by electrochemical etching for 30 s in concentrated nitric acid at 1 V that additions promote intercrystalline corrosion of the steels Ch18N10T (1.4510, AISI 439) and Ch18N40T (1.4541, AISI 321) in that they migrate to the grain boundaries under mechanical stress [434].

Weld seams of type X 12 CrNi 18 8 (1.4300) steels with a high degree of scaling are freed completely from the scale layer, to give a silky matt surface, after dipping in a 30% aqueous solution of 1.3 mol/l nitric acid + 0.85 mol/l hydrofluoric acid + 1.0 mol/l sulfuric acid + 0.001% fluorosurfactant, based on the total amount of the pickling solution, at 333 K (60 °C) for 5 min [447].

Tanks made of AISI 304 containing 0.5 boron (e.g. (%) Fe-0.068C-17.98Cr-8.04Ni-1.39Mn-0.02Al-0.65Si-0.024P-0.022S-0.53B) are used for storage of the spent fuel from nuclear reactors. The boride precipitates which form are more resistant than the steel matrix to etching solutions of 10 ml HNO_3 + 20 ml HCl + 30 ml water [588].

Pure nitric acid is not suitable for etching stainless steels. Hence the steel Ch18N10T is not etched in non-agitated 12% acid even after 240 min. At an acid flow rate of 0.4 m/s, etching occurs after 20 min, with a deposit of pickling sludge [1003].

A solution of 30% HNO_3 + 40% HCl + 8% H_2SO_4 + 12% water is recommended for removing the scale formed on the steel SUS-310 Nb ((%) Fe-<0.1C-25Cr-20Ni-1Nb) during rolling or after heat treatment. This solution also generates an excellent surface, as well as removing the oxide layer [541].

The rate of etching of a stainless steel at 353 K (80 °C) in an etching solution of (parts by volume) 1 conc. HNO_3 + 1 conc. HCl + 3 water was about 4.8 mm/h [443].

The most favorable concentration range for chemical etching of 12Ch18N10T (AISI 321) is shown in triangular diagrams with HNO_3, H_2SO_4 and HCl [287].

A pickling solution of 8 percent by volume each of concentrated nitric acid + hydrochloric acid + sulfuric acid + water as the remainder is used, inter alia, at room temperature for removal of the annealing colors after welding of stainless steels [521]. Passivating after-etching in dilute nitric acid is advisable in order to obtain the thin protective chromium oxide layer.

A mixed solution of (parts by volume) 3 HCl + 2 HNO_3 + 2 CH_3COOH is also used for etching chromium-nickel steels [398].

A pickling solution containing 12 – 15 percent by volume nitric acid (d = 1.38) + 1 – 3% hydrofluoric acid (d = 1.17) + 82 – 87% water with 0.1 g/l wetting agent and barium sulfate concentrate as a carrier has proved to be the most favorable pickling paste for chromium-nickel steels [226].

Steels of the type AISI 304 corrode at a material consumption rate of 0.15 or 0.13 g/m^2 h in boiling 65% nitric acid after mechanical or electrolytic polishing [366]. In the case of solution-annealed material, the mechanically polished specimens are attacked more severely than those polished electrolytically. Sensitized material also behaves in a similar way, the attack also being intercrystalline.

The etching time for stainless steels is shortened from 240 to 20 min by flowing 12 mol/l nitric acid (0.4 m/s). Furthermore, no rhythmic variations in potential, which are found in stationary acid, occur [266].

To protect the CrNi steel 18 8 (V2A, cf. AISI 321) during removal of deposits by nitric acid, 0.12 mol/l thiourea is added to the 10% nitric acid [410].

At certain proportions of nitric and hydrochloric acid, nitrosyl chloride is formed. This attacks the steel AISI 304 severely and with vigorous evolution of heat. Corrosion can be inhibited by additions of thiourea [582, 583].

The corrosion rate of the steel X8CrNiTi18-10 (1.4878, AISI 321 H) in 96 % sulfuric acid at 393 K (120 °C) is reduced to 0.1 mm/a (3.94 mpy) by addition of 0.03 – 0.04 % N_2O_3 as a result of passivation [465].

The material consumption rates of the steel 12Ch18N10T in solutions of 0.5 and 10 mol/l nitric acid containing 0.05 mol/l sodium fluoride at 353 K (80 °C) of about 0.6 and 20 g/m^2 h respectively are reduced by additions of 0.3 and 0.6 mol/l UO_2^{2+} to 0.03 and 0.7 g/m^2 h respectively. The corrosion decreases even further as a result of aluminium ions in the nitric acid [510].

A 21 Austenitic chromium-nickel-molybdenum steels

According to Figure 20, the material consumption rate of the steel X1CrNiMoN25-25-2 (AISI 310 MoLN, 1.4465) of 0.01 g/m^2 h in boiling 10 % nitric acid increases constantly as the acid concentration increases and reaches a constant value of about 0.8 g/m^2 h at 65 % and above [349].

Three isocorrosion curves of the niobium-stabilized steel 000Ch21N21M4B (Russ. type; (%) Fe-0.03C-(20-22)Cr-(20-21)Ni-(3.4-3.7)Mo-<0.6Mn-<0.6Si-<0.03P-<0.02S-(0.45-0.8)Nb) are shown in the temperature-concentration graph in Figure 35 [469].

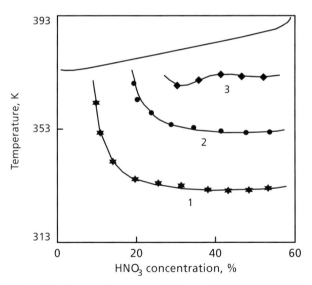

Figure 35: Isocorrosion curves of the steel 000Ch21N21M4B in a temperature/concentration graph at (1) 0.01, (2) 0.02 and (3) 0.3 g/m^2 h [469]

The solution-annealed and water-quenched steels GX5CrNiMo19-11-2 (UNS J92900, 1.4408; (%) Fe-(0.03-0.07)C-19.1Cr-(9.5-9.7)Ni-(2.08-2.35)Mo-(0.03-0.04)Nb-(0.59-0.68)Mn-(0.38-0.66)Si with a δ-ferrite phase of 8.5 – 11.5 %) corrode in 5 % nitric acid with a corrosion rate of < 0.01 mm/a (< 0.4 mpy) at both 293 and 343 K

(20 and 70 °C) [576]. Comparable CrNi steels with a low molybdenum content of 0.14 to 0.18 % of the type GX5CrNi19-10 (UNS J92600, 1.4308) show a similarly good corrosion resistance under the same test conditions.

As can be seen from Table 72, the susceptibility of the steel EI-943 to intercrystalline corrosion (0Ch23N28M3D3T) in nitric acid is determined by the Ti/C ratio [418]. At Ti/C values above 15, i.e. at a very low carbon content, no intercrystalline corrosion occurs.

Chemical composition, %						Ratio Ti/C	Observations
C	Cr	Ni	Ti	Cu	Mo		
0.05	23.42	20.2	0.76	2.88	2.56	15.2	–
0.06	22.87	27.17	0.5	2.77	2.5	8.3	intercrystalline corrosion
0.038	23.71	27.13	0.77	2.69	2.77	20.0	–
0.03	23.01	27.55	0.71	2.7	2.88	23.6	–
0.04	23.4	27.0	0.33	–	3.45	7.5	intercrystalline corrosion

Table 72: Chemical composition and intercrystalline corrosion of the steel EI-943 (after tempering at 973 K (700 °C)) in 5 % HNO_3 [418]

The corrosion-related weight loss of the steel AISI 316 L, which was produced by powder metallurgy, depends greatly on the rate of cooling after heat treatment. After 24 h tests at room temperature in 10 % nitric acid, the weight loss was about 0.01 % at a nitrogen content of < 0.001 % and about 2.5 % with 0.8 % nitrogen [537].

The chromium-nickel-molybdenum steels 06ChN28MDT (Russ. type; (%) Fe-0.06C-25Cr-28Ni-(1-3)Mo-(2-3)Cu; Ti-stabilized) and 08Ch21N6M2T ((%) Fe-0.08C-21Cr-6Ni-2Mo; Ti-stabilized) corrode in 10 and 15 % nitric acid containing 1 % H_2SiF_6 at room temperature at rates of about 0.1, 0.2, 0.3 and 0.7 mm/a (3.9, 7.9, 11.8 and 27.6 mpy) respectively after 48 h tests [470].

MnS impurities, which are the starting points for pitting corrosion in other media, are eliminated during passivation of the surface of the steel AISI 316 ((%) Fe-0.047C-17.8Cr-10.0Ni-1.72Mo-2.06Mn-0.22Cu-0.88Al-0.067W-0.040Ti-0.96Si-0.067P-0.017S) for 30 min in 20 % nitric acid at 323 K (50 °C) [477].

After a 144 d test in boiling 4.75 mol/l nitric acid, the material consumption rate of Incoloy® 825 was 0.0146 g/m² h [430].

The chromium-nickel-molybdenum steel FMN (1.4462, UNS S39209; (%) Fe-0.05C-25Cr-5Ni-2Mo-0.8Si-0.2N) corrodes in boiling 30 % nitric acid at a material consumption rate of 0.21 g/m² h [650].

Other material consumption rates for the low-carbon steels Fe21Cr24Ni3Mo ((%) Fe-0.03C-21.26Cr-24.31Ni-3.30Mo-0.62Mn-1.25Si-0.011P-0.010S) and Fe20Cr26Ni4-Mo1Cu ((%) Fe-0.03C-20.41Cr-25.52Ni-4.50Mo-0.67Mn-1.1Cu-0.15Ti-1.11Si-0.017P-0.020S) in boiling nitric acid can be seen from Table 73 [310]. The molybdenum-free steel (%) Fe-0.02C-23.8Cr-20.67Ni-1.10Mn-0.60Si-0.015P-0.014S shows a better corrosion behavior, with a corrosion rate of 0.028 g/m² h in 35 % HNO_3 at 382 K

(109 °C) and 0.270 g/m² h in 65 % HNO₃ at 393 K (120 °C). This is also confirmed generally elsewhere [330].

HNO$_3$ %	Temperature K (°C)	Fe21Cr24Ni3Mo	Fe20Cr26Ni4Mo1Cu
		Material consumption rate, g/m² h	
35	382 (109)	0.130	0.125
45	386 (113)	0.310	0.170
55	390 (117)	0.610	0.460
65	393 (120)	0.850	0.780

Table 73: Material consumption rates of two steels in HNO₃ at various concentrations and temperatures [310]

Cermets of stainless steel and UO₂ (steel/UO₂) from spent fuel elements are placed in a basket made of perforated niobium sheets with end holders made of fluorinated plastic in a titanium tank (as the cathode) and dissolved, for uranium processing, in 6.5 mol/l nitric acid at a rate of 0.66 g A/h and a current density of 1.59 A/cm². The potential difference between the titanium tank and the fuel element operating as the anode (cermet of steel/UO₂) is 20 V. Less than 0.05 % uranium remains in undissolved form in the total undissolved reside of 1 – 3 % [580].

The CrNiMo steels Sanicro® 28 ((%) Fe-0.011C-26.6Cr-31.1Ni-3.43Mo-1.0Cu-0.039N) and Carpenter® 7-Mo (cf. AISI 329; (%) Fe-0.054C-27.0Cr-4.2Ni-1.44Mo) shown in Table 74 corrode in nitric acid, especially at relatively high temperatures and concentrations, faster than the CrMo steel E-Brite® ((%) Fe-0.002C-26.1Cr-0.1Ni-1.00Mo-0.1Nb-0.010N) [356]. However, the corrosion of E-Brite® proceeds faster than that of Carpenter® 7-Mo or Sanicro® 28 in 60 % nitric acid at 422 K (149 °C) and in 70 % nitric acid at 366 K (93 °C).

HNO$_3$ %	Temperature K (°C)	Sanicro® 28	Carpenter® 7-Mo	E-Brite®
		Corrosion rate, mm/a (mpy)		
40	366 (93)	0.020 (0.79)	–	–
	384 (111)	0.223 (8.78)	0.067 (2.64)	0.021 (0.83)
	408 (135)	–	0.176 (6.93)	0.146 (5.75)
	422 (149)	–	0.750 (29.53)	0.226 (8.9)
60	366 (93)	0.026 (1.02)	–	0.024 (0.94)
	393 (120)	0.409 (16.10)	0.166 (6.54)	0.097 (3.82)
	422 (149)	–	2.87 (112.99)	3.52 (139)
70	366 (93)	0.049 (1.93)	–	0.119 (4.69)
	394* (121)	0.642 (25.28)	0.251 (9.88)	0.146 (5.75)

* boiling point

Table 74: Corrosion of some steels in nitric acid [356]

Tests under evaporation conditions with heat transfer in the evaporator from a titanium tank with heating steam coils made of steel containing 55 % nitric acid at 403 K (130 °C) showed that, according to Table 49, the chromium-nickel-molybdenum-copper steel EI-943 ((%) Fe-<0.01C-23Cr-28Ni-3Mo-3Cu) is superior in its corrosion behavior to normal chromium-nickel steel [261].

However, the greatest advantage of the molybdenum-containing steels is their good resistance to intercrystalline corrosion in nitric acid, such as, for example, the Swedish ELC steel Sandvik® 2RE10 (UNS S31002, 1.4335) with about 4.5 % molybdenum [487].

Figure 36 shows the dependence of the corrosion rate of the steel X 2 CrNiMo 18 12 ((%) Fe-0.03C-17.5Cr-12.4Ni-2.83Mo-0.97Mn-0.66Si-0.016N-0.012P-0.012S) in boiling 50 and 65 % nitric acid on the immersion time. While it has reached a stationary value of about 0.2 mm/a (7.9 mpy) after 200 h in 50 % acid, in 65 % acid it is still rising considerably even after 250 h. After 100 % cold-working, corrosion of the steel after 250 h tests in boiling 65 % nitric acid drops from 1.35 to 0.62 mm/a (53.15 to 24.41 mpy), and in 50 % acid from 0.24 to 0.12 mm/a (9.45 to 4.72 mpy) [480].

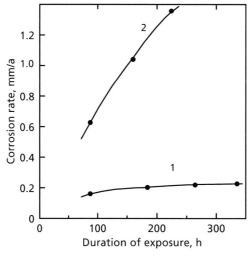

Figure 36: Corrosion rate of the solution-annealed steel X 2 CrNiMo 18 12 in (1) boiling 50 % and (2) 65 % HNO$_3$ as a function of the exposure time [480]

Pipe specimens of AISI 316 L ((%) Fe-0.07C-16.5Cr-11.8Ni-2.45Mo-0.6Mn-0.6Sr-0.002P-0.009S), produced from powder or cast ingots, corrode in boiling 65 % nitric acid at material consumption rates of 0.41 and 0.37 g/m^2 h respectively [577].

On the basis of studies in the Huey test, CrNiMo steels of the type (%) Fe-(0.034-0.043)S-(12.7-16.4)Cr-(3.9-5.9)Ni-(1.1-1.6)Mo-(0.28-0.039)Si-(0.030-0.037)N-(0.009-0.022)P-(0.007-0.020)S are not recommended for use in nitric acid [479, 503].

In agreement with the above values, steels of the type AISI 316 L heat-treated at 950 K (677 °C) for 20 min showed a corrosion rate of between 0.64 and 0.71 mm/a

(25.20 and 27.95 mpy) in the Huey test [354]. The steel AISI 316 gave a comparable corrosion behavior.

The corrosion-increasing effect of molybdenum in boiling 65 % nitric acid is also found in the Huey test (5 × 48 h in boiling 65 % HNO_3) on the steel Fe-0.1C-18Cr-16Ni-Mo, which corrodes with a material consumption rate of 0.0871 g/m^2 h without molybdenum and 0.0879 and 0.120 g/m^2 h with 2.5 and 5 % molybdenum respectively [482]. According to the Huey test, the steel AISI 316 L is attacked by boiling nitric acid, especially in the region of the σ-phase [395].

Even expensive high-alloy steels, such as, for example, Incoloy® 825 ((%) Fe-0.03C-20.8Cr-42.0Ni-3.0Mo-1.0Ti-1.74Cu) and Incoloy® 901 ((%) Fe-0.03C-13.1Cr-42.2Ni-6.2Mo-2.5Ti) are attacked with corrosion rates of 0.2 and 3.4 mm/a (7.87 and 134 mpy) in boiling 65 % nitric acid, since the chromium content is too low, especially in the latter, and the molybdenum content too high [499].

Figure 37 shows the tempering time and temperature of a CrNiMo steel ((%) Fe-0.020C-17.35Cr-13.70Ni-2.69Mo-1.67Mn-0.53Si-0.018P-0.008S) in the form of an isocorrosion curve at 0.25 and 0.35 g/m^2 h, determined in the Huey test after solution-annealing at 1573 K (1300 °C) for 15 min [330].

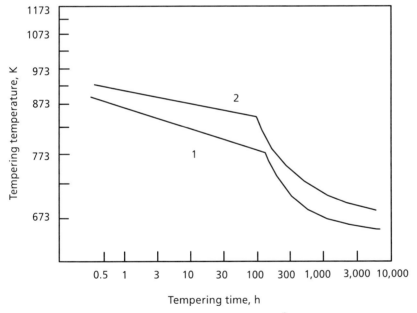

Figure 37: Isocorrosion curves for (1) 0.25 and (2) 0.35 g/m^2 h in the Huey test on the CrNiMo steel after solution-annealing at 1573 K (1300 °C) for 15 minutes [330]

Table 75 shows the dependence of the material consumption rate of some specimens of the steel 1.4435 (AISI 316 L) in boiling 65 % nitric acid (Huey test) on the heat treatment. The relatively large differences in corrosion which occur on the basis of uncontrollable changes within the tolerance range are remarkable [471].

The austenitic CrNiMo steel VEW A-963 with 6.3 % molybdenum and a low carbon content ((%) Fe-<0.03C-17.0Cr-16.0Ni-6.3Mo-0.15N) corrodes in the Huey test (5 × 48 h in boiling 65 % HNO_3) with a material consumption rate of < 0.20 g/m^2 h [472]. Comparable values are obtained with the quenched and sensitized steel 05Ch16N15M3 (Russian type; Fe-0.05C-16Cr-15Ni-3Mo) with material consumption rates of 0.16 and 0.32 g/m^2 h at polarization potentials of 1.2 and 1.1 V respectively [473]. The intercrystalline corrosion which is simultaneously observed results from the precipitation of the carbides (Cr, Fe, Mo)$_{23}$C$_6$ at the grain boundaries. The carbide $Cr_{15.5}Fe_{6.1}Mo_{1.4}C_6$ corrodes at 1.2 and 1.1 V with material consumption rates of 16 and 5 g/m^2 h respectively.

Welded pipes of CrNiMo steel ((%) Fe-<0.1C-20.25Cr-24.5Ni-6.25Mo-1.5Mn-0.50Si) available under the name AL-6X for cooling plants are also resistant to 65 % nitric acid [474].

Figure 38 shows the change with respect to time in the corrosion rate of the steel 05Ch16N15M3 (Fe-0.05C-16Cr-15Ni-3Mo) and the carbides $Cr_{16.9}Fe_{6.1}C_6$ and $Cr_{15.5}Fe_{6.1}C_6$ which precipitate at the grain boundaries during heat treatment in boiling 65 % nitric acid at a polarization potential of 1.2 V_{SHE} [476]. Here as well, the corrosion of the carbides is greater than that of the steel by more than a power of ten.

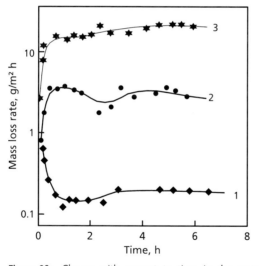

Figure 38: Change with respect to time in the material consumption rate (1) of the steel 05Ch16N15M3 and (2) of the carbides $Cr_{16.9}Fe_{6.1}C_6$ and (3) $Cr_{15.5}Fe_{6.1}Mo_{1.4}C_6$ in boiling 65 % HNO_3 at 1.2 V_{SHE} [476]

The influence of the 5 test periods of 48 h each in the Huey test on the material consumption rates of some chromium-nickel-molybdenum steels with the composition shown in Table 76 can be seen from Table 77. On some steels, the corrosion rate rises considerably with the number of weighings [399]. According to these tables, steels 5 and 6 show the greatest resistance in boiling 65 % nitric acid.

Chemical composition, %						Material consumption rate, g/m² h			
C	Si	Mn	Cr	Mo	Ni	a	b	c	d
0.024	0.64	0.86	17.40	2.81	14.47	0.29	1.39	3.8	7.3
0.020	0.70	0.92	17.99	2.75	14.53	0.26	4.59	13.7	18.9
0.024	0.72	0.89	17.67	3.02	14.77	0.24	8.1	51.2	33.3
0.015	0.65	0.88	17.68	2.82	14.59	0.33	0.31	0.39	1.5

a quenched
b quenched + heat treatment at 1073 K (800 °C) for 3 min
c quenched + heat treatment at 1073 K (800 °C) for 15 min and
d quenched + heat treatment at 1073 K (800 °C) for 30 min

Table 75: Composition of the steel 1.4435 (AISI 316 L) and corrosion behavior in the Huey test after various heat treatments at 1073 K (800 °C) [471]

No.	Chemical composition, %								Ferrite content, %
	C	Si	Mn	S	Cr	Ni	Mo	N	
1	0.029	0.34	0.70	0.016	15.02	14.02	3.12	–	0
2	0.045	–	4.10	–	16.98	15.68	2.13	–	0
3	0.020	0.34	0.14	0.033	17.52	11.22	2.05	0.16	0.2
4	0.03	0.25	1.24	0.017	18.06	13.40	2.96	–	2.3
5	0.025	0.30	1.89	0.020	18.17	14.81	2.61	–	0
6	0.01	0.36	1.60	0.014	18.98	16.48	3.80	–	0

Table 76: Chemical composition of the steels shown in Table 77 [399]

No.	Test period					Mean
	1	2	3	4	5	
	Material consumption rate, g/m² h					
1	0.703	0.889	1.040	–	–	0.887*
2	0.426	0.334	0.339	0.385	0.416	0.382
3	0.233	0.227	0.297	0.378	0.215	0.278
4	0.260	0.240	0.252	0.551	0.833	0.428*
5	0.212	0.127	0.130	0.228	0.126	0.143
6	0.391	0.207	0.272	0.387	0.215	0.278

* corrosion rate increasing with the number of periods

Table 77: Material consumption rates of the steels shown in Table 76 in boiling 65 % HNO₃ after 5 test periods of 48 h each [399]

These results are confirmed by earlier studies on the low-carbon austenitic steels 000Ch16N(13-16)M(2-4) (identical to AISI 316), where the material consumption rates of the quenched (hardened) and welded specimens in the Huey test were 0.20 – 0.33 and 0.25 – 0.58 g/m² h. Two ferritic-austenitic steels 000Ch21N10M2 ((%) Fe-0.020C-19.8Cr-10.5Ni-2.1Mo) and 000Ch21N6M2 ((%) Fe-0.036C-21.1Cr-6.5Ni-2.4Mo) have also been shown in Table 78 for comparison [492].

Steel	Specimen 1[1]	Specimen 2[2]	Susceptibility to intercrystalline corrosion (IC)
	Material consumption rate, g/m² h		
000Ch16N13M2	0.22 – 0.25	0.25 – 0.27	no or weak IC
000Ch16N13M3	0.26	0.35	none, only in HZ*
000Ch16N16M2	0.20	0.25	none, only weak IC in HZ
000Ch16N16M3	0.28	0.36	none, only weak IC in HZ
000Ch16N16M4	0.37	0.77	IC in HZ
000Ch18N13M2	0.25	0.19	none
000Ch18N16M2	0.17	0.25	none
000Ch18N16M4	0.34	0.82	IC in HZ and on the surface
000Ch21N10M2	0.28	0.22	none
000Ch21N6M2	0.43	0.48	none

HZ – heat-affected zone
[1] specimen 1 quenched
[2] specimen 2 welded

Table 78: Corrosion behavior of some CrNiMo steels in boiling 65 % HNO₃ in the Huey test [492]

CrNiMo steels with a ferritic-austenitic structure are used for special applications in the presence of nitric acid. The material consumption rate of the steel X2CrNi-MoN22-5-3 (UNS S39209, 1.4462; (%) Fe-0.020C-22.47Cr-5.21Ni-3.08Mo-1.82Mn-0.43Si-0.165N-0.031P-0.006S) in the Huey test (3 periods of 48 h in boiling 65 % nitric acid) is about 50 to 70 g/m² h after an annealing time of between 1 and 5 h at 973 K (700 °C), and about 0.6 to 0.8 g/m² h at 1073 K (800 °C). In the delivery state, the steel corroded at a material consumption rate of 0.45 g/m² h. According to Figure 39, relatively high annealing temperatures are more favorable for the corrosion behavior in nitric acid. The rate of corrosion of the steel is independent of the annealing time at 1423 K (1150 °C) [468].

The decisive influence of the correct heat treatment on the corrosion of stainless steels can be seen from Table 79 with the example of the niobium-containing steel ITM-43 ((%) Fe-0.074C-24.8Cr-3.93Ni-4.05Mo-0.51Nb-0.12Si-0.013P-0.002S-0.0117N) [545].

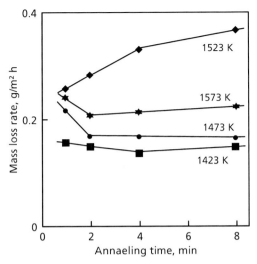

Figure 39: Material consumption rates of the steel X2CrNiMoN22-5-3 (UNS S39209) in the Huey test after short annealing times at high temperatures [468]

Heat treatment temperature K (°C)	Test cycles	Heat treatment duration			
		5 min	20 min	1 h	5 h
		Material consumption rate g/m² h			
923 (650)	3	25.0	–	–	59.1
1023 (750)	3	33.5	31.0	31.0	29.0
1073 (800)	5	18.8	23.1	15.6	2.11
1123 (850)	5	10.01	0.36	0.35	0.24
1173 (900)	5	0.20	0.20	0.26	0.15

Table 79: Corrosion of the steel ITM-43 in the Huey test after 3 and 5 cycles (48 h each) and various heat treatment durations [545]

If the heat treatment temperature is below 1100 K (827 °C), even a period of 5 h is too short to achieve a sufficiently low material consumption rate.

The influence of correct heat treatment on the corrosion of the steels AF-22 (UNS S39209, 1.4462; (%) Fe-0.028C-21.8Cr-5.00Ni-3.12Mo-1.63Mn-0.45Si-0.031P-0.012S-0.113N) and AF-22 + Mo ((%) Fe-0.027C-21.5Cr-7.21Ni-4.88Mo-1.67Mn-0.49Cu-0.29Si-0.021P-0.009S-0.140N) in boiling 65 % nitric acid can be seen in Figures 40 and 41. According to Figure 40, the corrosion rate of the steel AF-22 reaches the lowest values after heat treatment times of 20 min or 30 h between 573 and 673 K (300 and 400 °C) and above 1100 K (827 °C) respectively, and the highest value at about 900 K (627 °C).

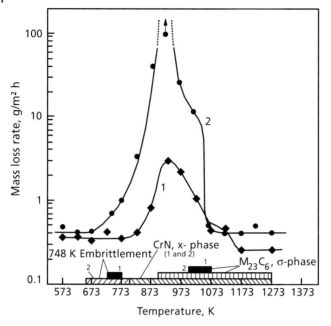

Figure 40: Influence of precipitates (Cr$_2$N, χ-phase, M$_{23}$C$_6$, σ-phase) on the material consumption rate reached in the Huey test (ASTM A 262, Practice C) on the steel AF-22, annealed at 1323 K (1050 °C) for 30 min and quenched in water, as a result of subsequent heat treatment between 573 and 1373 K (300 and 1100 °C) for (1) 20 min and (2) 30 h [608]

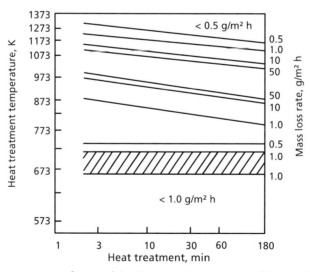

Figure 41: Influence of the duration and temperature of heat treatment after prior annealing at 1373 K (1100 °C) for 20 min and quenching in water on the corrosion of the steel AF-22 + Mo in the Huey test (ASTM A 262, Practice C) [608]

The steel AF-22 + Mo also shows a decrease in the material consumption rate in boiling acid of < 0.5 g/m² h after an annealing time of only 10 min above 1270 K (997 °C) [608]. Here also, annealing temperatures between 670 and 1070 K (397 and 797 °C) are unsuitable.

As a further example, Figure 42 shows the marked influence of the heat treatment (temperature and time) on the corrosion rate of the steel AISI 316 L in boiling 65 % nitric acid after exposure for 4 × 48 h periods. In all cases, the attack was intercrystalline. The corrosive attack on a steel specimen which had been heat-treated at 948 K (675 °C) for longer than 2 h was particularly intense [527].

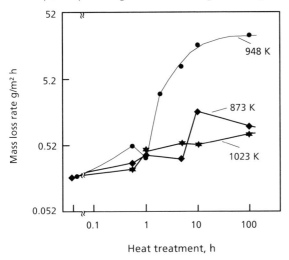

Figure 42: Material consumption rate of the steel AISI 316 L in boiling 65 % HNO₃ as a function of the temperature and duration of heat treatment [527]

To improve the corrosion resistance of the steel AISI 316 L ((%) Fe-0.02C-17.37Cr-15.45Ni-2.45Mo-1.73Mn-0.029Al-0.35Si-0.024P-0.014S-0.015N) in nitric acid, the occurrence of the ferrite and σ-phase should be prevented during heat treatment. A low carbon content is also necessary to avoid carbide precipitates. Similarly, the silicon, phosphorus and sulfur content should be kept low to prevent precipitates at the grain boundaries (intercrystalline corrosion) [525].

The influence of delayed cooling of the steel X2CrNiMoN17-13-5 (AISI 317 LMN, 1.4439) after solution-annealing at 1343 K (1070 °C) for 15 min on the corrosion rate in the Huey test after 96 h (in this case only 2 × 48 h) is informative in this connection. Figure 43 shows the influence of delayed cooling in the maximum precipitation range on corrosion [333].

Figure 44 also shows the marked influence of the annealing temperature and time on the corrosion of chromium-nickel-molybdenum steels in boiling 65 % nitric acid (Huey test). Below the solution-annealing temperature, the influence of precipitates becomes particularly critical as a result of the various periods of time at the susceptibility maximum of the steel Remanit® 4462 (UNS S39209, X2CrNiMoN22-5-3; (%) Fe-0.025C-22.42Cr-5.30Ni-3.06Mo-1.83Mn-0.40Si-0.022P-0.004S-0.110N) at 970 K

(697 °C) [731]. The tests on the steel Remanit® 4362 (UNS S32304, X2CrNiN23-4, (%) Fe-0.010C-23.56Cr-3.97Ni-<0.01Mo-1.15Mn-0.15Si-<0.005P-<0.003S-0.110N) performed for comparison show a considerably lower susceptibility to corrosion in the Huey test as a result of the annealing temperature and time.

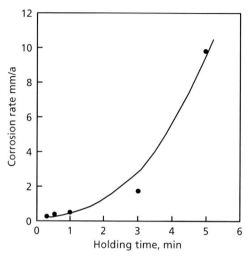

Figure 43: Influence of the holding time at 1073 K (800 °C) with subsequent quenching after solution-annealing of 15 min at 1343 K (1070 °C) on the corrosion rate of the steel X2CrNiMoN17-13-5 (AISI 317 LMN) in the Huey test after 96 h [333].

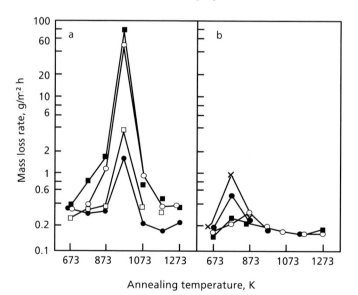

Figure 44: Influence of the annealing temperature and time on the material consumption rate in the Huey test on a) Remanit® 4462 (UNS S39209) and b) Remanit® 4362 (UNS S32304) [731]
a): ● = 5 min; □ = 15 min; ○ = 1 h; ■ = 10 h
b): ○ = 1 h; ■ = 10 h; ● = 30 h; x = 100 h

While addition of up to 2 % niobium to an untreated and solution-annealed CrNi-MoCu steel ((%) Fe-0.09C-20.0Cr-29.78Ni-2.31Mo-3.20Cu-0.91Mn-0.64Si-0.95N-0.018P-0.0 21S) caused no change in the corrosion rate of 0.18 mm/a (7.09 mpy) in boiling 65 % nitric acid, corrosion of sensitized, and of solution-annealed + stabilized + sensitized steel, which corroded more severely (1.25 and 0.75 mm/a (49.21 and 29.53 mpy)), was reduced to 0.13 mm/a (5.12 mpy) by addition of 1.4 % Nb [486]. This value should not be exceeded.

After 240 h tests in boiling 65 % nitric acid, a rapidly quenched steel AISI 316 ((%) Fe-0.05C-17.27Cr-13.09Ni-2.56Mo-0.24Cu-1.76Mn-0.13Co-0.04V-0.45Si-0.033P-0.015S) corroded with a corrosion rate of only 0.5 mm/a (19.69 mpy), and after only 30 % cold-working reached a rate of 0.23 mm/a (9.06 mpy), which remained constant with further cold-working [302].

Figure 45 shows particularly clearly the change with respect to time in the material consumption rate of the steel AISI 316 LN (1.4429, X2CrNiMoN17-13-3) in boiling 65 % nitric acid in the Huey test due to the degree of deformation, or the number of deformation stages. After passing through a corrosion maximum of > 12 g/m² h at a degree of deformation of about 40 %, the material consumption rate drops to about 0.3 g/m² h at about 50 % deformation [483]. It reaches a minimum of < 0.1 g/m² h after a degree of deformation of only 4 %.

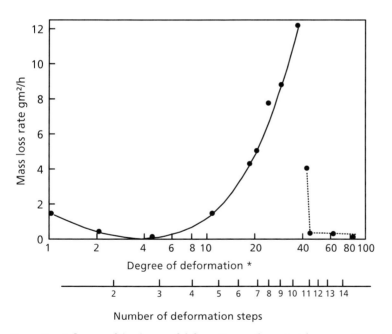

Figure 45: Influence of the degree of deformation on the material consumption rate of AISI 316 LN after 5 × 48 h periods in boiling 65 % HNO$_3$ [483].
* starting cross-section / deformed cross-section

Studies of damage to acid pumps made of Alloy-20 (Fe-(0.05-0.07)C-(19.5-20.5)Cr-(29-30)Ni-2.5Mo-3.2Cu-(0.5-1.0)Si) showed that the behavior under erosion corrosion conditions cannot be predicted by the Huey test in the laboratory [536].

The corrosion behavior and chemical composition of some Remanit steels in hot 86 % nitric acid at 333 K (60 °C) after 3 × 24 h tests can be seen from Table 80 [728]. The corresponding values in the Huey test (5 × 48 h in boiling 65 % nitric acid) are higher and are: 0.45, 0.40, 0.35 and 0.30 g/m^2 h. The steels Remanit® 4438 (AISI 317 L) and Remanit® 4565 (UNS S34565) are accordingly the most corrosion-resistant.

Remanit® Steel	No.*	Chemical composition, %								Material consumption rate, g/m^2 h
		C	Si	Mn	Cr	Mo	Ni	N	Nb	
4462	1	0.028	0.49	1.71	22.20	3.06	5.39	0.12	–	0.125
4429	2	0.022	0.30	1.31	17.26	2.68	13.36	0.15	–	0.091
4438	3	0.022	0.15	1.82	18.32	3.34	14.77	0.14	–	0.032
4565	4	0.020	0.62	5.79	23.14	3.27	16.53	0.39	0.22	0.032

* 1: 2205, UNS S39209, X2CrNiMoN22-5-3,
2: AISI 316 LN, X2CrNiMoN17-13-3,
3: AISI 317 L, X2CrNiMoN18-15-4,
4: UNS S34565X3CrNiMnMoNbN25-18-5-4

Table 80: Chemical composition and material consumption rates of the Remanit® steels investigated [728]

A valve made of the steel Alloy-20 CN-7M ((%) Fe-0.1C-20Cr-30Ni-2.3Mo-3.2Cu-0.9Mn) and also a pump of the same material were subjected to corrosion in 99 % nitric acid. The Huey test gave a corrosion rate of only 0.36 mm/a (14.17 mpy) on the base steel, while the casting surface corroded at a rate of 0.78 mm/a (30.71 mpy) [485]. Intercrystalline corrosion was found on the propeller of the acid pump.

The structure of the corrosion layer on the steels AISI 316 Ti (1.4571, X6CrNiMoTi17-12-2) and AISI 310 MoLN (1.4465, X1CrNiMoN25-25-2) after exposure to concentrated nitric acid for 3 years was investigated by Auger and X-ray fluorescence measurements [347]. Similar studies were also carried out on the oxidized surface layers on the austenitic-ferritic cast material Noridur® 1.4593 (GX3CrNiMoCuN24-6-2-3; (%) Fe-0.036C-24.73Cr-7.01Ni-2.46Mo-3.21Cu-1.11Mn-1.14Si-0.144N-0.021P-0.010S) [481].

The austenitic steel Sanicro® 28 (UNS N08028, 1.4563; (%) Fe-0.020C-27Cr-31Ni-3.5Mo-1.0Cu) has been developed for very severe corrosion conditions, such as, for example, for highly concentrated, boiling nitric acid [739].

The high-alloy steels which have already been mentioned several times, such as, for example, 03ChN28MDT and 03Ch21N21M4GB, are recommended for installations in fertilizer factories where highly concentrated nitric acid solutions occur [554]. The corrosion rate under the conditions which arise here is ≤ 0.1 mm/a (≤ 4 mpy).

Remanit® 4465 (1.4465, X1CrNiMoN25-25-2) and Remanit® 4575 (1.4575, 25-4-4, UNS S44635) still have an acceptable resistance to a hot solution of 10% HNO₃ + 2% HF at 343 K (70 °C) with material consumption rates of 0.6 and 0.2 g/m² h respectively [500]. This is confirmed for Remanit® 4575 under the same conditions with a material consumption rate of 0.18 g/m² h [1016].

In a solution of 10% nitric acid + 3% hydrofluoric acid at 323 K (50 °C), the steel EI-943 (06ChN28MDT) corrodes between + 0.3 and + 0.6 V_{SHE} with a corrosion rate of < 1 mm/a (< 40 mpy). This figure is < 0.1 mm/a (< 4 mpy) with an addition of 1% hydrofluoric acid at + 0.5 V_{SHE} [489].

Weld joints of the steel 06ChN28MDT corrode in 14 to 16% nitric acid containing 3.8% hydrogen fluoride between 328 and 333 K (55 and 60 °C) at a rate of 4.5 to 4.7 mm/a (177 mpy) after 336 h tests. The Nb-stabilized steel 04Ch21N21M4B has a higher corrosion resistance with a corrosion rate of 1.1 mm/a (43.31 mpy), but is unusable in this medium [416].

Even the high-molybdenum steel AISI 317 LMN (1.4439, X2CrNiMoN17-13-5) is noticeably attacked by nitric acid containing hydrogen fluoride. Figure 46 shows the dependence of the corrosion of welded specimens of the steel in 20% nitric acid at 303 K (30 °C) on the hydrofluoric acid content after 70 h tests. On the other hand, the material consumption rate of the steel AISI 316 also plotted in Figure 46 increases greatly as the concentration of HF increases [484].

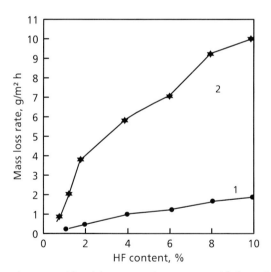

Figure 46: Material consumption rate on welded specimens of the steels (1) AISI 317 LMN (1.4439) and (2) AISI 316 (1.4401, X5CrNiMo17-12-2) in HF-containing 20% HNO₃ at 303 K (30 °C) as a function of the HF concentration, test duration 70 h [484]

The resistance of steels of the type ChN28MDT ((%) Fe-(0.03-0.046)C-(22.2-23.5)Cr-(26.55-27.88)Ni-(2.55-3.06)Mo-(2.68-3.38)Cu-(0.54-0.76)Ti-(0.15-0.30)Mn-(0.39-0.69)Si-(0.021-0.43)P-(0.008-0.017)S) to nitric acid containing hydrogen fluoride depends on the carbide precipitates caused by heat treatment and on the occurrence of the σ-phase, which increases corrosion. The risk can certainly be reduced by reducing the carbon content, but can still not be adequately eliminated. It is advisable for the steels to be additionally stabilized with niobium or zirconium [488].

The influence of a carbon content of 0.06 and 0.03%, shown in Figure 47, on corrosion of the steels 03ChN28MDT (Russ. types; (%) Fe-(0.03-0.06)C-22.3Cr-27Ni-2.3Mo-3.0Cu-(0.55-0.85)Ti-0.40Si-(0.006-0.010)N-0.010S-(0.018-0.033)P) in hot 10 mol/l nitric acid + 0.1 mol/l hydrogen fluoride at 363 K (90 °C) is thus to be expected. According to Figure 47, movement of the acid, expressed as the rotational velocity of the specimen, also has a corrosion-increasing influence [478].

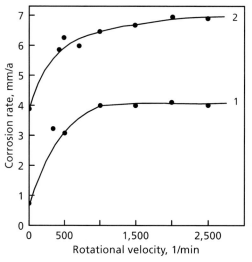

Figure 47: Influence of the rotational velocity and carbon content on the corrosion rate of the steels (1) 03ChN28MDT with 0.03 % C and (2) 06ChN28MDT with 0.06 % C in 10 mol/l HNO$_3$ + 0.1 mol/l HF at 363 K (90 °C) [478]

According to ESCA analyses of the surface, the chromium content in the oxide layer on the steel AISI 316 increases in the sequence of treatments in dry air, in nitric acid + hydrofluoric acid and by passivation in 30 % nitric acid, with the potential simultaneously rising from + 0.4 to + 1.05 V$_{SCE}$ [33, 490].

According to Table 81, corrosion of the steel 03Ch21N21M4B in nitric acid with and without hydrogen fluoride is increased both by additions of niobium and hydrofluoric acid [491]. The corrosion rate of the steel 06ChN28MDT in 40 % nitric acid containing 2 % hydrofluoric acid at 368 K (95 °C) is also increased from 11.4 to 38.5 mm/a (449 to 1515 mpy) by titanium additions of 0.43 to 0.86 %.

According to [334] the steel 08Ch21N6M2T has a superior resistance to the molybdenum-free chromium-nickel steels in nitric acid containing fluoride (8 to 12 mol/l HNO$_3$ + 0.02 to 0.12 mol/l NaF) between 323 and 373 K (50 and 100 °C).

Nb Content %	Medium	Temperature K (°C)	Duration h	Corrosion rate mm/a (mpy)
–	65 % HNO$_3$	boiling	240	0.13 (5.12)
0.45				0.22 (8.66)
0.90				0.32 (12.60)
–	40 % HNO$_3$ + 2 % HF	368 (95 °C)	48	14.2 (559)
0.45				15.9 (625)
0.90				27.6 (1087)

Table 81: Influence of the niobium content on the corrosion resistance of the steel 03Ch21N21M4B in nitric acid with and without HF [491]

To simulate the influence of flowing acid on the corrosion rate of the steels 03ChN28MDT and 06ChN28MDT, rotating steel specimens were exposed to attack by 10 mol/l nitric acid + 0.1 mol/l hydrogen fluoride at 363 K (90 °C) at an increasing rotational velocity. While the corrosion rates in the stationary state were 0.6 and 3.8 mm/a (23.6 and 149.6 mpy), they rose to 3.4, 6, 4.0 and 6.7 mm/a (134, 236, 157 and 264 mpy) at 500 and 2500 rpm respectively [548].

Carpenter® 20Cb-3 (cf. 2.4660; (%) Fe-0.036C-19.76Cr-33.70Ni-2.25Mo-3.14Cu-0.79Nb-0.23Mn-0.38Si-0.020P-0.04S), which has been annealed at 1477 K (1204 °C) for 1 h and quenched in water, corrodes in the Huey test (in boiling 65 % HNO$_3$) at a rate of 0.18 mm/a (7.09 mpy). Intercrystalline corrosion was observed neither in the Huey test nor in the test with 10 % HNO$_3$ + 3 % HF at 343 K (70 °C) (in accordance with ASTM A 262-D).

The corrosion rates of the steels 08Ch21N6M2T ((%) Fe-0.05C-20.6Cr-5.6Ni-2.4Mo-0.52Mn-0.25Ti) and 08Ch18G8N2M2T ((%) Fe-0.08C-18.2Cr-3.42Ni-8.9Mn-2.32Mo-0.22Ti) in 0.5 and 10 mol/l nitric acid with and without additions of 0.01 and 0.1 mol/l hydrofluoric acid or 5 g/l chloride at various temperatures are summarized in Table 82 [285, 408].

The welding material Ch17N18M2T ((%) Fe-0.09C-16.9Cr-12.3Ni-1.9Mo-1.4Mn-0.6Si) corrodes in fluoride-containing nitric acid (500 g/l HNO$_3$ + 600 g/l NaNO$_3$ + 50 g/l (Fe(NO$_3$)$_3$ + Cu(NO$_3$)$_2$) + 5 g/l NaF) at a material consumption rate of 1.97 g/m^2 h (about 2.4 mm/a (95 mpy)), knife-line corrosion and pitting corrosion also occurring [502].

The steels AISI 316, Carpenter® 20 Cb-3 and Hastelloy® C corrode in 25 % sulfuric acid containing 4 % nitric acid at 343 K (70 °C) at rates of 0.008, 0.005 and 0.08 mm/a (0.31, 0.20 and 3.15 mpy) respectively [493]. In acid mixtures containing 5.8 mol/l (HCl + HNO$_3$), the material consumption rate of AISI 316 is about 2 powers of ten higher, at about 10 g/m^2 h, at a mixing ratio of 50 percent by volume compared with that in mixtures with < 30 percent by volume HNO$_3$ [509]. The material consumption rate at about 80 percent by volume nitric acid is < 0.1 g/m^2 h.

The steel 08Ch21N6M2T corroded in the two mixtures of 36.6 % H$_2$SO$_4$ + 23.3 % HNO$_3$ + 39.1 % SO$_3$ and 28 % H$_2$SO$_4$ + 12 % HNO$_3$ + 60 % SO$_3$ with material consumption rates of 0.61 g/m^2 h (0.69 mm/a) and 1.30 g/m^2 h (1.46 mm/a) after 250 h tests [386].

Steel	Addition mol/l	0.5 mol/l HNO$_3$				10 mol/l HNO$_3$	
		293 K (20 °C)	323 K (50 °C)	373 K (100 °C)	Boiling point	293 K (20 °C)	323 K (50 °C)
		Corrosion rate, mm/a (mpy)					
08Ch21N6M2T	–	0.011 (0.43)	0.020 (0.79)	0.025 (0.98)	0.041 (1.61)	0.008 (0.31)	0.030 (1.18)
	0.01 HF	0.010 (0.39)	0.018 (0.71)	0.046 (1.81)	0.067 (2.64)	0.060 (2.36)	0.39 (15.35)
	0.1 HF	0.20 (7.87)	0.062 (2.44)	0.41 (16.14)	0.53 (20.87)	0.42 (16.54)	1.18 (46.46)
	0.01 HF + 5 g/l Cl$^-$	0.022 (0.87)	0.025 (0.98)	0.062 (2.44)	0.040 (1.57)	0.038 (1.5)	0.14 (5.51)
	5 g/l Cl$^-$	0.023 (0.91)	0.012 (0.47)	0.018 (0.71)	0.048 (1.89)	0.016 (0.63)	0.042 (1.65)
08Ch18G8N2M2T	–	0.016 (0.63)	0.016 (0.63)	0.030 (1.18)	0.053 (2.09)	0.013 (0.51)	0.019 (0.75)
	0.01 HF	0.036 (1.42)	0.025 (0.98)	0.075 (2.95)	0.096 (3.78)	0.080 (3.15)	0.31 (12.20)
	0.1 HF	0.73 (28.74)	0.90 (35.43)	2.11 (83.07)	11 (433)	0.38 (14.96)	1.42 (55.91)
	0.01 HF + 5 g/l Cl$^-$	0.22 (8.66)	0.47 (18.5)	0.19 (7.48)	5.69 (224)	0.036 (1.42)	0.16 (6.3)
	5 g/l Cl$^-$	0.028 (1.10)	0.019 (0.75)	0.018 (0.71)	0.058 (2.28)	0.025 (0.98)	0.044 (1.73)

Table 82: Corrosion of two CrNiMo steels in 0.5 and 10 mol/l HNO$_3$ with and without 0.01 or 0.1 mol/l HF or Cl$^-$ ions at various temperatures [285, 408]

The corrosion rate of the steels 10Ch17N13M2T (AISI 316 Ti, 1.4571) and 08Ch21N6M2T in 50 % sulfuric acid at 333 K (60 °C) is reduced from 171 and 361 mm/a to 0.02 mm/a (6,733 and 14,213 mpy to 0.79 mpy) by additions of ≥ 0.050 % nitric acid. A further increase in the nitric acid addition to 0.5 and 1 % caused no further reduction in the corrosion rate [441].

An austenitic steel of relatively high nickel and molybdenum content, Sandvik® 2RK65 (AISI 904 L, 1.4539; (%) Fe-0.020C-19.5Cr-25.0Ni-4.5Mo-1.5Cu), is used in the explosives industry for the reaction chamber for the production of TNT (trinitrotoluene) in an acid mixture of 25 % nitric acid and 75 % sulfuric acid which heats up during the course of the reaction [508].

The corrosion rate of the steels 06ChN28MDT, 03Ch21N21M4GB and 10Ch17N13M2T (AISI 316 Ti) increases with nitric acid-containing sulfuric acid (0.03 – 0.045 % HNO$_3$ in 75.3 to 75.6 % H$_2$SO$_4$) from < 0.05 at 333 K (60 °C) to 0.4, 0.45 and 0.85 mm/a (15.75, 17.72 and 33.46 mpy) respectively at 393 K (120 °C) [494].

According to Figure 48, the corrosion rate of the steels X 8 CrNiMoTi 18 11 (cf. 1.4541, AISI 321) and X5NiCrMoCuTi20-18 (1.4506) in hot 96 % sulfuric acid at

393 K (120 °C) is reduced from 1.2 and 0.6 mm/a (47.24 and 23.62 mpy) respectively to about 0.15 mm/a (5.91 mpy) by small amounts of nitric acid, expressed as N_2O_3, e.g. with 0.03 % N_2O_3. At the same time, instead of the crater-like local attack, general corrosion occurs. For the steels to be used in plants for further processing of concentrated sulfuric acid at 390 K (117 °C) (Müller-Kühne process), the acid must have the necessary concentration of 0.03 % N_2O_3 [465].

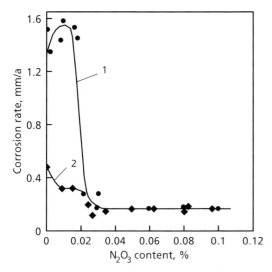

Figure 48: Dependence of the corrosion rate of austenitic steels in 96 % H_2SO_4 on the HNO_3 content (as N_2O_3) at 393 K (120 °C) [465]
(1) X 8 CrNiMoTi 18 11 (cf. 1.4541, AISI 321)
(2) X5NiCrMoCuTi20-18 (1.4506)

For surface pretreatment of implant steels used for replacement joints based on chromium-nickel-molybdenum steels, such as, for example, URX 2 CrNiMoN 18 ((18.12)) 12(0), the steels can be pickled either for 30 min at 293 K (20 °C) with 20 or 65 % nitric acid or for 30 min at 333 K (60 °C) with a solution of 20 % HNO_3 + 5 % H_2SO_4 + 4 % NaF. No increase in the pitting corrosion potential occurs with the latter solution, and the resistance to pitting corrosion is retained [547].

Table 83 shows the corrosion of the CrNiMo steels 10Ch17N13M3T (AISI 316 Ti, 1.4571; (%) Fe-0.10C-17Cr-13Ni-3Mo; Ti-stabilized) and 06ChN28MDT ((%) Fe-0.06C-(22-25)Cr-(26-29)Ni-(2.5-3.5)Mo-(2.5-3.5)Cu; Ti-stabilized) in hot nitric acid with various amounts of sodium fluoride and hydrazine (N_2H_4). If in each case 2 g/l sodium fluoride and hydrazine are simultaneously present, the material consumption rate of both steels increases by a factor of about 20, regardless of the concentration of nitric acid (75 – 100 g/l) [283].

The steel 0Ch23N28M3D3T ((%) Fe-<0.01C-23Cr-28Ni-3Mo-3Cu; Ti-stabilized) corrodes in 5 % sulfuric acid containing 1.2 % HCl + 2 % HNO_3 at 333 K (60 °C) with a corrosion rate of 0.3 mm/a (11.81 mpy), which corresponds to a current density of about 0.03 mA/cm^2 [371].

	Concentration		10Ch17N13M3T	06ChN28MDT
HNO_3	NaF	N_2H_4	Material consumption rate $g/m^2\,h$	
	g/l			
75	0	2	0.2 ± 0.1	0.08 ± 0.05
75	0.05	2	0.17 ± 0.13	0.2 ± 0.1
75	0.5	2	0.41 ± 0.16	0.16 ± 0.07
75	5.0	2	0.70 ± 0.1	0.11 ± 0.04
75	2	2	4.8 ± 0.5	0.3 ± 0.1
75	2	10	5.2 ± 0.3	0.2 ± 0.1
100	2	2	4.7 ± 0.7	0.27 ± 0.14
500	2	2	4.7 ± 1.2	1.7 ± 0.4

Table 83: Material consumption rate of CrNiMo steels in nitric acid with sodium fluoride and hydrazine at 368 K (95 °C) [283]

The steels summarized in Table 84 were developed based on a given corrosion problem for the purpose of discovering a steel resistant to a solution of 25 % HNO_3 + 10 % HCl + 65 % water at 303 and 333 K (60 °C). The additional cobalt content in some steels was added to improve the forming and mechanical properties. At a low silicon content, the corrosion resistance increases as the nickel content decreases.

Steel composition, %	Average material consumption rate $g/m^2\,h$
Fe-0.052C-0.33Si-1.54Mn-28.56Cr-9.03Ni-3.08Mo-0.40N	0.07
Fe-0.048C-0.37Si-1.40Mn-27.25Cr-8.94Ni-3.32Mo-0.042N	0.02
Fe-0.023C-0.43Si-1.52Mn-27.97Cr-5.27Ni-5.16Mo-3.29Co	0.08
Fe-0.018C-0.35Si-0.84Mn-27.19Cr-5.14Ni-3.62Mo	0.02
Fe-0.028C-0.44Si-1.62Mn-28.36Cr-2.98Ni-4.88Mo-3.29Co	0.03
Fe-0.026C-0.47Si-1.54Mn-28.75Cr-0.15Ni-4.36Mo-3.27Co	0.01

Table 84: Composition and average weight loss of specimens in 25 % HNO_3 + 10 % HCl + 65 % H_2O at 333 K (60 °C) [539]

Comparable tests on steels with the composition (%) Fe-0.020C-(24-29)Cr-24Ni-4Mo resulted in corrosion rates which were already higher by more than 2 powers of ten at 303 K (30 °C) [539].

The appearance of passivity on the steel Ch23N28M3D3T ((%) Fe-<0.1C-23Cr-28Ni-3Mo-3Cu; Ti-stabilized) is made difficult by the simultaneous presence of potassium chloride in nitric acid, so that a high degree of dissolution already occurs at room temperature [540].

The steel 08Ch21N6M2T corroded at 373 K (100 °C) in a solution of 40.4 g/l HNO_3 + 18.2 g/l oxalic acid at a corrosion rate of 0.004 mm/a (0.16 mpy) [412].

The corrosion rate of three steels used in a reactor for the synthesis of 3-chloro-2-hydroxypropanoic acid (RP) is shown in Table 85 [18].

Reaction medium, %	Steel	Corrosion rate mm/a (mpy)
Liquid phase		
* 45 – 46 HNO$_3$ + 20 – 21 epichlorohydrin + 28 – 29 H$_2$O + 5.3 – 5.4 HCHO	08Ch22N6M2T	1.7 – 3.45 (67 – 136)
	08Ch17N13M2T	0.85 – 1.24 (33 – 49)
** 29 – 30 RP + 9 – 12 H$_2$C$_2$O$_4$ + 55 – 56 H$_2$O	06ChN28MDT	0.29 – 0.78 (11 – 31)

* starting composition
** reaction products

Table 85: Corrosion rate of steels in HNO$_3$ containing process solution after 1270 h tests at 330 K (57 °C) [18]

While phosphoric acid, even in high concentrations, does not influence the corrosion of the steels 06ChN28MDT and 08Ch21N6M2T in 15 % nitric acid, according to Table 86 the addition of only 1 % H$_2$SiF$_6$ causes a considerable increase in the corrosion rate [470].

The steels already mentioned in Table 86 are recommended for the reactor containing nitric and phosphoric acid as well as small amounts of hydrogen fluoride in the production of Nitrophoska. Figure 49 shows the corrosion behavior of the steels in various production solutions of HNO$_3$ + H$_3$PO$_4$ + H$_2$SiF$_6$ + HF [497].

Composition of solution, %	06ChN28MDT	08Ch21N6M2T
	Corrosion rate, mm/a (mpy)	
15 HNO$_3$	0.007 (0.28)	0.001 (0.04)
15 HNO$_3$ + 15 H$_3$PO$_4$	0.013 (0.51)	0.001 (0.04)
15 HNO$_3$ + 15 H$_3$PO$_4$ + 1 H$_2$SiF$_6$	0.32 (12.60)	0.24 (9.45)
15 HNO$_3$ + 1 H$_2$SiF$_6$	0.32 (12.60)	0.62 (24.41)
15 HNO$_3$ + 1 H$_2$SiF$_6$ + 0.1 HF	0.54 (21.26)	0.72 (28.35)

Table 86: Corrosion of the steels 06ChN28MDT and 08Ch21N6M2T in 15 % HNO$_3$ with and without additions of H$_3$PO$_4$, HF and H$_2$SiF$_6$ at room temperature [470]

For the steel proposed for urea production to pass the Huey test, it must have a carbon content of < 0.03 % (for the welded regions). To obtain a ferrite-free structure at chromium contents of 17 to 18 % together with 2.6 to 3.0 % molybdenum, the nickel content should be not less than 15 %. The acceptable corrosion value in the Huey test is given as 0.54 g/m^2 h. As has already been pointed out several times, attention should also be paid to the correct heat treatment [523].

From their successful corrosion resistance in the Huey test, the steels UHB-724L (AISI 316 L, 1.4435; (%) Fe-<0.03C-17.4Cr-14.4Ni-2.6Mo) and UHB-725LN (AISI 310 MoLN, 1.4466; (%) Fe-0.020C-25.0Cr-22.0Ni-2.1Mo; nitrogen-containing) can be recommended for urea plants (specific corrosion rates for the urea region: 0.60 mm/a (23.62 mpy) for UHB-724L and 0.27 mm/a (10.63 mpy) for UHB-725LN) [498].

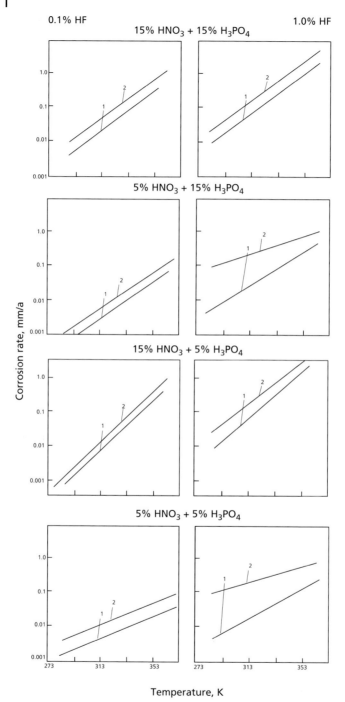

Figure 49: Temperature-dependence of the corrosion of (1) 06ChN28MDT and (2) 08Ch21N6M2T in (5 – 15) % HNO_3 + 1 % H_2SiF_6 + (0.1 – 1.0) % HF + (5 – 15) % H_3PO_4 [497]

If > 2 % molybdenum is added to chromium-nickel steel, the nitrogen content should be increased, this preventing the formation of intermetallic compounds as well as improving the structure. If the processing conditions and heat treatment of the steel are carefully observed, the resistance to nitric acid, which can also contain halide ions, is increased by the higher molybdenum content [544].

To evaluate the corrosion resistance of steels of the type AISI 316 L in urea synthesis solutions at 468 K (195 °C), the corrosion behavior of the steel in the Huey test (48 h in boiling 65 % HNO_3) was determined for comparison. The corresponding corrosion rates are compared in Table 87 [812]. According to these, steels with corrosion rates in the Huey test of between 0.27 and 0.65 mm/a (10.63 and 25.59 mpy) can be recommended as materials for a urea synthesis plant.

Steel composition, %	Huey test	Urea solution
	Corrosion rate, mm/a (mpy)	
Fe-0.019C-17.01Cr-12.04Ni-2.51Mo-0.93Mn-0.50Si	0.274 (10.79)	0.086 (3.39)
Fe-0.021C-17.24Cr-12.32Ni-2.11Mo-0.83Mn-0.45Si	0.365 (14.37)	0.080 (3.15)
Fe-0.010C-17.49Cr-12.75Ni-2.25Mo-1.73Mn-0.66Si	0.602 (23.70)	0.062 (2.44)
Fe-0.013C-17.12Cr-12.16Ni-2.09Mo-1.28Mn-0.67Si	0.274 (10.79)	0.079 (3.11)
Fe-0.014C-17.28Cr-13.12Ni-2.26Mo-1.41Mn-0.52Si	0.274 (10.79)	0.073 (2.87)
Fe-0.021C-16.40Cr-12.68Ni-2.31Mo-0.99Mn-0.53Si	0.657 (25.87)	0.099 (3.9)
Fe-0.014C-17.58Cr-12.48Ni-2.40Mo-1.00Mn-0.72Si	0.402 (15.83)	0.063 (2.48)

Table 87: Comparison of the corrosion rates of the steel AISI 316 L in a urea synthesis solution at 468 K (195 °C) and in the Huey test as a function of the composition [812]

The corrosion which occurs on the steels Ch21N6M2T and 0Ch23N28M3D3T under the conditions which prevail during production of methacrylic acid is shown in Table 88, using the following artificially prepared reaction mixtures (solution 1: 4 mol/l N_2O_4 + 1 mol/l HNO_3 + isobutylene, solution 2: 4 mol/l N_2O_4 + 1 mol/l HNO_3 + 0.5 mol/l isobutylene and solution 3: 4 mol/l N_2O_4 + 1 mol/l HNO_3) [538]. The normal CrNi steel 18 8 (cf. 1.4541, AISI 321) has also been shown for comparison. Relatively severe pitting corrosion was observed on the steels during continuous operation, with the exception of 0Ch23N28M3D3T.

Steel	Solution 1	Solution 2	Solution 3
	Material consumption rate g/m^2 h		
Ch21N6M2T	0.00065	0.0011	0.0007
0Ch23N28M3D3T	–	0.00007	0.00082
Ch18N10T	0.0002	0.00047	0.00133

Table 88: Corrosion of steels in 3 artificially prepared reaction solutions under the operating conditions in methacrylic acid production [538]

The sliding gates made of the steel (%) Fe-<0.1C-18Cr-10Ni-3Mo in sliding gate housings are also resistant to nitric acid at low temperatures [543].

The ferritic-austenitic steel Sandvik® 3RE60 (UNS S31500, 1.4417; (%) Fe-0.030C-18.5Cr-4.9Ni-2.7Mo-1.5Mn-1.7Si<0.03P-<0.03S) which was used for coolers in petroleum production in the North Sea and was in contact with gaseous nitric acid at 373 K (100 °C) at the intake and 313 K (40 °C) at the discharge for 2.5 years showed stress corrosion cracking [535].

A 22   Austenitic chromium-nickel steels with special alloying additions

The complex and costly CrNiMoCu steels X1NiCrMoCu31-27-4 (UNS N08028, 1.4563) often bring no improvement under the influence of nitric acid compared with CrNi steels with 18 to 20 % chromium, 10 to 14 % nickel, a low carbon content (< 0.05 % C) and the correct heat treatment, and in some cases they even worsen the corrosion resistance. Figure 50 shows the temperature dependence of corrosion of this steel in azeotropic nitric acid. The corrosion rate k in mm/a measured at 373 K (100 °C) for the range 20 to 75 % HNO_3 can be described by the equation

$$k = 1.95 \times 10^{-4} + 4.0 \times 10^{-6} [HNO_3]$$

where $[HNO_3]$ denotes the concentration of nitric acid in % [429].

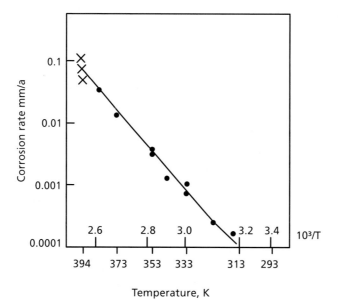

Figure 50:   Temperature dependence of the corrosion of X1NiCrMoCu31-27-4 in azeotropic nitric acid [429]
(x = values from the Huey test)

Table 89 summarizes steels which are particularly suitable for use in the presence of nitric acid. Of the steels listed, AISI 304 L (1.4306) gives the best results in the cost and behavior comparison if high purity (low C content, for example) and good homogeneity of composition are ensured [429].

DIN-Mat. No.	Composition, %				Huey test[1] Azeotr. HNO_3[2]		Relative cost[3]	Area of application
	Cr	Mo	Ni	Others	Material consumption rate, $g/m^2\,h$			
1.4306 (AISI 304 L)	19	–	11	–	0.20 – 0.50	–	0.7	–
1.4306S	20	–	12	[4]	0.07 – 0.09	0.09	1.0[5]	–
1.4335 (UNS S31002)	25	–	21	–	0.05 – 0.07	0.07	1.36[5]	–
1.4547 (UNS S31254)	24	–	20	Nb	0.07 – 0.11	–	–	–
1.4435 (UNS S31603)	17.5	2.6	14	–	0.13 – 0.30	–	1.15	[6]
1.4466 (AISI 310 MoLN)	25	2.3	22	N	0.04 – 0.08	0.09	1.27	[6]
1.4563 (UNS N08028)	27	3.5	31	Cu	0.07 – 0.08	0.08	1.72	[7]
2.4858 (UNS N08825)	21	3	40	Cu, Ti	0.15 – 0.20	0.25	1.9	[7]

1) determined on seamless pipes, 5 test periods of 48 h each
2) 50 test periods of 48 h each in boiling, azeotropic HNO_3 with continuous removal of the corrosion products (distillation process)
3) for hot-rolled sheets 2 – 3 mm thick; acceptance about 3 t, position: March 1985 in the FRG
4) with low C, Si and Mo contents and somewhat increased Cr and Ni contents
5) ESR material
6) urea industry
7) for mixed acids

Table 89: Steels for use in nitric acid of approximately azeotropic composition with their material consumption rate, relative cost and main area of application [429]

Ferralium® 255, a CrNiMoCu steel 25 5 3 2, is also resistant to stress corrosion cracking in nitric acid at high concentrations and temperatures [579].

According to Figure 51, the Si-containing steel 1815-LCSi (UNS S30600, 1.4361) of very low carbon content ((%) Fe-0.006C-18.3Cr-15.1Ni-1.5Mn-4.1Si-0.005S-0.010P-0.010N) has an adequate resistance in boiling nitric acid of between 28 and 95 %, apart from in the range 40 to 90 %. Hot-rolled steel specimens heat-treated at 1423 K (1150 °C) for 10 min, quenched in water and annealed again at 1423 K (1150 °C) for 30 min after welding corroded in boiling 95 % nitric acid with a corrosion rate of 0.11 to 0.12 mm/a (4.33 to 4.72 mpy) after 1296 h [634].

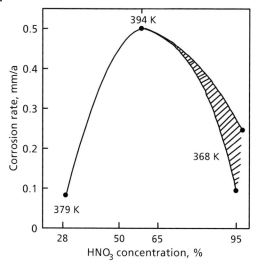

Figure 51: Corrosion rate of the steel 1815-LCSi (UNS S30600) in boiling nitric acid (28 to 95%) [634]

Steels of the type 10Ch14G14N4T ((%) Fe-<0.1C-(13-15)Cr-(3.5-4.5)Ni-(13-15)Mn-0.65Ti) can also be used for production of nitric acid and ammonium nitrate in the presence of 56% nitric acid below 320 K (47 °C) [387]. The corrosion rate of CrMnNi steels with 20% manganese and (2 – 5) % nickel (see section A 20) at room temperature and 363 K (90 °C) in 60% nitric acid is about 0.002 and 0.68 mm/a (0.08 and 26.77 mpy) respectively, regardless of the nickel content of between 2 and 5% [435].

Alloying up to 2% germanium with 25Cr6Ni duplex steel for the purpose of increasing the ferrite content has no influence on corrosion in boiling 65% nitric acid [609].

While addition of 0.042% cerium to the steel 02Ch18N11 (quenched in water after heating for 30 min and tempering at 923 K (650 °C) for 1 h) causes no noticeable increase in corrosion rates in the Huey test (5 × 48 h in boiling 65% HNO_3), the effect of boron is considerable even at 0.0015% (e.g. an increase from 0.26 to 5.38 mm/a (10.24 to 211 mpy), or from 0.23 to 11.6 mm/a (9.06 to 457 mpy) after tempering at 1353 and 1423 K (1150 °C) respectively) [813].

A protective SiO_2 layer is formed from the high silicon content in the steel. It should be mentioned that the corrosion rate of the steels AISI 304 L and 310 is lower up to about 70% HNO_3. Only above 80% HNO_3 is the steel 1815-LCSi (UNS S30600) superior to the other two in respect of corrosion behavior [635].

The material consumption rate of the steel KV-80 ((%) Fe-0.018C-17.7Cr-14.3Ni-0.19Cu-1.22Mn-0.05Si-0.006P-0.011S) and KV-81 ((%) Fe-0.016C-24.7Cr-19.8Ni-1.31Mn-0.17Cu-0.10Si-0.006P-0.016S) in boiling 98% nitric acid could be reduced from 7 and 5.5 g/m² h respectively to 0.07 and 0.31 g/m² h respectively by addition of 4% aluminium. While corrosion of the steel KV-81 is not improved sufficiently, a considerable improvement comparable to that on the steel KV-80 of 0.16 g/m² h is achieved on the steel KV-82 ((%) Fe-0.022C-18.4Cr-9.4Ni-0.25Mo-0.22Cu-1.18Mn-0.06Si-0.006P-0.016S) [578].

Welds produced with welding wires of the CrNiSi steels 02Ch17N(10-18)S6 ((%) Fe-0.02C-(4-6.5)Si-(0.43-0.52)Mn-(16.3-18.0)Cr-(10.5-18.2)Ni-(0.005-0.008)S-(0.012-0.014)P) and 02Ch8N(22-60)S6 ((%) Fe-0.02C-(5.5-6.5)Si-(0.004-0.005)Mn-(7.4-9.0)Cr-(21.1-61.0)Ni-(0.006-0.016)S-(0.013-0.014)P) show only a low corrosion rate, which does not exceed 0.1 mm/a (4 mpy), in boiling 98% nitric acid in the entire alloy range after 1000 h tests [603]. In practical tests in apparatuses for dissolution of Zr-plated uranium, Carpenter® 20 (cf. 2.4660), Durco® D-10, Durimet® 20 (CN-7M) and AISI 309 Nb (cf. 1.4828; (%) Fe-0.20-C(22-24)Cr-(12-15)Ni-2Nb) showed corrosion rates of 0.38 – 0.75 mm/a (14.96 – 29.53 mpy) during two batches (a year is calculated with 180 batches) in (mol/l) 10 nitric acid + 4 hydrogen fluoride. The stainless steel AISI 309 Nb had a higher corrosion resistance than Carpenter® 20. The corrosion rates were about 2 and 3.1 mm/a (79 and 122 mpy) at 313 and 353 K (40 and 80 °C) respectively. No galvanic action occurs when the steel is brought into contact with zirconium in the acid mixture [555]. Lining the steel tank with Teflon® (PTFE) had no effect, since the coverings are not sufficiently impermeable to acid.

The weld seam of the chromium-nickel steel 1Ch18N9T ((%) Fe-0.1C-19.0Cr-9.9Ni-0.68Mn-1.18V-0.76Nb-0.22Ti-1.24Si) shows pitting corrosion and intercrystalline corrosion at 303 K (30 °C) in a solution of 10% HCl + 5% HNO$_3$ [778].

The CrNiTi steel X6CrNiTi18-10 (AISI 321, 1.4541; (%) Fe-0.1C-18.0Cr-10.0Ni-(5 × C)Ti) corrodes in 0.1, 0.5 and 1.0 mol/l nitric acid at 353 K (80 °C) with a material consumption rate of 0.03, 0.07 and 0.09 g/m^2 h respectively, and if 0.05 mol/l ammonium cerium(IV)nitrate is also present in the acid, with rates of 0.16, 6.5 and 44 g/m^2 h [918].

The material consumption rate of chromium-nickel steels with additions of 4% silicon and niobium in boiling 56% nitric acid + 0.2% ammonium bichromate ($(NH_4)_2Cr_2O_7$) is shown in Table 90 [522]. The corrosion-reducing effect of silicon on the high-nickel CrNi steel which has already been mentioned above is also confirmed here.

Steel	Composition of steel, %						Material consumption rate g/m^2 h	
	C	Cr	Ni	Si	Mn	Nb	(a)	(b)
ChN40SB	0.040	18.8	39.4	4.3	0.06	0.63	0.25	0.30
ChN40S	0.031	20.0	38.9	4.2	0.05	0.13	0.27	0.32
Ch14N40SB	0.034	14.4	38.9	4.0	0.05	0.63	0.48	0.45
ChN40B	0.032	18.2	40.4	0.08	0.05	0.49	2.60	10.3

Table 90: Corrosion rate of steels after (a) quenching and (b) quenching and tempering at 923 K (650 °C) for 2 h in 56% HNO$_3$ + 0.2% $(NH_4)_2Cr_3O_7$ [522]

Solutions of 150 to 250 g/l HNO_3 + 75 to 100 g/l NH_4 HF have proved suitable for descaling the CrNiTiAl steel 36NChTJu (EI-702; (%) Fe-0.03C-12.1Cr-35.95Ni-3.03Ti-1.13Al-1.03Mn-0.03Si) after dispersion annealing and quenching from 1263 K (990 °C), excellent, silver-looking surfaces being produced. The scale is removed within 3 min. The steel itself is attacked in such a pickling solution at a rate of about 1 g/m² min, which corresponds to one tenth of the rate of dissolution of the scale [524].

A 23   Special iron-based alloys

Iron of high silicon content (10 to 18 % Si) is also resistant to nitric acid and can be cast into complicated shapes. The standard grade (Tantiron® N), which complies with the British Standard BS 1591/49, has a silicon content of 14.3 to 15.3 %, good physical properties and an excellent resistance to nitric acid. Tantiron® E with 16 to 18 % Si is recommended under severe corrosion conditions if mechanical properties are of secondary importance [556, 585].

The manganese-aluminium steels ((%) Fe-(0.1-1)C-(20-32.5)Mn-(7-10)Al) are not resistant in nitric acid at room temperature. Only the steel Fe-0.95C-23.3Mn-7.4Al, which is not resistant in nitric acid below 50 %, can be used in industry at higher concentrations. The material consumption rate of the steel at room temperature, e.g. in 65 % nitric acid, is thus about 0.21 g/m² h [557]. The corrosion potential here is -0.59 V_{SCE}.

Figure 52 shows the corrosion of the chromium-manganese steel AS-43 (Russ. type; (%) Fe-0.05C-16.5Mn-18.4Cr-1.60Ni-0.31Nb-0.01Ce-0.26N-0.004B) in nitric acid containing 0.1 and 0.01 mol/l hydrofluoric acid at 293 and 313 K (20 and 40 °C). The corrosion rate minimum is at 313 K (40 °C) at about 0.3 mm/a (11.81 mpy) in 5 mol/l HNO_3 + 0.1 mol/l HF [413].

Sintered stainless steels with an excellent absorption capacity for neutrons ((%) Fe-≤0.15C-(0.5-4)Si-≤0.6Mn-(7-50)Cr-≤25Ni-≤5Mo-≤5Cu-≤5Ti with and without ≤4Nb, mixed with 1 to 55 percent by volume metal boride) corrode in boiling 65 % nitric acid at a material consumption rate of 38 g/m² h [568].

Current density/potential curves using the iron alloy (atomic percent) Fe-40Ni-16P-6B in 15 % nitric acid show a higher corrosion rate for the crystalline alloy than for the amorphous alloy. However, at a polarization potential of about -0.2 V_{SHE}, a corrosion current density of 0.2 mA/cm² occurs on both, i.e. the corrosion rate is too high for industrial use [575].

Iron-nickel alloys with 10 to 30 % nickel are susceptible to corrosion in nitric oxide (NO) under 0.0027 MPa between 498 and 548 K (225 and 275 °C), iron oxide coatings being formed. The corrosion decreases as the nickel content increases [601].

For chemical surface treatment of the iron-nickel-cobalt alloy Kovar® (UNS K94610, 1.3981, Fe-29Ni-18Co), it is pickled in a solution of 2 H_2SO_4 : 1 HCl : 2 H_2O (percent by volume) at 343 K (70 °C) for 2 min and then polished in a solution of 30 HNO_3 : 1 HCl : 70 CH_3COOH (percent by volume) at the same temperature for

7 seconds. A layer of 2.5 μm is corroded here during the pickling operation, with a higher loss of nickel than cobalt. During the polishing operation, a 7.8 μm layer dissolves, but the ratio of the dissolved metals corresponds to that of the alloy [534].

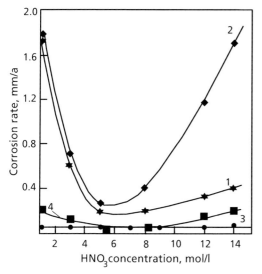

Figure 52: Dependence of the corrosion rate of the steel AS-43 on the HNO_3 concentration in HNO_3 containing 0.1 mol/l HF at (1) 293 K (20 °C) and (2) 313 K (40 °C) and in HNO_3 containing 0.01 mol/l HF at (3) 293 K (20 °C) and (4) 313 K (40 °C) [413]

An aqueous solution of 70 to 130 g tartaric acid + 80 to 120 ml hydrochloric acid + 30 to 70 ml nitric acid + 0.5 to 2 g/l of a non-ionic wetting agent is also proposed for polishing Kovar® [584].

For analyzing the structure of high-speed steels ((%) Fe-(0.59-1.10)C-(3.85-4.17)Cr-(0.25-0.32)Mn-(0.34-9.21)Mo-(1.1-1.9)V-(6.2-17.6)W-(0-17)Co-(0.23-0.30) Si), the best results in respect of ease of differentiation between austenite grain boundaries and the basic structure were achieved by pickling in a solution of 10 ml conc. HCl + 3 ml conc. HNO_3 + 80 ml methyl alcohol [442].

A 24 Magnesium and magnesium alloys

Magnesium and magnesium alloys are not resistant to nitric acid.

Shiny surfaces are obtained by dipping magnesium (99.96 %) and the alloys Mg-3.0Zn-0.68Zr-0.08Mn and Mg-5.0Zn-0.88Zr-0.06Mn, as well as Mg-0.8Zn-8.0Al-0.27Mn in 15 % nitric acid at temperatures of < 320 K (47 °C) for 5 min. The corrosion rate of magnesium and its alloys in 5, 10 and 15 % acid is about 0.3 mm/h, 1.2 to 4.2 mm/h and 9.6 to 12 mm/h [600].

A bath containing 8.5 % HNO_3 + 2.6 % HF + 4.4 % NH_4F + 0.03 % $Pb(NO_3)_2$ + 2.6 % gum arabic at 333 to 343 K (70 °C) is suitable for polishing magnesium and its alloys. 40 to 60 seconds are sufficient to produce a shiny surface [106].

To apply firmly adhering epoxy resin or polyethylene coatings to magnesium, the metal is degreased in a solution of 0.183 mol/l Na_3PO_4 + 0.28 mol/l Na_2CO_3 + 0.0024 mol/l soap at 373 K (100 °C) for 10 min, subsequently etched in nitric acid (d = 1.42 g/ml) at 298 K (25 °C) for 1 min and finally polarized anodically in a solution of 1.4 mol/l KOH + 5.2 mol/l KF at 323 K (50 °C) [548].

A 25 Molybdenum and molybdenum alloys

Molybdenum is severely attacked in 70 % nitric acid. At higher concentrations, however, it displays a passivity which is similar to that of iron [735].

Amorphous Mo-50Ni films 145 – 290 nm thick produced on nickel by mixed ionic beams irradiated with 1 and 2 Me V Au^+ ions give more uniform layers which are also more resistant to nitric acid (300 ml HNO_3 + 700 ml CH_3OH) than polycrystalline layers of the same composition. The amorphous films obtained by ion bombardment are resistant below 820 K (547 °C), but converted into polycrystalline material above 870 K (597 °C), this dissolving in nitric acid [599].

Industrial molybdenum TZM (Mo-0.5Ti-0.08Zr, arc-smelted) corrodes in 0.1, 0.5 and 1.0 mol/l nitric acid at 353 K (80 °C) with a material consumption rate of 0.05, 0.35 and 6 g/m² h. If 0.05 mol/l ammonium cerium(IV) nitrate is simultaneously present, the material consumption rates are 7.55 and 130 g/m² h [918].

Fractures caused on molybdenum coated with $MoSi_2$ layers by impact, bending and tensile tests are etched in a solution of 50 % HNO_3 + 50 % H_3PO_4 [574].

Molybdenum is unsuitable as material because of its high susceptibility to corrosion.

A 26 Nickel

Nickel and nickel alloys have little resistance to oxidizing acids such as nitric acid [562, 586].

The material consumption rate and the corrosion potential of nickel are shown in Figure 53 as a function of the concentration of nitric acid at room temperature [564]. At a concentration of about 8 mol/l, a sharp decrease in the material consumption rate from about 500 to 15 g/m² h occurs, coupled with a rise in the corrosion potential from about 0.2 to 0.85 V_{SHE}. While ammonia is the main product in dilute nitric acid, the reduction in concentrated nitric acid proceeds only to nitrous acid. By polarization in the potential range between + 0.6 and + 1.0 V_{SHE}, it was possible to keep the corrosion of nickel in 1 and 4 mol/l nitric acid at 303 K (30 °C) below 1 g/m² h [564, 565].

While nickel (99.99 %) corrodes with a material consumption rate of 114 g/m² h at 298 K (25 °C) in pure 8 mol/l nitric acid, the same acid saturated with nickel nitrate ($Ni(NO_3)_2 \times 6H_2O$) causes a material consumption rate of only 7.69 g/m² h, although this is still too high for industrial purposes [566]. The inhibitor p-nitroaniline has proved unusable.

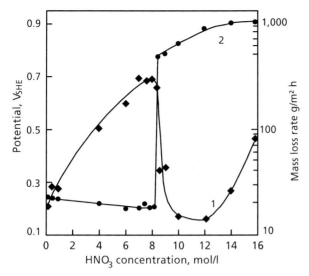

Figure 53: Dependence of (1) the material consumption rate and (2) the potential of nickel in nitric acid at room temperature on the acid concentration [564]

The behavior of nickel in nitric acid solutions depends greatly on the acid concentration. At concentrations below 40 %, Ni dissolves rapidly, while it is passivated in a stable form above 50 %. If chlorides are simultaneously present, pitting corrosion can be expected. Passivity is also achieved in dilute acid by oxidizing agents, such as, for example potassium dichromate [563].

The potentiodynamic polarization curves of nickel in 1 mol/l nitric acid with 2 % aqueous methanol as a solvent are characteristic of pitting corrosion. Nitrate ions can greatly accelerate dissolution of the metal in the passive state [567].

Thick nickel layers are attacked by nitric acid [569].

To detach a nickel layer from chromium-nickel steels, the layer is dipped in triethanolamine-containing nitric acid with 5 g/l bromide at 333 K (60 °C) for 15 min in order to keep the attack on the steel as low as possible [114].

According to current density/potential curves in an electrolyte consisting of 6 or 0.9 mol/l sodium nitrate + 0.1 mol/l nitric acid flowing at a rate of 10 m/s, anodic dissolution of nickel takes place under industrial conditions (15 to 20 A/cm^2) at a 100 % current yield in the transpassive potential range (>> 1 V_{SHE}). Evolution of oxygen is no longer observed under these conditions [570].

Layers of nickel about 500 μm thick precipitated on copper from a bath of Ni(CF$_3$COO)$_2$ halide in methanol at 303 K (30 °C) and 1.5 A/dm^2 corrode in 1.5 mol/l nitric acid + 20 g/l NH$_4$Cl or 8 g/l potassium bromide at material consumption rates of 1068 and 143 g/m^2 h respectively after short-term tests (10 min) at 298 K (25 °C). The corrosion rates did not change after additions of AlCl$_3$, NH$_4$Br and KCl. A commercially available rolled nickel specimen corroded in 1.5 mol/l nitric acid at 323 K (50 °C) at a rate of 116 g/m^2 h [571].

The following pickling solution, which does not attack steel, of 500 ml nitric acid + 5.7 g/l copper sulfate + 3.3 g/l sodium chloride + 3.0 g/l iron chloride, with and without Na_2TeO_3, is used to strip nickel layers applied to stainless steel in cladding devices. The pickling time is 90 to 140 s [572].

31 and 103 min respectively are needed to remove a 12 or 52 μm thick nickel layer on steel using the following pickling solution (100 ml/l HNO_3 + 200 ml/l ethylene-diamine + 50 g/l sodium m-nitrobenzenesulfonate; pH = 10) at 338 K (65 °C). The steel is not attacked during this process [573].

According to current density/potential curves on nickel in a solution containing 5 mol/l sodium nitrate + 1 mol/l nitric acid at 298 K (25 °C), agitated at a rate of 100 to 1000 cm/s, the current density starts to rise from < 0.1 to 4 to 15 A/cm^2 during polishing of steel above + 18 V_{SCE} at a running speed of 0.5 V/s. Almost 100 % polishing of the surface is achieved at about 5 A/cm^2 and an electrolyte agitation of 100 cm/s [592]. These test results are of importance for electrolytic processing of nickel.

A 27 Nickel-chromium alloys

According to Table 91, the nickel-chromium alloy Chromel® (UNS N06010) heat-treated at 1373 K (1100 °C) is severely attacked by nitric acid. The corrosion cannot be reduced sufficiently by addition of 5.0 g/l oxalic acid alone. Only if hydrochloric acid is present at the same time is a significant reduction in the material consumption rate achieved [587].

Solution, g/l	Without addition	With addition
	Material consumption rate g/m^2 h	
50 HNO_3	120	22
100 HNO_3	270	52
100 HNO_3 + 50 HCl	5.3	5
50 HNO_3 + 50 HCl	3	2
200 HNO_3 + 50 HCl	11	9

Table 91: Influence of 5.0 g/l oxalic acid on the corrosion of Chromel® in HNO_3 and HNO_3 + HCl at room temperature [587]

According to Figure 54, the material consumption rate of nickel-chromium alloys with increasing chromium content in 16 to 18 % nitric acid + 3 to 4 % hydrogen fluoride at 333 to 343 K (60 to 70 °C) drops from > > 1 to about 0.15 g/m^2 h with 50 % chromium. Specimens which had been heat-treated at 1373 K (1100 °C) and then quenched in water were analysed [598]. According to Table 92, the influence of the alloying additions – with the exception of 0.17 % carbon and 1.7 % niobium – on the corrosion of the alloy ChN60V in the solution can be ignored. The intercrystalline corrosion depends on the temperature and duration of the heat treatment.

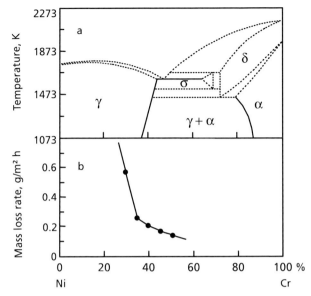

Figure 54: Phase diagram (a) and dependence of the material consumption rate (b) of NiCr alloys on the chromium content in 16 – 18 % HNO$_3$ + 3 – 4 % HF at 333 – 343 K (60 – 70 °C) after 500 h tests [598]

According to Figure 55, corrosion of the alloy Ni-40Cr in the above HNO$_3$/HF solution at 333 to 343 K (60 to 70 °C) is altered by addition of 1 % tungsten or 3 % molybdenum, depending on the temperature and duration of the heat treatment [598].

Alloying elements, %							Material consumption rate g/m^2 h	IC* of ≥ 15 µm	
C	N	Zr	Ti	Nb	Ce	B		Duration h	Temperature K (°C)
0.01	0.09	–	0.1	–	0.015	–	0.17	20	923 (650)
0.01	0.09	0.05	0.1	–	0.015	–	0.17	6	903 (630)
0.17	0.09	–	0.1	–	0.015	–	0.4	1	873 (600)
0.01	0.09	–	0.35	–	0.015	–	0.25	1	1003 (730)
0.02	0.09	–	0.1	1.7	0.015	–	0.3	2	973 (700)
0.01	0.09	–	0.1	–	0.05	–	0.17	30	1003 (730)
0.01	0.09	–	0.1	–	0.015	0.009	0.19	20	923 (650)

* intercrystalline corrosion

Table 92: Influence of alloying elements on the material consumption rate of ChN60V (quenched) with 40 % chromium in 16 – 18 % HNO$_3$ + (3 – 4) % HF at 333 to 343 K (60 to 70 °C), test duration 500 h [598]

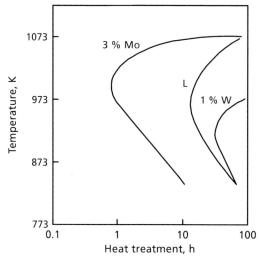

Figure 55: Influence of molybdenum and tungsten on the course of the isocorrosion curve (0.3 g/m² h) of ChN60 in 16 – 18 % HNO$_3$ + 3 – 4 % HF at 333 to 343 K (60 to 70 °C) in the time/-temperature diagram of the heat treatment [598]
L = ChN60 without additions

According to Figure 56, above 23 % chromium the corrosion rate of the nickel alloys (Ni-xCr-10W-2C) in 10 % nitric acid at room temperature reaches values of < 0.2 g/m² h, which are of industrial interest, in 48 h tests [611].

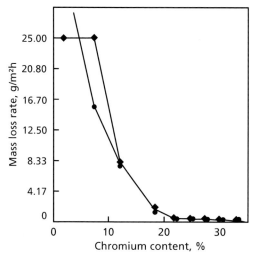

Figure 56: Corrosion of Ni-xCr-10W-2C in the cast state (●) and homogenized at 1323 K (1050 °C) for 180 h in 10 % HNO$_3$ at room temperature (♦) after 48 h tests, as a function of the chromium content [611]

In the modified Huey test (48 h in boiling 25 % nitric acid), the alloy I-82 ((%) Ni-0.007C-20.48Cr-3.04Mn-0.75Fe-0.38Ti-0.06Al-0.05Co-0.02Cu-0.13Si-0.007S) corrodes at a material consumption rate of 0.053 g/m^2 h [614].

For effective removal of the scale layer on the nickel-chromium alloy EI-345 ((%) Ni-0.12C-19.22Cr-1.0Fe-0.7Mn-0.2Cu-0.15Al-0.15Ti-0.8Si-0.01P-0.01S), it is immersed in a melt of 75 % NaOH + 25 % NaNO$_3$ at 673 K (400 °C) for 1 h and then treated in a solution of 10 % HNO$_3$ + 4 % HF at 328 K (55 °C). The material consumption rate of the alloy was determined as about 110 and 100 g/m^2 h, depending on the heat treatment temperature of 1273 and 1423 K (1000 and 1150 °C) respectively, with quenching in air. If treatment of the alloy in the melt was carried out at 773 K (500 °C), the rate was about 40 and 45 g/m^2 h respectively, under otherwise identical conditions [589].

The HNO$_3$/HF mixtures at which the corrosion rate does not exceed 0.5 mm/a (20 mpy) can be seen from the isocorrosion curves (0.5 mm/a (19.7 mpy)) shown in Figure 57 for the nickel-chromium alloy 03ChN60V (Russ. type) at various temperatures [478]. At a rotational velocity of the specimen of 500 or 2500 revolutions min^{-1}, the corrosion rate of the alloy in 10 mol/l HNO$_3$ + 0.1 mol/l HF at 363 K (90 °C) rises from 1.7 to 2 or 2.5 mm/a (67 to 79 or 98 mpy) respectively.

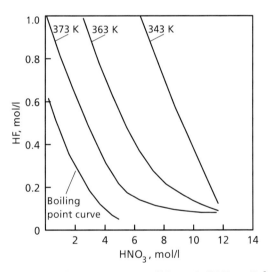

Figure 57: Isocorrosion curves (0.5 mm/a (19.7 mpy)) for 03ChN60V at various temperatures in HNO$_3$/HF solutions [478]

The alloy (%) N-(39-42)Cr-1Fe-1Mn-(0.5-2)W-(0.05-0.15)Mg-(0.01-0.02)Ce-0.15S-0.03C-0.02S-0.02P is said to be particularly resistant to nitric acid containing fluoride; no intercrystalline corrosion occurring even on welded joints [590].

The alloy Ni-(39-41)Cr-1W has proved to be suitable as a tank material for the etching solution consisting of nitric acid and hydrogen fluoride [591]. A solution of 250 g/l HNO$_3$ + 100 g/l NH$_4$F × HF is the best etching solution for the alloy 36NChTJu [610].

The rate of dissolution of the nickel-chromium alloy 36NCh (measured as the anodic corrosion current density) at 298 K (25 °C) in 1 % nitric acid was increased in the presence of a magnetic field [606].

The main advantage of nickel-chromium alloys in comparison with steels, such as, for example, 12Ch18N10T (AISI 321, 1.4541), is their resistance in nitric acid solutions containing fluoride ions [605].

Regardless of the method of smelting of the alloy EI-437B (Nimonic® alloy 80 A, 2.4631; (%) Ni-0.05C-0.27Si-0.16Mn-20.45Cr-0.70Al-0.48Fe-2.69Ti-0.11W-0.006S-0.006P), the rate of etching in 15 % nitric acid + 4 % hydrogen fluoride at 333 K (60 °C) after annealing at 1223 or 1373 K (950 or 1100 °C) is about 150 to 160 or 260 g/m^2 h respectively [612]. However, the vacuum-smelted alloy shows an etching rate of 380 g/m^2 h after annealing at 1373 K (1100 °C).

The nickel-chromium alloy (%) Ni-0.02C-34Cr-1.0Fe-0.6Mn-3.7Si with good forgeability and rollability is resistant to 65 % boiling nitric acid and recommended as the material for boiling Purex-IWW solution containing fluoride ions and Al(NO$_3$)$_3$ for reprocessing fuel rods [911].

A 28   Nickel-chromium-iron alloys

In the modified Huey test (48 h in boiling 25 % nitric acid), Inconel® 690 (2.4642) is more resistant than Inconel® 600 by a factor of 100. In contrast to Inconel® 600 (2.4816), no intercrystalline corrosion occurs on this alloy [614]. Some test values taking into account the penetration depth of intercrystalline corrosion are shown in Table 93 [607, 614].

Alloy	Heat treatment	Material consumption rate g/m^2 h	Depth of attack* mm
600	(a)	0.883	0.18
	(b)	2.03	0.20
	(c)	1.23	0.18
	(b) + (c)	2.89	0.48
690	(a)	0.0083	–
	(b)	< 0.001	–

* intercrystalline corrosion

Table 93: Corrosion of Inconel® 600 and 690 in boiling 25 % HNO$_3$ (a) in the delivery state, (b) after sensitization at low temperature (LTS: 24 h at 722 K (449 °C)) and (c) after sensitization in an annealing furnace (SA: 24 h at 894 K (621 °C)) [607, 614]

The corrosion rate of Inconel® 600 in boiling 25 % nitric acid is increased from 0.12 to 8.8 or 2.1 mm/a (4.72 to 346 or 83 mpy) by subsequent heat treatment, e.g. 4 h at 866 K (593 °C) or 15 h at 977 K (704 °C) respectively. However, the intercrystalline fractures drop from 62.2 to 18.6 % by this procedure [615].

According to Figure 58, the material consumption rate of the alloy 600 ((%) Ni-0.05C-16.32Cr-8.68Fe-0.24Mn-0.26Ti-0.04Nb-0.007S-0.009P-0.015N) in boiling 25 % nitric acid reaches values of 0.35 g/m² h after only 20 h by heat treatment at 973 K (700 °C). After treatment of the specimen at 873 K (600 °C), a period of 100 h was not sufficient to bring the material consumption rate to below 50 g/m² h [616]. As a result of the high intercrystalline corrosion, the partly oxidized specimens could be broken by hand.

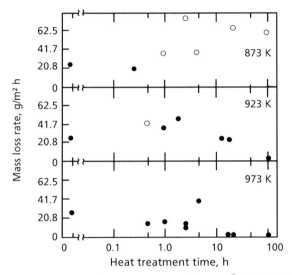

Figure 58: Material consumption rate of Inconel® 600 in boiling 25 % HNO₃ as a function of the heat treatment time at 873, 923 and 973 K (600, 650 and 700 °C). The open circles indicate specimens which could be broken by hand as a result of intercrystalline corrosion [616]

The corrosion behavior of various hard nickel-based alloys in hot 30 % nitric acid at 323 K (50 °C) is shown in Table 94 [127]. According to this table, the corrosion resistance of the alloys increases with increasing chromium content, decreasing molybdenum content and the addition of tungsten.

Alloy	Chemical composition, %*								Material consumption rate g/m² h
	C	Cr	Mo	Fe	W	Nb	Mn	Si	
Akrit® Ni 40	1.44	24.85	13.09	22.9	0.02	0.42	0.11	0.95	0.68
Alloy A	1.31	24.13	8.20	23.1	4.30	0.45	0.24	1.35	0.22
Alloy B	1.37	30.06	6.37	3.02	4.68	–	0.03	1.42	0.06

* balance Ni

Table 94: Composition and corrosion behavior of some Nickel alloys in 30 % HNO₃ at 323 K (50 °C) [127]

There is a direct relationship between the magnetic susceptibility, the Curie point and the corrosion rate of Alloy 600 in 40 % nitric acid via the chromium content of the alloy. The Curie point thus drops as the chromium content increases, from 680 at 2 % to 180 K (-93 °C) at 15 % chromium. The susceptibility and corrosion of the alloy decrease accordingly as the chromium content increases [617].

While the corrosion rate of the nickel-chromium-iron alloy (0.04 – 0.07C and 7 – 10 % Fe) in boiling 65 % nitric acid was still 2.3 mm/a (90.55 mpy) at 24 % chromium, it was reduced to 0.12 mm/a (4.72 mpy) by a chromium content of 30 % [675].

Table 95 shows the dependence of intercrystalline corrosion on Inconel® 600 ((%) Ni-0.030C-15.7Cr-7.63Fe-0.33Mn-0.21Ti-0.12Al-0.02Cu-0.005P-0.004S) in boiling 10 mol/l nitric acid on the heat treatment [674].

Heat treatment		Corrosion rate	$d\sigma_B/dt$
Temperature, K (°C)	Duration, h	mm/a (mpy)	MPa/h
623 (350)	200	4.4 (173)	–
	2000	17.4 (685)	–
873 (600)	0.2	2.6 (102)	5
	1.0	20.1 (791)	23
	10	84 (3307)	97
	500	10.4 (410)	12
973 (700)	0.2	34.8 (1370)	40
	1.0	82 (3228)	95
	10	30 (1181)	35
	500	13 (512)	15

Table 95: Dependence of the intercrystalline corrosion and rate of decrease of the tensile strength on the heat treatment of the alloy Inconel® 600 in boiling 10 mol/l HNO_3 [674]

Of nine heat-treated alloys of Inconel® 600 with different compositions, the following showed the lowest material consumption rates in the Huey test: (%) Ni-0.016C-15.4Cr-7.0Fe-0.20Mn-0.008Mg-0.37Al-0.33Ti-0.02Cu-0.11Si-0.014S-0.012P-0.008N and (%) Ni-0.016C-15.5Cr-7.1Fe-0.19Mn-<0.005Mg-0.38Al-0.33Ti-0.01Cu-0.11Si-0.0002S-0.032P-0.007N. The material consumption rate of both alloys, which were annealed at 1393 K (1120 °C) for 1 h and then quenched in water, in boiling 65 % nitric acid was 0.70 and 0.52 g/m² h, with corresponding depths of intercrystalline corrosion of 0.15 and 0.0 mm [594]. In boiling nitric acid, the corrosion rate of Inconel® 600 increases as the carbon and sulfur content increase.

In Figure 59, the curves (1) (solution-annealed at 1366 K (1093 °C)) and (2) (solution-annealed at 1366 K (1093 °C) and 25 % cold-worked) enclose corrosion regions of the alloy Inconel® 600 (2.4816) in which the corrosion rate in boiling 70 % nitric acid is > 30 mm/a (> 1180 mpy). In the regions outside the curves, the corrosion rate of the alloy is < 3 mm/a (< 118 mpy) [593].

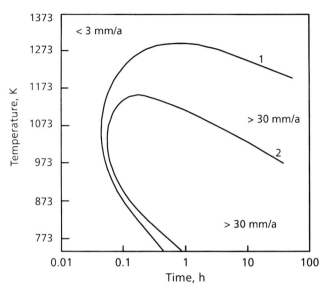

Figure 59: Time/temperature sensitization diagram for Inconel® 600 during corrosion in boiling 70% HNO$_3$ [593]
(1) solution-annealed at 1366 K (1093 °C), (2) as (1) with 70% cold working
< 3 mm/a: < 118 mpy   > 30 mm/a: > 1180 mpy

The corrosion rate of the complex alloy Hastelloy® G (cf. 2.4816; (%) Ni-0.05C-(21-23.5)Cr-(18-21)Fe-(5.5-7.5)Mo-(1-2)Mn-(1.5-2.5)Cu-(1.75-2.5)(Nb+Ta)-<1.0W-1.0Si-0.030S-0.040P) in boiling 70% nitric acid at 397 K (124 °C) is not low enough, at < 0.5 mm/a (< 20 mpy) [597]. Below 360 K (87 °C), the corrosion rate is < 0.1 mm/a (< 4 mpy).

As soon as boiling 65% nitric acid contains 0.1% hydrogen fluoride, the material consumption rate of the alloy 690 solution-annealed at 1373 K (1100 °C) for 0.5 h also reaches values of about 6 to 7 g/m^2 h with 30% chromium [617]. Addition of chromium salts to the nitric acid causes no change in the corrosion.

Inconel® 625 (1.4856) is also mentioned as a suitable material for processing fuel rods from a pressurized water reactor in solutions of 0.5 mol/l nitric acid and hydrofluoric acid (see Section A 29), while the NiCrFe alloys are unsuitable because their corrosion rates are too high [619]. Corrosion preferentially occurs in particular at welding joints.

Inconel® 600 is recommended for sprung, elastic pipe connections for turbine pipes operated with superheated steam. An alloy structure obtained by solution-annealing at 1173 K (900 °C) with subsequent etching in HNO$_3$/HF solutions leads to no intercrystalline corrosion during long-term turbine operation at 603 K (330 °C) [595].

In boiling 2 and 6 mol/l nitric acid + 0.5 mol/l hydrogen fluoride, Inconel® 690 (2.4816) corroded at a corrosion rate of 1.8 and 3.8 mm/a (71 and 150 mpy) respectively [340]. In spite of this, it is one of the materials which is used during processing of the fuel elements from nuclear reactors in nitric acid solutions containing hydrogen fluoride. In addition to Inconel® 690, the corrosion behavior of some other alloys is summarized in Table 96 [613].

Figure 60 shows that the intercrystalline corrosion of solution-annealed Inconel® 600 (2.4816), determined as the fracture stress after exposure to a solution of 300 ml 70% HNO_3 + 14 g $K_2Cr_2O_7$ + 700 ml water decreases with the contents of phosphorus and silicon in the alloy. However, a higher silicon content is far less hazardous than a higher phosphorus content [596].

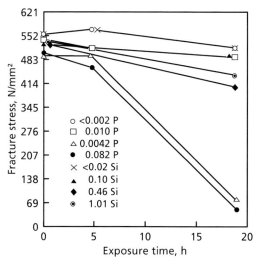

Figure 60: Influence of the exposure time in boiling HNO_3 + Cr^{6+} on the fracture stress of Inconel® 600 with various P and Si contents [596]

An electrolyte of 3% nitric acid + 2% perchloric acid in methanol was used at a polarization potential of + 1.30 V_{SCE} for quantitative extraction of the γ-phase ($Ni_{2.88}Al_{0.37}Ti_{0.51}Cr_{0.12}Fe_{0.12}$) in the NiCrFe alloy ((%) Ni-0.056C-14.68Cr-8.90Fe-2.25Ti-1.12Al-0.02Mn-0.004P-0.005S). The percentage content of the γ-phase was determined as 15.5% after heat treatment at 973 K (700 °C) for 1 hour [618].

Alloy	Forged		Welded
	HF addition, mol/l		
	0.01	0.1	0.1
	Corrosion rate, mm/a (mpy)		
Inconel® 690 (2.4642)	0.23 (9.1)	1.60 (63)	1.73 (68)
Inconel® 671 (52Ni-48Cr)	0.20 (7.9)	1.04 (41)	1.14 (45)
Hastelloy® G (cf. 2.4618, Ni-22Cr-18Fe-6Mo)	0.46 (18.1)	3.51 (138)	5.59 (220)
Inconel® 625 (1.4856, Ni-21Cr-2.5Fe-9Mo)	0.46 (18.1)	3.73 (147)	4.32 (170)

Table 96: Corrosion of some NiCrFe alloys in 10 mol/l HNO_3 with HF additions at 368 K (95 °C), test duration 120 h [613]

A 29 Nickel-chromium-molybdenum alloys

The good corrosion resistance of Hastelloy® C-4 alloy ((%) Ni-0.015C-(14-18)Cr-(14-17)Mo-<3Fe-<1Mn-<0.7Ti<2.8Co-0.08Si-<0.040P-<0.03S; cf. 2.4610) in nitric acid can be seen from the isocorrosion curves in the temperature/concentration graph in Figure 61 [597]. The corrosion rate, e.g. in 10% nitric acid at 373 K (100°C), is thus 0.2 mm/a (7.87 mpy) [620].

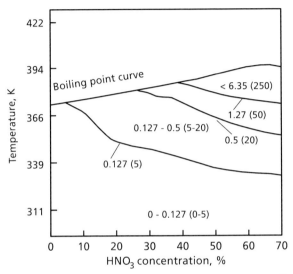

Figure 61: Isocorrosion curves in mm/a (mpy) of Hastelloy® C-4 in nitric acid in the temperature/concentration graph [597]

A comparable course of the isocorrosion curves is also obtained with Hastelloy® C-276 (cf. 2.4819, 2.4886) [597].

While the corrosion rate of Hastelloy® C-276 in boiling 10% nitric acid is 0.43 mm/a (16.93 mpy), the corrosion resistance of the Hastelloy® G-3 (cf. 2.4619; (%) Ni-0.015C-(21-23.5)Cr-(6.0-8.0)Mo-1.5W-(18-21)Fe-(1.5-2.5)Cu-0.50(Nb+Ta)) of 0.02 mm/a (0.79 mpy) is considerably better [624, 631]. For comparison, the corrosion rate of Hastelloy® C-4 (cf. 2.4610) under comparable test conditions (10% HNO_3 at 373 K (100°C)) is 0.20 mm/a (7.87 mpy). These results have also been confirmed elsewhere [635].

Alloys of the type NiMoCr 55 19 17 have only a limited resistance (0.1 to 1 g/m² h) in 5 to 65% nitric acid at 293 K (20°C) and are not resistant at 373 K [231].

The alloy ChN60MB (Russ. type; (%) Ni-0.015C-0.25Mn-0.1Si-23.8Cr-16.9Mo-0.18Nb-0.005S-0.007P) corrodes in boiling nitric acid (10, 20, 30 and 40%) with rates of 0.02, 0.055, 0.20 and about 1 mm/a (39.4 mpy) respectively [726].

The corrosion behavior of some nickel-chromium-molybdenum alloys in mixed acids containing nitric acid can be seen in Table 97 [621]. The approximate composition of the alloys tested in Table 97 is shown in Table 98. Hastelloy® C-22 is accordingly the most resistant to nitric acid electrolytes.

Medium %	Temperature K (C°)	Hastelloy® C-22	Hastelloy® C-4	Hastelloy® C-276	Inconel® 625
		2.4602	2.4610	2.4819	1.4856
		Corrosion rate, mm/a (mpy)			
5 HNO_3 + 15.8 HCl	boiling	0.01 (0.39)	0.28 (11)	0.20 (7.9)	0.03 (1.18)
8.8 HNO_3 + 15.8 HCl	325 (52)	< 0.10 (3.94)	2.90 (114)	0.84 (33)	> 250 (9843)
5 HNO_3 + 25 H_2SO_4 + 4 NaCl	boiling	0.03 (1.18)	2.46 (97)	1.63 (64)	18.1 (713)

Table 97: Corrosion of some NiCrMo alloys in mixed acids [621]

Alloy	Chemical composition, %					
	Cr	Mo	W	Fe	Nb	C
Hastelloy® C-22	22	13	3	3	–	≤ 0.010
Hastelloy® C-276	16	16	4	5	–	≤ 0.010
Hastelloy® C-4	16	16	–	≤ 3	–	≤ 0.010
Inconel® 625	21.5	9	–	≤ 5	3.5	≤ 0.10

Table 98: Chemical composition of the NiCrMo alloys shown in Table 97 [621]

Inconel® 625 ((%) Ni-<0.1C-(20-23)Cr-(8-10)Mo-<1Co-<5Fe-(3.15-4.15)Nb/Ta-<0.04Al-<0.5Si) corrodes in 10 mol/l nitric acid + 0.01 mol/l and 0.1 mol/l hydrogen fluoride at a corrosion rate of 0.46 and 3.7 mm/a (18.11 and 146 mpy) respectively [613]. That of the welded alloy of about 4.3 mm/a (169 mpy) is even higher.

Hastelloy® C-276 can be used for processing the spent fuel elements from pressurized water reactors in solutions of HNO_3 + HF + dissolved substances. However, the welding joints of the alloy are preferentially attacked. Hastelloy® C-4 would perhaps be preferable [619]. With a total concentration of the process solution, containing Zr salts, of 2 mol/l, including 0.5 mol/l HNO_3, a corrosion rate on the alloy C-276 of 3 mm/a (118 mpy) results after exposure at the boiling point for 168 h. At a total concentration of the process solution of 4 mol/l, the corrosion rate approximately doubles.

The corrosion behavior of the high-chromium NiCrMo alloy 47ChNM-2 in nitric acid/hydrogen fluoride solutions can be seen from the 100 h tests in Table 99 [496].

The alloy Ni-15Cr-15Mo used as a sintered compact is resistant to nitric acid containing hydrofluoric acid at room temperature [623].

The high-chromium alloy ChN58W (Russ. type EP-795) corrodes in a solution of 18 % nitric acid + 4 % hydrogen fluoride at 333 K (60 °C) at a material consumption rate of about 0.1 g/m^2 h [605].

Concentration, mol/l		Temperature	Corrosion rate
HNO$_3$	HF	K (°C)	mm/a (mpy)
1.5	0.02	323 (50)	0.021 (0.83)
1.5	1.0	323 (50)	0.046 (1.81)
1.5	0.02	367 (94)	0.012 (0.47)
1.5	1.0	367 (94)	0.372 (14.65)
4.0	0.2	323 (50)	0.011 (0.43)
4.0	1.0	323 (50)	0.083 (3.27)
4.0	0.2	367 (94)	0.285 (11.22)
6.0	0.2	343 (70)	0.103 (4.06)
6.0	1.0	343 (70)	0.286 (11.26)
10.0	0.6	355 (82)	0.656 (25.83)
14.0	1.0	323 (50)	0.457 (17.99)
14.0	1.0	353 (80)	2.00 (78.74)
14.0	1.0	367 (94)	4.35 (171.26)

Table 99: Corrosion rates of the NiCrMo alloy 47ChNM-2 in HNO$_3$ with additions of HF [496]

A nickel-chromium-tungsten-molybdenum alloy EP-567 (Russ. type; (%) Ni-0.03C-(14.5-16.5)Cr-(15-17)Mo-(3.0-4.5)W-1.5Fe-1.0Mn-<0.12Si-0.02S-0.025P) shows neither intercrystalline nor pitting corrosion after 96 h in a solution of 30% H$_2$SO$_4$ + 10% HNO$_3$ + 60% water at 353 K (80°C) if the silicon content is < 0.12%. The corrosion rate is < 0.1 mm/a (< 4 mpy) [626].

The nickel-chromium-molybdenum-niobium alloy JS-625 shows not only high strength but also a good corrosion resistance, also in contaminated nitric acid [649].

A 30 Nickel-copper alloys

Nickel-copper alloys are not recommended as materials in the presence of oxidizing acids, such as, for example, nitric acid, because their corrosion rates are too high [627, 628, 661].

For rapid and reliable differentiation between Monel® 400 (2.4360; (%) Ni-0.13C-33.7Cu-1.20Fe-0.98Mn-0.20Si-0.01S) and Monel® K 500 (2.4375; (%) Ni-0.16C-29.86Cu-2.81Al-0.52Ti-1.19Fe-0.62Mn-0.20Si-0.005S), the U.S. Navy utilizes rapid dissolution in 50% nitric acid at room temperature. The titanium salt which occurs in the solution is detected by a red-violet coloration with chromotropic acid [629].

A 31 Nickel-molybdenum alloys

The nickel-molybdenum alloys Ni28Mo (for example, Hastelloy® B 2, 2.4617; (%) Ni-<0.02C-(26-30)Mo-<2Fe-<1Mn-<1Co) and Ni30Mo (e.g. Hastelloy® B, 2.4810; (%)

Ni-0.05C-(26-30)Mo-(4-7)Fe-<1Cr-1.0Mn, with up to 2.5% Co and 0.6% V, Cu as admissible admixed components) are not resistant in nitric acid [628].

Hastelloy® B 2 corrodes in boiling 10% nitric acid with a corrosion rate of 483 mm/a (19,016 mpy) [625]. Hastelloy® B also suffers catastrophic corrosion in nitric acid even at room temperature [231, 597, 623, 630].

As can be seen from the comparison of the relative values of the material consumption of the nickel-molybdenum alloys with phosphorus and boron additions in Table 100, only corrosion of the alloy Ni-17Mo-0.3B in 7.5 mol/l nitric acid at 298 K (25 °C) is significantly lower than that of pure nickel-phosphorus Ni-11.0P [662].

Alloy	HNO_3 concentration, mol/l	
	7.5	15
	Relative dissolution rate*	
Ni-11.0P	1	1
Ni-11.0Mo-0.2P	2	2
Ni-7.3Mo-3.3P	20	55
Ni-17Mo-0.3B	0.07	0.9
Ni-1.5Mo-6.0B	2.25	90

* determined in mg/min

Table 100: Relative dissolution rate of some NiMo alloys with respect to the corrosion rate of Ni-11.0P in nitric acid at 298 K (25 °C), test duration 5 min [662]

A 32 Other nickel alloys

The nickel-cobalt-chromium-molybdenum alloy MP35N (2.4999; 35Ni-35Co-20Cr-10Mo) corrodes in boiling 65% nitric acid with a corrosion rate of 0.93 mm/a (36.61 mpy) [636].

The nickel alloy MP35N corrodes in 10 and 65% nitric acid at 323 K (50 °C) at a corrosion rate of 0.0025 and 0.0125 mm/a (0.10 and 0.49 mpy). Hastelloy® F (Ni-22Cr-19Fe-6Mo-3Co-2Mn-1Si) is comparably good [718].

The nickel alloy 617 (UNS N06617, 2.4663; (%) Ni-0.07C-22Cr-12.5Co-9Mo-1Al), which is an excellent material for gas turbines because of its high strength and heat stability, is also resistant to 20 and 70% nitric acid at room temperature with a corrosion rate of 0.025 and 0.5 mm/a (0.98 and 19.69 mpy) respectively [1017].

The nickel-chromium-iron-molybdenum-cobalt alloy ((%) Ni-(0.05-0.15)C-22Cr-18Fe-9Mo-2.5Co-1.0W-1.0Mn-1.0Si) corrodes in boiling 5 and 25% nitric acid at rates of 0.14 and 0.27 mm/a (5.51 and 10.63 mpy) respectively [656].

The alloy IN-504 with the preferred composition (%) Ni-9.5Si-2.85Ti-3.0Mo-2.5Cu (structure: β-phase with Ni_3Si precipitate) corrodes in 65% nitric acid at 433 K (160 °C) at a rate of 1.0 mm/a (39.37 mpy) [1073].

According to Table 101, corrosion of the nickel-phosphorus alloys in 1 mol/l nitric acid at 293 K (20 °C) decreases as the phosphorus content increases and reaches a value of < 0.1 g/m² h at 16% phosphorus [663]. Only uniform corrosion occurs, without fractures caused by intercrystalline corrosion.

Phosphorus content %	Material consumption rate g/m² h
–	15.8
4	9.54
6.6	4.43
10.0	3.60
15.0	0.55
16.0	< 0.1

Table 101: Average material consumption rates of NiP alloys in 1 mol/l HNO₃ at 293 K (20 °C), test duration 120 d [663]

The hardenable spring alloy Beryvac® 520, a nickel-beryllium alloy NiBe2, is not resistant in 10 and 65% nitric acid even at room temperature. Its corrosion rate in 10% nitric acid is > 10 mm/a (> 390 mpy) [632].

Nickel-phosphorus coatings containing 14% phosphorus are severely attacked or destroyed in 78% nitric acid even at room temperature (8 d tests) [633].

According to Figure 62, the relatively high chromium content in the alloys Ni-25Cr-20W-2C-xSi and Ni-30Cr-20W-2C-xSi has a decisive influence on the extent of corrosion in 20% nitric acid containing 4% hydrogen fluoride with an increasing silicon content at room temperature [611].

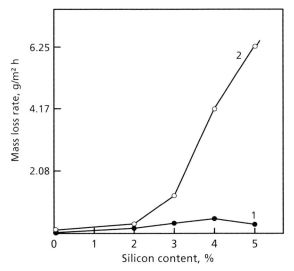

Figure 62: Corrosion of the nickel alloys (1) Ni-30Cr-20W-2C-xSi and (2) Ni-25Cr-20W-2C-xSi in 20% HNO₃ + 4% HF at room temperature as a function of the silicon content, test duration 48 h [611]

In chemical surface treatment of the nickel-iron alloy Permalloy® 52N (Ni-38.4Fe and Ni-49.5Fe) it is pickled in a solution of H_2SO_4:HCl:H_2O of 2:1:2 percent by volume at 343 K (70 °C) for 2 min, a layer of 2.5 µm being removed. This is followed by polishing in a solution of HNO_3:CH_3COOH:HCl of 30:70:1 percent by volume for 6 – 9 s at the same temperature, a layer of 3.39 µm being removed [534].

The intermetallic γ-phase (Ni_3Al) precipitated in a nickel-chromium-aluminium alloy can be rendered visible by pickling in an electrolyte of 43 % nitric acid + 12 % phosphoric acid + 45 % sulfuric acid at room temperature at a potential of 5.5 V in about 20 s using two platinum needle electrodes [637].

Because of the dissolution of the γ-phase and the resistance of the γ-phase (carbides) of the nickel superalloy ((%) Ni-0.19C-8.5Cr-4.4Al-3.7Ti-4.8Mo-5.2W-8.4Co-2.0Nb) in a solution of 5 g $(NH_4)_2SO_4$ + 15 ml conc. HNO_3 + 35 g citric acid + 1000 ml water at 288 – 293 K (15 – 20 °C) and anodic polarization with 1.02 – 1.06 V_{SHE} and 30 mA/cm^2, this solution is used for deep electrochemical etching of the alloy [672].

The following solution is suitable for etching the nickel alloy (%) Ni-0.18C-10Cr-15Co-5.7Al-5Ti-3Mo-0.9V-0.05Cu-0.05Zr-0.015B-0.015P-0.015S: (percent by volume) 40 % HNO_3 (65 %) + 50 % H_2SO_4 (97 %) + 10 % H_3PO_4 (85 %) at a direct voltage of 4 – 5 V overlapped by an alternating voltage [673].

In addition to the customary etching agent 5 ml HNO_3 + 2 ml HF + 93 ml H_2O and 6 ml HNO_3 + 14 ml HF + 60 ml H_2O for contrasting nickel alloys of high heat resistance, such as, for example, Inconel® 100 ((%) Ni-14.75Co-9.68Cr-5.55Al-3.01Mo-4.73Ti-0.77V-0.17Fe-0.060Zr-0.05Mn-0.05Cu-0.04Si) and Inconel® 939 ((%) Ni-19.11Co-22.67Cr-3.70Ti-2.04W-1.37Al-1.05Nb-0.45Mo-0.070Fe-0.10Zr-0.05Cu-0.03Mn-0.05Si), by visualizing the dendritic structure, the following are also suitable: 50 ml HNO_3 + 50 ml HCl + 50 ml glycerol and 41 ml HNO_3 + 47 ml H_2SO_4 + 12 ml H_3PO_4 (the latter at a polarization potential of 3 V) and 3 ml HNO_3 + 10 ml HCl + 100 ml ethanol at 323 – 333 K (50 – 60 °C) for a period of 15 s [734, 738].

A 33 Lead and lead alloys

Lead is severely attacked by nitric acid. Pitting corrosion occurs even in the presence of nitrate ions [639]. In 30 percent by volume nitric acid, lead is dissolved (at room temperature) at a corrosion rate of 217 mm/a (8,543 mpy) [676].

The further increase in the dissolution of lead in 2 mol/l nitric acid by alloying with sodium and potassium can be seen in Table 102 [640].

According to anodic polarization measurements with lead anodes in 1 mol/l nitric acid at room temperature between 0 and 2 V_{SHE}, an oxide layer consisting of PbO and β-PbO_2 forms [647].

The corrosion of lead in 1 mol/l nitric acid can be reduced by organic sulfonate compounds, but not to the extent that industrial use becomes possible [646].

As a result of the considerable solubility of lead in nitric acid, this solution is used with additions of ammonium molybdate for structure analysis and for improving the weldability of lead. A mixture of solution A: 45 g $(NH_4)_2MoO_4$ in 300 ml water

+ solution B: 80 ml conc. nitric acid in 220 ml water is recommended for this purpose, with an etching time of 10s [642]. Similar etching solutions (e.g. 16 g $(NH_4)_2MoO_4$ + 35 ml conc. HNO_3 + 200 ml water) with an etching time of 1 s to 1 min can also be used [643].

Addition		Na	K
Na %*	K %*	Material consumption rate $g/m^2\ h$	
–	–	160	160
0.9	5.0	175	193
5.0	15.0	165	243
10.0	–	200	–
20.0	–	223	–

*atomic percent

Table 102: Material consumption rate of lead with and without Na or K additions in 2 mol/l HNO_3 at room temperature [640]

The aim of the aqua regia test is to evaluate the susceptibility of lead in dilute aqua regia. The acid mixture used is: 100 ml conc. HCl + 285 ml conc. HNO_3 + 205 ml water nitric acid with hydrochloric acid. The acid mixture is preheated to 343 K (70 °C) and heated up to 382 K (109 °C) with a temperature increase of 3 K (3 °C) 1/min during the etching operation on the lead. The time between the first deposition of lead chloride ($PbCl_2$) and vigorous reaction is determined. The results of the aqua regia test are summarized in Table 103 [644].

Evaluation	Start of $PbCl_2$ deposition	Start of vigorous reaction	Appearance of specimen
Very good	≥ 16 min	none	no attack
Good	not < 14 min	none	slight attack
Limited suitability	not < 12 min	not < 18 min	moderate attack + weak pitting corrosion
Not usable	< 12 min	< 18 min	severe attack + pronounced pitting corrosion

Table 103: Aqua regia test for evaluating the corrosion behavior of lead [644]

According to Figure 63, the corrosion of lead-tin alloys in dilute nitric acid (0.02 to 0.07 mol/l) at room temperature decreases as the tin content increases [645]. The inhibition efficiency of thiourea regarding corrosion of the alloys in 0.1 mol/l nitric acid increases with an increase in the tin content from 60 % at 10 % tin to 85 % at 85 % tin.

The lead-indium alloy containing 50 % indium corrodes in 0.1 mol/l nitric acid at room temperature at a material consumption rate of 4.74 $g/m^2\ h$ (4.7 mm/a) [641]. The corrosion decreases as the indium content decreases, and reaches a value of 1.96 mm/a (77.17 mpy) at 10 % indium.

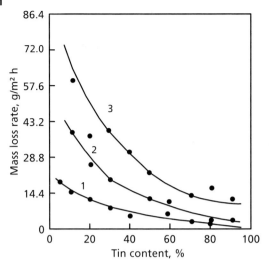

Figure 63: Influence of tin on the corrosion of PbSn alloys at room temperature in nitric acid as a function of the concentration [644]
1) 0.02; 2) 0.05; 3) 0.07 mol/l

Dipping in nitric acid (d = 1.33 g/ml) at room temperature is suitable for removing a layer of lead on aluminium [115].

A 34 Platinum and platinum alloys

Platinum is absolutely resistant to nitric acid. In the first 48 h, a platinum foil of $1 \times 1 \times 0.0127$ cm lost 8×10^{-7} g in conc. nitric acid. This corresponds to a material consumption rate of 0.000167 g/m² h. On continued exposure to nitric acid, no further weight loss was found [652].

Platinum is also resistant in 62% nitric acid at 373 K (100°C), but is severely attacked by aqua regia even at room temperature [727].

Thin platinum layers on steel or nickel can be removed by dipping in a solution of (percent by volume) 40% nitric acid (d = 1.33 g/ml) + 25% hydrochloric acid (d = 1.19 g/ml) + 25% water at room temperature [115].

Stainless steel and Nichrome® cermet nuclear fuels containing highly enriched uranium are dissolved in niobium or plastic baskets in a V-shaped solution tank made of titanium as the cathode and a corrosion-resistant platinum anode in 6.5 mol/l nitric acid or 3 mol/l nitric acid or 3 mol/l nitric acid with a 90 g/l metal content, which is circulated while cooling, at 322 K (49°C), 15 to 25 V and an anode current density of 1 to 2 A/cm². The material consumption rate which occurs under these conditions is about 1 g/m² h [653].

Alloys of platinum with 10 – 20 atomic percent ruthenium, rhodium, osmium or iridium are virtually no longer attacked by aqua regia. The resistance is attributed to the resistance of these metals to aqua regia [727].

A 35 Platinum metals (Ir, Os, Pd, Ru) and their alloys

Palladium is resistant in 5 percent by volume nitric acid at room temperature (at 288 K (15 °C) < 0.001 and at 298 K (25 °C) 0.0156 g/m^2 h). A content of just 1000 ppm copper ions in the acid increases the corrosion rate of palladium by 1.5 powers of ten [655].

The following palladium and ruthenium alloys show no noticeable attack after an immersion time of 1 d at 293 K (20 °C) in 50 % and concentrated nitric acid: $Pd_{75}Sn_{25}$, $Pd_{66.7}Sn_{33.3}$, $Pd_{50}Sn_{50}$, $Ru_{50}Al_{50}$, $Ru_{36}Cr_{64}$, $Ru_{33.3}Ta_{66.7}$, $Ru_{50}Ti_{50}$ and $Ru_{50}Zr_{50}$. Only the ruthenium alloys listed are resistant in aqua regia, while the palladium alloys mentioned disintegrate or are dissolved [654].

The alloy LiPdH produced from lithium hydride and palladium in vacuo at 1040 K (767 °C) reacts vigorously in concentrated nitric acid, hydrogen being released [657].

The 5 to 10 μm thick contact layers of palladium comparable to hard gold (Au + 0.5 % Co) may crack due to uncontrolled uptake of hydrogen. To detect this cracking, the layer is subjected to a nitric acid vapor test at room temperature in a desiccator, only the carrier metal, and not the palladium layer, being attacked [677].

Iridium, osmium and ruthenium are absolutely resistant to hot, concentrated nitric acid at 373 K (100 °C) [727].

A 36 Tin and tin alloys

Tin is dissolved in 5 mol/l nitric acid at 298 K (25 °C), and a tin-nickel alloy (Sn-(28-30)Ni) is also severely attacked [805].

The dissolution of tin in nitric acid, which has been investigated on several occasions, has unfortunately been investigated only using the increase in temperature with respect to time during corrosion caused by the heat of reaction, from which no quantitative calculation, but only an estimation of the corrosion, is possible [658 – 660, 664].

To estimate the dependence of the corrosion of tin on the concentration of nitric acid at room temperature, the increase in temperature with respect to time $T/\Delta t = (T_R-T_o)/\Delta t$ (T_R = temperature during the course of the reaction and T_o = starting temperature) in K/min is shown in Figure 64 [658]. According to this figure, a corrosion maximum results at 6 mol/l nitric acid.

Corrosion of tin in nitric acid is inhibited by alkylamines. Their effectiveness increases in the following order: $CH_3NH_2 < C_5H_{15}NH_2 < C_{12}H_{25}NH_2 < C_{18}H_{37}NH_2$ [664]. Measurement values for calculation of the material consumption rate are also lacking here.

The corrosion rate of tin-lead alloys in nitric acid can be seen from Figure 63. According to this figure, the material consumption rate increases as the lead content increases [645].

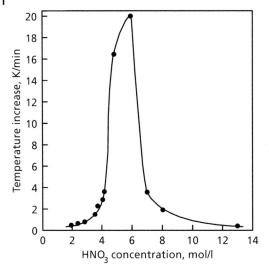

Figure 64: Dependence of the increase in temperature as a measure of the corrosion of tin in HNO_3 on the concentration of HNO_3 [658]

The tin-nickel alloy Sn-28Ni corrodes in 1 mol/l, 5 mol/l and concentrated nitric acid at 298 K (25 °C) at material consumption rates of 0.625, 7.25 and 6.0 g/m² h respectively. Both tin and nickel are dissolved in 5 mol/l HNO_3 [952].

Nitric acid with a density of 1.33 g/m can be used for detinning aluminium [115].

A 37 Tantalum, niobium and their alloys

Tantalum and niobium are resistant both to 5 % nitric acid at room temperature and to boiling, concentrated nitric acid [665]. Tantalum and niobium are also said to be still completely resistant to concentrated nitric acid at 453 K (180 °C). The corrosion resistance of niobium and in particular tantalum is based on the formation of a stable, dense, firmly adhering and self-regenerating oxide layer, which is insoluble in nitric acid [667, 668].

It is the outstanding corrosion resistance and the good thermal conductivity which makes these relatively expensive materials so attractive.

Tantalum is also used for heat exchangers and condensers in plants operated with nitric acid [751].

It is resistant to nitric acid up to the boiling point; but is destroyed, however, in mixtures of nitric acid and hydrogen fluoride [1004].

As determined by the radiotracer method, Ta and the alloy Ta-40Nb are attacked between 293 and 394 K (20 and 121 °C) with corrosion rates of between 10^{-8} and 4×10^{-6} or 2×10^{-7} and 8×10^{-6} mm/a respectively. The corrosion rates of Ta and Ta-40Nb were increased exponentially by fluoride additions above 10 ppm (NaF). The maximum corrosion rates were between 1.4×10^{-2} mm/a at 383 K (110 °C) and

320 ppm F⁻ and 8×10^{-3} mm/a at 323 K (50 °C) and 280 ppm F⁻. Ta proved to be ten times more resistant than the alloy.

Figure 65 shows the influence of F⁻ additions to azeotropic nitric acid on the corrosion rates of Ta and Ta-40Nb. The starting points of the curves correspond in practice to the corrosion rates of the materials which occur without F⁻ additions [21].

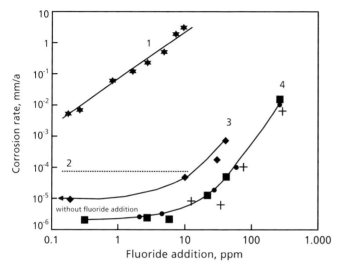

Figure 65: Influence of fluoride additions in the form of NaF on the corrosion rates of Zr, Ta and Ta-40Nb in azeotropic nitric acid [21]
1) Zr, 381 K (108 °C)
2) Zr without F⁻ addition
3) Ta-40Nb, 383 K (110 °C)
4) ■ Ta (500 A), 383 K (110 °C)
● Ta (1000 A), 383 K (110 °C)
+ Ta, 323 K (50 °C)

The corrosion resistance of the alloy KBI-40® Ta-40Nb in boiling nitric acid can be seen from Table 104. The corrosion rates of tantalum and niobium are also shown for comparison [753].

HNO₃ concentration %	Ta	Nb	Ta-40Nb
	Corrosion rate, mm/a (mpy)		
10	< 0.003 (0.12)	< 0.003 (0.12)	< 0.003 (0.12)
70	< 0.003 (0.12)	0.01 (0.39)	< 0.003 (0.12)

Table 104: Corrosion rates of tantalum, niobium and Ta-40Nb in boiling nitric acid [753]

The excellent resistance of tantalum in boiling nitric acid throughout the entire concentration range and in particular to fuming nitric acid up to pressures of 552 MPa and up to 590 K (317 °C) is also confirmed elsewhere. The tantalum-tungsten alloy Fansteel® 61 (Ta-7.5W) corrodes in 70% nitric acid at 472 K (199 °C) at 0.0025 mm/a (0.098 mpy) [750].

According to Table 105, nitration of niobium for 30 min in 0.005% oxygen-containing nitrogen at 2073 K (1800 °C) and of tantalum at 2123 K (1850 °C) has no noticeable influence on the corrosion rate of the metals in nitric acid and in mixtures of nitric and hydrochloric acid [671].

Solution	Temperature K (°C)	Nb	Nb, nitrated	Ta	Ta, nitrated
		Material consumption rate, g/m² h			
67% HNO_3	393 (120)	0.4	0.4	0.3	0.4
HNO_3/HCl (1:3)	353 (80)	0.4 – 0.5	0.3	0.35	0.35

Table 105: Corrosion rate of niobium and tantalum in HNO_3 containing solutions without and after nitration [671]

A solution of 33% HNO_3 + 33% HF + 34% H_2SO_4 is used for chemical etching of the niobium alloys Nb-0.24Zn-0.34N and Nb-3.06Zr-0.45N [814]. To remove the scaling from the alloy Nb-1Zr-0.1C, it is dipped in a solution of nitric and hydrofluoric acid [817].

Tantalloy® 63 (Ta-2.5W-0.15Nb) shows an excellent resistance to boiling nitric acid, coupled with a better strength than tantalum. The alloy is recommended as a material for evaporators, heat exchangers, condenser pipes and valves in the presence of nitric acid [907].

The alloy Tribocor®-532 N (Nb-30Ti-20W) has, inter alia, an excellent resistance to nitric acid [754, 1005].

Small amounts of nitric acid and nitrates, such as, for example, $Ni(NO_3)_2$, $Cu(NO_3)_2$ and $Ba(NO_3)_2$, present in sulfuric acid reduce the corrosion rate of tantalum and also the hydrogen absorption at 523 K (250 °C) after 100 h (Table 106) [666]. Inhibition of the corrosion and hydrogen uptake is evidently caused by the nitrate ions, regardless of whether they are in acid or salt form.

Addition %	Corrosion rate mm/a (mpy)	H absorption ppm
–	1.91 (75)	34.7
1 HNO_3	0.41 (16)	0.4
3 HNO_3	0.37 (14.6)	0.3
5 HNO_3	0.37 (14.6)	0.4
1 $Cu(NO_3)_2$	0.34 (13.4)	0.4
1 $Ba(NO_3)_2$	0.31 (12.2)	0.3
1 $Ni(NO_3)_2$	0.30 (11.8)	0.4

Table 106: Corrosion rate and hydrogen absorption and tantalum at 523 K (250 °C) in 95% H_2SO_4 with and without NO^{3-} additions [666]

In agreement with Table 106, the corrosion rate of pure tantalum between 423 and 543 K (150 and 270 °C), which is about 0.06, 0.7 and 5 mm/a (2.36, 27.56 and 197 mpy) in 98 % sulfuric acid at 473, 513 and 543 K (200, 240 and 270 °C) respectively, is reduced by a factor of 3 by a 10 % addition of 65 % nitric acid [752].

To produce highly concentrated nitric acid with sulfuric acid as the extraction agent, the heater in the forced circulation evaporator, where wall temperatures of 483 K (210 °C) occur, should be made of tantalum [669].

Tantalum is used as the tank material for wet combustion of organic α-emitting waste in recovery plants. In hot industrial 96 % sulfuric acid with 9 parts by weight 65 % nitric acid and 1 part by weight 96 % hydrochloric acid at 420 to 540 K (147 to 267 °C), the corrosion at < 520 K (247 °C) was sufficiently low if a decrease in wall thickness of 0.5 mm/a (19.69 mpy) can be tolerated [908]. Provisions should be made for the continuous presence of HNO_3.

To be able to use tantalum as a material for decomposition of plutonium-contaminated organic waste with 80 to 98 % sulfuric acid between 478 and 573 K (205 and 300 °C), the corrosion and hydrogen uptake must be reduced. The inhibiting effect needed for this could be achieved by additions of nitric acid, nitrate and nitrite salts [843].

A 38 Titanium and titanium alloys

Titanium has an excellent resistance to nitric acid in the entire concentration range up to 450 K (177 °C). The corrosion rate in 20 % nitric acid at 560 K (287 °C) is only 0.3 mm/a (11.81 mpy). Only in anhydrous, fuming nitric acid is caution required, since a pyrophoric reaction course occurs here and may proceed explosively [682, 688, 694]. Titanium is also resistant to aqua regia.

Titanium VT1-0 (similar to ASTM Grade 2), which corrodes at a rate of 0.006 g/m^2 h in hot 3.5 – 4 % nitric acid at 368 K (95 °C), becomes more corrosion-resistant in contact with lead (0.002 g/m^2 h) [381].

According to Table 107, the titanium alloys AT-3 (2.0-3.5Al; 0.2-0.5Cr; 0.2-0.5 Fe; 0.2-0.4 Si) and PT-3V (TiAl4V2) are no better than the commercially available pure titanium VT1-0 in their corrosion behavior in nitric acid as non-welded material and weld joints [725].

Amorphous phosphorus-containing titanium has a higher corrosion resistance than crystalline titanium in 1 mol/l nitric acid at room temperature. The same also applies to amorphous titanium-nickel-phosphorus alloys, such as, for example, Ti-27Ni-13P or Ti-27Ni-10P, in 1 mol/l nitric acid with a corrosion rate of < 0.01 mm/a (< 0.39 mpy) compared with the crystalline alloys of the same composition with rates of about 0.03 mm/a (1.18 mpy) [816]. At a titanium content of only 40 atomic percent, the corrosion rate of the amorphous alloy is 0.01 mm/a (0.39 mpy) and that of the crystalline alloy about 0.1 mm/a (3.94 mpy).

On the basis of the good corrosion resistance of titanium and titanium-tantalum alloys in nitric acid between 293 and 353 K (20 and 80 °C) under an anodic loading of between 0.20 and 0.40 A/cm^2 (see Table 108), titanium and the alloys mentioned are suitable as electrodes for anodic dissolution of fuel element sheaths made of stainless steel in nitric acid. Titanium and its alloys experience hardly any attack under these conditions [680].

According to Table 109, the corrosion of cathodically polarized titanium in dilute nitric acid at 353 K (80 °C) is increased by alloying with tantalum [684].

HNO_3 concentration %	Temperature K (°C)	VT1-0		AT3		PT-3V	
		(a)	(b)	(a)	(b)	(a)	(b)
		Corrosion rate, mm/a (mpy)					
5	293 (20)	0.0006 (0.02)	0.0007 (0.03)	0.0008 (0.03)	0.0010 (0.04)	0.0015 (0.053)	–
20		0.0010 (0.04)	0.0008 (0.03)	0.0013 (0.051)	0.0017 (0.067)	0.0025 (0.098)	0.0022 (0.087)
65		0.0008 (0.03)	0.0007 (0.03)	0.0023 (0.091)	0.0021 (0.083)	0.0031 (0.122)	0.0030 (0.118)
98		0.0006 (0.02)	0.0008 (0.03)	0.0005 (0.020)	0.0012 (0.047)	0.0008 (0.03)	0.0008 (0.03)
5	343 (70)	0.0105 (0.41)	0.0084 (0.331)	0.0125 (0.492)	0.0094 (0.370)	0.0102 (0.402)	0.0115 (0.453)
20		0.0232 (0.91)	0.0236 (0.929)	0.0333 (1.311)	0.0380 (1.496)	0.0420 (1.654)	0.0400 (1.575)
65		0.0158 (0.62)	0.0178 (0.701)	0.0596 (2.346)	0.0602 (2.370)	0.0910 (3.583)	0.0925 (3.642)
98		0.0019 (0.07)	0.0025 (0.098)	0.0010 (0.04)	0.0011 (0.043)	0.0020 (0.0079)	0.0018 (0.071)
5	boiling	0.0133 (0.52)	0.0140 (0.551)	0.0141 (0.555)	0.0282 (1.110)	0.0257 (1.012)	0.0300 (1.181)
20		0.0560 (2.2)	0.0475 (1.870)	0.290 (11.417)	0.190 (7.480)	0.194 (7.638)	0.180 (7.087)
65		0.0327 (1.29)	0.0375 (1.476)	0.0372 (1.465)	0.0425 (1.673)	0.0412 (1.622)	0.0700 (2.756)
80		0.0153 (0.60)	0.0185 (0.728)	0.0300 (1.181)	0.0355 (1.398)	0.0382 (1.504)	0.0472 (1.858)

Table 107: Corrosion of titanium VT1-0 and its alloys AT-3 and PT-3V in nitric acid as (a) non-welded material and (b) weld joints [725]

HNO$_3$ mol/l	Anodic current density A/cm^2	Temperature K (°C)	VT1-0	Ti-8Ta	Ti-10Ta	Ti-15Ta
			\multicolumn{4}{c}{Material consumption rate, g/m^2 h}			
2	0.20	293 (20)	0.012	0.180	< 0.01	0.260
	0.40		0.763	< 0.01	0.044	< 0.01
4	0.20		0.038	< 0.01	< 0.01	< 0.01
	0.40		< 0.01	0.033	0.044	0.033
8	0.20		< 0.01	< 0.01	< 0.01	0.20
	0.40		0.996	0.066	< 0.01	< 0.01
2	0.40	353 (80)	0.306	0.126	0.013	< 0.01
4	0.20		< 0.01	–	–	–
	0.40		0.038	0.110	0.100	0.070
8	0.40		0.114	< 0.01	< 0.01	0.190

Table 108: Corrosion rate of titanium VT1-0 (similar to ASTM Grade 2) and titanium-tantalum alloys in HNO$_3$ in contact with the steel EI-847, test duration 19 h [680]

HNO$_3$ mol/l	Current density A/cm^2	Potential V_{SHE}	Ti-8Ta	Ti-10Ta	Ti-15Ta
			\multicolumn{3}{c}{Material consumption rate, g/m^2 h}		
2	0.20	from – 0.132 to – 0.264	< 0.01	< 0.01	0.03
	0.60	from – 0.342 to – 0.454	< 0.01	–	0.16
4	0.20	from + 0.035 to – 0.084	0.03	< 0.01	0.07
	0.60	from – 0.144 to – 0.335	0.13	–	0.26
6	0.20	from + 0.307 to + 0.227	0.03	< 0.01	0.03
	0.60	from + 0.157 to + 0.087	< 0.01	0.16	0.06
8	0.20	from + 0.210 to + 0.150	< 0.01	< 0.01	0.07
	0.60	from + 0.084 to + 0.036	< 0.01	0.06	0.10

Table 109: Corrosion rate of titanium-tantalum alloys under cathodic polarization in dilute HNO$_3$ at 353 K (80 °C) [684]

At 473 K (200 °C), titanium VT1-0 (cf. Ti grade 2; UNS R50400) and Ti2Pd (cf. Ti grade 7, UNS R52400) corrode in 35 and 56 % nitric acid with rates of 0.06 and 0.054 mm/a (2.32 and 2.13 mpy), and 0.17 and 0.20 mm/a (6.69 and 7.87 mpy) respectively [256, 690]. Titanium with about 0.15 % palladium is said to display a better corrosion resistance than pure titanium [704].

According to Table 110, the corrosion rates of titanium and the titanium-palladium alloy Ti2Pd in boiling nitric acid are the same. Alloying with 30 % molybdenum leads to a marked increase in corrosion. All the tests were carried out in 5 × 48 h cycles. The good corrosion resistance of titanium was maintained by small additions of nickel [689].

HNO$_3$ %	Ti	Ti2Pd	Ti-1Ni	Ti-1.5Ni	Ti-30Mo
	Corrosion rate, mm/a (mpy)				
35	0.15 (5.9)	0.15 (5.9)	0.13 (5.1)	0.15 (5.9)	8.75 (344)
50	0.2 (7.9)	0.2 (7.9)	0.2 (7.9)	0.23 (9.1)	5.30 (209)
68	0.05 (2.0)	0.05 (2.0)	0.05 (2.0)	0.075 (2.9)	1.33 (52.3)

Table 110: Corrosion rate of titanium and titanium alloys in boiling nitric acid after 240 h tests in 48 h cycles [689]

The high-strength alloys VT-14 (Ti-(2.5-3.5)Mo-(0.1-1.5)V-(3.5-4.5)Al) and VT-15(Ti-(3-4)Al-(10-11.5)Cr(7-8)Mo) have a comparable resistance in 40% nitric acid at 353 K (80°C) in the presence of air in the passive state under industrial conditions, regardless of structure [743].

The corrosion rate of the titanium alloy VT-14 ((%) Ti-5.61Al-3.24Mo-1.78V-0.22Fe-0.09C-0.012H-0.03N-0.10O) in 40% nitric acid at 353 K (80°C) was reduced from 0.61 to 0.46 mm/a (18.11 mpy) by addition of 0.17% palladium [708]. At the same time, the strength rose from 1050 to 1090 N/mm^2.

The titanium evaporators which vaporize radioactive waste obtained in nitric acid by means of tributyl phosphate extraction by the Purex process are superior to the evaporators made of AISI 304 L, Sandvik® 2R12 and Nitronic® 50 in terms of corrosion resistance [746]. On the basis of corrosion tests with boiling 40% nitric acid (50 to 3500 h tests), the alloys VT5-1 (cf. Ti grade 6, UNS R54520) and AT-3 (2.0-3.5Al; 0.2-0.5Cr; 0.2-0.5 Fe; 0.2-0.4 Si) are also recommended for nitric acid evaporators [747].

The temperature dependence of the corrosion rate of industrially pure titanium VT1-1 and the alloy AT-3 in 40 and 80% nitric acid can be seen from Table 111 [781].

Corrosion proceeds uniformly. The breaking strength of the alloy AT-3 remains unchanged at 4.7 N/mm^2 after exposure for 24 h to 80% nitric acid at 723 K (450°C). A corresponding behavior is shown by VT1-1, with 1.45 N/mm^2.

Material	Temperature K (°C)	40% HNO$_3$		80% HNO$_3$	
		Test duration h	Material consumption rate g/m^2 h	Test duration h	Material consumption rate g/m^2 h
VT1-1	623 (350)	720	0.01	–	–
	673 (400)	1000	0.03	–	–
	723 (450)	1000	0.01	500	0.003
AT-3	623 (350)	720	0.02	–	–
	673 (400)	1000	0.03	–	–
	723 (450)	1000	0.05	500	0.0065

Table 111: Temperature dependence of the corrosion of VT1-1 (pure titanium) and AT-3 (2.0-3.5Al; 0.2-0.5Cr; 0.2-0.5 Fe; 0.2-0.4 Si) in nitric acid [781]

Regardless of the area ratio of titanium/steel (n = 10:1 to 1:10), corrosion in up to 50% nitric acid solutions at 343 K (70 °C) remains unchanged when titanium comes into contact with chromium-nickel steels of the type 03Ch18N11 (cf. AISI 304 L, 1.4306) and 12Ch18N10T (cf. AISI 321, 1.4541), e.g. in installations made of a mixture of titanium and chromium-nickel steels. In no instance does the material consumption rate exceed 0.02 g/m² h [811].

Titanium pipes where 56% nitric acid must be heated up to about 398 K (125 °C) are used in heat exchangers in ammonium nitrate production. They are also used to introduce the acid into the reactor. Pipes and pumps for handling 99.5% nitric acid between 293 and 353 K (20 and 80 °C) in the explosives industry are also made of titanium [695].

The titanium-molybdenum alloys Ti-10Mo, Ti-20Mo and Ti-30Mo corrode in hot 57% nitric acid at 373 K (100 °C) at material consumption rates of 0.7, 0.3 and 1.5 g/m² h respectively [256].

According to Figure 66, the material consumption rate of titanium-molybdenum-zirconium-niobium alloys in hot 57% nitric acid at 373 K (100 °C) already reaches about 1 g/m² h above 10% molybdenum + 2.5% zirconium and niobium. Such alloys are unusable as materials in production plants where nitric acid is processed [683].

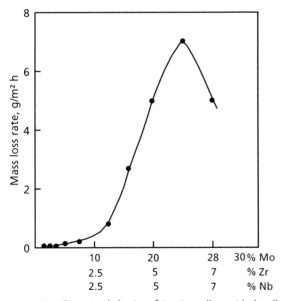

Figure 66: Corrosion behavior of titanium alloys with the alloy ratio 4 Mo:1 Nb:1 Zr as a function of the level of alloying additions in 57% HNO₃ at 373 K (100 °C) after 100 h [683]

The titanium alloys shown in Table 112 are absolutely resistant in 65% nitric acid at room temperature, but have only a satisfactory resistance at the boiling point [780]. The values deviate from those in Table 110.

Alloy	Titanium	Ti2Pd	Ti-15Mo	Ti-0.2Pd-15Mo	Ti-30Mo	Ti-50Ta
Corrosion rate, mm/a (mpy)	0.05 – 0.5 (1.9 – 19)	0.63 (24.8)	1.0 (39.4)	1.0 (39.4)	1.3 (51.2)	0.1 (3.9)

Table 112: Corrosion rates of some Ti alloys in boiling 65 % HNO$_3$, test duration 48 h [780]

The material consumption rates of titanium VT1-0 (cf. Ti grade 2; UNS R50400), VT5-1 (cf. Ti grade 6, UNS R54520) and its alloys Ti2Pd and Ti-33Mo, including in the welded state, in 70 % nitric acid at 453 K (180 °C) after 220 h are summarized in Table 113 [242].

Material	Non-welded	Welded	Observations
	Material consumption rate, g/m^2 h		
VT1-0	0.019	0.02	
VT5-1	0.018	0.02	uniform corrosion
Ti-5Ta	0.058	0.006	
Ti2Pd	2.8	3.9	
Ti-33Mo	19.0	21.0	strong attack

Table 113: Corrosion rate of titanium and titanium alloys in 70 % HNO$_3$ at 453 K (180 °C), test duration 220 h [242]

According to Table 114, the alloys AK-1 (Ti-2.0V-0.23O-0.005H) and AK-2 (Ti-2.9Al-2.0V-0.28O-0.005H) corrode in the gas phase above 99 % nitric acid at 323 K (50 °C) faster than in the liquid phase [691]. The welded joints show practically the same corrosion behavior. Because of their very low corrosion, they can be recommended as materials in nitric acid factories.

Alloy	Test duration h	Liquid phase	Gas phase
		Corrosion rate, mm/a (mpy)	
AK-1	1000	0.001 (0.04)	–
	600	–	0.003 (0.12)
AK-2	1000	0.003 (0.12)	–
	600	–	0.066 (2.6)

Table 114: Corrosion rate of AK-1 and AK-2 above and in 99 % HNO$_3$ at 323 K (50 °C) [691]

After a test duration of 3030 h, the weld joints of the titanium alloys VT1-0 (cf. Ti grade 2; UNS R50400), OT4-1 (TiAl2Mn1.5), PT-3V (TiAl4V2), AT-3 (2.0-3.5Al; 0.2-0.5Cr; 0.2-0.5 Fe; 0.2-0.4 Si) and AT-6 (Ti-6Al-1.5(Cr,Fe,Si)) corroded in boiling 99 % nitric acid at rates of 0.0003, 0.0011, 0.0002, 0.0007 and 0.0003 mm/a (all rates lower than 0.03 mpy) respectively [807]. According to corrosion tests, both the automatically welded and the manually welded joints of AT-3 are more resistant in 99 % nitric acid at 353 K (80 °C) than those of OT4-1 by a factor of 3 [810].

Titanium with small additions of palladium (0.2 – 0.5 %) improves the corrosion resistance, but also tends to undergo pyrophoric reaction in concentrated nitric acid if this contains < 2 % water or > 6 % nitrogen oxides [1002].

Heavy plate with a titanium deposit obtained by explosively applied cladding or roll bonding cladding from pre-primed semifinished material corrodes in boiling concentrated nitric acid at a corrosion rate of 0.034 and 0.032 mm/a (1.34 and 1.26 mpy) respectively [699].

The alloy TiAl6V4 (cf. Ti grade 5; (%) Ti-(5.5-6.5)Al-(3.5-4.5)V-<0.1C-<0.3Fe-<0.015H-<0.04N) differs hardly at all from unalloyed titanium in its corrosion behavior in nitric acid [736].

Urea synthesis and the production and processing of nitric acid as well as the production of fertilizers and processing of fuel rods from nuclear reactors may be mentioned as fields of use for apparatus and installations made of titanium and titanium alloys. Plant is chiefly produced as a welded construction, titanium claddings and linings of steel tanks with titanium also being used, alongside solid wall constructions [716].

The material consumption rates of the titanium alloys OT4-1 (TiAl2Mn1.5) and VT6-S (Ti-5.6Al-4.4V; cf. Ti grade 5) in 50 % nitric acid at 293 and 353 K (20 and 80 °C) are 0.001 and 0.0015, and 0.1 and 0.15 g/m^2 h respectively. Addition of up to 15 % sulfuric acid has virtually no influence on the corrosion rate. Conversely, the material consumption rate of the two alloys in 70 % sulfuric acid, which is about 3 or 5 g/m^2 h respectively at 293 K (20 °C) and about 100 or 150 g/m^2 h respectively at 353 K (80 °C), is reduced to 0.02 and 1 g/m^2 h by addition of 5 % nitric acid. Further addition of up to 17 % HNO_3 produces no additional reduction in corrosion rate [809].

Table 115 shows the high inhibition efficiency of nitric acid towards the corrosion of titanium (Ti + (ppm) 280C+(40-60)Cr+(50-70)Ni+(900-1800)O+(150-250)Fe+(10-200)Al+(15-60)V+(1-3)Cu+(10-40)Zr) in 5 and 12.5 mol/l sulfuric acid at room temperature. A comparably good inhibition is achieved with potassium dichromate ($K_2Cr_2O_7$) [711]. The limited duration of action of the inhibitor, due to its limited life (reduction), should also be referred to here.

Addition	Potential V_{SCE}	1 h	24 h	30 d	30 d*
		Material consumption rate, g/m^2 h			
–	– 0.79 (- 0.66*)	8.95	10.4	–	1.04
2 ml/l HNO_3	+ 0.58 (+ 0.61*)	0.14	0.010	0.083	0.00153
41 ml/l HNO_3	+ 0.055 (+ 0.765*)	< 0.01	< 0.01	0.0097	< 0.001
1 g/l $K_2Cr_2O_7$	+ 0.43 (+ 0.55*)	< 0.01	0.023	0.0097	0.00069

* 5 mol/l H_2SO_4

Table 115: Material consumption rate of titanium electrodes in 12.5 mol/l H_2SO_4 with additions of HNO_3 and $K_2Cr_2O_7$ at room temperature as a function of the test duration [711]

In 14 mol/l sulfuric acid at room temperature, the material consumption rate of titanium could be reduced to < 1 g/m² h by addition of 0.01 – 0.06 mol/l nitric acid, and in an acid mixture of 14 mol/l sulfuric acid + 1.6 mol/l phosphoric acid (H_3PO_4), it could even be reduced to 0.03 g/m² h by addition of 0.0025 mol/l nitric acid [681]. Complete passivity is achieved.

According to Table 116, the corrosion of titanium in sulfuric and hydrochloric acid can already be reduced drastically by small additions of nitric acid [689]. Nitric acid acts as an inhibitor here.

Acid %	Nitric acid %	Temperature K (°C)	Corrosion rate mm/a (mpy)
20 HCl	–	308 (35)	3.4 (134)
	1	308 (35)	< 0.025 (< 1)
65 H_2SO_4	–	311 (38)	35.8 (1400)
	5	308 (35)	< 0.025 (< 1)

Table 116: Influence of HNO_3 additions on the corrosion of titanium in H_2SO_4 and HCl [689]

Table 117 shows the influence of cathodic polarization on the crack propagation rate of the titanium alloy VT-20 (Ti-7Al-2Zr-1.5Mo-1V) in nitric acid solutions. A delay in crack growth is found both at – 0.4 and – 0.6 V_{SHE} [703].

Electrolyte composition %	Corrosion potential V_{SCE}	Cracking rate mm/h	Stress intensity factor kg/mm$^{3/2}$	Polarization potential V_{SCE}	
				– 0.4	– 0.6
4 HNO_3	0.66	0.1 – 1.0	250*	DCG	DCG
4 HNO_3 + 0.1 HCl	0.6	0.1 – 1.0	250*	DCG	DCG
4 HNO_3 + 3.6 HCl	0.54	0.01	240*	**	**
15 HNO_3 + 3.6 HCl	0.55	0.1 – 1.0	200	DCG	DCG
0.6 HNO_3 + 13 HCl	0.55	0.01 – 0.1	160	–	**
1 HNO_3 + 5 CH_3COOH	0.20	0.1 – 1.0	180	DCG	DCG

DCG: delayed crack growth
* crack growth at the start of the plastic deformation range
** cathodic polarization accelerates crack growth

Table 117: Influence of cathodic polarization on the crack propagation rate of the titanium alloy VT-20 in HNO_3 containing solutions [703]

In a fertilizer production plant, a flange made of the steel 08Ch17T (1.4510, AISI 439) was accidentally used alongside titanium VT1-0 and was severely attacked, leading to a pyrophoric reaction producing flames with the chloride ions in the nitric acid (aqua regia) in contact with titanium [273].

While 99.5 % titanium ((%) Ti-(0.02-0.15)Fe-(0.02-0.10)Si-(0.02-0.03)C-(0.01-0.03)N-(0.01-0.10)-O) corroded in an acid mixture (1 HNO_3 : 3 HCl) at 353 K (80 °C) at a rate of 0.4 g/m² h, the dissolution of titanium nitrated on the surface (α-phase) reached about 10 g/m² h [687].

Even the tiniest amounts of nitric acid, e.g. 0.013 % HNO_3 to 20 % HCl, cause a noticeable decrease in the material consumption rate of unalloyed titanium and the alloy Ti-0.1Pd at room temperature (e.g. from 0.33 to 0.13 g/m² h) [713].

On the basis of electrochemical studies, addition of 7 and 18 g/l HNO_3 is necessary to protect titanium from corrosion in 30 and 35 % hydrochloric acid respectively. However, since nitric acid is reduced to NOCl by hydrochloric acid, the protective action of the nitric acid is of limited duration. The protection time is thus stated as 60 or 80 d, for example, on addition of 10 and 20 g/l nitric acid respectively to 30 % hydrochloric acid [706].

Titanium VT1-0 (cf. Ti grade 2; UNS R50400) corrodes in 0.5 or 5 % sulfuric acid with a mixed addition of 8 % NaCl + 2 % HNO_3 + Na_2SO_3 at a corrosion rate of < 0.1 mm/a (< 4 mpy) at 313 K (40 °C) and 0.2 mm/a (7.87 mpy) at 353 K (80 °C). The corrosion rate of the welded titanium specimen in the more dilute acid at 353 K (80 °C) was about 0.5 mm/a (19.7 mpy) [290].

The titanium alloys Ti-35Nb, Ti-5Ta, Ti-6Si and Ti2Pd corrode at room temperature in 70 % nitric acid + 1 mol/l potassium chloride at ≤ 0.15 – 0.22 mm/a (5.91 – 8.66 mpy) [540].

As well as a corrosion rate of VT1-0 in 70 % HNO_3 + 3 % Cl^- + 0.1 % F^- at 363 K (90 °C) of ≤ 20 mm/a (≤ 790 mpy), cracking also occurs, reaching an average depth of 0.5 mm after exposure for 300 h [806].

Titanium passivated in a solution of potassium dichromate or chromium trioxide in hydrochloric or sulfuric acid corroded in 10 % nitric acid + 2 % HF + organic substances at 0.02 mm/a (0.79 mpy) [748].

According to Figure 67, the increase in the corrosion rate of titanium VT1-0 (cf. Ti grade 2; UNS R50400) and its alloy with molybdenum 4201 ((%) Ti-(29-33)Mo-(0.4-0.6)Zr-(0.2-0.8)Fe-(0.03-0.08)Si-(0.007-0.017)C-(0.06-0.11)O-(0.05-0.08)N-0.0017H) caused by hydrogen fluoride in nitric acid is reduced again with an increasing concentration of nitric acid [685]. According to this figure, for example, the corrosion rate of titanium at 298 K (25 °C) is 0.06 mm/h in 6 mol/l HNO_3 + 2 mol/l HF at a potential of + 0.1 V_{SHE} and 0.375 mm/h in 2 mol/l HNO_3 + 2 mol/l HF at – 0.5 V_{SHE}.

The corrosion rate of the alloy TiAl6V4 (cf. Ti grade 5) in nitric acid/hydrogen fluoride solutions at 319 K (46 °C) is shown in Table 118 [710].

Etch solution, %*	Corrosion rate, mm/h
40 HNO_3 + 4.9 HF	0.090
80 HNO_3 + 4.9 HF	0.150
80 HNO_3 + 7.35 HF	0.228

*percent by volume

Table 118: Corrosion rate of TiAl6V4 (cf. Ti grade 5) at 319 K (46 °C) in various HNO_3/HF solutions [710]

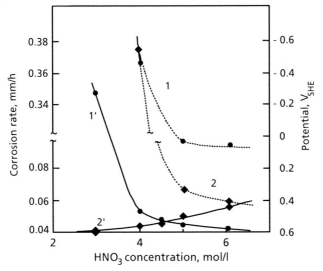

Figure 67: Influence of the HNO$_3$ concentration in HNO$_3$ with 2 mol/l HF at 298 K (25 °C) on the potential 1 – 1' and the corrosion rate 2 – 2' of titanium 1 – 2 and the TiMo alloy 1' – 2' [685]

According to current density/potential measurements, attack on VT1-0 in 70% nitric acid at 453 K (180 °C) is increased by more than 3 powers of ten by addition of 0.1% hydrogen fluoride. The increase in corrosion is cancelled out again by a further addition of 0.1% sulfuric acid [686].

A solution of 400 g/l HNO$_3$ + 5 g/l HF is suitable for chemical descaling and milling (controlled removal of metal by etching) of titanium materials (e.g. TiAl6V4 or Ti-2Cu) at room temperature [714]. During this operation, the hydrogen concentration rises from 560 to 585 ppm in the alloy Ti-2Cu or from 575 to 585 ppm in the alloy TiAl6V4V.

As may be assumed on the basis of the rate of dissolution, which is independent of the potential (from – 1.6 to + 0.8 V$_{SHE}$), of the alloy TiAl6V4 in (percent by volume) 80% HNO$_3$ + 4.9% HF at 319 K (46 °C), dissolution proceeds according to a chemical mechanism. The dissolution was 1590 g/m^2 h, or 0.168 mm/h [710].

The best polishing results are obtained on titanium in a vigorously agitated solution of 16 ml HNO$_3$ + 16 ml HF + 68 ml glycerol. After a certain start-up time, until the reaction between the specimen and polishing solution becomes vigorous, polishing should be carried out for only 10 s [701].

For effective pickling of titanium alloys while reducing the hydrogen uptake, the alloy VT-15 (Ti-(3-4)Al-(10-10.5)Cr-(7-8)Mo) was immersed in various nitric acid solutions at room temperature. According to Table 119, pickling solution 1 causes a low hydrogen uptake at an acceptable pickling rate [709].

On exposure to a solution of 70% nitric acid + 0.1% sodium fluoride at 453 K (180 °C), the oxide layer formed on VT1-0 (cf. Ti grade 2; UNS R50400) consists of rutile + TiOF$_2$ [693]. When the same titanium surface is dipped in 50% nitric acid at

388 K (115 °C) for 5 h, a thin anatase layer 0.15 μm thick is formed with a breakdown potential of 8 V and a field strength of 8×10^5 V/cm.

Pickling solution No.	Composition of electrolyte %	Pickling rate mm/h	Hydrogen uptake %
1	14 HCl + 4 HNO$_3$ + 4.4 NaF	0.33	0.11
2	12 HNO$_3$ + 2.5 HF	0.07	0.15
3	20 HNO$_3$ + 5 HF	0.32	0.17

Table 119: Influence of the electrolyte composition on the pickling rate and hydrogen uptake of the alloy VT-15 [709]

As can be seen from Table 120, the corrosion rate of titanium VT1-0 in nitric acid increases as the concentration of sodium fluoride increases. The increase in corrosion due to NaF cannot be reduced significantly by addition of hydrazine [283].

Composition of solution, g/l			Material consumption rate, g/m^2 h	Corrosion rate, mm/a (mpy)
HNO$_3$	NaF	N$_2$H$_4$		
75	0	2	0.2 ± 0.1	0.4 ± 0.2 (15.75 ± 7.87)
75	0.05	2	3.2 ± 0.6	6.3 ± 1.2 (248 ± 47)
75	0.5	2	41 ± 10	81 ± 20 (3,189 ± 787)
75	5.0	2	220 ± 20	432 ± 40 (17,008 ± 1,575)
75	2	0	42 ± 3	83 ± 6 (3,268 ± 236)
75	2	2	138 ± 9	271 ± 18 (10,669 ± 708)
75	2	10	80 ± 16	157 ± 32 (6,181 ± 1,260)
100	2	2	140 ± 16	275 ± 30 (10,826 ± 1,181)
500	2	2	62 ± 6	122 ± 12 (4,803 ± 472)

Table 120: Corrosion rate of VT1-0 in HNO$_3$ with NaF and hydrazine at 368 K (95 °C) [283]

The titanium-vanadium alloy PT-3V (TiAl4V2) corrodes in a solution of 10 % HNO$_3$ + 4 % NH$_4$F × NH$_4$HF$_2$ at 298 K (25 °C) at a rate of 85.5 g/m^2 h, which can be reduced to 2.5 g/m^2 h by 2.0 g/l CuSO$_4$. The hydrogen content in the alloy remained constant at 0.0043 % [702].

A hot solution of 30 % HNO$_3$ + 2 % NH$_4$F at 338 K (65 °C) is suitable for removing oxide layers on titanium and its alloys. The etching time is 10 to 100 min, depending on the heat treatment used on the specimen (920 – 1120 K (647 – 847 °C)) [707].

Two-stage titanium pumps are recommended for conveying mixtures of nitric acid with CO$_2$ at 366 K (93 °C) [745].

The influence of hydrogen peroxide in nitric acid and the temperature on corrosion of VT1-0 (cf. Ti grade 2; UNS R50400) can be seen in Figure 68. According to this figure, the accelerating influence of H$_2$O$_2$ on the corrosion of titanium decreases with increasing concentration of nitric acid and increasing temperature [700].

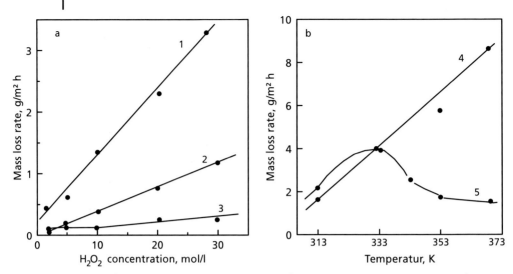

Figure 68: Influence of the H_2O_2 concentration on corrosion of (a) VT1-0 at 368 K (95 °C) in HNO_3 with 1) 5, 2) 7 and 3) 9 mol/l H_2O_2 and of the temperature on corrosion of (b) in 4) 5 and 5) 8 mol/l HNO_3 in the presence of 30 g/l H_2O_2 [700]

Under cathodic polarization, titanium is resistant in 2 to 6 mol/l nitric acid if this contains 6 to 8 mol/l nitrate ions, the current density is between 0.2 and 0.6 A/cm^2 and the temperature does not exceed 353 K (80 °C). Some test results showing the range of potential are summarized in Table 121 [684].

HNO_3 mol/l	Current density A/cm^2	Potential V_{SHE}	293 K (20 °C)	313 K (40 °C)	333 K (60 °C)	353 K (80 °C)
			Material consumption rate, g/m^2 h			
2	0.20	from − 0.084 to − 0.384	1.50	2.86	3.83	0.30
	0.60	from − 0.180 to − 0.407	−	−	−	5.70
4	0.20	from + 0.130 to − 0.308	4.46	1.33	0.10	0.20
	0.60	From + 0.027 to − 0.377	13.46	−	−	3.90
6	0.20	from + 0.110 to + 0.070	0.033	0.10	0.00*	0.00*
	0.60	from + 0.066 to − 0.072	10.3	−	−	0.00*

* no measurable corrosion

Table 121: Corrosion rate of titanium under cathodic polarization in nitric acid as a function of the temperature [684]

Titanium VT1-0 which is used in a production plant for methacrylic acid corrodes at a material consumption rate of 0.0020 g/m^2 h (0.0023 mm/a) after 1600 h. These data agree with the corrosion values shown in Table 122, which were obtained with VT1-1 (pure titanium) in the three solutions prepared.

	Reaction solutions mol/l	Material consumption rate g/m² h
1)	4 N_2O_4 + 1 HNO_3 + 1 isobutylene	0.00032
2)	4 N_2O_4 + 1 HNO_3 + 0.5 isobutylene	0.0005
3)	4 N_2O_4 + 1 HNO_3	0.0024

Table 122: Corrosion of VT1-1 in 3 synthetically prepared reaction solutions under the operating conditions of methacrylic acid production [538]

Titanium is also suitable for the reactor in a production plant for 3-chloro-2-hydroxypropanoic acid. Its corrosion rate in three regions of the production plant is shown in Table 123 [18].

Production region and reaction solution %	Temperature K (°C)	Test duration h	Corrosion rate mm/a (mpy)
Synthesis reactor; liquid phase 45 – 46 HNO_3 + (20 – 21) ECH + (28 – 29) H_2O + (5.3 – 5.4) HCHO + (29 – 30) RP + (9 – 12) $H_2C_2O_4$ + other starting materials	333 – 341 (60 – 68)	1270	0.0035 – 0.0038 (0.14 – 0.15)
Synthesis reactor; gas phase with Ti-0.2Pd	333 – 341 (60 – 68)	1270	0.013 – 0.018 (0.5 – 0.7)
Condensation column after reactor: 0.3 – 0.5 HNO_3 + (77 – 89) H_2O + 2 HCHO + (6 – 20) HCl	328 (55)	106	0.01 – 0.2 (0.4 – 7.9)
Recipe for RP: (65 – 68) RP + 2 HNO_3 + (1 – 4) HCl + (0.5 – 1) $H_2C_2O_4$ + 1 additions + 0.2 resin + (20 – 25) H_2O	293 – 298 (20 – 25)	2460	0.001 (0.04)

RP reaction product
ECH epichlorohydrin

Table 123: Corrosion rate of VT1-1 (pure titanium) in 3 regions of 3-chloro-2-hydroxypropanoic acid production [18]

In conclusion, Table 124 gives a few examples of the use of titanium and its alloys for plants and apparatus which have proven sufficiently resistant to process solutions containing nitric acid [808].

As already mentioned, titanium is suitable as the material for solution tanks for spent nuclear fuels from a light water reactor, where nitric acid occurs together with oxidizing substances, such as $Cr_2O_7^{2-}$, Fe^{3+}, VO_2^{2+} and cerium ions. Figure 69 shows the dependence of the corrosion rate and the thickness of the passive layer of industrially pure titanium on the concentration of nitric acid at the boiling point. In spite of an increasing thickness of the passive layer, the corrosion rate increases initially, and decreases again when a maximum is exceeded at 8 mol/l HNO_3 [733, 773].

Plant component	Reaction medium	Temperature K (°C)	Duration of use years
Heat exchanger	87 % HNO₃ + contamination with 20 % Na₂CO₃ + 80 % H₂SO₄ with sodium hypochlorite	303 – 318 (30 – 45)	1
	40 % HNO₃ with hydrocarbon contaminants	493 – 533 (220 – 260)	3.5
	30 % HNO₃ + nitrogen	353 – 553 (80 – 280)	3
Column	31 % HCl + up to 5 g/l HNO₃	393 (120)	6
Sewage filter	mixture of 36 % HCl + 68 % HNO₃ in the ratio 1:3, containing non-ferrous salts	353	1

Table 124: Use of plant components made of titanium showing the conditions in HNO₃ containing reaction media [808]

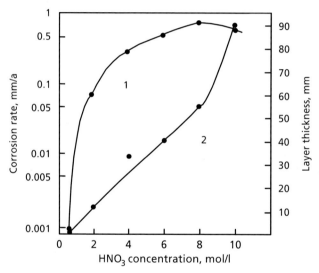

Figure 69: Influence of the HNO₃ concentration on 1) the corrosion rate and 2) the thickness of the passive titanium layer at the boiling point, test duration 65 h [733]

Table 125 shows the marked decrease in the corrosion rate of titanium in boiling 6 mol/l nitric acid due to 0.1 % oxidizing ions [733]. In critical cases, $Cr_2O_7^{2-}$ ions will therefore be added if the nitric acid is contaminated.

The electrochemical and embrittlement properties of titanium electrodes were investigated in pure nitric acid and $HNO_3/UO_2^{2+}/N_2H_5^+$ solutions such as occur during reprocessing of nuclear fuels. Specimens of Ti (3.7035, Ti grade 2, UNS R50400) and Pd-containing Ti (3.7235, Ti grade 7, UNS R52400) with 0.15 % Pd were used. Compared with hydride formation in pure HNO_3, this is reduced in the solution containing UO_2^{2+} and $N_2H_5^+$. In polarization tests, the penetration depth of hydride formation remained limited to < 100 μm in the immediate region of the electrode surface as a result of a parabolic growth law. No significant embrittlement phenomena were detectable on tensile specimens. More precise data on the test parameters are given in [81].

Addition of 0.1 % ions	Redox system	Potential V_{SHE}	Corrosion rate mm/a (mpy)
–	–	–	0.43 (16.9)
Fe^{2+}	Fe^{2+}/Fe	– 0.44	0.44 (17.3)
Cr^{3+}	Cr^{3+}/Cr	– 0.34	0.45 (17.7)
Cu^{2+}	Cu^{2+}/Cu^{+}	+ 0.34	0.38 (14.9)
Fe^{3+}	Fe^{3+}/Fe^{2+}	+ 0.77	0.14 (5.5)
VO_2^{+}	VO_2^{+}/VO_2^{2+}	+ 1.00	< 0.01 (< 3.9)
$Cr_2O_7^{2-}$	$Cr_2O_7^{2-}/Cr^{3+}$	+ 1.33	< 0.01 (< 3.9)
Ce^{4+}	Ce^{4+}/Ce^{3+}	+ 1.70	< 0.01 (< 3.9)

Table 125: Influence of oxidizing ions on corrosion of industrially pure titanium in boiling 6 mol/l HNO_3, test duration 65 h [733]

In nitric acid, colored oxide films produced anodically on titanium VT1-1 (pure titanium) and the alloy OT4-0 (Ti-(0.2-1.4)Al-(0.2-1.3)Mn-≤0.3Fe-≤0.1C-≤0.15Si-<0.8 other elements) tend to suffer a breakdown in the film above 4 % HNO_3 at a potential of > 14 V, depending on the film thickness, followed by rapid anodic dissolution with formation of a violet film [732]. Anodic dissolution of VT1-1 takes place faster than that of OT4-0. Oxide layers generated by heat are more resistant to nitric acid.

Of the oxide layers produced by oxidation in air at 873 K (600 °C), the dark yellow layer (~ 35.8 nm) is particularly resistant to nitric acid, compared with the violet (~ 49.2 nm) and blue (~ 51.5 nm) layers. The material consumption rate of the alloy OT4-0 provided with a dark yellow oxide layer was 0.00011 and 0.00023 g/m² h after 1500 h in 30 and 55 % nitric acid respectively at room temperature. Table 126 shows the dependence of corrosion of the alloy OT4-0 in nitric acid at 293 K (20 °C) on the concentration [782, 785]. According to this table, corrosion of the alloy provided with a yellow oxide layer in 50 % nitric acid is less than that of the untreated alloy by a factor of about 6.

HNO_3 concentration, %	20	30	40	45	50
Material consumption rate, g/m² h	0.000238	0.000475	0.000713	0.00095	0.00143
Corrosion rate, mm/a	0.0005	0.0009	0.0014	0.0018	0.0028
Corrosion rate, mpy	0.02	0.04	0.06	0.07	0.11

Table 126: Corrosion of OT4-0 in nitric acid at 293 K (20 °C), test duration 720 h [782]

Oxide layers generated on titanium by heat provide better protection than those which have been obtained in nitric acid, because of the more stable crystal structure and the more uniform composition [697].

A 39 Zinc, cadmium and their alloys

Zinc is not resistant to nitric acid at room temperature, even at low concentrations (0.1 %). Zinc-containing alloys with more than 15 % zinc are also severely attacked by dilute 0.1 % nitric acid [755]. Dezincification starts above 15 % zinc. This phenomenon has in particular been found with high-zinc brass (see Section A 12).

Reference is made to an item on the interrelationship between the adsorption of surface-active substances and their protective action against dissolution of zinc in nitric acid [760].

In surface treatment of zinc-based printing plates, the plates are given a one-stage treatment with a pickling solution of 8 to 10 % nitric acid + 10 g/l surface-active substances with an emulsion and diethylbenzene [756]. The structure and extent of selective protection is determined by capacitance and resistance measurements.

Before chromization of zinc surfaces, the zinc specimen is electropolished in ethanol/phosphoric acid mixtures and then activated in 0.1 mol/l nitric acid [757].

The zinc alloy (%) Zn-1.3Pb-0.35Cd-0.09Fe-0.01Cu can be polished chemically in an aqueous solution of HCl, HNO_3 and HF with anionic and cationic wetting agents [758].

The following two solutions, inter alia, are also mentioned as chemical polishing agents for preparation of the cadmium-zinc eutectic ((percent by volume) 20.4 % Zn + 79.6 % Cd): (1) 75 ml fuming nitric acid + 25 ml water and (2) 50 ml nitric acid + 950 ml water + 200 g CrO_3 + 15 g Na_2SO_4. While a high corrosion rate of 60 µm/min is achieved in solution 1 with virtually uniform corrosion of both phases, the corrosion rate in solution 2 is only 7 µm/min [759].

A 40 Zirconium and zirconium alloys

Like tantalum, niobium and titanium, zirconium is absolutely resistant both in dilute and concentrated boiling nitric acid [665]. Nitric acid-resistant superficially oxidized zirconium and Zircaloy® 2 (Zr-1.5Sn-0.35(Fe-Cr-Ni)-0.15O) are the preferred materials in nuclear reactor construction [761].

Powdery zirconium, which is used for the production of carbide, corrodes in 12 mol/l nitric acid at 380 K (107 °C) at a rate of 0.075 g/m² h [871].

Up to 525 K (252 °C), zirconium corrodes in all concentrations of nitric acid at a rate of < 0.02 mm/a (< 0.8 mpy) [762]. The metal is used as a material for nuclear fuel reprocessing plants in Oak Ridge National Laboratories.

The resistance to hot concentrated nitric acid reaches that of platinum and is based on the formation of a dense, firmly adhering layer of disordered zirconium dioxide which protects the metal from chemical attack by the acid up to temperatures of about 570 K (297 °C) as well as from mechanical attack. Zirconium is also resistant to crevice corrosion and stress corrosion cracking, as well as uniform corrosion [763]. Galvanic corrosion can occur in contact with more noble metals if the protective layer is destroyed, i.e. in the active state.

Zirconium and its alloys Zr-1.5Sn and Zr-2.5Nb exhibit a high resistance to stress corrosion cracking in up to 70 % nitric acid up to the boiling point [960].

The dependence of the corrosion rate of Zr10NbTi, Zr10MoTi and ZrTi alloys in the β-annealed state in boiling 20 % nitric acid can be seen in Figure 70. The lowest material consumption rate here is achieved by the alloy Zr5Ti10Nb at about 0.006 g/m² d or 0.00025 g/m² h [764].

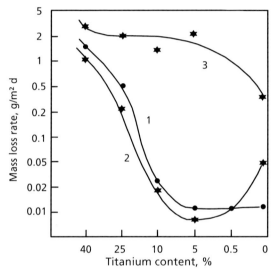

Figure 70: Material consumption rates of various zirconium alloys as β-annealed specimens in boiling 20 % HNO$_3$ as a function of the titanium content [764]
1) ZrTi; 2) Zr10NbTi; 3) Zr10MoTi

The material consumption rates of some zirconium-titanium alloys in boiling 20 % nitric acid are summarized in Table 127. The best corrosion behavior is shown by the alloy Zr10Ti10Ta in the α-annealed state [764].

The corrosion cracking rate of U-shaped specimens of the zirconium-hafnium alloys Zr-702 (ASTM Grade R60702; (%) Zr-4.5Hf-<0.2(Fe+Cr)-0.005H-0.025N-0.05C-0.16O), Zr-704 (R60704: (%) Zr-4.5Hf-(0.2-0.4)(Fe+Cr)-(1-2)Sn-0.005H-0.025N-0.05C-0.18O) and Zr-705 (R60705; (%) Zr-4.5Hf-(2-3)Nb-<0.2(Fe+Cr)-0.005H-0.025N-0.05C-0.18O) with tensile strengths of 450, 485 and 585 MPa and yield strengths of 310, 380 and 450 MPa was determined in nitric acid. The times to failure are summarized in Table 128, differently worked specimens of each alloy being used. The figures in parentheses denote the hours after which still no cracking was found. Only the U-shaped and welded band of the alloy Zr-705 showed corrosion cracking in red, fuming nitric acid after only 118 h, while the other alloys were still free from cracks after 1440 h. The cracking on the welded alloys Zr-704 and Zr-705, which starts after only 2 h on both longitudinally and transversely cut specimens in 80 % nitric acid at 303 K (30 °C), is also worth pointing out [765].

Alloy	α-annealed	β-annealed
	Material consumption rate, $g/m^2\,h$	
Zr	0.00050	–
Zr-10Ti	0.00071	–
Zr-10Nb	0.0019	0.00083
Zr-5Ti-10Nb	0.00029	0.00083
Zr-7.5Ti-15Nb	0.00029	0.0010
Zr-10Ti-5Nb	0.0012	0.00096
Zr-10Ti-7.5Nb	0.0012	0.0033
Zr-10Ti-15Nb	0.0026	0.0014
Zr-15Ti-5Nb	0.0018	0.0015
Zr-15Ti-7.5Nb	0.0041	0.0019
Zr-15Ti-15Nb	0.0058	0.0067
Zr-10Ti-7.5Ta	0.0011	0.0050
Zr-10Ti-10Ta	0.00025	0.0050
Zr-10Ti-5Ta-5Nb	0.0023	0.0015
Zr-5Ti-0.1Fe-0.3Cr-0.1Ni	0.0021	0.0033
Zr-5Ti-0.5Cr	0.0033	0.0042
Zr-5Ti-10Mo	0.0083	0.0042

Table 127: Material consumption rates of ZrTi alloys of various composition in boiling 20 % HNO_3 as a function of the structure [764]

Vapor oxidation of zirconium is greatly accelerated if the metal is washed with tapwater (60 ppm Ca) after pickling in HNO_3/HF solutions. No impurities which dampen the corrosion are known [763].

Zircaloy® 4 (UNS R60804; (%) Zr-0.0095C-0.120Cr-0.210Fe-0.008Hf-<0.0035Ni-1.54Sn) exhibits cracking (> 3.6×10^{-4} mm/h) at 298 K (25 °C) in nitric acid above 10 % under a strain rate of 2.5×10^{-6} s^{-1}. The cracking rate of Zircaloy® 4 under polarization-free conditions is shown in Figure 71, while the cracking rate of the alloy in 90 % nitric acid at 298 K (25 °C) is given as a function of the potential applied in Figure 72 [766]. According to this figure, cracking is prevented above + 1.2 V_{SCE}.

Alloy	Working*	70% HNO₃ boiling	80% HNO₃ 303 K (30 °C)	Red, fuming HNO₃ 297 K (24 °C)
		Time to failure, h		
Zr-702	LCNW	(1440)	(1440)	(1440)
	TCNW	(1440)	(1440)	(1440)
	LCWW	(1440)	23.5	(1440)
	TCWW	(1440)	23.5	(1440)
Zr-704	LCNW	(1440)	(1440)	(1440)
	TCNW	(1440)	307 and (1440)	(1440)
	LCWW	(1440)	2	(1440)
	TCWW	(1440)	2	(1440)
Zr-705	LCNW	(1440)	(1440)	(1440)
	TCNW	(1440)	980 and (1440)	(1440)
	LCWW	(1440)	2	118
	TCWW	17	2	118

* LCNW: longitudinally cut, non-welded
TCNW: transversely cut, non-welded
LCWW: longitudinally cut, welded
TCWW: transversely cut, welded

Table 128: Times to failure of the ZrHf alloys Zr-702, Zr-704 and Zr-705 in various states of working in HNO₃ [765]

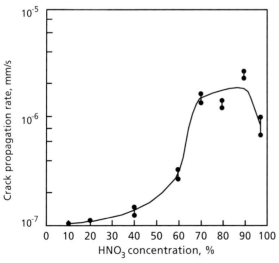

Figure 71: Influence of the HNO₃ concentration on the crack propagation rate of Zircaloy® 4 (UNS R60804) at 298 K (25 °C) without an external electrical field at a strain rate of 2.5×10^{-6} s^{-1} [766]

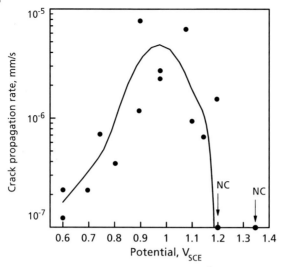

Figure 72: Crack propagation rate of Zircaloy® 4 (UNS R60804) in 90% HNO$_3$ at 298 K (25 °C) as a function of the potential [766]
NC: no crack formation

To improve the coefficient of friction of Zircaloy® 2 (UNS R60802; (%) Zr-1.2Sn-0.16Fe-0.06Mg-0.10N-0.12O), it is polarized anodically in concentrated nitric acid. A 65% acid gave compact, hard and sufficiently thick (up to 30 µm) oxide layers with abrasion-resistant properties at 273 K (0 °C) and low current densities of 1.5 to 3.1 mA/cm^2 [767].

According to Table 129, zirconium alloys annealed at 1473 K (1200 °C) for 50 h in purified argon are considerably weakened in their corrosion behavior in 50% nitric acid + 0.1% hydrogen fluoride by, in each case, two metals, such as Ti + Hf, Y + La and Co + Pd [768].

The Alniflex process uses a solution of 2 mol/l HF + 1 mol/l HNO$_3$ + 1 mol/l Al(NO$_3$)$_3$ + 0.1 mol/l (NH$_4$)$_2$Cr$_2$O$_7$ for processing spent nuclear fuel rods coated with Zircaloy® 2 (UNS R60802), because this alloy is attacked far more severely than the tank steel AISI 304 L SS because of the complex aluminium fluorides which form (especially at higher temperatures). The dissolution of Zircaloy® 2 thus corresponds, for example, to 0.5 mm/h at 363 K (90 °C), while the steel corrodes at a rate of about 0.5 mm/a (20 mpy). A fuel rod made of Zircaloy® 2 + 6% natural uranium used in pressurized water reactors shows a dissolution of zirconium higher than that of uranium by a factor of about 3 after 15 h [769].

As has been determined with the radiotracer method, zirconium corrodes in azeotropic nitric acid between 293 and 394 K (20 and 121 °C) at rates of between 7×10^{-6} and 5×10^{-4} mm/a (0.0003 and 0.0197 mpy). The corrosion rate is increased exponentially with even very small additions of 10 ppm fluoride (NaF). With fluoride additions of 0.15 to 10 ppm, the corrosion rates for Zr at 381 K (108 °C) were between 5×10^{-3} and 3.1 mm/a (0.19 and 122.1 mpy). No F$^-$ threshold can be specified. F$^-$ contents in the nitric acid should be avoided as far as possible (see Figure 65) [21].

Foreign metal additions %	Material consumption rate g/m² h	Corrosion rate mm/a (mpy)	Quality rating
–	< 0.001	< 0.001 (0.04)	1
0.0149 Ti + 0.036 Hf	< 0.001	< 0.001 (0.04)	1
0.089 Ti + 0.189 Hf	0.0048	0.0064 (0.25)	2
0.309 Ti + 0.764 Hf	0.0093	0.0125 (0.49)	3
0.0292 Y + 0.0305 La	0.0011	0.0014 (0.06)	2
0.1949 Y + 0.153 La	0.0247	0.0335 (1.32)	4
0.5937 Y + 0.608 La	0.318	0.428 (16.85)	6
0.011 Cr + 0.017 Mo	< 0.001	< 0.001 (0.04)	1
0.306 Cr + 0.417 Mo	0.0038	0.0051 (0.20)	2
0.104 Ni + 0.107 Pd	< 0.001	< 0.001 (0.04)	1
0.317 Ni + 0.456 Pd	0.0019	0.0025 (0.10)	2
0.117 Fe + 0.061 Co	0.0016	0.0021 (0.08)	2
0.319 Fe + 0.241 Co	0.0043	0.0057 (0.22)	2
0.008 Fe + 0.017 Pd	< 0.001	< 0.001 (0.04)	1
0.341 Fe + 0.437 Pd	0.003	0.004 (0.16)	2
0.013 Co + 0.208 Pd	< 0.001	< 0.001 (0.04)	1
0.118 Co + 0.093 Pd	0.0014	0.0018 (0.07)	2
0.371 Co + 0.451 Pd	0.0053	0.0071 (0.28)	3

Table 129: Corrosion rates of zirconium alloys with small additions of two metals in 50% HNO_3 + 0.1% HF at room temperature [768]

If zirconium is exposed to 70% nitric acid at higher temperatures, tensile stresses originating from the processing must be eliminated. In spite of its excellent resistance to nitric acid, higher corrosion rates can be expected in fluoride-containing acid and in vapors of chloride-containing nitric acid [772].

Oxidizing metal ions are also used to inhibit the corrosion of zirconium in boiling nitric acid (65%), and in this case, for example, ruthenium ions in a concentration of 0.001 mol/l. Rhodium, palladium, osmium, iridium, platinum, chromium, vanadium and cerium ions are also mentioned [770]. A uniformly protecting oxide layer is said to be generated by the presence of these ions.

Powder metallurgy zirconium and Zr-2.5Nb, which has a density of between 98 and 99% of the compact material, have the same corrosion behavior in boiling 70% nitric acid as the cast alloy. Welded material behaves similarly [771].

In addition to the concentration of nitric acid, the potential (1 to 12 V) and temperature have a large influence on the extent of the anodic current density on zirconium as an electrode, which can be varied in this way, for example, between 5 and 45 mA/cm² in 6 mol/l nitric acid [783].

A 41 Other metals and alloys

Because of its tendency to form oxyfluorides, tungsten is already noticeably attacked at room temperature in a solution of 200 g/l HNO_3 + 174 g/l HF (about 0.7 mm/h) [735].

Since the weight increase determined for tungsten was proportional to the charges which flowed through, the current density of 8 mA/cm^2 at about 80 V between 298 and 323 K (25 and 50 °C) results in a corrosion rate of about 65 to 70 mm/a (2,560 to 2,756 mpy) [784]. The surface layer which forms consists mainly of WO_3.

If tungsten is not contaminated with low-melting metals, such as iron, nickel and chromium, it can be used for nozzles and other rocket components which are subjected to high temperatures with high-energy fuels, such as nitric acid + ammonia (2600 K (2327 °C)) and nitric acid + gasoline (3120 K (2847 °C)) [1001].

Hafnium shows approximately the same corrosion behavior in nitric acid as zirconium. The corrosion rate in 12 mol/l nitric acid at the boiling point is thus ≤ 0.0003 mm/a (≤ 0.01mpy), and up to 333 K (60 °C) is only ≤ 0.00005 mm/a (≤ 0.002 mpy) [786]. A loop specimen showed no stress corrosion cracking after 5000 h in 17 % nitric acid between 293 and 333 K (20 and 60 °C). Since hafnium also displays a high neutron absorption, it is used as an absolutely safe material for the construction of plants for reprocessing nuclear fuels [787].

From electrochemical measurements with hafnium in 1 mol/l nitric acid, the corrosion current density is < 1.1 µA/cm^2, from which a material consumption rate of < 0.0013 g/m^2 h is calculated [789].

Hafnium-niobium alloys with 10 to 70 atomic percent niobium, which are used at high temperatures (~ 2000 K (1727 °C)), are etched in 4 HNO_3 + 2 HF + 4 glycerine for determination of the microstructure at up to 40 atomic percent Nb, and in 4.5 HCl + 2.5 HNO_3 + 1.5 HF with higher Nb contents [788].

Unalloyed vanadium is already non-resistant in nitric acid at room temperature. Table 130 shows some vanadium alloys which are in some cases quite resistant in 20 % nitric acid at room temperature [790].

Alloy composition %*	Material consumption rate g/m^2 h
V	> 200
V-50Ti	0.63
V-20Cr	0.033
V-30Cr	0.029
V-16.6Cr-33.4Ti	0.042
V-13.3Cr-6.7Ti	0.33
V-1Zr-20Cr	0.040
V-10Cr-10Al	0.21

*atomic percent

Table 130: Material consumption rates of vanadium and some vanadium alloys in 20 % HNO_3 at 293 K (20 °C) [790]

According to Table 131, the corrosion resistance of high-alloyed vanadium-tantalum and in particular vanadium-niobium alloys in 10 to 40 % boiling nitric acid is considerably better than that of the alloys listed in Table 130 [791].

In agreement with these results, the alloy V-40Nb-10Ti corrodes in boiling 65 % nitric acid at a rate of 0.45 mm/a (17.72 mpy) [792].

Above 22 atomic percent Ta, the vanadium-tantalum alloys achieve a good corrosion resistance in nitric acid which is comparable to that of the vanadium-niobium alloys. The alloys V-30Ta and V-40Ta thus corrode in hot 56 % nitric acid at 393 K (120 °C), for example, at a rate of 0.08 and 0.006 mm/a (3.15 and 0.24 mpy). The resistance of the alloys is said to be based on the formation of an oxide layer consisting mainly of Ta_2O_5, as is the case with pure tantalum [793].

Rhenium corrodes in 40 % nitric acid at room temperature at a rate of about 200 g/m^2 h [794].

The solder PInK-24 (Russ. type; In-24Cd) used in semi-conductor technology is severely attacked by 0.1 mol/l nitric acid at room temperature after 40 d (12.3 mm/a (484 mpy)) [795].

Uranium (U-3Fs and U-10Fs) irradiated with neutrons corrodes in nitric acid faster than non-irradiated uranium. This is shown by some of the results in Table 132. The large difference in the corrosion rates of non-irradiated uranium and irradiated uranium is based on the occurrence of fission products, in particular noble ruthenium and rhodium (U-3Fs; (%) U-1.44Mo-1.29Ru-0.179Rh-0.126Pd-0.019Zr-0.006Nb) and U-10Fs; (%) U-4.91Mo-4.33Ru-1.56Rh-0.419Pd-0.097Zr-0.020Nb)) [796].

HNO_3 concentration %	V-9Ta	V-13Ta	V-12Nb	V-21Nb	V-26Nb	V-31Nb
	Corrosion rate, mm/a (mpy)					
5	0.15 (5.91)	~0.01 (0.39)	0.7 (27.5)	~0.01 (0.39)	~0.01 (0.39)	~0.01 (0.39)
10	4 (157.48)	0.5 (19.7)	6 (236)	0.1 (3.94)	0.01 (0.39)	<0.01 (0.39)
20	40 (1574.8)	4 (157)	100 (3,937)	1 (39.4)	0.05 (1.97)	0.01 (0.39)
40	–	–	800 (31,496)	7 (275.6)	0.2 (7.87)	0.015 (0.59)

Table 131: Corrosion rates of vanadium-tantalum and vanadium-niobium alloys in boiling nitric acid [791]

Potentiostatic and galvanostatic anodic polarization tests show dilute nitric acid to be a suitable electrolyte for the anodic dissolution of uranium. The addition of an oxidizing agent such as iron(II) salts allows the dissolution to be made more effective [798].

The etching rate of the uranium alloys U-0.75Ti and U-2.3Nb in nitric acid solution (HNO_3 + $ZnCl_2$) can be seen in Table 133 [797].

Material	Period 1			Period 2			Period 3		
	Time h	Corrosion rate mm/a (mpy)	Current density* A/m^2	Time h	Corrosion rate mm/a (mpy)	Current density* A/m^2	Time h	Corrosion rate mm/a (mpy)	Current density* A/m^2
U	45	0.05 (1.97)	0.71	90	0.33 (12.9)	0.48	102	0.38 (14.9)	0.56
U-3Fs	2	1100 (43,307)	1600	2	1100 (43,307)	1600	–	–	–
U-10Fs	19	**	**	–	–	–	–	–	–

* current density A/m^2 calculated on the basis of the reaction U $\rightarrow$ UO$_2^{2+}$ during the dissolution $\rightarrow$ U^{6+}
** specimens were destroyed

Table 132: Corrosion rates of pure, non-irradiated uranium and neutron-irradiated uranium U-3Fs and U-10Fs in 3.5 mol/l HNO$_3$ at 310 K (37 °C) [796]

Material	Composition of solution	Etching conditions		Corrosion rate mm/h
		Duration, min	Temperature, K (°C)	
U-0.75Ti	100 ml/l HNO$_3$ + 1350 g/l ZnCl$_2$	10	295 (22)	0.06
	200 ml/l HNO$_3$ + 900 g/l ZnCl$_2$	10	295 (22)	0.2
	500 ml/l HNO$_3$ + 20 g/l ZnCl$_2$	10	295 (22)	0.56
	200 ml/l HNO$_3$ + 1350 g/l ZnCl$_2$	20	322 (49)	0.48
U-2.3Nb	100 ml/l HNO$_3$ + 1350 g/l ZnCl$_2$	10	295 (22)	0.27
	400 ml/l HNO$_3$ + 450 g/l ZnCl$_2$	10	295 (22)	0.75
	200 ml/l HNO$_3$ + 1350 g/l ZnCl$_2$	20	322 (49)	0.36
	450 g/l ZnCl$_2$	10	295 (22)	0.00

Table 133: Corrosion rate of U-0.75Ti and U-2.3Nb in HNO$_3$/ZnCl$_2$ solutions [797]

After a certain heat treatment, during which a finely divided two-phase microstructure of high carbon content (> 150 ppm) is formed, a violent explosion may occur when uranium-niobium alloys are immersed in oxidizing acids, such as nitric acid. This can be initiated either by sparks or mechanical shock, or even by friction. The risk of explosion is prevented by addition of fluoride ions [797, 799].

Plutonium is dissolved in 10 mol/l nitric acid + 0.05 mol/l hydrogen fluoride by anodic polarization at $\leq$ 14 V and a current density of 1.86 A/cm^2 at room temperature with an average material consumption rate of 3.75 g/cm^2 h or 37,500 g/m^2 h (see the structure of the electrolyte cell in [800]).

In 1 and 10 % nitric acid + 200 g/l CrO$_3$, the corrosion rate of beryllium at room temperature and an anodic polarization potential of 3 V is about 0.6 and 1.2 mm/h respectively at a final current density of 20 and 30 mA/cm^2 respectively. A crystalline BeO layer forms [801].

An acid mixture of 1 part fuming nitric acid + 5 parts glacial acetic acid diluted with water in a ratio of 8:2 to 2:8 parts, depending on the aim, is used as the standard etching agent for bismuth [802].

According to Table 134, gallium is already not resistant in nitric acid at 298 K (25 °C). Furthermore, it also exhibits local attack [803].

HNO_3 %	Potential U_{st}* V_{SHE}	Potential U_R** V_{SHE}	Material consumption rate $g/m^2\ h$
1	− 0.32	− 0.61	3.3
5	− 0.27	− 0.60	17.2

* U_{st} stationary potential
** U_R reversible potential of the anodic process

Table 134: Material consumption rates and electrochemical behavior of gallium in nitric acid at 298 K (25 °C), test duration 160 h [803]

B
Non-Metallic Inorganic Materials

B 2   Natural rock

According to Table 135, basalt, a high-silicon stone containing oxides of manganese and iron which occurs widely in nature, is quite resistant to nitric acid at room temperature [844].

HNO_3 concentration %	Static	Agitated	Static	Agitated
	291 K (18 °C)		373 K (100 °C)	
	Corrosion rate, mm/a (mpy)			
10	0.30 (11.81)	0.048 (1.89)	–	–
35	0.14 (5.51)	0.033 (1.3)	3.33 (131)	1.27 (50)
50	0.066 (2.60)	0.015 (0.59)	–	–
60	0.051 (2.01)	0.010 (0.39)	–	–
Concentrated	–	–	14.50 (571)	8.23 (324)

Table 135: Resistance of recrystallized cast basalt (melting point about 1520 K (1247 °C)) to nitric acid [844]

B 3   Carbon and graphite

Dense graphite is resistant in up to 20 % nitric acid, but not at higher concentrations [958]. Graphite impregnated with furylphenol-formaldehyde resin is resistant, inter alia, to 5 % nitric acid at 313 K (40 °C) and 10 % acid at 298 K (25 °C) [827].

Heat exchangers made of graphite (high thermal conductivity) with a special impregnation of polytetrafluoroethylene should guarantee problem-free operation even in very aggressive media, such as mixtures of nitric acid and hydrogen fluoride [818, 1008].

Sigraflex®, a sealing material made of flexible graphite which is also used as a graphite film or laminate and has a high thermal conductivity, forms intercalation compounds with nitric acid which diffuses in. Its use in concentrated nitric acid is critical [847]. Its resistance can be improved by impregnation with corrosion-resistant resins, such as polytetrafluoroethylene.

Caution is required when using normal graphite seals, such as TBA-700, for sealing off from highly oxidizing media, such as fuming nitric acid. TBA-26 is more suitable for this [846].

Grafilex® 6501, a graphited cord, is resistant to 10 and 20% nitric acid up to 358 and 333 K (85 and 60 °C) respectively [823].

Highly orientated pyrolytic graphite incorporates HNO_3 molecules when in contact with nitric acid vapors in the absence of air and water vapor. After an initial superficial oxidation of the graphite, red-brown vapors of NO_2 arise. Prior treatment with nitrogen reduces the inward diffusion of HNO_3 from about 45% to 10%. In the presence of oxygen, no incorporation of HNO_3 was found even after 24 h [819].

The chemisorption of ammonia (NH_3) is increased several-fold and becomes more stable on oxidative treatment of carbon films with nitric acid, so that desorption of ammonia at room temperature is impossible. This does not take place until the temperature reaches 470 K (197 °C) [820].

The thermodynamic state and the composition of the surface of graphite fibers after treatment in nitric acid are being investigated by the IDS and SIMS methods (ion diffraction and secondary ion mass spectroscopy) [822].

Polyethylene-coated carbon fibers in epoxy-phenolic resin guarantee the shear strength of the plastic even after prolonged exposure to a medium containing nitric acid (Table 136). The shear strength of the C fiber-reinforced plastic is still 95% of the starting value even after exposure for 360 d [824].

Exposure time d	Fiber content %*	Porosity %	Density g/ml	Shear strength, R_τ N/mm²	R_τ/R_0**
0	58	4.8	1.45	62	1.0
42	59	5.4	1.45	59	0.95
90	56	1.3	1.48	59	0.95
180	57	0.0	1.47	61	0.98
270	58	2.7	1.46	58	0.94
360	59	3.2	1.46	59	0.95

* percent by volume
** starting value

Table 136: Influence of the exposure time of an HNO_3 containing medium on the porosity, density and shear strength of C fiber-containing epoxy-phenolic resin at room temperature [824]

The inward diffusion or diffusion rate of gaseous nitric acid above 98% into highly orientated pyrolytic graphite can be calculated using the known diffusion equation. Some values are summarized in Table 137 [821]. The HNO_3 molecules in the graphite display a behavior like that of liquid nitric acid.

Graphite compacted by deposition of pyrocarbon is suitable as a material for heat exchangers in numerous chemical processes. The weight increase in sulfuric acid containing 1 to 16% nitric acid at 398 K (125 °C) in the course of 960 h could thus be reduced from about 2% for untreated graphite to 0.6% for compacted graphite [828].

HNO₃ pressure kPa	Temperature K (°C)	Diffusion rate cm²/h
10.8	303 (30)	0.012
10.8	313 (40)	0.0080
4.9	303 (30)	0.0051

Table 137: Diffusion of HNO_3 into highly orientated pyrolytic graphite [821]

For processing the graphite which holds back the fission gases in gas-cooled high temperature reactors, it was comminuted electrolytically in 4 mol/l nitric acid at 312 K (39 °C) and 14 V in a suitable electrochemical cell. While the corrosion rate of irradiated graphite was 1.9 mm/h, that of non-irradiated graphite reached about 3.4 mm/h [951]. By this process, up to 90 % of the graphite can be removed from the fuel to be processed.

The infiltration of nitric acid into graphite results in a very low specific electrical resistance parallel to the layer planes of about 3×10^{-6} Ω cm [829].

Of the non-metallic materials, vitreous carbon is particularly suitable for media such as aqua regia (HNO_3 + HCl) [391].

The adsorption isotherm of water vapor on synthetic diamond powder treated with 12 % nitric acid differs only slightly from that of the untreated powder at 298 K (25 °C) [953].

Materials consisting of graphite carbide, e.g. graphite silicon carbide (doped with boron), are resistant to nitric acid. The compressive strength of graphite silicon carbide of 150 N/mm² determined at the start was thus virtually retained, at 140 N/mm², after exposure for 720 h to hot and cold (343 K (70 °C)) concentrated nitric acid [869].

B 4 Binders for building materials (e.g. mortar and concrete)

Concrete made from Portland cement is not resistant to nitric acid even at only low concentrations. It is already rapidly destroyed in 2 % nitric acid at room temperature [832].

Urethane concrete, a mixture of polyurethane and cement which combines the strength of concrete with the chemical resistance of polyurethane, is resistant to 10 % nitric acid at room temperature, but attacked by 70 % nitric acid [833]. Urethane concrete is suitable for floors in chemical factories.

Ucrete®, a coating material made of polyurethane, is suitable for coating concrete floors which suffer severe exposure to chemicals, including up to 50 % nitric acid [845].

Polymer concrete based on polymethyl methacrylate, which is used for shaped articles and coatings, is resistant to 10 and 30 % nitric acid, but has only a limited resistance to concentrated acid [937].

Concrete impregnated with methyl methacrylate displays the properties shown in Table 138 in nitric acid (pH 2) at room temperature [957]. After exposure for 360 d

(8640 h), protection provided by impregnation with polymethyl methacrylate is inadequate [957].

Test duration h	Weight loss %	Compressive strength MPa
720	3	34
8640	15	16

Table 138: Weight loss and compressive strength of impregnated concrete in HNO_3 at room temperature [957]

B 5   Acid-resistant building materials and binders (putties)

Acid-resistant stones have a relatively good resistance in 50, 55 % and 10 mol/l nitric acid at 323 K (50 °C) with a relative change in weight of + 0.01, − 0.05 and − 0.19 % respectively. In acid mixtures of 18 mol/l HNO_3 + 2 mol/l H_3PO_4, 6 mol/l HNO_3 + 4 mol/l H_3PO_4 and 2 mol/l HNO_3 + 6 mol/l H_3PO_4, the corrosion resistance becomes considerably poorer with a relative weight change of + 0.80, + 1.01 and 1.32 % respectively [986]. Acid putties have a somewhat better behavior with values of − 1.05, − 0.98 and + 0.97 %.

B 6   Glass

Glass is absolutely resistant to nitric acid. Because of this resistance, borosilicate glass 3.3 (80 % SiO_2 + (12 − 13) % B_2O_3 + 2 % Al_2O_3 + 4 % alkali metals) is used in concentrating plants for nitric acid and in the nitration of organic substances. If no particular thermal conduction problems have to be taken into consideration, borosilicate glass 3.3 can replace tantalum, which is similarly resistant to nitric acid at high temperatures [834].

Borosilicate glass, which has become well known as Duranglas®, can be combined with other materials, such as graphite in circulation apparatus. Downstream evaporators made of Duranglas® in denitration plants render re-use of the acid possible by concentrating the nitric acid to at least 98 % [669, 850, 854]. Glass apparatus components are sealed with metallic materials, such as titanium and tantalum, using PTFE (see Section C 19). The use of Duranglas® apparatus in waste water lines from nitric acid plants and in liquid/liquid extraction in the presence of nitric acid should also be mentioned [848, 852, 853].

Duranglas® is resistant in the entire concentration range of the system $HNO_3/H_2SO_4/H_2O$, even at higher temperatures [837].

Foamed borosilicate glass is used to protect chimneys and washers. It is not attacked by 20 % nitric acid at 376 K (103 °C) [835].

The corrosion resistance of glass-ceramic to nitric acid is somewhat lower than that of glass, and can be increased by application of a suitable layer of enamel [837].

Glass-ceramic building materials are also resistant to acid mixtures containing nitric acid. As well as water glass, epoxy resin and polyester are also recommended as extremely durable putties for glass-ceramic tiles [851].

Corrosion of lead silicate glass in nitric acid increases with additions of up to 0.5 mol/l lead oxide and decreases again with higher additions [849].

B 7 Quartz ware and quartz glass

Quartz ware and quartz glass are resistant to nitric acid even at high temperatures.

Acid mixtures of nitric and sulfuric acid containing hydrogen fluoride are used as etching solutions for quartz sheets in piezoelectric resonators [855].

B 8 Enamel

In view of the use of enamelled apparatus and reaction boilers, resistance to highly oxidizing acids, such as nitric acid, is of decisive importance. Enamel 7701, with a corrosion rate of 0.1 mm/a (3.94 mpy) in all concentrations of nitric acid up to 423 K (150 °C) is superior in corrosion behavior to the high-alloy steels, such as Hastelloy® C-276 (cf. 2.4819, 2.4886), apart from tantalum [838].

The high resistance of chemical service enamels to nitric acid can be seen from the isocorrosion curve of 0.1 mm/a (3.94 mpy) in the temperature/concentration graph in Figure 73. In contact with aqueous acids, a silicate skeleton with water added on is formed on the enamel surface. The resulting dense gel layer is largely insoluble in nitric acid [837].

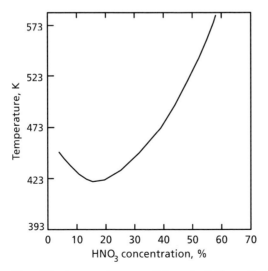

Figure 73: Isocorrosion curve (0.1 mm/a (3.94 mpy)) of chemical service enamel in nitric acid [837]

The attack by aqueous nitric acid on chemical service enamel at 433 K (160 °C) is greatly inhibited by silicon dioxide dissolved in the acid (Table 139). The following silicon-containing substances have proven suitable for use on enamelled apparatuses in practice: finely divided, solid silicic acid (e.g. Aerosil®), aqueous solutions of sodium silicate or water glass, and colloidal silicic acid solution [839].

The dependence of the corrosion of chemical service enamel 1 (volume-surface ratio V/S of 45 ml/cm^2) in nitric acid at 413 K (140 °C) on the concentration is shown in Figure 74. The dependence in 20 % nitric acid on the temperature under comparable experimental conditions can be seen from Table 140 [840].

HNO$_3$ concentration %	SiO$_2$ content ppm	Corrosion rate mm/a (mpy)
20	0.8	0.23 (9.06)
	150	0.12 (4.72)
30	1.2	0.23 (9.06)
	100	0.08 (3.15)

Table 139: Corrosion rates of the enamel C-1 (containing about 65 % SiO$_2$) in HNO$_3$ with and without dissolved SiO$_2$ [839]

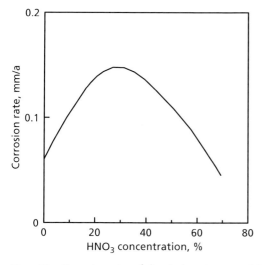

Figure 74: Corrosion rates of chemical service enamel 1 as a function of the HNO$_3$ concentration at 413 K (140 °C) after 24 h [840]

Highly resistant chemical service enamels of average composition 65 % SiO$_2$ + 17 % Me$_2$O + 5 % MeO + 3 % Me$_2$O$_3$ + 10 % MeO$_2$ are applied to tantalum-lined autoclaves to guarantee good protection against nitric acid at higher temperatures [906].

Temperature, K (°C)	Corrosion rate, mm/a (mpy)
373 (100)	0.04* (1.57)
413 (140)	0.13 (5.12)
434 (161)	0.22 (8.66)
455 (182)	0.35* (13.78)

*extrapolated

Table 140: Dependence of the corrosion rate of chemical service enamel 1 in 20% HNO_3 on the temperature, test duration 24 h [840]

The material consumption rate of the Russian enamel 143-V passes through a maximum of 18 and 5 g/m^2 h in 10 and 20% nitric acid respectively at 533 K (260 °C), measured in an autoclave under a nitrogen atmosphere. The corrosion rate decreases below and above this temperature. It is about 1 and 0.5 g/m^2 h at 493 and 558 K (220 and 285 °C) respectively. In 50% nitric acid, the corrosion at 493 and 533 K (220 and 260 °C) reaches about < 0.5 – 0.3 g/m^2 h, regardless of the temperature [841].

Enamel-lined pulsed absorption tray columns are resistant to nitric acid and its vapors up to 470 K (197 °C) [418].

The enamel Lampart Univer-80® is said to provide effective protection against nitric acid and its vapors [856].

A layer of enamel 10-1 (Russ. type) applied to the steel 12Ch18N10T (AISI 321, 1.4541) corroded in 20% nitric acid at 377 K (104 °C) at a rate of 0.06 mm/a (2.36 mpy) [954].

B 9 Porcelain

Porcelain is absolutely resistant to nitric acid at all concentrations, even at the boiling point. Porcelain dishes made of chemical porcelain (about 54% kaolin + 28% feldspar + 18% quartz) are used for evaporation of concentrated nitric acid and aqua regia (see analytical chemistry textbooks).

Porcelain with the composition 70.67% SiO_2 + 23.18% Al_2O_3 + 0.51% Fe_2O_3 + 0.16% TiO_2 + 0.75% CaO + 0.43% MgO + 3.78% alkali metal oxides corrodes in 10% nitric acid at 373 K (100 °C) in 32 h with a weight loss per unit time of 0.0063% per hour. Over longer periods, further weight loss Δm becomes very small and can be calculated as follows:

$$-\Delta m = 0.21 \{1 - \exp(0.07 \times t)\} \text{ with t in h.}$$

The dependence of the weight loss on the concentration is obtained analogously:

$$-\Delta m = C_{HNO_3} \times \exp(0.0075) - 3.33$$

with C_{HNO_3} in % [870].

Because of the resistance of porcelain to nitric acid, it can be used to remove an undesirable black glaze caused by finely divided silicon, by oxidation of the silicon [950].

B 10 Stoneware

In contrast to basalt, terracotta is resistant both to dilute and concentrated nitric acid [1010]. Siliceous stones also exhibit a high resistance to nitric acid.

Coarsely crystalline cast stones of hornblende, a silicate rock which approximately corresponds to the general formula $Ca_2Mg_5(OH)_2Si_8O_{22}$, corrode in 20% nitric acid at room temperature at a rate of 0.0013 mm/a (0.05 mpy) [1012].

The essential elements of a corrosion-resistant pump for handling nitric acid and other acids with a control system for preventing penetration into the motor and the current-carrying components consist of the ceramic types SK-1 and SK-2 (Russ. types) [873].

B 11 Refractory materials

Refractory materials with a high aluminium oxide content are resistant to nitric acid, even at higher temperatures [870].

B 12 Oxide ceramic materials

Pipe sections made of the aluminium oxide Rapal® are a recommended solution for a packing with a high corrosion resistance. No noticeable attack by nitrogen oxides and nitric acid is said to take place at temperatures below 1273 K (1000 °C) [898].

Building stones with a high aluminium oxide content, good strength and adequate corrosion resistance decrease in strength in nitric acid at 363 K (90 °C). The greatest corrosion is said to occur at 363 K (90 °C) in 40% nitric acid [897].

Because the attack by nitric acid on ceramics of aluminium oxide is low, this acid is not suitable as an etching agent for roughening ceramic for the purpose of improving the adhesion of vapor-deposited copper [860].

Uranium dioxide (UO_2) is dissolved by nitric acid [899]. It is therefore possible to dissolve it out of comminuted fuel rods.

Plutonium oxide is dissolved in an aqueous solution of nitric acid and hydrogen fluoride more and more rapidly as the concentration of either acid increases. The dissolved Pu^{4+} causes fluoride ions to be obtained as a type of catalyst. On the other hand, because of the low solubility of the Pu^{4+} ions in the acid solution, the rate of dissolution of the oxide is reduced by foreign metal ions as an impurity [867, 1011].

PbO_2 layers precipitated on tin dioxide (SnO_2) by anodic polarization in 1 mol/l nitric acid + 0.25 mol/l $Pb(NO_3)_2$ at 0.6 V can be redissolved by cathodic polarization [868].

Andesite (silicate rock) and to an even greater extent corundum, inter alia, are also resistant in the mixtures of phosphoric and nitric acid found in the fertilizer industry. While andesite corrodes at 0.8 to 1.0 mm/a (31.5 to 39.4 mpy) between 298 and 383 K (25 and 110 °C), the corresponding values for corundum are 0.01 to 0.06 mm/a (0.39 to 2.36 mpy) [872].

The corrosion current density (anodic dissolution) of uranium dioxide from spent fuel rods reaches a maximum of about 3.5 A/cm^2 at 325 K (52 °C) and 20 V in 6 mol/l nitric acid in the absence of iron salts [580].

Zirconium dioxide with small amounts of the carbide $Zr(O_2C)$ exhibits no corrosion at all after exposure to mixtures of $H_2SO_4 + HNO_3$ (1:4) (d = 1.37; 1 ml) at 373 K (100 °C) for 1 hour [879].

The microhardness of the (001) face of magnesium oxide monocrystals (99.99 %) of 810 Vickers microhardness remained unchanged after exposure to aqueous nitric acid at pH 1 and room temperature for 28 h. If the pH changes, only a relative small change occurs [874, 875].

B 13   Metallo-ceramic materials (carbides, nitrides, borides)

The carbides hitherto chiefly used for cutting tools are distinguished above all by a high compressive strength, rigidity and hardness as well as wear resistance, and also have a high corrosion resistance in the presence of a suitable binder. Titanium carbide with a curable binder, for example, thus exhibits an excellent resistance to nitric acid [865]. Tungsten carbide and tantalum carbide are not so resistant to nitric acid. While tungsten carbide (TiC) is attacked in boiling 65 % nitric acid at 0.6 – 0.7 V, niobium carbide (NbC) remains absolutely resistant [861].

A material consisting of in each case 50 % TiC and the chromium steel Ch25 (with 25 % chromium) corrodes in 10 % nitric acid at 293 and 353 K (20 and 80 °C) at rates of 0.45 and 1.91 mm/a (17.72 and 75.2 mpy). Corresponding values of 0.50 and 0.66 mm/a (19.69 and 25.98 mpy) are found in 20 and 60 % nitric acid at 293 K (20 °C) [863].

According to Table 141, the carbides belonging to the sintered composite materials, such as Cr_3C_2 and WC, with a binder phase such as nickel and cobalt, are attacked to different degrees by concentrated nitric acid at room temperature [876].

The corrosion of a chromium-nickel steel is shown for comparison.

Material	10 % HNO_3	HNO_3, conc.
	Material consumption rate g/m^2 h	
Cr_3C_2 + 12 % Ni	0.83	0.33
WC + 8 % Ni-Cr	0.0042	0.0042
WC + 6 % Co	6.46	0.75
CrNi steel 18 8 (1.4301)	0.15	0.19

Table 141:   Comparison of the material consumption rates of some carbides with metals (hard metals) in HNO_3 at room temperature [876]

Table 142 summarizes the resistance to nitric acid of other hard metals with a high TiC content and two different binder metal contents, based on WC and Cr_3C_2 with cobalt and nickel as binder metals [877, 878].

Material composition %	Material consumption rate g/m^2 h
80 TiC – 5 NbC – 10 Ni – 5 Mo	6.33
60 TiC – 5 NbC – 30 Ni – 5 Mo	2.92
94 WC – 6 Co	0.23
88 Cr_3C_2 – 12 Ni	0.083

Table 142: Material consumption rates of some hard metals in 65 % HNO_3 at 298 K (25 °C), test duration 720 h [877, 878]

A solution of 15 ml nitric acid + 60 ml hydrochloric acid + 15 ml acetic acid + 15 ml water for up to 0.5 min or 25 ml nitric acid + 75 ml hydrochloric acid for up to 5 min is used to identify the β-phase in hard metals of the type 62WC-20TiC-TaC-Co or 62WC-20TiC-NbC-Co respectively [1013].

Tungsten carbide (WC) with 30 % cobalt as the binder metal is not resistant to 65 % nitric acid [126]. Tungsten carbide-cobalt alloys can be rendered resistant to 10 % nitric acid by chromizing treatment [148].

Tantalum carbide (TaC), which is resistant to hydrofluoric acid, is not resistant to oxidizing acid mixtures, such as HNO_3 + HF [884].

Lanthanum carbide powder (LaC_2) is attacked by both dilute and concentrated nitric acid at room temperature [892].

The material HTi-10 (92 % WC + 6 % Co + 2 % (TiC + TaC)) corrodes in 10 % nitric acid at room temperature at a rate of 4.62 g/m^2 h. This could be reduced to 0.94 or 0.32 g/m^2 h respectively by application of a surface coating of NbC or TaC [956].

The carbide materials WC-(5.5-15)Co described as Igetalloy® are severely attacked in boiling 5 % nitric acid. Only Igetalloy® G-1 (WC-5.5Co) corrodes at a material consumption rate of < 0.3 g/m^2 h [1009].

Zirconium carbide and zirconium oxycarbide powder (particle size 15 – 20 μm) is noticeably attacked by nitric acid solutions at 373 K (100 °C) (Table 143) [879].

Solution	$ZrC_{0.94}$	$ZrC_{0.82}O_{0.12}$	$ZrC_{0.77}O_{0.16}$	$ZrC_{0.7}O_{0.24}$
	Residue of powder which remains, %			
HNO_3 (d = 1.37 g/ml)	44	83	88	96
HNO_3, diluted (1:1)	82	90	94	97
HCl + HNO_3 (3:1)	0	0	56	84
H_2SO_4 + HNO_3 (1:4)	0	18	97	97

Table 143: Weight loss of $ZrC_{0.94}$ and ZrC_xO_y powder in HNO_3 containing solutions at 373 K (100 °C) after 1 h [879]

The corrosion and electrochemical properties of hot-pressed Mo_2C powder (with a residual porosity of 2 – 5 %) were investigated potentiostatically in nitric acid between 298 and 358 K (25 and 85 °C) and the gases CO and CO_2 formed were analyzed. The current density (as a measure of the dissolution) of 100 mA/cm^2 which occurs at 1 and 3 mol/l nitric acid appears at 298 K (25 °C) and + 1.2 or 0.65 V_{SHE} respectively. At 358 K (85 °C), the corresponding value is + 0.85 or + 0.7 V_{SHE} respectively, 1 mA/cm^2 corresponding to about 7.6 g/m^2 h [880].

Silicon nitride Si_3N_4 is not attacked by concentrated nitric acid [881]. Fibers of Si_3N_4 are resistant in fuming nitric acid at room temperature [955].

Silicon nitride with 10 % magnesium oxide and 1 to 20 % calcium fluoride (to reduce the friction) is severely attacked in concentrated nitric acid at room temperature, in contrast to pure Si_3N_4, since hydrogen fluoride is liberated from the CaF_2 and attacks the Si_3N_4. Silicon nitride without any addition corrodes with a weight loss per unit time of 0.0126 % per hour [885].

0.1 g silicon oxynitride Si_2ON_2 is dissolved completely in HNO_3 (d = 1.4) + HF (d = 1.15) after 2 h, while in the case of Si_3N_4, 88.4 % of the starting amount was still present.

Table 144 shows the weight loss of hot-pressed specimens of Si_3N_4 + 10 % MgO with and without additions of CaF_2 in concentrated nitric acid at room temperature [883].

CaF_2 addition %	Weight loss %	Dissolution per h %
–	0.033	0.00017
1	0.133	0.00067
5	1.074	0.00537
10	3.112	0.0156
20	4.17*	0.104

* after 40 h

Table 144: Weight loss of hot-pressed specimens of silicon nitride + 10 % MgO with and without additions of CaF_2 in conc. HNO_3 at room temperature, test duration 200 h [883]

Titanium nitride TiN corrodes in aqua regia (HNO_3 + HCl = 1:3) at 353 K (80 °C) at a rate of 10 g/m^2 h, i.e. 25 times faster than titanium at 0.4 g/m^2 h [687].

Table 145 summarizes the corrosion behavior of nitrides of some transition metals in nitric acid at 368, 378 and 393 K (95, 105 and 120 °C) [886]. According to this table, the nitrides of titanium and chromium are the most resistant. However, the resistance of these nitrides in nitric acid shows a high dependence on the temperature.

In spite of the inhibiting effect of sulfamic acid and hydrazine sulfate on the corrosion of titanium nitride powder, these are unusable because they are oxidized by nitric acid in the course of exposure. Phosphoric acid (H_3PO_4) and sodium phosphate are industrially usable inhibitors because of their resistance to nitric acid (Table 146) [887].

Material	Temperature K (°C)	Test duration h	Undissolved residue %	Dissolution per h	
				g	%
TiN	368 (95)	1	50	–	–
Ti	378 (105)	1	99.1	0.0036	0.9
ZrN	368 (95)	1	87.1	0.0516	12.9
	368 (95)	1	91.8	0.0328	8.2
Zr	368 (95)	1	97.5	0.010	2.5
HfN	368 (95)	1	58.9	0.164	41
Hf	368 (95)	1	99.6	0.0016	0.4
VN	368 (95)	1	0	0.400	100
V	393 (120)	1	0	rapid dissolution	
CrN_2	378 (105)	1	99.9	0.0004	0.1
	393 (120)	1	99.2	0.0032	0.8
CrN	378 (105)	4	98.9	0.0044	1.1
	393 (120)	4	80.0	0.080	20
Cr	393 (120)	1	99.5	0.0020	0.5

Table 145: Resistance of the nitrides of some transition metals to pure HNO_3 (d = 1.4), and for comparison also the metal powders. Weight of the material 0.4 g and solution volume 100 ml [886]

Nitrated titanium VT1-0 showed no change in appearance in the presence of 0.03 mol/l Na_3PO_4 in 6 mol/l nitric acid after exposure at 298 K (25 °C) for 480 h. The corresponding gravimetric determination showed a material consumption rate of 0.0024 g/m² h [887].

Abrasion-inhibiting layers of $TiN_{0.95}$ and $TiN_{0.84}O_{0.11}$ are resistant to concentrated nitric acid at room temperature. The layer mentioned last can also be used in hot acid [904].

Duration h	HNO_3 mol/l	Without addition	With addition, mol/l			
			0.01 NH_2SO_3H	0.01 $(N_2H_6)SO_4$	0.01 H_3PO_4	0.001 Na_3PO_4
			Degree of dissolution			
1	6	0.38	0.24	0.04	0.08	0.10
6	6	0.88	0.81	0.30	0.15	0.19
1	1	0.05	0.02	0.00	0.01	–
6	1	0.28	0.04	0.03	0.03	0.05
1	3	0.24	0.02	0.01	0.05	0.04 (0.02)*
6	3	0.22	0.14	0.06	0.12	0.14 (0.07)*

* at 0.06 Na_3PO_4

Table 146: Influence of additions on the extent of dissolution of TiN in HNO_3 at 353 K (80 °C) [887]

The manganese nitrides Mn_3N_2 and Mn_4N are not resistant in hot nitric acid. 0.2 g Mn_3N_2 powder, for example, is dissolved in HNO_3 (d = 1.4) at 393 K (120 °C) within 1 minute, molecular nitrogen being formed. Dissolution of 0.2 g Mn_4N powder (13 g/min) takes place even more rapidly under the same reaction conditions [888].

Indium nitride films (InN) are dissolved in boiling concentrated nitric acid at a material consumption rate of 23.8 g/m^2 h. The rate of dissolution is even greater in a boiling acid mixture of HNO_3 = CH_3COOH (1:1) or HNO_3 + oxalic acid (1:1), at 35 and 38 g/m^2 h respectively [890].

According to Table 147, the nitrides of lanthanum (LaN), cerium (CeN), praseodymium (PrN) and neodymium (NdN) are not resistant to nitric acid [889].

Nitride	Temperature K (°C)	Relative dissolution rate g/min
LaN	298 (25)	0.0041
	393 (120)	0.216
CeN	298 (25)	0.0021
	393 (120)	0.102
PrN	298 (25)	0.0087
	393 (120)	0.333
NdN	298 (25)	0.0287
	393 (120)	0.209

Table 147: Dissolution of some rare earth metal nitrides (0.2 g) in 50 ml HNO_3 (d = 1.4) [889]

While the borides TiB_2, ZrB_2, HfB_2 and VB_2 are destroyed in dilute nitric acid (1:1), those of niobium (NbB_2) and especially tantalum (TaB_2) are resistant [891].

Molybdenum boride Mo_2B_5 corrodes in 1.5 mol/l nitric acid even at room temperature, the material consumption rate of molybdenum being 0.120 and that of boron 0.900 g/m^2 h [893]. The stationary corrosion potential was determined as + 0.23 V_{SHE}.

The average corrosion rate of Mo_2B_5 compact or in the form of a coating material in 1.5 mol/l nitric acid at 298 K (25 °C) and a corrosion potential of 0.28 V_{SHE} is 1.24 and 1.05 mm/a (48.82 and 41.34 mpy) respectively [1041].

The protective action of a passive layer which forms on ternary borides (germanoborides) in 1 mol/l nitric acid decreases in the following order: Nb_5Ge_2B > Nb_5Ge_3B > Ta_5Ge_2B > Ta_5Ge_3B. Quantitative corrosion data cannot be obtained from the current density/potential measurements [894].

Similar circumstances have also been found on the germanoborides $Mo_{1.7}Ge_{0.3}B$, $Co_{21}Ge_2B_6$ and $Ni_{12}Ge_3B_5$ in 1 mol/l nitric acid at room temperature. In the potential range – 0.4 to – 0.5 $V_{Ag/AgCl}$, the corrosion current density is < 10^{-5} A/cm^2 [895].

The cubic hexaborides, such as CaB_6, which belong to the refractory hard substances and have a hardness in the region of that of corundum, are rapidly dissolved by oxidizing acids, such as nitric acid [896].

B 14 Other inorganic materials

The phosphate formed on phosphation of cobalt is converted to the phosphide Co_2P between 770 and 1270 K (497 and 997 °C), the product corroding in 1 mol/l nitric acid at 303 K (30 °C) at a rate of 0.219 g/m² h [900].

Nickel phosphide is severely destroyed in 78 % nitric acid at room temperature within 8 h [959].

The resistance of boron phosphide to hydrofluoric acid and mixtures thereof with nitric acid or aqua regia largely depends on the size of the crystallites. Coarsely crystalline boron phosphide is considerably more resistant [916].

According to Table 148, rhenium silicide and germanides are not resistant in hot nitric acid solutions [902].

Solution	Concentration %	Temperature K (°C)	Re	Si	Ge	$ReGe_2$	Re_3Si	ReSi	$ReSi_2$
			\multicolumn{7}{c}{Undissolved portion, %}						
HNO_3	66	383 (110)	0.0	100	31	< 1	0.0	0.0	50
	33	378 (105)	0.0	100	*	0	0.0	0.0	48
HNO_3 + 3 HCl	66 and 38	383 (110)	0	–	0	–	0.0	0	54
HNO_3 + HF	66 and 48	378 (105)	0	–	–	0	0	0	0

* mostly dissolved

Table 148: Corrosion of Ge, Re, Si and compounds thereof in HNO_3 containing solutions after a duration of exposure of 1 h [902]

The following silicides are resistant in dilute nitric acid between 373 and 388 K (100 and 115 °C): (Ti, Zr)Si_2, Zr_2Si, (V, Nb, Ta)Si_2, Cr_3Si, CrSi, $CrSi_2$, Cr_5Si_3, WSi_2, $MoSi_2$, $MnSi_2$, Mn_3Si, Mn_5Si_3, (Fe, Co, Ni)Si_2, Co_3Si, Fe_3Si and FeSi. Only the silicides Mg_2Si and $ReSi_2$ and also Re_3Si are soluble in nitric acid. All the silicides mentioned dissolve in HF-containing nitric acid [901].

The following germanides are dissolved in nitric acid (in 1:1, 1:4 and 1:10) at 373 K (100 °C) after only 1 h: La_5Ge_3, LaGe, $LaGe_2$, Ce_5Ge_3, CeGe, $CeGe_2$, Pr_5Ge_3, PrGe, $PrGe_2$, Nd_5Ge_3, NdGe and $NdGe_2$ [903].

The phosphate layers on metal surfaces are converted to soluble nitrate compounds by vapors of both 5 and 30 % nitric acid. This process proceeds over a period of months, while complete conversion takes place within only a few seconds in nitric acid solution [905].

While the corrosion of silicon in pure nitric acid is very low, a solution of (amounts by weight) 8 HNO_3 + 1 HF + 1 H_2O has proven suitable for defined removal of p-n layers in silicon in the presence of aluminium contacts for semi-conductor components [912].

Silicon may become discolored in nitric acid/hydrogen fluoride solutions (12:1 to 14:1) alone and in the presence of Cr^{6+}, Fe^{3+} and Cu^{2+} ions as a result of the formation of a suboxide (SiO_x where $x < 2$). $KMnO_4$ containing dilute hydrofluoric acid (2 parts HF + 1 part 0.038 mol/l $KMnO_4$), which causes no discoloration, dissolves the discolored suboxide layer without noticeable attack on the silicon [913]. Other etching processes for silicon are found elsewhere [914, 915].

Treatment in aqua regia (68 % HNO_3 + 32 % HCl) at 473 K (200 °C) and 1.0 MPa in a Teflon®–lined autoclave has proven suitable for freeing metallurgically produced silicon from metallic impurities. Corrosion of the pulverulent silicon can be ignored [995].

The corrosion rate of cadmium telluride CdTe in (parts by volume) 2 HNO_3 (70 %) + 2 HCl (37 %) + 1 water at room temperature (~ 300 K (27 °C)) was 0.75 mm/h [917].

The selenides of scandium, gadolinium and erbium ((γ'-) Sc_2Se_3, (γ-)Gd_2Se_3 and ξ-Er_2Se_3) are not resistant in nitric acid. While the selenides preferentially form NO_2 in concentrated nitric acid at 293 K (20 °C) and thereby dissolve as $Me(NO_3)_3$, only NO is formed at the boiling point, with total dissolution of the selenides. Dissolution takes place considerably more slowly in 0.1 mol/l nitric acid, hydrogen selenide (H_2Se) and NO being formed [985].

C
Organic Materials

C 4 Wood

Wood is destroyed even at room temperature in a mixture of nitric and sulfuric acid. This is used for the analysis of wood, e.g. eucalyptus, for metals (Cu, Cr, As and Zn) [938].

C 6 Furan resins

The laminates based on furan resin (furanoplasts) Quacorr® 1200 FR are also resistant, inter alia, to 5 % nitric acid [920].

The good resistance of moldings based on polydifurfurylidene acetone with 65 % filler material (kaolin, n-bentonite or graphite) to nitric acid can be seen in Table 149 [919].

Plastics based on resol-formaldehyde resin and furfurolacetone monomer in a ratio of 1:3 have a high breaking strength (10 – 15 N/mm^2) and compressive strength (80 N/mm^2) and are resistant to 40 % nitric acid [966].

Furan resin-glass fiber laminates (Quacorr®-RP 100A/RP 104B) retain 80.5 % of their original flexural strength of 109.8 N/mm^2 and 89.4 % of their flexural E modulus of 426 N/mm^2 after exposure to 5 % nitric acid at room temperature for 12 months [1082].

Glass fiber-reinforced furan resins (Rigidon®-4889) are used in contact with nitric acid. They are also said to be resistant, inter alia, to nitric acid vapors [1083].

C 7 Polyolefins and their copolymers

Polyethylene (without more detailed identification) is completely resistant to 10 % nitric acid at room temperature [19].

According to Figure 76, the time to fracture of high-density polyethylene (HDPE) in 30 % nitric acid at 353 K (80 °C) decreases as the stress increases to a considerably greater degree than in water. The running times and resistance factors both of low density polyethylene (LDPE) and high-density polyethylene (HDPE) in the nitric acid creep test at 353 K (80 °C) are summarized in Table 150 [922]. The kink in the water curve in Figure 75 is at 5 N/mm² and 2000 h for HDPE and 1.8 N/mm² and 10 000 h for LDPE.

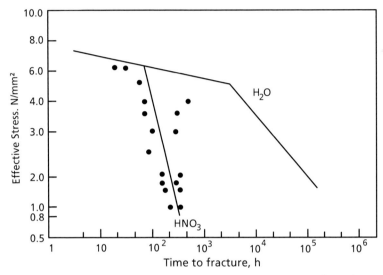

Figure 75: Dependence of the time to fracture on the effective stress of high-density PE (HDPE) in 30 % HNO$_3$ at 353 K (80 °C) [922]

HNO$_3$ %	Test conditions		Change in weight %	Swelling %	Flexural strength N/mm²
	Temperature, K (°C)	Duration, h			
5	293 (20)	1000	0.56	0.24	31.5
10	293 (20)	1000	1.24	0.43	30.0
20	293 (20)	1000	2.12	0.81	29.0
5	373 (100)	24	0.92	0.31	32.0

Table 149: Behavior of moldings based on polydifurfurylidene acetone with 65 % filler material in nitric acid [919]

Polyethylene is inadequately resistant in concentrated nitric acid [947].

Because of their resistance to acid, polyolefins can also be used for heat exchangers by increasing the exchange surface and reducing the wall thickness, and by optimizing the flow rate of the cooling medium and the liquid to be cooled, such as nitric acid [1080].

For pipe specimens of HDPE with a hot-gas and hot plate butt welded joint in accordance with DIN 16 960 or supplement DVS 2207, the resistance factor f_{CR} in 53 % nitric acid at a test stress of 5 to 6.5 N/mm² at 293 and 313 K (20 and 40 °C) is

0.46 and 0.5 respectively. The weld joints are not attacked by nitric acid more severely than the basic material [1079].

Plastic	HNO$_3$ %	Stress N/mm^2	Time to fracture h	Resistance factor	Observations
HDPE	5	2	10000	0.1	–
		1.2	> 20000	–	–
	30	4	100	–	–
		1	100	–	oxidation
LDPE	5	2	2000	–	flat branch of curve
		1.3	2500	0.2	flat branch of curve
	30	2	80	–	flat branch of curve
		1.8	100	–	flat branch of curve
		1.3	150	0.01	–
		0.8	150	0.005	–

Table 150: Running times and resistance factors f_R in the HNO$_3$ creep test at 353 K (80 °C) with HDPE and LDPE under various stresses [922]

With a carbonyl group, C = O content in polyethylene of 0.8×10^{-4} mol/ml, the penetration depth of 58 % nitric acid at 363 K (90 °C) within 20 and 34 h is 195 and 519 μm respectively, which corresponds to a penetration rate of about 9.75 and 15.3 μm/h respectively [921]. If the C = O group content is only 0.1×10^{-4} mol/ml, the penetration rate calculated for HNO$_3$ is 23 μm/h (test duration 20 h).

Table 151 summarizes the permeability of an HDPE film to nitric acid as well as the diffusion rate [923].

HNO$_3$ %	Temp. K (°C)	Test duration h	Diffusion coefficient 10^9, cm^2/s	Permeability coefficient 10^{12}, g/cm s
40	343 (70)	16.5	0.52	10.53
	323 (50)	42.0	0.31	2.78
	293 (20)	380	0.02	0.95
25	343 (70)	31.0	0.27	2.84
	323 (50)	162	0.06	1.31
	293 (20)	620	0.01	0.76
5	323 (50)	195	0.04	0.63
	293 (20)	870	0.008	0.12

Table 151: Diffusion rate and permeability coefficient of HNO$_3$ in an HDPE film of the Russ. grade 102-44 [923]

The promoting influence of gasoline on the diffusion of nitric acid in distilled water at room temperature from a mixture of 10% nitric acid and gasoline in a volume ratio of 4:1 in a polyethylene specimen (90 × 60 × 10) is remarkable (Figure 76) [924]. According to this figure, the passage of HNO_3 through a polyethylene film in the presence of gasoline is so fast that it cannot be used as a protective coating for materials sensitive to HNO_3. Only if gasoline is absent does the passage of HNO_3 through the film take place so slowly that the film provides good protection.

Low-pressure polyethylene (HDPE) shows no change in nitric acid at 353 K (80 °C) after exposure for 350 h, so that it can also be recommended as effective protection against nitric acid under these conditions [925].

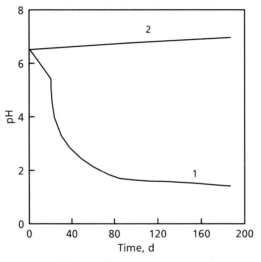

Figure 76: Change in the pH in dist. water with respect to time after immersion of a PE specimen (90 × 60 × 10 mm) in a mixture of 10% HNO_3 ① with and ② without gasoline in a volume ratio of 4:1 at room temperature [924]

HDPE film of the type 102-44, which is used for protecting concrete from corrosion by acid – including HNO_3 – shows a weight increase after exposure to warm 5 and 25% nitric acid at 293 and 323 K (20 and 50 °C) for 12 months of 0.19 and 0.42% and 0.2 and 0.63% respectively [926]. Service lives of 20 to 25 years are given.

Low-density polyethylene (LDPE) containing 30% modified glass wool SV (modified with γ-aminopropyltriethoxysilane (AGM-9) and epoxy resin) and with a tensile strength improved to 32 N/mm² still retains a value of 31.5 N/mm² after exposure to 5% nitric acid at 293 K (20 °C) for 1000 h. The swelling after this treatment is 0.9% [927].

Flamulit® WS PE-25, a high-pressure polyethylene of low-density (LDPE) stabilized to heat and stress cracking, can be recommended as corrosion protection for coating galvanizing frames. It is resistant to 10% nitric acid [932].

The creep strength of pipes of polyethylene (HDPE) and polypropylene is not impaired by a mixture of nitric acid and hydrogen fluoride (4 and 15%) [939, 940].

The resistance of polyethylene and polyisobutylene (PIB) to concentrated nitric acid at 323 K (50 °C) is also of interest in the fertilizer industry where liquid fertilizers are also produced in addition to ammonium phosphate. The change in weight of these plastics after exposure to 40 and 58 % nitric acid at 323 K (50 °C) for 12 h is summarized in Table 152 [986].

HNO_3 concentration %	PE	PIB
	Change in weight, %	
40	− 0.23	+ 0.18
50	− 0.26	+ 0.29
55	− 0.30	+ 0.40
58	− 0.45	+ 0.68

Table 152: Relative change in weight of PE and PIB in nitric acid at 323 K (50 °C) after exposure for 12 h [986]

Solution, %	Chemical resistance			Observations
	293 K (20 °C)	333 K (60 °C)	373 K (100 °C)	
10 HNO_3	A	A	A	–
60 HNO_3	A	D (313 K (40 °C))	–	–
50 HNO_3 + 50 HCl	A	D (353 K (80 °C))	–	crack formation possible under stress
50 HNO_3 + 50 H_2SO_4	C	D	–	crack formation possible under stress

A negligible attack
B slight attack
C severe absorption and/or rapid permeation
D severe attack

Table 153: Behavior of PP in HNO_3 containing solutions [930]

A polyethylene (LDPE) layer modified by grafting on N-vinylpyrrolidone and about 70 μm thick is resistant to boiling concentrated nitric acid. The steel St-3 coated with this material is not attacked by nitric acid [929].

The behavior of polypropylene (PP) in solutions containing nitric acid is shown in Table 153 [930].

In agreement with Table 153, PP is recommended for 10 % nitric acid both at 295 and 369 K (22 and 96 °C) [994]. It is similarly suitable for nitric acid washers because of its low specific gravity and resistance [931].

Tanks resistant to nitric acid at 373 K (100 °C) can still be produced from polypropylene sheets provided with several layers of glass fiber fabric and hardened at 353 K (80 °C) for 3 to 4 h [933, 996].

Polypropylene filters are used for removing nitric acid vapor and mist from waste gases (capacity: 0.09 m³/m² h at a linear speed of 0.24 m/s) [934]. About 98 % of the HNO$_3$ present in the waste gas can be removed. Polypropylene is also resistant to nitrous waste gases (0.5 g/m³) up to 380 K (107 °C) [935].

The action of 65 % nitric acid at 323 K (50 °C) on polypropylene can be determined from its microhardness. Figure 78 shows the microhardness of a polypropylene film as a function of the thickness of the specimen after various exposure times in nitric acid. Above a thickness of 0.5 or 1 mm, the microhardness remains constant as the thickness increases [1033].

Since crystalline regions in the plastic are attacked more slowly by etching solutions than the amorphous regions, they can be rendered visible by etching. Exposure to an etching agent of 100 ml HNO$_3$ + 100 ml chromic/sulfuric acid at 343 K (70 °C) for 120 s is needed for development of the structure in polypropylene [936].

Ethylene/propylene rubber containing carbon black is resistant to 15 % nitric acid [967].

C 8 Polyvinyl chloride and its copolymers

Polyvinyl chloride (PVC), which is also used, inter alia, for coatings, has an excellent resistance to 10 % nitric acid at room temperature (according to customer tests with WS-PVC-22).

It is recommended for 10 % nitric acid both at 295 and 369 K (22 and 96 °C) [19, 994].

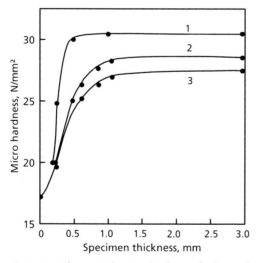

Figure 77: Change in the microhardness of polypropylene as a function of the specimen thickness in 65 % HNO$_3$ at 323 K (50 °C) over various exposure times [1033]

Table 154 summarizes the change in weight and tensile strength of a Russ. PVC in 35 % nitric acid at 293 K (20 °C) as a function of the test duration [941]. While a considerable change in weight of –21.3 % occurs after exposure for 250 h, the tensile strength rises to about 31.5 N/mm².

Test duration h	Change in weight %	Tensile strength N/mm²
10	– 8.5	22
50	–	25
100	– 20.7	28
200	– 21.6	30.5
250	– 21.3	31.5

Table 154: Change in weight and tensile strength of PVC in 35 % HNO_3 at 293 K (20 °C) with the test duration [941]

The resistance under stress of PVC pipes on exposure to nitric acid at 333 K (60 °C) and various loads can be seen in Table 155 [922].

PVC pastes (plastisols) with and without pigments for coloring are used for corrosion protection on galvanizing frames. They are absolutely resistant to 10 % nitric acid at room temperature, but attacked by concentrated nitric acid [942].

HNO_3 concentration %	Load N/mm²	Time to failure h
5	8	1000
	7	4000
30	9	20
	6	100

Table 155: Time to failure of PVC in HNO_3 at 333 K (60 °C) under various loads [922]

PVC adhesive joints, e.g. for gluing concrete, are being used more and more frequently in concrete construction and system building, alongside other plastic adhesives. While resistant to 20 and 50 % nitric acid at 293 K (20 °C), they are attacked in 50 % nitric acid at 313 and 333 K (40 and 60 °C) [943].

Sleeve joints on commercially available pipes made of rigid PVC for which low-dissolving adhesive was used remained tight at 293 K (20 °C) under an internal pressure of 1.0 MPa for more than 1000 h both in 20 and 50 % nitric acid. The shear strength also remained constant (about 23.5 N/mm² in 20 % and 22 N/mm² in 50 % nitric acid) [1076].

The effect of nitric acid on the physico-mechanical values of chlorinated polyvinyl chloride (CPVC) is summarized in Table 156. According to this table, the CPVC tested can be used as a material for industrial reaction solutions for the production of nitrogen fertilizers in the presence of nitric acid [944].

The resistance of a coating based on CPVC ChV-785 in 25 % nitric acid at 293 K (20 °C) could be increased from 167 to about 250 h by addition of surface-active substances, such as 0.4 % ODA (octadecylamine), and priming with ChS-010 (Russ. type) [982].

PVC copolymers, such as Hostaflex® (75 parts vinyl chloride + 25 parts vinyl isobutyl ether), as versatile binders for protecting metal, concrete and plastic during use, proved to be resistant after exposure to 10 to 40 % nitric acid for 3 months. They have only a limited resistance to 50 % nitric acid, however, and are not resistant in 60 % acid [946].

Flexible PVC (containing plasticizer) is completely resistant in 10 % nitric acid at room temperature [19].

After exposure to solutions of 48 parts 96 % H_2SO_4 + 49 parts 53 % HNO_3 + 3 parts water at 293 K (20 °C) for 3 days, stress corrosion cracking occurs on specimens of Trovidur® NL (rigid PVC) under a bending stress of > 30 N/mm². The resistance factor of pipes made of rigid PVC during exposure to 30 % nitric acid at 313 K (40 °C) is 1. This material should not be exposed to 50 % nitric acid in a temperature range above 293 K (20 °C).

C 9 Polyvinyl esters and their copolymers

Polyvinyl ester is recommended quite generally as a material for 10 % nitric acid at 295 to 369 K (22 to 96 °C) [994]. If 338 K (65 °C) is not exceeded, vinyl resin coatings about 500 µm thick for exposure to the atmosphere are also resistant to vapors of 10 % nitric acid. Vapors of concentrated nitric acid attack vinyl resin.

Coatings of polyvinyl ester reinforced with glass spangles are also resistant to dilute nitric acid at room temperature [278].

Levasint®, an ethylene/vinyl acetate copolymer used for corrosion protection as a whirl-sintered coating, is also resistant to 10 % nitric acid at 293 K (20 °C), i.e. the weight increase after exposure for 30 d is still < 0.3 %. At higher acid concentrations and temperatures, however, Levasint® is attacked [945].

Polyvinyl ester (of 78 – 86 % vinyl chloride + 13 – 16 % vinyl acetate; marketed in the USA as Bakelite Vinyl Ester) is resistant in 10 % nitric acid, but attacked in 70 % acid [1077].

C 10 Phenolic resins

Reinforcing asbestos with epoxy and phenol-formaldehyde resin gives a material which, inter alia, is also resistant to 10 % nitric acid between 245 and 440 K (– 28 and 167 °C) [931].

Phenol-formaldehyde resin alone, however, is inadequately resistant to 10 % nitric acid [19, 1084].

According to Table 157, cast phenol-formaldehyde resin filled with 38.6 % synthetic graphite and 3.6 % sulfosalicylic acid as a catalyst for the hardening becomes more and more corrosion-resistant in 10 % nitric acid at 293 K (20 °C) as the heat treatment temperature increases [999]. Only after treatment at about 1000 K

(727 °C) in an inert atmosphere at a heating up rate of 20 K (-253 °C) h^{-1} can the resistance of the material be described as good.

HNO$_3$ %	Tensile strength MPa	Elongation at break %	Softening point K (°C)	Notched impact strength kJ/m^2	Permeability coefficient 10^{12}, g/cm s	Diffusion rate 10^8, cm^2/s
–*	64.4	56	391 (118)	2.5	–	–
10	68.8	39	390 (117)	2.9	6.43	0.33
40	69.5	27	391 (118)	2.9	3.81	0.32

* values in air

Table 156: Dependence of the physico-mechanical values of CPVC on the exposure to HNO$_3$ after 2352 h at 293 K (20 °C) [944]

Heat treatment temperature, K (°C)	293 (20)	373 (100)	473 (200)	573 (300)	673 (400)	773 (500)	973 (700)	1173 (900)
Material consumption rate, g/m^2 h	3.9	1.8	1.1	0.74	0.46	0.35	0.13	0.018

Table 157: Material consumption rates of heat-treated material consisting of phenol-formaldehyde resin + graphite in 10 % HNO$_3$ at 293 K (20 °C) [999]

C 11   Acrylic resins

The time taken for the first change to occur in acrylic resin coatings at room temperature is 6 d in 10 % nitric acid and 2 d in 20 % and 30 % nitric acid [1047].

Weakly acidic cation exchangers based on acrylic acid are relatively resistant to 20 % nitric acid. No precise measurements on the consequences of exposure are known [949].

The acrylic resin coating powder P-AK-1138 (Russ. type) showed no change after exposure to 15 % nitric acid at room temperature for 360 h [965].

Varnish coatings applied by the two-coat method using Glassodur® Acryl-Autolack 21 ((parts by volume) 100 Glassodur® Acryl-Autolack 21 + 50 Glassodur® curing agent 929-28 or 929-29) showed a change in color after 240 h in 65 % nitric acid [964].

C 12   Polyamides

The highly abrasion-resistant Flamulit® WS PA-12, a heatstable polyamide with a melting point of 450 to 457 K (177 to 184 °C), a Shore D hardness of 77 and a resistance to heat/low temperatures from 213 to 353 K (– 60 to 80 °C), is resistant to 10 % nitric acid [932].

PRD®-49, an aromatic polyamide, is used as a synthetic fiber. The fiber is readily hygroscopic, sensitive to the ultraviolet fraction of sunlight, but quite resistant to nitric acid. The change in weight compared with the dry fiber in 65 % or concentrated nitric acid at room temperature is + 0.80 and + 0.60 % respectively [968].

The aromatic polyamides used for the production of aramide fibers of high strength (2500 to 3000 N/mm^2) show a relative residual tensile strength of 29, 23 and 17 % after exposure to 10 % nitric acid for 12, 40 and 120 d at room temperature [969].

The polyimides which are counted as heterocyclic polymers and are used long-term in air up to about 530 K (257 °C) lose about 60 % of their initial strength and 75 % of their elongation after exposure to concentrated nitric acid at room temperature for 120 h [970].

Imilon® S, a polyimide molding compound, is attacked in a manner comparable to above in concentrated nitric acid at 293 K (20 °C) [971].

After exposure to 10 % nitric acid at room temperature for 24 h, 98 % of the adhesive strength of polyimide on metal, such as silver, copper and iron, was retained [972].

Elastomeric aromatic polyimides which contain perfluoroalkylene for the purpose of reducing the glass transition temperature or increasing internal plastification without changing the thermal and chemical stability, are attacked by 35 % nitric acid. The tensile strength of the copolyimide after exposure at 296 K (23 °C) for 170 h was thus 77 % of the starting value. The tensile strength of fluorosilicone rubber under the same test conditions was only 85 % [973].

C 13 Polyacetals

Both the acetal homo- and copolymer are attacked by nitric acid [976].

Oilon® PV-80 consists of an acetal resin, a filler and a dispersed lubricant. It is attacked by 10 % nitric acid [1018].

C 14 Polyesters

Polyesters based either on isophthalic acid or bisphenol are absolutely resistant to 2 and 5 % nitric acid at 302 K (29 °C). The two polyesters have a satisfactory resistance to 30 % nitric acid only with frequent rinsing in water, and are no longer recommended for 56 % nitric acid [974]. Glass fibers (also in the form of glass fabric) can be used as fillers.

The polyester Atlac® 382 and the glass fiber-reinforced grade Atlac® 711-05A are recommended in the presence of nitric acid at pH 1 – 2 [931].

The critical temperatures for using Atlac® 382 are 377 and 353 K (104 and 80 °C) for 2 and 5 % nitric acid and 339 and 295 K (66 and 22 °C) for 15 and 50 % acid [975].

The polyester polybutene terephthalate is attacked by nitric acid even at room temperature [976].

Xydar®, a self-reinforcing aromatic polyester which forms fiber-like polymer chains like wood (nematic structure) is resistant to aggressive media even at elevated temperature. Xydar® thus retains, for example, 98 % of its tensile strength after exposure to 70 % nitric acid for 11 d [977].

The hydrolysis resistance and protective properties of complex polyester urethanes in 20 % nitric acid at room temperature can be seen from Table 158. According to this table, the most resistant is PU – ODDF, which contains oligodiethylene diphenate, while the polyester urethane made up of oligodiethylene adipate is less resistant, especially as regards the extent of sorption. The best resistance to hydrolysis in 20 % nitric acid is shown by PU – ODAF, containing oligodiethylene adipate phthalate, as can be seen from a weight increase of only 1 % after 2 h which remains unchanged after 24 h [979].

The chemical resistance of copolyesters of terephthalate and α-hydroxy acids to 5 % nitric acid at room temperature is shown by the small change in weight of 0.009 to 0.037 % [980].

Polyester urethane	Sorption 10^2, g/cm^3	Diffusion rate 10^{10}, cm^2/s	Permeability coefficient 10^{11}, g/cm s
PU – ODA	5.80	3.57	2.07
PU – ODAF	3.75	1.74	0.65
PU – ODADF	3.60	1.23	0.44
PU – ODDF	1.32	2.00	0.26

Table 158: Sorption and diffusion properties of complex polyester urethanes in 20 % HNO$_3$ as a function of the structure of the oligoester blocks [979]

Polyesters based on isophthalate or bisphenol A are recommended as materials for 10 % nitric acid at 295 K (22 °C), but only polyesters based on bisphenol A are recommended at 369 K (96 °C) [994].

Increasing the strength of polyesters by glass fiber reinforcement has no influence on the resistance in nitric acid in comparison with the non-reinforced polyesters [978].

The breakdown and nitration of the unsaturated polyester resins PN-15 and PN-16 (Russ. types) in 60 % nitric acid at 293 K (20 °C) could be determined from infrared measurements after exposure for up to 1000 h. While nitric acid penetrates into PN-16 down to a depth of 300 μm after 250 h, the penetration depth in PN-15 is only 20 μm. The tear strength also decreases accordingly [981].

The protective action of coatings of unsaturated polyester on metals against nitric acid can be ascertained from the diffusion coefficients determined for nitric acid through the coating. Figure 79 shows the diffusion rate of HNO$_3$ through unsaturated polyester as a function of the temperature and the acid concentration [990].

C 15 Polycarbonates

According to Table 159, test specimens of polycarbonate (PC) kept in 65% nitric acid at 293 K (20 °C) for 672 h show a marked decrease in elongation at break and notched impact strength under a load of 30 N/mm^2 after drying at 313 K (40 °C). The nitric acid which has penetrated into the PC attacks the plastic at higher temperatures. In contrast, PC is resistant to 10% nitric acid under otherwise identical test conditions [983]. Because of its high impact strength and high permeability to light, it can be used for tanks and inspection windows in chemical plants in the presence of dilute nitric acid.

Figure 79 shows the course of penetration of concentrated HNO_3 into the polycarbonate Diflon® at various temperatures as a function of time. According to this graph, the penetration depth of HNO_3 in Diflon® at 303 K (30 °C) after 100 h is about 1.35 mm [984]. The deformation work of Diflon® in nitric acid at 323 K (50 °C) is independent of time within 100 h and is about 3 N/mm^2.

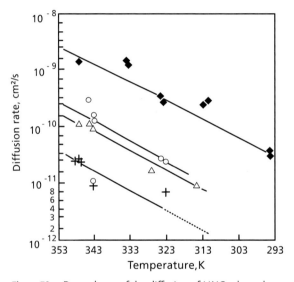

Figure 78: Dependence of the diffusion of HNO_3 through unsaturated polyester without pigment additions on the temperature and concentration [990]
● 5, + 15, △ 30, ○ 40 and ◆ 50% HNO_3

The low-dissolving adhesive made of polycarbonate is resistant to 20 and 50% nitric acid at 293 K (20 °C). At 313 and 333 K (40 and 60 °C), it has only a limited resistance in 20% acid, and in 50% acid is non-resistant [943].

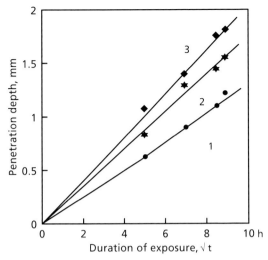

Figure 79: Dependence of the penetration depth ($\Delta\zeta$) of concentrated HNO_3 into Diflon® at ① 303 (30), ② 313 (40) and ③ 323 K (50 °C) on the duration of exposure [984]

HNO_3 %	Change in weight %	Limiting flexural stress %*	Tensile strength %*	Elongation at break %*	Hardness** 358/30 %*	Notched impact strength %*
10	+ 0.8	97 (91)	100 (102)	97 (95)	100 (100)	130 (100)
65	+ 2.9	84 (91)	90 (98)	94 (64)	95 (100)	108 (6)

* based on the starting values
** ball indentation hardness

Table 159: Change in the mechanical properties of PC after exposure to HNO_3 at 293 K (20 °C) for 672 h. The values in parentheses are values after exposure and drying at 313 K (40 °C) [983]

C 16 Polyurethanes

Solvent-free polyurethane raw materials for coatings and seals are resistant to 10 % nitric acid [988]. The good resistance of polyurethanes to nitric acid is mentioned quite generally [987].

Polyurethane used for coating and modified by 5 different curing processes is not recommended as protection against 70 % nitric acid, and is only of limited use even against 10 % acid [989]. These statements are confirmed by the concentration dependence of the diffusion of HNO_3 through polyurethane at 298, 323 and 343 K (25, 50 and 70 °C) shown in Figure 81. According to this figure, the polyurethane coating should protect the carrier material (e.g. metal) only against < 15 % nitric acid at room temperature [990].

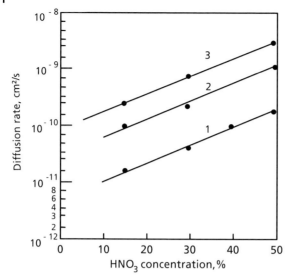

Figure 80: Concentration dependence of the diffusion of HNO_3 through polyurethane at ① 298 (25), ② 323 (50) and ③ 343 K (70 °C) [990]

Aluminium is protected against 10 % nitric acid, even at higher temperatures, by a polyurethane primer coating about 100 µm thick and a 400 µm thick coating of pitch polyurethane on top. It suffers massive attack without this protective coating [991].

Weathering-resistant polyurethane varnish based on a mixture of aliphatic and aromatic polyisocyanates with and without pigments of TiO_2 is also resistant to 10 % nitric acid at room temperature [1019].

Glycidyl urethane oligomers as plastic coatings are frost-resistant and have high chemical resistance in 5 and 10 % nitric acid at 313 K (40 °C) [1020].

The values of the diffusion coefficient of nitric acid in a coating of polyurethane with and without additions of bis-epoxypropylbenzimidazole in 20 % nitric acid at room temperature are summarized in Table 160. According to this table, the diffusion rate and therefore also the corrosion rate of the material to be protected is increased drastically by addition of the epoxide compound [993].

Addition of bis-epoxypropylbenzimidazole %	20 % HNO_3	H_2O
	Diffusion rate 10^{12}, cm^2/s	
–	0.04	0.078
5	67.0	0.027
10	52.8	0.025
15	39.1	0.023
25	59.0	0.209

Table 160: Diffusion rate of HNO_3 and, for comparison, H_2O through polyurethane with and without additions of the epoxide compound at room temperature [993]

Polyurethanes as polyaddition products of β-hydroxyethyl esters of piperazine-N,N-thiolthionic acid and isocyanates are resistant to dilute nitric acid at room temperature, but severely attacked by concentrated acid, especially at higher temperatures [1093].

The mechanical properties of polysiloxane urethanes (PSU) and polyurethanes (PU) after exposure to 5 % nitric acid can be seen in Table 161. According to this table, compounds I and II are more resistant than III and IV ((molar ratio) PSU I = 1 polyethylene glycol adipate: 2 tolylene diisocyanate: 1.1 $HO(CH_2)_3Si(CH_3)_2O(CH_3)_2$-$Si(CH_2)_3OH$, PSU II = 1:2:1.05, PU III = 1 polyethylene glycol adipate:2 tolylene diisocyanate:1.1 butanediol and PU IV = 1:2:1.05) [1100].

Resin	Starting value	After 312 h	Starting value	After 312 h
	Tear strength MPa		Elongation at break %	
PSU I	28.3	20.0	310	41
PSU II	31.2	22.0	330	40
PU III	30.5	2.0	320	20
PU IV	29.0	2.0	300	0

Table 161: Influence of HNO_3 on the mechanical properties of PSU and PU after exposure to 5 % HNO_3 at room temperature for 312 h [1100]

A mixture of polyurethane and concrete has also proven itself against 10 % nitric acid as chemically resistant floors at room temperature. 70 % nitric acid attacks urethane concrete [833].

Polyurethane has an excellent resistance to 5 % nitric acid, and a satisfactory resistance to 30 % acid only if it is washed with water. It is unsuitable at higher acid concentrations [974].

C 18 Epoxy resins

Epoxy resin is recommended in the presence of 10 % nitric acid at room temperature, but classified as unusable at 340 K (67 °C) [994].

The time which elapses before an initial change takes place in an epoxy resin coating at room temperature is 200 d in 10 % nitric acid and 6 and 30 d in 20 and 30 % acid respectively [1047].

Grilonit® G 16.05, a liquid, non-modified epoxy resin which, in combination with suitable curing agents (polyamines), combines the excellent properties of epoxy resins based on bisphenol A and epichlorohydrin with a low viscosity and absence of crystallization, is unchanged, apart from a slight discoloration, after exposure to 10 % nitric acid at 293 K (20 °C) for 3 months [1022].

Filler-containing epoxy resin consisting of (parts by weight) 200 parts Araldite® CT-200 + 50 parts curing agent HY-923 + 235 parts silica powder can be recommended in the presence of 5 % nitric acid. However, it is not resistant to 20 % acid [1023].

At a temperature of 353 K (80 °C), the curing time of the epoxy resin oligomer is shortened by half and the resistance to 10 % nitric acid is greatly improved [1024].

The epoxy resin EP-793 (Russ. type; from a low-viscosity epoxy composition) was destroyed after exposure to 20 % nitric acid at 293 K (20 °C) for 2.5 months. In nitric acid vapors above 5 to 25 % HNO_3 solutions, the resin specimen remained unchanged at 293 K (20 °C) for a maximum of 8 months [1025].

Filler-containing epoxy resin coatings (Asodur® HCB) 800 to 850 μm thick suffer a loss in gloss, without blistering, after exposure to 20 % nitric acid at room temperature for 3 months [1026].

According to Figure 81, the creep elongation of the epoxy resin ED-6 under a given load of 40 N/mm^2 in nitric acid decreases as the concentration and duration of exposure increase. The strain increases rapidly with respect to time in 30 % nitric acid after about 800 h [1028]. When $\sigma/\sigma\sigma_0$ reaches ≈ 0.33, creep is no longer observed in the epoxy resin. The life is thus shortened considerably under the simultaneous action of tensile stress and a corrosive acid.

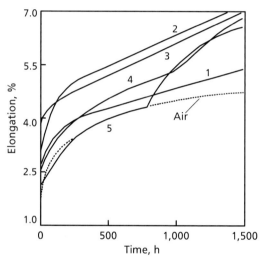

Figure 81: Creep of epoxy resin specimens under σ = 40 N/mm^2 in aqueous HNO_3 solutions of increasing concentration [1028]
① 3, ② 5, ③ 7, ④ 15 and ⑤ 30 % HNO_3

The creep capacity of an epoxy graft polymer (ED-6 as the base), expressed in ∈ (%), changes only from 2.73 to 2.38 % after exposure to 3 to 30 % nitric acid under a load of 40 N/mm^2 for 1500 h. The comparable value for the specimen after exposure to water or air is 2.66 and 2.34 % respectively [1035].

Epoxy resins based on arenephenol-formaldehyde (polyphenols with aromatic hydrocarbon radicals without hydroxyl groups) also show resistance to dilute nitric acid, as well as improved mechanical properties. The maximum swelling of the resin in 5 and 10 % acid is, for example, 1.5 to 2 %. After exposure to 20 and 30 %

nitric acid for 90 d, the swelling is 8.3 and 8.9 % respectively. The resin is destroyed in 40 % acid [992].

Epoxide oligomers filled with residue tar from coke production ((parts by weight) 10 to 70 parts tar per 100 parts resin) are said to be still resistant in 25 % nitric acid [1029].

In contrast, according to another paper, epoxy resin containing coal tar is attacked after 140 h in 10 % nitric acid at room temperature [991].

The resin mixture of epoxy resin EP-00-10 + fluorocarbon epoxy resin varnish LFE-32 + fluorocarbon resin F-32 L used in radiation protection withstands exposure to 12 mol/l nitric acid at 293 K (20 °C) for 1 month [1021].

Asbestos-reinforced epoxy resin and phenol-formaldehyde resin exhibits a high corrosion resistance to 10 % nitric acid [931].

A solvent-free composition of 73.5 % of the low molecular weight epoxy resin ED-5 + 15.4 % diglycidyl ether (DGE) + 11.1 % polyethylenepolyamide PO-300 with curing of 70 % exhibits a resistance of > 140 d in 10 % nitric acid [1034]. If the composition of the resin is changed, with the same components, selective corrosive attack occurs after only 20 d.

The life of polyamine-cured epoxy resins in nitric acid at 328 K (55 °C) ($L = A\,c^{-m}$, where L is the life, c the concentration and A and m are constants) decreases from about 100 to 2 h, i.e. by a factor of 50, in the concentration range 5 to 50 %. The diffusion coefficients follow the same dependence on the concentration of nitric acid [1030].

The epoxide-ester resins obtained by reaction of epoxides with fatty acids and polyhydric alcohols are attacked even in 10 % nitric acid [1031].

A material based on epoxy resin with inert fillers protects the concrete floor in an electroplating plant even from contamination (e.g. splashes) with 70 % nitric acid [1032].

The protective coatings based on epoxide-novolak block copolymers, which are quite resistant to mineral acids, have only a limited resistance after a test duration of 720 h in 40 % nitric acid at 293 K (20 °C), with swelling of 5.3 %. In spite of swelling and blistering, however, no corrosion was found on the metal underneath the coating. Nevertheless, at 373 K (100 °C) the coating flaked off, with destruction of the resin and corrosion of the underlying metal [1036].

The resistance of the protective coating of the epoxide-novolak block copolymer (ENBS) with a low drying temperature in 10 % nitric acid at room temperature can be seen in Table 162 [1037].

To evaluate the protection of these coatings against attack by nitric acid on the underlying metal, the diffusion of the acid into the layer is also determined from the swellability. According to this method, the extent of swelling of the coating in 15 % nitric acid at 293 K (20 °C) reaches a stationary value of 0.36 % after about 35 d. The corresponding value for water is 0.33 %. These block copolymers are accordingly more resistant to nitric acid than the normal epoxy resins [1038].

Epoxy resin containing 80 % 1,7-bis(α-furyl)-1,3,6-heptatrien-5-one showed a weight increase of 2.2 % after exposure to nitric acid at 313 K (40 °C) for 2400 h. A coating of this resin could not prevent corrosion on the underlying metal [1088].

Coatings of epoxy-furfurol resin EF on concrete surfaces have proven suitable corrosion protection against 10% nitric acid. The impact strength, for example, of the coating after exposure to nitric acid for one year is virtually unchanged. The change in weight is only 1.5% and the swelling 2.2%. According to Figure 82, adhesion of the resin to the concrete surface decreases relatively little after 12 months, in contrast to the epoxy resin EATK based on ED-16 with a readily volatile solvent (acetone, toluene or xylene) and the plasticizing agent dibutyl phthalate [1040].

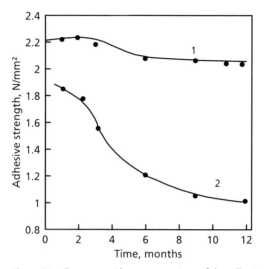

Figure 82: Decrease with respect to time of the adhesive strength of ① the resin EF and ② EATK on concrete in 10% HNO_3 at 293 K (20 °C) [1040]

The latex Ebolat® based on the epoxy resin EP-00-10, vulcanized at 423 K (150 °C) for 2 h, can also be used as a protective coating for carbon steels after exposure to 10% nitric acid at room temperature for 1000 h. It is attacked in the presence of 30% acid [1039].

In contrast to polyamide-containing epoxy resins, amine-containing epoxy resins are excellently suited as flooring in the presence of 5% nitric acid. If frequent washing of the floor can be guaranteed, the amine-containing resin can also be used when the floor is contaminated with 55% acid.

Addition, %	Swelling, %	Adhesion*	Appearance of coating	Resistance
1 HMTA	0.17	satisfactory	loss in gloss and crack formation	limited
1 TEA	0.15	good	somewhat darkened	resistant

* adhesion by net-like notches in the layer

Table 162: Behavior of the protective coating of the block copolymer ENBS with additions of HMTA (hexamethylenetetraamine) and TEA (triethanolamine) in 10% HNO_3 at room temperature, test duration 500 h [1037]

No noticeable corrosion, in addition to NO_2 formation, is observed as a result of the action of γ-rays on exposure of epoxy resin to dilute nitric acid [1042].

The resistance of epoxy rubber with 19 percent by volume acid or alkaline filler, such as diabase and fine quartz sand, in 10 % nitric acid at 293 and 313 K (20 and 40 °C) as a function of the duration of exposure is shown in Table 163. R_B/R_o gives the ratio of the breaking strength under static bending of the specimen in nitric acid compared with that of an unstressed control specimen [1043].

Temperature K (°C)	Duration of exposure, d		
	7	14	21
	R_B/R_o		
293 (20)	1.25	1.1	1.0
313 (40)	0.99	0.93	0.85

Table 163: Resistance of filler-containing epoxy rubber in 10 % HNO_3 as a function of the duration of exposure [1043]

Films based on epoxy resins (ED-20, E-40, E-41, E-49, SNS-epoxide-2000) cured with polyethylene polyamide are attacked by nitric acid on an iron carrier more severely than in the free state without a carrier. In 15 % nitric acid at 328 K (55 °C), the diffusion coefficient is 0.21×10^{-9} cm^2/s and the permeability coefficient 0.52×10^{-7} g/cm s [1044].

C 19 Fluorocarbon resins

The fluorocarbon resin polytetrafluoroethylene (PTFE) (Hostaflon®, Teflon® etc.) is absolutely resistant in 10 and 50 % as well as concentrated nitric acid in the temperature range 293 to 423 K (20 to 150 °C). The same behavior applies to tetrafluoroethylene/hexafluoropropylene FEP (Teflon® FEP) and polytetrafluoroethylene/perfluoroalkyl vinyl ether PFA/TFA (Teflon® PFA, Hostaflon® TFA) [1045].

According to Table 164, the following fluorocarbon resins polyethylene/polytetrafluoroethylene E/TFE (Tefzel®, Hostaflon® ET), polytrifluorochloroethylene PCTFE (Voltalef®, Kel-F), polyvinylidene fluoride PVDF (Kynar®, Solef®, Foraflon®, Dyflor®), trifluorochloroethylene/ethylene copolymer ECTFE (Halar®) and polyvinyl fluoride PVF (Tedlar®) also show a good corrosion resistance to nitric acid [1045].

In agreement with the data in Table 164, all the fluorocarbon resins show only a small weight increase of about 0.1 % after exposure to 10 % nitric acid at 343 K (70 °C) for 12 months [1055].

The Teflon® distillation column (height 3 m, external diameter 35 cm with a wall thickness of 3 cm) in a plant for removing nitric acid from a mixture of $HNO_3/HF/H_2SO_4$ showed a service life several times longer than that of high-quality nickel alloys, which lowers costs [1058].

Electrically insulating varnishes and layers based on PTFE provide the required protection against nitric acid even at higher concentrations and temperatures [1056].

Since PTFE is a thermoplastic which cannot be injection molded, in contrast to the other fluorocarbon resins, its use is limited.

An immersed circulation pump for conveying concentrated nitric acid solution at temperatures up to 353 K (80 °C) is made of PTFE components. The shaft is also coated with PTFE, rendering it corrosion-resistant, and has excellent running properties [1050].

HNO_3, %	Temperature, K (°C)	PTFE	PFA/TFA	FEP	E/TFE	PVDF	ECTFE	PCTFE
10	293 (20)	1	1	1	1	1	1	1
	333 (60)	1	1	1	1	1	1	1
	373 (100)	1	1	1	1	–	1	1
	423 (150)	1	1	1	1	–	–	–
50	293 (20)	1	1	1	1	1	1	1
	333 (60)	1	1	1	1	2	1	1
	373 (100)	1	1	1	2	–	2	1
	423	1	1	1	–	–	2	1
Concentrated	293	1	1	1	1	2	1	1
	333	1	1	1	1	2	1	1
	373	1	1	1	3	–	2	1
	423	1	1	1	–	–	2	2

1 resistant
2 limited resistance
3 not resistant

Table 164: Chemical resistance of fluorocarbon resins in nitric acid [1045]

A porous filter body made of PTFE obtained by pressing PTFE powder together with NaCl crystals and then washing out the sodium chloride is used to remove impurities from nitric acid [1096].

FEP, which, according to Table 164, is absolutely resistant to nitric acid, also withstands exposure to heat and can easily be processed as a thermoplastic. This also applies to PVDF. Unfortunately, a temperature of 295 K (22 °C) should not be exceeded in 50 % nitric acid with this fluorocarbon resin [1051].

PVDF can be recommended as a coating material 1 mm thick in up to 10 % nitric acid below 330 K (57 °C) for galvanizing frames [932, 994].

According to Figure 83, stabilization occurs after the action of 65 % nitric acid at 343 K (70 °C) on PVDF after 14 and 28 d, i.e. no further change in weight, tensile strength or elongation at break occurs. In spite of the initial weight increase of 1.5 %, no damage to PVDF is observed [1053].

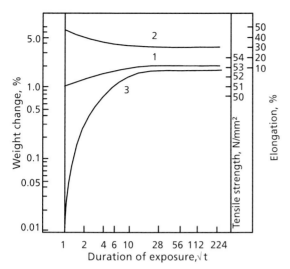

Figure 83: Change with time of ① the tensile strength, ② the elongation and ③ the weight of PVDF under the action of 65 % HNO_3 at 343 K (70 °C) [1053]

Pore-free coatings of ECTFE protect steel tanks from attack by pickling solutions of nitric acid and hydrogen fluoride at 333 K (60 °C). Because of the good lubricating properties coupled with a high corrosion resistance, valves with a fluorocarbon resin coating are also proposed for these baths [1052].

The fluorocarbon resin Kynar® (PVDF) has proven to be particularly suitable for increasing safety in plutonium-processing plants. Coating with Kynar®, which has an average life of about 2 years, can be recommended for the dissolution reactor with a mixture of nitric acid and hydrogen fluoride between 353 and 403 K (80 and 130 °C) under radiolytic radiation. A special long-life flourel mixture has been developed for such solutions. PVDF linings in pipelines of such plants have proven themselves over 10 years [1057].

After 7000 h, pipes made of Solef® (PVDF) are resistant in hot 85 % nitric acid at 358 K (85 °C) under a test pressure of 12.5 MPa. In the plant, these pipes have been operating trouble-free for 7 years with a mixture of one third each of 96 % sulfuric acid + 63 % nitric acid + water under 0.4 MPa at an average operating temperature of 358 K (85 °C) [1059].

Tefzel® or Hostaflon® ET (E/TFE) are resistant to fuming nitric acid at room temperature and therefore used for caps for closing vessels [1060].

Pump components which come into contact with nitric acid or mixtures of 50 % nitric acid and 50 % sulfuric acid between 294 and 408 K (21 and 135 °C) can be made of Kynar® [1061, 1078]. However, since it is very expensive, it can be used only where cheaper fluorocarbon resins fail.

According to Table 165, cold-cured fluorocarbon resins based on polyfluoroperchlorovinyl SP-FP-7 and fluoroepoxy resin SP-FE-8 are noticeably attacked by nitric acid only above 320 K (47 °C) under simultaneous irradiation [1062].

HNO₃ concentration %	Temperature K (°C)	SP-FP-7	SP-FE-8
		Maximum operating life months	
5	293 (20)	42	48
17	293 (20)	42	48
	313 (40)	8	24
	333 (60)	3	–
56	293 (20)	36	30

Table 165:   Maximum operating life of fluorocarbon resins in HNO_3 [1062]

Iron beneath a fluoron-3 coating (Russ. type) 1 μm thick corrodes in an HNO_3 containing atmosphere at a material consumption rate of 0.017 g/m² h [1104].

The perfluoropolyether Fomblin® Y-N VAC 140/13, a resistant lubricant in high vacuum technology, remains unchanged in 60% nitric acid at 343 K (70 °C), in contrast to silicone and petroleum-based lubricants [1063]. It is also usable at 473 K (200 °C) for up to 48 h.

Teflon®, Hostaflon® and Kynar® are corrosion-resistant in the entire concentration range of the system $HNO_3/H_2SO_4/H_2O$ up to 390 K (117 °C) [830].

Of the sulfocationic membranes based on copolymers of vinylidene fluoride/hexafluoropropylene with grafted-on styrene (MPFS-26) and α, β, β'-trifluorostyrene (MRF-26), and a copolymer of tetrafluoroethylene/hexafluoropropylene with grafted-on α, β, β-trifluorostyrene (MRF-4MB), the last two show the lowest change in specific electrical resistance (> 500 h up to doubling) as an indirect measure of the chemical resistance in 100% nitric acid at 293 K (20 °C) and in 64% acid at 373 K (100 °C) [1064]. Attack by nitric acid is accompanied by evolution of CO_2.

Coatings based on perfluoroalkoxyalkane (PFA) are also said to have an excellent resistance to nitric acid [1066].

Because of their suitability for injection molding, PFA plastics are appropriate materials for construction elements with a complicated structure in the presence of nitric acid, e.g. for pumps, etching trays, membrane valves and also for lining [1048].

Polyvinylidene fluoride PVDF (Foraflon®) is resistant to dilute nitric acid, but attacked by concentrated acid (above 50%) [1098].

C 22   Silicones

Silicones with a tensile strength at room temperature and 395 K (122 °C) of 2.7 to 10 and 5.8 MPa respectively have a satisfactory resistance to dilute nitric acid. After exposure to 7% nitric acid at 293 K (20 °C) for 168 h, the change in weight and volume was < 1%. Comparable values in concentrated nitric acid were + 10% [1065]. Fluorosilicone rubber showed a change in volume of +5% in 70% nitric acid at 294 K (21 °C) after 168 h.

Organosilicon elastomers are resistant in 7% nitric acid at room temperature [1086]. Polysilicones of the type VN-30 (Russ.) are not attacked by 1% aqueous nitric acid at room temperature [1087]. 130 μm thick coatings are used for protection purposes.

C 23 Other plastics

Styrene/acrylonitrile copolymer, which can be printed on with sharp contours, suffers a weight increase of 0.37 and 0.29% respectively in 10 and 20% nitric acid at room temperature after 240 h. In contrast, attack in 66% acid causes a weight change of about 12% combined with a considerable change in color [1067].

No change in shape and color was caused to a coating of modified polystyrene exposed to 10% nitric acid for 3 months at room temperature [1068].

Polystyrene from the rectification residues of styrene shows no change in 20% nitric acid at room temperature after 48 h in the absence and presence of UV radiation [1069].

After exposure of 75/25 and 50/50 polyethylene/butyl rubber of high and low density to nitric acid at room temperature for 96 h, no change in weight, tested in accordance with ASTM D-543, was found [1070]. The plastic can be easily pigmented by carbon black and processed.

Pentaplast® (poly-3,3-bis(chloromethyl)oxacyclobutane) is also resistant to dilute nitric acid (5%). Monitoring of the change in density in the wavelength range 1720 to 1570 cm^{-1} is suitable for predicting the life of Pentaplast® in concentrated nitric acid. Figure 84 shows the course with time of the penetration depth of nitric acid and the tensile strength of Pentaplast® in 5 to 60% nitric acid at 353 K (80°C) [1072].

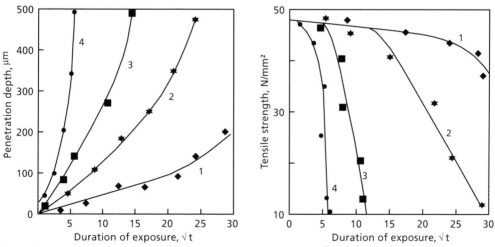

Figure 84: Dependence of the penetration depth of HNO$_3$ in Pentaplast® and the tensile strength on the duration of exposure in h and the HNO$_3$ concentration at 353 K (80°C) [1072]
① 5, ② 15, ③ 30, ④ 60% HNO$_3$

The behavior of Pentaplast® in HNO₃ containing solutions can be seen from Table 166 [1071].

Composition of solution	Temperature K (°C)	Time h
15 % HCl + 30 g/l HNO₃	293 – 295 (20 – 22)	1200
40 % HNO₃ + 5 g/l HF	293 – 295 (20 – 22)	1464
500 g/l HNO₃ + 10 g/l HF	293 – 295 (20 – 22)	5000
500 g/l HNO₃ + 100 g/l HF	293 – 295 (20 – 22)	5000
10 mol/l HNO₃ + 2 mol/l NaF	293 – 295 (20 – 22)	624
600 g/l HNO₃ + 200 g/l HF	333 – 338 (60 – 65)	144

Table 166: Duration of exposure in HNO₃ containing solutions during which no change in the appearance of Pentaplast® occurs [1071]

In agreement with this table, Pentaplast® PL-2 (Russ. type) is said to have a high resistance in solutions containing 10 % HNO₃ [1103].

The swelling of a coating of 1,2-polybutadiene in 10 % nitric acid of 1.2 % achieved after exposure at room temperature for 10 d also remained unchanged on the 15th day. Nevertheless, the plastic is said to be unsuitable as a protective coating for nitric acid without a special primer [1089].

While polystyrenesulfonic acid suffers scarcely any attack in 20 % nitric acid at room temperature even over a relatively long period of time, it is significantly attacked in 40 % acid. If the temperature is increased to 353 K (80 °C), noticeable attack occurs already in 12 % acid [1090].

During processing of nuclear fuels, in addition to accelerated attack, some explosions also occurred on cation exchangers based on strongly basic polystyrene resins in solutions containing 7 to 9 mol/l nitric acid in the presence of radiation and catalytic ions. Because of this it is not possible to use these ion exchangers [1090].

Polysulfone, which is used for components in milk-measuring machines, is also resistant to dilute nitric acid at 353 K (80 °C), which occurs, for example, during the cleaning process [1092].

Ethylene polysulfide is already destroyed by dilute nitric acid at room temperature [1093].

Protective coatings of epoxy coumarone indole resin containing (parts by weight) 50 – 70 parts coumarone indole resin + 30 parts epoxy resin ED-16 + 15 parts chloroparaffin ChP-470 or + 5 – 6 parts dibutyl phthalate are not resistant in concentrated nitric acid [1094].

Polychloral, which does not melt at higher temperatures, but decomposes into gaseous products, is resistant to nitric acid, even when fuming [1095].

According to Table 167, cyclorubber has only a limited resistance to nitric acid at room temperature. It is severely attacked at higher concentrations and temperatures [1101].

HNO₃ %	Temperature K (°C)	Time to fracture of the layer
10	313 (40)	2 months
	333 (60)	5 d
	353 (80)	15 h
	373 (100)	2 h
30	293 (20)	4 months
	313 (40)	8 d
	333 (60)	10 h
	343 (70)	2 h
56	293 (20)	2 d
	373 (100)	1 h

Table 167: Dependence of the life of coatings of cyclorubber in nitric acid on the concentration and temperature [1101]

Because of the good resistance of soft rubber coatings based on butyl rubber to dilute nitric acid, they are suitable as protective coatings of 1.5 to 2 mm thickness [932, 998].

Under certain conditions, nitrile rubber can also be used in the presence of up to 20% nitric acid. However, the manufacturer should be consulted before its use [1102].

A soft rubber lining based on Baypren® is superior in its mechanical (tensile strength, impact resilience) and corrosion properties to linings with hard and natural rubber as well as styrene/butadiene rubber vulcanizates. After 56 d, the change in weight in 30% nitric acid at room temperature was between 2 and 5%. However, Baypren® is not resistant at higher temperatures [1105].

D
Materials with special properties

D 1 Coatings and linings

Brass tubular conductors silvered on the inside, in which stresses occur during soldering, may be attacked by HNO_3 containing air, formed as a result of electrical discharges, with stress corrosion cracking occurring [3]. The silver layer can be protected by rhodium.

Coatings of nickel-phosphorus alloys with 14% phosphorus are attacked after only 8 h in 78% nitric acid at room temperature [633]. In 1 mol/l nitric acid at room temperature, the material consumption rate is < 0.1 g/m² h [663].

By applying coatings of titanium carbide and zirconium carbide or zirconium carbonitride to steel, as well as improving the wear properties, the corrosion resistance in 4.6% nitric acid is also increased [866]. Mechanically resistant coatings of $TiN_{0.58}O_{0.39}$ are not noticeably attacked by hot and cold nitric acid [904].

According to Table 17, diffusion layers of chromium and chromium-titanium on carbon steel greatly reduce the material consumption rate [140, 153, 156, 1106].

Hard gold layers are resistant to nitric acid vapors, but attacked by aqua regia [83].

Thin platinum coatings can be removed from steel and nickel, in spite of attack to the carrier material, if they are exposed to the following solution: (percent by volume) 40% HNO_3 (d = 1.33 g/ml) + 25% HCl (d = 1.19 g/ml) + remainder water. Since the solution has a low storage life, it must be freshly prepared each time [115].

The material consumption rate of carbon steel could be reduced from 72 to 0.57 and 0.8 g/m² h by diffusion coatings with CrAlTi and CrAl respectively (prepared in 4 h at 1373 K (1100 °C) with a mixture of 40% Al_2O_3 + 60% (45% Al + 55% (70% Cr_2O_3 + 30% TiO_2)) and from 30% Al_2O_3 + 65% (70% Cr_2O_3 + 30% Al) + 5% AlF_3 [140].

Nickel coatings on machinery and apparatus components are attacked by nitric acid [567].

A hot solution, at 333 – 344 K (60 – 71 °C), of (percent by volume) 43 – 48% HNO_3 + 7 – 12% HCl + balance water with additions of 0.008 – 0.025 mol/l $FeCl_3$ and > 0.016 mol/l $CuSO_4$ is used to remove nickel aluminide coatings on nickel-based superalloys without noticeably attacking them [1108].

Heavy plate with a titanium deposit as explosion cladding and roll-bonded cladding corrodes in boiling concentrated nitric acid at a corrosion rate of 0.045 and

0.032 mm/a (1.77 and 1.26 mpy) respectively [699]. Amorphous TiNi films generated by ionic beams (by alternating irradiation with titanium and nickel under MeV) are more corrosion-resistant than normal TiNi coatings [749, 804].

It was possible for destruction of the carbon steel St-3 which takes place at room temperature in 50 % nitric acid to be prevented by a titanium diffusion coating about 0.25 mm thick. A corrosion rate of only 0.01 mm/a (0.39 mpy) was found after 450 h [692].

During processing of nitric acid and in the production of fertilizers, as well as solid wall constructions, titanium claddings and linings of steel tanks are also used [716].

As well as improving the wear resistance of copper and bronze by diffusion coating with aluminium and chromium (Al at 1223 K (950 °C) and Cr at 1173 – 1303 K (900 – 1030 °C)), the resistance to concentrated nitric acid is also increased drastically [775].

A solid solution zone based on chromium is formed on the surface of the bronzes Br-Ch08 and SP 19 by alitochroming, the material consumption rate in concentrated nitric acid being reduced from 7650 and 4500 to 75 and 15 g/m^2 h [559].

According to Table 168, chromium diffusion coatings of varying chromium content applied to the low-alloy steels 08 KP and 09 G2 (Russ. types) result in a corrosion resistance comparable to that of chromium-nickel steel [156].

An apparatus made of aluminium alloy clad with chromium-nickel steel is proposed for concentrating nitric acid by absorption of NO_2 in dilute acid [1109].

Solution %	Cr-content of layer, %			Ch18N10T
	20 – 22	30 – 32	38 – 41	
	Corrosion rate, mm/a (mpy)			
2 – 5 HNO_3	0.00082 (0.03)	0.00025 (0.01)	0.00029 (0.01)	0.0004 (0.02)
30 – 40 HNO_3	0.0017 (0.07)	0.0009 (0.04)	0.00065 (0.03)	0.0008 (0.03)

Table 168: Corrosion rates of low-alloy steels provided with Cr diffusion coatings in HNO_3 solutions at 303 – 313 K (30 – 40 °C), test duration 2100 h [156]

The material 1.4571 which is used for the pump housing with lid and impeller and which comes into contact with HNO_3 containing solutions with iron sludge is severely attacked. The service life can be increased eight to ten-fold by thermal spraying with the alloy $W_2C/Ni/Cr/B/Si$ and subsequent compaction [1110].

Under no circumstances was it possible to reduce the material consumption rate of the steel St-50 in 50 % nitric acid from about 1 g/m^2 h to < 0.2 g/m^2 h by alitizing under various conditions after 172 h tests [1107].

An amorphous film of Mo-50Ni produced by the mixed ion beam technique is resistant to 70 % nitric acid at room temperature, in contrast to the polycrystalline film [1108].

For processing fuel rods of zirconium-clad uranium, these are dissolved in a solution of 10 mol/l nitric acid + 4 mol/l hydrogen fluoride between 340 and 360 K (67 and 87 °C) [555].

To detach a nickel layer from chromium-nickel steels without noticeable corrosion of the latter, the layer is immersed in triethanolamine-containing nitric acid with 5 g/l bromide at 333 K (60 °C) for 15 min [114].

For pickling nickel layers on stainless steel, a pickling solution which does not attack the steel of 500 ml nitric acid + 5.7 g/l copper sulfate + 3.3 g/l sodium chloride + 3.0 g/l iron chloride with and without small additions of Na_2TeO_3 is used [572].

To remove a lead layer on aluminium, dipping in nitric acid (d = 1.33 g/ml) at room temperature is recommended [115].

Silicon nitride layers about 100 μm thick sprayed cathodically onto copper can protect the metal from attack by 10 % nitric acid for about 3 h [882]. Corrosion is observed in concentrated acid after only 5 min.

Coatings of chemical service enamel are superior to high-alloy steels in all concentrations of nitric acid up to 423 K (150 °C), with a corrosion rate of about 0.1 mm/a (3.94 mpy) (see Figure 74) [837, 838, 841, 954].

Glass enamel coatings protect steels of the 12Ch18N10T type against 20 % nitric acid at 377 K (104 °C). The corrosion rate here is 0.06 mm/a (2.36 mpy) [954].

A coating of polyurethane is suitable for protecting concrete floors against nitric acid (up to 50 %) [845].

Solvent-free polyurethanes for coatings are resistant to 10 % nitric acid [988]. Pigmented polyurethanes are also resistant to 10 % nitric acid, but not to 2 % acid. The resistance to nitric acid vapors is also said to be inadequate [111].

Aluminium is protected against 10 % nitric acid, even at higher temperatures, by a polyurethane primer coating about 100 μm thick, on top of which is a 400 μm thick layer of pitch polyurethane [991]. Further data on the behavior of polyurethane coatings of other compositions in nitric acid are found in Section C 16.

Since the permeability of polyethylene films to HNO_3 is decisive for their protective action against nitric acid on metals, Table 151 summarizes the diffusion and permeability of HNO_3 as a function of the temperature and acid concentration [923]. According to this table, a good protective action by PE films is observed. The promoting effect of gasoline on the diffusion of HNO_3 through polyethylene is worth noting [924].

HDPE films of the type 102-44 show a weight increase of 0.2 and 0.4 % and 0.2 and 0.6 % after exposure to 5 and 25 % nitric acid at 293 and 323 K (20 and 50 °C) respectively for 12 months.

Flamulit® WS PE-25, a high-pressure polyethylene stabilized to heat and stress cracking, is recommended as corrosion protection for coating galvanic frames [932].

Polyvinyl chloride (PVC) coatings have an excellent resistance to 10 % nitric acid at room temperature [932, 945].

Coatings of Levasint® (ethylene/vinyl acetate copolymer) produced by a whirl sintering method, are resistant to 10 % nitric acid. After exposure for 30 d, the weight increase was < 0.3 % [945].

Vinyl resin coatings 250 to 500 μm thick are also resistant to HNO_3 vapors [1085].

Acrylic resin coatings have only a limited resistance to nitric acid [965, 1047].

The protective action of coatings of unsaturated polyester on metals against nitric acid can be estimated from the diffusion coefficients of HNO_3 plotted in Figure 79 [990].

Filler-containing epoxy resin coatings 800 to 850 µm thick certainly show a loss in gloss, but no blistering after exposure to 20% nitric acid at room temperature for 3 months [1026].

The resistance of a protective coating of the 1% triethanolamine-containing epoxide-novolak block copolymer of low drying temperature in 10% nitric acid at room temperature is worth mentioning [1037].

Coatings of epoxide-furfurol resin have proven suitable as protection for concrete surfaces against 10% nitric acid [1040]. The latex based on the epoxy resin EP-00-10 can also be used as a protective coating for carbon steels against 10% acid at room temperature [1039].

Linings of pipelines with the fluorocarbon resin Kynar® are said to have been in operation for 10 years in plutonium-processing plants using mixtures of nitric acid and hydrogen fluoride [1057].

The dependence of the service life of coatings of cyclo-rubber in 10, 30 and 56% nitric acid between 293 and 373 K (20 and 100 °C) can be seen from Table 167 [1101].

Coatings of butyl rubber are not resistant to 10% nitric acid [932].

A soft rubber lining based on Baypren® is superior to linings with hard and natural rubber not only in respect of its mechanical properties, but also in terms of its corrosion behavior in 30% nitric acid [1105].

Coatings of fluoroepoxy resin with permitted operating times in 17% nitric acid at 293 and 313 K (20 and 40 °C) of 48 and 24 months respectively are superior in resistance to those of polyfluoroperchlorovinyl at 42 and 8 months respectively.

Epoxy resin coatings of the type EP-1155 are destroyed in 25% nitric acid at room temperature only after about 500 h [1112].

Coatings of perfluoroalkoxy resins (PFA) are also resistant to nitric acid at room temperature [1066].

Pore-free coatings of the fluorocarbon resin ECTFE are an effective protection against pickling solutions of nitric acid and hydrogen fluoride at 333 K (60 °C) [1113].

After exposure of a coating of modified polystyrene to 10% nitric acid at room temperature for 3 months, no change was found [1068].

D 2 Seals and packings

Sigraflex®, a sealing material made of flexible graphite, certainly forms intercalation compounds with nitric acid, but is resistant nevertheless [847]. Its resistance can be increased further by impregnation with HNO_3 resistant resins, such as polytetrafluoroethylene.

Seals for water taps made of Al_2O_3 ceramic which have been used in installations for a number of years have also found acceptance in the nitric acid industry, since aluminium oxide is absolutely resistant to nitric acid, even at high temperatures.

Sleeve connections on rigid PVC pipes still remain tight even above 1000 h in the presence of nitric acid (20 to 50%) at 293 K (20 °C) under an internal pressure of 1.0 MPa [1076].

D 3 Composite materials

Hard metals based on WC or TiC + Mo$_2$C with binder phases of nickel and cobalt belong to the two-phase composite materials. The corrosion behavior of some hard metals in 6 mol/l nitric acid is summarized in Table 169 [119].

The WC hard metals with an Ni5Cr alloy as the binder phase show the greatest resistance. Further data on the corrosion behavior of other hard metals can be found in Tables 141 and 142 [876 – 878].

According to Table 170, the chromium steel 30Ch13 is superior to the composite materials of TiC-chromium steel in terms of resistance to nitric acid [420].

The corrosion resistance of steel-glass composite is always higher than that of the pure steels, the degree of improvement depending on the composition and amount of the glass phase and on the production conditions. The material consumption rate of the composite system Ch23N18-VVS in boiling 30% nitric acid from 95 h tests is shown in Table 171 [401]. The composition of the two components of the composite for Ch23N18 is (%) Fe-0.12C-25.57Cr-17.1Ni, and for VVS 71.4% SiO$_2$ + 7.90% CaO + 16.20% Na$_2$O + 3.0% MgO + 1.2% Al$_2$O$_3$.

Aluminium reinforced with graphite fibers increases in strength from about 540 MPa at 20 percent by volume fibers to 810 MPa at 34 percent by volume. It can be treated with concentrated nitric acid for the purpose of cleaning. Apart from zinc, other alloying components, such as titanium and boron, are not attacked. A solution of 50% HNO$_3$ + 2% HF used at room temperature causes blackening of the surface of the specimens (e.g. of wires) [38].

Hard metal	Composition %	Density g/ml	Hardness HV 30	298 K (25 °C)	373 K (100 °C)
				Material consumption rate, g/m^2 h	
WC6Ni	94 WC + 6 Ni	15.0	1400	0.04 – 0.4	> 4
WC9Ni	91 WC + 9 Ni	14.6	1150	0.04 – 0.4	> 4
TCR 10	94 WC + 6 NiCr (5 Cr)	14.8	1520	< 0.04	> 4
TCR 30	91 WC + 9 NiCr (5 Cr)	14.4	1420	< 0.04	> 4
H 10 T	94.5 WC + 5.5 Co	15.0	1730	0.04 – 0.4	> 4
H 20 T	94 WC + 6 Co	14.9	1610	0.04 – 0.4	> 4
H 40 T	88 WC + 12 Co	14.3	1340	0.4 – 4	> 4
TWF 18	66 TiC + 16 Mo$_2$C + 18 Ni	6.0	1470	> 4	> 4

Table 169: Material consumption rates of some hard metals in 6 mol/l nitric acid [119]

Material composition %	HNO₃ concentration, %			
	10	10	20	60
	273 K (0 °C)	353 K (80 °C)	273 K (0 °C)	273 K (0 °C)
	Corrosion rate, mm/a (mpy)			
35 TiC + 65 19 Cr steel	0.26 (10.2)	1.41 (55.5)	0.303 (11.9)	0.25 (9.8)
40 TiC + 60 21 Cr steel	0.30 (11.8)	1.82 (71.7)	0.274 (10.8)	0.42 (16.5)
50 TiC + 50 25 Cr steel	0.45 (17.7)	1.91 (75.2)	0.500 (16.7)	0.66 (25.9)
Chromium steel 30Ch13, 0.3 C	0.003 (0.12)	0.26 (10.2)	0.003 (0.12)	< 0.001 (0.04)

Table 170: Corrosion rates of the hard metals TiC-chromium steel and the 13 Cr steel 30Ch13 in HNO₃ [420]

Glass content %	Material consumption rate g/m² h
–	1.49
3	1.51
5	1.35
7	0.85
10	0.84

Table 171: Corrosion of the sintered and tempered steel-glass composite Ch23N18-WS in boiling 30 % HNO₃ as a function of glass content [401]

Films of aluminium with fibers of indium or bismuth can be used for production of filters with micropores (~ 10 μm) if the fibers are dissolved out selectively with dilute nitric acid (1 – 20 percent by volume) in the case of indium, or 70 percent by volume in the case of bismuth. The average diameter of the pores after an action time of < 20 h is 8 to 10 μm [314].

Copper or tantalum in a composite of tantalum fibers and a copper matrix is attacked, depending on the choice of the corrosive agent, such as 70 % nitric acid or 45 % hydrofluoric acid. Non-metallic fiber reinforcements are – apart from graphite and boron – non-conductors and are therefore not subject to galvanic corrosion [1114].

The quite good resistance of the composite material consisting of hot-pressed powder mixtures of silicon nitride with 10 % magnesium oxide to concentrated nitric acid is cancelled out by calcium fluoride additions (see Table 144) [883].

Materials consisting of a mixture of graphite and silicon carbide with and without boron doping are resistant to nitric acid. The compressive strength of 150 N/mm² determined at the start is retained approximately, at 140 N/mm², after exposure to hot and cold (343 K (70 °C)) concentrated nitric acid for 720 h [869].

Furan resin glass fiber laminates (see Section C 6) retain 80.5 % of their initial flexural strength and 89.4 % of their flexural E modulus after exposure to 5 % nitric acid at room temperature for 12 months [1082].

Urethane concrete, a mixture of polyurethane and cement, combines the strength of concrete and the chemical resistance of polyurethane. Because of its good resistance to 10 % nitric acid, it is used as a concrete floor which also tolerates brief contamination with concentrated acid [833].

To improve the resistance of heat exchangers made of graphite (high thermal conductivity) to nitric acid and mixed acids of nitric acid and hydrogen fluoride, the graphite is given a special impregnation of PTFE. This material guarantees problem-free operation [818, 1008].

Polypropylene sheets with glass fiber woven layers, which are still resistant to nitric acid at 373 K (100 °C) and hardened at 353 K (80 °C) for 3 to 4 h, are recommended for nitric acid tanks [933, 996].

Polyethylene-coated carbon fibers in epoxy-phenol resin guarantee that the compressive strength and shear strength of the plastic remain constant even after prolonged exposure to nitric acid solutions [824].

Ethylene/propylene rubber containing carbon black is resistant to 15 % nitric acid [967].

According to Table 157, cast phenol-formaldehyde resin filled with 36.6 % synthetic graphite and 3.6 % sulfosalicylic acid, as a curing catalyst, becomes more and more resistant in 10 % nitric acid at 293 K (20 °C) as the heat treatment temperature increases [999].

Incorporation of glass fibers to increase the strength of polyesters causes no decrease in their resistance to nitric acid [978].

Filler-containing epoxy resin ((parts by weight) 200 parts Araldite® CT-200 + 50 parts curing agent HY-923 + 235 parts silica powder) can be recommended in the presence of 5 % nitric acid. It is attacked too severely by 20 % acid [1023].

Epoxide oligomers filled with residue tar ((parts by weight) about 10 – 70 parts tar per 100 parts resin) are also resistant in 25 % nitric acid [1029].

D 4 Heat-resistant and scaling-resistant alloys

Scaling-resistant chromium steels with 14 % chromium corrode in boiling 65 % nitric acid at a rate of < 0.015 g/m^2 h [344]. Ferritic chromium steels with > 12 % chromium have an inadequate resistance to boiling 65 % nitric acid (see Section A 18).

The chromium-nickel steels X 3 CrNi 18 10 and X 2 CrNiSi 18 15 show an excellent resistance even in 98 % nitric acid on addition of 4 % silicon (~ 0.04 g/m^2 h) [259].

The scaling-resistant steel AISI 304 also has an adequate resistance to nitric acid (see Section A 20). The chromium-nickel-molybdenum steels are not as corrosion-resistant and often exhibit intercrystalline corrosion (A 21). Generally, however, the Huey test indicates acceptable corrosion, as can be seen in the example of corrosion on the steel AISI 316 L in the Huey test in Table 87 [812]. Corrosion is reduced if occurrence of the ferrite and σ-phase can largely be prevented during heat treatment. The carbon content should be as low as possible to avoid carbide precipitates [525].

The complex and costly chromium-nickel-molybdenum-copper steels (e.g. X1NiCrMoCu31-27-4, 1.4563) provide no improvement in corrosion behavior towards nitric acid in comparison with the chromium-nickel steels of low carbon content (< 0.05 % C) and the correct heat treatment, and in some cases even cause a deterioration (see Figure 50) [429].

The scaling-resistant nickel-chromium alloys are noticeably attacked by nitric acid, especially at higher temperatures (see Section A 27). Nickel-chromium-iron alloys with a high chromium content (30 %) and 7 to 10 % iron are attacked only slightly in boiling 65 % nitric acid (0.12 mm/a (4.72 mpy)) [675]. Inconel® 600 and 690 are recommended here [595, 607, 614].

The heat-resistant and scale-resistant alloy Hastelloy® G-3 ((%) Ni-0.015C-(21-23.5)Cr-(6.0-8.0)Mo-1.5W-(18-21)Fe-(1.5-2.5)Cu-0.50(Nb+Ta) corrodes in boiling 10 % nitric acid at a rate of 0.02 mm/a (0.79 mpy) [624, 631].

Figure 71 shows the material consumption rate of the zirconium alloy Zr-5Ti-10Nb in the β-annealed state in boiling 20 % nitric acid as being about 0.00025 g/m^2 h [764].

According to Table 131, the corrosion rate of vanadium-niobium alloys in up to 40 % boiling nitric acid becomes sufficiently low only above 26 % niobium (about 0.01 mm/a (0.4 mpy)) [791].

D 5   Natural and synthetic elastomers

Ethylene/propylene rubber containing carbon black is also resistant to 15 % nitric acid [967].

Perfluoroalkylene-containing elastomers and aromatic polyimides are attacked by 35 % nitric acid [973].

Ebolat®, which is based on epoxy resin and vulcanized at 423 K (150 °C) for 2 h, protects carbon steels from corrosion after exposure to 10 % nitric acid at room temperature for 1000 h. The protective action fails in 30 % acid [1039].

Epoxy rubber filled with 19 % by volume diabase or fine quartz sand is not noticeably attacked in 10 % nitric acid at room temperature. Its flexural strength also remains unchanged after exposure to the acid for 21 d [1043].

After exposure to 70 % nitric acid at 294 K (21 °C) for 168 h, fluorosilicone rubber shows a change in volume of + 5 % [1065]. Organosilicon elastomers are resistant to 7 % nitric acid at room temperature [1086].

Cyclorubber is attacked in nitric acid (> 10 %) even at room temperature [1101].

Soft rubber coatings based on butyl rubber are resistant to dilute HNO_3 [932, 998]. Nitrile rubber can also be used in the presence of 20 % nitric acid [1102].

Linings with Baypren® are superior to those of hard rubber, natural rubber and styrene/butadiene rubber vulcanizates in respect of resistance to nitric acid. The change in weight which occurs at room temperature after the action of 30% HNO_3 on Baypren® for 56 d is 2 to 5% [1105].

D 6 Powder metallurgical materials

Powder metallurgical chromium (99.96%) can be shaped to produce semi-finished and finished products of limited size. A corrosion test over a period of one week showed corrosion rates of 0.09 mm/a (3.54 mpy) in 67% HNO_3 at a temperature of 393 K (120 °C). For media mixtures such as $H_2SO_4/HNO_3/H_2O$ (55/30/15) and $HNO_3/HF/H_2O$ (10/6/84), rates of 0.028 mm/a (1.10 mpy) were determined in the first case at 393 K (120 °C) and 0.011 mm/a (0.43 mpy) in the second case at 313 K (40 °C) [863].

A specimen pressed from chromium steel powder (Fe-30Cr) and sintered at 1573 K (1300 °C) for 5 h (d = 7.144, 93.2% of the compact steel) corroded in 58% nitric acid at room temperature at a rate of about 8 g/m^2 h. If the sintering was carried out at 1573 K (1300 °C) for 10 h and the specimen was then forged and rolled, a density of 7.66 g/ml (= 100% of the compact steel) being achieved, the material consumption rate was 1.9 g/m^2 h [415].

The corrosion of the powder metallurgical steel AISI 316 L in nitric acid depends greatly on the cooling rate after heat treatment (see Section A 21) [537].

Prealloyed powder of the steel 304 LSC (= 304 L with 1% Zn and 2% Cu), sintered at high temperatures, has an excellent resistance at 298 K (25 °C) in 10% nitric acid. The resistance of the sintered steel increases with an increasing sintering temperature and subsequent rapid cooling and a low concentration of metal ions at interstitial sites [436].

Specimens which had been pressed from the steel powder Ch18N15, sintered at 1673 K (1400 °C) for 10 h and subsequently forged and rolled and had been quenched in water after a heat treatment at 1373 K (1100 °C) corroded in 58% nitric acid at a rate of 0.88 g/m^2 h. The sintered steel had no pores and the density of compact steel. At a density of 81% of that of compact steel, the material consumption rate was 120 g/m^2 h [415].

A test specimen produced from pulverulent Ch18N15 or Ch23N18 by sintering at high temperatures and having 6.5, 14 and 1% pores corroded in boiling 60% nitric acid at rates of 0.055, 0.64 and 0.045 mm/a (2.17, 25.2 and 1.77 mpy) respectively [250].

The corrosion behavior of specimens of the steel powder Ch23N18 and a premix powder of equivalent composition of 59% iron + 18% nickel + 23% chromium, obtained by thermal pressure working, is identical in nitric acid at room temperature and at the boiling point (Table 172). There is the possibility of producing complex moldings by thermal pressure working of a powder alloy, or a powder mixture of the metals from which the alloy is composed [325].

The titanium alloys Ti-10Cr and Ti-6Al produced by electrical sintering corrode in 10% nitric acid at approximately the same rate as unalloyed titanium [741].

Pulverulent zirconium used for the preparation of carbide corrodes in 12 mol/l HNO_3 at 380 K (107 °C) at a rate of 0.075 g/m² h [871]. Powder metallurgical zirconium or Zr-2.5Nb, which has a density of 98 – 99 % of the compact material, shows an equally good resistance to boiling 70 % nitric acid as a cast alloy of the same composition [771].

Material	Density	Tensile strength MPa	20% HNO_3		60% HNO_3	
			293 K (20 °C)*	Boiling**	293 K (20 °C)*	Boiling**
			Corrosion rate, mm/a (mpy)			
St-P	7.75	638	0.003 (0.12)	0.088 (3.46)	0.002 (0.08)	0.045 (1.77)
FeCrNi-P	7.75	485	0.003 (0.12)	0.050 (1.97)	0.003 (0.12)	0.050 (1.97)

* 264 h tests
** 100 h tests

Table 172: The corrosion behavior of vacuum-pressed and diffusion-annealed sintered specimens of steel powder (St-P) and (Fe-Cr-Ni) powder FeCrNi-P in HNO_3 [325]

Bibliography

[1] Horn, E.-M.; Schoeller, K.
Temperatur- und Konzentrationsabhängigkeit der Korrosion nichtrostender austenitischer und ferritischer Werkstoffe in 20- bis 75 %iger Salpetersäure
(Corrosion resistance of austenitic and ferritic stainless alloys in 20 to 75 % nitric acid as a function of temperature and concentration) (in German)
Werkst. Korros. *41* (1990) 3, p. 97

[2] Forssell, H. W.
Vergolden – Schichtdickenmessung
(Photometric determination of gold for coating – Thickness measurement) (in German)
Galvanotechnik *67* (1976) 2, p. 106

[3] Raffalovich, A. J.
Waveguide corrosion
Mater. Performance *13* (1974) 11, p. 9

[4] Landolt, D.; Mathieu, H. J.
Application des méthodes de spectroscopie électronique à l'étude de problèmes électrochimiques
(On the influence of ion energy and incidence angle on auger depth profiles of binary alloys) (in French)
Oberfläche Surf. *21* (1980) 1, p. 8

[5] Gräf, I.; Lange, H.; Röschenbleck, B.
Contribution to the preparation and microstructure display of joints of silver solder and CrNi steel
Prakt. Metallogr. *20* (1983) p. 280

[6] Forty, A. J.
Corrosion micromorphology of noble metal alloys and depletion gilding
Nature *282* (1979) Dec., p. 597

[7] Forty, A. J.; Rowlands, G.
A possible model for corrosion pitting and tunnelling in noble-metal alloys
Philosophical Magazine A *43* (1981) 1, p. 171

[8] Forty, A. J.; Durkin, P.
A micromorphological study of the dissolution of silver-gold alloys in nitric acid
Philosophical Magazine A *42* (1980) 3, p. 295

[9] Anonymous
Edelmetall-Taschenbuch „Chemische Reaktionen mit wäßrigen Lösungen"
(Pocket book of noble metals, "Chemical reactions in aqueous solutions") (in German), p. 103
Degussa AG, Frankfurt am Main, 1967

[10] Singh, D. D. N.; Singh, M. M.; Chaudhary, R. S.; Agarwal, C. V.
Inhibition and polarization studies of some substituted urea compounds for corrosion of aluminium in nitric acid
Electrochim. Acta *26* (1981) 8, p. 1051

[11] Singh, D. D. N.; Chaudhary, R. S.; Agarwal, C. V.
Corrosion characteristics of some aluminium alloys in nitric acid
J. Electrochem. Soc. *129* (1982) 9, p. 1869

[12] Turashev, A. I.; Belyaeva, Z. G.
Effect of temperature on the electrochemical reactions of aluminium with chemical polishing electrolyte (in Russian)
Zashch. Met. *18* (1982) 1, p. 136

[13] Heibrock, W.
Organisatorische Sofortmaßnahmen nach einem Brand und Sanierungsmöglichkeiten an Gebäuden, maschinellen sowie elektrotechnischen Einrichtungen
(Immediate organizational measures after a fire and sanitation possibilities in buildings, automated as well as electrical facilities) (in German)
VGB Kraftwerkstech. *62* (1982) p. 775

[14] Horner, L.; Meisel, K.
Inhibitoren der Korrosion 22 (1) – Einfluß

von pH, Sauerstoff, Fremdionen, Art und Konzentration von Säuren auf den Korrosionsverlauf von Aluminium
(Corrosion inhibitors 22 (1) – Influence of pH, oxygen, added ions, type and concentration of acids on the corrosion of aluminium) (in German)
Werkst. Korros. 29 (1978) 8; p. 511

[15] Makwana, S. C.; Patel, N. K.; Vora, J. C.
Corrosion of 3S aluminium in the mixture of acid solutions
J. Indian Chem. Soc. 51 (1974) p. 1051

[16] Singh, D. D. N.; Chaudhary, R. S.; Prakash, B.; Agarwal, C. V.
Inhibitive efficiency of some substituted thioureas for the corrosion of aluminium in nitric acid
Br. Corros. J. 14 (1979) 4, p. 235

[17] Mohler, J. B.
Finishing pointers. Acid dipping
Met. Finishing 70 (1972) 11, p. 48

[18] Ponomareva, O. V.; Khramushina, M. I.; Etlis, V. S.
Corrosion resistance of steel and alloys and the effect of some metal cations on the synthesis of 3-chloro-2-hydroxypropanoic acid (in Russian)
Khim. Prom. (1983) 10, p. 600

[19] Stahl, R.; Kiefer, P.
Korrosion und Korrosionsschutz in der Lebensmittelindustrie
(Corrosion and corrosion prevention in the food industry) (in German)
Werkst. Korros. 24 (1973) 6, p. 513

[20] Nényei, Z.
Passivierung von Halbleiterbauelementen durch anodisiertes Aluminium
(Passivation of semiconductor devices by anodized aluminium) (in German)
Metalloberfläche 31 (1977) 3, p. 104

[21] Klas, W.; Herpers, U.; Michel, R.; Reich, M.; Droste, R.; Holm, R.; Horn, E.-M.; Müller, G.
Bestimmung der Abtragungsraten von Zirconium, Tantal und der Legierung Tantal-40Niob in fluoridhaltiger azeotroper Salpetersäure mit Hilfe der Radiotracer-Methode
(The radiotracer technique as a means to investigate the corrosion of zirconium, tantalum, and a Ta-40Nb alloy in fluoride containing azeotropic nitric acid) (in German)
Werkst. Korros. 42 (1991) 11, p. 570

[22] Chakrabarty, C.; Singh, M. M.; Agarwal, C. V.
New class of inhibitor for 1060 aluminium in nitric acid: Aryl substituted-3-formamidino thiocarbamides
Br. Corros. J. 18 (1983) 2, p. 107

[23] Chakrabarty, C.; Singh, M. M.; Agarwal, C. V.
Inhibitive action of benzoic acid and its derivatives on dissolution of aluminium alloys in nitric acid
Corrosion 39 (1983) 12, p. 481

[24] Campanella, L.
Low voltage anodizing of super-purity aluminium in molten KNO_3 at 450 °C
Annali di Chimica 63 (1973) p. 485

[25] Yukhas-Kish, Yu.; Lipovets, I.; Lokhonyai, N.; Shekhter, K.
Corrosion of aluminium in nitric acid solution. Degree of development of metallic corrosion protection (in Russian)
Period. Polytech. Chem. Eng. 17 (1973) 2, p. 197

[26] Malakhova, E. K.; Zinchenko, R. G.
Crevice corrosion of aluminium in strong nitric acid (in Russian)
Khim. Prom. (1980) p. 358

[27] Ivanov, E. S.; Komolova, L. F.; Makovskii, R. D.
Investigation of the surface of 36 alloys after etching in nitric acid fluoride solutions (in Russian)
Zashch. Met. 19 (1983) 6, p. 935

[28] Bouraya, T. A.; Chak, R. O.; Adler, Yu. A.; Turkovskaya, A. V.
Corrosion of aluminium and steel Kh18N10T (in Russian)
Zashch. Met. 11 (1975) 3, p. 329

[29] Singh, D. D. N.; Chaudhary, R. S.; Agarwal, C. V.
Inhibitive effects of some thioureas towards corrosion of aluminium in nitric acid
Indian J. Technol. 18 (1980) p. 167

[30] Herbsleb, G.; Jäkel, U.; Schwaab, P.
Einfluß von Verformung und spannungsinduzierten Ausscheidungen auf die Korrosionsbeständigkeit von siliciumlegiertem nichtrostendem Stahl X2CrNiSi 18 15 in Salpetersäure
(Effect of deformation and stress-induced precipitations on the corrosion resistance of silicon alloyed stainless steel X2CrNiSi 18 15 in nitric acid) (in German)
Werkst. Korros. 41 (1990) 4, p. 170

[31] Roßmann, Ch.
Einflüsse der Vorbehandlung. Die Bedeutung

der Reinigung und Vorbehandlung für die Qualität der galvanischen Beschichtung. (Influence of pretreatment. The importance of cleaning and pretreatment in the quality of electroplated coatings is examined. The condition of the material and the nature of the surface treatment have a great effect) (in German)
Galvanotechnik 69 (1978) p. 784

[32] Isano, T.; Takano, M.; Shimodaira, S.
On the relationship between grainboundary corrosion and stress corrosion cracking of AlZnMg alloys (in Japanese)
J. Jpn. Inst. Met. 37 (1973) p. 110

[33] Storp. S.; Holm, R.
ESCA investigation of the oxide layers on some Cr containing alloys
Surf. Sci. 68 (1977) p. 10

[34] Wälchli, W.
Touching precious metals
Gold Bull. 14 (1981) 4, p. 154

[35] Holtzer, M.
Effect of carbon and silicon on the structure and corrosion resistance of 17Cr-8Ni cast steel in concentrated nitric acid solutions
Werkst. Korros. 41 (1990) 1, p. 25

[36] Gindin, L. G.; Dmitrenko, V. E.; Zherdeva, T. I.; Smirenkina, I. P.; Fedorov, V. V.
Kinetics of the reaction of nitric acid with aluminium. I. Kinetics of the dissolution of aluminium in nitric acid and of the reduction of the latter
Russian J. Phys. Chem. 45 (1971) 2, p. 740

[37] Gindin, L. G.; Dmitrenko, V. E.; Zherdeva, T. I.; Smirenkina, I. P.
Kinetics of the reaction of nitric acid with aluminium. II. Effect of ammonium, chloride and fluoride ions
Russian J. Phys. Chem. 45 (1971) 3, p. 330

[38] Harrigan, jr., W. C.; Goddard, D. M.
Aluminium graphite composites: Effect of processing on mechanical properties
J. Met. 27 (1975) 5, p. 20

[39] Ammar, I. A.; Darwish, S.; Khalil, M. W.
Galvanostatic formation of barrier-type anodic oxides
Z. Werkstofftech. 12 (1981) p. 309

[40] Furneaux, R. C.; Wood, G. S.
Assessment of acid immersion test for the sealing of anodized aluminium
Trans. Inst. Metal Finishing 60 (1982) 1, p. 14

[41] Levy, G. G.
Sealing of aluminium surfaces
Brit. Pat. 1 232 693 (Chrysler Corp.; May 19, 1971)

[42] Fouda, A. S.; Mohamed, A. K.
Substituted phenols as corrosion inhibitors for copper in nitric acid
Werkst. Korros. 39 (1988) 1, p. 23

[43] Staronka, A.; Holtzer, M.
Korrosionsbeständigkeit von CrNi-Stahlguß (Typ 18/9 bzw. 18/13) mit erhöhtem Siliciumgehalt in konzentrierten Salpetersäurelösungen
(Corrosion resistance of chromium nickel cast steel (types 18/9 and 18/13) with increased silicon content in concentrated nitric acid solutions) (in German)
Werkst. Korros. 38 (1987) 8, p. 431

[44] Nguyen, T. H.; Randall, J. J.; McPherson, A. B.
Etching aluminium capacitor foil
Brit. Pat. 2 127 435 (Sprague Electric Comp.; August 25, 1983)

[45] Anonymous
Rapid anodization of aluminium (alloys) in aqueous solution of organic acids
Jap. Pat. 1066 243 and 1066 244 (Fuji Sash Ltd.; Dec. 5, 1974)

[46] Sato, E.
Aluminium anode for cathodic protection. III. Corrosion tests and anodic and cathodic polarization curves of aluminium in hydrochloric, sulfuric, nitric and oxalic acids (in Japanese)
Kinzoku Hyomen Gijutsu 22 (1971) p. 449

[47] Zaitseva, L. V.; Kuzyukov, A. N.; Yuferev, A. S.; Abramtsov, I. I.
Solution for testing metals for intergranular corrosion
USSR Pat. 597 950 (March 15, 1978)

[48] Endo, H.; Kitani, M.
Electropolishing and coating the surface of aluminium or aluminium alloy products
Jap. Pat. 7251 928 (Honny Chemicals Co. Ltd.; Dec. 27, 1972); (C.A. 79 (1973) 152286s)

[49] Logan, N.; Dove, M. F. A.
Corrosion chemistry in inhibited HDA
Report AFRPL-TR-81-81 (1980) No. AD-A104772 NTIS, Gov. Rep. Announce Index (U.S.) 82 (1982) 3, p. 634

[50] Lommel, H.
Oberflächenbehandlung hochglänzender Aluminiumwerkstoffe, chemische und elektrochemische Glänzverfahren

(Surface treatment of specular aluminium materials chemical and electrochemical brightening processes) (in German)
Metall *32* (1978) 10, p. 1005

[51] Läser, L.; Meyers, Chr.
Galvanisieren von Aluminium
(Electroplating of aluminium) (in German)
Galvanotechnik *69* (1978) 4, p. 306

[52] Dölling, H.
Übersicht über die Vorbehandlungsverfahren für das Beschichten von Aluminiumoberflächen
(Survey of pretreatment processes for the surface coating of aluminium) (in German)
Aluminium *54* (1978) 11, p. 708

[53] Turashev, A. L.; Belayeva, Z. G.
Effect of nitric acid concentration on coupled electrochemical reactions during chemical polishing of aluminium (in Russian)
Zashch. Met. *13* (1977) 4, p. 441

[54] Niculescu, I.
The influence of various parameters on the electrographical solving of some metals (in Rumanian)
Metalurgia (Bucharest) *29* (1977) 5, p. 277

[55] Friedemann, W.; Göhausen, H. J.; Puderbach, H.
Untersuchungen zum Bewitterungsverhalten anodisch erzeugter Aluminiumoxidschichten
(Investigation into the weathering behavior of anodized aluminium) (in German)
Aluminium *53* (1977) 8, p. 471

[56] Mohler, J. B.
Aluminum bright dip – Specific gravity as a control factor
Met. Finish. *80* (1982) 3, p. 35

[57] Endtinger, F.
Neues Schwarzfärbe-Verfahren für Aluminium
(New blackening processes for aluminium) (in German)
Metalloberfläche *32* (1978) 4, p. 157

[58] Simon, H.
Vorbehandeln metallischer Oberflächen für das Beschichten mit Lack oder Trockenschmierstoffen
(Pretreatment of metallic surfaces for coating with paint and dry lubricants) (in German)
Galvanotechnik *73* (1982) p. 1320

[59] Subramanyam N. C.; Mayanna, S. M.
Effect of haloacetic acids on corrosion of aluminium in nitric acid
J. Electrochem. Soc. India *33* (1984) 3, p. 273

[60] Tonk, E.
Corrosion of welded aluminium transport containers (in Hungarian)
Magy. Alum. *13* (1976) 1, p. 9

[61] Schenkel, H.
Hydratisierung und Aktivierung der auf Aluminium anodisch erzeugten Oxidschicht
(Hydration and activation of anodically produced oxide films on aluminium) (in German)
Aluminium *53* (1977) 11, p. 681

[62] Vitkovskij, M. N.; Maslov, V. A.
Corrosion stability of Al in 98% HNO_3
Zavod. Lab. *24* (1958) 4, p. 429

[63] Sato, E.
Anode current-time curves of aluminium in oxalic and nitric acid under controlled potential electrolysis. Studies on aluminium anode for cathodic protection (Part 1) (in Japanese)
J. Met. Finish. Soc. Jpn. *22* (1971) 6, p. 284

[64] Devereux, O. F.; Libsch, T. A.; Choo, Y. H.; Bambri, A. K.
The effect of environmental pH on the failure strain of anodic films formed on the aluminium alloy 2024
Corros. Sci. *15* (1975) p. 361

[65] Mercer, A. D.; Butler, G.; Warren, G. M.
Cleaning of corroded test specimens
Br. Corros. J. *12* (1977) 2, p. 122

[66] Sinyavski, V. S.; Kalinin, V. D.; Golyakov, G. M.; Dorokhina, V. E.; Ivanenko, N. I.; Kozanova, V. V.
Method for determining the susceptibility of highly alloyed aluminium alloys to corrosion (in Russian)
Zashch. Met. *16* (1980) 4, p. 422

[67] Horn, E.; Schoeller, K.
Korrosion nichtrostender austenitischer Stähle in (kondensierender) chloridhaltiger Salpetersäure
(Corrosion of austenitic stainless steels in (condensing) nitric acid containing chlorides) (in German)
Werkst. Korros. *42* (1991) 11, p. 569

[68] Bakulin, A. V.; Malyshev, V. N.
Acoustic emission during the layer corrosion of aluminium alloys (in Russian)
Zashch. Met. *14* (1978) 2, p. 197

[69] Zhuravleva, L. V.; Rebrunov, V. P.
Aluminium alloys for nitric acid cisterns (in Russian)
Zashch. Met. *6* (1970) 2, p. 224

[70] Horn, E.-M.; Schoeller, K.; Dölling, H.
Zur Korrosion von Aluminium-Werkstoffen in Salpetersäure
(Corrosion of aluminium and aluminium alloys in nitric acid) (in German)
Werkst. Korros. *41* (1990) 6, p. 308

[71] Kon, K.; Ohtani, N.
Crack growth in stress corrosion cracking of Al-4% Cu single crystals (in Japanese)
J. Jpn. Inst. Met. *40* (1976) 5, p. 498

[72] Kon, K.; Sato, A.; Ohtani, N.
Dissolution current at pits in the stress corrosion cracking of aluminium alloy single crystals (in Japanese)
J. Jpn. Inst. Met. *39* (1975) 11, p. 1182

[73] Lifka, B. W.
SCC resistant aluminum alloy 7075-T73 performance in various environments
Aluminium *53* (1977) 12, p. 750

[74] Fredriksson, H.; Hillert, M.; Lange, N.
The modification of aluminium-silicon alloys by sodium
J. Inst. Met. *101* (1973) Nov./Dec., p. 285

[75] Singh, D. D. N.; Singh, M. M.; Chaudhary, R. S.; Agarwal, C. V.
Inhibitive effects of isatin, thiosemicarbazide and isatin-3-(3-thiosemicarbazone) on the corrosion of aluminium alloys in nitric acid
J. Appl. Electrochem. *10* (1980) p. 587

[76] Soni, K. P.; Bhatt, I. M.
Corrosion and inhibition of copper, brass and aluminium in nitric acid, sulphuric acid and trichloroacetic acid
J. Electrochem. Soc. India *34* (1985) 1, p. 76

[77] Cordier, H.; Schippers, M.; Polmear, I.
Microstructure and intercrystalline fracture in a weldable AlZnMg alloy
Z. Metallkde. *68* (1977) 4, p. 280

[78] Paul, J.; Bauer, B.
Contrast techniques for phase separation in the scanning electron microscope
Prakt. Metallogr. *60* (1983) p. 213

[79] Sklenička, V.; Oliveriová, A.
Dislocation structure revealed by etch pits on an Al-20% Zn alloy after high temperature creep
Prakt. Metallogr. *12* (1975) p. 372

[80] Klemm, H.
Susceptibility of materials to intergranular corrosion
Prakt. Metallogr. *9* (1972) p. 441

[81] Fiebiger, C.; Kilian, R.; Degelmann, E.; Robisch, O.; Kaiser, H.; Kaesche, H.
Electrochemical behavior and hydride embrittlement evaluation of titanium during cathodic polarization in $HNO_3/UO_2^{2+}/N_2H_5^+$ solutions
Werkst. Korros. *40* (1989) 12, p. 695

[82] Ganaha, T.; Pearce, B. P.; Kerr, H. W.
Grain structures in aluminium alloy GTA welds
Metall. Trans. S *11A* (1980) p. 1351

[83] Graf, L.
Was spricht gegen das Aufreißen von Deckschichten und gegen eine Versprödung als Ursachen der Spannungsrißkorrosion?
(Stress corrosion cracking and the hypotheses of film rupture and embrittlement)
(in German)
Z. Metallkd. *66* (1975) 12, p. 749

[84] Dotsenko, V. K.; Marchenko, V. A.; Polguev, Yu. V.; Uglyanskaya, R. A.; Sevruk, O. K.
Removal of shell remnants from aluminium castings in an alkali melt (in Russian)
Liteinoe Proizvod. (1979) 4; p. 26

[85] Durkin, P.; Forty, A. J.
Oxide formation during the selective dissolution of silver from silver-gold alloys in nitric acid
Philos. Mag. A *45* (1982) 1, p. 95

[86] Forty, A. J.
A microscopic study of elastic deformation in an oxide layer on a plastically deformed metal
Philos. Mag. A *46* (1982) 3, p. 521

[87] Vvedenskii, A. V.; Stekol'nikov, Yu. A.; Tutukina, N. M.; Marshakov, I. K.
Electrochemical oxidation kinetics for silver in a nitrate solution
Sov. Electrochem. *18* (1982) p. 1468

[88] Pilyankevich, A. N.; Mel'nikova, V. A.
Structural morphological transformations in metallic films under the effect of electrolytes
Sov. Progress Chem. *48* (1982) 12, p. 39

[89] Chaudhary, R. S.; Singh, D. D. N.; Yadava, P. N. S.; Agarwal, C. V.
Inhibition of corrosion of aluminium in nitric acid by tolyl thioureas
J. Electrochem. Soc. India *28* (1979) 3, p. 169

[90] Arrowsmith, D. J.; Clifford, A. W.; Crowther, J. C.
The control of transfer etch in chemical polishing of aluminium
Trans. Inst. Metal Finish. *57* (1979) 2, p. 89

[91] Ortí, S.; Cruz, A.; Lizarbe, R.
Nuevas posibilidades del ensayo de la "gota de colorante" en el control de calidad del sell-

ado de los recubrimientos anódicos de óxido de aluminio
(New possibilities on the use of the "dye absorption test" for sealing quality control in aluminium oxide anodic coatings) (in Spanish)
Rev. Metal. *13* (1977) 1; p. 18

[92] Smith, H. V.
Aluminum deoxidizing and desmutting
Plat. Surf. Finish. *62* (1975) p. 870

[93] Choudhary, R. S.; Singh, D. D. N.; Agarwal, C. V.
Inhibitive action of some para-substituted aromatic amines towards corrosion of 1060 aluminium in nitric acid solution
J. Electrochem. Soc. India *27* (1978) 2, p. 91

[94] Takahashi, H.; Nagayama, M.
Surface films formed on aluminium by different pretreatments. I. XPS analysis of thickness and chemical composition
J. Met. Finish. Soc. Jpn. *36* (1985) 3, p. 96 (in Japanese)

[95] Singh, D. D. N.; Chaudhary, R. S.; Agarwal, C. V.
A thiourea-based inhibitor for the corrosion of aluminium in nitric acid
J. Electrochem. Soc. India *28* (1979) 4, p. 241

[96] Zajtseva, L. V.; Levchenko, V. A.; Kuzyukov, A. N.; Romaniv, V. I.; Mashchenko, N. S.
Effect of the stress state on the corrosion resistance of technical-grade aluminium (in Russian)
Fiz.-Khim. Mekh. Mater. *16* (1980) 6, p. 99

[97] Everhart, J. L.
Aluminium alloy castings
Mater. Eng. *47* (1958) 2, p. 125

[98] Kesui, K.; Yamamoto, E.
A preparation for the chemical polishing of aluminium (in Japanese)
Japan Pat. 54 32255 (Sept. 26, 1980)

[99] Hasegawa, A. Tanigawa, K.
Method for the surface treatment of aluminium and its alloys (in Japanese)
Japan Pat. 55 168 034 (June 8, 1982)

[100] Moriyasu, S.
Cleaning of the surfaces of aluminium and its alloys after electrolytic deposition of reinforced coatings (in Japanese)
Japan Pat. 51 5133 (July 25, 1977)

[101] Cork, F.
Method of and mixture for alloy coating removal
Brit. Pat. 3 169 677 (Rolls Royce Ltd.; April 16, 1980)

[102] Zajtseva, L. V.; Kuzyukov, A. N.; Yuferev, A. A.; Abramtsov, I. I.
Solution for examining metals for intercrystalline corrosion (in Russian)
USSR Pat. 2 315 197 (February 22, 1978)

[103] Ullmann, P.; Clausnitzer, P.
Etching agent for structurizing Al layers (in German)
DDR Pat. 90 681 (June 12, 1972)

[104] Taccani, G.
Pretreatment for chromating and phosphochromating aluminium parts (in Italian)
Riv. Colore-Verniciat. Ind. *11* (1978) p. 123, 265

[105] Isano, T.; Takano, M.; Shimodaira, S.
On the relationship between grainboundary corrosion and stress corrosion cracking of Al-Zn-Mg alloys (in Japanese)
J. Jpn. Inst. Met. *37* (1973) 1, p. 110

[106] Bacquias, G.
Le brillantage chimique des métaux communs et des métaux précieux
(Chemical brightening of common metals and precious metals) (in French)
Oberfläche Surf. *20* (1979) p. 178

[107] Sklenička, V.; Oliveriová, A.
Dislocation structure revealed by etch pits on an Al-20 % Zn alloy after high temperature creep
Prakt. Metallogr. *12* (1975) p. 372

[108] Mamiya, F.
Performance test on acidic cleaners for aluminum alloys
J. Jpn. Inst. Light Met. *29* (1979) 7, p. 273

[109] Horn, E.-M.; Schoeller, K.; Storp, S.; Dölling, H.
Zur Korrosion der Legierung AlMg 2 Mn 0,8 in hochkonzentrierter Salpetersäure
(Corrosion of the alloy AlMg 2 Mn 0.8 in highly-concentrated nitric acid) (in German)
Werkst. Korros. *37* (1986) 2, p. 74

[110] Stanaland, V. A. et al.
Evaluation of the corrosion properties of aluminium alloy 5083 weldments
Oak Ridge Y-12 Plant TN (1984) 17 p.; NTIS GPO Dep. File No. DE-8500 7729

[111] Cherenkova, Z. V.
Bestimmung von Korrosionsschäden und Vorhersage der Korrosionsanfälligkeit von Aluminiumlegierungen durch elektroinduktiven Leitfähigkeitsprüfer

(Detection of corrosion damage and prediction of the susceptibility to corrosion of aluminium alloys by means of an electroinductive conductivity tester) (in Russian)
Prom. Primenenie Elektromagn. Metodov Kontrolysa (1974) p. 49

[112] Andrew, A.
Improvements in or relating to the sealing of anodized aluminium
Brit. Pat. 1 350 292 (April 18, 1974)

[113] Kay, C. J.
A method of etching aluminium alloys
Brit. Pat. 1 371 035 (Rolls Royce Ltd.; October 23, 1974)

[114] Metzger, W.; Pappe, G.; Schmitz, M.
Electrolytic metal stripping in solutions containing carboxylic acids
Trans. Inst. Met. Finish. 53 (1975) p. 184

[115] Hammel, B.
Entmetallisieren
(Stripping) (in German)
Oberfläche Surf. 23 (1982) 7, p. 238

[116] Chakrabarty, C.; Singh, M. M.; Agarwal, C. V.
Preventing corrosion of aluminium in nitric acid by tolyl guanidine derivatives
Corros. Prev. Control 33 (1986) 3, p. 72

[117] Uhlig, H. H.; Asphanani, A. I.
Corrosion behavior of cobalt base alloys in aqueous media
Mater. Performance 18 (1979) 11, p. 9

[118] Pitlenko, V. I.; Proskurykova, L. A.; Sukhotin, A. M.
Pitting and activation of cobalt under the action of nitrate ions (in Russian)
Zashch. Met. 20 (1984) 2, p. 254

[119] Kny, E.; Bader, T.; Hohenrainer, C.; Schmid, L.
Korrosionsresistente, hochverschleißfeste Hartmetalle
(Corrosion and highly wear resistant hard metals) (in German)
Werkst. Korros. 37 (1986) 5, p. 230

[120] Baldi, A. L.
Stripping aluminide coatings from cobalt- and nickelbase superalloys
US Pat. 3 622 391 (Alloy Surfaces Co., Inc.; Nov. 23, 1971)

[121] Ivanov, A. F.; Sudakov, V. S.
Effect of melting methods on the corrosion resistance of nonmagnetic 40KChNM alloy (in Russian)
Metalloved. Term. Obrab. Met. (1976) 9, p. 62

[122] Stupnisek-Lisac, L.; Karsulin, M.; Takenouti, H.
Passivation of iron, nickel and cobalt in concentrated nitric acid solutions
Passivity of Metals and Semiconductors, Bombanne (France), (Proc. Conf.), ed. by M. Froment (1983) p. 327
Elsevier Science Publishers B. V., Amsterdam

[123] Demo jr., J. J.; Ferriss, D. P.
Verschleiß- und korrosionsbeständige Kobalt- und Nickellegierungen
(Abrasion and corrosion-resistant cobalt and nickel alloys) (in German)
DOS 2 348 703 (August 22, 1974)

[124] Feiler, C. E.; Morrell, G.
Effect of fluoride on the corrosion of the aluminium grade 2S-0 and the stainless steel grade 347 in fuming nitric acid at 170°F
Natl. Advisory Comm. Aeronautic Res. Memo E-53L-176 (1954) 22 p.

[125] Chine, C.
Soft magnetic platinum-cobalt products
US Pat. 4 221 615 (Sept. 9, 1980)

[126] Weith, W.
Verschleißteile aus Hartmetall für die chemische Industrie
(Wearing parts made of hard metal in the chemical industry) (in German)
Chem.- Anl. + Verfahren 12 (1979) 8, p. 81

[127] Spyra, W.; Klupsch, H.
Über die Entwicklung einer Nickelhartlegierung zum Auftragsschweißen
(The development of a hard nickel alloy for deposition welding) (in German)
Thyssen Edelst. Tech. Ber. 8 (1982) 2, p. 162

[128] McCurrie, R. A.; Carswell, G. P.
Effects of etching on magnetization reversal in SmCo$_5$ particles
Philos. Mag. 28 (1973) p. 611

[129] Kolew, A. S.; Bratoewa, M.; Muleschkow, N.
Einfluß der Glanzbildner auf die Struktur und die Korrosionsbeständigkeit von Silberüberzügen
(The influence of brighteners upon the structure and corrosion resistance of silver electrodeposits) (in German)
8th Congress Intern. Union Electrodeposit. & Surface Finishing, Basel, 1972 (Proc. Conf.), publ. 1973, p. 127

[130] Kodolov, V. P.; Murav'ev
Preparation of specimens of aluminum alloys for electron microscopy (exchange of experience)
Ind. Lab. (U.S.S.R.) 44 (1978) p. 1713

[131] Gindin, L. G.; Dimitrenko, V. E.; Zherdeva, T. I.; Smirenkina, I. P.
Kinetics of the reaction of nitric acid with aluminium. II. Effect of ammonium chloride and fluoride ions (in Russian)
Russ. J. Phys. Chem. *45* (1971) p. 330

[132] Kuzyukov, A. N.; Zaitseva, L. V.
Intergranular corrosion of stressed aluminium (in Russian)
Fiz.-Khim. Mekh. Mater. *16* (1982) 6, p. 97

[133] Zaitseva, L. V.; Malakhova, E. K.; Romaniv, V. I.; Mashchenko, N. S.
The corrosion stability of aluminium and AlMg alloys in 98 % nitric acid (in Russian)
Khim. Neft. Mashinostr. (1985) 9, p. 26

[134] Stahl, R.; Kiefer, P.
Korrosion und Korrosionsschutz in der Lebensmittelindustrie
(Corrosion and corrosion prevention in the food industry) (in German)
Werkst. Korros. *24* (1973) p. 513

[135] Zaitseva, L. V.; Kuzyukov, A. N.; Malakhova, E. K.; Mashchenko, N. S.
Corrosion of welds of technical-grade aluminium in nitric acid (in Russian)
Khim. Neft. Maschinostr. (1981) 8, p. 27

[136] Chakrabarty, C.; Singh, M. M.; Yadav, P. N. S.; Agarwal, C. V.
Inhibition of corrosion of aluminium (1060) by benzoic acid and its derivatives in nitric acid
Transactions SAEST *18* (1983) 1, p. 13

[137] Anonymous
Some features of the dissolution of aluminium in nitric acid
Sb. Tr. Vses. Zaochn. Politekh. Inst. *100* (1976) p. 171

[138] Uchiyama, I.; Sato, E.
Electrochemical properties of boehmite film in sodium chloride solution
J. Met. Finish. Soc. Jpn. *28* (1977) 7, p. 374

[139] Zaitseva, L. V.; Kuzyukov, A. N.
Assessment of the intercrystalline corrosion tendency of aluminium (in Russian)
Khim. Neft. Mashinostr. (1979) 5, p. 20

[140] Voroshnin, L. G.
Corrosion-resistant diffusion coatings
Met. Sci. Heat Treat. (1984) p. 764

[141] Kutsova, V. Z.; Borshchevskaya, D. G.; Evina, T. Y.; Goncharova, T. A.
The fine structure of an AMg6-AD1-AMg6 trimetallic composite
Russ. Metall. (1980) p. 168

[142] Mohler, J. B.
Deoxidizing and brightening aluminum alloys
Met. Finish. *70* (1972) 10, p. 43

[143] Weber, H.; Hönig, A.
Duratherm® aushärtbare Federwerkstoffe für vielseitige Anwendungen
(Duratherm® heat treatable spring materials for versatile applications) (in German)
Metall *30* (1976) 11, p. 1041

[144] Vakulenko, L. I.; Kozlovskaya, N. A.; Shedenko, L. I.; Vdovenko, I. D.
The effect of polyvinyl alcohol on the chemical polishing of copper in electrolytes based on orthophosphoric and nitric acids
Sov. Progress Chem. *47* (1981) 11, p. 81

[145] Naka, M.; Miyake, M.; Maeda, M.; Okamoto, I.; Arata, Y.
Effect of chromium addition on corrosion resistance of amorphous Co-Zr alloys
Trans. JWRI Jpn. *11* (1982) 2, p. 171

[146] Pakhomov, V. S.
Nitric acid as a corrosion medium (in Russian)
Tr. Mosk. Inst. Khim. Mashinostr. *67* (1975) p. 102

[147] Sayed, S. M.; Gouda, V. K.
Korrosionsinhibierung von Weichstahl in verschiedenen Säurelösungen durch Petrolsulfonate
(Corrosion inhibition of mild steel in various acid solutions using petroleum sulfonate) (in German)
Metalloberfläche *32* (1978) p. 298

[148] Ito, H.; Mihashi, Y.
Corrosion-resistant hard alloys containing tungsten carbide
Japan Pat. 72 38 511 (Nippon Tungsten Co. Ltd.; Dec. 5, 1972)

[149] Storp, S.; Holm, R.
ESCA investigation of the oxide layers on some Cr containing alloys
Surface Sci. *68* (1977) p. 10

[150] Mirolyubov, E. N.; Merendi, Yu. Ya.
Self-dissolution of chromium in diluted nitric acid (in Russian)
Zashch. Met. *11* (1975) 4, p. 420

[151] Mirolyubov, E. N.; Merendi, Yu. Ya.
Effect of temperature on the corrosion and electrochemical behavior of chromium in nitric acid solutions (in Russian)
Zashch. Met. *11* (1975) 5, p. 599

[152] Konstantinova, E. V.; Muravyev, L. L.
Time factor in corrosion research (in Russian)
Zashch. Met. *12* (1975) 5, p. 599

[153] Jostan, J. L.; Schwing, C.
Dünne Chrom- und Chrom-Nickelschichten in der Elektronik
(Thin chromium and chrome-nickel films in the electronic industry) (in German)
Galvanotechnik *64* (1973) 1, p. 3

[154] Kanetake, N.
Surface treatment of powder metallurgy products
Japan Pat. 73 39 309 (Kito Co. Ltd.; June 9, 1973)

[155] Evtukhov, V. I.; Suprun, A. K.; Goryachev, P. T.
Improvement of mechanical properties by chromium diffusion coating (in Russian)
Tekhnol. Organ. Proizvod., Kiev (1983) 1, p. 53

[156] Ponomarenko, E. P.; Filina, L. I.; Rebrunov, V. P.; Shatalov, N. I.; Brusentzova, V. M.
Corrosion-resistant carbon steel with vacuum-diffused chrome coating (in Russian)
Zashch. Met. *9* (1973) 5, p. 161

[157] Hirschfeld, D.; Ostermann, G.
Interessante Eigenschaften von Hartmetallen für die Chemische Industrie
(Properties of hardmetals of interest to the chemical industry) (in German)
Z. Werkstofftech. *4* (1973) p. 367

[158] Antropov, L. I.; Donchenko, M. I.; Saenko, T. V.
Dissolution of copper in dilute nitric acid (in Russian)
Zashch. Met. *14* (1978) 6, p. 657

[159] Chen, N. G.; Kiryukha, A. S.; Chen, L. N.; Sokolyan, L. N.; Panfilova, Z. V.
Inhibitors of acid corrosion of metals EK and EK-1 (in Russian)
Zashch. Met. *9* (1973) 2, p. 211

[160] Donchenko, M. I.; Sayenko, T. V.
Inhibition of copper corrosion by dimethylolthiourea in dilute nitric acid (in Russian)
Zashch. Met. *15* (1979) 1, p. 96

[161] Könneke, D.; Lacmann, R.
The dissolution of copper single crystal spheres
J. Crystal Growth *46* (1979) p. 15

[162] Schmitt, G
Application of inhibitors for acid media
Report prepared for the European Federation of Corrosion working party on inhibitors
Brit. Corros. J. *19* (1984) p. 165

[163] Walker, R.
Corrosion inhibition of copper by tolyltriazole
Corrosion *32* (1976) 8, p. 339

[164] Beccaria, A. M.; Mor, E. D.
Inhibitive effect of tannic acid on the corrosion of copper in acid solutions
Br. Corros. J. *11* (1976) 3, p. 156

[165] Walker, R.
Benzotriazole as a corrosion inhibitor for immersed copper
Corrosion *29* (1973) 7, p. 290

[166] Desai, M. N.; Rana, S. S.
Inhibition of the corrosion of the copper in nitric acid
Anticorros. Methods Mater. *20* (1973) 2, p. 8

[167] Beccaria, A. M.; Mor, E. D.
Inhibiting effect of some monosaccharides on the corrosion of copper in nitric acid
Br. Corros. J. *13* (1978) 4, p. 186

[168] Joshi, K. M.; Kamat, P. V.
Dissolution of metals in magnetized water at various pH
Indian J. Appl. Chem. *33* (1970) p. 376

[169] Shams El Din, A. M.; El Hosary, A. A.; Saleh, R. M.; Abd El Kader, J. M.
Peculiarities in the behaviour of thiourea as corrosion-inhibitor
Werkst. Korros. *28* (1977) 1, p. 26

[170] Desai, M. N.; Rana, S. S.
Inhibition of the corrosion of copper in nitric acid aniline and other substituted aromatic amines
Anticorros. Methods Mater. *20* (1973) 6, p. 16

[171] Blahnik, R.; Kosobud, J.
The corrosion resistance of wrapped and soldered elec. connections in corrosive atmospheres (in Czech)
Koroze Ochr. Mater. *21* (1977) 1, p. 3

[172] Winkley, D. C.; Grand, A. F.
Hydrogen peroxide pickling of copper rod
Wire J. *6* (1973) p. 64

[173] Martschewska, M.; Penew, P.; Janatschkowa, I.; Spassowa, E.
Oxidation von Kupfer durch Wechselstrom mit technischer Frequenz. I. In wäßrigen Natriumnitrat- und Natriumkarbonat-Lösungen
(Oxidation of copper by alternating current with a technical frequency. I. In aqueous

solutions of sodium nitrate and sodium carbonate) (in German)
Metalloberfläche 27 (1973) 11, p. 416

[174] Anonymous
Pretreatment of copper surfaces before coating with aluminium
UdSSR Pat. 1802058; (Galvanotechnik 70 (1979) 7, p. 685)

[175] Costas, L. P.
Some chemical aspects of the mercuric nitrate test used for copper alloys
Corrosion 22 (1966) 3, p. 74

[176] Ahmad, Z.; Ghafelehbashi, M.; Nategh, S.
Corrosion resistance of high manganese copper aluminium alloys in sea water
Anticorros. Methods Mater. 21 (1974) 8, p. 13

[177] Zhuravlev, B. L.; Nazmutdinova, A. S.; Dresvyannikov, A. F.
Coulometric control of the corrosion stability of copper-nickel-chrome coatings (in Russian)
Zashch. Met. 20 (1984) 2, p. 319

[178] Dubinin, G. N.; Sokolov, V. S.
Improvement in the scaling and corrosion resistance of copper-based alloys (in Russian)
Izv. V. U. Z. Tsvet. Metall. (1974) 5, p. 133

[179] Patel, M. M.; Patel, N. K.; Vora, J. C.
Azoles as corrosion inhibitors for 63/37 brass in acidic media
Metals & Minerals Review (1975) 8, p. 4

[180] Desai, M. N.; Thakar, B. C.; Shah, D. K.; Gandhi, M. H.
Cathodic protection of 70/30 brass in 2.0 M nitric acid in the presence of organic corrosion inhibitors
Br. Corros. J. 10 (1975) 1, p. 41

[181] Desai, M. N.; Shah, V. K.
Aromatic amines as corrosion inhibitors for 70/30 brass in nitric acid
Corros. Sci. 12 (1972) p. 725

[182] Desai, M. N.
Corrosion inhibitors for brasses
Werkst. Korros. 24 (1973) 8, p. 707

[183] Patel, N. K.; Patel, B. B.
Influence of toluidines and acridine derivatives on corrosion of 63/37 brass in nitric acid solution
Werkst. Korros. 26 (1975) 2, p. 126

[184] Desai, M. N.; Shah, Y. C.; Punjani, B. K.
Inhibition of the corrosion of 63/37 brass in nitric acid
Br. Corros. J. 4 (1969) Nov., p. 309

[185] Walker, R.
Triazole, benzotriazole, and naphthotriazole as corrosion inhibitors for brass
Corrosion 32 (1976) 10, p. 414

[186] Dinnappa, R. K.; Mayanna, S. M.
Haloacetic acids as corrosion inhibitors for brass in nitric acid
Corrosion 38 (1982) 10, p. 525

[187] Desai, M. N.; Shah, Y. C.
Thioureas as corrosion inhibitors for 63/67 brass in nitric acid
Trans. SAEST 6 (1971) 3, p. 81

[188] Patel, M. M.; Patel, N. K.; Vora, J. C.
Thiocresol as a corrosion inhibitor for brass
J. Electrochem. Soc. India 27 (1978) 3, p. 171

[189] Desai, M. N.; Joshi, J. S.
Alkyl anilines as corrosion inhibitors for 63/37 brass in nitric acid
J. Indian Chem. Soc. 52 (1975) p. 884

[190] Dinnappa, R. K.; Mayanna, S. M.
Effect of thioureas on corrosion of brass in nitric acid
Trans. SAEST 19 (1984) 2, p. 93

[191] Desai, M. N.; Joshi, J. S.
p-Substituted aromatic amines as corrosion inhibitors for 63/37 brass in nitric acid
J. Indian Chem. Soc. 52 (1975) Sept., p. 878

[192] Desai, M. N.; Patel, B. M.; Thakker, B. C.
Meta substituted aromatic amines as corrosion inhibitors for 70/30 brass in nitric acid
J. Indian Chem. Soc. 52 (1975) June, p. 554

[193] Desai, M. N.; Thaker, B. C.; Patel, B. M.
Heterocyclic amines as corrosion inhibitors for 70/30 brass in nitric acid
J. Electrochem. Soc. India 24 (1975) 4, p. 184

[194] Shams El Din, A. M.; El Hosary, A. A.; Gawish, M. M.
A thermometric study of the dissolution of some Cu-Zn alloys in acid solutions
Corros. Sci. 16 (1976) p. 485

[195] Boron, K.; Dabrowiecki, K.
Corrosion resistance of brass pipes with varying inner surface coarseness (in Polish)
Rudy Met. Niezelaz. 27 (1982) 1, p. 36

[196] Leonard, R. B.
Corrosion of metals by aliphatic organic acids
Mater. Performance 19 (1980) 9, p. 65

[197] Mai, Y. W.
Crack propagation resistance of alpha-brass subjected to prior-exposure in corrosive environments
Corrosion 32 (1976) 8, p. 336

[198] Dönnges, E.
Ausmaß und Rhythmus der Auflösungsgeschwindigkeit von Metallhalbzeug in

angreifenden Agenzien in Abhängigkeit vom Grade der Kaltverformung
(Extent and rhythm of the dissolution rate of semifinished metal products in corrosive media independent of the degree of cold working) (in German)
Metall 33 (1979) 12, p. 1269

[199] Graf, L.; Ata, H. O. K.
Untersuchung der Spannungskorrosionsempfindlichkeit von Ein- und Vielkristallen aus Messing und Kupfer-Gold-Legierungen
(Investigation of stress corrosion sensitivity of single and polycrystalline specimens of brass and copper-gold alloys) (in German)
Z. Metallkde. 64 (1973) 5, p. 366

[200] Wawra, H.
Verarbeitung von Messingblechen unter dem Gesichtspunkt hoher Spannungskorrosionsbeständigkeit, Teil I
(Processing tin sheets with the aim of achieving high stress corrosion cracking resistance, Part I) (in German)
Blech 19 (1972) 11, p. 591

[201] Wawra, H.
Verarbeitung von Messingblechen unter dem Gesichtspunkt hoher Spannungskorrosionsbeständigkeit, Teil II
(Processing tin sheets with the aim of achieving high stress corrosion cracking resistance, Part II) (in German)
Blech 19 (1972) 12, p. 631

[202] Altgeld, W.; Läser, L.
Chemisches Glänzen von Kupfer-Zink-Legierungen
(Chemical polishing of copper-zinc alloys) (in German)
Metalloberfläche 36 (1982) 4, p. 177

[203] Srivastava, R. D.; Mukerjee, R. C.; Agarwal, A. K.
Corrosion of electroplated Cu-Cd alloys and its inhibition
Corros. Sci. 19 (1979) 1, p. 27

[204] Srivastava, R. D.; Agarwal, A. K.
Corrosion inhibition of copper-cadmium alloy electroplates in nitric acid solutions
Anticorros. Methods Mater. 24 (1977) 11, p. 6

[205] Naka, M.; Hashimoto, K.; Masumoto, T.
Corrosion behavior of amorphous and crystalline $Cu_{50}Ti_{50}$ and $Cu_{50}Zr_{50}$ alloys
J. Non-Crystalline Solids 30 (1978) p. 29

[206] Afendik, K. F.; Krichmar, S. I.
Model of a viscous electrolyte for electropolishing of copper
Sov. Electrochem. 6 (1970) p. 122

[207] Komp, M. E.; Mathay, W. L.
Steels for the pulp and paper industry
Mater. Performance 16 (1977) 6, p. 22

[208] Vdovenko, I. D.; Vakulenko, L. I.; Kozlovskaya, N. A.
Effect of catapin A on corrosive and electrochemical behavior of iron in hydrochloric and nitrogen acid mixtures (in Russian)
Ukr. Khim. Zh. 46 (1980) 4, p. 360

[209] Vdovenko, I. D.; Vakulenko, L. I.; Shedenko, L. I.
Corrosion and electrochemical behaviour of iron in electrolytes on the basis of nitric and orthophosphoric acids (in Russian)
Ukr. Khim. Zh. 49 (1983) 10, p. 1077

[210] Vakulenko, L. I.; Kozlovskaya, N. A.; Shedenko, L. I.
The effect of sulfamic acid on the corrosion and electrochemical behavior of iron in nitric acid solutions
Sov. Progress Chem. 48 (1982) 9, p. 62

[211] Srivastava, R. D.; Mukerjee, R. C.; Tripathi, D.
Bath characteristics and corrosion nature of electrodeposited brass
Indian J. Technol. 15 (1977) p. 446

[212] Gavrilenko, A. G.
Dissolution of iron in nitric acid (in Russian)
Voronezh. Thekhnol. Inst., Voronezh, USSR, Deposited Doc. (1974), VINITI 1527-74, 8 p.

[213] Lavrishin, B. N.; Krilyuk, S. S.; Fedin, S. P.; Miskidzhyan, S. P.
Cleansing and protection from corrosion of plates of rectification columns (in Russian)
Zashch. Met. 18 (1982) 5, p. 823

[214] Eisenkolb, F.; Kühnel, M.
Korrosion von Metallpulvern
(Corrosion of metal powders) (in German)
3rd International Powder Metallurgy Conference, Eisenach, 1965, publ. Berlin 1966, p. 49

[215] Fedorov, Yu. V.; Uzlyuk, M. V.
Effect of mixture of indole and ammonium thiocyanate on the corrosion of steel in acid solutions
Sov. Progress Chem. 37 (1971) 1, p. 75

[216] Uzlyuk; M. V.; Fedorov, Yu. V.; Pinus, A. M.; Tolstykh, V. F.; Panfilova, Z. V.; Shatukhina, L. I.; Miskidzhyan, S. P.
Inhibiting influence of thiocyanate compounds of pyridine during dissolution of steel in nitric acid (in Russian)
Zashch. Met. 13 (1977) 2, p. 212

[217] Sayed, S. M.; Gouda, V. K.
Korrosionsinhibierung von Weichstahl in verschiedenen Säurelösungen durch Petrolsulfonate
(Corrosion inhibition of mild steel in various acid solutions using petroleum sulfonate) (in German)
Metalloberfläche 32 (1978) 7, p. 298

[218] Krotil, B.
Oceli pro difúzní chromování
(Methods of chromium diffusion) (in Czech)
Koroze Ochr. Mater. 18 (1974) 4, p. 53

[219] Sathianandhan, B.; Balakrishnan, K.; Subramanyan, N.
Triazoles as inhibitors of corrosion of mild steel in acids
Br. Corros. J. 5 (1970) 4, p. 270

[220] Abd El Haleem, S. M.; Khedr, M. G. A.; Killa, H. M.
Corrosion behaviour of metals in HNO_3. I. Contribution to the study of the dissolution of steel in HNO_3
Br. Corros. J. 16 (1981) 1, p. 42

[221] El-Basiouny, M. S.; Elnagdi, M. H.; Ismail, A. R.
Einfluß von Benzolsulfonsäure-Derivaten auf die Auflösungsgeschwindigkeit von Eisen in Salpetersäure. Grenzen der Anwendbarkeit thermometrischer Verfahren zum Messen von Korrosionsgeschwindigkeiten
(Effect of benzenesulfonic acid derivatives on the dissolution rate of iron in nitric acid. Limits of applying thermometric methods for measurement of corrosion rates) (in German)
Metalloberfläche 35 (1981) 12, p. 482

[222] Donchenko, M. I.; Sribnaya, O. G.; Zheleznyak, Yu. Yu.
Retarding iron corrosion in nitric acid by means of organic and inorganic additives (in Russian)
Zashch. Met. 17 (1981) 2, p. 156

[223] Uzlyuk, M. V.; Fedorov, Yu. B.; Zelenin, B. M.
Effect of mixtures of sodium sulfide with indole and thiourea on steel corrosion in nitric acid (in Russian)
Zashch. Met. 10 (1974) 4, p. 482

[224] Gerasimov, V. V.; Gromova, A. I.; Samarin, O. P.; Belous, V. N.
Kinetics of the anodic process on steel 20 at high temperatures in water media
(in Russian)
Zashch. Met. 11 (1975) 4, p. 480

[225] Baumann, W.; Fritz, H.; Steinbrecher, M.
Untersuchungen der Gasatmosphäre von Winderhitzern zur Klärung der interkristallinen Spannungsrißkorrosion
(Investigation of the gaseous atmosphere of hot-blast stoves to elucidate intercrystalline stress-corrosion cracking) (in German)
Stahl Eisen 97 (1977) p. 633

[226] Steinhäuser, S.; Graubner, E.; Goldschmidt, V.
Oberflächenbehandlung im Schweißnahtbereich nichtrostender und säurebeständiger Stähle mit Beizpasten
(Surface treatment in the weld seam region of stainless and acid-resistant steels with pickling pastes) (in German)
Schweisstechnik (Berlin) 20 (1970) 6, p. 266

[227] Hasegawa, M.; Osawa, M.
Anomalous corrosion of hydrogen-containing ferritic steels in aqueous acid solution
Corrosion 39 (1983) 4, p. 115

[228] Vivet, F.
Décapage des cuivreux procédé non polluant au peroxyde d'hydrogène
(Copper pickling, nonpolluting hydrogen peroxide process) (in French)
Galvano-Organo 45 (1976) p. 875

[229] Vaccari, J. A.
Cast irons
Mater. Eng. 79 (1974) 6, p. 52

[230] Tsejtlin, Kh. L.; Sorokin, Yu. I.; Ryzhkova, Zh. S.
Corrosion of equipment and problems of accident prevention (in Russian)
Khim. Prom. (1977) 3, p. 179

[231] Dilthey, U.; Wanke, R.
Hochleistungs-Schweißplattierverfahren für den Chemie-Apparatebau
(Highly efficient plating in chemical plant construction) (in German)
Chem.-Ing.-Tech. 46 (1974) 11, p. 467

[232] Ford, E.
Korrosionsbeständige Metalle
(Corrosion-resistant metals) (in German)
Metall 28 (1974) 5, p. 459

[233] Saldanha, B. J.; Streicher, M. A.
Effect of silicon on the corrosion resistance of iron in sulfuric acid
Mater. Perform. 25 (1986) 1, p. 37

[234] Salminkeit, V.; von Plessen, H.; Vollmüller, H.
Die Korrosion der Gußeisenkessel beim Pauling-Verfahren
(Corrosion of cast iron boiler in the Pauling process) (in German)
Chem.-Ing.-Tech. 53 (1981) p. 822

[235] Poulson, B.
The sensitization of ferritic steels containing less than 12 % Cr
Corros. Sci. *18* (1978) p. 371

[236] Luklinska, Z. B.; Castle, J. E.
Microstructural study of initial corrosion product of aluminium-brass alloy after exposure to natural seawater
Corros. Sci. *23* (1983) 11, p. 1163

[237] Miska, K. H.
Wrought brass
Mater. Eng. *80* (1974) 5, p. 65

[238] Charbonnier, J. C.; Lena, M.; Thomas, B. J.
The influence of molybdenum on the intergranular corrosion of a pure ferritic steel with 17 % chromium
Corros. Sci. *19* (1979) p. 23

[239] Knoth, J.
High purity ferritic Cr-Mo stainless steel – Five years' successful fight against corrosion in the process industry
Werkst. Korros. *28* (1977) p. 409

[240] Fortunati, S.; Tamba, A.
Mechanical and corrosion properties of welded ELI 18-2 CrMo(Ti) steel pipes
Arch. Eisenhüttenwes. *53* (1982) 3, p. 105

[241] Moller, G. E.; Franson, I. A.; Nichol, T. J.
Experience with ferritic stainless steel in petroleum refinery heat exchangers
Mater. Performance *20* (1981) 4, p. 41

[242] Sedenko, A. M.
Corrosion stability of stainless steels and titanium alloys in nitric acid (in Russian)
Ukrain. Khim. Zh. *51* (1985) 11, p. 1226

[243] Nelzina, I. V.; Zaplatina, F. V.; Radomyselskij, I. D.
Properties of powder stainless Kh25 steel produced by swaging (in Russian)
Poroshk. Metall. *24* (1984) 1, p. 35

[244] Vaccari, J. A.
New ferritic stainless steels beat stress corrosion, ease fabrication
Mater. Eng. *82* (1975) 7, p. 24

[245] Brown, R. S.
The three-way trade off in stainless-steel selection
Mater. Eng. *96* (1982) 5, p. 58

[246] Troselius, L.; Andersson, I.; Andersson, T.; Bernhardsson, S. O.; Degerbeck, J.; Henrikson, S.; Karlsson, A.
Corrosion resistance of type 18Cr2MoTi stainless steel
Br. Corros. J. *10* (1975) 4, p. 174

[247] Vehlow, J.
Eine Methode zur kontinuierlichen und elementspezifischen Messung des korrosiven Abtrags von metallischen Werkstoffen (A method of continuous and element specific measurement of corrosive abrasion in metallic materials) (in German)
Z. Werkstofftech. *12* (1981) 9, p. 324

[248] Miska, K. H.
Ferritic stainless steels
Mater. Eng. *85* (1977) 4, p. 69

[249] Demo, J. J.
Weldable and corrosion-resistant ferritic stainless steels
Metallurg. Trans. *5* (1974) Nov., p. 2253

[250] Napara-Volgina, S. G.
Properties of materials based on alloy powders prepared by diffusion saturation from point sources, and areas of their use (in Russian)
Poroshk. Metall. *24* (1984) 11; p. 1

[251] Glazkova, S. A.; Karasyuk, T. N.; Bukanova, G. S.; Zheltova, G. A.; Lejbzon, V. M.; Kalinin, B. P.
Corrosion resistance of weld joints of the steel 08Ch13AG20 (in Russian)
Khim. Neft. Mashinostr. (1985) 2, p. 28

[252] Konstantinova, E. V.; Ryabova, N. I.; Lomovtsev, V. I.
Corrosion resistance of welded joints in low-nickel and nickel-free steels
Weld. Prod. (USSR) *29* (1982) 2, S. 8

[253] Tomashov, N. D.; Chernova, G. P.; Rutten, M. Y.
Investigation of the electrochemical behavior of stainless steels in solutions of nitric acid with additions of iron chloride and hydrochloric acid (in Russian)
Zashch. Met. *18* (1982) 6, p. 850

[254] Dimov, I.; Rashev, T.; Dzhambazova, L.; Andreev, Ch.
Corrosion of austenitic chromium-manganese steels with increased nitrogen content (in Bulgarian)
Materialoznanie Tekhnol. (1976) 2, p. 20

[255] Razygraev, V. P.; Mirolubov, E. N.
Influence of Cr(VI) on the kinetics of anodic and cathodic processes during corrosion of stainless steels and alloys in nitric acid solutions (in Russian)
Zashch. Met. *9* (1973) 1, p. 44

[256] Tomashov, N. D.
Corrosion resistant alloys and the prospects for their development (in Russian)
Zashch. Met. 17 (1981) 1, p. 16

[257] Anonymous
Vallourec® VLX 562 high strength austenitic-ferritic alloys with superior resistance to stress corrosion cracking
Hydrocarbon Processing 65 (1986) 11, p. 39

[258] Komp, M. E.; Mathay, W. L.
Steels for the pulp and paper industry
Mater. Performance 16 (1977) 6, p. 22

[259] Bäumel, A.
Korrosion in der Wärmeeinflußzone geschweißter chemisch beständiger Stähle und Legierungen und ihre Verhütung (Corrosion in the heat-affected zone of welds in chemically resistant steels and alloys, and respective preventive measures) (in German)
Werkst. Korros. 26 (1975) 6, p. 433

[260] Razygraev, V. P.; Lebedeva, M. V.
Effect of carbon on the electrochemical and corrosion behavior of Ch18N10 stainless steel in nitric acid media containing fluoride (in Russian)
Zashch. Met. 11 (1975) 4, p. 469

[261] Zilberman, B. Y.; Kotlyar, N. Z.; Sakulin, S. V.
Effect of the temperature of the heat-transfer surface of an evaporation apparatus on the corrosion rate (in Russian)
Zashch. Met. 11 (1975) 3, p. 354

[262] Lebedev, A. N.; Blinova, V. A.
Effect of ammonia on the corrosion of carbon steel in solutions containing copper ions (in Russian)
Zashch. Met. 15 (1979) 4, p. 446

[263] Lukin, B. V.; Kurtepov, M. M.; Kazarin, V. I.
Electrochemical corrosion behavior of stainless steel in nitrogen fluoride solutions (in Russian)
Zashch. Met. 15 (1979) 4, p. 445

[264] Volikova, I. G.; Petrova, T. V.; Lebedev, B. V.
Quick electrochemical method for determining the inclination of low-carbon chromium-nickel steels to intercrystalline corrosion (in Russian)
Zashch. Met. 15 (1979) 4, p. 502

[265] Razygraev, V. P.; Yegorikhin, E. V.
Self-passivation of Ch18N10 steel in boiling, dilute nitrogen chloride solutions (in Russian)
Zashch. Met. 15 (1979) 1, p. 67

[266] Devyatkina, T. S.; Lipkin, Ya. N.; Korosteleva, T. K.
Effect of circulation of nitric acid solution on the kinetics of etching of stainless steels (in Russian)
Zashch. Met. 14 (1978) 1, p. 92

[267] Kopeliovich, D. H.; Anashkin, R. D.; Ivlev, V. I.
Inhibition of intercrystalline corrosion on the steel 12Ch18N10T by fluorides in nitric acid solutions containing other oxidizers (in Russian)
Zashch. Met. 14 (1978) 4, p. 457

[268] Glukhova, A. I.
Effect of temperature and the nature of corrosive destruction on stainless steel in acid solutions (in Russian)
Zh. Prikl. Khim. 33 (1960) 8, p. 1853

[269] Razygraev, V. P.; Ponomareva, E. Y.
Behavior of the steel 12Ch18N10T in nitric acid solutions containing fluoride during heat emission (in Russian)
Zashch. Met. 14 (1978) 4, p. 461

[270] Lozovatskaya, L. P.
Effect of silicon on the stability of grain boundaries in chromium-nickel steels in strong oxidizing media (in Russian)
Zashch. Met. 19 (1983) 6, p. 923

[271] Makarov, V. A.; Stoyanovskaya, T. N.; Koloskova, E. F.
Special features of flame atom-absorption determination of iron, chromium and nickel during corrosion-electrochemical investigations (in Russian)
Zashch. Met. 19 (1983) 4, p. 579

[272] Razygraev, V. P.; Lebedeva, M. V.
Influence of some secondary reactions on the redox-potential and corrosion process in nitric acid solutions (in Russian)
Zashch. Met. 19 (1983) 5, p. 771

[273] Ponomarenko, T. T.; Shilov, V. R.; Sheremetikov, Yu. P.; Kolesnik, V. P.
Pyrophoric effect at point of contact of stainless steel and titanium in nitric acid media (in Russian)
Zashch. Met. 18 (1982) 6, p. 985

[274] Blom, U.; Kvarnback, B.
The importance of high purity in stainless steels for nitric acid service – Experience from plant service
Mater. Performance 14 (1975) 7, p. 43

[275] Osawa, M.
Effect of hydrogenation on the stress corrosion cracking of pre-stressed type 304 steel in

boiling 45 % MgCl$_2$ solution (in Japanese)
J. Jpn. Inst. Met. *45* (1981) 10, p. 1056

[276] Ishigami, I.; Tsunasawa, E.; Yamanaka, K.
Carburization during vacuum oil-quenching of SUS 304 and a proposal of prevention of it (in Japanese)
J. Jpn. Inst. Met. *46* (1982) 12, p. 1168

[277] Kumada, M.
Intergranular corrosion and fatigue strength of nonsensitized austenitic stainless steels (in Japanese)
J. Jpn. Inst. Met. *44* (1980) 5, p. 562

[278] Anonymous
Progress in fighting corrosion
Processing *31* (1985) 2, p. 22

[279] Razygraev, V. P.; Lebedeva, M. V.; Egorikhin, E. V.; Ponomareva, E. Y.; Lobanova, L. P.
Safe ion concentration of Ch18N10 type steel in boiling solutions of dilute nitric acid (in Russian)
Zashch. Met. *16* (1980) 5, p. 586

[280] Kolotyrkin, Ya. M.
The development of the corrosion theory its successes and tasks (in Russian)
Zashch. Met. *16* (1980) 6, p. 660

[281] Razygrayev, V. P.; Lobanova, L. P.; Lebedeva, M. V.
Electrochemical behavior of carbide inclusions during the corrosion of Ch18N10 type steel in boiling nitro-fluoride solutions (in Russian)
Zashch. Met. *16* (1980) 6, p. 728

[282] Razygraev, V. P.; Lebedeva, M. V.
Inhibition of corrosion and cathodic reactions on stainless steels in nitric acid media by fluoride ions (in Russian)
Zashch. Met. *18* (1980) 2, p. 227

[283] Kopeliovich, D. K.; Saprykina, N. P.; Glagolenko, Yu. V.
Corrosion of certain materials in nitro-fluoride media containing hydrazine (in Russian)
Zashch. Met. *18* (1982) 2, p. 232

[284] Kasparova, O. V.; Bogolyubski, S. D.; Kolotyrkin, Ya. M.; Milman, V. M.; Mekhryusheva, L. I.; Smakhtin, L. A.; Yudina, N. S.
Effect of carbon and phosphorus on the intergranular corrosion of tempered steel Ch20N20 in nitric acid solutions (in Russian)
Zashch. Met. *18* (1982) 1, p. 18

[285] Konstantinova, E. V.; Lomovtsev, V. M.
Corrosion behavior of economically alloyed steels in dilute nitric acid solutions with activating additives (in Russian)
Zashch. Met. *18* (1982) 1, p. 82

[286] Merendi, Yu. Ya.; Metsik, R. E.
Behavior of stainless steels in models of the production process of saturated dicarbonic acids by the destructive oxidation of nitric acid kerogene (in Russian)
Zashch. Met. *19* (1983) 3, p. 444

[287] Lipkin, Ya. N.; Bershadskaya, T. M.; Albrut, V. M.; Pugacheva, L. V.; Chepurova, L. G.
Special features of chemical polishing of steels (in Russian)
Zashch. Met. *20* (1984) 6, p. 908

[288] Kopeliovich, D. K.; Ivlev, V. I.
Corrosion of structural materials in nitrate solutions of thiourea (in Russian)
Zashch. Met. *22* (1986) p. 421

[289] Konstantinova, E. V.; Mikhlynina, T. A.; Feldgandler, E. G.
Comparative investigation of corrosion of the sheet and bar rolled steels 12Ch18N10T, 08Ch18N10T and 03Ch23N6 in boiling nitric acid (in Russian)
Zashch. Met. *20* (1984) 4, p. 622

[290] Sorokin, Yu. I.; Zeitlin, Kh. L.; Gleizer, M. M.; Potapova, K. V.; Isayenko, G. I.
Effect of certain salts on the corrosion of metals in diluted sulfuric acid (in Russian)
Zashch. Met. *20* (1984) 5, p. 780

[291] Lozovatskaya, L. P.; Levin, I. A.; Burtseva, I. K.; Goldshstejn, Ya. E.; Shmatko, M. N.; Piskunova, A. I.
Increasing the resistance of the steel 03Ch18N11 against ICC by means of correcting its chemical composition (in Russian)
Zashch. Met. *20* (1984) 3, p. 411

[292] Kurtepov, M. M.
Effect of uranium on the corrosion of stainless steel in nitric fluorine solutions (in Russian)
Zashch. Met. *21* (1985) 1, p. 104

[293] Chronister, D. J.; Spence, T. C.
Influence of higher silicon levels on the corrosion resistance of modified CF-type cast stainless steels
Corrosion in Sulfuric Acid, Boston (USA), March 1985 (Proc. Conf.), p. 75
NACE, Houston, Tx., 1985

[294] Naumann, G.
Korrosionsverhalten des austenitischen Chrom-Nickel-Stahles X5CrNiN 19.7 in Sal-

petersäure bei erhöhten Temperaturen (Corrosion behaviour of the austenitic chromium nickel steel X5CrNiN 19.7) (in German)
Chem. Tech. *30* (1978) 2, p. 94

[295] Bodine jr., G. C.; Sump, C. H.
Development of wrought austenitic-ferritic stress corrosion resistant alloys with less than 40 % ferrite for pipe and tube
Mater. Performance *16* (1977) 6, p. 13

[296] Zhadan, T. A.; Babakov, A. A.; Sharonova, T. N.; Vasilieva, N. M.
Inclination of steel 000Ch20N20S5 (ZI-52) to intercrystalline corrosion (in Russian)
Zashch. Met. *9* (1973) 1, p. 42

[297] Andersen, T. N.; Vanorden, N.; Schlitt, W. J.
Effects of nitrogen oxides, sulfur dioxide, and ferric ions on the corrosion of mild steel in concentrated sulfuric acid
Metall. Trans. A *11A* (1980) Aug., p. 1421

[298] Steigerwald, R. F.
The effects of metallic second phases in stainless steels
Corrosion *33* (1977) 9, p. 338

[299] Streicher, M. A.
The role of carbon, nitrogen, and heat treatment in the dissolution of iron-chromium alloys in acids
Corrosion *29* (1973) 9, p. 337

[300] Anonymous
The effect of heat treatments in the prevention of intergranular corrosion failures of AISI 321 stainless steel
Mater. Performance *22* (1983) 9, p. 22

[301] Smallwood, R. E.
Heat exchanger tubing reliability
Mater. Performance *16* (1977) 2, p. 27

[302] Hahin, C.; Stoss, R. M.; Nelson, B. H.; Reucroft, P. J.
Effect of cold work on the corrosion resistance of nonsensitized austenitic stainless steels in nitric acid
Corrosion *32* (1976) 6, p. 229

[303] Hahin, C.
Effect of cooling rate and subsequent plastic deformation on the corrosion rate of AISI 304 in boiling nitric acid
Corrosion *38* (1982) 2, p. 116

[304] Briant, C. L.
The effects of sulfur and phosphorus on the intergranular corrosion of 304 stainless steel
Corrosion *36* (1980) 9, p. 497

[305] Babakov, A. A.; Ostapenko, T. V.
Effect of rare-earth metals on the inclination of steel 00Ch18N11 to intercrystalline corrosion (in Russian)
Zashch. Met. *10* (1974) 3, p. 282

[306] Slate, S. C.; Maness, R. F.
Corrosion experience in nuclear waste processing at Battelle-Northwest
Mater. Performance *17* (1978) 6, p. 13

[307] Henthorne, M.
Corrosion testing of weldments
Corrosion *30* (1974) 2, p. 39

[308] Devine, T. M.; Drummond, B. J.
Use of accelerated intergranular corrosion tests and pitting corrosion tests to detect sensitization and susceptibility to intergranular stress corrosion cracking in high temperature water of duplex 308 stainless steel
Corrosion *37* (1981) 2, p. 104

[309] Robinson, F. P. A.; Scurr, W. G.
The effect of boron on the corrosion resistance of austenitic stainless steels
Corrosion *33* (1977) 11, p. 408

[310] Eremias, B.; Prazak, M.
Polarization resistance measurements of high alloyed austenitic steels with low carbon content in boiling solutions of concentrated nitric acid
Corrosion *35* (1979) 5, p. 216

[311] Chung, P.; Szklarska-Smialowska, S.
The effect of heat treatment on the degree of sensitization of Type 304 stainless steel
Corrosion *37* (1981) 1, p. 39

[312] Razygraev, V. P.; Lebedeva, M. V.; Panova, O. A.
Corrosion of 12Ch18N10T steel in hot nitric acid solutions during cathodic polarization (in Russian)
Zashch. Met. *14* (1978) 6, p. 704

[313] Kasparova, O. V.; Bogolyubski, S. D.; Kolotyrkin, Ya. M.; Milman, V. M.; Lubnin, E. N.; Shapovalov, E. T.; Yudina, N. S.
Role of silicon in the intergranular corrosion of phosphorus steel Ch20N20 (in Russian)
Zashch. Met. *18* (1982) 3, p. 336

[314] Angers, L. M.; Grugel, R. N.; Hellawell, A.; Draper, C. W.
Selective etching and laser melting studies of monotectic composite structures
In Situ Composites IV, Boston (USA), 1981 (Proc. Conf.), p. 205
Elsevier Science Publishing Co., Inc., New York, 1982

[315] Patel, M. M.; Patel, N. K.; Vora, J. C.
Azoles as corrosion inhibitors for 63/37 brass in acidic media
Vishwakarma *16* (1975) 3, p. 4

[316] Peapell, P. N.; Walkley, E. C.
Acoustic emission during stress corrosion cracking
Metall. Mater. Technol. *11* (1979) 2, p. 95

[317] Cepero, A. E.; Guedes, J.
Influencia de los iones cloruros en la corrosividad del acido nitrico
(Influence of chloride ions on the corrosiveness of nitric acid) (in Spanish)
R. Tecnológica *13* (1975) 4, p. 80

[318] Piguzov, Yu. P.; Drapkin, B. M.; Ivanov, Yu. N.
Effect of regulation on the corrosion resistance of cast iron (in Russian)
Zashch. Met. *16* (1980) 3, p. 307

[319] Telmanova, O. N.; Karyazin, P. P.; Shtanko, V. M.
Behavior of pig iron with lamellar and globular-shaped graphite in dilute acids (in Russian)
Zashch. Met. *16* (1980) 1, p. 57

[320] Dmitrieva, E. S.
Study of the corrosion resistance of parts made from the steel Ch17N2 (in Russian)
Vopr. Optim. Rezaniya Met. (1978) 1, p. 160

[321] Sotnichenko, A. L.; Agapov, G. N.; Yarkovoj, V. S.
Corrosion resistance of the nickel-free chromium manganese steel AS-43 in nitric acid (in Russian)
Khim. Neft. Mashinostr. (1974) 10, p. 23

[322] Zhuravleva, L. V.; Kasinskaya, L. K.; Nosivets, L. A.
Corrosion resistance of steel 08Ch18G8N2T in sulfuric acid (in Russian)
Khim. Neft. Mashinostr. (1979) 12, p. 21

[323] Süry, P.; Brezina, P.
Kurzzeitprüfmethoden zur Untersuchung des Einflusses von Wärmebehandlung und chemischer Zusammensetzung auf die Korrosionsbeständigkeit martensitischer Chrom-Nickel-(Molybdän)-Stähle mit tiefem Kohlenstoff-Gehalt
(Short-duration test for evaluating the influence of heat-treatment and chemical composition on the corrosion resistance of low carbon martensitic chromium nickel (molybdenum) steels) (in German)
Werkst. Korros. *30* (1979) 5. p. 341

[324] Babakov, A. A.; Novokschenova, S. M.; Levin, F. L.; Zhadan, T. A.; Sharonova, T. N.
Silicon as an alloying element in 000Ch20N20 steel (in Russian)
Zashch. Met. *10* (1974) 5, p. 552

[325] Radomyselskij, I. D.; Napara-Volgina, S. G.; Orlova, L. N.; Apininskaya, L. M.; Grabchak, A. K.; Vergeles, N.M.G.
Structure, mechanical and corrosion properties of powder stainless steel Ch23N18 (in Russian)
Poroshk. Metall. *23* (1983) 1, p. 43

[326] Abd El Haleem, S. M.; Khedr, M. G. A.; Killa, H. M.
Corrosion behavior of metals in HNO_3. I. Contribution to the study of the dissolution of steel in HNO_3
Br. Corros. J. *16* (1981) 1, p. 42

[327] Melkumov, S. B.
Effect of a welding thermal cycle on the corrosion resistance welding zone of the ferritic-austenitic steel 0Ch21N5T in a 65 % solution of HNO_3 (in Russian)
Zashch. Met. *9* (1973) 2, p. 185

[328] Kuzyukov, A. N.
Effect of welding seam tensions on the intercrystalline corrosion of 08Ch18N10T and 03Ch18N11 steels (in Russian)
Zashch. Met. *14* (1978) 4, p. 430

[329] Chernova, G. P.; Rutten, M. Y.; Tomashov, N. D.
Accelerated electrochemical method for determining the susceptibility of stainless steels to intercrystalline corrosion (in Russian)
Korroz. Zashch. (1979) 8, p. 3

[330] Herbsleb, G.; Westerfeld, K.-J.
Der Einfluß von Stickstoff auf die korrosionschemischen Eigenschaften lösungsgeglühter und angelassener austenitischer 18/10 Chrom-Nickel- und 18/12 Chrom-Nickel-Molybdän-Stähle – II. Interkristalline Korrosion in Kupfersulfat-Schwefelsäure-Lösungen und in siedender 65 %iger Salpetersäure
(Influence of nitrogen on the corrosion behavior of solution treated and annealed austenitic 18/10 chromium-nickel and 18/12 chromium-nickel-molybdenum steels – II. Intercrystalline corrosion in copper sulfate sulfuric acid solution and in boiling 65 % nitric acid) (in German)
Werkst. Korros. *27* (1976) p. 404

[331] Herbsleb, G.; Schüller, H.-J.; Schwaab, P.
Ausscheidungs- und Korrosionsverhalten unstabilisierter und stabilisierter 18/10 Chrom-Nickel-Stähle nach kurzzeitigem sensibilisierendem Glühen
(Precipitation and corrosion behavior of unstabilized and stabilized 18 10 CrNi steels after short term sensibilizing annealing) (in German)
Werkst. Korros. 27 (1976) p. 560

[332] Zhuravlev, V. K.; Kurtepov, M. M.; Bochkareva, E. F.
Effect of cerium and silver ions on the corrosion and electrochemical behavior of stainless steel in ozonized nitric acid solutions (in Russian)
Zashch. Met. 10 (1975) 3, p. 294

[333] Stanz, A.; Schäfer, K.
Einfluß einer Wärmenachbehandlung auf die mechanischen Eigenschaften und das Korrosionsverhalten nichtrostender Stähle
(Influence of an ultimate heat treatment on the mechanical properties and the corrosion behavior of stainless steels) (in German)
Werkst. Korros. 27 (1976) p. 701

[334] Lebedev, A. N.; Blinova, V. A.
The use of low nickel content stainless steel in nitrofluorine solutions (in Russian)
Zashch. Met. 13 (1977) 6, p. 710

[335] Konstantinova, E. V.; Muravyev, L. L.
Time factor in corrosion research (in Russian)
Zashch. Met. 12 (1976) 5, p. 599

[336] Sumner, W. B.
Corrosion tests in boiling, simulated radioactive wastes utilizing neutron activation
Mater. Performance 18 (1979) 9, p. 39

[337] Clark, A. T.
Processing of power reactor fuels
AEC Report DP-740 (1962) Jan.-April

[338] Maness, R. F.
Effects of Cr(VI) and Fe(III) on purex plant corrosion
AEC Report HW-72076 (1962) Jan.

[339] Votinov, S. N.; Kazennov, Yu. I.; Bogoyavlenskij, V. L.; Belokopytov, V. S.; Krylov, E. A.; Klestova, L. M.; Reviznikov, L. I.
The effect of reactor irradiation on the tendency of austenitic steels to intercrystalline corrosion (in Russian)
Atomnaya Energiya 41 (1976) 6, p. 405

[340] Mah, R.; Terada, K.; Cash, D. L.
Corrosion of distillation equipment by HNO_3-HF solutions during HNO_3 recovery
Mater. Performance 14 (1975) 11, p. 28

[341] Zamiryakin, L. K.; Chikunov, V. K.
Corrosion stability of welded joints of austenitic steels in nitric acid solutions (in Russian)
Zashch. Met. 5 (1969) 4, p. 422

[342] Uzlyuk, M. V.; Zimina, V. M.; Fedorov, Yu. V.; Pinus, A. M.
A corrosion inhibitor for copper and brass in nitric acid solutions (in Russian)
Vopr. Khim. i Khim. Technol. 62 (1981) p. 17

[343] Sukhodolskaya, E. A.; Odarchenko, V. V.
Improvement in the properties of corrosion-resistant cast iron (in Russian)
Liteinoe Proizvod. (1980) 10, p. 8

[344] Truman, J. E.
Corrosion resistance of 13 % chromium steels as influenced by tempering treatments
Br. Corros. J. 11 (1976) 2, p. 92

[345] Williams, D. E.; Asher, J.
Measurement of low corrosion rates: Comparison of a.c. impedance and thin layer activation methods
Corros. Sci. 24 (1984) 3, p. 185

[346] Covino jr., B. S.; Scalera, J. V.; Driscoll, T. J.; Carter, J. P.
Dissolution behavior of 304 stainless steel in HNO_3/HF mixtures
Metall. Trans A 17A (1986) Jan., p. 137

[347] Diekmann, H.; Gräfen, H.; Holm, R.; Horn, E.-M.; Storp, S.
ESCA – Untersuchungen zum Aufbau von Oxidschichten auf austenitischen Stählen nach Salpetersäurebeanspruchung
(ESCA – Investigations of the passive films formed on austenitic stainless steels in nitric acid) (in German)
Z. Werkstofftech. 9 (1978) 2, p. 37

[348] Horn, E.-M.; Kilian, R.; Schoeller, K.
Deckschichtbildung und Korrosionsverhalten des hochsiliciumlegierten austenitischen Sonderstahles X2CrNiSi 18 15 in Salpetersäuren
(The formation of surface films and the corrosion resistance of the silicon containing austenitic steel X2CrNiSi 18 15 in nitric acid) (in German)
Z. Werstofftech. 13 (1982) 8, p. 274

[349] Gräfen, H.
Die Bedeutung der Schadensanalyse für Werkstoffentwicklung, Konstruktion und Fertigung im Chemieapparatebau
(The importance of case histories with respect

[350] Herbsleb, G.
to materials, development, design and construction of chemical plant) (in German)
Werkst. Korros. 26 (1975) 9, p. 675

[350] Herbsleb, G.
Werkstoff- und Korrosionsprobleme in wäßrigen Lösungen bei hohen Temperaturen und Drücken
(Material and corrosion problems in aqueous solutions at high temperatures and pressures) (in German)
Werkst. Korros. 27 (1976) 3, p. 145

[351] Sydberger, T.
Influence of the surface state on the initiation of crevice corrosion on stainless steels
Werkst. Korros. 32 (1981) 3, p. 119

[352] Kügler, A.
Die Wahl der Edelstähle bei aggressiven Medien
(The selection of high-grade steels in aggressive media) (in German)
VDI Z. 119 (1977) 8, p. 411

[353] Drehsen, H.; Strassburg, F. W.
Neue austenitische Chrom-Nickel-Stähle und Nickel-Chrom-Eisen-Legierungen für den Chemieanlagenbau und ihr Schweißverhalten
(New austenitic chromium-nickel steels and nickel-chromium-iron alloys for chemical plant construction and their welding behavior) (in German)
Schweissen Schneiden 27 (1975) 1, p. 11

[354] Brown, M. H.
Behavior of austenitic stainless steels in evaluation tests for the detection of susceptibility to intergranular corrosion
Corrosion 30 (1974) 1, p. 1

[355] Kabayashi, M.; Miki, M.; Ohkubo, K.
Nitric acid resistant stainless steels
(in Japanese)
Bull. Jpn. Inst. Met. 22 (1983) p. 320

[356] Johnson, M. J.; Kearns, J. R.; Deverell, H. E.
The corrosion of the new ferritic stainless steels in nitric acid
Corrosion '84, International Corrosion Forum Devoted Exclusively to the Protection and Performance of Materials, New Orleans, April 1984 (Proc. Conf.), Paper No. 144

[357] Pakhomov, V. S.
Self dissolution and corrosion behavior of stainless 17 % chromium steel in nitric acid
(in Russian)
Tr. Moskov. Inst. Khim. Mashinostr. 67 (1975) p. 59

[358] Bejan, F.; Mercea, V.; Balaban, L.
Several views on the corrosion behavior of thin sheets of the stainless steel 18 8, stabilized with titanium (in Rumanian)
Revista de Chimie 28 (1977) 3, p. 254

[359] Dzhambazova, L.; Zlateva, G.; Kamenova, Ts.
Effect of the characteristics of the nitride phase on the corrosion behavior of nitrogen chromium-manganese steels (in Bulgarian)
Tekh. Mis'l 15 (1978) 4, p. 93

[360] Kuzyukov, A. N.; Zajtseva, L. V.; Khanzadeev, I. V.
Effect of stress on intergranular corrosion of austenitic steels in the production of weak nitric acid (in Russian)
Zashch. Met. 18 (1982) 3, p. 413

[361] Donat, H.; Schäfer, K.
Stand und Entwicklungstendenzen des Schweißens von korrosionsbeständigen Stählen
(Current state and trends in development in the welding of corrosion-resistant steels) (in German)
Schweissen Schneiden 27 (1975) 9, p. 343

[362] Dorn, L.; Anik, S.; Günaltan, A.
Untersuchungen zur Beständigkeit widerstandspunktgeschweißter Verbindungen aus X 12 CrNi 18 8 gegenüber interkristalliner Korrosion
(Investigations relating to the intergranular corrosion resistance of resistance spot welded joints in X 12 CrNi 18 8) (in German)
Schweissen Schneiden 32 (1980) 1, p. 5

[363] Gulyaev, V. M.; Chulkova, V. M.
Effect of phosphorus on the corrosion properties of stainless steel (in Russian)
Zashch. Met. 12 (1976) 3, p. 287

[364] Kachanov, V. A.; Nikitin, D. G.; Klyushnikova, L. A.; Kabashnyj, A. I.; Ponomarenko, V. I.
Corrosion of welded joints in derivatives of adipic acid (in Russian)
Zashch. Met. 17 (1981) 6, p. 739

[365] Mirolyubov, E. N.; Kazakov, A. P.; Kurtepov, M. M.
The influence of chlorides on the corrosion resistance of stainless steels in nitric acid
in: Corrosion of Metals and Alloys (Izdatel'stvo Metallurgiya), Ed. Tomashow, N. D.; Miralyubov, E. N., Part 2 (1965) p. 101

[366] Buttinelli, D.; Capotorto, C.; Memmi, M.
Corrosion behavior of austenitic chromium-nickel steels in boiling nitric acid (Huey test).

[367] Tomashov, N. D.; Chernova, G. P.; Rutten, M. Y.; Radetskaya, G. K.; Yakovleva, L. F.
III. Effect of polishing the surface on the corrosion rate (in Italian)
Ann. Chimica 63 (1973) p. 181

[367] Tomashov, N. D.; Chernova, G. P.; Rutten, M. Y.; Radetskaya, G. K.; Yakovleva, L. F.
Testing the tendency of stainless steels to undergo intercrystalline corrosion when exposed to drops of electrolyte (in Russian)
Korroz. Zashch. 12 (1979) 10, p. 3

[368] Low, A. J.
Plutonium recovery from scrap
Chem. Eng. 74 (1967) 11, p. 132

[369] Bingham jr., E. C.
Compact design pays off at new nitric acid plant
Chem. Eng. 73 (1966) 11, p. 116

[370] Vasileva, V. A.; Dzutsev, V. T.; Moskvichev, I. F.; Ulyanin, E. A.; Feldgandler, E. G.; Lobova, M. V.
Nitric acid effect on steels corrosion resistance in diluted sulphuric-nitric acid mixtures (in Russian)
Khim. Prom. (1982) p. 165

[371] Shulgat, G. A.; Pakhomov, V. S.; Klinov, I. Y.
Local fracture of steel 12Ch18N10T in nitric acid containing chlorides (in Russian)
Khim. Neft. Mash. (1976) 9, p. 21

[372] Glejzer, M. M.; Tsejtlin, Kh. L.; Sorokin, Yu. N.; Isaenko, G. I.; Babitskaya, S. M.
The effect of organic and inorganic oxidizers on the corrosion of the steel 12Ch18N10T and VT1-0 titanium in sulfuric acid (in Russian)
Zashch. Met. 13 (1977) 6, p. 684

[373] Anonymous
Lagertank aus austenitischem CrNiSi-Stahl (Storage tank made of austenitic CrNiSi steel) (in German)
Nickel-Ber. 38 (1973) p. 6

[374] Zhadan, T. A. et al.
Corrosion resistant steel alloy used for weld seams
USSR Pat. 377 404 (April 17, 1973); (C.A. 79 (1973) 9559s)

[375] Kobayashi, M.; Fujiyama, S.; Araya, Y.; Wada, S.; Sunayama, Y.
Study on concentrated nitric acid-resistant stainless steel
Nippon Sutenreso Giho (1976) 12, p. 1

[376] Lipodaev, V. N.; Yushenko, K. A.; Shulskij, V. Y. et al.
Effect of stabilizers and the ferrite phase on the corrosion resistance and ductility of welded joints of austenitic steels with a high silicon content (in Russian)
Avtom. Svarka (1986) 6, p. 9

[377] Horn, E.-M.; Kügler, A.
Development, properties, processing and applications of high-silicon steel grade X 2 CrNiSi 18 15
Z. Werkstofftech. 8 (1977) p. 362

[378] Vehlow, J.
Kontinuierliche Verfolgung der Korrosion des Stahls X 2 CrNiSi 18 15 in 98 %iger Salpetersäure
(Continuous measurement of corrosion on steel X 2 CrNiSi 18 15 in 98 % nitric acid) (in German)
Z. Werkstofftech. 13 (1982) p. 286

[379] Zitter, H.
Prüfung geschweißter und ungeschweißter austenitischer Chrom-Nickel-Stähle auf interkristalline Korrosion
(Testing of welded and non-welded austenitic chromium-nickel steels for intercrystalline corrosion) (in German)
Arch. Eisenhüttenwes. 28 (1957) 7, p. 401

[380] Chak, R. O.; Turkovskaya, A. V.
Effect of redox potential and contact with precious metals on behavior of the steel Ch18N10T in mixtures of sulfuric and nitric acids (in Russian)
Zashch. Met. 9 (1973) 6, p. 705

[381] Gorelik, G. N.; Nikonova, E. A.
Influence of contact with lead on the behavior of stainless steels and titanium (in Russian)
Zashch. Met. 6 (1970) 4, p. 416

[382] Oknin, I. V.; Gabrielyan, S. G.
Corrosion of the steel Ch18N10T in a solution of nitric acid with sulphite additions (in Russian)
Zashch. Met. 6 (1970) 3, p. 274

[383] Kasparova, O. V.; Bogolyubskij, S. D.; Kolotyrkin, Ya. M.; Ulyanin, E. A.; Vasyukov, A. B.; Yudina, N. S.; Kostromina, S. V.
Investigation of the effect of additions of phosphorus, sulphur and molybdenum on the corrosion-electrochemical behavior of the steel Ch20N20 in acid media in the over-passivation area (in Russian)
Zashch. Met. 15 (1979) 2, p. 147

[384] Demyanets, A. A.; Ezau, Ya. Ya.
The effect of chloride ions on the corrosion and electrochemical behavior of iron and car-

bon steels in nitric acid (in Russian)
Khim. i Khim. Tekhnol., Minsk 15 (1980) p. 21

[385] Zavrazhina, V. I.; Artemenko, A. I.
The effect of the potassium salts of some hydroxamic acids on the corrosion of steel in aqueous solutions (in Russian)
Sb. Tr. Belgor. Tekhnol. Inst. Stroit. Mater. 22 (1976) p. 67

[386] Sidorenko, N. R.; Chekhovskii, A. V.; Prokhorov, V. M.; Kuznetsov, V. N.; Makartsev, V. V.
Study of corrosion resistance of structural materials in sulfur-nitrogen mixtures with high free sulfur trioxide content (in Russian)
Khim. Mashinostr. 12 (1980) p. 160

[387] Vazhenin, S. F.; Remashevskaya, Z. V.; Bojko, A. Z.; Korotenko, N. D.
Possible use of nickel-free and economically alloyed, low-nickel stainless steels in the production of nitric acid and ammonium nitrate (in Russian)
Vopr. Khim. i Khim. Tekhnol., Kharkov 67 (1982) p. 99

[388] Agranat, B. A.; Bashkirov, V. I.; Vedeneeva, M. A.; Semenovykh, N. V.; Mironov, I. I.
The effect of ultrasound and cementation on the pickling of scale on the steel 18ChGT (in Russian)
Sb. Mosk. Inst. Stali Splavu 77 (1974) p. 108

[389] Warwick, R. G.
An inspection authority's experience of difficulties met in the operation of steam plant
Anticorros. Methods Mater. 25 (1978) 4, p. 8

[390] Voroshilov, I. P.; Shilov, V. R.; Voroshilova, E. P.; Shapovalova, L. P.
Corrosion behavior of certain structural materials in a mixture of nitric acid and potassium chloride (in Russian)
Zashch. Met. 11 (1975) 5, p. 602

[391] Čihal, V.; Lehká, N.; Malík, J. K.
Probleme der interkristallinen und der Messerlinienkorrosion von Schweißverbindungen der mit Niob stabilisierten korrosionsbeständigen Stähle
(Intercrystalline and knife corrosion of welded joints of chromium-nickel niobium-stabilized corrosion resisting steels) (in German)
Metalloberfläche 26 (1972) 2, p. 453

[392] Yudina, N. S.; Shapovalov, E. T.
Preferential dissolution of high-chrome carbides of the chomium-nickel steels (in Russian)
Zashch. Met. 18 (1982) 5, p. 686

[393] Azzerri, N.; Tamba, A.
Potentiostatic pickling: a new technique for improving stainless steel processing
J. Appl. Electrochem. 6 (1976) p. 347

[394] Sukhotin, A. M.; Lantratova, N. Y.; Trubnikov, V. P.; Atroshenko, E. I.
The corrosion resistance of structural materials in N_2O_4 and coolant technology (in Russian)
Atomnaya Energiya 36 (1974) p. 496

[395] Leymonie, C.
Étude de l'influence de la structure sur la résistance à la corrosion des soudures d'acier 18-10 au Mo à bas carbone (316 L)
(Study of the influence of the structure of welds of the steel 18-10 (316 L) with Mo and low in carbon on the corrosion resistance) (in French)
Rev. Metall. (1971) p. 289

[396] Kilman, Ya. I.; Zaichko, N. D.
Structural materials selection in ammonium nitrate production by single-stage non-steaming method and production safety questions (in Russian)
Khim. Prom. (1983) 1, p. 28

[397] Volikova, I. G.; Shapiro, M. B.; Subbotina, L. A.
Effect of ferrite on the corrosion resistance of the chromium-nickel steel Ch18N10T (in Russian)
Zashch. Met. 8 (1972) 5, p. 555

[398] Searson, P. C.; Latanision, R. M.
A comparison of the general and localized corrosion resistance of conventional and rapidly solidified AISI 303 stainless steel
Corrosion 42 (1986) 3, p. 161

[399] Vyklick, M.; Brenner, O.; Hamouz, E.; Podhradsk, M.
Langzeit-Korrosionsuntersuchungen an austenitischen Stählen im Hochdruckteil einer Harnstoffanlage
(Longterm corrosion field tests with austenitic steels in the high pressure parts of a urea plant) (in German)
Werkst. Korros. 32 (1981) 2, p. 85

[400] Anonymous
Anlagen und Verfahren zur elektrochemischen Metallreinigung und -vorbehandlung
(Plants and processes for the electrochemical cleaning of metals and treatment)

(in German)
Ingenieur-Digest *11* (1972) 11, p. 55

[401] Yagupolskaya, L. N.; Ivanova, S. V.; Lysenko, E. V.; Shcherban, N. I.
Corrosion resistance of metal-glass sintered materials (in Russian)
Poroshk. Metall. (1976) 11, p. 54

[402] Yudina, N. S.; Shapovalov, E. T.; Bortsov, A. N.
Intercrystalline corrosion of austenitic chromium-nickel steel in a strongly oxidizing medium
Prot. Met. (U.S.S.R.) *20* (1984) 1, p. 58

[403] Khokhlova, P. M.; Levin, I. A.
Selective structural corrosion of the ferritic austenitic chromium-nickel steel Ch22N5 (in Russian)
Zashch. Met. *10* (1974) 6, p. 674

[404] Pavlova, F. S.; Kuznetsova, V. N.; Gerasimov, V. V.
Perfection of the cleaning technology of pipe lines made from the steel 0Ch18N10T (in Russian)
Zashch. Met. *5* (1969) 4, p. 451

[405] Drehsen, H.; Strassburg, F. W.
Das Schweißverhalten neuer austenitischer Chrom-Nickel-Stähle und Nickel-Chrom-Eisen-Legierungen
(The welding behavior of new austenitic chromium-nickel steels and nickel-chromium-iron alloys) (in German)
Praktiker *27* (1975) 8, p. 146

[406] Kuzyukov, A. N.; Zaytseva, L. V.; Khanzadeev, I. V.
Effect of stress on intergranular corrosion of austenite steels in the production of weak nitric acid (in Russian)
Zashch. Met. *18* (1982) 3, p. 413

[407] Schmidt, L.
(The corrosion resistance of chromium steels) (in Rumanian)
Revista de Chimie *37* (1986) 1, p. 68

[408] Konstantinova, E. V.; Lomovtsev, V. I.
Corrosion of stainless steels containing little or no nickel in concentrated nitric acid with hydrofluoric acid or chloride ions (in Russian)
Zashch. Met. *19* (1983) 6, p. 933

[409] Koren, M.; Hochörtler, G.
Eigenschaften des ferritisch-austenitischen Stahles X 3 CrMnNiMoN 25 6 4
(Properties of the ferritic-austenitic steel X 3 CrMnNiMoN 25 6 4) (in German)
Stahl Eisen *102* (1982) 10, p. 509

[410] Cojocaru, V.
Inhibitors for fighting corrosion during removal of deposits by chemical treatment of steels (in Rumanian)
Revista de Chimie *25* (1974) 11, p. 908

[411] Mertins, K.
Metallographic investigations as a supplement to short-time corrosion tests on high alloy chromium and chromium-nickel steels
Prakt. Metallogr. *10* (1973) 2, p. 75

[412] Migaj, L. L.; Malchevskij, E. G.
Stability of sparingly alloyed stainless steels in rare metal industrial media (in Russian)
Zashch. Met. *16* (1980) 2, p. 143

[413] Lomovtsev, V. I.; Konstantinova, E. V.
Corrosion of austenitic chromium-manganese steel AS-43 in nitrofluoride media (in Russian)
Zashch. Met. *14* (1978) 60, p. 702

[414] Charbonnier, J. C.; Lena, M.; Thomas, B. J.
Influence du molybdène sur la corrosion intergranulaire d'un acier pur à 17 % de chrome
(Effect of molydenum on the intergranular corrosion of a pure 17 % chromium steel) (in French)
Rev. Métall. *76* (1979) p. 469

[415] Nelzina, I. V.; Radomyselskij, I. D.
Production technique and properties of dense powder stainless steels (in Russian)
Poroshk. Metall. (1981) 12, p. 31

[416] Kakhovskij, N. I.; Lipodaev, V. N.; Deryugin, B. G.
Corrosion resistance of high-alloyed steels and their weld joints (in Russian)
Avtom. Svarka (1980) 2, p. 72

[417] Lazebnov, P. P.; Savonov, Yu. N.; Aleksandrov, A. G.
Influence of titanium nitride, cerium and vanadium on the corrosion resistance of welded chromium-nickel metals (in Russian)
Avtom. Svarka (1981) 8, p. 69

[418] Tomashov, N. D.; Chernova, G. P.; Rutten, M. Y.
Determination of the susceptibility of stainless steels to intercrystalline corrosion (in Russian)
Korroz. Zashch. (1978) 4, p. 3

[419] Kudryavtseva, E. F.; Dolinkin, V. N.
Chemical methods for removing high-temperature oxide films (in Russian)
Zashch. Met. *13* (1977) 5, p. 567

[420] Kyubarsepp, Ya. P.; Valdma, L. E.; Kallast, V. A.
Corrosive resistance of TiC steel powder alloys (in Russian)
Poroshk. Metall. (1980) 4, p. 99

[421] Tokareva, T. B.; Kolyada, A. A.; Smolin, V. V.; Medvedev, E. A.
Corrosion mechanical properties of high-chromium content ferritic steels produced by vacuum melting (in Russian)
Zashch. Met. 13 (1977) 5, p. 529

[422] Leali, C.; Montagna, G.; Fortunati, S.; Tamba, A.
Caratteristiche meccaniche e di resistenza alla corrosione di tubi saldati in acciaio ELI 18-2 Cr Mo (Ti)
(Mechanical properties and corrosion resistance of welded ELI 18-2 titanium-stabilized chromium-molybdenum steel tubes)
(in Italian)
Metall. Ital. 71 (1979) 7/8, p. 331

[423] Dzhambazova, L. D.; Zlateva, G.; Kamenova, Ts.
Effect of nitride phase on the corrosion behavior of chrome manganese steels alloyed with nitrogen (in Russian)
Zashch. Met. 15 (1979) 2, p. 202

[424] Dzhambazova, L.; Rashev, Ts.; Zlateva, G.
Corrosion of nitrogenized CrMn steel in nitric acid (in Russian)
Zashch. Met. 14 (1978) 4, p. 465

[425] Kraft, R.; Leistikow, S.; Pott. E.
Untersuchungen zur Korrosion des austenitischen 25Cr20Ni-Stahls UHB 25 L in Salpetersäure und salpetersauren Lösungen oxidierender Metallionen
(Investigation of the corrosion of the austenitic steel 25Cr20Ni UHB 25 L in nitric acid and its solutions containing metal ions)
(in German)
Kernforschungszentrum Karlsruhe, ISSN 0303-4003, 1983, 40 p.

[426] van Maaren, P. W.
Possible applications of optical methods in failure
Prakt. Metallogr. 18 (1981) p. 494

[427] Droste, R.; Marx, G.
Nachweis selektiver Korrosion am Edelstahl 1.4306 mit Hilfe der Radioisotopenmethode
(Detection of selective corrosion on stainless steel material no 1.4306 by radioisotope methods) (in German)
Werkst. Korros. 37 (1986) 2, p. 101

[428] Tischner, H.; Pieger, B.; Kratzer, A.; Horn, E.-M.
A cast low carbon stainless steel for use in nitric acid at higher concentrations and temperatures
Werkst. Korros. 37 (1986) 3, p. 119

[429] Horn, E.-M.; Kohl, H.
Werkstoffe für die Salpetersäure-Industrie
(Materials for the nitric acid industry)
(in German)
Werkst. Korros. 37 (1986) 2, p. 57

[430] Vogg, H.; Braun, H.; Löffel, R.; Lubecki, A.; Merz, A.; Schmitz, J.; Schneider, J.; Vehlow, J.
Anwendung der Radionuklidtechnik in Chemie und Verfahrenstechnik
(Use of radionuclide techniques in chemistry and chemical engineering) (in German)
J. Radioanal. Chem. 32 (1976) p. 495

[431] Kuzyukov, A. N.; Levchenko, V. A.
Influence of substructure on intergranular corrosion of 08Ch18N10T and 03Ch18N11 steels
Sov. Mater. Sci. 16 (1980) p. 329

[432] Eremenko, A. S.; Shchesno, L. P.; Taraban, A. I.; Severina, L. S.
Effect of alloying with nitrogen on the properties of low-carbon stainless steels
Sov. Mater. Sci. 11 (1975) p. 672

[433] Furman, A. Y.; Romanov, V. V.
Effect of intercrystalline corrosion on the cyclic strength of steel Ch18N9
Sov. Mater. Sci. 11 (1975) 1, p. 96

[434] Shuvalov, V. A.; Emel'yantseva, Z. I.; Gerasimov, V. V.
Relationship of the character of distribution of dislocations in austenitic steels to corrosion under stress
Sov. Mater. Sci. 12 (1976) 3, p. 313

[435] Storchaj, E. I.; Kuznetsova, M. V.; Elchinova, L. N.; Ezhov, N. V.
Corrosion Electrochemical behavior of steel 03Ch13AG19 and 07Ch13N4AG20
(in Russian)
Khim. Neft. Mashinostr. (1984) 3, p. 37

[436] Lei, G.; German, R. M.; Nayar, H. S.
Corrosion control in sintered austenitic stainless steels
Progress in Powder Metallurgy, New Orleans (Proc. Conf.) 39 (1983) p. 391

[437] Shiganova, L. N.; Pakhomova, N. M.; Obnosova, V. A.
Corrosion resistance of some steels in superphosphoric acid (in Russian)

Khim. Prom., Ser.: Azotn. Prom-st (1979) 5, p. 68

[438] Maslov, V. A.; Semenova, A. A.
Corrosion resistance of the steel 03Ch18N11 and its weld joints in nitric acid solutions (in Russian)
Svar. Proizvod. (1980) 11, p. 29

[439] Yamamoto, K.; Koumi, N.
Corrosion of 25Cr-20Ni-Nb used in nuclear plant handling nitric acid
Internat. Conf. on World Nuclear Energy Accomplishments and Perspectives, Illinois (USA), 1980 (Proc. Conf.), p. 231

[440] Kato, T.; Fujikura, M.; Ichikawa, J.
Some properties of boron-containing 18Cr9Ni stainless steel for nuclear engineering
Denki Seiko 49 (1978) p. 108

[441] Fajngold, L. L.; Shatova, T. Y.
The corrosion and electrochemical behavior of stainless steels in sulfuric acid (in Russian)
Zashch. Met. 17 (1981) 3, p. 312

[442] Leitner, S.; Köstler, H. J.
Investigations into the application of etching reagents for austenite grain boundaries in tempered high-speed steels
Prakt. Metallogr. 15 (1978) p. 66

[443] Anonymous
Formteilätzen
(Etching of molded parts) (in German)
Oberfläche Surf. 19 (1978) p. 36

[444] Fleer, R.; Rickel, J.; Draugelates, U.
Metallographic detection of carbides in the steel X 41 CrMoV 5 1 after different austenizing processes
Prakt. Metallogr. 16 (1979) p. 105

[445] Ozawa, R.
Grain boundary corrosion-resistant austenitic stainless steels
Jap. Pat. 78 19 915 (February 23, 1978)

[446] Ito, N.; Kobayashi, M.; Okubo, K.; Miki, M.
Stainless steel for use in concentrated nitric acid
Jap. Pat. 77 156 122 (December 26, 1977)

[447] Schmidt, M.; Kowall, D.; Schöler, W.
Mittel zum chemischen Entzundern, Beizen und Glänzen von Edelstahl
(Agent used for chemical descaling, pickling and shining) (in German)
Galvanotechnik 72 (1981) p. 664

[448] Maness, R. F.
Inhibition corrosion of stainless steel by ruthenium-containing nitric acid solution
US Pat. 4 111 831 (Sept. 5, 1978)

[449] Ondrejcin, R. S.; McLaughlin, B. D.
Corrosion of high NiCr-alloys and type 304 L stainless steel in HNO_3-HF
Du Pont de Nemours & Co., DP-1550, April 1980, 46 p.

[450] Biefer, G. J.; Garrison, J. G.
Comparison of the effects of uranium and molybdenum on the corrosion resistance of the AISI 430 stainless steel grade
Can. Dept. Mines. Tech. Surv., Mines Branch, Tech. Bull. TB-74 (1965) 18 p.

[451] Kobayashi, M.; Yoshida, T.; Aoki, M.; Ikeda, N.; Takahashi, M.
Development of nitric acid corrosion resistant duplex stainless steel NAR-SN-5
Nippon Stainless Tech. Rep. 17 (1982) 23 p.

[452] Denhard jr., E. E.
Austenitic stainless steel combining strength and resistance to intergranular corrosion
US Pat. 3 645 725 (Febr. 1972)

[453] Culling, J. H.
Corrosion resistant chromium-nickel alloys
US Pat. 3 759 704 (Sept. 18, 1973)

[454] Shapovalov, E. T.; Kazakova, G. V.; Andrushova, N. V.
An investigation of the processes of pickling and passivation in a non-contact electrochemical pickling bath (in Russian)
Stal USSR (1983) 1, p. 39

[455] Gramberg, U.; Günther, T.; Palla, H.
Erfahrungen beim Einsatz der Rasterelektronenmikroskopie im Rahmen der Untersuchung von Schadensfällen unter korrosiven Bedingungen
(Experience gained with the scanning electron microscopy during the evaluation of case histories with corrosive conditions) (in German)
Werkst. Korros. 26 (1975) 6, p. 461

[456] Stefec, R.; Protiva, K.
Improvement of the corrosion resistance of type FeCr21Ni-32TiAl steel by a cold work and anneal combination treatment (in Czech)
Kovove Mater. 20 (1977) p. 225

[457] Malakhova, E. K.; Kuzyukov, A. N.
Plant experience with equipment in chemicals that cause intercrystalline corrosion and corrosion cracking (in Russian)
Tr. Vses. Konstr. Inst. Khim. Mashinostr 78 (1977) p. 110

[458] Ruf, R. R.; Tsuei, C. C.
Extremely high corrosion resistance in amorphous Cr-B alloys
J. Appl. Phys. *54* (1983) 10, p. 5705

[459] Vajpeyi, M.; Gupta, S.; Dhirendra; Pandey, G. N.
Corrosion of stainless steel (AISI 304) in H_2SO_4 contaminated with HCl and HNO_3
Corros. Prev. Control *32* (1985) 5, p. 102

[460] Kowaka, M.; Yamanaka, K.
The effect of alloying elements and testing methods on intergranular fracture in chloride stress corrosion cracking of austenitic stainless steels (in Japanese)
Corros. Eng. (Boshoku Gijutsu) *32* (1983) 9, p. 449

[461] Tokunaga, K.; Sakitani, K.
Thin oxide film on stainless steel in 30% nitric acid solution (in Japanese)
Corros. Eng. (Boshoku Gijutsu) *32* (1983) 4, p. 221

[462] Sinigaglia, D.; Re, G.
Intergranular corrosion (in Italian)
Metall. Ital. *67* (1975) 11, p. 660

[463] Gräfen, H.; Gerischer, K.; Gramberg, U.
Schadensverhütung in Chemiebetrieben durch systematische Schadensuntersuchung (Prevention of damage in chemical plants through systematic failure analysis) (in German)
Z. Werkstofftech. *8* (1977) 11, p. 357

[464] Houdremont, E.
Handbuch der Sonderstahlkunde (Handbook of special steels) (in German), 3rd edition
Springer-Verlag, Berlin; Verlag Stahleisen, Düsseldorf, 1956

[465] Niendorf, K.; Peuker, R.
Zum Einfluß des Nitrosegehaltes konzentrierter Schwefelsäure auf das Korrosionsverhalten austenitischer Chrom-Nickel-Stähle bei hohen Temperaturen (The influence of the nitrogen content of sulfuric acid on the corrosion behavior of austenitic chromium-nickel steels at elevated temperatures) (in German)
Chem. Tech. *28* (1976) 4, p. 236

[466] Naumann, G.
Korrosionsverhalten des austenitischen Chrom-Nickel-Stahles X 5 CrNiN 19.7 in Salpetersäure bei erhöhten Temperaturen (Corrosion behaviour of the austenitic chromium nickel steel X 5 CrNiN 19.7 in nitric acid at elevated temperatures) (in German)
Chem. Tech. *30* (1978) 2, p. 94

[467] Casarini, G.; Colonna, C.; Songa, T.
Comportamento elettrochimico dell'acciaio inossidabile AISI 304 accoppiato con Pt, Au e grafite in soluzioni concentrate di HNO_3 (65% in peso) (Electrochemical behavior of the stainless steel AISI 304 in contact with Pt, Au and graphite in concentrated nitric acid (65 percent by weight)) (in Italian)
Metall. Ital. *62* (1970) p. 183

[468] Wehner, H.; Speckhardt, H.
Zum Ausscheidungs- und Korrosionsverhalten eines ferritisch-austenitischen Chrom-Nickel-Molybdän-Stahles nach kurzzeitiger Glühbehandlung unter besonderer Berücksichtigung des Schweißens (Structural changes and corrosion behaviour of a ferritic-austenitic chromium-nickel-molybdenum steel after short time heat treatments with special consideration of welding) (in German)
Z. Werkstofftech. *10* (1979) 9, p. 317

[469] Babakov, A. A.; Posysaeva, L. I.; Petrovskaya, V. A.; Sidorkina, Yu. S.
New, high-alloyed anti-corrosion steel 000Ch21N21M4B (in Russian)
Zashch. Met. *7* (1972) 2, p. 99

[470] Ogneva, V. K.; Kiselev, V. D.; Chernova, T. V.
Stainless steels corrosion resistance in media of nitric acid decomposition of apatite (in Russian)
Khim. Prom. (1980) p. 480

[471] Weingerl, H.; Straube, H.; Blöch, R.
Über die Auswirkung von Seigerungen auf das Korrosionsverhalten austenitischer Cr-Ni-Mo-Stähle (Contribution to the effect of segregation on the corrosion resistance of austenitic Cr-Ni-Mo-steels) (in German)
Werkst. Korros. *27* (1976) 2, p. 69

[472] Kohl, H.; Hochörtler, G.; Kriszt, K.; Koren, M.
Sonderstähle für den Apparatebau mit sehr guter Korrosionsbeständigkeit und erhöhter Festigkeit (Special steels with superior corrosion resistance and strength for chemical equipment manufacture) (in German)
Werkst. Korros. *34* (1983) 1, p. 1

[473] Plaskeev, A. V.; Knyazheva, V. M.; Dergach, T. A.; Dembrovskij, M. A.
The corrosion behavior of CrNiMo steels in

nitric acid (in Russian)
Zashch. Met. *14* (1978) 4, p. 393

[474] Anonymous
New stainless tubing for salt water service
Mater. Eng. *84* (1976) 3, p. 20

[475] Kolotij, A. A.; Ivanova, E. G.
Corrosion of certain steels in caustic alkali melts (in Russian)
Zashch. Met. *14* (1978) 2, p. 190

[476] Knyazheva, V. M.; Chigal, V.; Kolotyrkin, Ya. M.
Role of surplus phases in the corrosion resistance of stainless steels (in Russian)
Zashch. Met. *11* (1975) 5, p. 531

[477] Barbosa, M. A.
The pitting resistance of AISI 316 stainless steel passivated in diluted nitric acid
Corros. Sci. *23* (1983) 12, p. 1293

[478] Konstantinova, E. V.; Chechetina, N. A.; Lomovtsev, V. I.
Construction materials for hot nitrofluoride solutions (in Russian)
Zashch. Met. *16* (1980) 1, p. 66

[479] Süry, P.; Brezina, P.
Kurzzeitprüfmethoden zur Untersuchung des Einflusses von Wärmebehandlung und chemischer Zusammensetzung auf die Korrosionsbeständigkeit martensitischer Chrom-Nickel-(Molybdän)-Stähle mit tiefem Kohlenstoffgehalt
(Short-duration test for evaluating the influence of heat-treatment and chemical composition on the corrosion resistance of low carbon martensitic chromium nickel (molybdenum) steels) (in German)
Werkst. Korros. *30* (1979) 5; p. 341

[480] Süry, P.
Untersuchungen zum Einfluss der Kaltverformung auf die Korrosionseigenschaften des rostfreien Stahles X 2 CrNiMo 18 12
(Studies on the effect of cold deformation on corrosion properties of the stainless steel X 2 CrNiMo 18 12) (in German)
Material u. Technik *8* (1980) 4; p. 163

[481] Holm. R.; Horn, E.-M.; Storp, S.
Oxidische Deckschichten auf dem austenitisch-ferritischen Gußwerkstoff G-X 3 CrNiMoCuN 24 6
(Oxide surface layers on the austenitic-ferritic cast iron material G-X 3 CrNiMoCuN 24 6) (in German)
VDI Z. *124* (1982) 23/24, p. 917

[482] Charbonnier, J.-C.
Influence du molybdène sur la résistance des aciers inoxydables à différents types de corrosion, en présence de solutions aqueuses minérales ou de milieux organiques (étude bibliographique)
(Effect of molybdenum on the resistance of stainless steels to different types of corrosion in aqueous mineral solutions or organic media. Bibliographic study) (in French)
Métaux-Corros.-Ind. (1975) Nr. 598, p. 201

[483] Böhm, K.; Frohberg, M. G.
Korrosionsverhalten austenitischer Stähle bei der Prüfung nach Huey in Abhängigkeit vom Verformungsgrad
(Corrosion behaviour of austenitic steels in the Huey test as a function of the degree of deformation) (in German)
Arch. Eisenhüttenwes. *44* (1973) 7, p. 553

[484] Bäumel, A.; Horn, E. M.; Siebers, G.
Entwicklung, Verarbeitung und Einsatz des stickstofflegierten, hochmolybdänhaltigen Stahles X 3 CrNiMoN 17 13 5
(Development, processing and use of nitrogen alloyed high molybdenum steel X 3 CrNiMoN 17 13 5) (in German)
Werkst. Korros. *23* (1972) 11, p. 973

[485] Herrnstein, W. H.; Cangi, J. W.; Fontana, M. G.
Effect of carbon pickup on the serviceability of stainless steel alloy castings
Mater. Performance *14* (1975) 10, p. 21

[486] Minick, G. A.; Wilfley, A. R.; Olson, D. L.
Effect of columbium additions in austenitic stainless steel castings to inhibit intergranular corrosion
Mater. Performance *14* (1975) 9, p. 41

[487] Edström, J. O.; Carlén, J. C.; Kämpinge, S.
Anforderungen an Stähle für die chemische Industrie
(Properties required for steels in the chemical industry) (in German)
Werkst. Korros. *21* (1970) 10, p. 812

[488] Sidorkina, Yu. S.; Zhadan, T. A.; Khakhlova, N. V.; Shapiro, M. B.; Shetkov, A. Y.; Larinova, R. F.
Effect of carbon and of stabilizing elements on the corrosion stability of the ChN28MDT alloy (in Russian)
Zashch. Met. *18* (1982) 1, p. 71

[489] Bershadskaya, T. M.; Makarov, V. A. Lipkin, Ya. N.; Artamonova, N. M.; Albrut, V. M.
Analysis of corrosion conditions in the opera-

tion of pickling baths made of EI-943 alloy (in Russian)
Zashch. Met. *17* (1981) 5, p. 565

[490] Asami, K.; Hashimoto, K.
An X-ray photo-electron spectroscopic study of surface treatments of stainless steels
Corros. Sci. *19* (1979) p. 1017

[491] Sidorkina, Yu. S.; Khakhlova, N. V.; Bekoyeva, G. P.; Sergeyeva, G. V.
Corrosion of the steels 03Ch21N21M4B and 06ChN28MDT in nitric acid containing fluoride ions (in Russian)
Zashch. Met. *13* (1977) 1, p. 39

[492] Feldgandler, E. G.; Kareva, E. N.; Brusentsova, V. M.
Corrosion resistance of low-carbon stainless steels under urea synthesis conditions (in Russian)
Zashch. Met. *6* (1970) 5, p. 540

[493] Liening, E. L.
Practical applications of electrochemical techniques to plant localized corrosion problems
Mater. Performance *19* (1980) 2, p. 35

[494] Poluboyartseva, L. A.; Rejfer, A. A.
Steel and alloy corrosion resistance in production sulfuric acid (in Russian)
Khim. Prom. (1980) p. 481

[495] Fokin, M. N.; Petrovskaya, V. N.; Nikolayeva, G. N.
Evaluating the inclination of the steel 03Ch21N21M4B (ZI-35) to intercrystalline corrosion by potentiostatic etching (in Russian)
Zashch. Met. *13* (1977) 1, p. 71

[496] Konstantinova, E. V.; Zazarova, N. P.
Corrosion of the 47ChNM-2 alloy in nitric fluoride media (in Russian)
Zashch. Met. *14* (1978) 2, p. 187

[497] Ogneva, V. K.; Sidlin, Z. A.; Stroev, V. S.; Dobrolyubov, V. V.; Chernova, T. V.
Economically alloyed corrosion-resisting steels used in Nitrophoska® production (in Russian)
Khim. Prom. (1982) p. 605

[498] Nordin, S.
Rostfreie Spezialstähle für die chemische Industrie. Teil I: Korrosionseigenschaften und Anwendungsbereiche (Stainless special steels in the chemical industry. Part I: Corrosion properties and fields of use) (in German)
Chem.-Anl. + Verfahren (1979) 5, p. 67

[499] Scarberry, R. C.; Graver, D. L.; Stephens, C. D.
Alloying for corrosion control. Properties and benefits of alloy materials
Mater. Protection 6 (1967) 6, p. 54

[500] Anonymous
Korrosionsbeständigkeit von Superferrit (Remanit® 4575) – 6. Korrosionsbeständigkeit
(Corrosion resistance of superferrite (Remanit® 4575) – 6. Corrosion resistance) (in German)
Thyssen Edelst. Tech. Ber. *5* (1979) 1, p. 48

[501] Ohkubo, M.; Takekawa, T.; Miki, M.
Corrosion prevention for stainless steel
Jap. Pat. 73 47 453 (Sumitomo Chemical Co., Ltd.; July 5, 1973)

[502] Zamiryakin, L. K.; Chikunov, V. K.
Corrosion stability of welded joints made of austenitic steels in nitric acid solutions (in Russian)
Zashch. Met. *5* (1969) 4, p. 422

[503] Shapiro, M. B.; Gorlenko, A. P.; Adugina, N. A.; Shadrukhina, E. I.
A new higher strength corrosion-resistant steel with nitrogen (in Russian)
Khim. Neft. Mashinostr. (1977) 9, p. 26

[504] Tokareva, T. B.; Ershova, N. I.; Zubchenko, A. S.
Intercrystalline corrosion of low-carbon, high-chromium ferritic steels (in Russian)
Metalloved. Term. Obrab. Met. (1974) 12, p. 33

[505] Kasparova, O. V.; Bogolyubski, S. D.; Kolotyrkin, Ya. M.; Milman, V. M.
Enhancing the stability of austenitic stainless steels to intergranular corrosion in strongly oxidizing media by regulating the composition of impurities (in Russian)
Zashch. Met. *20* (1984) 6, p. 844

[506] Dundas, H. J.; Bond, A. P.
Niobium and titanium requirements for stabilization of ferritic stainless steels
Intergranular Corrosion of Stainless Alloys, Toronto (Canada), 1977 (Proc. Conf.)
American Society for Testing and Materials (1978), p. 154

[507] Kiesheyer, H.; Lennartz, G.; Brandis, H.
Korrosionsverhalten hochchromhaltiger, ferritischer, chemisch beständiger Stähle
(Corrosion behaviour of high-chromium ferritic stainless steels) (in German)
Werkst. Korros. *27* (1976) p. 416

[508] Edström, J. O.; Carlén, J. C.; Kampinge, S.
Nouveaux matériaux de construction pour le

[508] génie chimique
(New construction materials for chemical engineering) (in French)
La Métallurgie 101 (1969) 8/9, p. 391

[509] Dhirendra, Pandey, G. N.; Sanyal, B.
Comparative corrosion of unalloyed and alloyed steels in single and mixed mineral acids
Corros. Prev. Control 28 (1981) 1, p. 19

[510] Kurtepov, M. M.
Effect of uranium on the corrosion of stainless steel in nitric fluoride solutions (in Russian)
Zashch. Met. 21 (1985) 1, p. 104

[511] Weingerl, H.; Diebold, A.
Surface treatment of corrosion-resistant components for boiling water reactors
Kerntechnik 18 (1976) 3, p. 105

[512] Ponomarenko, V. I.; Novachek, L. A.; Fedorov, Yu. V.
The effect of N-decyl and N-dodecylquinoline iodide on the acid corrosion of carbon steels (in Russian)
Vopr. Khim. i. Khim. Tekhnol. 71 (1983) p. 14

[513] Demyanets, A. A.; Ezau, Ya. Ya.
Effect of chloride ions on the anodic behavior of carbon steels in nitric acid (in Russian)
Khim. i Khim. Tekhnol., Minsk 18 (1983) p. 3

[514] Tolstykh, V. F.; Fedorov, Yu. V.; Uzlyuk, M. V.
Combined inhibitor of acid corrosion on S-5 metals (in Russian)
Zashch. Met. 18 (1982) 2, p. 272

[515] Oknin, I. V.
Iron corrosion in nitric acid solutions (in Russian)
Zashch. Met. 14 (1978) 1, p. 25

[516] Apostolache, S.
Influence of temperature on corrosion inhibitors in the system Fe-HNO_3 (in Rumanian)
Revista de Chimie 30 (1979) 7, p. 693

[517] Milyan, V. V.; Petrivskaya, M. A.; Esip, M. P.; Kuzma, Yu. B.
Einfluß von Zirkoniumdiborid auf die Korrosionsbeständigkeit von Eisen in sauren Medien
(Influence of zirconium diboride on the corrosion resistance of iron in acid media) (in Russian)
Fiz.-Khim. Mekh. Mater. 18 (1982) 5, p. 112

[518] Miskidzhyan, S. P.; Kirilyuk, S. S.; Lavrishin, B. N.; Fedin, R. M.
Cleaning of heat exchangers with mineral acids (in Russian)
Korroz. Zashch. (1982) 8, p. 7

[519] Miller, R. F.; Rhodes, P. R.
Nitric acid as corrosion inhibitor for carbon steel in sulfuric acid
Mater. Protection and Performance 9 (1970) 10, p. 33

[520] Levyanto, S. I.; Putilova, I. N.
Inhibitoren zum Schutz von verchromten Eisenblechen gegen Korrosion
(Inhibitors used to protect chrome-plated iron sheets against corrosion) (in Russian)
Konserv. Ovoshchesus. Promst. 26 (1971) 9, p. 13

[521] Killing, R.
Beseitigung der Anlauffarben beim Schweißen nichtrostender Stähle
(Removal of annealing colours from the welding of stainless steels) (in German)
Praktiker 24 (1972) 11, p. 224

[522] Grishin, A. M.; Kondrashin, Yu. V.; Sentyurev, V. P.
Effect of silicon on the corrosion properties of ChN40B alloy (in Russian)
Zashch. Met. 16 (1980) 6, p. 718

[523] Weingerl, H.; Kahler, E.; Kriszt, K.; Kühnelt, G.; Plessing, R.
Auswirkung der chemischen Zusammensetzung und der Erschmelzung auf die Eigenschaften von Stählen für Harnstoffreaktoren
(Effects of the chemical composition and smelting on the properties of steels for urea-reactors) (in German)
Berg Hüttenmann. Monatsh. 122 (1977) 9, p. 388

[524] Ivanov, E. S.; Makovetski, R. D.
Investigation of the etching of 36NChTYu alloy after thermal treatment (in Russian)
Zashch. Met. 17 (1981) 2, p. 195

[525] Hooper, R. A. E.; Honess, C. V.
Corrosion of type 316 L in nitric acid
Stainless Steel, London, Sept. 1977 (Proc. Conf.)
Climax Molybdenum Co., New York (1978), p. 247

[526] Lang, C.; Teindl, J.
Spannungsrißkorrosion bei MnSiCr-Stählen
(Stress corrosion cracking in MnSiCr steels) (in German)
Neue Hütte 18 (1973) 4, p. 240

[527] Devine, T. M.; Briant, C. L.; Drummond, B. J.
Mechanism of intergranular corrosion of 316

L stainless steel in oxidizing acids
Scr. Metall. *14* (1980) p. 1175

[528] Srivastava, P.; Srivastava, K.
Inhibition of corrosion of mild steel in nitric acid by garlic
Corrosion and Maintenance *6* (1983) 2, p. 149

[529] Gupta, S.; Vajpeyi, Dhirendra; Pandey, G. N.
Determination of corrosion of stainless steel (AISI 304) by mixed vapours of HCl and HNO_3
Corros. Prev. Control *33* (1986) 4, p. 47

[530] Lavrishin, B. N.; Isaev, N. I.; Kvokova, T. M.; Kirilyuk, S. S.
Optimization of the composition of the wash solution used for cleaning heat exchangers in the chemical industry (in Russian)
Fiz.-Khim. Mekh. Mater. *19* (1983) 5, p. 101

[531] Lavrishin, B. N.; Kirilyuk, S. S.; Isaev, N. I.; Fedin, I. M.
Effect of cleaning the floors of rectification columns with inhibiting acid on their coarseness and corrosion resistance (in Russian)
Fiz.-Khim. Mekh. Mater. *18* (1982) 5, p. 80

[532] Cihal, V.; Kubelka, J.
Corrosion cracks on steel in nitrate solutions (in Czech)
Strojirenstvi *13* (1963) 11, p. 837

[533] Dhirendra; Pandey, G. N.; Sanyal, B.
Comparative corrosion of iron and iron alloys by vapours of single and mixed acids
Corros. Prev. Control *29* (1982) 1, p. 10

[534] Musina, A. S.; Lange, A. A.; Bukhman, S. P.; Semashko, T. S.
Change in the surface state of the alloys Permalloy® and Kovar® after chemical treatment (in Russian)
Izv. Akad. Nauk Kaz. SSR. Ser. Khim. (1984) 3, p. 12

[535] Eden, A.
Recent advances in the use of stainless steel for offshore and seawater applications
Anticorros. Methods Mater. *26* (1979) 11, p. 7

[536] Klodt, D. T.; Minick, G. A.
Acid pump impeller corrosion
Mater. Protection and Performance *12* (1973) 6, p. 28

[537] Nayar, H. S.; German, R. M.; Johnson, W. R.
Effect of sintering on the corrosion resistance of 316L stainless steel
Ind. Heat. *48* (1981) 12, p. 23

[538] Ponomareva, O. V.; Shvarts, G. L.; Sycheva, V. N.
Corrosion stability of stainless steel and titanium under conditions of methacrylic acid production (in Russian)
Khim. Prom. (1972) p. 668

[539] Buchner, K. H.
Versuch zur Optimierung von Werkstoffeigenschaften durch gezielte Variation der Legierungszusätze und Anwendung der Multiregressionsanalyse
(An attempt to optimize material by deliberate variation of alloying additions and application of multi-regression analysis) (in German)
Z. Metallkd. *65* (1974) 11, p. 667

[540] Voroshilov, I. P.; Shilov, V. R.; Voroshilova, E. P.; Shapovalova, L. P.
Corrosion behavior of certain structural materials in a mixture of nitric acid and potassium chloride (in Russian)
Zashch. Met. *11* (1975) 5, p. 602

[541] Konda, K.
Descaling of stainless steels
Japan Pat. 77 44 743 (April 8, 1977)

[542] Lang, G. P.; Weidman, S. W.; Armstrong, W. P.; Martin, G. L.; Bradford, W. G.
Removal of chloride from nitric acid by oxidation with permanganate
U.S. At. Energy Comm. MCW 1436 (1958) 28 p.

[543] Kemplay, J., Schieber
Chem. Process Eng. *47* (1966) 7, p. 53

[544] Anonymous
Influence of molybdenum on austenitic chromium-nickel steels (in Bulgarian)
Metallurgiya (Sofia) *39* (1984) 12, p. 12

[545] Rondelli, G.; Vicentini, B.; Sinigaglia, D.
Investigation into precipitation phenomena following heat treatment of ELI ferritic stainless steels and their influence on mechanical and corrosion properties
Microstructural Sci. *12* (1985) p. 73

[546] Krutikov, P. G.; Nemirov, N. V.; Papurin, N. M.
State of the surface of the steel 16GS after various chemical actions (in Russian)
Zashch. Met. *19* (1983) 3, p. 455

[547] Werner, H.; Erkel, K.-P.
Der Einfluß der Oberflächenvorbehandlung auf die Lochfraßkorrosionsbeständigkeit des austenitischen Implantatstahles UR X 2 CrNiMoN 18.12(0)
(The effect of surface treatment on the pitting corrosion resistance of the austenitic implant steel UR X 2 CrNiMoN 18 12(0)) (in German)
Korrosion *16* (1985) 3, p. 131

[548] Arslanov, V. V.; Gevorkyan, O. M.; Ogarev, V. A.
Strength and stability of adhesion compounds formed between anodized magnesium and polymers
Colloid J. of the USA, New York; (Kolloidny Zh. *43* (1981) 5, p. 952)

[549] Apparao, B. V.
Pickling treatment of hot rolled stainless steel (304) problem of pitting corrosion in local industry
J. Electrochem. Soc. India *30* (1981) 3, p. 213

[550] Risch, K.; Althen, W.
Gezieltes Beizen von Apparaten aus chemisch beständigen Stählen
(Controlled pickling of apparatus made of chemical-resistant steels) (in German)
Z. Werkstofftech. *12* (1981) 1, p. 23

[551] Anonymous
Cronifer® 2522 LCN (austenitic stainless steel)
Alloy Dig. (1983) Aug., SS-430, p. 2

[552] Čihal, V.; Kaspar, M.
Trends in requirements for steel quality and steel treatment for urea synthesis (in Czech)
Hutn. Listy *38* (1983) p. 783

[553] Pavlović Milovanovi, J.
Corrosion of stainless steels in nitric acid production plants (in Serbo-Croatian)
Hemijska Industrija, Beograd *27* (1973) p. 203

[554] Sidorkina, Yu. S.; Larionova, R. F.; Bekoeva, G. P.; Khakhlova, N. V.
Corrosion resistance of high-alloyed steels and alloys based on iron and nickel in reaction media used in fertilizer production (in Russian)
Tr. Vses. Nauchno-Issled. Konstrukt. Inst. Khim. Mashinostr. (1977) No. 78, p. 61

[555] Shuler, W. E.
Corrosion by fluoride solutions
U.S. At. Energy Comm. DP-348 (1959) 15 p.

[556] Ford, E.
Silizium macht Eisen säurefest
(Silicon makes iron acid-resistant) (in German)
Technica *21* (1962) 16, p. 1381

[557] Cavallini, M.; Felli, F.; Fratesi, R.; Veniali, F.
Aqueous solution corrosion behaviour of "poor man" high manganese-aluminum steels
Werkst. Korros. *33* (1982) 5; p. 281

[558] Tarasenko, V. A.; Belousova, L. Y.
Reinigen der Rohroberflächen nach dem Pressen von Glasschmiermitteln und Zunder
(Cleaning of pipe surfaces after the pressing of glass lubricants and scale) (in German)
Stal' *9* (1969) 8; p. 821

[559] Dubinin, G. N.; Sokolov, V. S.
Improving the heat and corrosion resistance of copper-base alloys
Sov. Non-Ferrous Metals Res. *5* (1974) p. 302

[560] Staronka, A.; Holtzer, M.
Korrosionsbeständigkeit von CrNi-Stahlguß (Typ 18/9 bzw. 18/13) mit erhöhtem Siliciumgehalt in konzentrierten Salpetersäurelösungen
(Corrosion resistance of chromium nickel cast steel (types 18/9 and 18/13) with increased silicon content in concentrated nitric acid solutions) (in German)
Werkst. Korros. *38* (1987) p. 431

[561] Horn, E.-M.; Schoeller, K.; Storp. S: Dölling, H.
Zur Korrosion der Legierung AlMg2MnO,8 in hochkonzentrierter Salpetersäure
(Corrosion of the alloy AlMg2MnO.8 in highly concentrated nitric acid) (in German)
Werkst. Korros. *37* (1986) p. 74

[562] Schulz, W.-D.
Einige Fragen des Korrosionsschutzes durch Spritzmetallschichten bei vorwiegend medialer Korrosion
(Some questions concerning corrosion protection with sprayed metal layers against predominantly medial corrosion) (in German)
Schweisstechnik (Berlin) *30* (1980) 12, p. 539

[563] Shams El Din, A. M.; Al-Kharafi, F. M.; Al-Fahd, Z.; El-Tantawy, Y. A.
Corrosion behaviour of nickel in nitric acid solutions
Corros. Prev. Control *32* (1985) 5, p. 92

[564] Razygraev, V. P.; Mirolyubov, E. N.; Pisarenko, T. A.
Mechanism of nickel corrosion in nitric acid solutions (in Russian)
Zashch. Met. *9* (1973) 1, p. 48

[565] Khedr, M. G. A.; Mabrouk, H. M.; Abd El Haleem, S. M.
Autocatalytic dissolution of Ni in HNO_3
Corros. Prov. Control *30* (1983) 2, p. 17

[566] Stupniek-Lisac, E.; Karšulin, M.
Electrochemical behaviour of nickel in nitric acid
Electrochim. Acta *29* (1984) 10, p. 1339

[567] Tousek, J.
Die aktivierende Wirkung der Nitrationen auf die Geschwindigkeit der Metallauflösung im passiven Zustand
(The conditions of activation of pit corrosion of metals by nitrate ions were investigated) (in German)
Corros. Sci. *12* (1972) p. 799

[568] Anonymous
Sintered stainless steels with excellent neutron absorbability
Japan Pat. 81 23 256 (Daido Steel Co., Ltd.; March 5, 1981)

[569] Anonymous
Erfolgreich im Kampf gegen Korrosion und Verschleiß: Dicknickelschichten
(Successful in the fight against corrosion and wear: thick nickel layers) (in German)
Metalloberfläche *33* (1979) 11, p. 521

[570] Landolt, D.
Mechanistische Gesichtspunkte der elektrochemischen Metallbearbeitung
(Views on the mechanism of electrochemical treatment of metals) (in German)
Chem.-Ing. Tech. *45* (1973) 4, p. 188

[571] Takei, T.
The effect of various acids on the behavior of nickel electrodeposited from Ni $(CF_3COO)_2$-halide-MeOH bath (in Japanese)
J. Met. Finish Soc. Jpn *36* (1985) 4, p. 166

[572] Anonymous
Stripping nickel layers
USA Pat. 3 856 694 (Oxy Metal Finishing Corp., Warren (Mich.))

[573] Sriveeraraghavan, S.; Krishnan, R. M.; Natarajan, S. R.; Parthasaradhy, N. V.; Udupa, H. V. K.
Immersion stripping of nickel deposits
Met. Finish. *77* (1979) 12, p. 67

[574] Sivakova, E. V.; Lazarev, E. M.; Kornilova, Z. I.; Suchkova, E. M.; Vojtekhova, E. A.
Fractographic examination of disilicide coatings on molybdenum alloys (in Russian)
Fiz. Khim. Obrab. Mater. (1982) 3, p. 70

[575] Melnik, P. I.; Shumilov, V. N.; Rimskaya, E. A.; Sokolovskij, M. F.
Electrochemical behavior of alloys based on nickel and iron (in Russian)
Fiz. Khim. Mekh. Mater. *20* (1984) 2, p. 33

[576] Süry, P.
Korrosionseigenschaften gegossener Chrom-Nickel- und Chrom-Nickel-Molybdänstähle in aggressiven Medien
(Corrosion properties of cast chromium-nickel and chromium-nickel-molybdenum steels in aggressive media) (in German)
Chemie-Technik *4* (1975) 7, p. 241

[577] Åslund, C.; Gemmel, G.; Andersson, T.
Stranggepreßte Rohre auf pulvermetallurgischer Basis
(Extruded tubes based on powder metallurgy) (in German)
Bänder Bleche Rohre *22* (1981) 9, p. 223

[578] Miki, M.
Studies on cathodic protection of stainless steel by contact of aluminum metal and aluminum bearing stainless steel in strong concentrated nitric acid solution (in Japanese)
Corros. Eng. (Boshoku Gijutsu) *32* (1983) 12, p. 701

[579] Product Information (High Technology Materials Division, Kokomo)
Testing will prove Ferralium® alloy 255 outperforms ordinary stainless steels
Chem. Eng. Progr. *77* (1981) 3, p. 29

[580] Lakey, L. T.; Kerr, W. B.
Pilot plant development of an electrolytic dissolver for stainless steel alloy nuclear fuels
Ind. Eng. Chem. Proc. Des. Develop. *6* (1967) 2, p. 174

[581] Asher, J.; Carney, R. F. A.; Conlon, T. W.; Wilkins, N. J. M.; Shaw, R. D.
The detection of pitting corrosion in stainless steel using double layer activation
Corros. Sci. *24* (1984) 5, p. 411

[582] Gupta, S.; Kumar, Y.; Sanyal, B.; Pandey, G. N.
The exothermic reaction of stainless steel (AISI 304) in mixtures of HCl and HNO_3 and the inhibition of its corrosion
Corros. Prev. Control *30* (1983) 1, p. 11

[583] Gupta, S.; Vajpeyi, M.; Dhirendra, N. I.; Pandey, G. N.
Determination of corrosion of stainless steel (AISI 304) by mixed vapours of HCl and HNO_3
Corros. Prev. Control *33* (1986) 2, p. 47

[584] Nakamura, M.; Ito, H.
Polishing solution for iron-nickel-cobalt alloy
Japan Pat. 78 47 335 (April 27, 1978)

[585] Ford, E.
Tantiron® (in German)
Oberflächentechnik *50* (1973), Metall-Journal 2/73, p. A 3

[586] Feldstein, N.
Composite coatings: plating plus particles. Tiny particles in a nickel-alloy matrix stand

up to tough service environments
Mater. Eng. *94* (1981) 1, p. 38

[587] Vdovenko, I. D.; Vakulenko, L. I.;
Shedenko, L. I.; Vakulenko, L. V.
Effect of oxalic acid on the pickling of Chromel (in Russian)
Zh. Prikl. Khim. *57* (1984) p. 2624

[588] Loria, E. A.; Isaacs, H. S.
Type 304 stainless steel with 0.5 % boron for storage of spent nuclear fuel
J. Met. *32* (1980) 12; p. 10

[589] Solok, A. M.; Babchenko, V. A.; Vydra, E. I.; Kuznetsov, V. M.
Use of molten sodium hydroxide for removing high-temperature scale from certain heat-resistant alloys
Sov. Appl. Chem. *46* (1973) p. 2044

[590] Svistunova, T. V.
Nickel alloy
UdSSR Pat. 450 844 (Nov. 25, 1974)

[591] Kireeva, T. S.; Svistunova, T. V.; Runova, Z. K.
A corrosion-resistant alloy used for work with hydrogen fluoride-containing nitric acid etching solutions
Stal' (1979) 2, p. 141

[592] Datta, M.; Landolts, D.
Surface brightening during high rate nickel dissolution in nitrate electrolytes
J. Electrochem. Soc. *122* (1975) p. 1466

[593] Scarberry, R. C.; Pearman, S. C.; Crum, J. R.
Precipitation reactions in Inconel® alloy 600 and their effect on corrosion behavior
Corrosion *32* (1976) 10, p. 401

[594] McIlree, A. R.; Michels, H. T.; Morris, P. E.
Effects of variations of carbon, sulfur and phosphorus on the corrosion behavior of Alloy 600
Corrosion *31* (1975) 12, p. 441

[595] Kramer, L. D.; Michael, S. T.; Pement, F. W.
Service experience and stress corrosion of Inconel® 600 bellows expansion joints in turbine steam environments
Mater. Performance *14* (1975) 9, p. 15

[596] Vermilyea, D. A.; Tedmon jr., C. S.; Broecker, D. E.
Effect of phosphorus and silicon on the intergranular corrosion of a nickel-base alloy
Corrosion *31* (1975) 6, p. 222

[597] Lee, III, T. S.; Hodge, F. G.
Resistance of Hastelloy® alloys to corrosion by inorganic acids
Mater. Performance *15* (1976) 9, p. 29

[598] Svistunova, T. V.; Kireyeva, T. S.; Runova, Z. K.
Structure and corrosion behavior of ChN 60 type in a nitrofluoride solution (in Russian)
Zashch. Met. *19* (1983) 2, p. 212

[599] Rai, A. K.; Bhattacharya, R. S.; McCormick, A. W.; Pronko, P. P.; Khobaib, M.
Corrosion resistant behavior of amorphous Mo-Ni films formed by ion beam mixing
Appl. Surface Sci. *21* (1985) p. 95

[600] Narayana, A.; Nair, K.; Chatterji, B.; Srinivasan, S. S.
Studies in chemical milling of magnesium alloys
J. Electrochem. Soc. India *27* (1978) 3, p. 149

[601] Roque, R.
Mecanismo de la oxidación de las aleaciones Fe-Ni (de 5 a 30 %)
(Oxidation mechanism of FeNi alloys with 5 to 30 % Ni) (in Spanish)
Revista Cenic *11* (1980) 1 – 2, p. 79

[602] Kvärnbach, B.
Sandvik® 2R12 AISI 304 L steel for nitric acid
Chem. Age India *25* (1974) 1, p. 49

[603] Lipodaev, V. N.; Yushchenko, K. A.; Skulskij, V. Y.; Tikhonovskaya, L. D.; Dzykovich, I. Y.
Effect of nickel on the structure and properties of corrosion-resistant welds with high silicon content (in Russian)
Avtom. Svarka (1985) 9, p. 9

[604] Harrison, J. M.; Shaw, R. D.; Worthington, S. E.; Thomas, J. G. N.; Andon, R. J. L.; Pemberton, R. C.
Localised corrosion of stainless steels in nitric acid
UK Corrosion, Wembley, Middlesex (GB) (Proc. Conf.) *2* (1984) p. 180

[605] Svistunova, T. V.
Corrosion-resistant alloys based on nickel
Met. Sci. Heat Trat. (USSR) (1980) p. 483

[606] Kravchinskii, A. P.; Isaev, N. I.; Baluev, V. N.; Zakharov, A. I.; Shumilov, V. N.; Revyakin, A. V.
The anodic behavior of various metals in a magnetic field
Russ. Metall. *25* (1984) 6, p. 30

[607] Page, R. A.; McMinn, A.
Relative stress corrosion susceptibilities of Alloys 690 and 600 in simulated boiling water reactor environments
Metall. Trans. A *17A* (1986) May, p. 877

[608] Herbsleb, G.; Schwaab, P.
Precipitation of intermetallic compounds, nitrides and carbides in AF 22 duplex steel and their influence on corrosion behavior in acids
Mannesmann Forschungsber. (1983) No. 957, 26 p.

[609] Imai, H.; Fukumoto, I.; Masuko, N.
The influence of germanium on the corrosion resistance of 25Cr-6Ni duplex steel (in Japanese)
Corros. Eng. (Boshoku Gijutsu) 34 (1985) p. 339

[610] Ivanov, E. S.; Komolova, L. F.; Makovskii, R. D.
Investigation of the surface of 36 alloys after etching in nitric acid/fluoride solutions (in Russian)
Zashch. Met. 19 (1983) 6, p. 935

[611] Knotek, O.; Lugscheider, E.; Wichert, W.
On the structure and properties of wear- and corrosion-resistant nickel-chromium-tungsten-carbon-(silicon) alloys
Thin Solid Films 53 (1978) p. 303

[612] Babchenko, A. M.; Khodos, R. S.; Kistanov, A. I.; Maximenko, A. S.; Nozdrachev, A. V.
Effect of the smelting method on the etching of alloy EI437B in a mixture of nitric and hydrofluoric acids (in Russian)
Zashch. Met. 14 (1978) 4, p. 463

[613] Crum, J. R.; Scarberry, R. C.
Corrosion testing of Inconel® alloy 690 for PWR steam generators
J. Mater. Energy Systems 4 (1982) 3, p. 125

[614] Page, R. A.
Stress corrosion cracking of Alloys 600 and 690 and Nos. 82 and 182 weld metals in high temperature water
Corrosion 39 (1983) 10, p. 409

[615] Crum, J. R.
Effect of composition and heat treatment on stress corrosion cracking of Alloy 600 steam generator tubes in sodium hydroxide
Corrosion 38 (1982) 1, p. 40

[616] Briant, C. L.; O'Toole, C. S.; Hall, E. L.
The effect of microstructure on the corrosion cracking of Alloy 600 in acidic and neutral environments
Corrosion 42 (1986) 1, p. 15

[617] Okada, Y.; Yoshikawa, K.; Yukitoshi, T.
Measurement of sensitization behavior in Ni-based Alloy 600 by means of magnetic property (in Japanese)
J. Jpn. Inst. Met. 45 (1981) 5, p. 496

[618] Kurosawa, F.; Taguchi, I.; Matsumoto, R.
Determination of gamma prime in Ni-base by non-aqueous electrolyte-potentiostatic method (in Japanese)
J. Jpn. Inst. Met. 44 (1980) 10, p. 1187

[619] Paige, B. E.
Corrosion of nickel alloys in nuclear fuel reprocessing
Mater. Performance 15 (1975) 12, p. 22

[620] Hodge, F. G.; Kirchner, R. W.
An improved Ni-Cr-Mo alloy for corrosion service
Corrosion 32 (1976) 8, p. 332

[621] Manning, P. E.; Schöbel, J.-D.
Hastelloy® Alloy C 22 – ein neuer und vielseitig anwendbarer Werkstoff für die chemische Industrie
(Hastelloy® alloy C 22 – a new and versatile material for the chemical process industries) (in German)
Werkst. Korros. 37 (1986) p. 137

[622] Sidorkina, Yu. S.; Khakhlova, N. V.; Snetkov, A. Y.; Sergeeva, G. V.; Larionova, R. F.; Akshentseva, A. P.
Investigation of the effect of the α-phase on the corrosion stability of the alloys 06ChN28MDT, 03ChN28MDT and the steels 03Ch21N21M4B and 10Ch17N13M2T (in Russian)
Zashch. Met. 16 (1980) 5, p. 589

[623] Solonin, S. M.; Chernyshev, L. I.; Fedorchenko, I. M.
Investigation of the properties of porous materials made of nickel-molybdenum and nickel-chromium-molybdenum alloys (in Russian)
Poroshk. Metall. 12 (1972) 6, p. 36

[624] Yoshida, T.; Takizawa, Y.; Sekine, I.
The characteristics and applications of recently improved Hastelloy® alloys (in Japanese)
Corros. Eng. (Boshoku Gijutsu) 31 (1982) 2, p. 81

[625] Asphahani, A. I.
Corrosion resistance of high performance alloys
Mater. Performance 19 (1980) 12, p. 33

[626] Abelev, M. M.; Shvarts, G. L.; Gerasimenko, G. I.; Belinkij, A. L.; Besednyi, V. A.; Kostyuchenko, V. A.
Fabrication of fractionating columns of

[626] nickel-chromium-molybdenum alloy EP567
Chem. & Petroleum Eng., New York (1971) p. 699

[627] Nickel, O.
Eigenschaften und Anwendungen von Reinnickel, Nickel-Basislegierungen und nickelhaltigen NE-Werkstoffen im Rohrleitungsbau
(Properties and uses of pure nickel, nickel-base alloys and nickel-bearing non-ferrous metals in piping) (in German)
3R international 14 (1975) 8, p. 419

[628] Nickel, O.
Eigenschaften und neuere Anwendungen nickelhaltiger Werkstoffe
(Properties and new applications of nickel-containing materials) (in German)
VDI Z 121 (1979) 7, p. 313

[629] Zall, D. M.; Bolander, E. H.
In situ identification of alloys. Spot testing alloys for submarine construction
Mater. Protection 6 (1967) 7, p. 37

[630] Anonymous
Improved version of Hastelloy® alloy B
Anticorros. Methods Mater. 22 (1975) 11, p. 19

[631] Anonymous
Nickel alloy for use in as-welded condition
Mater. Eng. 90 (1979) 1, p. 22

[632] Schneider, W.; Weinlich, W.
Nickel-Beryllium aushärtbare Federlegierung mit besonderen Eigenschaften
(Nickel-beryllium age-hardable spring alloy with special properties) (in German)
Metall 36 (1982) 11, p. 21

[633] Volokhova, V. I.; Vakhidov, R. S.; Lukyanitsa, A. I.
Corrosion stability of nickel-phosphorus galvanic coatings (in Russian)
Zashch. Met. 11 (1975) 3, p. 370

[634] Kirchheiner, R. R.; Hofmann, F.; Hoffmann, T.; Rudolph, G.
A silicon-alloyed stainless steel for highly oxidizing conditions
Mater. Performance 26 (1987) 1, p. 49

[635] Kirchheiner, R. R.; Heubner, U.; Hofmann, F.
Increasing the life-time of nitric acid equipment using improved stainless steels and nickel alloys
Corrosion 88, St. Louis-Missouri, 1988, Nr. 318, 22 p.

[636] Steigerwald, R. F.
New molybdenum stainless steels for corrosion resistance: A review of recent developments
Mater. Performance 13 (1974) 9, p. 9

[637] Bartosiewicz, L.
Preferential electrolytic etching technique to reveal the phase in nickel base alloys
Prakt. Metallogr. 10 (1973) p. 450

[638] Horn, E.-M.; Storp, S.; Holm, R.
Korrosion und Deckschichtenaufbau von X 5 CrTi 12, Werkstoff-Nr. 1.4512, bei Beanspruchung in Salpetersäure
(Corrosion resistance and formation of protective surface films on X 5 CrTi 12, material no. 1.4512, in nitric acids) (in German)
Werkst. Korros. 37 (1986) p. 69

[639] Miska, K. H.
Take a look at lead to lock out corrosion, radiation and noise
Mater. Eng. 81 (1975) 3, p. 22

[640] Sokolova, I. V.; Kiryakov, G. Z.; Dunaev, Yu. D.
Corrosion of lead-alkaline metal alloys (in Russian)
Zashch. Met. 9 (1973) 3, p. 332

[641] Kovriga, Yu. P.; Yartsev, M. G.; Bardin, V. A. Corrosion of indium-lead alloys in solutions of HCl, HNO_3, NaOH and NaCl (in Russian)
Zashch. Met. 18 (1982) 3, p. 433

[642] Reinert, M.
Bleiqualität aus technologischer und anwendungstechnischer Sicht
(Lead grade seen from a technological and application viewpoint) (in German)
Metall 28 (1974) 2, p. 135

[643] Müller, I.
Metallographie von Bleiwerkstoffen
(Metallography of lead materials) (in German)
Metall 33 (1979) 9, p. 929

[644] Kraft, G.
Bleiqualität aus analytischer Sicht
(Lead quality from an analytical point of view) (in German)
Metall 27 (1973) 12, p. 1171

[645] Fawzy, M. A.; Sedahmed, G. H.; Mohamed, A. A.
Corrosion behaviour of Pb-Sn binary alloys in acid solutions
Surf. Tech. 14 (1981) p. 257

[646] Lehonyai, N.; Juhasz-Kis, J.; Lipovetz, I.
Acidic lead corrosion in the presence of inhibitors 41st Corrosion Week, Manifestation Eur. Fed. Corrosion,
1968, publ. 1970, p. 659

[647] Vaidyanathan, H.; Narasagoudar, R. A.; O'Keefe, T. J.
Film formation on anodically polarized lead
J. Electrochem. Soc. *121* (1974) p. 876

[648] Horinchi, K. et al.
Chromium-Molybdenum-Nickel alloy
Japan Pat. 70 36 659 (Nov. 20, 1970)

[649] Jesso Steel Comp., Washington, Pa
When you need strength plus corrosion resistance, specify Jesso JS 625
Mater. Eng. *99* (1984) 5, p. 17

[650] Anonymous
FMN duplex steel for chemical process plant
Anticorros. Methods Mater. *26* (1979) 10, p. 10

[651] Tikhonenko, A. D.; Arustamyan, E. S.; Olevskij, V. M.
Reduction of platinum losses in weak nitric acid production (in Russian)
Khim. Prom. (1980) p. 608

[652] Hoare, J. P.
A determination of the oxygen content of platinum-oxygen alloys
J. Electrochem. Soc. *127* (1980) p. 1758

[653] Kerr, W. B.; Lakey, L. T.; Denney, R. G.
Development of a continuous electrolytic process for recovery of uranium from high-enriched stainless steel and Nichrome fuels
US At. Energy Comm. IDO-14643 (1965) 37 p.

[654] Wopersnow, W.; Raub, Ch. J.
Eigenschaften einiger binärer intermetallischer Phasen des Palladiums und Rutheniums mit anderen Metallen
(Properties of some binary, intermetallic phases of palladium and ruthenium with other metals) (in German)
Metall *33* (1979) 7, p. 736

[655] Szaplonczay, A. M.
Effect of copper on the solubility of palladium in nitric acid solutions in analysis of electroplated palladium layers
Anal. Chem. *55* (1983) 13, p. 2202

[656] Spendelow jr., H. R.; Crafts, W.
Nickel-base alloy for high temperature service
US Pat. 2 703 277 (Union Carbide & Carbon Corp.; March 1, 1955)

[657] Liu, J.
Synthesis and the crystal structure of lithium-palladium hydride
Ph. D. Thesis Denver Univ., Colorado, Order No. 74-24661 (1974) 87 p.

[658] Abd El Haleem, S. M.; Khedr, M. G. A.; El Kot, A. M.
The dissolution of tin in HNO_3
Corros. Prev. Control *28* (1981) 2, p. 5

[659] Baraka, A.; Ibrahim, M. E.; Al-Abdallah, M. M.
Thermometric study of the dissolution of tin in acid solutions
Br. Corros. J. *15* (1980) 4, p. 212

[660] Ibrahim, M. E.; El-Khrisy, E. A. M.; Al-Abdallah, M. M.; Baraka, A.
Bewertung der Inhibitorwirkung einiger als Korrosionsinhibitoren verwendeter Naturstoffe. 1. Bei der Auflösung von Zinn in Salpetersäure
(Assessment of the inhibition efficiency of some natural substances used as inhibitors) (in German)
Metalloberfläche *35* (1981) 4, p. 134

[661] Anonymous
Le leghe Wiggin di nichel negli impianti chimici ed industriali
(The Wiggin nickel alloy in chemical and industrial plants) (in Italian)
Ingegneria Chim. *27* (1978) 2, p. 12

[662] Mallory, G. O.
Properties of electroless nickel-molybdenum alloys
Plat. Surf. Finish. *63* (1976) 6, p. 34

[663] Masui, K.; Yamada, T.; Hisamatsu, Y.
Effects of phosphorus content and heat-induced structural change of electro-deposited Ni-P alloys on acid corrosion resistance (in Japanese)
J. Met. Finsih. Soc. Jpn. *32* (1981) 8, p. 410

[664] Baraka, A.; Ibrahim, M. E.; Al-Abdallah, M. M.
Die Verzögerungswirkung von n-Alkylaminen bei der Auflösung von Zinn in Salpetersäurelösung
(The delay effect of n-alkylamines on the dissolution of tin in nitric acid solutions) (in German)
Metalloberfläche *35* (1981) 7, p. 263

[665] Shuker, F. S.
When to use refractory metals and alloys in the plant
Chem. Eng. (New York) *90* (1983) 9, p. 81

[666] Fukuzuka, T.; Shimogori, K.; Satoh, H.; Kamikubo, F.
Inhibition of corrosion and hydrogen embrittlement of tantalum in concentrated sulfuric acid solutions at elevated temperatures by addition of N-O compounds (in Japanese)
Corros. Eng. (Boshoku Gijutsu) *30* (1981) 6, p. 327

[667] Schreiber, F.
Herstellung von Tantal, Niob und ihre Eigenschaften
(Production of tantalum, niobium and their properties) (in German)
Chemie-Technik 11 (1982) 6, p. 685

[668] Brandner, R.
Reaktoren aus Sondermetall
(Reactors made of special metal) (in German)
Chemie-Technik 12 (1983) 2, p. 72

[669] Anonymous
Konzentrieren von Salpetersäure mit Zusatz von Schwefelsäure. Teil 2
(Concentration of nitric acid with addition of sulfuric acid. Part 2) (in German)
Maschine + Werkzeug 79 (1978) 19, p. 56

[670] Ammar, I. A.; Darwish, S.; Ammar, E. A.
Parameter des anodischen Oxidwachstums auf Niob
(Parameters of anodic oxide growth on niobium) (in German)
Metalloberfläche 27 (1973) 5, p. 163

[671] Kiparisov, S. S.; Levinskij, Yu. V.; Stroganov, Yu. D.; Kruchkova, N. A.
Effect of nitriding on the corrosion resistance of niobium and tantalum (in Russian)
Zashch. Met. 7 (1971) 4, p. 464

[672] Zhu Hun-De; Wang Jing-Yun
The morphology of carbides and α-phase in cast-nickel-base superalloys
Prakt. Metallogr. 17 (1980) p. 680

[673] Meisel, H.; Johner, G.; Scholz, A.
Metallographic development of the structure of nickel based-alloys
Prakt. Metallogr. 17 (1980) p. 261

[674] Sato, A.; Kon, K.; Tsujikawa, S.; Hisamatsu, Y.
Effects of aging and misorientation on intergranular corrosion of Inconel® Alloy 600 (in Japanese)
J. Jpn. Inst. Met. 43 (1979) 7. p. 664

[675] Copson, H. R.; van Rooyen, D.; McIlree, A. R.
Stress corrosion behavior of Ni-Cr-Fe alloys in high temperature aqueous solutions
5th Intern. Congress Metallic Corrosion, Tokyo 1972, publ.
NACE, Houston (1974) p. 376

[676] Pini, G. C.; Weber, J.
Die Abscheidung von dispersionsgehärteten Blei-Überzügen
(The deposition of dispersion-hardened lead coatings) (in German)
Oberfläche Surf. 19 (1978) 3, p. 54

[677] Simon, F.; Zilske, W.
Abscheidung von Palladium aus einem sauren Elektrolyten
(Deposition of palladium from an acid electrolyte) (in German)
Galvanotechnik 73 (1982) 9, p. 981

[678] Hochörtler, G.; Horn, E.-M.
Austenitic stainless steel with approximately 5.3 % silicon
8th Internat. Congress Metallic Corrosion, Mainz, 1981
(Proc. Conf.) Vol. 2, p. 1447

[679] Horn, E.-M.; Kügler, A.
Entwicklung, Eigenschaften, Verarbeitung und Einsatz des hochsiliciumhaltigen austenitischen Stahls X 2 CrNiSi 18 15
(Development, properties, processing and applications of high-silicon steel grade X 2 CrNiSi 18 15) (in German)
Z. Werkstofftech. 8 (1977) p. 410

[680] Karasev, A. F.; Stabrovskij, A. I.
Behavior of titanium and its tantalum and niobium alloys during anodic polarization in nitric acid (in Russian)
Zashch. Met. 11 (1975) 4, p. 461

[681] Zhivotovskij, E. A.; Zhivotovskaya, G. P.; Brynza, A. P.; Levin, A. I.
Corrosion and passivation of titanium in mixtures of sulfuric, phosphoric and nitric acids (in Russian)
Zashch. Met. 14 (1978) 1, p. 72

[682] Rüdinger, K.
Moderne Werkstoffe Auswahl Prüfung Anwendung. Übersichten über Sondergebiete der Werkstofftechnik für Studium und Praxis. Titan und Titanlegierungen
(Modern materials selection testing application. Review of special areas of material technology for study and practice. Titanium and its alloys.) (in German)
Z. Werkstofftech. 9 (1978) p. 181

[683] Tomashov, N. D.; Kazarin, V. I.; Mikheev, V. S.; Goncharenko, B. A.; Sigalovskaya, M. P.; Kalyanova, M. P.
Investigation of titanium-molybdenum-zirconium alloys with increased corrosion stability in acid solutions (in Russian)
Zashch. Met. 13 (1977) 1, p. 3

[684] Karasev, A. F.; Stabrovskij, A. I.
Behavior of titanium and its tantalum alloys during cathodic polarization in nitric acid solutions (in Russian)
Zashch. Met. 6 (1970) 3, p. 324

[685] Ruskol, Yu. S.; Tomashov, N. D.; Modestova, V. N.; Estrina, N. D.; Shamis, N. V.; Lobanova, L. P.
Hydrogenization and dissolution of the alloy Ti-Mo (33 %) during contact with titanium in acid solutions (in Russian)
Zashch. Met. 13 (1977) 2, p. 154

[686] Sedenkov, A. M.; Berezovskij, L. R.
Inhibition of titanium corrosion in hot nitro-nitrate solutions by sulfate ions (in Russian)
Zashch. Met. 21 (1985) 1, p. 111

[687] Kiparisov, S. S.; Levinskii, Yu. V.; Strogonov, Yu. D.; Kryuchkova, N. A.
Corrosion of nitrided titanium in certain corrosive media
Russ. Appl. Chem. 45 (1972) p. 1794

[688] Anonymous
Titanium and its alloys
Mater. Eng. 80 (1974) 7, p. 61

[689] Bomberger, H. B.; Plock, L. F.
Methods used to improve corrosion resistance of titanium
Mater. Protection 8 (1969) 6, p. 45

[690] Tomashow, N. D.
Untersuchungen zur Passivität und Korrosionsbeständigkeit von Legierungen
(Tests on the passivity and corrosion resistance of alloys) (in German)
Chem. Tech. 30 (1978) 1, p. 6

[691] Zotova, L. M.; Blashchuk, V. E.; Maximov, Yu. A.; Vavilova, V. V.
Stress corrosion of welded seams on alloys AK1 and AK2 (in Russian)
Zashch. Met. 9 (1973) 6, p. 707

[692] Shapovalov, V. P.; Gornunov, N. S.; Brynza, A. P.; Fedash, V. P.; Legashova, T. P.
Corrosion stability and electrochemical behavior of titanium coatings (in Russian)
Zashch. Met. 9 (1973) 4, p. 465

[693] Sedenkov, A. M.; Kuzub, V. S.; Berezovskij, L. R.
On the nature of the oxide film on titanium in nitric acid containing fluorides (in Russian)
Zashch. Met. 19 (1983) 4, p. 589

[694] Rüdinger, K.
Beeinflussung des Korrosionsverhaltens von Titan durch konstruktive und fertigungstechnische, insbesondere schweißtechnische Maßnahmen
(Influencing the corrosion behavior of titanium by constructional, manufacturing and in particular welding measures) (in German)
Schweissen Schneiden 27 (1975) p. 436

[695] Anonymous
Properties and applications of titanium
Corros. Prev. Control 24 (1977) 4, p. 11

[696] Okhramovich. L. N.; Tomashov, N. D.; Burchuk, L. I.; Onasenko, V. K.; Kovalenko, A. K.
Acid solutions for the removal of scale from titanium surfaces (in Russian)
Zashch. Met. 18 (1982) 1, p. 66

[697] Peksheva, N. P.; Vorontsov, E. S.
Chemical stability of coloured films on titanium OT4-0 obtained by anodic oxidation (in Russian)
Zashch. Met. 10 (1974) 3, p. 334

[698] Stöckel, D.
Herstellung und Eigenschaften von Spezialrohren aus Titan
(Manufacture and characteristics of special titanium tubes) (in German)
Bänder-Bleche-Rohre 15 (1974) 11, p. 444

[699] Pircher, H.; Pennenkamp, R.; Sussek, G.
Plattiertes Grobblech mit Titan-Auflage
(Titanium-clad plates) (in German)
Z. Werkstofftech. 13 (1982) 11, p. 371

[700] Kopeliovich, D. K.; Ivlev, V. I.
Corrosion of titanium and its alloys in nitric acid solutions with additions of hydrogen peroxide (in Russian)
Zashch. Met. 13 (1977) 4, p. 439

[701] Simmen, B.; Schmalfuss, D.
Comments on the preparation of metallographic specimens of titanium with particular reference to chemical polishing
Prakt. Metallogr. 15 (1978) p. 78

[702] Tyr, S. G.; Boboshko, Z. A.; Glushko, I. D.; Kechedzhi, T. V.; Figurnaya, S. V.; Belchikov, V. I.; Tembel, V. P.
Comparison of corrosion inhibitors developed and produced in the USSR and the socialist countries. On the attempt to use inhibitors in the etching of titanium alloy PT-3V (in Russian)
Zashch. Met. 19 (1983) 4, p. 629

[703] Marichev, V. A.; Lunin, V. V.
The non-singular effect of cathodic polarization on the corrosion cracking of titanium alloys (in Russian)
Zashch. Met. 16 (1980) 6, p. 674

[704] Tostmann, K.-H.
Korrosion von Titan im galvanotechnischen Einsatz
(Corrosion of titanium in electroplating prac-

tice) (in German)
Galvanotechnik *71* (1980) 12, p. 1310

[705] Kossyj, G. G.; Novakovskij, V. M.; Kolotyrkin, Ya. M.
Anodic dissolution of passive titanium in the presence of hydrofluoric acid (in Russian)
Zashch. Met. *5* (1969) 2, p. 210

[706] Kopeliovich, D. H.; Ivlev, V. I.
Resistance of the passive state of titanium in concentrated nitric acid-containing hydrochloric acid (in Russian)
Zashch. Met. *15* (1979) 6, p. 684

[707] Tomashov, N. D.; Okhramovich, L. N.; Burchuk, L. I.
Etching of oxidized titanium in acid solutions (in Russian)
Zashch. Met. *11* (1975) 1, p. 10

[708] Tomashov, N. D.; Anoshkin, N. F.; Moroznikova, S. V.; Oginskaya, E. I.; Ruskol, Yu. S.; Chernova, G. P.
Effect of palladium on the technological, mechanical and corrosion properties of titanium alloys (OT4 and VT14) (in Russian)
Zashch. Met. *9* (1973) 6, p. 672

[709] Rozenfeld, I. L.; Babkin, Yu. A.; Alekseeva, E. I.
Etching of high-strength titanium steels without hydrogenizing (in Russian)
Zashch. Met. *6* (1970) 4, p. 410

[710] Mansfeld, F.; Kenkel, J. V.
An example of chemical corrosion
Corros. Sci. *16* (1976) p. 653

[711] Abdel Hady, Z.; Pagetti, J.
Anodic behaviour of titanium in concentrated sulphuric acid solutions. Influence of some oxidizing inhibitors
J. Appl. Electrochem. *6* (1976) p. 333

[712] Savochkin, V. R.; Nagaj, I. N.
Gap voltage and protective properties of oxide films on titanium (in Russian)
Zashch. Met. *17* (1981) 3, p. 318

[713] Fischer, W. R.; Ilschner-Gensch, C.; Knorr, W.
Über den Einfluß von Legierungszusätzen auf das Korrosionsverhalten von Titan
(The effect of alloying additions on the corrosion behavior of titanium) (in German)
Werkst. Korros. *12* (1961) 10, p. 597

[714] Simon, H.
Oberflächenreaktionen an Titanwerkstoffen. Probleme und Problemlösungen
(Surface reactions on titanium materials. Problems and solutions) (in German)
Metalloberfläche *36* (1982) 5, p. 211

[715] Ageev, N. V.; Rubina, E. B.; Babareko, A. A.; Khorev, A. I.
Pickling effects on some titanium alloys as a function of the structure (in Russian)
Fiz. Khim. Obrab. Mater. (1977) 4, p. 99

[716] Köcher, R.
Anwendung von hochkorrosionsbeständigen Werkstoffen im Chemie-Apparatebau
(Application of high corrosion resistance materials in chemical apparatus construction) (in German)
VDI Z. *115* (1973) 8, p. 649

[717] Krivenko, M. P.; Rozenkova, V. A.; Solntsev, S. S.; Arzhakov, V. M.; Zhukov, N. D.
Effect of technical coatings on the fatigue strength of titanium alloy pressed parts (in Russian)
Fiz. Khim. Mekh. Mater. *16* (1980) 1, p. 106

[718] Taylor, E.
Corrosion resistance of multiphase alloys
Mater. Protection *9* (1970) 3, p. 29

[719] Ezau, Ya. Ya.
Protective effect of cathodic currents and sulfur-containing compounds in nitric acid on iron (in Russian)
Khim. i Khim. Tekhnol., Minsk *12* (1977) p. 12

[720] Ezau, Ya. Ya.
Effect of mixtures of tin(IV) chloride and sulfur-containing compounds on the corrosion of iron and carbon steel in nitric acid (in Russian)
Khim. i Khim. Tekhnol., Minsk *10* (1976) p. 72

[721] Lyons, V. E.
Corrosion problems in fertilizer plants
Chem. Processing. Eng., India *5* (1971) 4, p. 11

[722] Shibuya, Y.
Corrosion resistivity of boronized steel (in Japanese)
J. Soc. Mater. Sci. Jpn. *26* (1977) p. 120

[723] Demyanets, A. A.; Ezau, Ya. Ya.
Effect of chloride ions on the corrosion and electrochemical behavior of iron in carbon steels in nitric acid (in Russian)
Khim. i. Khim. Tekhnol., Minsk *15* (1980) p. 21

[724] Ulyanin, E. A.
New corrosion-resistant steels and alloys (in Russian)
Metalloved. Term. Obrab. Met. (1972) 4, p. 14

[725] Maslov, V. A.; Semenova, L. A.
Evaluation of the corrosion resistance of

welded joints in titanium alloys in nitric acid solutions
Weld. Prod. (USSR) (1982) 6, p. 29

[726] Shvarts, G. L.; Moroz, V. A.; Gerasimenko, G. I.; Akshentseva, A. P.; Afanasenko, E. A.
Fields of use of the nickel-chromium-molybdenum alloy ChN60MB (in Russian)
Khim. Neft. Mashinostr. (1975) 8, p. 20

[727] Anonymous
in: Edelmetall-Taschenbuch „Chemische und elektrochemische Eigenschaften"
(Noble metal pocket book, "Chemical and electrochemical properties" (in German) p. 46)
Degussa AG, Frankfurt, 1967

[728] Grundmann, R.; Gümpel, P.; Michel, E.
Betrachtungen über die Einsatzmöglichkeiten eines ferritisch-austenitischen und eines hochfesten austenitischen Stahles im Chemikalientankerbau
(Considerations regarding the application possibilities of a ferritic austenitic, high-strength steel in chemical tanker construction) (in German)
Thyssen Edelst. Tech. Ber. 14 (1988) 1, p. 49

[729] Arlt, N.; Gümpel, P.; Michel, E.
Korrosionsverhalten von X 2 CrNi 19 11 (Remanit® 4306) in Salpetersäure mit Zusatz von sechswertigem Chrom
(Corrosion behavior of X 2 CrNi 19 11 (Remanit® 4306) in nitric acid with the addition of hexavalent chromium) (in German)
Thyssen Edelst. Tech. Ber. 14 (1988) 1, p. 26

[730] Kiesheyer, H.
Korrosionsverhalten von betrieblich erzeugtem chemisch beständigen Stahl X CrNi 19 11 (Remanit® 4306) in Salpetersäure
(Corrosion behavior of the works-produced, chemically resistant steel X 2 CrNi 19 11 (Remanit® 4306) in nitric acid) (in German)
Thyssen Edelst. Tech. Ber. 14 (1988) 1, p. 35

[731] Chlibec, G.; Gümpel, P.; Ladwein, T.
Einsatzmöglichkeiten verschiedener Werkstoffe in salpetersäurehaltigen Medien
(Application possibilities of various materials in nitric acid-containing media) (in German)
Thyssen Edelst. Tech. Ber. 14 (1988) 1, p. 39

[732] Peksheva, N. P.; Vorontsov, E. S.
Thermal and anodic oxidation of titanium and the chemical stability of interference-colored films on it
Russ. J. Appl. Chem. 46 (1973) p. 1798

[733] Satoh, H.; Kamikubo, F.; Shimogori, K.; Fukuzuka, T.
Effect of oxidizing agents on corrosion resistance of commercially pure titanium in nitric acid solution (in Japanese)
Corros. Eng. (Boshoku Gijutsu) 31 (1982) 12, p. 769

[734] Kohlhaas, E.; Fischer, A.
The use of the carbon-extraction technique in investigations of superalloys
Prakt. Metallogr. 6 (1969) S. 291

[735] Stamets, R. J.
Chemical cleaning of tungsten and molybdenum
Met. Finish. 73 (1975) 7, p. 29

[736] Knorr, W.
Die technische Titan-Legierung Ti-6Al-4V, ihre Eigenschaften und Wärmebehandlung
(The technical titanium alloy Ti-6Al-4V, its properties and heat treatment) (in German)
Tech. Mitt. Krupp 14 (1956) 4, p. 88

[737] Schulze, B.
Planung und Konstruktion hochkorrosionsfester Anlagen für die thermische Verfahrenstechnik
(Design and construction of highly corrosion-resistant plants for thermal process technology) (in German)
Chem.-Ing.-Tech. 52 (1980) p. 298

[738] Rickes, M.
The contrasting of high strength, high temperature nickel alloys with protective coatings by etching
Prakt. Metallogr. 22 (1985) p. 430

[739] Hummel, O. H.
Werkstoffe im Behälter- und Apparatebau
(Materials in container and apparatus construction) (in German)
Chem. Tech. 11 (1982) 8, p. 930

[740] Hauffe, K.
Über das Korrosionsverhalten von Titan und Titanlegierungen gegen Chlor, Salzsäure und chlorionenhaltige Lösungen
(The corrosion behavior of titanium and its alloys towards chlorine, hydrochloric acid and solutions containing chloride ions) (in German)
Metalloberfläche 36 (1982) p. 594

[741] Hara, Z.; Akechi, K.
Flash resistance sintering of titanium alloy
Funtai Oyobi Funmatsuyakin 24 (1977) 3, p. 71

[742] Karyazin., P. P.; Khukhareva, N. N.; Perimov, Yu. A.; Shtanko, V. M. et al.
Electrolyte for electrochemical polishing of articles made of titanium and titanium alloys
US Pat. 4 220 509 (Sept. 2, 1980)

[743] Ruskol, Yu. S.; Fokin, M. N.; Mochaev, A. S.
Behavior of high-strength titanium alloys in their passivated state
Titanium and Titanium Alloys, Sci. and Technol. Aspects, Moscow, 1976 (Proc. Conf.) Vol. 2, p. 967
Plenum Press, New York

[744] Dobrunov, Yu. V.; Volynskii, V. V.; Kolobov, G. A.; Kuznetsov, S. I.
Effect of inhibitors on the corrosion of titanium alloys in hydrochloric acid solutions in titanium/magnesium production (in Russian)
Tsvetn. Met. (1977) 1, p. 60

[745] Anonymous
New two-stage titanium pump
Brit. Chem. Eng. 15 (1970) 1, S. 17

[746] Villemez, R.; Millet, C.
Evaluation of alloys for nuclear waste evaporators
Corrosion 78, Idaho Chem. Processing Plant, Idaho Falls (Proc. Conf.) Paper No. 114
NACE, Houston (Texas), 1978

[747] Antonovskaya, E. I.; Pozdeeva, A. A.; Ovchinnikov, P. N.
Using titanium equipment in nitric acid at high temperatures
Titan Nar. Khoz. (1976) p. 136

[748] Fukui, S.; Utsu, Y.; Yamazaki, T.; Arakawa, T.; Kai, T.
Corrosion prevention of apparatus and parts made of titanium
Japan Pat. 73 09, 702 (Mitsubishi Heavy Industries, Ltd.; March 27, 1973)

[749] Bhattacharyan, R. S.; Rai, A. K.; Pronko, P. P.
Corrosion behavior of amorphous TiNi films fabricated by ion beam mixing
Mater. Letter 2 (1984) p. 483

[750] Schussler, R.
Tantalum immune to most corrosive attack
Mater. Eng. 77 (1973) 4, p. 24

[751] Rees, T. W.
Metals for small diameter tubing
Mater. Eng. 76 (1972) 5, p. 24

[752] Vehlow, J.
Corrosion of tantalum in sulfuric acid between 150 and 270 °C with nitric acid and hydrochloric acid metered in: 8th Intern. Congress on Metallic Corrosion, Mainz 1981 (Proc. Conf.) Vol. 2, p. 1436

[753] Burns, R. H.; Schwartz, H. D.
Corrosion resistance and mechanical properties of a tantalum-40 weight percent niobium alloy
Corrosion 84, New Orleans, April 1984 (Proc. Conf.), Paper No. 148, 13 p.
NACE, Houston (Texas)

[754] Lambert, J. B.; Ziegler, P. F.; Rausch, J. J.
Tribocor® 532 N a new material for environments involving wear and corrosion
11th Inter. Plansee Seminar 185, Reutte (Österreich), 1985 (Proc. Conf.) Vol. 2, p. 181

[755] Miska, K. H.
Wrought brass
Mater. Eng. 80 (1974) 5, p. 65

[756] Petrov, L. N.
Physical chemistry of selective protection of printing elements during one process emulsion etching (in Russian)
Zashch. Met. 13 (1977) 2, p. 216

[757] Williams, L. F. G.
The mechanism of formation of chromate conversion films on zinc
Surf. Tech. 4 (1976) p. 355

[758] Oda, N.; Morioka, N.
Chemical polishing of zinc alloys
Japan Pat. 78 52,253 (May 12, 1978)

[759] Rost, J.; Schneider, H. G.
The preparation of specimens of directionally solidified lamellar eutectics for transmission electron microscopy
Prakt. Metallogr. 15 (1978) p. 171

[760] Buryanenko, A. F.; Pashulya, P. L.; Khelevina, O. G.
Correlation between the adsorption of surface-active substances and their protective action in the dissolution of zinc in nitric acid (in Russian)
Tekhn., Tekhnol. Ekon. Poligrafni., Kiew, Nauka Dumka (1976) p. 53

[761] Wehrenberg, R. H.
Oxidized zirconium not only for nuclear applications. Corrosion, wear and high-temperature resistance promise a bright future for element No. 40
Mater. Eng. 98 (1983) 1, p. 46

[762] Knittel, D. R.
Zirconium: a corrosion-resistant material for industrial applications
Chem. Eng. 87 (1980) 11, p. 95

[763] Demant, J. T.; Wanklyn, J. N.
Effects of contamination on oxidation of Zr in steam
Corrosion 22 (1966) 3, p. 60

[764] Jangg, G.; Baroch, E. F.; Kieffer, R.; Watti, A.
Untersuchungen über das Korrosionsverhalten von Zirkoniumlegierungen – III. Untersuchungen an Zirkonium-Titan-Legierungen (Investigations into the corrosion behavior of zirconium alloys – III. Investigations on zirconium titanium alloys) (in German)
Werkst. Korros. 24 (1973) 10, p. 845

[765] Te-Lin Yau
SCC of zirconium and its alloys in nitric acid
Corrosion 39 (1983) 5, p. 167

[766] Beavers, J. A.; Griess, J. C.; Boyd, W. K.
Stress corrosion cracking of zirconium in nitric acid
Corrosion 37 (1981) 5, p. 292

[767] Conte, A.; Borello, A.; Cabrini, A.
Anodic oxidation of Zircaloy® 2
J. Appl. Electrochem. 6 (1976) p. 293

[768] Rogov, M. I.; Slobodskij, M. I.; Filippov, V. F.
Corrosion resistance of zirconium with complex additions of transition metals (in Russian)
Fiz.-Khim. Mekh. Mater. 15 (1979) 3, p. 121

[769] Caracciolo, V. P.; Rust, F. G.
Alniflex process for dissolving zirconium-uranium alloy in 304 L stainless steel vessels
Ind. Eng. Chem. Process Des. Develop. 5 (1966) 4, p. 364

[770] Sasaki, Y.; Suzuki, K.; Minato, A.; Yoshida, T.
Method of inhibiting corrosion of zirconium or its alloy
Europe-Patent 158 177 (March 20, 1985)

[771] Wojcik, C. C.
Corrosion resistant zirconium alloys prepared by powder metallurgy
Modern Dev. Powder Metallurgy, Toronto (Canada), 1984
(Proc. Conf.), Vol. 17, p. 145

[772] Yau, T. L.
Zirconium for nitric acid solutions
Industr. Appl. Titanium and Zirconium, Philadelphia (Pa.),
1984 (Proc. Conf.), Vol. 4, p. 57

[773] Satoh, H.; Kamikubo, F.; Shimogori, K.
Effect of oxidizing agents on corrosion resistance of commercially pure titanium in nitric acid solution
Titanium Science and Technology, München, 1984 (Proc. Conf.), Vol 4, p. 2649

[774] Uzlyuk, M. V.; Fedorov, Yu. V.
Effect of pyridine thiocyanates on the corrosion of iron in nitric acid (in Russian)
Okislit. Vosstanov, i Adsorbts. Protsessy na Poverkhn. Tverd. Metallov, Izhevsk (1980) 2, p. 29

[775] Dubinin, G. N.; Sokolov, V. S.
Heat and corrosion resistance of copper and bronze after diffusion-coating with aluminium and chromium (in Russian)
Korroz. Zashch. (1979) 13, p. 79

[776] Antropov, L. I.; Donchenko, M. I.; Saenko, T. V.
Dissolution of copper in dilute nitric acid (in Russian)
Zashch. Met. 14 (1978) 6, p. 657

[777] Kufa, F.; Polachova, J.
Properties of steels used for urea and nitric acid production plants (in Czech)
Hutn. Listy 34 (1979) 9, p. 618

[778] Chen, N. G.; Bocharov, V. A.; Fursov, P. F.; Shust, T. F.; Dektyareva, V. K.; Borozdina, R. R.; Yudina, S. M.
Delaying the corrosion of weld seams in carbon and stainless steels in acid solutions (in Russian)
Zashch. Met. 1 (1965) p. 726

[779] Kasinskaya, L. L.; Yudina, S. M.; Murashkina, A. A.
Comparison of the corrosion resistance of aluminium weld joints (in Russian)
Tekhnol. Organ. Proizvod. (1971) 4, p. 71

[780] Seraphin, M. L.
Caractères principaux de la résistance à la corrosion du titane et de ses alliages
(Main characteristics of the corrosion resistance of titanium and its alloys) (in French)
Corros.-Trait.-Prot.-Finition 18 (1970) 4, p. 237

[781] Antonovskaya, É. I.; Pozdeeva, A. A.
Corrosion of titanium alloy AT-3 in nitric acid at high temperatures
Russ. J. Appl. Chem. 48 (1975) p. 2572

[782] Peksheva, N. P.; Vorontsov, E. S.
Influence of nitric acid on colored layers obtained by thermal and anodic oxidation of titanium alloy OT4-0
Russ. J. Appl. Chem. 48 (1975) p. 1356

[783] Stabrovskij, A. I.; Karasev, A. F.
Electrochemical behavior of zirconium during anodic polarization in nitrate solutions (in Russian)
Elektrokhimiya 19 (1983) 5, p. 594

[784] Di Paola, A.; Di Quarto, F.; Sunseri, C.
Anodic oxide films on tungsten. I. The influence of anodizing parameters on charging curves and film composition
Corros. Sci. *20* (1980) p. 1067

[785] Vorontsov, E. S.; Peksheva, N. P.; Peshkov, V. V.
Interference colours of oxide films on titanium as an indication of heterogeneous processes on its surface
Russ. J. Phys. Chem. *48* (1974) 4, p. 558

[786] Rühle, M.
Rohstoffprofil Hafnium
(Raw material profile) (in German)
Metall *37* (1983) 2, p. 171

[787] Anonymous
Hafnium ein neuer Werkstoff in der Kerntechnik
(Hafnium a new material in nuclear technology) (in German)
Chemiker-Ztg. *104* (1980) 10, p. 298

[788] Ruda, G. I.; Samgina, O. Y.; Smirnov, V. P.
Phase composition and refractoriness of Hf-Nb alloys (in Russian)
Zashch. Met. *19* (1983) 6, p. 984

[789] Brauer, E.; Piroth, J.; Streb, B.
Zur Korrosion des Hafniums Oxide und Hydride
(Corrosion of hafnium oxides and hydrides) (in German)
Werkst. Korros. *36* (1985) p. 511

[790] Jangg, G.; Kieffer, R.; Retelsdorf, H. J.; Prem, F.
Untersuchungen über Vanadinbasislegierungen
(Tests on alloys based on vanadium) (in German)
Metall *26* (1972) 7, p. 720

[791] Vorobyeva, L. P.; Gulyaev, A. P.; Druzhinina, I. P.
Corrosion resistance of vanadium alloys in boiling acids (in Russian)
Zashch. Met. *6* (1970) 5, p. 537

[792] Wlodek, S. T.
Vanadium-Niobium alloys
US Pat. 3 136 631 (June 9, 1964)

[793] Andreeva, V. V.; Stepanova, T. P.; Druzhinina, I. P.; Vladimirskaya, T. M.
Effect of tantalum on the corrosion resistance, electrochemical and physical-mechanical properties of vanadium (in Russian)
Zashch. Met. *9* (1973) 2, p. 181

[794] Allen, B. C.
Corrosion of refractory metals in aqueous and gaseous media
React. Met. *13* (1970) 1, p. 47

[795] Kovriga, Yu. P.; Yartsev, M. G.; Bardin, V. A.; Matveev, V. S.
Corrosion resistance of PInK-24 alloy in acids, caustic soda and sodium chloride (in Russian)
Zashch. Met. *15* (1979) 2, p. 224

[796] Kindlimann, L. E.; Greene, N. D.
Mechanism for the acid corrosion behavior of neutron-irradiated uranium
Corrosion *26* (1970) 7, p. 189

[797] Johnson, H. R.; Dini, J. W.
Etching and plating of uranium alloys
Met. Finish. *74* (1976) 3, p. 37

[798] Greene, N. D.; Kindlimann, L. E.
Anodic dissolution of uranium
U.S. Atom Energy Comm. RPI-2714-4 (1965) 10 p.

[799] Johnson, H. R. et al.
Etching and plating of nominal U-6 wt% Nb alloys
High-density Penetration Materials, Charlottesville (Va), 1980 (Proc. Conf.) (SAND-80-8033). 26 p.

[800] Wheelwright, E. J.; Fox, R. D.
Development of an electrolyte dissolver for plutonium metal
Ind. Eng. Chem. Proc. Des. Dev. *16* (1977) 3, p. 297

[801] Shehata, M. T.; Kelly, R.
The formation and structure of anodic films on beryllium
J. Electrochem. Soc. *122* (1975) p. 1359

[802] Maniar, L. K.; Mehta, B. J.; Shah, B. S.
Etching characteristics of bismuth single crystals in nitric acid-acetic acid solutions
Surf. Tech. *15* (1982) p. 287

[803] Malchevskij, E. G.; Migaj, L. L.; Vedeneeva, M. A.
Corrosion and electrochemical behavior of gallium in acids (in Russian)
Zashch. Met. *10* (1974) 6, p. 727

[804] Bhattachary, R. S.; Rai, A. K.; Raffoul, C. N.; Pronko, P. P.; Khobaib, M.
Corrosion behavior of amorphous Ni based alloy coatings fabricated by ion beam mixing
J. Vac. Sci. Technol. A *3* (1985) 6, p. 2680

[805] Enomoto, H.; Ishikawa, M.
Corrosion resistance of tin-nickel alloys deposited from pyrophosphate bath

(in Japanese)
J. Met. Finsih. Sco. Jpn. *30* (1979) 6, p. 284

[806] Sedenkov, A. M.; Berezovskij, L. R.; Kuzub, V. S.
Breakdown of the passivity of titanium in nitric acid due to fluoride and chloride ions (in Russian)
Elektrokhimiya *20* (1984) p. 1004

[807] Shelenkov, G. M.; Troyanovskii, V. E.; Blashchuk, V. E.; Onoprienko, L. M.
The corrosion resistance of welded joints in certain titanium alloys
Automatic Welding (1985) 4, p. 21

[808] Vinogradov, Yu. M.; Shvarts, G. L.; Moroz, V. A.; Makarova, L. S.
Experience in the operation of titanium equipment in corrosive media
Chem. & Petroleum Eng., New York (1973) p. 1094

[809] Klinov, I. Y.; Zaretskii, E. M.; Zakharchuk, E. A.
Corrosion and electrochemical behavior of titanium alloys in mixtures of sulfuric and nitric acids
Chem. & Petroleum Eng., New York (1972) p. 913

[810] Blashchuk, V. E.; Gurevich, S. M.; Onoprienko, L. M.; Shelenkov, G. M.
Properties of AT3 titanium alloy and welded joints
Chem. & Petroleum Eng., New York (1979) p. 778

[811] Shvarts, G. L.; Glazkova, S. A.; Moroz, V. A.
Effect of contact titanium on the resistance of structural materials to general and pitting corrosion in corrosive media
Chem. & Petroleum Eng., New York (1986) p. 26

[812] Matsumoto, K.; Shinohara, T.; Kasamatsu, A.
Effects of process conditions on corrosion of 316 L in urea reactor and evaluation of corrosion resistance (in Japanese)
Corros. Eng. (Boshoku Gijutsu) *33* (1984) 11, p. 643

[813] Lozovatskaya, L. P.; Shmatko, M. N.; Goldshtejn, Ya. N.; Piskunova, A. E.
Effect of micro-alloying with boron and cerium on the corrosion stability and hot ductility of the steel 02Ch18N11 (in Russian)
Zashch. Met. *21* (1985) 1, p. 98

[814] Sheftel, E. N.; Usmanova, G. Sh.; Grigorovich, V. K.
Ageing niobium alloys strengthened with zirconium nitrides
Russ. Metall. (1980) 5, p. 149

[815] Mushagi, A.; Rayevskaya, M. V.; Loboda, T. P.; Bodak, O. I.
Phase equilibria and corrosion properties of alloys of the Pd-Y-Cu system
Russ. Metall. (1983) 1, p. 169

[816] Naka, M.; Asami, K.; Hashimoto, K.; Masumoto, T.
Corrosion behavior of amorphous titanium alloys
Intern. Conf. on Titanium, Kyoto (Japan), 1980 (Proc. Conf.) Vol. 4, p. 2677

[817] Glazunov, M. G.; Minakov, V. N.; Turtsevich, E. V.; Yakushina, A. I.
Structural sensitivity of the mechanical properties of deformed Nb-1 % Zr-0.1 % C alloy
Russ. Metall. (1978) 5, p. 98

[818] Product Information (Deutsche Carbone AG, Frankfurt)
Wärmeaustauscher aus imprägniertem Graphit
(Information heat exchangers made of impregnated graphite) (in German)
Werkst. Korros. *31* (1980) p. 301

[819] Forsman, W. C.; Vogel, F. L.; Carl, D. E.; Hoffman, J.
Chemistry of graphite intercalation by nitric acid
Carbon *16* (1978) p. 269

[820] Zawadzki, J.
IR spectroscopy investigations of acidic character of carbonaceous films oxidized with HNO_3 solution
Carbon *19* (1981) p. 19

[821] Dowell, M. B.; Badorrek, D. S.
Diffusion coefficients of Br_2, HNO_3 and $PdCl_2$ in graphite
Carbon *16* (1978) p. 241

[822] Lawrence, T. D.
The surface composition and energetics of type A graphite fibers
Carbon *15* (1977) p. 129

[823] Anonymous
Grafilex® 6501 nastro per baderne in grafite pura
(Grafilex® 6501 tape for pure graphite packing) (in Italian)
Ingegneria Chim. *26* (1977) 1, p. 22

[824] Kobets, L. P.; Golikova, L. A.; Strebkova, T. S.; Chubarova, M. A.; Frolov, F. I.
Investigation of the service life of activated

carbon fibers (in Russian)
Fiz. Khim. Obrab. Mater. (1977) 6, p. 129

[825] Antonov, A. N.; Fokin, M. N.; Vorob'eva, N. F.
Corrosion resistance of graphitized materials in concentrated phosphoric acid
Russ. Appl. Chem. *50* (1977) p. 2175

[826] Antonov, A. N.; Fokin, M. N.; Vorob'eva, N. F.
Resistance of carbon-graphite materials to the action of sulfuric acid
Russ. Appl. Chem. *50* (1977) p. 1847

[827] Kralin, A. A.; Fokin, V. P.; Kanevskij, L. S.; Tashchilova, L. P.; Zarechenskij, E. T.
Graphite for chemical apparatus (in Russian)
Plast. Massy *9* (1983) p. 51

[828] Maryasin, I. L.; Vlasov, E. G.; Lukina, T. V.; Ivanova, L. I.; Vinokurov, Yu. V.
Effectiveness of protection of graphite with pyrocarbon against aggressive media (in Russian)
Khim. Prom. (1977) p. 451

[829] Wege, E.
Kohlenstoff und Graphit – Stand der Technik und Perspektiven
(Carbon and graphite – current state of the technology and perspectives) (in German)
Sprechsaal *113* (1980) 6, p. 432

[830] McDowell jr., D. W.
Handling mixed nitric and sulfuric acid
Chem. Eng. *81* (1974) 23, p. 133

[831] Kowatschewa, R.; Djambazowa, L.
Precipitation and corrosion processes in ferrite containing nitrogen-chromium-manganese steels after isothermal annealing
Prakt. Metallogr. *17* (1980) p. 560

[832] Lankard, D. R.
Cement and concrete technology for the corrosion engineer
Mater. Performance *15* (1976) 8, p. 24

[833] Dadson, L. M.; Shearing, H. J.
Urethane concretes in the chemical industry
Z. Werkstofftech. *6* (1974) 2, p. 77

[834] Pilhofer, T.
Einsatzbereich des Werkstoffes Glas in der Anlagentechnik
(Scope of glass as material of construction in plant engineering) (in German)
Chem.-Ing.-Tech. *56* (1984) 4, p. 299

[835] Anonymous
High-temperature masonry shuns acids, salts, solvents
Chem. Eng. *82* (1975) 13, p. 96

[836] Anonymous
Nonmetallic coating takes on hot acids
Mater. Eng. *75* (1972) 3, p. 50

[837] Gräfen, H.; Gramberg, U.; Schindler, H.
Verwendung nichtmetallischer Werkstoffe zur Verringerung von Korrosionsschäden
(Use of non-metallic materials for reducing corrosion damage) (in German)
Werkst. Korros. *30* (1979) 5, p. 297

[838] Hohenhinnebusch, W.
Email im Apparatebau. Fortschritte und Möglichkeiten
(Enamel in apparatus construction. Progress and possibilities) (in German)
Chemie Technik *11* (1982) 6, p. 645

[839] Lorentz, R.
Inhibition des Säureangriffs auf Chemieemail
(Inhibition of the acid attack on chemical service glass-enamel) (in German)
Werkst. Korros. *34* (1983) 9, p. 437

[840] Lorentz, R.
Angriff wäßriger Säuren auf Chemieemail
(Attack of aqueous acids on chemical service glass-enamel) (in German)
Werkst. Korros. *34* (1983) 5, p. 219

[841] Alikina, I. B.; Bobovich, O. V.; Smirnov, N. S.; Ivakhin, S. I.; Taran, N. G.
Corrosion resistance of silicate enamels at increased temperatures and pressure (in Russian)
Zashch. Met. *9* (1973) 1, p. 89

[842] Sawin, L. S.; Barinow, J. D.; Petzold, A.
Die Rolle oberflächenaktiver Stoffe beim Emaillierprozeß. Wirkung auf Säurebeize, Wasserstoffverhalten und Fischschuppenbildung aluminiumhaltiger Emaillierbleche
(The role of surface-active substances in the enamelling process. Effect of acid pickling, hydrogen behavior and fish scale formation of aluminium-containing enamel sheets) (in German)
Silikattechnik *25* (1974) 4, p. 120

[843] Fukuzuka, T.; Shimogori, K.; Satoh, H.; Kamikubo, F.
Inhibition of corrosion and hydrogen embrittlement of tantalum in concentrated sulfuric acid solutions at elevated temperatures by addition of N-O-compounds (in Japanese)
Corros. Eng. (Boshoku Gijutsu) *30* (1981) p. 327

[844] Crook, A.; Beal, R. E.
Basalt: Properties and applications in corrosion-erosion problems
Mater. Performance 20 (1981) 1, p. 45

[845] Subat, G.
Spezielle Kunstharz-Beschichtungssysteme zum Schutz von stark beanspruchten Betonflächen
(Special synthetic resin coating systems for protecting concrete surfaces under extreme stress) (in German)
Sprechsaal 113 (1980) 3, p. 189

[846] Anonymous
Asbestfreie Stopfbuchspackungen
(Asbestos-free stuffing box packing) (in German)
Chem.-Anl. + Verfahren (1979) 7, p. 42

[847] Engelmann, A.; Hochegger, G.; Eicher, H.
Sigraflex® ein neuer Dichtungswerkstoff aus Graphit. Flexibler Graphit ein Widerspruch in sich oder Tatsache?
(Sigraflex® a new sealing material made of graphite. Flexible graphite a contradiction in terms or a fact?) (in German)
Chem.-Anl. + Verfahren (1973) 2, p. 80

[848] Wittenberger, W.
Glasapparaturen im Chemiebetrieb
(Glass apparatus in the chemical plant) (in German)
Chem. Labor Betr. 31 (1980) 4, p. 138

[849] El-Shamy, T. M.; Taki-Eldin, H. D.
The chemical durability of $PbO-SiO_2$ glasses
Glass Technol. 15 (1974) 2, p. 48

[850] Schulze, B.
Planung und Konstruktion hochkorrosionsfester Anlagen für die thermische Verfahrenstechnik
(Design and construction of highly corrosion-resistant plant for thermal processes) (in German)
Chem.-Ing.-Tech. 52 (1980) 4, p. 298

[851] Chekhov, A. P.; Vinarskij, V. L.
Glass-ceramic slags in corrosion protection (in Russian)
Budivelni Mater. Konstr. (1970) 1, p. 5

[852] Gericke, D.
Glasapparate und Rohrleitungen in der chemischen Industrie
(Glass apparatus and pipelines in the chemical industry) (in German)
Tech. Mitt. 62 (1969) p. 234

[853] Schober, S.
Technisches Glas für den Rohrleitungs- und Apparatebau. Derzeitiger Stand und Perspektive
(Technical glass for pipeline and apparatus construction. Current state and perspectives) (in German)
Mitt. Inst. Leichtbau, Dresden 9 (1970) p. 328

[854] Anonymous
Warum niemand Schwefelsäure-, Salz- und Salpetersäureanlagen korrosionssicherer baut als Schott
(Why no one builds sulfuric, hydrochloric and nitric acid plants more corrosion-resistant than Schott) (in German)
Chem.-Anl. + Verfahren (1969) 9

[855] Vondeling, J. K.
Fluoride-based etchants for quartz
J. Mater. Sci. 18 (1983) p. 304

[856] Lachowszky, K.
Der Email Lampart Univer® 80 ein wirksamer Schutz in der chemischen Industrie
(The enamel Lampart Univer 80 effective protection in the chemical industry) (in Czech)
Koroze Ochr. Mater. 14 (1970) 4, p. 74

[857] Lavrishin, B. N.; Kirilyuk, S. S.; Isaev, N. I.; Fedin, I. M.
The effect of cleaning with inhibited acids on the surface roughness and corrosion stability of the bubble-cap trays of fractionating towers (in Russian)
Fiz. Khim. Mekh. Mater. (1982) 5, p. 80

[858] Gajduchok, V. M.; Kasperskij, G. A.
Effect of plastic working on the oxidation of steel (in Russian)
Probl. Treniya Iznashivaniya (1972) 2, p. 118

[859] Anonymous
Lexikon der Korrosion, Abschnitt „Salpetersäure"
(Corrosion lexicon, section "Nitric acid") (in German)
Mannesmann-Werke 2 (1970) p. 150

[860] Ameen, J. G.; McBride, D. G.; Phillips, G. C.
Etching of high alumina ceramics to promote copper adhesion
J. Electrochem. Soc. 120 (1973) p. 1518

[861] Cihal, V.; Kashova, I.; Kubelka, J.
Effect of titanium and niobium carbides on the corrosion resistance of stainless steels in oxidizing media (in French)
Métaux-Corros.-Ind. (1969) No. 529, p. 281

[862] Samsonov, G. V.; Zhunkovskij, G. L.; Luchka, M. V.
Studies on the formation process for composite coatings on the titanium carbide base

(in Russian)
Poroshk. Metall. *16* (1976) 7, p. 53

[863] Martinz, H.-P.; Eck, R.; Eiter, J.; Sakaki, T.; Kato, M.
Das Korrosionsverhalten von pulvermetallurgischem Chrom
(Corrosion behaviour of powder metallurgically produced chromium) (in German)
Werkst. Korros. *40* (1989) 12, p. 715

[864] Brynza, A. P.; Buglakova, V. S.; Khmelovskaya, S. A.
Inhibitors of titanium carbide dissolution in nitric acid (in Russian)
Zashch. Met. *17* (1981) 6, p. 729

[865] Miska, K. H.
Refractory metal carbides fight wear, heat and corrosion
Mater. Eng. *82* (1975) 6, p. 67

[866] Sintsova, I. T.; Kozlovski, L. V.
Protective properties of titanium and zirconium carbonitride coatings on steel (in Russian)
Zashch. Met. *6* (1970) 5, p. 616

[867] Tallent, O. K.; Mailen, J. C.
Study of the dissolution of refractory Pu-oxide in nitric-hydrofluoric acid dissolvents at 100 °C
Oak Ridge National Laboratory, Tenn, ORNL/TM-5181 (1976) 31 p.

[868] Laitinen, H. A.; Watkins, N. H.
Mechanism of anodic deposition and cathodic stripping of PbO_2 on conductive tin oxide
J. Electrochem. Soc. *123* (1976) p. 804

[869] Shurshakov, A. N.; Posos'eva, G. D.; Mashkovich, L. A.; Subbotina, I. G.; Dergunova, V. S.
Chemical resistance of graphite-carbide composite materials
Inorg. Mater. USSR *12* (1976) p. 470

[870] Avgustinik, A. I.; Mironov, I. M.
Action of nitric and phosphoric acid solutions on chemical porcelain
Russ. Appl. Chem. *46* (1973) p. 2536

[871] Kopylova, V. P.; Kornilova, V. I.; Hazarchuk, T. N.; Fedorus, V. B.
Combined effect of powdery zirconium and niobium carbides with some acids in the homogeneous range (in Russian)
Poroshk. Metall. *20* (1980) 1, p. 12

[872] Pctrascu, E.; Visa, C.
Studiul rezisten ei la coroziune a unor materiale in medii acide
(Investigation of the corrosion resistance of some materials in acid media) (in Rumanian)
Revista de Chimie *27* (1976) p. 786

[873] Zabrodskij, A. G.; Kononyuk, A. E.
Pumps with controllable output for acids and other aggressive liquids (in Russian)
Fermentnaya Spirtovaya Prom. *35* (1969) 6, p. 25

[874] Miyoshi, K.; Buckley, D. H.; Rengstorff, G. W. P.; Ishigaki, H.
Surface effects of corrosive media on hardness, friction and wear of materials
Intern. Conf. on Wear of Materials, Vancouver, British Columbia, 1985 (Proc. Conf.), p. 302

[875] Miyoshi, K.; Buckley, D. H.; Rengstorff, G. W. P.; Ishigaki, H.
Surface effects of corrosive media on hardness, friction and wear of materials
Ind. Eng. Chem. Prod. Res. Dev. *24* (1985) 3, p. 425

[876] Grewe, H.
Einfluß der Hartstoffkomponenten auf einige Eigenschaften von Hartmetallen unter besonderer Berücksichtigung neuerer Werkstoffentwicklungen
(Influence of the hard material components on some properties of hard metals respecting especially modern developments of working materials) (in German)
Z. Werkstofftech. *4* (1973) 4, p. 209

[877] Hirschfeld, D.; Ostermann, G.
Interessante Eigenschaften von Hartmetallen für die Chemische Industrie
(Properties of hard metals of interest to the chemical industry) (in German)
Z. Werkstofftech. *4* (1973) 7, p. 367

[878] Weith, W.
Verschleißteile aus Hartmetall für die chemische Industrie
(Working parts made of hard metal in the chemical industry) (in German)
Chem.-Anl. + Verfahren (1979) 8, p. 81

[879] Kutysheva, E. V.; Kosolapova, T. Y.
Chemical stability of zirconium carbide and oxycarbides and chemical phase analysis of their mixtures
Sov. Progress Chem. *41* (1975) 1, p. 84

[880] Ilicheva, L. A.; Frejd, M. K.; Suprunov, V. A.; Tyurina, N. A.
Corrosion and electrochemical properties of molybdenum carbide in acid and alkali solutions (in Russian)
Poroshk. Metall. *18* (1978) 8, p. 27

[881] Felten, R. P.
Siliziumnitrid – eine Literaturübersicht

(Silicon nitride – a review of the literature) (in German)
Sprechsaal *107* (1974) 3, p. 92

[882] Leidheiser jr., H.; Audisio, S. C.
Projection cathodique de nitrure de silicium sur cuivre. Propriétés protectrices du revêtement
(Cathodic projection of silicon nitride on copper. Protective properties of coatings) (in French)
Trait. Protection, Finition *20* (1972) 3, p. 200

[883] Postogvard, G. I.; Makarenko, A. E.; Ryzhova, T. P.; Ostapenko, I. T.
Resistance of silicon nitride base materials in solutions of certain acids and alkalis (in Russian)
Poroshk. Metall. *20* (1980) 1, p. 70

[884] Binder, F.
Die Technologie des Tantalcarbids
(Tantalum carbide technology) (in German)
Radex Rundsch. (1978) 2, p. 507

[885] Kopylova, V. P.; Nazarchuk, T. N.
Chemical stability of silicon nitride and oxynitride powders (in Russian)
Poroshk. Metall. *15* (1975) 10, p. 38

[886] Lyutaya, M. D.; Kulik, O. P.
Chemical properties of nitrides of some transient metals (in Russian)
Poroshk. Metall. *10* (1970) 10, p. 48

[887] Brynza, A. P.; Krivonogova, O. M.
Inhibitors of dissolution of titanium nitride in nitric acid (in Russian)
Zashch. Met. *14* (1978) 6, p. 721

[888] Ljutaja, M. D.; Goncharuk, A. B.
On manganese nitrides (in Russian)
Poroshk. Metall. *17* (1977) 3, p. 65

[889] Lyutaya, M. D.; Goncharuk, A. B.; Gordienko, S. P.
Chemical properties of nitrides of cerium subgroup lanthanides
Sov. Progress Chem. *40* (1974) 11, p. 18

[890] Andreeva, A. F.; Chernysh, I. G.
Chemical stability of indium nitride films
Russ. Appl. Chem. *44* (1971) p. 417

[891] Kugaj, L. N.; Nazarchuk, T. N.
On chemical stability of diborides of transient metals of groups IV – V of periodic system (in Russian)
Poroshk. Metall. *11* (1971) 3, p. 51

[892] Kosolapova, T. Y.; Gordienko, S. P.; Domasevich, L. T.
Reaction of lanthanum carbide, boride and silicide with acids and water
Sov. Progress Chem. *37* (1971) 1, p. 1

[893] Lesnikova, K. P.; Frejd, M. K.
Influence of the anionic composition of electrolytes on the corrosion behavior of Mo_2B_5 (in Russian)
Izv. Vuzov. Khim. i Khim. Tekhnol. *21* (1978) p. 1015

[894] Marko, M. A.; Dikii, I. I.; Petrivskii, R. I.; Kuzma, Yu. B.
Corrosion properties of niobium and tantalum germanoborides
Sov. Progress Chem. *46* (1980) 1, p. 26

[895] Marko, M. A.; Dikii, I. I.; Petrivskii, R. I.; Kuzma, Yu. B.
Corrosion properties of Mo, Co and Ni germanobromides
Sov. Progress Chem. *46* (1980) 5, p. 29

[896] Binder, F.
Ein Beitrag zur Kenntnis der kubischen Hexaboride
(A contribution to knowledge on cubic hexaboride) (in German)
Radex Rdsch. (1977) 1, p. 52

[897] Bennett, J. P.
Corrosion resistance of selected ceramic materials to nitric acid (Pamphlet)
US Dept. of Interior, Bureau of Mines, Washington. D. C., 1984, Rep. Invest. No. 8851, 12 p.

[898] Anonymous
Hochkorrosionsfeste Füllkörper Rapal®
(Highly corrosion-resistant packing Rapal®) (in German)
Chemie-Technik *11* (1982) p. 1268

[899] Issel, W.; Schüller, W.
Kerntechnische Besonderheiten bei der apparativen Ausstattung der Wiederaufarbeitungsanlage Karlsruhe
(Characteristics of nuclear technology in the apparatus set-up of the reprocessing plant in Karlsruhe) (in German)
Kerntechnik *9* (1968) p. 238

[900] Motojima, S.; Nakayama, Y.
Phosphidation of cobalt plate and some of its properties
J. Less-Common Met. *118* (1986) p. 109

[901] Popova, O. I.
Regularities of variations in chemical properties of silicide powders (in Russian)
Poroshk. Metall. *22* (1982) 3, p. 59

[902] Popova, O. I.
Chemical properties of rhenium silicides and

germanide (in Russian)
Poroskh. Metall. *17* (1977) 7, p. 1

[903] Lynchak, K. A.; Kosolapova, T. Y.
Certain chemical properties of the germanides of lanthanum, cerium, praseodymium and neodymium
Sov. Progress Chem. *42* (1976) 7, p. 8

[904] Hintermann, H. E.; Boving, H.
Verschleißfeste dünne Schichten
(Abrasion-resistant thin layers) (in German)
Technik *33* (1978) 7, p. 387

[905] Reka, B. A.; Vojnov, I. D.; Doroshenko, V. G.
Investigation of interaction of phosphate layers with acid pairs (in Russian)
Lakokras. Mater. Primen. *24* (1983) 4, p. 18

[906] Lorenz, R.
Emailkorrosion durch Säuren bei höheren Temperaturen und Druck. Aufklärung des Korrosionsmechanismus und Entwicklung von praxisgerechten Prüfmethoden
(Enamel corrosion due to acids at high temperature and pressure. Elucidation of the corrosion mechanism and development of testing methods suited to practice) (in German)
Werkst. Korros. *34* (1983) p. 573

[907] Product Information (Fansteel Metals, Chicago, Ill. (USA)
Tantaloy® "63" Metal: the corrosion fighter (Pamphlet), 1981, 13 p.
Werkst. Korros. *24* (1983) 8, p. R 190 (83-2510).

[908] Vehloh, J.
Corrosion testing using radionuclides – tantalum as a material in wet ashing
KFK-Nachr. *15* (12983) p. 31

[909] Bachmann, W. T.
Die Benutzung hitzebeständiger Metalle bei Korrosionsbeanspruchung
(The use of heat-resistant metals against corrosion stress) (in German)
Mater. Design. Eng. *64* (1966) 6, p. 106

[910] Baranova, I. K.; Fionova, L. K.
Etching technique for determination of crystallographic orientation in niobium
Scr. Metall. *14* (1980) p. 167

[911] Maness, R. F.
Korrosionsbeständige Nickellegierung
(Corrosion-resistant nickel alloy) (in German)
DOS 2 002 652 (United States Atomic Energy Commission; Nov. 19, 1970)

[912] Schwartz, B.; Robbins, H.
Chemical etching of silicon. IV. Etching technology
J. Electrochem. Soc. *123* (1976) p. 1903

[913] Schimmel, D. G.; Elkind, M. J.
An examination of the chemical staining of silicon
J. Electrochem. Soc. *125* (1978) p. 152

[914] Weidner, G.
Application of hillock etching of microdefects to the investigation of Czochralski silicon by transmission electron microscopy
Phys. Stat. Sol. (a) *48* (1978) K 105

[915] Sopori, B. L.
A new defect etch for polycrystalline silicon
J. Electrochem. Soc. *131* (1984) p. 667

[916] Andreeva, I. A.; Efremov, G. V.
Chemical properties of boron phosphide (in Russian)
Vestn. Leningr. Univ. *19* (1964) 2, p. 130 (Ser. Fiz. i Khim.)

[917] Gangash, P.; Milnes, A. G.
Etching of cadmium telluride
J. Electrochem. Soc. *128* (1981) p. 924

[918] Wiedemann, K. H.
Korrosionsuntersuchungen zur Entwicklung von Dekontaminationslösungen für Nuklearanlagen. Teil 1: Das Korrosionsverhalten verschiedener Werkstoffe der Kerntechnik in sauren und alkalischen Redoxlösungen
(Investigation of corrosion in the development of decontamination solutions for nuclear systems. Part I: The corrosion behavior of several nuclear materials in acid and alkaline redox solutions) (in German)
Werkst. Korros. *39* (1988) 5, p. 185

[919] Davlyatshaev, A.; Kamenenskij, I. V.
Properties of molded parts based on difurfurylidene acetone (in Russian)
Plast. Massy (1974) 7, p. 37

[920] Kasthuri, R.
Storage and handling of some difficult chemicals
Anticorros. Methods Mater. *27* (1980) 10, p. 14

[921] Murov, V. A.; Radaev, A. N.; Musaelyan, I. N.
Properties of polyethylene in nitric acid (in Russian)
Plast. Massy (1978) 2, p. 16

[922] Ehrbar, J.; v. Meysenbug, C.-M.
Untersuchungen zur Zeitstandfestigkeit von Thermoplast-Rohren unter Chemikalieneinwirkung
(Investigations on long term strength of thermoplastic tubes in chemical media)

(in German)
Z. Werkstofftech. 7 (1976) p. 429

[923] Frolova, M. K.; Shevchenko, E. N.; Shnejderova, V. V.
Acceleration methods for testing the permeability of polymer films to aggressive liquid media (in Russian)
Plast. Massy (1983) 5, p 46

[924] Plisov, V. G.; Zelenev, Yu. V.
Permeability of low-molecular weight compounds through a polyethylene film (in Russian)
Khim. Prom. 53 (1975) p 913

[925] Mulin, Yu. A.; Yakovlev, A. D.
Use of polyethylene powder in protective coatings (in Russian)
Lakokr. Mater. Primen. 15 (1974) 2, p. 30

[926] Frolova, M. K.; Shevchenko, E. N.
Determination of the protective action of polyethylene sheath (in Russian)
Plast. Massy (1983) 4, p. 55

[927] Mzuykantova, A. I.; Egorova, T. A.; Konoval, I. V.; Evdokimov, E. I.; Zlobina, V. A.; Listkov, V. T.; Ershov, L. A.
Properties of polyethylene filled with modified glass wool (in Russian)
Plast. Massy (1975) 3, p. 69

[928] Vlasova, S. V.
Effect of electrolytes on the electrical behavior of HDPE (in Russian)
Plast. Massy (1982) 4, p. 27

[929] Yulchibaev, A. A.; Latypov, T.; Usmanov, Kh. U.; Bakhramov, N.
Modication of polyethylene by grafting with N-vinylpyrrolidone (in Russian)
Plast. Massy (1977) 8, p. 53

[930] Boova, A. A.
Covers for continuous strip pickling
Met. Finish. 80 (1982) 9, p. 57

[931] Kotlobaj, A. P.; Zemlyanichenko, M. A.
Use of glass-reinforced plastic materials in chemical industry abroad (in Russian)
Khim. Prom. 48 (1972) p. 387

[932] Schötz, H.
Beschichten von Galvanogestellen
(Coatings for electroplating racks) (in German)
Galvanotechnik 72 (1981) 3, p. 268

[933] Zudilin, E. V.; Kupkina, N. P.; Volchek, A. M.; Kryshchenko, K. I.; Goryainova, A. V.
Polypropylene composite container for chemical reagents (in Russian)
Khim. Prom. 47 (1971) p. 709

[934] Kuryaeva, R. I.; Makhotkin, A. F.; Sapozhnikov, A. D.
Waste gases from nitric acid vapor and mist with the aid of polypropylene filters (in Russian)
Khim. Prom. 55 (1979) p. 485

[935] Dobrolyubov, V. V.; Dashevskij, S. L.; Efimova, A. R.; Shmelev, V. K.; Gorshenina, G. I.; Mozhukhin, V. G.
Polymeric materials used in construction of chemical equipment for inorganic processes (in Russian)
Khim. Prom. 48 (1972) p. 764

[936] Linke, U.; Kopp, W.-U.
Preparation of polished specimens and thin sections of plastics
Prakt. Metallogr. 17 (1980) p. 479

[937] Buchholz, W.
Anwendung von Polymerbeton im Maschinenbau
(Use of polymer concrete in machine construction) (in German)
Fachber. f. Metallbearbeitung 60 (1983) p. 214

[938] Johanson, R.
Determination of Cu-Cr-As and Zn in sulfuric/nitric acid digested extracts of preserved wood by atomic absorption spectroscopy with reference to As in Karri Rail sleepers
Holzforschung 28 (1974) p. 117

[939] Diedrich, G.; Kempe, B.; Graf, K.
Zeitstandfestigkeit von Rohren aus Polyethylen (HDPE) und Polypropylen (PP) unter Chemikalieneinwirkung
(Creep strength of polyethylene pipes (HDPE) and polypropylene (PP) under the action of chemicals) (in German)
Kunststoffe 69 (1979) p. 470

[940] Gaube, E.; Diedrich, G.
Kunststoffrohre im Rohrleitungs- und Apparatebau
(Plastic pipes in pipeline and apparatus construction) (in German)
Chem.-Ing.-Tech. 46 (1974) p. 273

[941] Obryadchikova, K. N.; Pavlov, N. N.
Chemical resistance of some organic glasses and materials based on polyvinylchloride (in Russian)
Plast. Massy (1976) 2, p. 72

[942] Berger, H.
Korrosionsschutz von Galvanik-Gestellen und Anlageteilen in der Galvanik mit PVC-Pasten
(Protection of electroplating jigs and plant

components against corrosion in the plating industry with PVC pastes) (in German)
Galvanotechnik 69 (1978) p. 210

[943] Woebcken, W.
Langzeitverhalten der Kunststoffe als Baustoffe
(Long-term behavior of plastics as construction materials) (in German)
VDI Z. 116 (1974) 10, p. 829

[944] Artyushenko, N. K.; Maznitsina, N. N.; Stulova, L. I.; Arkhipova, L. I.; Glazkova, N. V.; Savelev, A. P.
Resistance of CPVC to aggressive media (in Russian)
Plast. Massy (1983) 3, p. 23

[945] Anonymous
Erfolgreich gegen Korrosion: Wirbelgesinterte Kunststoffschicht
(Successful against corrosion: whirl-sintered plastic coating) (in German)
Werkst. u. ihre Veredlung 2 (1980) 12, p. 557

[946] Angelmayer, K.-H.
PVC-Polymerisate – vielseitige Bindemittel für den Schutz von Metall, Beton und Kunststoff
(PVC polymers – versatile binders for protection of metal, concrete and plastic) (in German)
Nachrichten der Hoechst AG 41 (1982) 18, p. 1

[947] Reichert, O.
Korrosionsschutz in Anlagen, Teil II
(Corrosion protection in plants, Part II) (in German)
Chem.-Anl. + Verfahren (1977) 11, p. 126

[948] Ioshpe, M. L.
Surface staining of some plastics with varnishes (in Russian)
Lakokras. Mater. 26 (1985) 1, p. 40

[949] Kühne, G.; Martinola, F.
Ionenaustauscher – ihre Beständigkeit gegen chemische und physikalische Einwirkungen
(Ion exchangers – their resistance to chemical and physical loads) (in German)
VGB Kraftwerkstech. 57 (1977) p. 173

[950] Reumann, O.
Fehlerursachen und Fehlerbeseitigung in der Keramik, insbesondere bei der Porzellanherstellung
(Causes of defects and their removal in ceramics, particularly in porcelain production) (in German)
Glas-Email-Keramo-Technik 24 (1973) p. 325

[951] Beone, G.; Lazzaretto, G.; Moccia, A.
Electrolytical disgregation tests of irradiated and non-irradiated graphite
Kerntechnik 15 (1973) 6, p. 270

[952] Enomoto, H.; Ishikawa, M.
Corrosion resistance of tin-nickel alloys deposited from pyrophosphate bath (in Japanese)
J. Met. Finish. Soc. Jpn. 30 (1979) p. 284

[953] Gordeev, S. K.; Smirnov, E. P.; Koltsov, S. I.; Nikitin, Yu. I.
Effect of liquid phase oxidation on the surface properties of synthetic diamonds (in Russian)
Sverkhtverdye Materialy (1979) 3, p. 27

[954] Bednosheya, V. Y.; Olejnik, M. I.; Ivanov, V. A.; Bobovich, O. V. et al.
One-coat enameling of stainless steel (in Russian)
Khim. Neft. Mashinostr. (1982) 2, p. 25

[955] Ivanov, A. V.; Keshishyan, T. N.; Vlasov, A. S.
Untersuchung des Mikrogefüges und einiger Eigenschaften von Fasern aus Siliciumnitrid und kombinierten Fasern aus Siliciumnitrid und Siliciumdioxid
(Examination of the microstructure and some properties of silicon nitride fibers and combined silicon nitride/silicon dioxide fibers) (in Russian)
Trudy Instituta, Moscow, Nr. 98, p. 44

[956] Arai, T.; Sugimoto, Y.; Komatsu, N.
Vanadium-, niobium- and tantalum-carbide on cemented carbides by immersion process in fused borax bath (in Japanese)
J. Met. Finish. Soc. Jpn. 28 (1977) p. 107

[957] Bazhenov, Yu. M.
Verbesserung des Korrosionswiderstandes von Beton durch Polymer-Imprägnierung
(Improvement in the corrosion resistance of concrete by means of polymer impregnation) (in Russian)
Prom. Stroitelstvo. USSR (1978) 8, p. 37

[958] Schley, J. R.
Imperious graphite for process equipment, Part 1
Chem. Eng. 81 (1974) 4, p. 144

[959] Volokhova, V. I.; Vakhidov, R. S.; Lukyanitsa, A. I.
Corrosion stability of nickel-phosphorus galvanic coatings (in Russian)
Zashch. Met. 11 (1975) 3, p. 370

[960] Yau, T. L.
SCC of zirconium and its alloys in nitric acid

Corrosion, Toronto, Canada, 1981 (Proc. Conf.) p. 26

[961] Vrabely, E.
Effect of the corrosion product on the intergranular corrosion tendency of austenitic corrosion-resistant steel in nitric acid tests (in Hungarian)
Banyasz. Kohasz. Lapok Kohasz. *107* (1974) p. 507

[962] Babchuk, G. M. et al.
Selection of a material for equipment in the production of chemically pure silver nitrate for use in the ciné industry (in Russian)
Vses. Nauchno-Issled. Inst. Khim. Reakt. Osob. Chist. Khim. Veshchestv. *44* (1978) p. 166

[963] Schmitz, D.
Werkstoffe für Kreiselpumpen (Materials for centrifugal pumps) (in German)
3R International *22* (1983) p. 276

[964] Anonymous
System 21 – Rationelles Zweischicht-Verfahren für die Lackierung von Großfahrzeugen (System 21 – rational two-layer process for the lacquering of large vehicles) (in German)
Glasurit-Rdsch. *23* (1973) Nr. 51, p. 17

[965] Plyplina, A. I.; Smekhov, F. M.; Amfiteatrova, T. A.; Fartunin, V. I.; Barenbaum, A. L. et al.
Development of recipes for acrylic resin coating powder (in Russian)
Lakokras. Mater. Primen. *17* (1976) 1, p. 9

[966] Klauzner, G. M.; Murasev, N. V.
Development of the production technique of a synthetic resin based on resol formaldehyde resin and furfural-acetone monomer used as a binder for producing effective anticorrosion materials (in Russian)
Mater. Nauch.-Tekh. Konf. Molodykh Uch. Spets. Tyumeni, 1967 (publ. 1968) p. 498

[967] Kopylov, V. T.; Konova, V. I.; Kabanov, E. N.; Zhukov, V. V.; Furnichenko, V. V.
Chemical stability of some brands of rubbers (in Russian)
Khim. Neft. Mashinostr. (1973) 7, p. 21

[968] Däppen, W.
Untersuchungen an PRD 49-Fasern und -Laminaten
(Tests on PRD-49 fibers and laminates) (in German)
Kunststoffe *63* (1973) p. 919

[969] Hillermeier, K. H.
Aramide Verstärkungsfasern für Kunststoffe (Aramides reinforcing fibers for plastics) (in German)
Kunststoffe *66* (1976) p. 802

[970] Anonymous
Polyimide plastics
Mater. Eng. *79* (1974) 2, p. 69

[971] Levshanov, V. S.; Volkov, V. S.; Sholokhova, L. A.; Bojko, L. I.; Dolmatov, S. A.
Effect of test conditions on the properties of fiber glass-reinforced Imilon-S (in Russian)
Plast. Massy (1982) 5, p. 35

[972] Rajsin, I. B.; Basin, V. E.
Adhesion strength of coating systems (in Russian)
Plast. Massy (1977) 10, p. 23

[973] Strepparola, E.; Caporiccio, G.; Monza, E.
Elastomeric polyimides from α,ω-bis(aminomethyl)polyoxyperfluoroalkylenes and tetracarboxylic acids
Ind. Eng. Chem. Prod. Res. Dev. *23* (1984) 4, p. 600

[974] Anonymous
Recommended practice: Monolithic organic corrosion resistant floor surfacings
Mater. Performance *15* (1976) 10; NACE Standard RP-03-76, p. 1

[975] Burbridge, J. F.
Corrosion-resistant reinforced polyester for process plant
Anticorros. Methods Mater. *23* (1976) 2, p. 7

[976] Braun, K.
Polybutenterephthalat und Polyoxymethylen im Spritzgussverfahren (Polybutene terephthalate and polyoxymethylene in the injection molding process) (in German)
Kunststoffe-Plastics *21* (1974) 12, p. 18

[977] Kubel jr., E. J.
Gallery of accomplishments in plastics
Mater. Eng. *101* (1985) 1, p. 41

[978] English, L. K.
Liquid-crystal polymers: In a class of their own
Mater. Eng. *103* (1986) 3, p. 36

[979] Kadurina, T. I.; Omelchenko, S. I.
Hydrolytic stability and protective properties of polyester-based polyurethanes (in Russian)
Lakokras. Mater. Primen. *21* (1980) 3, p. 4

[980] Moshchinskaya, N. K.; Olifer, C. S.; Rakova, T. A.; Kunitsa, T. S.; Klimenko, L. K.; Majofis, I. M.; Chebotareva, N. A.; Lebedeva, N. I.
Enamel varnish based on copolyesters of tere-

[981] Knopova, S. I.; Kajser, M. F.; Gorbunova, V. V.; Mukhaeva, R. K.; Mukhajlova, Z. V.; Orlova, G. A.
The chemical resistance of unsaturated polyester resins (in Russian)
Plast. Massy (1976) 7, p. 65

[982] Shcherbakova, L. S.; Shabanova, S. A.; Tolstaya, S. N.
Improving the protective action of coatings by adding surface-active substances (in Russian)
Lakokras. Mater. Primen. 18 (1977) 2, p. 37

[983] Gnauck, B.
Chemikalien- und Spannungsrißbeständigkeit von Polycarbonat
(On the resistance of polycarbonate to chemicals and environmental stress cracking) (in German)
Kunststoffe 7 (1981) 1, p. 51

[984] Glukhov, L. V.; Vasileva, Yu. M.; Kostrov, V. I.; Sayushkina, O. D.; Vavilova, V. P.; Balusova, N. P.
Interaction of polycarbonate with concentrated nitric acid (in Russian)
Plast. Massy (1976) 2, p. 66

[985] Obolonchik, V. A.; Skripka, I. P.
The chemical stability of scandium, gadolinium and erbium selenides in various corrosive media
Sov. Progress Chem. 43 (1977) 7, p. 5

[986] Nabiev, M. N.; Li, M. G.; Kamalov, K. M.; Radzhabov, R.; Grinenko, G. G.
Corrosion resistance of amorphous complex equipment materials in the transitions to SUM-V-zh production (in Russian)
Uzbek. Khim. Zh. (1980) 2, p. 56

[987] Anonymous
Scuffs and scratches snuffed with polyurethane coatings
Mater. Eng. 77 (1973) 4, p. 67

[988] Product Information
Desmodur®/Desmophen® Lösungsmittelfreie Polyurethanrohstoffe für Beschichtungen und Abdichtungen
(Desmodur®/Desmophen® solvent-free polyurethane raw materials for coatings and sealings) (in German)
Bayer AG, Leverkusen, 1983

[989] Wyatt, C. H.; Montle, J. F.
Urethane protective coatings for atmospheric exposures
phthalate and α-oxalic acids (in Russian)
Lakokras. Mater. Primen. 17 (1976) 6, p. 12
Mater. Protection and Performance 12 (1973) 10, p. 30

[990] Menges, S.; Stoff, F.
Haftung und Versagen organischer Oberflächenschutzschichten
(Adhesion and failure of protective organic coatings) (in German)
Farbe + Lack 81 (1975) p. 204

[991] Wiktorek, S.
Evaluation of coating systems for the internal protection of blast furnace stoves
Anticorros. Methods Mater. 27 (1980) 12, p. 4

[992] Vinarskij, V. L.; Vorobev, I. E.; Popkova, E. E.; Vishnevetskij, V. M.; Sukhovaya, E. Y.; Moshchinskaya, N. K.; Kulev, G. B.
Chemical resistant paints based on new epoxy resins (in Russian)
Lakokras. Mater. Primen. 14 (1973) 5, p. 28

[993] Kuznetsova, V. P.; Omelchenko, S. I.; Zapunnaya, K. V.; Dobrovolskaya, A. N.
Modification of polyurethane binders with bis-epoxypropylbenzimidazole (in Russian)
Plast. Massy (1980) 11, p. 43

[994] Rolston, J. A.
Fiberglass composites in filter applications
Chem. Eng. Progr. 78 (1982) 10, p. 75

[995] Chu, T. L.; Chu, S. S.
Partial purification of metallurgical silicon by acid extraction
J. Electrochem. Soc. 130 (1983) p. 455

[996] Anonymous
Polypropylenes are better than ever
Mater. Eng. 75 (1972) 4, p. 30

[997] Anonymous
Improved cost performance of two-pack systems
Paint Manufacture 50 (1980) 1, p. 19

[998] Margus, E. A.
Plastic pumps for corrosive services
Pumps-Pompes-Pumpen (1980) Nr. 164, p. 202

[999] Antonov, A. N.; Zhdanov, V. K.; Fokin, M. N.; Khananashvili, L. M.; Andrianov, K., A.
Influence of heat treatment on the physical and mechanical properties and corrosion resistance of graphite-polymer molding material
J. Appl. Chem. USSR 51 (1978) p. 1502

[1000] Anonymous
Asbestfreie Dichtungsplatten
(Asbestos-free sealing plates) (in German)
Werkst. Korros. 38 (1987) p. 203

[1001] Kampmann, Ch.; Kirner, K.,
Investigation of the failure of a tungsten turbulence plate
Prakt. Metallogr. 9 (1972) p. 139

[1002] Cole, S. A.
Materials for tubular filters
Chem. Eng. Progr. 78 (1982) 10, p. 70

[1003] Devyatkina, T. S.; Linkin, Ya. N.; Korosteleva, T. K.
Effect of the circulation of nitric acid solution on the etching kinetics of stainless steels (in Russian)
Zashch. Met. 14 (1798) 1, p. 92

[1004] Cox, F. G.
Tantalum (2)
Corros. Prev. Control 10 (1963) 2, P. 56

[1005] Ziegler, P. F.; Rausch, J. J.
Tribocor® 532 N a new material for environments involving wear and corrosion
Surface modifications and coatings, Toronto (Canada), Oktober 1985 (Proc. Conf.) ASM Metal Park, Ohio (1986) p. 301

[1006] Box, S. M.; Wilson, F. G.
Effect of carbide morphology and composition on the intergranular corrosion of titanium stabilized austenitic stainless steels
J. Iron Steel Inst. 210 (1972) p. 71

[1007] Furman, A. Y.; Romanov, V. V.
Effect of intercrystalline corrosion on the cyclic strength of steel Ch18N9
Sov. Materials Sci. 11 (1975) p. 96

[1008] Dittmer, H.
Anorganische Werkstoffe
(Inorganic materials) (in German)
Chem.-Ing.-Tech. 51 (1979) p. 1216

[1009] Hara, A.; Saito, Y.
Corrosion and oxidation resistance of Igetalloy®
Sumitomo Electr. Techn. Rev. 13 (1970) p. 40

[1010] Georgieva, Z.; Ivanova, A.; Dimitrova, E.
Resistance of floor tiles in acid solutions and alkalis (in Bulgarian)
Stroit. Mater. i Silikatna Prom., Sofia 10 (1969) 8, p. 9

[1011] Tallent, O. K.; Mailen, J. C.
Study of the dissolution of refractory PuO_2 in nitric-hydrofluoric acid dissolvents at 100 °C
Nuclear Technol. 32 (1977) p. 167

[1012] Chechulin, V. A.; Chunaev, V. V.; Novikov, A. I.; Sukhanov, A. I.; Karpov, V. M.
Investigation of the chemical resistance of cast stones made of hornblende in acid media and chloride melts (in Russian)
Khim. i Neft Mashinostr. (1976) 7, p. 17

[1013] Uygur, E. M.
Metallography and microstructural characterization of some hardmetal grades by optical and electron microscopy
Prakt. Metallogr. 19 (1982) p. 592

[1014] Henthorne, M.; DeBold, T. A.
Intergranular corrosion resistance of Carpenter 20Cb-3
Corrosion 27 (1971) p. 255

[1015] Brown, M. H.
Behavior of austenitic stainless steels in evaluation tests for the detection of susceptibility to intergranular corrosion
Corrosion 30 (1974) 1, p. 1

[1016] Anonymous
Korrosionsbeständigkeit von Superferrit (Remanit® 4575). 6. Korrosionsbeständigkeit
(Corrosion resistance of superferrit (Remanit® 4575). 6. Corrosion resistance)
(in German)
Thyssen Edelst. Tech. Ber. 5 (1979) 1, p. 48

[1017] Anonymous
A new nickel alloy for gas turbines
Mater. Eng. 75 (1972) 5, p. 28

[1018] Anonymous
Selbstschmierender Acetalwerkstoff von geringer Reibung und kleinem Abrieb bei hohen Belastungen
(Low-friction acetal composite takes high bearing loads) (in German)
Mater. Eng. 71 (1970) 4, p. 21

[1019] Golovko, L. I.; Rumyantsev, L. Y.; Kuznetsova, V. P.
Weather-resistant polyurethane varnish based on a mixture of aliphatic and aromatic polyisocyanates (in Russian)
Lakokras. Mater. Primen. 23 (1982) 2, p. 56

[1020] Sorokin, M. F.; Shode, L. G.; Klochkova, L. V.; Mirenskij, R. B.
Glycidylurethane oligomers and their use (in Russian)
Lakokras. Mater. Primen. 21 (1980) 5, p. 32

[1021] Shigorina, I. I.; Egorov, B. N.; Timofeeva, L. P.
Cold drying Ftorlon® coatings for radiation protection technology (in Russian)
Lakokras. Mater. Primen 18 (1977) 2, p. 45

[1022] Hoppe, M.
Grilonit® G 16.05 – Ein neues Epoxidharz

der Emser Werke AG (Grilonit® G 16.05 – a new epoxy resin from Emser Werke AG) (in German)
Oberfläche Surf. 13 (1972) 5, p. 108

[1023] Anonymous
Epoxy resins for chemical engineering
Processing 23 (1977) 7, p. 55

[1024] Sorokin, M. F.; Dzhurichina, T. N.
Benzoyl peroxide – effect on the hardening of epoxy resin oligomers; protective properties of the varnish coatings (in Russian)
Lakokras. Mater. Primen. (1977) 3, p. 12

[1025] Sorokin, M. F.; Shode, L. G.; Shigorin, V. G.; Bojtsova, L. S.; Lysov, V. P.
Low viscosity epoxide compositions (in Russian)
Plast. Massy (1979) 5, p. 3

[1026] Helmdach, V.; Adler, J.
Lösungsmittelbeständige Epoxidharzbeschichtungen für den Bautenschutz (Solvent-resistant epoxy resin coatings for building construction) (in German)
Kunststoffe 76 (1986) 9, p. 786

[1027] Lushchik, V. V.
Effect of the short time action of static loads and aggressive liquid media on the mechanical properties of three dimensional cross-linked epoxide polymers (in Russian)
Sov. Mater. Sci. 9 (1973) p. 413

[1028] Lushchik, V. V.
Creep of cross-linked epoxy polymers strained in tension under the influence of aggressive liquid media (in Russian)
Sov. Mater. Sci. 10 (1974) p. 420

[1029] Khozin, V. G.; Murafa, A. V.; Vinarskij, V. L.
Modification of epoxy oligomers with the sediments from the coke industry (in Russian)
Lakokras. Mater. Primen. 18 (1977) 4, p. 21

[1030] Tsingarelli, E. P.; Orzhakhovskij, M. L.
Comparison of the temperature and concentration dependence of the operating life of varnish coatings in aggressive gases and liquids (in Russian)
Lakokras. Mater. Primen. 18 (1977) 4, p. 40

[1031] Young, P.
Epoxy ester coatings for atmospheric service
Mater. Performance 24 (1985) 1, p. 54

[1032] Anonymous
Flooring protection – A maintenance-free floor for a pickling shop utilises epoxy resins
Anticorros. Methods Mater. 18 (1971) 3, p. 29

[1033] Lobanov, Yu. E.; Bogatyreva, E. A.; Subbotkina, A. A.; Shterenzon, A. L.
Assessment of the chemical resistance of polymer materials based on their microhardness (in Russian)
Plast. Massy (1983) 6, p. 28

[1034] Shode, L. G.; Sorokin, M. F.; Bojtsova, L. S.; Dobrovinskij, L. A.
Solvent-free epoxy combinations (in Russian)
Lakokras. Mater. Primen. 14 (1973) 1, p. 19

[1035] Lushchik, V. V.
Creep resistance of epoxy graft polymers under pressure in aggressive liquids (in Russian)
Fiz.-Khim. Mekh. Mater. 9 (1973) 5, p. 34

[1036] Trizno, M. S.; Apraskina, L. M.; Borobeva, G. Y.
Chemical resistance of protective coatings based on epoxy-novolak block copolymers (in Russian)
Lakokras. Mater. Primen. 15 (1974) 5, p. 31

[1037] Trizno, M. S.; Apraksina, L. M.
Chemically resistant coatings based on epoxy-novolak block copolymers with a low drying temperature (in Russian)
Lakokras. Mater. Primen. 16 (1975) 2, p. 48

[1038] Trizno, M. S.; Apraksina, L. M.; Eronko, O. N.
Assessment of the protective properties of epoxy-novolak block copolymers based on the aggressive liquids diffusing into the coatings (in Russian)
Lakokras. Mater. Primen. 19 (1978) 2, p. 40

[1039] Lebedeva, N. N.; Shitov, V. S.
Latex compound "Ebolats" for anticorrosion protection of the equipment (in Russian)
Lakokras. Mater. Primen. 22 (1981) 4, p. 10

[1040] Mantsevich, R. P.; Chernov, A. V.
Test on the use of epoxy-furfurol coatings as protection on concrete surfaces (in Russian)
Lakokras. Mater. Primen. 15 (1974) 1, p. 61

[1041] Lesnikova, K. P.; Freid, M. K.
Corrosion and electrochemical properties of a boride coating on molybdenum in acid and alkaline media
Prot. Metals 15 (1979) p. 579

[1042] Sidyakin, P. V.; Egorov, B. N.; Adamchikova, L. M.
Resistance of some epoxy resin coatings under the action of nuclear radiation and corrosive media (in Russian)
Lakokras. Mater. Primen. *15* (1974) 5, p. 37

[1043] Tuishev, Sh. M.; Gotlib, E. M.; Sokolova, Yu. A.; Figovskij, O. L.
The question of fillers for epoxy rubber binders in corrosions protection substances (in Russian)
Korroz. Zashch. (1982) 4, p. 16

[1044] Orzhakhovskij, M. L.; Tsingarelli, E. P.
Failure mechanism of the protective action of epoxy resin paints on metals towards organic acids (in Russian)
Lakokras. Mater. Primen. *20* (1979) 4, p. 20

[1045] Homann, J.
Fluorkunststoffe in der chemischen Verfahrenstechnik, insbesondere als Auskleidungs-/Beschichtungswerkstoffe (Fluoroplastics in chemical process engineering, especially as lining/coating materials) (in German)
Werkst. Korros. *37* (1986) p. 532

[1046] Kuzmenko, A. P.; Kolokolova, T. G.; Ogilets, L. I.; Doroshenko, I. D.; Kiseleva, L. K.; Klimenko, T. G.
Effect of the test conditions on the mechanical properties of fluoroplastics (in Russian)
Plast. Massy (1983) 11, p. 59

[1047] Kut, S.
Pulverlacke (Teil I) – Typen, Trends, Vorbehandlung
(Powder coatings (Part 1) – types, trends, pretreatment) (in German)
Metalloberfläche *28* (1974) 6, p. 201

[1048] Heusser, P.
Spritzgiessbare Fluorkunststoffe und ihre Anwendung
(Injection-moldable fluoroplastics and their application) (in German)
Kunststoffe Plastics *21* (1974) 3, p. 14

[1049] Rodchenko, D. A.; Parshikova, Z. V.
Modifications with Ftorlon® F-3 (in Russian)
Lakokras. Mater. Primen. *22* (1981) 4, p. 41

[1050] Anonymous
Tauch-Kreiselpumpe aus Stahl oder Kunststoff
(Immersed centrifugal pump made of steel or plastic) (in German)
Chem.-Anl. + Verfahren *9* (1976) 4, p. 42

[1051] Werthmüller, E.
Fluorkunststoffe im modernen Behälter-, Rohrleitungs- und Apparatebau
(Thermoplastic fluoropolymers in the construction of modern tanks, pipes and apparatus) (in German)
Werkst. Korros. *34* (1983) 1, p. 32

[1052] Anonymous
Coated valves resist acids, oxidizers
Mod. Plastics Intern. *10* (1980) 3, p. 17

[1053] Barth, E.; Schommer, R.
PVDF im Chemieanlagenbau – Erfahrungen mit einem neuen Werkstoff
(PVDF in chemical plant construction – experience with a new material) (in German)
Werkst. Korros. *31* (1980) p. 303

[1054] Vaccari, J. A.
New heat resistant elastomers are more versatile
Mater. Eng. *85* (1977) 5, p. 45

[1055] Houston, A. M.
Heat and stress key to chemical resistance of plastics
Mater. Eng. *84* (1976) 1, p. 36

[1056] Blahnik, R.; Hraničková
Resistance of electrically isolating varnishes and polymer films to the surrounding medium (in Czech)
Koroze Ochr. Mater. *18* (1974) 5, p. 79

[1057] Bruns, L. E.
Plastics in nuclear processing plants
Chem. Eng. Progr. *71* (1975) 1, p. 59

[1058] Anonymous
Chemical processing reactor
Corros. Prev. Control *25* (1978) 4, p. 17

[1059] Product Information (Deutsche Solvay-Werke GmbH, Solingen-Ohligs)
Rohre aus Solef®
(Pipes made of Solef®) (in German)
Werkst. Korros. *35* (1984) p. 463

[1060] Anonymous
For purity within a few parts per billion, mold parts without using processing aids
Mater. Eng. *92* (1980) 3, p. 9

[1061] Margus, E.
Polyvinylidene fluoride for corrosion-resistant pumps
Chem. Eng. *82* (1975) 23, p. 133

[1062] Shigorina, I. I.; Zvyagintseva, N. V.; Egorov, B. N.
Chemical resistance of cold-set fluorocarbon resin coatings for radiation protection facilities (in Russian)
Lakokras. Mater. Primen. *18* (1977) 6, p. 38

[1063] Caporiccio, G.; Corti, C.; Soldini, S.
A new perfluorinated grease for high-vacuum technology
Ind. Eng. Chem. Prod. Res. Dev. *21* (1982) 3, p. 520

[1064] Ryzhov, M. G.; Vauchskij, Yu. P.; Larin, A. M.; Velts, A. A.
Chemical resistance of membranes based on trifluorostyrene (in Russian)
Plast. Massy (1976) 2, p. 68

[1065] Vaccari, J. A.
Silicone rubbers resist heat, cold, weathering, chemicals
Mater. Eng. *77* (1973) 4, p. 40

[1066] Anonymous
Coating stands up well against acids, alkalis
Mater. Eng. *90* (1979) 1, p. 36

[1067] Shafir, K. F.; Bryskovskaya, A. V.; Lejtman, K. A.; Tkachenko, G. F. et al.
New sheet material with multicolored print reduces inflammability (in Russian)
Plast. Massy (1974) 9, p. 52

[1068] Kurbanova, R. A.; Alieva, D. N.; Ragimov, A. V.
Coatings based on modified polystyrene (in Russian)
Lakokras. Mater. Primen. *21* (1980) 1, p. 51

[1069] Yukelson, I. I.; Butenko, T. R.; Berdutin, A. Y.
Film former based on the sediments in the rectification of styrene (in Russian)
Lakokras. Mater. Primen. *20* (1979) 4, p. 9

[1070] Anonymous
Polyethylene-butyl rubber copolymer resists aging, ozone, stress cracking
Mater. Eng. *72* (1970) 1, p. 28

[1071] Medvedeva, T. I.; Yankin, G. D.
Auftragen von Schichten aus einer Pentaplast®-Suspension durch Tauchen
(Immersion deposition of layers from a Pentaplast® suspension) (in Russian)
Lakokras. Mater. Primen. *20* (1979) 5, p. 55

[1072] Kajser, M. F.; Tarasevich, B. N.
Resistance of Pentaplast® in nitric acid (in Russian)
Past. Massy (1978) 11, p. 29

[1073] Williams, K. J.; Evans, T. E.
Neue Legierung auf NiSi-Basis für Schwefelsäure
(New NiSi-based alloy for sulfuric acid) (in German)
Chem. Process Eng. *51* (1970) 9, p. 57

[1074] Margus, E. A.
Engineering plastics for pumps
Chem. Eng. Progr. *78* (1982) 12, p. 69

[1075] Barth, E.
Das Verhalten von PVC hart gegen Chemikalieneinwirkung
(The behavior of rigid PVC on exposure to chemicals) (in German)
Z. Werkstofftech. *17* (1986) 3, p. 98

[1076] Poschet, G.; Zöhren, J.
Chemikalienbeständigkeit von Klebverbindungen an Rohren aus Polyvinylchlorid hart unter Berücksichtigung von Fugenbreite und biaxialer Spannung
(Chemical resistance of adhesive joints on rigid polyvinylchloride pipes taking into account joint width and biaxial stress) (in German)
Schweissen Schneiden *31* (1979) 8, p. 332

[1077] Anonymous
Vinyl paints for maintenance systems
Anticorros. Methods. Mater. *20* (1973) 6, p. 4

[1078] Firmenmitteilung (Vanton Pump and Equipment Corp., Hillside, N.J.)
Case history library PVC, CPVC, polypropylene, Teflon®, Ryton®, Kynar® plastic pumps in critical service
Chem. Eng. Progr. *76* (1980) 7, p. 19

[1079] Kempe, B.; Hessel, J.
Zeitstandverhalten von Schweißverbindungen aus HDPE bei der Einwirkung von Chemikalien
(Long-term creep tests with HDPE pipes under the influence of several fluids) (in German)
Z. Werkstofftech. *14* (1983) 2, p. 37

[1080] Hapke, J.
Theoretische Grundlagen für den Einsatz von Wärmeaustauschern aus Polyolefin-Kunststoffen
(Theoretical fundamentals in the use of heat exchangers made of polyolefin plastics) (in German)
CZ-Chemie-Technik *3* (1974) 1, p. 15

[1081] Pauli, W. P.
Kunststoffarmaturen für die chemische

Industrie
(Plastic fittings in the chemical industry) (in German)
Chem.-Anl. + Verfahren *14* (1981) 7, p. 36

[1082] Radcliffe, A. T.; Lens, T. J.
Furanharz/Glasfaser-Laminate eine Kombination hervorragender Werkstoff-Eigenschaften
(Furan resin/glass fiber laminates a combination of outstanding material properties) (in German)
Kunststoffe *63* (1973) 12, p. 854

[1083] Anonymous
Chemical fume fans resist corrosion
Can. Chem. Process. *56* (1972) 11, p. 21

[1084] Barton, H. D.
Phenolics and furans in chemical process equipment
Mater. Protection and Performance *12* (1973) 6, p. 16

[1085] Gelfer, D. H.; Tator, K. B.
Vinyl coatings for resistance to atmospheric corrosion
Mater. Performance *16* (1977) 12, p. 9

[1086] Losev, V. B.
Chemical resistance of silico-organic elastomers (in Russian)
Korroz. Zashch. (1976) 9, p. 25

[1087] Repin, A. A.; Martynov, O. M.; Novgorodskij, V. I.
Testing the protective properties of organosilicates in the corrosion protection of metal constructions (in Russian)
Sb. Tr. Vses. N.-I. Proekt. Inst. Teploproekt (1977) No. 44, p. 96

[1088] Kondrashov, G. A.
Epoxyfuran compositions (in Russian)
Tr. Krasnodar. Politekh. Inst. *66* (1975) p. 104

[1089] Pushkarev, Yu. N.; Labutin, A. L.; Anosov, V. I.; Antonova, N. G.
Coatings based on low molecular weight 1,2-polybutadiene (in Russian)
Lakokras. Mater. Primen. *20* (1979) 4, p. 32

[1090] Kühne, G.; Martinola, F.
Ionenaustauscher – ihre Beständigkeit gegen chemische und physikalische Einwirkungen
(Ion exchanger – resistance to chemical and physical exposure) (in German)
VGB Kraftwerkstech. *57* (1977) 3, p. 173

[1091] Fenner, O. H.
Plastics testing
Chem. Eng. *77* (1970) 22, p. 53

[1092] Anonymous
Milchmeßgerät mit Teilen aus Polysulfon
(Milk measuring device with polysulfone components) (in German)
Kunststoffe *72* (1982) 8, p. 478

[1093] Sigwalt, P.
Polysulfures d'éthylène – Synthèse et propriétés
(Ethylene polysulfides – synthesis and properties) (in French)
Chimie et Industrie *104* (1971) 1, p. 47

[1094] Gluskin, V. M.; Borisova, S. V.; Varene, O. V.; Troyan, L. A.; Ademenko, A. A.; Solntseva, I. L.
Protective coatings based on coumarone-indol resins (in Russian)
Lakokras. Mater. Primen. *17* (1976) 3, p. 59

[1095] Kline, G. M.
Technical review. New nonflammable polymer: polychloral
Mod. Plastics Intern. *2* (1972) 6, p. 84

[1096] Rybakov, K. V.; Kovalenko, V. P.; Rozanova, L. M.
Filters for removing contaminants from corrosive liquids (in Russian)
Khim. Neft Mashinostr. (1971) 19, p. 39

[1097] Anonymous
Chemical processing reactor
Corros. Prev. Control *25* (1978) 4, p. 17

[1098] Anonymous
Polyvinylidene Fluoride (PVDF) "Foraflon®" (in German)
Werkst. Korros. *37* (1986) p. 565

[1099] Szymik, Z.; Borek, J.
Polyadditionsprodukte aus ß-Hydroxyäthylester der Piperazin-N,N-bis-thiolthionsäure und Isozyanaten
(Polyaddition products from ß-hydroxyethylesters of piperazine-N,N-bis-thiolthione acid and isocyanates) (in German)
Plaste und Kautschuk *25* (1978) 8, p. 448

[1100] Kotomkin, V. Y.; Baburina, V. A.; Lebedev, E. P.; Kercha, Yu. Yu.
Resistance of polysilane urethanes to solvents and aggressive media (in Russian)
Plast. Massy (1981) 2, p. 27

[1101] Egorov, B. N.; Shigorina, I. I.; Zvyagintseva, N. V.
Chemical resistance of some cyclorubber

coatings in aggressive liquids (in Russian)
Lakokras. Mater. Primen. *14* (1973) 3, p. 38

[1102] Houston, A. M.
Nitrile rubbers excel in oil over wide temperature range
Mater. Eng. *84* (1976) 3, p. 78

[1103] Sukhov, S. I.; Kajser, M. F.
Testing the corrosion resistance of Pentaplast® in mineral acid solutions
(in Russian)
Zash. Korr. Khim. Promst., Cherkassy (1975) p. 41

[1104] Zadorozhnyi, V. G.
Protective properties of Fluoron-3- and Fluoron-4-coatings in different aggressive media (in Russian)
Korroz. Zash. (1980) p. 387

[1105] Product Information
Korrosionsschutz durch Weichgummi-Auskleidung auf Basis Baypren®, Information No. KA 30515, Dec. 1972
(Corrosion protection by means of a soft rubber lining based on Baypren®)
(in German)
Bayer AG, Leverkusen

[1106] Petergerya, D. M.; Ponomarenko, E. P.; Domio, A. A.; Rastorgueva, V. G.
Fatigue strength of the steel St-3 with vacuum diffusion chromium layers in aggressive media (in Russian)
Fiz.-Khim. Mekh. Mater. *5* (1969) p. 637

[1107] Bogdanov, S. G.
Testing the corrosion and abrasion resistance of alloyed steels (in Russian)
Metalloved. Obrab. Met. (1955) 3, p. 25

[1108] Fishter, R. E.; Lada, H.
Stripping solution for nickel superalloys
Brit. Pat. 2 099 459 (June 1, 1982)

[1109] Anonymous
Concentration of nitric acid
(Pintsch-Bamag-Patent No. 37 222/67)
Brit. Chem. Eng. *15* (1970) p. 601

[1110] Anonymous
Thermische Spritztechnik 1977
(Thermal spraying technology 1977)
(in German)
Metalloberfläche *32* (1978) 1, p. 30

[1111] Product Information (Lord Hughson Chemicals, Erie, Pa. (USA))
Chemglaze® coatings: America's first family of single-package pigmented polyurethanes for maximum protection, durability and service
Mater. Eng. *76* (1972) 5, p. 58

[1112] Sokolova, T. G.; Susorova, V. I.
New varnish coating systems in mining
(in Russian)
Lakokras. Mater. Primen. *19* (1978) 4, p. 68

[1113] Anonymous
Coated valves resist acids, oxidizers
Mod. Plastics Int. *10* (1980) 3, p. 17

[1114] Greene jr., N. D.; Ahmed, N.
Performance in corrosive environment.
Filament reinforced metallic composites
Mater. Protection *9* (1970) 3, p. 16

Key to materials compositions

Table 1: Chemical compositions of alloys according to German and other standards

German Standard		Materials Compositions	US-Standard
Mat.-No.	DIN-Design	Percent in Weight	AISI/SAE/ASTM/UNS
0.6656	GGL-NiCuCr 15 6 3	Fe-max. 3.0C-1.0-2.8Si-0.5-1.5Mn-13.5-17.5Ni-2.5-3.5Cr-5.5-7.5Cu	A 436 Type 1b
0.6660	GGL-NiCr 20 2	Fe-max. 3.0C-1.0-2.8Si-0.5-1.5Mn-18.0-22.0Ni-1.0-2.5Cr	A 436 Type 2
0.6661	GGL-NiCr 20 3	Fe-max. 3.0C-1.0-2.8Si-0.5-1.5Mn-18.0-22.0Ni-2.5-3.5Cr	A 436 Type 2b
0.6655	GGL-NiCuCr 15 6 2	Fe-max. 3.0C-1.0-2.8Si-0.5-1.5Mn-13.5-17.5Ni-1.0-2.5Cr-5.5-7.5Cu	A 436 Type 1
0.6676	GGL-NiCr 30 3	Fe-max. 2.5C-1.0-2.8Si-0.5-0.8Mn-2.5-3.5Cr-28.0-32.0Ni	A 436 Type 3
1.0037	St 37-2; S235JR	Fe-$\leq$0.17C-$\leq$0.3Si-$\leq$1.4Mn-$\leq$0.045P-$\leq$0.045S-$\leq$0.009N	A 283; SAE 1015
1.0038	RSt 37-2	Fe-$\leq$0.17C-$\leq$1.4Mn-$\leq$0.045P-$\leq$0.045S-$\leq$0.009N	UNS K02502
1.0040	USt 42-2	Fe-$\leq$0.25C-$\leq$0.2-0,5Mn-$\leq$0.05P-$\leq$0.05S-$\leq$0.007N	
1.0050	E295; St 50-2	Fe-$\leq$0.045P-$\leq$0.045S-$\leq$0.009N	
1.0204	UQSt 36	Fe-$\leq$0.14C-$\leq$0.25-0.50Mn-$\leq$0.040P-0.040S	SAE 1008
1.0256	St 44.0	Fe-$\leq$0.21C-$\leq$0.55Si-$\leq$1.60Mn-$\leq$0.040P-$\leq$0.040S-$\leq$0.009N	A 106
1.0305	St 35.8	Fe-$\leq$0.17C-$\leq$0.10-0.35Si-$\leq$0.40-0.80Mn-$\leq$0.040P-$\leq$0.040S	UNS K01200
1.0308	St 35	Fe-$\leq$0.17C-$\leq$0.35Si-0.4Mn-$\leq$0.05P-$\leq$0.05S-$\leq$0.007N	SAE 1010
1.0309	DX55D	Fe-$\leq$0.16C-0.17-0.40Si-0.35-0.65Mn-$\leq$0.05P-$\leq$0.050S-$\leq$0.30Cr-$\leq$0.30Ni-$\leq$0.30Cu	UNS K02501
1.0330	St 2; St 12	Fe-$\leq$0.12C-$\leq$0.60Mn-$\leq$0.045P-$\leq$0.045S	A 366 (C)

German Standard		Materials Compositions	US-Standard
MatNo.	DIN-Design	Percent in Weight	AISI/SAE/ASTM/UNS
1.0333	USt 3; USt 13	Fe-≤0.08C-≤0.007N	
1.0336	USt 4	Fe-≤0.09C-≤0.25-0.50Mn-≤0.030P-≤0.030S-≤0.007N	
1.0345	P235GH; H I	Fe-≤0.16C-≤0.35Si-≤0.40-1.20Mn-≤0.030P-≤0.025S-0.02Al-≤0.30Cr-≤0.30Cu-≤0.08Mo-≤0.010Nb-≤0.30Ni-≤0.03Ti-≤0.02V	A 285; A 414
1.0375	TH57; T 57	Fe-≤0.1C-Traces Si-0.25-0.45Mn-≤0.04P-≤0.04S-0.007N	
1.0401	C 15	Fe-≤0.12-0.18C-≤0.40Si-≤0.3-0.6Mn-≤0.045P-≤0.045S	SAE 1015
1.0402	C 22	Fe-≤0.17-0.24C-≤0.4Si-≤0.4-0.7Mn-≤0.045P-≤0.045S-≤0.4Cr-≤0.1Mo-≤0.4Ni	SAE 1020
1.0405	St 45.8	Fe-≤0.21C-≤0.10-0.35Si-≤0.40-1.20Mn-≤0.040P-≤0.040S	A 106
1.0408	St 45	Fe-≤0.25C-≤0.035Si-0.40Mn-≤0.050P-≤0.050S	A 108; SAE 1020
1.0414	C20D; D 20-2	Fe-≤0.18-0.23C-≤0.30Si-≤0.3-0.6Mn-≤0.035P-≤0.035S-≤0.01Al-≤0.2Cr-≤0.3Cu-≤0.05Mo-≤0.25Ni	UNS G10200; SAE 1020
1.0425	P265GH; H II	Fe-≤0.20C-≤0.4Si-≤0.5-1.4Mn-≤0.030P-≤0.025S-0.02Al-≤0.3Cr-≤0.3Cu-≤0.08Mo-≤0.01Nb-≤0.3Ni-≤0.03Ti-≤0.02V	UNS K01701
1.0426	P280GH	Fe-≤0.08-0.20C-≤0.4Si-≤0.9-1.5Mn-≤0.025P-≤0.015S-≤0.30Cr	A 662 (A)
1.0473	P355GH	Fe-≤0.1-0.22C-≤0.6Si-≤1.0-1.7Mn-≤0.03P-≤0.025S-≤0.30Cr-≤0.3Cu-≤0.08Mo-≤0.3Ni	A 299
1.0481	17 Mn 4	Fe≤0.08-0.20C-≤0.4Si-≤0.90-1.50Mn-≤0.030P-≤0.025S-0.02Al-≤0.30Cr-≤0.30Cu-≤0.08Mo-≤0.010Nb-≤0.3Ni-≤0.03Ti-≤0.02V	A 414, 515
1.0501	C 35	Fe-≤0.32-0.39C-≤0.4Si-≤0.5-0.8Mn-≤0.045P-≤0.045S-≤0.4Cr-≤0.1Mo-≤0.4Ni	SAE 1035
1.0503	C 45	Fe-≤0.42-0.50C-≤0.4Si≤0.5-0.8Mn-≤0.045P-≤0.045S-≤0.4Cr-≤0.1Mo-≤0.4Ni	AISI 1045
1.0505	StE 315	Fe-≤0.18C-≤0.45Si-≤0.70-1.50Mn-≤0.035P-≤0.030S-≤0.30Cr-0.020Al-≤0.20Cu-≤0.020N-≤0.03Nb-≤0.08Mo-0.30Ni	A 573
1.0528	C 30	Fe-≤0.27-0.34C-≤0.4Si-≤0.5-0.8Mn-≤0.045P-0.045S-≤0.4Cr-≤0.1Mo≤0.4Ni	SAE 1030
1.0545	S355N	Fe-≤0.20C-≤0.50Si-≤0.90-1.65Mn-≤0.035P-0.030S-0.02Al-≤0.30Cr-≤0.35Cu-≤0.10Mo-≤0.015N-≤0.050Nb-≤0.50Ni-≤0.03Ti-≤0.12V	UNS K12709

German Standard		Materials Compositions	US-Standard
Mat.-No.	DIN-Design	Percent in Weight	AISI/SAE/ASTM/UNS
1.0562	StE 355	Fe-≤0.20C-≤0.50Si-≤0.90-1.70Mn-≤0.03P-≤0.025S-0.02Al-≤0.30Cu-≤0.30Cr-≤0.08Mo-≤0.02N-≤0.05Nb-≤0.50Ni-0.03Ti	A 633 (C)
1.0564	N-80	Fe-≤0.030P-≤0.030S	—
1.0570	St 52-3	Fe-≤0.20C-≤0.55Si-≤1.60Mn-≤0.035P-≤0.035S	SAE 1024
1.0605	C 75	Fe-≤0.7-0.8C-≤0.15-0.35Si-≤0.6-0.8Mn-≤0.045P-≤0.045S	SAE 1074
1.0616	C86D	Fe-≤0.83-0.88C-≤0.10-0.30Si-≤0.50-0.80Mn-≤0.035P-≤0.035S-≤0.01Al-≤0.15Cr-≤0.25Cu-≤0.05Mo-≤0.20Ni	SAE 1086
1.0619	GP240GH	Fe-≤0.18-0.23C-≤0.60Si-≤0.50-1.20Mn-≤0.03P-≤0.02S	A 216
1.0670	P-105	Fe-≤0.70C-≤0.03-0.30Si-1.0Mn-≤0.04P-≤0.04S-≤0.007N	—
1.0912	46Mn7	Fe-≤0.42-0.50C-≤0.15-0.35Si-≤1.6-1.9Mn-≤0.05P-≤0.05S-≤0.007N	SAE 1345
1.1104	EStE 285	Fe-≤0.16C-≤0.4Si-≤0.5-1.5Mn-≤0.025P-≤0.015S-0.02Al-≤0.30Cr-≤0.30Cu-≤0.02N-≤0.5Ni	P275NL2
1.1106	P355NL2	Fe-≤0.18C-≤0.5Si-≤0.9-1.7Mn-≤0.025P-≤0.015S-≤0.3Cr-≤0.3Cu-≤0.3Mo-0.5Ni-≤0.02N	A 707
1.1121	C10E	Fe-≤0.07-0.13C-≤0.40Si-≤0.30-0.60Mn-≤0.035P-≤0.035S	SAE 1010
1.1127	36Mn6	Fe-≤0.34-0.42C-≤0.15-0.35Si-≤1.4-1.65Mn-≤0.035P-≤0.035S	
1.1151	Ck 22	Fe-≤0.17-0.24C-≤0.4Si-≤0.4-0.7Mn-≤0.035P-≤0.035S-≤0.4Cr-≤0.1Mo-≤0.4Ni	SAE 1023
1.1186	C40E	Fe-≤0.37-0.44C-≤0.4Si-≤0.5-0.8Mn-≤0.035P-≤0.035S-≤0.4Cr-≤0.1Mo-≤0.4Ni	SAE 1040
1.1191	C45E; Ck 45	Fe-≤0.42-0.50C-≤0.4Si-≤0.50-0.80Mn-≤0.035P-≤0.035S-≤0.40Cr-≤0.10Mo-≤0.4Ni	SAE 1045
1.1520	C 70 W1	Fe-≤0.65-0.74C-≤0.10-0.30Si-≤0.10-0.35Mn-≤0.030P-≤0.030S	
1.1525	C 80 W1	Fe-≤0.75-0.85C-≤0.10-0.25Si-≤0.10-0.25Mn-≤0.020P-≤0.020S	AISI W 108
1.1545	C 105 W1	Fe-≤1.0-1.1C-≤0.10-0.25Si-≤0.10-0.25Mn-≤0.020P-≤0.020S	AISI W 110
1.2365	X 32 CrMoV 3 3	Fe-≤0.28-0.35C-≤0.10-0.40Si-≤0.15-0.45Mn-≤0.030P-≤0.030S-≤2.70-3.20Cr-≤2.60-3.00Mo-≤0.40-0.70V	AISI H 10

German Standard		Materials Compositions	US-Standard
Mat.-No.	DIN-Design	Percent in Weight	AISI/SAE/ASTM/UNS
1.2787	X23CrNi17	Fe-≤0.10-0.25C-≤1.00Si-≤1.00Mn-≤0.035P-≤0.035S≤15.5-18.0Cr-≤1.0-2.5Ni	
1.3355	S 18-0-1	Fe-≤0.70-0.78C-≤0.45Si-≤0.4Mn-≤0.030P-≤0.030S-≤3.8-4.5Cr-≤1.0-1.2V-≤17.5-18.5W	A 600
1.3505	100Cr6	Fe-≤0.93-1.05C-≤0.15-0.35Si-≤0.25-0.45Mn-≤0.025P-≤0.015S-≤0.05Al-≤1.35-1.60Cr-≤0.30Cu-≤0.10Mo	SAE 52100; A 29
1.3813	X40MnCrN19	Fe-≤0.30-0.50C-≤0.80Si-≤17.0-19.0Mn-≤0.10P-≤0.030S-≤3.0-5.0Cr-≤0.08-0.12N	
1.3974	X2CrNiMnMoNNb23-17-6-3	Fe-≤0.30C-≤1.0Si-≤4.5-6.5Mn-≤0.025P-≤0.010S-≤21.0-24.5Cr-≤2.8-3.4Mo-≤0.3-0.5N-≤0.1-0.3Nb-≤15.5-18.0Ni	—
1.3981	NiCo 29 18; (X3NiCo29-18)	Fe-≤0.050C-≤0.30Si-≤0.50Mn-≤17.0-18.0Co-≤28.0-30.0Ni	UNS K94610
1.4000	X6Cr13	Fe-≤0.08C-≤1.0Si-≤1.0Mn-≤0.04P-≤0.015S-≤12.0-14.0Cr	AISI 403, 410S
1.4002	X6CrAl13	Fe-≤0.08C-≤1.0Si-≤1.0Mn-≤0.04P-≤0.015S-≤0.10-0.3Al-≤12-14Cr	AISI 405
1.4003	X2CrNi12; X 2 Cr 11	Fe-≤0.03C-≤1.0Si-≤1.5Mn-≤0.04P-≤0.015S-≤10.5-12.5Cr-≤0.03N-≤0.3-1.0Ni	UNS S40977
1.4005	X12CrS13	Fe-≤0.08-0.15C-≤1.0Si-≤1.5Mn-≤0.04P-≤0.15-0.35S-≤12.0-14.0Cr-≤0.6Mo	AISI 416
1.4006	X12Cr13	Fe-≤0.08-0.15C-≤1.00Si-≤1.5Mn-≤0.04P-≤0.015S-≤11.0-13.5Cr-≤0.75Ni	AISI 410
1.4016	X6Cr17	Fe-≤0.08C-≤1.0Si-≤1.0Mn-≤0.04P-≤0.015S-≤16.0-18.0Cr	AISI 430
1.4021	X20Cr13	Fe-≤0.16-0.25C-≤1.0Si-≤1.5Mn-≤0.04P-≤0.015S-≤12.0-14.0Cr	AISI 420
1.4024	X15Cr13	Fe-≤0.12-0.17C-≤1.0Si-≤1.0Mn-≤0.045P-≤0.03S-≤12.0-14.0Cr	SAE 420
1.4028	X30Cr13	Fe-≤0.26-0.35C-≤1.0Si-≤1.5Mn-≤0.04P-≤0.015S-≤12.0-14.0Cr	A 743 UNS J91153
1.4031	X39Cr13	Fe-≤0.36-0.42C-≤1.00Si-≤1.00Mn-≤0.04P-≤0.015S-≤12.5-14.5Cr	UNS S42080
1.4034	X46Cr13	Fe-≤0.43-0.5C-≤1.0Si-≤1.0Mn-≤0.04P-≤0.015S-≤12.5-14.5Cr	
1.4057	X17CrNi16-2	Fe-≤0.12-0.22C-≤1.0Si-≤1.5Mn-≤0.04P-≤0.015S-≤15.0-17.0Cr-≤1.5-2.5Ni	AISI 431
1.4104	X14CrMoS17	Fe-≤0.10-0.17C-≤1.0Si-≤1.5Mn-≤0.04P-≤0.15-0.35S-≤15.5-17.5Cr-≤0.2-0.6Mo	AISI 430 F
1.4110	X55CrMo14	Fe-≤0.48-0.60C-≤1.0Si-≤1.0Mn-≤0.04P-≤0.015S-≤13.0-15.0Cr-≤0.5-0.8Mo-≤0.15V	

Key to materials compositions

German Standard		Materials Compositions	US-Standard
Mat.-No.	DIN-Design	Percent in Weight	AISI/SAE/ASTM/UNS
1.4112	X90CrMoV18	Fe-≤0.85-0.95C-≤1.0Si-≤1.0Mn-≤0.04P-≤0.015S-≤17.0-19.0Cr-≤0.9-1.3Mo-≤0.07-0.12V	AISI 440 B
1.4113	X6CrMo17-1	Fe-≤0.80C-≤1.0Si-≤1.00Mn-≤0.04P-≤0.03S-≤16.0-18.0Cr-≤0.90-1.40Mo	AISI 434
1.4120	X20CrMo13	Fe-≤0.17-0.22C-≤1.0Si-≤1.0Mn-≤0.04P-≤0.015S-≤12-14Cr-≤0.9-1.3Mo-≤1.0Ni	
1.4122	X39CrMo17-1; GX35CrMo17	Fe-≤0.33-0.45C-≤1.0Si-≤1.5Mn-≤0.04P-≤0.015S-≤15.5-17.5Cr-≤0.8-1.3Mo-≤1.0Ni	
1.4126	X 110 CrMo 13	Fe-≤1.05-1.15C-≤1.0Si-≤1.0Mn-≤0.040P-≤0.030S- ≤17.0-18.0Cr-≤0.8-1.0Mo	
1.4131	X 1 CrMo 26 1	Fe-≤0.010C-≤0.40Si-≤0.40Mn-≤0.020P-≤0.020S-≤0.015N-≤25.0-27.5Cr-≤0.75-1.50Mo-≤0.50Ni	
1.4300	X 12 CrNi 18 8	Fe-≤0.12C-≤1.0Si-≤2.0Mn-≤0.045P-≤0.030S-≤17.0-19.0Cr-≤8.0-10.0Ni	
1.4301	X5CrNi18-10	Fe-≤0.07C-≤1.0Si-≤2.0Mn-≤0.045P-≤0.015S-≤17-19.5Cr-≤0.11N-≤8.0-10.5Ni	AISI 304
1.4302	X5CrNi19-9	Fe-≤0.05C-≤1.40Si-≤1.90Mn-≤0.025P-≤0.015S-≤18.2-19.8Cr-≤8.70-10.30Ni	UNS S30888
1.4303	X4CrNi18-12	Fe-≤0.06C-≤1.0Si-≤2.0Mn-≤0.045P-≤0.015S-≤17-19Cr-≤0.11N-≤11.0-13.0Ni	AISI 305/308
1.4304	X5CrNi18-12E	Fe-≤0.12C-≤1.0Si-≤2.0Mn-≤0.045P-≤0.030S-≤17-19Cr-≤8.0-10.0Ni	
1.4305	X8CrNiS18-9	Fe-≤0.10C-≤1.0Si-≤2.0Mn-≤0.045P-≤0.15-0.35S-≤17-19Cr-≤1.0Cu-≤0.11N-≤8.0-10.0Ni	AISI 303
1.4306	X2CrNi19-11	Fe-≤0.03C-≤1.0Si-≤2.0Mn-≤0.045P-≤0.015S-≤18-20Cr-≤0.11N-≤10.0-12.0Ni	AISI 304 L
1.4307	X2CrNi18-9	Fe-≤0.03C-≤1.0Si-≤2.0Mn-≤0.045P-≤0.015S-≤17.5-19.5Cr-≤0.11N-≤8.0-10.0Ni	
1.4308	GX5CrNi19-10	Fe-≤0.07C-≤1.5Si-≤1.5Mn-≤0.04P-≤0.030S-≤18.0-20.0Cr-≤8.0-11.0Ni	SAE 304 H
1.4310	X 12 CrNi 17 7	Fe-≤0.05-0.15C-≤2.0Si-≤2.0Mn-≤0.045P-≤0.015S-≤16.0-19.0Cr-≤0.8Mo-≤0.11N≤6.0-9.5Ni	AISI 301
1.4311	X2CrNiN18-10	Fe-≤0.03C-≤1.0Si-≤2.0Mn-≤0.045P-≤0.015S-≤17.0-19.5Cr-≤8.5-11.5Ni-≤0.12-0.22N	AISI 304 LN
1.4312	GX10CrNi18-8	Fe-≤0.12C-≤2.0Si-≤1.5Mn-≤0.045P-≤0.03S-≤17.0-19.5Cr-≤8.0-10.0Ni	A 743

German Standard		Materials Compositions	US-Standard
MatNo.	DIN-Design	Percent in Weight	AISI/SAE/ASTM/UNS
1.4313	X3CrNiMo13-4	Fe-≤0.05C-≤0.7Si-≤1.5Mn-≤0.04P-≤0.015S-≤12.0-14.0Cr-≤0.3-0.7Mo-≤0.02N-≤3.5-4.5Ni	UNS J91540
1.4315	X5CrNiN19-9	Fe-≤0.06C-≤1.0Si-≤2.0Mn-≤0.045P-≤0.03S-≤18.0-20.0Cr-≤0.12-0.22N-≤8.0-11.0Ni	
1.4318	X2CrNiN18-7	Fe-≤0.03C-≤1.0Si-≤2.0Mn-≤0.045P-≤0.015S-≤16.5-18.5Cr-≤0.1-0.2N-≤6.0-8.0Ni	
1.4330	X 2 CrNi 25 20	Fe-≤0.03C-≤1.0Si-≤1.5Mn-≤0.045P-≤0.035S-≤18.0-22.0Cr-≤23.0-27.0Ni	
1.4335	X1CrNi25-21	Fe-≤0.02C-≤0.25Si-≤2.0Mn-≤0.025P-≤0.010S-≤24.0-26.0Cr-≤0.20Mo-≤0.110N-≤20.0-22.0Ni	
1.4340	GX40CrNi27-4	Fe-≤0.30-0.50C-≤2.00Si-≤1.50Mn-≤0.045P-≤0.030S-≤26.0-28.0Cr-≤3.5-5.5Ni	A 743
1.4361	X1CrNiSi18-15-4	Fe-≤0.015C-≤3.70-4.50Si-≤2.00Mn-≤0.025P-≤0.010S-≤16.5-18.5Cr-≤0.20Mo-≤0.110N-≤14.0-16.0Ni	A 336
1.4362	X2CrNiN23-4	Fe-≤0.03C-≤1.00Si-≤2.0Mn-≤0.035P-≤0.015S-≤22.0-24.0Cr-≤0.10-0.60Cu-≤0.10-0.60Mo-≤0.050-0.200N-≤3.5-5.5Ni	
1.4371	X2CrMnNiN17-7-5	Fe-≤0.30C-≤1.0Si-≤6.0-8.0Mn-≤0.045P-≤0.015S-≤16.0-17.0Cr-≤0.15-0.20N-≤3.5-5.5Ni	AISI 202
1.4401	X5CrNiMo17-12-2	Fe-≤0.07C-≤1.0Si-≤2.0Mn-≤0.045P-≤0.015S-≤16.5-18.5Cr-≤2.0-2.5Mo-≤0.110N-≤10.0-13.0Ni	AISI 316
1.4404	X2CrNiMo17-12-2; X2CrNiMo17-13-2	Fe-≤0.03C-≤1.0Si-≤2.0Mn-≤0.045P-≤0.015S-≤16.5-18.5Cr-≤2.0-2.5Mo-≤10.0-13.0Ni≤0.110N	AISI 316 L
1.4405	X 5 CrNiMo 16 5; GX4CrNiMo16-5-1	Fe-≤0.07C-≤1.0Si-≤1.0Mn-≤0.035P-≤0.025S-≤15.0-16.5Cr-≤0.5-2.0Mo-≤4.5-6.0Ni	
1.4406	X2CrNiMoN17-12-2	Fe-≤0.03C-≤1.0Si-≤2.0Mn-≤0.045P-≤0.015S-≤16.5-18.5Cr-≤2.0-2.5Mo-≤0.12-0.22N-≤10.0-12.0Ni	AISI 316 LN
1.4408	GX5CrNiMo19-11-2	Fe-≤0.07C-≤1.50Si-≤1.50Mn-≤0.04P-≤0.030S-≤18.0-20.0Cr-≤2.0-2.5Mo-≤9.0-12.0Ni	CF-8M
1.4410	X2CrNiMoN25-7-4	Fe-≤0.03C-≤1.0Si-≤2.0Mn-≤0.035P-0.015S-≤24.0-26.0Cr-≤3.0-4.5Mo-≤0.200-0.350N-≤6.0-8.0Ni	2507; A 182

German Standard		Materials Compositions	US-Standard
MatNo.	DIN-Design	Percent in Weight	AISI/SAE/ASTM/UNS
1.4417	GX2CrNiMoN25-7-3	Fe-≤0.03C-≤1.0Si-≤1.5Mn-≤0.030P-≤0.020S-≤24.0-26.0Cr-≤1.0Cu-≤3.0-4.0Mo-≤0.150-0.250N-≤6.0-8.5Ni-≤1.0W	3RE60; A 789
1.4418	X4CrNiMo16-5-1	Fe-≤0.06C-≤0.7Si-≤1.5Mn-≤0.040P-≤0.015S-≤15.0-17.0Cr-≤0.8-1.5Mo-≤0.020N-≤4.0-6.0Ni	
1.4427	X12CrNiMoS18-11	Fe-≤0.12C-≤1.0Si-≤2.0Mn-≤0.06P-≤0.15-0.35S-≤16.5-18.5Cr-≤2-2.5Mo-≤10.5-13.5Ni	
1.4429	X2CrNiMoN17-13-3	Fe-≤0.03C-≤1.0Si-≤2.0Mn-≤0.045P-≤0.015S-≤16.5-18.5Cr-≤2.5-3.0Mo-≤11.0-14.0Ni-≤0.12-0.22N	AISI 316 LN
1.4435	X2CrNiMo18-14-3	Fe-≤0.03C-≤1.0Si-≤2.0Mn-≤0.045P-≤0.015S-≤17.0-19.0Cr-≤2.5-3.0Mo-≤0.110N-≤12.5-15.0Ni	AISI 316 L
1.4436	X5CrNiMo17-13-3	Fe-≤0.05C-≤1.0Si-≤2.0Mn-≤0.045P-≤0.015S-≤16.5-18.5Cr-≤2.5-3.0Mo-≤0.110N-≤10.5-13.0Ni	AISI 316
1.4438	X2CrNiMo18-15-4	Fe-≤0.03C-≤1.0Si-≤2.0Mn-≤0.045P-≤0.015S-≤17.5-19.5Cr-≤3.0-4.0Mo-≤0.110N-≤13.0-16.0Ni	AISI 317 L
1.4439	X2CrNiMoN17-13-5	Fe-≤0.030C-≤1.00Si-≤2.00Mn-≤0.045P-≤0.015S-≤16.5-18.5Cr-≤4.00-5.00Mo-≤12.5-14.5Ni-≤0.12-0.22N	
1.4449	X3CrNiMo18-12-3	Fe-≤0.035C-≤1.00Si-≤2.00Mn-≤0.045P-≤0.015S-≤17.0-18.2Cr-≤1.0Cu-≤2.25-2.75Mo-≤0.08N-≤11.5-12.5Ni	AISI 317
1.4460	X3CrNiMoN27-5-2	Fe-≤0.05C-≤1.0Si-≤2.0Mn-≤0.035P-≤0.015S-≤25.0-28.0Cr-≤1.3-2.0Mo-≤0.05-0.2N-≤4.5-6.5Ni	AISI 329
1.4462	X2CrNiMoN22-5-3	Fe-≤0.03C-≤1.00Si-≤2.00Mn-≤0.035P-≤0.015S-≤21.0-23.0Cr-≤2.50-3.50Mo-≤4.50-6.50Ni-≤0.1-0.22N	2205; A 182
1.4465	X1CrNiMoN25-25-2	Fe-≤0.02C-≤0.70Si-≤2.0Mn-≤0.020P-≤0.015S-≤24.0-26.0Cr-≤2.0-2.5Mo-≤22.0-25.0Ni-≤0.08-0.16N	AISI 310 MoLN
1.4466	X1CrNiMoN25-22-2	Fe-≤0.02C-≤0.7Si-≤2.0Mn-≤0.025P-≤0.010S-≤24.0-26.0Cr-≤2.0-2.5Mo-≤21.0-23.0Ni-≤0.1-0.16N	
1.4467	X2CrMnNiMoN26-5-4	Fe-≤0.03C-≤0.8Si-≤4.0-6.0Mn-≤0.03P-≤0.015S-≤24.5-26.5Cr-≤2.0-3.0Mo-≤0.3-0.45N-≤3.5-4.5Ni	

German Standard		Materials Compositions	US-Standard
MatNo.	DIN-Design	Percent in Weight	AISI/SAE/ASTM/UNS
1.4492	X 8 CrNiMoN 17 5	Fe-≤0.07-0.11C-≤0.5Si-≤0.5-1.25Mn-≤0.04P-≤0.03S-≤16.0-17.0Cr-≤2.5-3.25Mo-≤4.0-5.0Ni	
1.4500	GX7NiCrMoCuNb25-20	Fe-≤0.08C-≤1.50Si-≤2.0Mn-≤0.045P-≤0.03S-≤19.0-21.0Cr-≤1.5-2.5Cu-≤2.5-3.5Mo-≤24.0-26.0Ni	A 351
1.4501	X2CrNiMoCuWN25-7-4	Fe-≤0.03C-≤1.0Si-≤1.0Mn-≤0.035P-≤0.015S-≤24.0-26.0Cr-≤0.5-1.0Cu-≤3.0-4.0Mo-≤0.2-0.3N-≤6.0-8.0Ni-≤0.5-1.0W	UNS S32760
1.4505	X4NiCrMoCuNb20-18-2	Fe-≤0.05C-≤1.0Si-≤2.0Mn-≤0.045P-≤0.015S-≤16.5-18.5Cr-≤2.0-2.5Mo-≤19.0-21.0Ni-≤1.8-2.2Cu	
1.4506	X5NiCrMoCuTi20-18	Fe-≤0.07C-≤1.0Si-≤2.0Mn-≤0.045P-≤0.030S-≤16.5-18.5Cr-≤2.0-2.5Mo-≤19.0-21.0Ni-≤1.8-2.2Cu	
1.4507	X2CrNiMoCuN25-6-3	Fe-≤0.03C-≤0.7Si-≤2.0Mn-≤0.035P-≤0.015S-≤24.0-26.0Cr-≤1.0-2.5Cu-≤0.15-0.30N-≤5.5-7.5Ni-≤2.7-4.0Mo	UNS S43940
1.4509	X2CrTiNb18	Fe-≤0.03C-≤1.0Si-≤1.0Mn-≤0.040P-≤0.015S-≤17.5-18.5Cr-≤0.10-0.60Ti	
1.4510	X3CrTi17	Fe-≤0.05C-≤1.0Si-≤1.0Mn-≤0.040P-≤0.015S-≤16.0-18.0Cr	AISI 430 Ti
1.4511	X3CrNb17	Fe-≤0.05C-≤1.0Si-≤1.0Mn-≤0.040P-≤0.015S-≤16.0-18.0Cr	
1.4512	X2CrTi12	Fe-≤0.03C-≤1.0Si-≤1.0Mn-≤0.040P-≤0.015S-≤10.5-12.5Cr	AISI 409
1.4515	GX2CrNiMoCuN26-6-3	Fe-≤0.03C-≤1.0Si-≤2.0Mn-≤0.030P-≤0.020S-≤0.12-0.25N-≤24.5-26.5Cr-≤2.5-3.5Mo-≤5.5-7.0Ni-≤0.8-1.3Cu	
1.4520	X2CrTi17	Fe-≤0.025C-≤0.5Si-≤0.5Mn-≤0.04P-≤0.015S-≤16.0-18.0Cr-≤0.015N-≤0.3-0.6Ti	
1.4521	X2CrMoTi18-2	Fe-≤0.025C-≤1.0Si-≤0.040P-≤0.015S-≤1.0Mn-≤17.0-20.0Cr-≤1.8-2.5Mo-≤0.030N	AISI 444
1.4522	X2CrMoNb18-2	Fe-≤0.025C-≤1.0Si-≤1.0Mn-≤0.040P-≤0.015S-≤17.0-19.0Cr-≤1.8-2.3Mo-≤0.25Ni	AISI 443
1.4523	X2CrMoTiS18-2	Fe-≤0.03C-≤1.0Si-≤0.5Mn-≤0.040P-≤0.15-0.35S-≤17.5-19.0Cr-≤2.0-2.5Mo-≤0.30-0.80Ti	
1.4528	X105CrCoMo18-2	Fe-≤1.0-1.1C-≤1.0Si-≤1.0Mn-≤0.045P-≤0.030S-≤16.5-18.5Cr-≤1.0-1.5Mo-≤0.30-0.80Ti-≤1.3-1.8Co≤0.07-0.12V	

German Standard		Materials Compositions	US-Standard
Mat.-No.	DIN-Design	Percent in Weight	AISI/SAE/ASTM/UNS
1.4529	X1NiCrMoCuN25-20-7	Fe-≤0.02C-≤0.50Si-≤1.00Mn-≤0.030P-≤0.010S-≤19.0-21.0Cr-≤6.0-7.0Mo-≤0.5-1.5Cu-≤0.15-0.25N-≤24.0-26.0Ni	A 249; ASTM N08926
1.4539	X1NiCrMoCu25-20-5	Fe-≤0.02C-≤0.70Si-≤2.0Mn-≤0.030P-≤0.010S-≤19.0-21.0Cr-≤4.0-5.0Mo-≤24.0-26.0Ni-≤1.2-2.0Cu-≤0.150N	AISI 904 L
1.4541	X6CrNiTi18-10	Fe-≤0.08C-≤1.0Si-≤2.0Mn-≤0.045P-≤0.015S-≤17.0-19.0Cr-≤9.0-12.0Ni	AISI 321
1.4542	X5CrNiCuNb16-4	Fe-≤0.07C-≤0.70Si-≤1.5Mn-≤0.040P-≤0.015S-≤15.0-17.0Cr-≤3.0-5.0Ni-≤3.0-5.0Cu-≤0.60Mo	AISI 630; 17-4 PH
1.4544	X 10 CrNiMnTi 18 10	Fe-≤0.08C-≤1.0Si-≤2.0Mn-≤0.035P-≤0.025S-≤17.0-19.0Cr-≤9.0-11.5Ni	SAE 321 UNS J92630
1.4546	X5CrNiNb18-10	Fe-≤0.08C-≤1.0Si-≤2.0Mn-≤0.045P-≤0.030S-≤17.0-19.0Cr-≤9.0-11.5Ni	SAE 347
1.4547	X1CrNiMoCuN20-18-7	Fe-≤0.02C-≤0.70Si-≤1.0Mn-≤0.030P-≤0.010S-≤19.5-20.5Cr-≤0.5-1.0Cu-≤6.00-7.00Mo-≤0.18-0.25N-≤17.5-18.5Ni	254 SMO; A 182
1.4548	X5CrNiCuNb17-4-4	Fe-≤0.07C-≤1.0Si-≤1.0Mn-≤0.025P-≤0.025S-≤15.0-17.5Cr-≤3.0-5.0Cu-≤0.15-0.45Nb-≤3.00-5.00Ni	17-4 PH; AISI 630
1.4550	X6CrNiNb18-10	Fe-≤0.08C-≤1.0Si-≤2.0Mn-≤0.045P-≤0.015S-≤17.0-19.0Cr≤9.0-12.0Ni	AISI 347
1.4558	X2NiCrAlTi32-20	Fe-≤0.03C-≤0.70Si-≤1.0Mn-≤0.020P-≤0.015S-≤0.15-0.45Al≤20.0-23.0Cr-≤32.0-35.0Ni	
1.4561	X1CrNiMoTi18-13-2	Fe-≤0.02C-≤0.50Si-≤2.0Mn-≤0.035P-≤0.015S-≤17.0-18.5Cr-≤2.0-2.5Mo-≤11.5-13.5Ni-≤0.4-0.6Ti	
1.4562	X1NiCrMoCu32-28-7	Fe-≤0.015C-≤0.30Si-≤2.0Mn-≤0.020P-≤0.010S-≤26.0-28-0Cr-≤1.0-1.4Cu-≤6.0-7.0Mo-≤0.15-0.25N-≤30.0-32.0Ni	Alloy 31
1.4563	X1NiCrMoCu31-27-4	Fe-≤0.02C-≤0.70Si-≤2.0Mn-≤0.030P-≤0.010S-≤26.0-28.0Cr-≤3.0-4.0Mo-≤30.0-32.0Ni-≤0.70-1.5Cu-≤0.11N	B 668
1.4565	X2CrNiMnMoNbN25-18-5-4	Fe-≤0.03C-≤1.0Si-≤3.5-6.5Mn-≤0.030P-≤0.015S-≤23.0-26.0Cr-≤3.0-5.0Mo-≤0.3-0.5N-≤0.15Nb-≤16.0-19.0Ni	UNS S34565
1.4568	X7CrNiAl17-7	Fe-≤0.09C-≤0.70Si-≤1.0Mn-≤0.040P-0.015S-≤16.0-18.0Cr-≤6.5-7.80Ni-≤0.70-1.5Al	17-7 PH; AISI 631

German Standard		Materials Compositions	US-Standard
Mat.-No.	DIN-Design	Percent in Weight	AISI/SAE/ASTM/UNS
1.4571	X6CrNiMoTi17-12-2	Fe-≤0.08C-≤1.0Si-≤2.0Mn-≤0.045P-≤0.015S-≤16.5-18.5Cr-≤2.0-2.5Mo-≤10.5-13.5Ni	AISI 316 Ti
1.4573	GX3CrNiMoCuN24-6-5	Fe-≤0.40C-≤1.0Si-≤1.0Mn-≤0.030P-≤0.020S-≤22.0-25.0Cr-≤1.5-2.5Cu-≤4.5-6.0Mo-≤0.15-0.25N-≤4.5-6.5Ni	AISI 316 Ti
1.4575	X1CrNiMoNb28-4-2	Fe-≤0.015C-≤1.0Si-≤1.0Mn-≤0.025P-≤0.015S-≤26.0-30.0Cr-≤1.8-2.5Mo-≤0.035N-≤3.0-4.5Ni	25-4-4; A 176
1.4577	X3CrNiMoTi25-25	Fe-≤0.04C-≤0.50Si-≤2.0Mn-≤0.030P-≤0.015S-≤24.0-26.0Cr-≤2.0-2.5Mo-≤24.0-26.0Ni	
1.4580	X6CrNiMoNb17-12-2	Fe-≤0.08C-≤1.0Si-≤2.0Mn-≤0.045P-≤0.015S-≤16.5-18.5Cr-≤2.0-2.5Mo-≤10.5-13.5Ni	AISI 316 Cb UNS J92971
1.4583	X10CrNiMoNb18-12	Fe-≤0.10C-≤1.00Si-≤2.00Mn-≤0.045P-≤0.030S-≤16.5-18.5Cr-≤2.5-3.0Mo-≤12.0-14.5Ni	318 (Spec)
1.4585	X6CrNiMoNb17-12-2; G-X 10 CrNiMoNb 18 10	Fe-≤0.080C-≤1.50Si-≤2.0Mn-≤0.045P-≤0.030S-≤16.5-18.5Cr-≤2.0-2.5Mo-≤19.0-21.0Ni-≤1.8-2.4Cu	
1.4589	X5CrNiMoTi15-2	Fe-≤0.080C-≤1.0Si-≤1.0Mn-≤0.045P-≤0.030S-≤13.0-15.5Cr-≤0.2-1.2Mo-≤1.0-2.5Ni-≤0.3-0.5Ti	UNS S42035
1.4591	X1CrNiMoCuN33-32-1	Fe-≤0.015C-≤0.5Si-≤2.0Mn-≤0.020P-≤0.010S-≤31.0-35.0Cr-≤0.3-1.2Cu-≤0.5-2.0Mo-≤0.35-0.6N-≤30.0-33.0Ni	
1.4652	X1CrNiMoCuN24-22-8	Fe-≤0.02C-≤0.5Si-≤2.0-4.0Mn-≤0.03P-≤0.005S-≤23.0-25.0Cr-≤0.3-0.6Cu-≤7.0-8.0Mo-≤0.45-0.55N-≤21.0-23.0Ni	
1.4712	X10CrSi6	Fe-≤0.12C-≤2.00-2.50Si-≤1.00Mn-≤0.045P-≤0.030S-≤5.50-6.50Cr	
1.4713	X10CrAlSi7	Fe-≤0.12C-≤0.50-1.00Si-≤1.00Mn-≤0.040P-≤0.015S-≤0.5-1.0Al-≤6.00-8.00Cr	
1.4722	X 10 CrSi 13	Fe-≤0.12C-≤1.90-2.40Si-≤1.00Mn-≤0.045P-≤0.030S-≤12.0-14.0Cr	
1.4724	X10CrAlSi13	Fe-≤0.12C-≤0.70-1.40Si-≤1.00Mn-≤0.040P-≤0.015S-≤0.70-1.20Al-≤12.0-14.0Cr	
1.4762	X10CrAlSi25	Fe-≤0.12C-≤0.70-1.40Si-≤1.00Mn-≤0.040P-≤0.015S-≤1.20-1.70Al-≤23.0-26.0Cr	AISI 446
1.4828	X15CrNiSi20-12	Fe-≤0.20C-≤1.50-2.50Si-≤2.0Mn-≤0.045P-≤0.015S-≤19.0-21.0Cr-≤0.11N-≤11.0-13.0Ni	AISI 309

German Standard		Materials Compositions	US-Standard
Mat.-No.	DIN-Design	Percent in Weight	AISI/SAE/ASTM/UNS
1.4833	X 7 CrNi 23 14	Fe-≤0.15C-≤1.00Si-≤2.00Mn-≤0.045P-≤0.015S-≤22.0-24.0Cr-≤0.11N-≤12.0-14.0Ni	AISI 309 S
1.4835	X9CrNiSiNCe21-11-2	Fe-≤0.05-0.12C-≤1.4-2.5Si-≤1.0Mn-≤0.045P-≤0.015S-≤0.030-0.080Ce-≤20.0-22.0Cr-≤0.12-0.20N-≤10.0-12.0Ni	253 MA; A 182
1.4841	X15CrNiSi25-20	Fe-≤0.20C-≤1.50-2.50Si-≤2.00Mn-≤0.045P-≤0.015S-≤24.0-26.0Cr-≤0.11N-≤19.0-22.0Ni	3RE60; AISI 310; AISI 314
1.4845	X12CrNi25-21	Fe-≤0.15C-≤1.50Si-≤2.00Mn-≤0.045P-≤0.015S-≤24.0-26.0Cr-≤0.11N-≤19.0-22.0Ni	AISI 310 S
1.4848	GX40CrNiSi25-20	Fe-≤0.30-0.50C-≤1.00-2.50Si-≤1.50Mn-≤0.035P-≤0.030S-≤24.0-26.0Cr-≤19.0-21.0Ni	A 297 (HK)
1.4856	GX40NiCrSiNbTi35-25	Fe-≤0.35-0.45C-≤1.00-1.50Si-≤0.5-1.50Mn-≤0.035P-≤0.030S-≤23.0-27.0Cr-≤0.9-1.5Nb-≤33.0-37.0Ni-≤0.10-0.25Ti	—
1.4857	GX40NiCrSi35-25	Fe-≤0.30-0.50C-≤1.00-2.50Si-≤1.50Mn-≤0.035P-≤0.030S-≤24.0-26.0Cr-≤34.0-36.0Ni	A 297 (HP)
1.4864	X12NiCrSi36-16	Fe-≤0.15C-≤1.0-2.0Si-≤2.0Mn-≤0.045P-≤0.015S-≤15.0-17.0Cr-≤0.11N-≤33.0-37.0Ni	AISI 330
1.4876	X10NiCrAlTi32-20	Fe-≤0.12C-≤1.00Si-≤2.00Mn-≤0.030P-≤0.015S-≤19.0-23.0Cr-≤30.0-34.0Ni-≤0.15-0.60Al-≤0.15-0.60Ti	B 163; Alloy 800
1.4878	X12CrNiTi18-9	Fe-≤0.10C-≤1.0Si-≤2.0Mn-≤0.045P-0.015S-≤17.0-19.0Cr-≤9.0-12.0Ni	
1.4903	X10CrMoVNb9-1	Fe-≤0.08-0.12C-≤0.20-0.50Si-≤0.30-0.60Mn-≤0.020P-≤0.010S-≤0.04Al-≤8.0-9.5Cr-≤0.85-1.05Mo-≤0.030-0.070N-≤0.06-0.1Nb-≤0.4Ni-≤0.18-0.25V	A 182
1.4919	X6CrNiMo17-13	Fe-≤0.04-0.08C-≤0.75Si-≤2.0Mn-≤0.035P-≤0.015S-≤0.0015-0.0050B-≤16.0-18.0Cr-≤2.0-2.5Mo-≤0.11N-≤12.0-14.0Ni	AISI 316 H
1.4922	X20CrMoV12-1	Fe-≤0.17-0.23C-≤0.50Si-≤1.00Mn-≤0.030P-≤0.030S-≤10.0-12.5Cr-≤0.80-1.20Mo-≤0.30-0.80Ni-≤0.25-0.35V	
1.4948	X6CrNi18-11	Fe-≤0.04-0.08C-≤0.75Si-≤2.0Mn-≤0.035P-≤0.015S-≤17.0-19.0Cr-≤10.0-12.0Ni	AISI 304 H
1.4961	X8CrNiNb16-13	Fe-≤0.04-0.1C-≤0.3-0.6Si-≤1.5Mn-≤0.035P-≤0.015S-≤15.0-17.0Cr-≤12.0-14.0Ni	

German Standard		Materials Compositions	US-Standard
Mat.-No.	DIN-Design	Percent in Weight	AISI/SAE/ASTM/UNS
1.4981	X8CrNiMoNb16-16	Fe-≤0.04-0.10C-≤0.30-0.60Si-≤1.50Mn-≤0.035P-≤0.015S-≤15.5-17.5Cr-≤1.60-2.00Mo-≤15.5-17.5Ni	
1.4986	X8CrNiMoBNb16-16	Fe-≤0.04-0.1C-≤0.3-0.6Si-≤1.5Mn-≤0.045P-≤0.030S-≤0.05-0.1B-≤15.5-17.5Cr-≤1.6-2.0Mo-≤15.5-17.5Ni	
1.4988	X8CrNiMoVNb16-13	Fe-≤0.04-0.1C-≤0.3-0.6Si-≤1.5Mn-≤0.035P-≤0.015S-≤15.5-17.5Cr-≤1.1-1.5Mo-≤12.5-14.5Ni-≤0.60-0.85V-≤0.06-0.14N	
1.5122	37MnSi5	Fe-≤0.33-0.41C-≤1.1-1.4Si-≤1.1-1.4Mn-≤0.035P-≤0.035S	
1.5415	15 Mo 3	Fe-≤0.12-0.2C-≤0.35Si-≤0.4-0.9Mn-≤0.030P-≤0.025S-≤0.30Cr-≤0.30Cu-≤0.25-0.35Mo-≤0.30Ni	A 204 (A)
1.5511	35B2	Fe-≤0.32-0.39C-≤0.4Si-≤0.5-0.8Mn-≤0.035P-≤0.035S-≤0.02Al-≤0.0008-0.005B	—
1.6511	36CrNiMo4	Fe-≤0.32-0.40C-≤0.4Si-≤0.5-0.8Mn-≤0.035P-≤0.035S-≤0.9-1.2Cr-≤0.15-0.3Mo-≤0.9-1.2Ni	SAE 9840
1.6545	30NiCrMo2-2	Fe-≤0.27-0.34C-≤0.15-0.4Si-≤0.7-1.0Mn-≤0.035P-≤0.035S-≤0.4-0.6Cr-≤0.15-0.3Mo-≤0.4-0.7Ni	SAE 8630
1.6565	40NiCrMo6	Fe-0.35-0.45C-≤0.15-0.35Si-≤0.50-0.70Mn-≤0.035P-≤0.035S-≤0.90-1.4Cr-≤0.20-0.30Mo-≤1.4-1.7Ni	SAE 4340
1.6580	30CrNiMo8	Fe-≤0.26-0.34C-≤0.4Si-≤0.3-0.6Mn-≤0.035P-≤0.035S-≤1.8-2.2Cr-≤0.3-0.5Mo-≤1.8-2.2Ni	
1.6582	34CrNiMo6	Fe-≤0.3-0.38C-≤0.4Si-≤0.5-0.8Mn-≤0.035P-≤0.035S-≤1.3-1.7Cr-≤0.15-0.3Mo-≤1.3-1.7Ni	
1.6751	22NiMoCr3-7	Fe-≤0.17-0.25C-≤0.35Si-≤0.5-1.0Mn-≤0.02P-≤0.02S-≤0.05Al-≤0.3-0.5Cr-≤0.18Cu-≤0.5-0.8Mo-≤0.6-1.2Ni-≤0.03V	A 508
1.6900	X 12 CrNi 18 9	Fe-≤0.12C-≤1.00Si-≤2.00Mn-≤0.045P-≤0.030S-≤17.00-19.00Cr-≤0.5Mo-≤8.00-10.00Ni	UNS J92801
1.6903	X 10 CrNiTi 18 10	Fe-≤0.10C-≤1.00Si-≤2.00Mn-≤0.045P-≤0.030S-≤17.0-19.0Cr-≤0.5Mo-≤10.0-12.0Ni	
1.6906	X 5 CrNi 18 10	Fe-≤0.07C-≤1.0Si-≤2.0Mn-≤0.045P-≤0.030S-≤17.0-19.0Cr-≤0.50Mo-≤9.0-11.5Ni	
1.6957	26NiCrMoV14-5	Fe-≤0.22-0.32C-≤0.3Si-≤0.15-0.40Mn-≤0.015P-≤0.018S-≤1.2-1.8Cr-≤0.25-0.45Mo-≤3.4-4.0Ni-≤0.05-0.15V	A 470 (9)

German Standard		Materials Compositions	US-Standard
Mat.-No.	DIN-Design	Percent in Weight	AISI/SAE/ASTM/UNS
1.7033	34Cr4	Fe-≤0.3-0.37C-≤0.4Si-≤0.6-0.9Mn-≤0.035P-≤0.035S-≤0.9-1.2Cr	UNS G51320
1.7035	41Cr4	Fe-≤0.38-0.45C-≤0.4Si-≤0.6-0.9Mn-≤0.035P-≤0.035S-≤0.9-1.2Cr	SAE 5140; UNS H51400
1.7218	25CrMo4	Fe-≤0.22-0.29C-≤0.40Si-≤0.60-0.90Mn-≤0.035P-≤0.035S-≤0.90-1.20Cr-≤0.15-0.30Mo	SAE 4130
1.7219	26CrMo4	Fe-≤0.22-0.29C-≤0.35Si-≤0.5-0.8Mn-≤0.03P-≤0.025S-≤0.9-1.2Cr-≤0.15-0.30Mo	A 372
1.7220	34CrMo4	Fe-≤0.3-0.37C-≤0.4Si-≤0.6-0.9Mn-≤0.035P-≤0.035S-≤0.9-1.2Cr-≤0.15-0.30Mo	SAE 4130
1.7259	26CrMo7	Fe-≤0.22-0.30C-≤0.15-0.35Si-≤0.50-0.70Mn-≤0.035P-≤0.035S-≤1.50-1.80Cr-≤0.20-0.25Mo	
1.7273	24CrMo10	Fe-≤0.20-0.28C-≤0.15-0.35Si-≤0.50-0.80Mn-≤0.035P-≤0.035S-≤2.30-2.60Cr-≤0.20-0.30Mo-≤0.80Ni	
1.7276	10CrMo11	Fe-≤0.08-0.12C-≤0.15-0.35Si-≤0.30-0.50Mn-≤0.035P-≤0.035S-≤2.70-3.00Cr-≤0.20-0.30Mo	
1.7281	16CrMo9-3	Fe-≤0.12-0.20C-≤0.15-0.35Si-≤0.30-0.50Mn-≤0.035P-≤0.035S-≤2.00-2.50Cr-≤0.30-0.40Mo	
1.7335	13CrMo4-4	Fe-≤0.08-0.18C-≤0.35Si-≤0.4-1.0Mn-≤0.030P-≤0.025S-≤0.7-1.15Cr-≤0.3Cu-≤0.4-0.6Mo	A 182
1.7357	G17CrMo5-5	Fe-≤0.15-0.20C-≤0.60Si-≤0.50-1.0Mn-≤0.020P-≤0.020S-≤1.00-1.50Cr-≤0.45-0.65Mo	A 217; UNS J11872
1.7362	X12CrMo5	Fe-≤0.08-0.15C-≤0.50Si-≤0.30-0.60Mn-≤0.025P-≤0.020S-≤4.00-6.00Cr-≤0.45-0.65Mo	AISI 501
1.7375	12CrMo9-10	Fe-≤0.10-0.15C-≤0.30Si-≤0.30-0.80Mn-≤0.015P-≤0.010S-≤0.01-0.04Al-≤2.00-2.50Cr-≤0.20Cu-≤0.9-1.10Mo-≤0.012N-≤0.30Ni	UNS K21590
1.7380	10CrMo9-10	Fe-≤0.08-0.14C-≤0.50Si-≤0.40-0.80Mn-≤0.030P-≤0.025S-≤2.00-2.50Cr-≤0.30Cu-≤0.90-1.10Mo	A 182 (F22); UNS J21890
1.7383	11CrMo9-10	Fe-≤0.08-0.15C-≤0.50Si-≤0.40-0.80Mn-≤0.030P-≤0.025S-≤2.00-2.50Cr-≤0.30Cu-≤0.90-1.10Mo	

German Standard		Materials Compositions	US-Standard
Mat.-No.	DIN-Design	Percent in Weight	AISI/SAE/ASTM/UNS
1.7386	X12CrMo9-1	Fe-≤0.07-0.15C-≤0.25-1.0Si-≤0.30-0.60Mn-≤0.025P-≤0.020S-≤8.0-10.0Cr-≤0.90-1.1Mo	AISI 504; UNS S50488
1.7715	14MoV6-3	Fe-≤0.1-0.18C-≤0.1-0.35Si-≤0.4-0.7Mn-≤0.035P-≤0.035S-≤0.3-0.6Cr-≤0.5-0.7Mo-≤0.22-0.32V	UNS K11591
1.7766	17CrMoV10	Fe-≤0.15-0.20C-≤0.15-0.35Si-≤0.30-0.50Mn-≤0.035P-≤0.035S-≤2.70-3.00Cr-≤0.20-0.30Mo-≤0.10-0.20V	
1.7779	20CrMoV13-5	Fe-≤0.17-0.23C-≤0.15-0.35Si-≤0.30-0.50Mn-≤0.025P-≤0.020S-≤3.00-3.30Cr-≤0.50-0.60Mo-≤0.45-0.55V	
1.7783	X41CrMoV5-1	Fe-≤0.38-0.43C-≤0.80-1.0Si-≤0.20-0.40Mn-≤0.015P-≤0.010S-≤4.75-5.25Cr-≤1.2-1.4Mo-≤0.40-0.60V	AISI 610
1.8070	21CrMoV5-11	Fe-≤0.17-0.25C-≤0.30-0.60Si-≤0.30-0.60Mn-≤0.035P-≤0.035S-≤1.20-1.50Cr-≤1.00-1.20Mo-≤0.60Ni-≤0.25-0.35V	
1.8075	10CrSiMoV7	Fe-≤0.12C-≤0.9-1.2Si-≤0.35-0.75Mn-≤0.035P-≤0.035S-≤1.6-2Cr-≤0.25-0.35Mo-≤0.25-0.35V	
1.8905	StE 460	Fe-≤0.20C-≤0.60Si-≤1.00-1.70Mn-≤0.030P-≤0.025S-≤0.02Al-≤0.30Cr-≤0.70Cu-≤0.10Mo-≤0.025N-≤0.050Nb-≤0.80Ni-≤0.03Ti-≤0.20V	A 633 (E)
1.8907	StE 500	Fe-≤0.21C-≤0.1-0.6Si-≤1.0-1.7Mn-≤0.035P-≤0.03S-≤0.02Al-≤0.30Cr-≤0.70Cu-≤0.10Mo-≤0.020N-≤0.05Nb-≤1.0Ni-≤0.2Ti-≤0.22V	6386 B; UNS K02001
1.8912	S420NL	Fe-≤0.2C-≤0.6Si-≤1.0-1.7Mn-≤0.03P-≤0.025S-≤0.02Al-≤0.3Cr-≤0.7Cu-≤0.1Mo-≤0.025N-≤0.050Nb-≤0.8Ni-≤0.03Ti-≤0.2V	A 737: UNS K02002
1.8935	WstE 460; P460NH	Fe-≤0.20C-≤0.60Si-≤1.0-1.70Mn-≤0.030P-≤0.025S-≤0.02Al-≤0.30Cr-≤0.70Cu-≤0.10Mo-≤0.025N-≤0.050Nb-≤0.8Ni-≤0.03Ti-≤0.2V	A 350; UNS K02900
1.8961	S235J2W	Fe-≤0.13C-≤0.4Si-≤0.2-0.6Mn-≤0.040P-≤0.035S-≤0.02Al-≤0.4-0.8Cr-≤0.25-0.55Cu-≤0.015-0.060Nb-≤0.65Ni-≤0.02-0.10Ti-≤0.02-0.10V	
1.8962	9CrNiCuP3-2-4	Fe-≤0.12C-≤0.25-0.75Si-≤0.2-0.5Mn-≤0.07-0.15P-≤0.035S-≤0.5-1.25Cr-≤0.25-0.55Cu-≤0.65Ni	A 242; UNS K11430

Key to materials compositions

German Standard		Materials Compositions	US-Standard
Mat.-No.	DIN-Design	Percent in Weight	AISI/SAE/ASTM/UNS
1.8963	S355J2G1W	Fe-≤0.16C-≤0.50Si-≤0.50-1.5Mn-≤0.035P-≤0.035S-≤0.02Al-≤0.40-0.80Cr-≤0.25-0.55Cu-≤0.3Mo-≤0.015-0.060Nb-≤0.65Ni-≤0.02-0.10Ti-≤0.02-0.12V-≤0.15Zr	A 588 (A)
1.8977	L485MB	Fe-≤0.16C-≤0.45Si-≤1.70Mn-≤0.025P-≤0.020S-≤0.015-0.06Al-≤0.30Cr-≤0.25Cu-≤0.10Mo-≤0.012N-≤0.06Nb-≤0.30Ni-≤0.06Ti-≤0.10V	
2.1525			
2.4060	Ni 99,6	≤99.60Ni-≤0.08C-≤0.15Si-≤0.35Mn-≤0.005S-≤0.15Cu-≤0.25Fe-≤0.15Mg-≤0.10Ti	UNS N02200
2.4066	Ni 99,2; S-Ni 99,2	≤99.20Ni-≤0.10C-≤0.25Si-≤0.35Mn-≤0.005S-≤0.25Cu-≤0.40Fe-≤0.15Mg-≤0.10Ti	UNS N02200
2.4068	LC-Ni 99	≤99.0Ni-≤0.02C-≤0.25Si-≤0.35Mn-≤0.005S-≤0.25Cu-≤0.40Fe-≤0.15Mg-≤0.10Ti	UNS N02201
2.4360	NiCu 30 Fe	≤63.0Ni-≤0.15C-≤0.50Si-≤2.0Mn-≤0.020S-≤0.5Al-≤28.0-34.0Cu-≤1.0-2.5Fe-≤0.3Ti	UNS N04400
2.4366	EL-NiCu 30 Mn	≤62.0Ni-≤0.15C-≤1.0Si-≤1.0-4.0Mn-≤0.030P≤0.015S-≤0.5Al-≤27.0-34.0Cu-≤0.5-2.5Fe-≤1.0Nb≤1.0Ti	B 127-98
2.4375	NiCu 30 Al	Ni-≤0.20C-≤0.50Si-≤1.5Mn-≤0.015S-≤2.2-3.5Al-≤27.0-34.0Cu-≤0.5-2.0Fe-≤63.0Ni-≤0.3-1.0Ti	UNS N05500
2.4600	NiMo29Cr	Ni-≤0.01C-≤0.1Si-≤3.0Mn-≤0.025P-≤0.015S-≤0.1-0.5Al-≤3.0Co-≤0.5-3.0Cr-≤0.5Cu-≤1.0-6.0Fe-≤26.0-32.0Mo-≤0.4Nb-≤0.2Ti-≤0.2V-≤3.0W	
2.4602	NiCr21Mo14W	Ni-≤0.01C-≤0.08Si-≤0.5Mn-≤0.025P-≤0.010S-≤2.5Co-≤20.0-22.5Cr-≤2.0-6.0Fe-≤12.5-14.5Mo-≤0.35V-≤2.5-3.5W	UNS N06022
2.4603	NiCr30FeMo	Ni-≤0.03C-≤0.08Si-≤2.0Mn-≤0.04P-≤0.02S-≤5.0Co-≤28.0-31.5Cr-≤1.0-2.4Cu-13.0-17.0Fe-≤4.0-6.0Mo-≤0.3-1.5Nb-≤1.5-4.0W	UNS N06002
2.4605	NiCr23Mo16Al	Ni-≤0.01C-≤0.10Si-≤0.5Mn-≤0.025P-≤0.015S-≤0.1-0.4Al-≤0.3Co-≤22.0-24.0Cr-≤0.5Cu-≤1.5Fe-≤15.0-16.5Mo	UNS N06059
2.4606	NiCr21Mo16W	Ni-≤0.01C-≤0.08Si-≤0.75Mn-≤0.025P-≤0.015S-≤0.5Al-≤1.0Co-≤19.0-23.0Cr-≤2.0Fe-≤15.0-17.0Mo-≤0.02-0.25Ti-≤0.2V-≤3.0-4.0W	UNS N06686

German Standard		Materials Compositions	US-Standard
MatNo.	DIN-Design	Percent in Weight	AISI/SAE/ASTM/UNS
2.4607	SG-NiCr23Mo16; UP-NiCr23Mo16	Ni-≤0.015C-≤0.08Si-≤0.50Mn-≤0.02P-≤0.015S-≤0.1-0.4Al-≤0.3Co-≤22.0-24.0Cr-≤1.5Fe-≤15.0-16.5Mo	
2.4610	NiMo16Cr16Ti	Ni-≤0.01C-≤0.08Si-≤1.0Mn-≤0.025P-≤0.015S-≤2.0Co-≤14.0-18.0Cr-≤0.5Cu-≤3.0Fe-≤14.0-18.0Mo-≤0.7Ti	UNS N06455
2.4617	NiMo28	Ni-≤0.01C-≤0.08Si-≤1.0Mn-≤0.025P-≤0.015S-≤1.0Co-≤1.0Cr-≤0.5Cu-≤2.0Fe-≤26.0-30.0Mo	UNS N10665
2.4618	NiCr22Mo6Cu	Ni-≤0.05C-≤1.0Si-≤1.0-2.0Mn-≤0.025P-≤0.015S-≤2.5Co-≤21.0-23.5Cr-≤1.5-2.5Cu-≤18.0-21.0Fe-≤5.5-7.5Mo-≤1.75-2.5Nb-≤1.0W	UNS N06007
2.4619	NiCr22Mo7Cu	Ni-≤0.015C-≤1.0Si-≤1.0Mn-≤0.025P-≤0.015S-≤5.0Co-≤21.0-23.5Cr-≤1.5-2.5Cu-≤18.0-21.0Fe-≤6.0-8.0Mo-≤0.5Nb-≤1.5W	UNS N06985
2.4631	NiCr20TiAl	Ni-≤0.04-0.10C-≤1.0Si-≤1.0Mn-≤0.03P-≤0.015S-≤1.0-1.8Al-≤0.008B-≤2.0Co-≤18.0-21.0Cr-≤0.2Cu-≤1.5Fe-≤1.8-2.7Ti	UNS N07080
2.4642	NiCr29Fe	≤58.0Ni-≤0.05C-≤0.5Si-≤0.5Mn-≤0.020P-≤0.015S-≤0.5Al-≤27.0-31.0Cr-≤0.5Cu-≤7.0-11.0Fe-≤0.5Ti	UNS N06690
2.4650	NiCo20Cr20MoTi	Ni-≤0.04-0.08C-≤0.4Si-≤0.6Mn-≤0.007S-≤0.3-0.6Al-≤0.005B-≤19.0-21.0Co-≤19.0-21.0Cr-≤0.2Cu-≤0.7Fe-≤5.6-6.1Mo-≤1.9-2.4Ti	UNS N07263
2.4652	EL-NiCr26Mo	≤37.0-42.0Ni-≤0.03C-≤0.7Si-≤1.0-3.0Mn-≤0.015S-≤0.1Al-≤23.0-27.0Cr-≤1.5-3.0Cu-≤30.0Fe-≤3.5-7.5Mo-≤37.0-42.00Ni-≤1.0Ti	UNS S32654
2.4660	NiCr20CuMo	≤32.0-38.0Ni-≤0.07C-≤1.0Si-≤2.0Mn-≤0.025P-≤0.015S-≤19.0-21.0Cr-≤3.0-4.0Cu-≤2.0-3.0Mo	UNS N08020
2.4663	NiCr23Co12Mo	Ni-≤0.05-0.10C-≤0.2Si-≤0.2Mn-≤0.01P-≤0.01S-≤0.7-1.4Al≤0.006B-≤11.0-14.0Co-≤20.0-23.0Cr-≤0.5Cu-≤2.0Fe-≤8.5-10.0Mo-≤0.2-0.6Ti	UNS N06617
2.4669	NiCr15Fe7TiAl; NiCr15Fe7Ti2Al	Ni-≤0.08C-≤0.5Si-≤1.0Mn-≤0.02P-≤0.015S-≤0.4-1.0Al≤1.0Co-≤14.0-17.0Cr-≤0.5Cu-≤5.0-9.0Fe-≤0.7-1.2Nb-≤2.25-2.75Ti	UNS N07750
2.4675	NiCr23Mo16Cu	Ni-≤0.01C-≤0.08Si-≤0.5Mn-≤0.025P-≤0.015S-≤0.5Al≤2.0Co-≤22.0-24.0Cr-≤1.3-1.9Cu-≤3.0Fe-≤15.0-17.0Mo	—

Key to materials compositions

German Standard		Materials Compositions	US-Standard
MatNo.	DIN-Design	Percent in Weight	AISI/SAE/ASTM/UNS
2.4681	CoCr26Ni9Mo5W	Ni-≤1.0C-≤1.0Si-≤1.5Mn-≤23.5-27.5Cr-≤1.0-3.0Fe-≤4.0-6.0Mo-≤0.12N-≤7.0-11.0Ni-≤1.0-3.0W	
2.4800	—	—≤60.0Ni-≤0.05C-≤1.0Si-≤1.0Mn-≤0.045P-≤0.025S-≤2.5Co-≤1.0Cr-≤4.0-7.0Fe-≤26.0-30.0Mo-≤0.2-0.4V	UNS N10001
2.4810	NiMo 30 †	≤62.0Ni-≤0.05C-≤0.5Si-≤1.0Mn-≤0.030P-≤0.015S-≤2.5Co-≤1.0Cr-≤0.5Cu-≤4.0-7.0Fe-≤26.0-30.0Mo-≤0.6V	UNS N10001
2.4816	NiCr15Fe	≤72.0Ni-≤0.05-0.1C-≤0.5Si-≤1.0Mn-≤0.020P-≤0.015S-≤0.3Al-≤0.0060B-≤1.0Co-≤14.0-17.0Cr-≤0.5Cu-≤6.0-10.0Fe-≤0.3Ti	UNS N06600
2.4817	LC-NiCr15Fe	≤72.0Ni-≤0.025C-≤0.50Si-≤1.0Mn-≤0.020P-≤0.015S-≤0.3Al-≤0.0060B-≤1.0Co-≤14.0-17.0-≤0.5Cu-≤6.0-10.0Fe-≤0.3Ti	
2.4819	NiMo16Cr15W	Ni-≤0.01C-≤0.08Si-≤1.0Mn-≤0.025P-≤0.015S-≤2.5Co-≤14.5-16.5Cr-≤0.5Cu-≤4.0-7.0Fe-≤15.0-17.0Mo-≤0.35V-≤3.0-4.5W	UNS N10276
2.4856	NiCr22Mo9Nb	Ni-≤0.03-0.10C-≤0.5Si-≤0.5Mn-≤0.020P-≤0.015S-≤0.4Al-≤1.0Co-≤20.0-23.0Cr-≤0.5Cu-≤5.0Fe-≤8.0-10.0Mo-≤3.15-4.15Nb-≤0.4Ti	UNS N06625
2.4858	NiCr21Mo	≤38.0-46.0Ni-≤0.025C-≤0.50Si-≤1.0Mn-≤0.025P-≤0.015S-≤0.20Al-≤1.0Co-≤19.5-23.5Cr-≤1.5-3.0Cu-≤2.5-3.5Mo-≤0.6-1.2Ti	B 163 UNS N08825
2.4886	SG-NiMo16Cr16W; UP-NiMo16Cr16W	≤50.0Ni-≤0.02C-≤0.08Si-≤1.0Mn-≤0.015S-≤14.5-16.5Cr-≤4.0-7.0Fe-≤15.0-17.0Mo-≤0.4V-≤3-4.5W	
2.4887	EL-NiMo15Cr15W	≤50.0Ni-≤0.02C-≤0.20Si-≤1.0Mn-≤0.015S-≤14.5-16.5Cr-≤4.0-7.0Fe-≤15.0-17.0Mo-≤0.4V-≤3-4.5W	—
2.4951	NiCr20Ti	Ni-≤0.08-0.15C-≤1.0Si-≤1.0Mn-≤0.020P-≤0.015S-≤0.3Al-≤0.0060B-≤5.0Co-≤18.0-21.0Cr-≤0.5Cu-≤5.0Fe-≤0.2-0.6Ti	UNS N06075
2.4964	CoCr20W15Ni	≤9.0-11.0Ni-≤0.05-0.15C-≤0.4Si-≤2.0Mn-≤0.020P-≤0.015S-≤19.0-21.0Cr-≤3.0Fe-≤9.0-11.0Ni-≤14.0-16.0W	UNS R30605
2.4999	MP35N†	35.0Ni-≤0.01C-≤20.0Cr-≤9.5Mo	—

Table 2: Chemical compositions of different American, CIS, Bulgarian and other steels

Steel	Materials Compositions, Percent in Weight	Note
000Ch16N13M2	Fe-≤0.07C-≤1.0Si-≤2.0Mn-16.5-18.5Cr-2-2.5Mo-10-13Ni-≤0.045P-≤0.015S-≤0.11N	CIS, formerly USSR, identical with AISI 316
000Ch16N13M3	Fe-≤0.07C-≤1.0Si-≤2.0Mn-16.5-18.5Cr-2-2.5Mo-10-13Ni-≤0.045P-≤0.015S-≤0.11N	CIS, formerly USSR, identical with AISI 316
000Ch16N16M4	Fe-≤0.07C-≤1.0Si-≤2.0Mn-16.5-18.5Cr-2-2.5Mo-10-13Ni-≤0.045P-≤0.015S-≤0.11N	CIS, formerly USSR, identical with AISI 316
000Ch18N10	Fe-≤0.03C-≤0.8Si-≤2.0Mn-17-19Cr-≤0.3Mo-9-11Ni	CIS, formerly USSR/Bulg.
000Ch18N11	Fe-≤0.03C-≤1.0Si-≤2.0Mn-18-20Cr-10-12Ni-≤0.045P-≤0.015S-≤0.11N	Bulg., comparable with 1.4306
000Ch20N20	Fe-≤0.03C-≤18.57Cr-≤19.40Ni-≤0.71Mn-≤0.26Si	CIS, formerly USSR
000Ch21N10M2	Fe-≤0.02C-≤19.8Cr-≤10.5Ni-≤2.1Mo	CIS, formerly USSR
000Ch21N21M4B	Fe-≤0.03C-≤20-22Cr-20-21Ni-3.4-3.7Mo-≤0.6Mn-≤0.6Si-≤0.03P-≤0.02S-0.45-0.8Nb	CIS, formerly USSR
000Ch21N6M2	Fe-≤0.036C-≤21.1Cr-6.5Ni-2.4Mo	CIS, formerly USSR
005Ch25B	Fe-0.005C-0.007N-25Cr	CIS, formerly USSR
00Ch18G8N2T	Fe-≤0.08C-≤0.8Si-7-9Mn-17-19Cr-≤0.3Mo-1.8-2.8Ni-≤0.2W-≤0.3Cu-≤0.2Ti-0.2-0.5Al	CIS, formerly USSR
00Ch18N10	Fe-≤0.015C-≤0.7Si-≤1.7Mn-≤17.3Cr-≤10.4Ni	CIS, formerly USSR
02Ch12N10S5	Fe-0.02C-12Cr-10Ni-5Si, Nb-stabilized	CIS, formerly USSR
02Ch12N10S5B	Fe-0.02C-12Cr-10Ni-5Si, Nb-stabilized	CIS, formerly USSR
02Ch12N10S5T	Fe-0.02C-12Cr-10Ni-5Si, Nb-stabilized	CIS, formerly USSR
02Ch17NS6	Fe-0.02C-4-6.5Si-0.43-0.52Mn-16.3-18Cr-10.5-18.2Ni-0.005-0.008S-0.012-0.014P-	CIS, formerly USSR
02Ch8N22S6	Fe-≤0.02C-≤5.4-6.7Si-0.6Mn-≤0.030P-≤0.020S-7.5-10Cr-0.3Mo-21-23Ni-≤0.2Ti-≤0.20W	CIS, formerly USSR
02Ch8N22S6B	Fe-0.02C-8Cr-122Ni-6Si, Nb stabilized	CIS, formerly USSR
02Ch8N22T	Fe-0.02C-8Cr-122Ni-6Si, Ti-stabilized	CIS, formerly USSR
03Ch18N11	Fe-0.03C-≤0.80Si-≤0.70-2.0Mn-≤0.035P-0.020S-17.0-19.0Cr-≤0.10Mo-10.5-12.5Ni-≤0.20W-0.30Cu-≤0.50Ti	CIS, formerly USSR, comparable with DIN-Mat.No. 1.4306
03Ch18N14	Fe-0.03C-18Cr-14Ni	CIS, formerly USSR
03Ch21N21M4GB	Fe-≤0.03C-≤0.60Si-1.8-2.5Mn-≤0.030P-≤0.020S-20-22Cr-3.4-3.7Mo-20-22Ni-≤0.2W-≤0.3Cu-≤0.2Ti, Nb 15 x C-0.80	CIS, formerly USSR
03Ch23N6	Fe-≤0.03C-≤0.40Si-1.0-2.0Mn-≤0.035P-≤0.020S-22-24Cr-≤5.3-6.3Ni	CIS, formerly USSR
03Ch25	Fe-about 0.03C-25Cr-0.6Ni	CIS, formerly USSR
03ChN28MDT	Fe-≤0.03C-≤0.80Si-≤0.80Mn-≤0.035P-≤0.020S-22.0-25.0Cr-2.5-3.0Mo-26.0-29.0Ni-0.50-0.90Ti-2.5-3.5Cu	CIS, formerly USSR

Steel	Materials Compositions, Percent in Weight	Note
04Ch18N10	Fe-≤0.04C-≤0.8Si-≤2.0Mn-17-19Cr-≤0.030P-≤0.02S-≤0.3Mo-9-11Ni-≤0.2W-≤0.3Cu-≤0.2Ti	CIS, formerly USSR
04Ch18N10T	Fe-≤0.04C-≤0.8Si-≤2.0Mn-17-19Cr-≤0.3Mo-9-11Ni-≤0.2W-≤0.3Cu-≤0.2Ti	CIS, formerly USSR
05Ch16N15M3	Fe-0.05C-16Cr-15Ni-3Mo	CIS, formerly USSR
06Ch17G15NAB	Fe-0.05C-18.36Cr-16.5Mn-1.6Ni-0.31Nb-ß.12Si-0.01Ce-0.017P-0.014S	CIS, formerly USSR
06Ch23N28M3D3T	Fe-≤0.06C-≤0.8Si-≤2.0Mn-22-25Cr-≤2.4-3Mo-26-29Ni-0.5-0.9Ti-2.5-3.5Cu	CIS, formerly USSR/Bulg.
06Ch28MDT	Fe-≤0.06C-≤0.8Si-≤0.80Mn-22.0-25.0Cr-2.5-3.0Mo-26.0-29.0Ni-0.50-0.90Ti-2.5-3.5Cu	CIS, formerly USSR
06ChN28MDT	Fe-≤0.06C-≤0.8Si-≤0.8Mn-≤0.035P-≤0.02S-22-25Cr-2.5-3.0Mo-26-29Ni-0.5-0.9Ti-2.5-3.5Cu	CIS, formerly USSR
06ChN40B	Fe-0.055C-17.01Cr-39.04Ni-1.99Mn-0.50Nb-0.60Si-0.013S-0.022P	CIS, formerly USSR
07Ch13AG20	Fe-≤0.07C-≤0.60Si-≤19-22Mn-≤0.035P-≤0.025S-≤0.0030B-≤0.1Ca-≤0.1Ce -12.-14.8Cr-≤0.30Cu≤0.1Mg-≤0.30Mo≤0.08-0.18N-≤1.0Ni-≤0.20W≤0.20Ti	CIS, formerly USSR
07Ch16N4B	Fe-0.05-0.10C-≤0.60Si-≤0.2-0.5Mn-≤0.025P-≤0.020S-15.16.5Cr-≤0.30Cu-≤0.30Mo-0.2-0.4Nb-3.5-4.5Ni-≤0.20W	CIS, formerly USSR
07Ch17G15NAB	Fe-0.05C-18.4Cr-16.5Mn-1.6Ni-0.01Ce-0.005B-0.32N	CIS, formerly USSR
07Ch17G17DAMB	Fe-0.06C-17.6Cr-15.2Mn-0.43Mo-0.3Nb-0.005B-0.38N	CIS, formerly USSR
08Ch17N15M3T	Fe-≤0.08C-≤0.80Si-≤2.0Mn-≤0.35P-≤0.020S-16.0-18.0Cr-3.00-4.00Mo-14.0-16.0Ni≤0.20W-0.30Cu-0.30-0.60Ti	CIS, formerly USSR
08Ch17N5M3	Fe-0.06C-0.10C-≤0.80Si-≤0.80Mn-≤0.035P-≤0.020S-16.0-17.5Cr-3.0-3.5Mo-4.5-5.5Ni-≤0.20W-≤0.30Cu-≤0.20Ti	CIS, formerly USSR
08Ch17T	Fe-≤0.08C-≤0.80Si-≤0.80Mn-≤0.035P-≤0.025S-16.0-18.0Cr-≤0.6Ni-≤0.30Cu	CIS, formerly USSR
08Ch18G8N2M2T	Fe-0.08C-18.2Cr-3.42Ni-8.9Mn-2.32Mo-0.22Ti	CIS, formerly USSR
08Ch18G8N2T	Fe-≤0.08C-≤0.80Si-7.0-9.0Mn-≤0.035P-17.0-19.0Cr-≤0.30Mo-1.80-2.80Ni-≤0.30Cu-≤0.035P-≤0.025S-≤0.20-0.50Ti-≤0.20W	CIS, formerly USSR
08Ch18N10	Fe-≤0.08C-≤0.8Si-≤2.0Mn-≤0.035P-≤0.020S-17-19Cr-0.3Mo-9.0-11.0Ni-≤0.2W-≤0.3Cu	CIS, formerly USSR
08Ch18N10T	Fe-≤0.08C-≤0.8Si-≤2.0Mn-≤0.035P-≤0.020S-17-19Cr-0.5Mo-9.0-11.0Ni-0.5Ti-≤0.2W-≤0.3Cu	CIS, formerly USSR
08Ch21N6M2T	Fe-≤0.08C-≤0.8Si-≤0.8Mn-≤0.035P-≤0.025S-20-22Cr-1.8-2.5Mo-5.5-6.5Ni-0.2-0.4Ti-≤0.2W-≤0.3Cu	CIS, formerly USSR

Steel	Materials Compositions, Percent in Weight	Note
08Ch22N6M2T	Fe-≤0.08C-≤0.80Si-≤0.80Mn-20.0-22.0Cr-1.80-2.50Mo-5.50-6.50Ni-≤50.20W-≤0.30Cu-≤0.035P-≤0.025S-0.20-0.40Ti	CIS, formerly USSR
08Ch22N6T	Fe-≤0.08C-≤0.8Si-≤0.8Mn-≤0.035P-≤0.025S-21-23Cr-≤0.3Mo-5.3-6.3Ni-≤0.2W-≤0.3Cu, 5x% C max. 0.65Ti	CIS, formerly USSR
08ChP	Fe-0.25Cr-0.25Ni-0.25Mo	CIS, formerly USSR
09G2S	Fe-≤0.12C-0.5-0.8Si-1.3-1.7Mn-≤0.035P-≤0.035S-≤0.3Cr-≤0.3Ni-≤0.3Cu	Bulg.
0Ch17N16M3T	Fe-≤0.080C-≤0.8Si-≤2.00Mn-16.0-18.0Cr-3.00-4.00Mo-14.0-16.0Ni-≤ 0.035P-≤0.025S-0.30-0.60Ti	CIS, formerty USSR
0Ch18G8N3M2T	Fe-about 18 Cr-8Mn-3Ni-2Mo, Ti	CIS, formerly USSR
0Ch18N10T	Fe-≤0.08C-≤1.0Si-≤2.0Mn-≤0.045P-≤0.015S-17-19Cr-9-12Ni, Ti 5xC-0.70	CIS, formerly USSR/Bulg.
0Ch18N12B	Fe-≤0.08C-≤1.0Si-≤2.0Mn-≤0.045P-≤0.015S-17-19Cr-≤9.0-12Ni, Nb 10xC-1.00	CIS, formerly USSR/Bulg., comparable with DIN-Mat. No. 1.4550
0Ch20N14S2	Fe- about 20Cr-14Ni-2Si	CIS, formerly USSR
0Ch21N5T	Fe-≤0.08Cr-21Cr-5Ni, Ti	CIS, formerly USSR s. text HNO3
0Ch23N18	Fe-≤0.20C-≤1.00Si-1.50Mn-22.0-25.0Cr-≤0.30Mo-17.0-20.0Ni-≤0.035P-≤0.025S	CIS, formerly USSR
0Ch23N28M3D3T	Fe-≤0.06C-≤0.8Si-≤2Mn-22-25Cr-2.4-3Mo-26-29Ni-0.5-0.9Ti-2.5-3.5Cu	CIS, formerly USSR/Bulg.
0Ch25T	Fe-≤0.01C-25Cr, Ti-stabilized	CIS, formerly USSR
0H17N12M2T	Fe-≤0.05C-≤1.0Si-≤2.0Mn-16-18Cr-2-3Mo-11-14Ni-<0.045P-≤0.030S, Ti 5xC-0.60	Poland
10Ch13 (1Ch13)	Fe-0.08-0.15C-≤1.0Si-≤1.5Mn-≤0.040P-≤0.015S-11.5-13.5Cr-≤0.75Ni	CIS, formerly USSR/Bulg.
10Ch14AG15	Fe-≤0.10C-≤0.80Si-14.5-16.5Mn-≤0.045P-≤0.030S-13.0-15.0Cr-<0.60Ni-≤0.60Cu-≤0.20Ti-0.15-0.25N	CIS, formerly USSR
10Ch14G14N4T	Fe-≤0.10C-≤0.80Si-13.0-15.0Mn-13.0-15.0Cr-≤0.30Mo-2.80-4.50Ni-≤0.20W-≤0.30Cu-≤0.035P-≤0.020S-5x% C max. 0.60Ti	CIS, formerly USSR
10Ch17	Fe-0.10C-17Cr	CIS, formerly USSR
10Ch17N13M2T	Fe-≤0.10C-≤0.80Si-≤2.0Mn-≤0.035P-≤0.020S-16.0-18.0Cr-≤0.30Cu-2.0-3.0Mo-12.0-14.0Ni-≤0.20W, Ti>-5x% C	USA, comparable with DIN-Mat. No. 1.4571
10Ch17N13M3T	Fe-≤0.10C-≤0.80Si-≤2.0Mn-≤0.035P-≤0.020S-16.0-18.0Cr-≤0.30Cu-3.0-4.0Mo-12.0-14.0Ni-≤0.20W-≤0.7Ti	USA, comparable with DIN-Mat. No. 1.4573
10Ch18N10M2T	Fe-≤0.10C-18Cr-10Ni-2Mo, Ti stabilized	CIS, formerly USSR

Key to materials compositions

Steel	Materials Compositions, Percent in Weight	Note
10Ch18N10T	Fe-≤0.10C-≤0.80Si-≤1.0-2.0Mn-≤0.035P-≤0.020S-≤17.0-19.0Cr-≤10.0-11.0Ni	CIS, formerly USSR/Bulg.
10Ch18N9T(Ch18N9T)	Fe-0.08C-1.0Si-≤2Mn-≤0.045P-≤0.015S-17-19Cr-≤0.3Mo-9-12Ni-Ti5xC-0.70	CIS, formerly USSR/Bulg.
12Ch13G18D	Fe-0.12C-13Cr-18Mn-Cu	CIS, formerly USSR
12Ch17G9AN4	Fe-≤0.12C-≤0.80Si-8.0-10.5Mn-≤0.035P-≤0.020S-16.0-18.0Cr-≤0.30Mo-3.5-4.5Ni-≤0.20W-≤0.30Cu-≤0.20Ti-0.15-0.25N	CIS, formerly USSR
12Ch18N10T	Fe-≤0.12C-≤0.8Si-≤2.0Mn-≤0.025P-≤0.020S-17.0-19.0Cr-≤0.30Cu-≤0.50Mo-9.0-11.0Ni-≤0.20W-≤0.70Ti	CIS, formerly USSR; comparable with DIN-Mat. No. 1.4878
12Ch18N9T	Fe-≤0.12C-≤0.80Si-≤2.0Mn-≤0.035P-≤0.020S-17.0-19.0Cr-≤0.50Mo-8.0-9.5Ni-≤0.20W-≤0.30Cu-Ti = 5x % C	CIS, formerly USSR
12Ch21N5T	Fe-0.09-0.14C-≤0.80Si-≤0.80Mn-≤0.035P-≤0.025S-20.0-22.0Cr-≤0.30Mo-4.80-5.80Ni-≤0.20W-≤0.30Cu-0.25-0.50Ti-≤0.08Al	CIS, formerly USSR
12Ch2M1	Fe-0.12C-2Cr-1Mo	CIS, formerly USSR/Bulg., comparable with DIN-Mat. No. 1.7380, A 182, F 22, B.S. 1501-622
12Ch2N4A	Fe-0.09-0.15C-0.17-0.37Si-0.30-0.60Mn-≤0.025P-≤0.025S-1.25-1.65Cr-3.25-3.65Ni-≤0.30Cu-≤0.15Mo-≤0,03Ti-≤0.05V-≤0.12W	CIS, formerly USSR
12ChN2	Fe-0.09-0.16C-0.17-0.37Si-0.30-0.60Mn-≤0.035P-≤0.035S-0.60-0.90Cr-1.50-1.90Ni-≤0.30Cu-≤0.15Mo-≤0.03Ti-≤0.05V-≤0.20W	CIS, formerly USSR/Bulg.
13-4-1	Fe-0.043C-12.7Cr-3.9Ni-1.5Mo-0.68Mn-0.39Si-0.009P-0.013S-0.030N	CIS, formerly USSR
14Ch17N2	Fe-≤0.11-0.17C-≤0.80Si-≤0.80Mn-16.0-18.0Cr-≤0.30Mo-1.50-2.50Ni-≤0.20W-≤0.30Cu-≤0.030P-≤0.025S-≤0.20Ti	CIS, formerly USSR
15Ch17N2	Fe-0.13C-0.49Si-0.52Mn-17.17Cr-1.75Ni-0.012P-0.09S	CIS, formerly USSR
15Ch25T	Fe-≤0.15C-≤1.00Si-≤0.80Mn-≤0.035P-≤0.025S-24.0-27.0Cr-≤0.30Cu-≤1.00Ni-0.09Ti	CIS, formerly USSR
15Ch28	Fe-≤0.15C-≤1.00Si-≤0.80Mn-27.0-30.0Cr-0.60Ni-≤0.30Cu-≤0.035P-≤0.025S-≤1.0Ni-≤0.20Ti	CIS, formerly USSR
15Ch2M2FBS	Fe-about 0.15C-2Cr-2Mo-V-Nb-Si	CIS, formerly USSR
15Ch5M	Fe-≤0.15C-≤0.50Si-≤0.50Mn-4.50-6.00Cr-0.40-0.60Mo-≤0.60Ni-≤0.03Ti-≤0.030P-≤0.025S-≤0.20Cu-≤0.05V-≤0.30W	CIS, formerly USSR

Steel	Materials Compositions, Percent in Weight	Note
16GS	Fe-≤0.12-0.18C-≤0.40-0.70Si-0.90-1.20Mn-≤0.30Cr-≤0.30Ni-≤0.30Cu-≤0.035P-≤0.040S-≤0.05Al-≤0.08As-≤0.012N-≤0.03Ti	CIS, formerly USSR/Bulg., comparable with DIN-Mat. No. 1.0481, A 414, A 515, A 516
18/8-CrNi-steel	Fe-≤0.12C-1.0Si-≤2.0Mn-≤0.045P-≤0.030S-≤17.0-19.0Cr≤-8.0-10.0Ni	–
1815-LCSi	Fe-0.006C-18.3Cr-15.1Ni-1.5Mn-4.1Si-0.005S-0.010P-0.010N	UNS S30600, comparable with DIN-Mat. No. 1.4361
18-18-2	Fe-≤0.08C-1.5-2.5Si-≤2.0Mn-≤0.030P-≤0.030S-17.0-19.0Cr-17.5-18.5Ni	USA
18G2A	Fe-≤0.20C-≤0.50Si-≤0.9-1.7Mn-≤0.025P-≤0.020S-≤0.0200Al-≤0.30Cr-≤0.50Ni-≤0.30Cu.-≤0.08Mo-≤0.020N-≤0.050Nb-≤0.03Ti-≤0.1V	Poland
20Ch13 (2Ch13)	Fe-0.16-0.25C-≤0.8Si-≤0.8Mn-12-14Cr-≤0.6Ni-≤0.030P-≤0.025S-≤0.30Cu-≤0.20Ti	CIS, formerly USSR/Bulg., comparable with DIN-Mat. No. 1.4021, AISI 420, 420 S 29
20Ch23N18	Fe-≤0.2C-≤1.0Si-≤2.0Mn-22-25Cr-≤0.3Mo-17-20Ni-≤0.2W-≤0.3Cu-≤0.2Ti-≤0.035P-≤0.025S	CIS, formerly USSR
20Ch2G2SR	Fe-0.16-0.26C-0.75-1.55Si-1.4-1.8Mn-≤0.040P-≤0.040S-1.4-1.8Cr-≤0.30Ni-≤0.30Cu-0.02-0.08Ti-0.015-0.050Al-0.001-0.007B	CIS, formerly USSR
2320	Fe-≤0.08-1.0Si-1.0Mn-≤0.040P-≤0.030S-16.0-18.0Cr-≤1.0Ni	Sweden, comparable with DIN-Mat. No. 1.4016; AISI 430, 10Ch17T
23Ch2G2T	Fe-0.19-0.26C-0.40-0.70Si-1.4-1.7Mn-≤0.045P-≤0.045S-1.35-1.70Cr-≤0.30Ni-≤0.30Cu-0.02-0.08Ti-0.015-0.05Al	CIS, formerly USSR
36NChTJu	Fe-≤0.05C-≤0.3-0.7Si-≤0.8-1.2Mn-≤0.020P-≤0.020S-11.5-13Cr-35-37Ni-0.9-1.2Al-2.7-3.2Ti	CIS, formerly USSR
40Ch13	Fe-0.36-0.45C-≤0.80Si-≤0.80Mn-≤0.030P-≤0.025S-12.0-14.0Cr-≤0.30Cu-≤0.60Ni-≤0.20Ti	CIS, formerly USSR, comparable with DIN-Mat. No. 1.4031
45 G2	Fe-0.41-0.49C-0.17-0.37Si-1.4-1.8Mn-0.035P-≤0.035S-≤0.30Cr-≤0.30Cu-≤0.15Mo-≤0.30Ni-≤0.03Ti-≤0.05V-≤0.20W	CIS, formerly USSR, comparable with DIN-Mat. No. 1.0912
50Ch	Fe-0.46-0.54C-0. 17-0.37Si-0.50-0.80Mn-≤0.035P-≤0.035S-0.80-1.10Cr-≤0.30Ni-≤0.30Cu-≤0.15Mo-≤0.03Ti-≤0.05V-≤0.20W	CIS, formerly USSR
70G	Fe-0.67-0.75C-0.17-0.37Si-0.90-1.20Mn-≤0.035P-≤0.035S-≤0.25Cr-≤0.25Ni- ≤0.20Cu	CIS, formerly USSR
80S	Fe-0.74-0.82C-0.60-1.10Si-0.50-0.90Mn-≤0.040P-≤0.045S-≤0.30Cr-≤0.30Ni-≤0.30Cu-0.015-0.040Ti	CIS, formerly USSR

Steel	Materials Compositions, Percent in Weight	Note
AISI 1008	Fe-0.10C-0.30-0.50Mn-≤0.030P-≤0.050S	USA, comparable with DIN-Mat. No. 1.0204
AISI 1018	Fe-0.15-0.20C-0.60-0.90Mn-≤0.030P-≤0.050S	USA
AISI C-1018	Fe-0.20C-0.25Si-0.58Mn-0.16Cr-0.04Mo-0.012-0.014P-0.02S	USA
ASTM A-159	Fe-3.1-3.4C-1.9-2.3Si-0.6-0.9Mn-0.15S-0.15P	USA
ASTM A-516 Gr. 70	Fe-0.27C-0.13-0.45Si-0.79-1.30Mn-≤0.035P-≤0.040S	USA, comparable with DIN-Mat. No. 1.0050 and No. 1.0481
ASTM XM-27	Fe-≤0.01C-≤0.40Si-≤0.40Mn-≤0.020P-≤0.020S-25.0-27.5Cr-≤0.20Cu-0.75-1.50Mo-≤0.015N-≤0.050-0.2Nb-0.50Ni, Ni+Cu≤0.50	USA, comparable with AISI XM-27
C 1204	Fe-0.20C-0.35Si-0.50Mn-0.050P-0.050S-0.30Cr	Yugoslavia, comparable with DIN-Mat. No. 1.0425, B.S. 1501 Gr. 161-400, 164-350, 164-400; 16 K
C 90	Fe-0.85-0.94C-≤-0.35Si-≤0.35Mn-≤0.03P-≤0.03S	Italy
Carpenter 20 Cb-3	Fe-≤0.06C-≤1.00Si-≤2.00Mn-19.0-21.0Cr-2.0-3.0Mo-32.5-35Ni-3.0-4.0Cu-≤0.035P-≤0.035S	USA
Ch12M	Fe-1.45-1.65C-0.15-0.35Si-0.15-0.4Mn-11-12.5Cr-0.4-0.6Mo-≤0.35Ni-15-0.3V-≤0.2W≤0.3Cu-≤0.03Ti	CIS, formerly USSR
Ch14N40SB	Fe-0.034C-4.0Si-0.05Mn-14.4Cr-38.9Ni-0.63Nb	CIS, formerly USSR
Ch15T	Fe-≤0.1C-≤0.8Si-≤0.8Mn-14-16Cr-≤0.3Mo-≤0.6Ni, 5x%C≤Ti≤0.8	CIS, formerly USSR/Bulg.
Ch17	Fe-≤0.08C-≤1Si-≤1Mn-≤0.040P-≤0.015S-16.0-18.0Cr	Bulg., comparable with DIN-Mat. No. 1.4016, AISI 430, 12Ch17T, X6Cr17
Ch17N12M3T	Fe-≤0.12C-≤1.5Si-≤2.0Mn-16-19Cr-3-4Mo-11-13Ni-0.3-0.6Ti	CIS, formerly USSR/Bulg.
Ch17N18M2T	Fe-0.09C-0.6Si-1.4Mn-16.9Cr-1.9Mo-12.3Ni, Ti stab. (p.a.)	CIS, formerly USSR/Bulg.
Ch17N2	Fe-17Cr-2Ni	CIS, formerly USSR
Ch17T	Fe-≤0.05C-≤1Si-≤1Mn-≤0.040P-≤<0-0.15S-16.0-18.0Cr, Ti4x(C+N)+0.15-0.80	Bulg., comparable with DIN-Mat. No. 1.4510, 08Ch17T, X3CrTi17
Ch18AG14	Fe-18Cr-14Mg-0.5N	
Ch18N9T Ch18N10T	Fe-≤0.08C-≤1Si-≤2Mn-≤0.045P-≤0.015S-17.0-19.0Cr-9.0-12.0Ni, Ti 5xC-0.70	CIS, formerly USSR/Bulg., comparable with DIN-Mat. No. 1.4541, X6CrNiTi18-10
Ch18N10	Fe-0.08C-18.4Cr-10.2Ni-1.08Mn-0.3Si-0.005P-0.014S-0.005N	
Ch18N12M2T	Fe-≤0.15C-≤5 1.5Si-≤2Mn-17-19Cr-2-2.5Mo-11-13Ni, 4x%C≤Ti≤0.8	CIS, formerly USSR/Bulg.

Steel	Materials Compositions, Percent in Weight	Note
Ch18N12T	Fe-0.08C-1.0Si-≤2.0Mn-≤0.045P-0.015S-17.0-19.0Cr-9.0-12.0Ni-Ti = 5x%C≥0.70	CIS, formerly USSR/Bulg.
Ch18N14	Fe-0.035C-18.8Cr-14.6Ni-0.35Mn-0.75Si-0.005P-0.03S-0.004N	
Ch18N40T	Fe-<0.08C-<1.0Si-<2.0Mn-<0.045P-<0.015S-17-19Cr-9-12Ni-<0.7Ti	Comparable with DIN-Mat. No. 1.4541, AISI 321
Ch20N20	Fe-0.004-0.015C-19.4-21.8Cr-19.3-20.8Ni-0.05-5.40Si-0.002-0.1P	CIS, formerly USSR/Bulg
Ch22N5	Fe-0.07C-21.54Cr-5.73Ni	CIS, formerly USSR
Ch23N18	Fe-≤0.20C-≤1.00Si-≤1.50Mn-22.0-25.0Cr-≤0.30Mo-17.0-20.0Ni-≤0.035P-≤0.025S	CIS, formerly USSR/Bulg.
Ch23N27M2T	Fe-27Ni-23Cr-2Mo-Ti	CIS, formerly USSR/Bulg.
Ch23N28M3D3T	Fe-28Ni-23Cr-3Mo-3Cu-Ti	CIS, formerly USSR
Ch25T	Fe-≤0.15C-≤1.00Si-≤0.80Mn-24.0-27.0Cr-≤0.30Mo-≤0.60Ni-≤0.035P-≤025S-5x%C≤Ti≤0.9	CIS, formerly USSR/Bulg.
Ch28N18	Fe-0.16C-1.7Mn-1.1Si-22.6Cr-18Ni-0.4Ti (p.a.)	CIS, formerly USSR/Bulg.
ChN28MDT	Fe-0.03-0.046C-22.2-23.5Cr-26.55-27.88Ni-2.55-3.06Mo-2.68-3.38Cu-0.54-0.76Ti-0.15-0.30Mn-0.39-0.69Si-0.021-0.43P-0.008-0.017S	CIS, formerly USS
ChN40B	Fe-0.032C-18.2Cr-40.4Ni-0.08Si-0.05Mn-0.49Nb	CIS, formerly USSR
ChN40S	Fe-0.031C-20.0Cr-38.9Ni-4.2Si-0.05Mn-0.13Nb	CIS, formerly USSR
ChN40SB	Fe-0.04C-18.8Cr-39.4Ni-4.3Si-0.06Mn-0.63Nb	CIS, formerly USSR
ChN58W	Ni-0.03C-14.5-16.5Cr-15-17Mo-3.0-4.5W-1.5Fe-1.0Mn-<0.12Si-0.02S-0.025P	CIS, formerly USSR
ChN60V	Fe-0.01-0.02C-0.09N-max.0.05Zr-0.1Ti-0.015Ce-max.1.7Nb-max0.009B	CIS, formerly USSR
FC 20	Fe-3.92C-1.12Si-0.63Mn-0.072P-0.012S	Japan
JS 700	Fe-0.04C-≤1.0Si-≤2.0Mn-≤0.040P-≤0.030S-19-23Cr-4.3-5.0Mo-24-26Ni, Nb≥8xC≤0.40	USA
OZL-17u	Fe-0.04C-0.32Si-1.5Mn-23.2Cr-0.2Mo-29.4Ni-0.01P-0.01S-0.1Ti	CIS, formerly USSR
S35C	Fe-0.32-0.38C-0.15-0.35Si-0.60-0.90Mn-≤0.030P-≤0.035S-≤0.20Cr-≤0.30Cu-≤0.20Ni	Japan, comparable with DIN-Mat. No. 1.0501
SIS 2333	Fe-≤0.05C-1.0Si-2.0Mn-≤0.045P-≤0.030S-17.0-19.0Cr-8.0-11.0Ni	Sweden, comparable with DIN-Mat. No. 1.4303; AISI 304, 03Ch18N11
SKH 2	Fe-0.73-0.83C-0.45Si-0.40Mn-≤0.030P-≤0.030S-3.8-4.5Cr-≤0.25Cu-17.0-19.0W-1.0-1.2V	Japan, comparable with DIN 1.3355
SKH-4A	Fe-0.80C-0.29Si-0.31Mn-4.16Cr-17.64W-9.3Co-1.1V	Japan
SKH-9	Fe-0.85C-0.16Si-0.31Mn-4.14Cr-4.97Mo-6.03W-1.88V	Japan
SS41	Fe-≤0.050P-≤0.050S	Japan, comparable with DIN-Mat. No. 1.0040

Steel	Materials Compositions, Percent in Weight	Note
St 38b-2	Fe-0.12-0.20C-0.17-0.37Si-0.40-0.65Mn-≤0.045P-≤0.050S-Cr+Cu+Ni≤0.70	Germany, formerly GDR, comparable with DIN-Mat. No. 1.0038, BS 4360-40C and A 570 Gr. 36
St35b-2	Fe-≤0.16C-0.17-0.40Si-0.35-0.65Mn-≤0.30Cr-≤0.30Ni-≤0.30Cu-0.050P-0.050S	Germany, formerly GDR, comparable with DIN-Mat. No. 1.0309
St35hb	Fe-≤0.18C-≤0.17Si-0.35-0.65Mn-0.050P-0.050S	Germany, formerly GDR
St38	Fe-≤0.20C-≤0.080-≤0.060S	Germany, formerly GDR, comparable with DIN-Mat. No. 1.0037
St5	Fe-≤0.045P-≤0.045S-≤0.009N	Poland, comparable with DIN-Mat. No. 1.0050
SUS 304	Fe-≤0.08C-≤1.0Si-≤2.0Mn-≤0.045P-≤0.030S-18.0-20.0Cr-8-10.5Ni	Japan, comparable with DIN-Mat. No. 1.4301
SUS 430	Fe-≤0.12C-≤0.75Si-≤1.00Mn-16.0-18.0Cr-≤0.60Ni-≤0.040P-≤0.030S	Japan, comparable with DIN-Mat. No. 1.4016, AISI 430, 430 S 15, 12Ch17
Sv-08	Fe-≤0.10C-≤0.03Si-0.35-0.60Mn-≤0.040P-≤0.040S-≤0.015Cr-≤0.30Ni-≤0.01Al	CIS, formerly USSR
TsL-17	Fe-0.1C-5Cr-1Mo	CIS, formerly USSR
TsL-9	Fe-0.07C-1.00Si-2.3Mn-24.1Cr-12.9Ni-1.1Nb-0.03P-0.02S	CIS, formerly USSR
U10A	Fe-0.96-1.03C-0.17-0.33Si-0.17-0.28Mn-≤0.025P-≤0.018S-≤0.20Cr-≤0.20Ni-≤0.20Cu	CIS, formerly USSR, comparable with DIN-Mat.No. 1.1545, AISI W 1 10
U7A	Fe-0.65-0.75C-0.10-0.30Si-0.10-0.40Mn-≤0.030P-≤0.030S	CIS, formerly USSR, comparable with DIN-Mat. No. 1.1520
U8A	Fe-0.75-0.85C-0.10-0.30Si-0.10-0.40Mn-≤0.030P-≤0.030S	CIS, formerly USSR, comparable with DIN-Mat. No. 1.1525
X5CrNiMoCuTi18-18	Fe-≤0.07C-≤0.80Si-≤2.0Mn-≤0.045P-≤0.030S-16.5-18.5Cr-2.0-2.5Mo-19.0-21.0Ni-1.8-2.2Cu-Ti≥7x%C	Germany, formerly GDR, comparable with DIN-Mat. No. 1.4506
X8CrNiTi18-10	Fe-≤0.10C-≤1.0Si-≤2.0Mn-≤0.045P-≤0.015S-17.0-19.0Cr-9.0-12.0Ni-Ti≥5 x C≤0.80	Germany, formerly GDR, comparable with DIN-Mat. No. 1.4541

Index of materials

0

0Ch18N10T 154, 374, 380
0Ch18N12B 374
0Ch21N5T 346–347
0Ch23N28M3D3T 352, 384, 401, 405
0Ch25T 341
00Ch18N10 374
000Ch16N13M2 390
000Ch16N13M3 390
000Ch16N16M4 390
000Ch18N11 352, 359
000Ch18N16M2 390
000Ch18N16M4 390
000Ch20N20 360
000Ch21N10M2 390
000Ch21N21M4B 383
000Ch21N6M2 390
005Ch25B 339
02Ch12N10S5 363
02Ch12N10S5B 363
02Ch12N10S5T 363
02Ch17NS6 409
02Ch19N9 352
02Ch8N22S6 363
02Ch8N22S6B 363
02Ch8N22T 363
03Ch18N11 348–349, 351–354, 358, 378, 439
03Ch18N14 364
03Ch21N21M4B 398–399
03Ch21N21M4GB 378, 396, 400
03Ch23N6 369, 372
03ChN28MDT 378, 396, 398
03ChN60V 417
04Ch18N10 368–369, 380
04Ch18N10T 373
04Ch18N27 380
04Ch18N40 380
05Ch16N15M3 388

06Ch17G15NAB 343–344
06ChN28MDT 367, 375–376, 384, 397–404
06ChN40B 379
07Ch13AG20 338
07Ch17G15NAB 342
07Ch17G17DAMB 342–344
08Ch17T 442
08Ch18G8N2M2T 399
08Ch18G8N2T 341, 344–345, 352
08Ch18N10 379
08Ch18N10T 352, 358–359, 373
08Ch21N6M2T 375–376, 384, 398–400, 402–404
08Ch22N6T 352, 359, 369, 375–376, 378–379

1

1Ch13 345
1Ch18N10T 349
1Ch18N9T 374, 409
1.0256 67
1.0308 92
1.0333 331
1.0402 59
1.0481 67
1.0503 198
1.0564 60
1.0670 60
1.1191 198
1.1520 332
1.1545 332
1.3974 84
1.3981 410
1.4006 343–345
1.4016 67, 92–93, 198, 341
1.4057 76
1.4300 87, 382

Index of materials

1.4301	72, 80, 83–85, 87–91, 93, 110, 177, 198, 350, 352–353, 358, 368
1.4303	84, 354
1.4306	88, 96, 348–349, 351–353, 355, 359, 364, 368, 378, 407, 439
1.4306S	407
1.4308	384
1.4310	343
1.4313	346
1.4315	352
1.4335	359, 386
1.4361	348, 350, 353, 359, 362, 407
1.4401	72, 80, 88, 90, 93, 96, 134, 177, 198–199, 397
1.4404	106, 108, 110
1.4406	88
1.4408	383
1.4410	254
1.4417	406
1.4429	395
1.4435	84, 387, 389, 403, 407
1.4436	93, 198
1.4439	84, 87, 90, 393, 397
1.4462	77, 84, 87, 345, 384, 390–391
1.4465	349, 383, 396–397
1.4466	347, 368, 403, 407
1.4467	346
1.4492	106
1.4501	254
1.4506	400–401
1.4510	381, 442
1.4512	336
1.4522	83–84
1.4529	87, 253
1.4539	84, 87, 254, 400
1.4541	92–93, 95–96, 154, 198, 315, 338–339, 348–352, 358–359, 365, 367–369, 372, 374–377, 379–381, 400–401, 405, 409, 418, 439, 466
1.4547	87, 97, 254
1.4558	90
1.4562	254
1.4563	254, 396, 406–407
1.4565	87
1.4571	84, 90, 343–344, 396, 400–401, 501
1.4575	339, 397
1.4762	341
1.4828	409
1.4845	365
1.4856	421–422
1.4878	352–353, 383
1.5511	69–70
1.6580	70
1.7033	70
1.7220	70
1.7362	67
1.7375	67
1.7380	67
1.7386	67
1.7783	345
10Ch13	343–344
10Ch14G14N4T	345, 348, 408
10Ch17	343–344
10Ch17N13M2T	400
10Ch17N13M3T	401–402
12Ch13G18D	345
12Ch17G9AN4	350
12Ch18N10T	348–352, 354, 365, 368–377, 379–382, 418, 439, 466
13-4	346
13-4-1	346
1815-LCSi	407–408
18-2	73–74
18/8-CrNi-steel	72
1010	92–93
1042	197–198
1045	197–198

2

2.0321	55
2.1525	56, 253
2.4060	99
2.4066	99
2.4360	117, 425
2.4366	117
2.4375	425
2.4600	253
2.4602	106, 129, 254
2.4603	106
2.4605	99, 129, 254
2.4606	129
2.4610	99, 254, 423
2.4617	106, 117, 253, 425
2.4618	88, 422
2.4619	99, 106, 423
2.4631	418
2.4642	372, 418, 422
2.4652	97, 254
2.4660	88, 399, 409
2.4663	426
2.4675	254
2.4800	117
2.4810	117, 425
2.4816	99, 418, 420–422
2.4819	88, 99, 129, 254, 423, 464
2.4856	99, 254
2.4858	88, 99, 407
2.4886	423, 464

Index of materials

2.4887   88
2.4964   42
2.4999   42, 426
2RE 69   347
20   396
20 CN-7M   396
21   313
25   313
25-4-4   339, 397
26-1S   73
28   254
29-4   73–74
29-4-2   73–74, 341
214   117
242   125–126
254 SMO   96, 254
276   103
2017   23
2205   77, 84, 87
2507   254

3
3.0525   305
3.2315   305
3.3207   305
3.3315   305
3.3523   305
3.3527   305
3.3537   305
3.7035   448
3.7235   448
30CrNiMo8   70
31   254
34Cr3   72
34Cr4   70
34CrMo4   70
35B2   69–71
36NCh   418
36NChTJu   410, 417
302 SS   343, 345
304   72–74, 80, 82–85, 87–91, 93–95, 110, 177, 197–198, 348, 350, 352–353, 355–358, 368, 375, 379, 381–382
304 L   88, 96, 297, 347–360, 365, 368, 372, 378–379, 407–408, 438–439, 454, 508
304 LSC   508
304 SS   370–371
308   84, 354
308 SS   354
309 L   87
309 Nb   409
310   347, 350
310 L   350, 374, 408
310 mod.   347

310 MoLN   349, 368, 383, 396, 403, 407
310 S   365–366
316   72–74, 80, 88, 90, 93–96, 114, 134, 177, 197–199, 348, 356, 384, 387, 390, 395, 397–399
316 L   79, 84, 106, 108, 110–111, 113, 345, 384, 386–387, 389, 393, 403, 405
316 LN   88, 395
316 Ti   84, 90, 343–344, 396, 400–401
317 L   396
317 LMN   84, 87, 90, 393–394, 397
321   92–93, 95, 154, 198, 315, 338–339, 348–352, 354, 358–359, 365, 367–377, 379–382, 400–401, 405, 409, 418, 439, 466
321 H   353, 383
321 SS   354
329   385
3931   330

4
47ChNM-2   424
400   99
409   336
410   343–345
430   67–68, 73–74, 92–95, 197–198, 341
431   74, 76
439   73, 381, 442
440   93–94
443   83–84
444   73
446   341
4200   144
4207   144–145
4238   330
4240   330

5
52Ni-48Cr   422
59   99–100, 102, 105, 129, 254

6
600   99, 420
602 CA   117
610   345
617   426
625   99–100, 103, 105, 115–117, 253
654 SMO   97, 254
686   117, 129
690   421

7
70Cr-10Ni-20P   173

Index of materials

8
80Cr-20P 173
800 177
825 99–100

9
904 L 84, 253, 400

a
A 419
A 106 67–68
A 269 253
A 335 (P 5) 67–68
A 335 (P 9) 67–68
A 335 (P 11) 67–68
A 335 (P 22) 67–68
A 336 348, 350, 353, 359, 362
A 355 (P 22) 68
A 436 67
A 905 346
A120Zn 309
AA 1050A 293–294
AA 1060 294–295, 298–300
AA 1100 298
AA 2017 23–24
AA 2024 21
AA 2090 21
AA 2091 21
AA 3003 21, 303, 306, 308
AA 3005 294, 305
AA 5005 305
AA 5005A 294
AA 5049 294, 305
AA 5052 294, 303, 305
AA 5056A 296, 303–304
AA 5083 309
AA 5454 294, 305
AA 5754 296, 303–304
AA 6060 294, 305
AA 6082 294, 305
AA 7004 309
AA 7020 307
AA 7075 307
AA 8090 21–23
α-Al$_2$O$_3$ 198
α-brass 55, 326–327
ABS 209, 230
acid-resistant building materials 463
acrylamide 182
acrylic resins 483
acrylonitrile 182
acrylonitrile-butadiene rubber 209, 237, 243
acrylonitrile/butadiene/styrene copolymer 209, 230
Admiralty brass 325
Aerosil® 465
AF-22 391–392
Aflon® 243
Ag19Cu21Zn20Cd 293
Ag-20Pd 292
Ag-25Au 291
AgAu alloys 292
AgPd alloys 293
AISI 302 SS 343, 345
AISI 304 72–74, 80, 82–85, 87–91, 93–95, 110, 177, 197–198, 348, 350, 352–353, 355–358, 368, 375, 379, 381–382
AISI 304 L 88, 96, 297, 347–359, 365, 368, 372, 378–379, 407–408, 439, 454
AISI 304 SS 370–371
AISI 308 84, 354
AISI 308 SS 354
AISI 309 L 87
AISI 309 Nb 409
AISI 310 347, 350
AISI 310 L 350, 374
AISI 310 mod. 347
AISI 310 MoLN 349, 368, 383, 396
AISI 310 MoLN, 403
AISI 310 MoLN 407
AISI 310 S 365–366
AISI 316 72–74, 80, 88, 90, 93–96, 114, 134, 177, 197–199, 348, 356, 384, 387, 390, 395, 397–399
AISI 316 L 79, 84, 106, 108, 110–111, 113, 345, 384, 386–387, 389, 393, 403, 405
AISI 316 LN 88, 395
AISI 316 Ti 84, 90, 343–344, 396, 400–401
AISI 317 L 396
AISI 317 LMN 84, 87, 90, 393–394, 397
AISI 321 92–93, 95, 154, 198, 315, 338–339, 348–352, 354, 358–359, 365, 367–377, 379–382, 400–401, 405, 409, 418, 439, 466
AISI 321 H 353, 383
AISI 321 SS 354
AISI 329 385
AISI 409 336
AISI 410 343–345
AISI 430 67–68, 73–74, 92–95, 197–198, 341
AISI 431 74, 76
AISI 439 73, 381, 442
AISI 440 93–94
AISI 443 83–84
AISI 444 73

Index of materials

AISI 446   341
AISI 610   345
AISI 904 L   84, 400
AISI 1010   92–93
AISI 1042   197–198
AISI 1045   197–198
AISI W1   332
Akrit® Ni 40   419
Al-0.1Si-0.11Fe   296
Al-0.30Si-0.14Fe-0.06Cu   296
Al-1.8Si-4.0Mg   303
Al-2.4Cu-1.5Mg-1.2Ni-1.0Fe   310
Al-3s   307
Al-4.43Zn-1.34Mg-0.13Cu   307
Al-4.57Zn-1.35Mg   307
Al-4.63Zn-2.46Mg   307
Al-4.86Zn-1.44Mg-0.3Cr   307
Al4Cu   307
Al-5.25Si   303
AL-6X   388
Al-6XN   254
Al-6XN PLUS   254
Al-7.0Si-0.3Mg   303
Al12Si casting alloy   309
Al-14Si   309
Al 99.5   293–294
Al 99.99   294
Al ADO   296
AlCuLi   21
AlCuLiMg   21
alkali-lime glass   250, 256
alkali silicate glass   189, 255
alkyd resin   241
ALLCORR®   115–116
AlLi alloys   21
AlLiCuMg   21
alloy-20   396
alloy-20 CN-7M   396
alloy 21   313
alloy 25   313
alloy 28   254
alloy 31   254
alloy 59   99–100, 102, 105, 129, 254
alloy 242   125–126
alloy 254 SMO   254
alloy 400   99
alloy 600   99, 420
alloy 617   426
alloy 625   99–100, 103, 115–116, 253
alloy 654 SMO   97, 254
alloy 686   129
alloy 690   421
alloy 800   177
alloy 825   99–100

alloy 904 L   254
alloy 2507   254
alloy A   419
alloy B   117, 119–120, 419
alloy B-2   106, 108, 117–120, 125, 128, 177, 253
alloy B-3   117, 120, 125–126
alloy B-4   117, 120
alloy C   73, 103
alloy C-4   99–100, 102–103, 177, 254
alloy C-22   103–106, 116, 177, 254
alloy C-276   100–101, 103–106, 108, 115–116, 177, 254, 313
alloy C-2000   125, 253
alloy G-3   99–100, 106
alloy G-30   100–101
alloy HC new   103
AlMg   16
AlMg1   294, 305
AlMg2.5   294, 305
AlMg2.7Mn   294, 305
AlMg2Mn0.8   294, 305–306
AlMg3   296, 303–304
AlMg5   296, 303–304
AlMg6   304–305, 310
AlMgSi0.5   294, 305
AlMgSi1   294, 305
AlMn   16, 294, 305, 307
AlMn1Cu   23
AlMn1Mg0.5   294, 305
Alpolit UP®   243
aluminium   15, 30, 38, 289, 293, 297, 302
aluminium alloys   18, 303
aluminium-boron-silicate glass   250
aluminium brass   328
aluminium bronze   48–51
aluminium casting alloys   303
aluminium-copper-magnesium alloys   310
aluminium-lithium alloy   22
aluminium-magnesium alloys   303, 310
aluminium-magnesium-manganese alloy   306
aluminium-manganese alloys   303
aluminium oxide   467
aluminium oxide ceramics   197
aluminium-silicate glass   250, 255
aluminium-silicon alloys   309
aluminosilicate glass   185
AlZnMg   307
AlZnMg1   307
aminoplasts   235
Amts   304
Andesite   468
Araldit®   243

Index of materials

Araldite® CT-200   489, 506
Aramid®   248
Aramide   232
Armco® iron   67–68, 331–333
AS-43   338, 342, 410
asbestos   232
Asodur® HCB   490
Asplit®   243
ASTM XM-27   341
AT 3   158, 435–436, 438, 440
AT 6   158, 440
Atlac® 382   484
Atlac® 711-05A   484
Au-0.2Co   311
Au-20Ag-21.5Cu   311
austenitic chromium-nickel-molybdenum steels   383
austenitic chromium-nickel steels   336, 347, 406
austenitic CrNiMo(N) and CrNiMoCu(N) steels   96
austenitic CrNi steel   74, 79
austenitic steel   83, 199
austentic cast iron   67

b

B   117, 119–120, 419
B-2   106, 108, 117–120, 125, 128, 177, 252
B-3   117, 120, 125–126, 253
B-4   117, 120
B-150   49
basalt   460, 467
Baypren®   243, 499, 503, 507–508
Beckopox®   243
beryllium   458
Beryvac® 520   427
binders   181, 462
bisphenol A-polyester   251
bisphenol resin   234
borides   468
Bornumharz®   243
boron phosphide   473
borosilicate glass   185, 187, 189–190, 256, 463
brass   54, 291, 321, 324–328
brass casting alloys   321
bromobutyl rubber   237, 244, 257
bronze   51, 54, 321
BS   209
building stones   467
Buna®   243
Buna Hüls®   243
butadiene/styrene copolymer   209

Butylkautschuk®   243
butyl rubber   237, 257, 263

c

C   73, 103
C-4   99–100, 102–103, 105, 177, 254
C-22   59, 103–106, 108, 116, 129, 177, 254
C70U   332
C-75   60
C-276   99, 100–101, 103–106, 108, 115–116, 129, 177, 254, 313
C-2000   125–127, 129, 253
C64900   56, 253
C65500   56, 253
CaB_6 472
cadmium   159, 450
cadmium telluride   474
carbide   198, 199, 468
carbon   178, 233, 255, 460
carbon fiber   248
carbon graphite   263
carbon steel   59, 63, 66, 80, 329, 333–334
Cariflex®   243
Cariflex IR®   243
Carpenter® 7-Mo   385
Carpenter® 20   409
Carpenter® 20 Cb-3   88, 399
cast alloy   97
cast aluminium   302
cast basalt   460
cast brass   325
cast iron   134, 253, 335–336
cast steel   56, 329
CdTe   474
CeGe   473
$CeGe_2$ 473
cement   181–183, 462
CeN   472
CERAFIL-S®   197
ceramic   201, 467
ceramic materials   289
ceramic types SK-1   467
cermets   385
Cer-Vit®   197
Ce_5Ge_3 473
C glass   250, 256
Ch14N40SB   409
Ch17N18M2T   399
Ch18AG14   340
Ch18N10   364
Ch18N10T   315–316, 338–339, 349, 352, 377–378, 380–382
Ch18N14   364
Ch18N40T   381

Index of materials

Ch18N9T 367
Ch20N20 364
Ch21N6M2 405
Ch22N5 345
Ch23N28M3D3T 402
Ch28N18 374
chemical porcelain 466
chemical-resistant enamel 189, 256
chemical service enamel 465
chemical service enamel 1 465–466
chemical service enamels 464–465
chlorobutyl rubber 237, 244, 257
chloroprene rubber 209, 238, 243–244
chlorosulfonylpolyethylene 209, 238, 263
chlorosulphonated polyethylene 243
chlorotrifluoroethylene 209, 243
chlorotrifluoroethylene copolymer 226
ChN28MDT 397
ChN40B 409
ChN40S 409
ChN40SB 409
ChN58W 424
ChN60 416
Chromel® 414
chromium 42, 198, 203, 288, 314
chromium alloys 42, 314
chromium carbide 198
chromium-free steel 68
chromium layer 43
chromium-manganese steel 341
chromium-manganese-nickel steels 350
chromium-manganese steels 338–340, 342–343
chromium-molybdenum steels 338, 340, 342
chromium-nickel-molybdenum-copper steels 386
chromium-nickel-molybdenum steels 289, 381, 384, 388, 393, 401
chromium-nickel steels 348, 358, 363, 365, 380, 468
chromium-nickel-titanium steels 348
chromium powder 43
chromium steels 83, 337, 340
chromized steels 330
ChS-13 336
CN-7M 409
Co3Si 473
Co-50Pt 314
coated aluminium 15
coating glass 187
coatings 239, 500
cobalt 42, 312
cobalt alloys 41, 312, 313

cobalt-chromium-molybdenum alloy 314
cobalt-platinum alloy 314
cobalt-zirconium-chromium alloys 313
CoCr20W15Ni 42
composite materials 504
concrete 181, 462
constructional steel 67, 69
copper 12, 46, 288, 292, 316, 316–319
copper alloys 56, 292
copper-aluminium alloys 48, 320
copper-cadmium alloys 328
copper-gold alloys 329
copper-nickel alloys 51, 320
copper-tin alloys 54, 321
copper-tin-zinc alloys 54
copper-zinc alloys 54, 321
corundum 198, 468
$Co_{21}Ge_2B_6$ 472
CR 209, 220, 238, 243–244, 247, 259
CrB alloy 174
CrMnNiN17-19-4 367
CrMn steels 344
CrMo26-1 339
CrMoTi steel 18 2 342
CrN 471
CrNi cast steel 366
CrNiMoCu28-28 367
CrNiMo steel 388
CrNiSi steel 17 14 4 353
CrNiSi steel 18 15 2 348
CrNiSi steel 20 15 359
CrNi steel 18 3 346
CrNi steel 18 8 343, 345
CrNi steel 18 9 348
CrNiTi18-10 368
CrNiTiAl steel 410
CrP-alloy 174
$Cr_{23}O_6$ 198
Cr_3O_2 198
Cr_3O_2 469
CrSi 473
$CrSi_2$ 473
Cr_3Si 473
Cr_5Si_3 473
Cr_7O_3 198
CrTi25 367
crystalline titanium 435
CSM 209, 220, 238, 243–244, 263
CTFE 209
Cu-15Zn 327
Cu-1Pd-1Y 329
Cu-2Pd-2Y 329
Cu-3Pd 329
Cu-3Y 329

Index of materials

Cu-6Cd 328
Cu-20Cd 328
Cu-20Zn-2Al 325
Cu-22Zn-2Al-0.03As 328
Cu-28Zn-1Sn 325
Cu-28Zn-1Sn-P 325
Cu-30Zn 321–325
Cu-36Zn 322, 324
Cu-37Zn 321–322, 324–325
Cu-40Zn 322–325
Cu-50Ti 328
Cu-50Zr 328
CuAl alloys 320
CuCd alloys 328
CuNi 70/30 51
CuNi alloy 51
CuSi3Mn 56, 252
CuZn37 55
CuZn alloy 54, 321–326
CW 116C 56, 253
CW 508L 55

d
Diabon® 179
dicobalt monophosphide 41
Diflon® 487
duplex steel 77, 79
Duranglas® 463
Duratherm® 313
Duratherm® 600 312
Durco® D-10 409
Durimet® 20 409
duroplast® 206–208, 213, 243
Dyflor® 493

e
Ebolat® 492, 507
E-Brite® 26-1 73, 337–338, 341, 385
ECFE 263
ECR glass 250, 256, 258
ECTFE 209, 226, 229, 240, 242, 243, 245, 258, 260–262, 494–495
E glass 250
EI-345 417
EI-702 410
EI-943 352, 384, 386, 397
EI-943 352
elastomer 206, 208, 213, 236, 243
Elgiloy® 312
ELI 18-2CrMo 340
enamel 189–190, 464, 466
enamel 10-1 466
enamel 143-V 466
enamel 7701 464

enamel coatings 289
EN AW-2017 23–24
EN AW-2024 21
EN AW-2090 21
EN AW-2091 21
EN AW-3003 23–24
EN AW-3005 305
EN AW-5005 305
EN AW-5049 305
EN AW-5052 305
EN AW-5454 305
EN AW-6060 305
EN AW-6082 305
EN AW-8090 21–23
EP 209, 219, 232, 240–241, 243, 264
EP-567 425
EP-795 424
EPDM 209, 220, 238, 247, 257, 259, 261, 263
EP-1155 503
EP-GF 209, 213
EP (Novolac) 219
epoxy 248
epoxy resin 207–209, 232, 240, 241, 243, 290, 489
epoxy-vinyl ester resin 258
EP-VE 219
ETFE 209, 219, 229, 243
E/TFE 494–495
ethylene/chlorotrifluoroethylene copolymer 209
ethylene/propylene copolymer 209
ethylene-propylene-diene rubber 209, 238
ethylene-propylene rubber 263
ethylene/tetrafluoroethylene copolymer 209

f
Fansteel® 61 434
Fe-13P-7C 177
Fe18Cr10NiTi 348
Fe-18Cr-13Ni-1Nb 351
Fe-20B 177
Fe20Cr26Ni4Mo1Cu 384–385
Fe-20P 177
Fe21Cr24Ni3Mo 384–385
Fe31Cr27Ni31Mo6.5 254
Fe-35Cr 339
Fe-36Ni-12Cr-14P-6B 177
Fe37Cr27Ni25Mo4.5 254
Fe-40Ni-14P-6B 170–171, 177
Fe42Cr24Ni22Mo7.3 254
Fe46Cr22Ni25Mo6.7 254
Fe-47Cr 341–342

Index of materials

Fe47Cr20Ni25Mo6.2   254
Fe47Cr21Ni25Mo4.5   254
Fe49Cr20Ni24Mo6.2   254
Fe55Cr20Ni18Mo6.1   254
FeCr21Ni32TiAl   348
FeCr25Ni7Mo3.5   254
feldspar   233
FEP   209, 213, 219, 226, 229, 243, 245, 247, 258, 259–260, 263, 493–494
Ferralium®   177
Ferralium® Alloy 255   106, 108, 110–111, 113, 407
ferritic-austenitic chromium steels   345
ferritic-austenitic steels   77, 87, 345–346
ferritic chromium steels   67, 72, 74, 336–337
ferritic/pearlitic-martensitic steels   74
ferritic steel   330
FeSi   473
Fe$_2$C   199
Fe$_3$Si   473
fiber-reinforced duroplast   199, 240
films   239
FKM   247
Flamulit® WS PA-12   483
Flamulit® WS PE-25   478, 502
flexible graphite   460
Fluorel®   243
fluorinated cyclic ether copolymer   229
fluorinated rubber   209, 243
fluorocarbon resins   290, 493
fluoroelastomers   238
fluoroplastics   226
fluoropolymers   11, 207, 216, 225, 240, 243, 246, 258, 259–261, 265
FMN   384
FN   243
foamed borosilicate glass   463
Fomblin® Y-N VAC 140/13   496
Foraflon®   243, 493
FPM   209, 220, 238, 243, 247
FRP   262
Fu   209, 233, 243
furan resins   178, 180, 207, 209, 223, 233, 241, 243, 258, 475

g

G-3   99–100, 106
G-30   100–101, 117
G34CrMo4   70
G10200   59
gallium   459
G-AlMg2Mn0.8   303
G-AlMg2.7Mn   303
G-AlMn   303
Genakor®   243
germanides   473
germanoborides   472
glass   184, 188, 223, 226, 234, 256, 289, 463
glass-ceramic   463
glass fiber   245, 248–249
glass fiber-reinforced plastic   209
Glassodur®   483
Glassodur® Acryl-Autolack 21   483
gold   12, 40, 148, 288, 311
gold alloys   40, 311
gold-copper alloys   311
gold-silver alloys   311
Grafilex® 6501   461
graphite   7, 80, 178, 180–181, 223, 226, 233, 246, 255, 460–462
graphite carbide   462
graphite materials   255
graphite silicon carbide   462
gravel   181
gray cast iron   336
Grilonit® G 16.05   489
GRP   207, 209, 257
GRP laminate   220, 258
GS-42A   303
GX2CrNiN18-9   96
GX3CrNiMoCuN24-6-2-3   396
GX5CrNi19-10   384
GX5CrNiMo19-11-2   383
gypsum   181

h

H 10 T   504
H 20 T   504
H 40 T   504
Hafnium   456
Hafnium-niobium alloys   456
Halar®   243
Halon®   243
hand-laid laminate   242
hardened chromium steels   337
hard gold   431
hard rubber   236, 242, 244, 257, 261
hard rubber NR   213
Hastelloy® alloy B   110
Hastelloy® alloy B-2   110–112
Hastelloy® alloy C-22   115
Hastelloy® alloy C-276   88, 110–111, 113, 114
Hastelloy® alloy G   111, 113, 114
Hastelloy® alloy G-30   106
Hastelloy® alloys   166

Hastelloy® B   120, 123, 425–426
Hastelloy® B-2   119–124
Hastelloy® B 2   425–426
Hastelloy® B-4   122
Hastelloy® C   263, 399
Hastelloy® C-4   99, 423–424
Hastelloy® C-22   103, 424
Hastelloy® C-276   99, 106–107, 423–424, 464
Hastelloy® F   426
Hastelloy® G   113, 421–422
Hastelloy® G-3   423, 507
Hastelloy® G 30 alloy   177
Havar®   312–313
Haynes® 20 Mod. alloy   88
Haynes® alloy 188   312
Haynes® alloy 25   42, 312
Haynes® alloy 625   114
Haynes® alloy 6B   312
Haynes® alloy No. 20-Mod.   110–111, 113
Haynes® alloy No. 625   110–111, 113
Haynes® No. 20-Mod.   110
HC-new   103
HDPE   475–478, 502
hexafluoropropylene   238, 243
hexafluoropropylene copolymer   226, 229
HfB_2 472
HfN   471
high-alloy cast iron   67, 335
high-chromium ferritic steel   339
high-performance alloys   110
high-performance plastics   230
high-silicon cast iron   335
high-silicon stone   460
high-strength tempering steel   69
hornblende   467
Hostaflex®   482
Hostaflon®   243, 289, 493, 496
Hostaflon ET®   243, 493, 495
Hostaflon FEP®   243
Hostaflon PFA®   243
Hostaflon TF®   493
Hostalen®   243
Hostalen PP®   243
Hostalit®   243
HTi-10   469
Hypalon®   243

i

I-82   417
Igetalloy®   469
Igetalloy® G-1   469
IIR   209, 243–244

IIR-PVC   220
IMI 25   146
Imilon® S   484
Incoloy® 825   88, 99, 112, 114, 384, 387
Incoloy® 901   387
Inconel® 100   428
Inconel® 600   99, 177, 418–422
Inconel® 625   73, 99, 421–422, 424
Inconel® 671   422
Inconel® 690   372, 418, 421–422
Inconel® 800   99
Inconel® 939   428
Inconel® G-3   99
Inconel® X 750   177
Indium nitride   472
industrial-grade aluminium   294
Industrial molybdenum   412
InN   472
IR   209, 243
iridium   129, 169, 431
iron   56, 288, 330–331, 334
iron-nickel-cobalt alloy   410
iron-silicon alloys   336
isobutylene-isoprene rubber   209, 243
isophthalate resin   234
isoprene rubber   236
ITM-43   390–391

j

J-55   60
J92600   384
J92900   383
JS-625   425

k

K-299   359
K94610   410
Kalrez®   238, 243, 263
KBI-40® Ta-40Nb   433
Kel-F®   243, 493
Kevlar®   180
killed steel   330
Kovar®   410–411
KV-80   408
KV-81   408
KV-82   409
Kynar®   243, 493, 495–496, 503

l

LaC_2 469
LaGe   473
$LaGe_2$ 473
La_5Ge_3 473

Index of materials | 605

Lampart Univer-80® 466
LaN 472
Lanthanum carbide 469
LC-20 356
LDPE 475, 477–479
lead 129, 428, 428–429, 435
lead alloys 129, 428
lead-indium alloy 429
lead silicate g 464
lead-tin alloys 429
Levasint® 482, 502
lime mortar 181
linings 239, 500
LiPdH 431
low-alloy aluminium 298, 308
low-alloy cast iron 66
low-alloyed steels 56
low-alloy steels 329–332
low-carbon steel 346
Lupolen® 243

m

magnesium 98, 411
magnesium alloys 98, 411
manganese nitride 472
marble 181
MBS 209
MC-20 356
melamine-formaldehyde plastics 209, 235
mercury 288
metal-ceramic material 204
Metglas 2826 170
methylmethacrylate/butadiene/styrene copolymer 209
MF 209, 235
Mg_2Si 473
$MnSi_2$ 473
Mn_3N_2 472
Mn_3Si 473
Mn_5Si_3 473
Mo-0.5Ti-0.08Zr 412
molybdenum 98, 130, 412
molybdenum alloys 98, 412
molybdenum boride 472
Monel® 400 425
Monel® alloy 400 117
Monel® K 500 425
mortar 181, 462
$MoSi_2$ 473
$Mo_{1.7}Ge_{0.3}B$ 472
Mo_2B_5 472
MP 35N 42, 312–313, 426
Ms-72 327

Ms-80 327
mullite 369

n

N02200 99
N04400 117
N06010 414
N06022 103, 106, 115, 129, 254
N06030 100, 106, 117
N06059 99, 129, 254
N06110 115
N06200 125, 129, 254
N06455 99, 254
N06600 99
N06617 426
N06625 99, 115, 117, 254
N06686 117, 129
N06985 99, 106
N07214 117
N08028 254, 396, 406–407
N08031 254
N08367 97, 254
N08825 99, 407
N08904 254
N08926 87, 97, 254
N10001 117, 119–120
N10242 124–125
N10276 99, 106, 115, 129, 254
N10629 117, 120, 122, 253
N10665 106, 117, 119–120, 253
N10675 117, 120, 125, 253
Natsyn® 243
N-80 60
natural fiber 248
natural rubber 208–209, 236–237, 243–244, 246, 263
natural stones 460
Nb-0.24Zn-0.34N 434
Nb-1Zr-0.1C 434
Nb-3.06Zr-0.45N 434
Nb-20Ta 138–139, 141
Nb-30Ti-20W 434
Nb-40Ta 138, 141–142
Nb-60Ta 138, 141–142
Nb-80Ta 138, 141
NbB_2 472
NbC 468
NBR 209, 237, 243–244, 247, 259, 261
Nb_5Ge_2B 472
Nb_5Ge_3B 472
NdGe 473
$NdGe_2$ 473
NdN 472
Nd_5Ge_3 473

606 | Index of materials

Neoflon® 243
Neopren® 243
Ni-10Cr-20P 177
Ni-20P 177
Ni-30Fe-14P-6B 177
Ni-38.4Fe 428
Ni-40Cr-20P 177
Ni-40Fe-20B 177
Ni-49.5Fe 428
Ni57Cr21Mo13Fe4W3 254
Ni59Cr16Mo16Fe6W4 254
Ni59Cr23Mo16Fe1 254
Ni59Cr23Mo16Fe1Cu1.6 254
Ni61Cr21Mo9Fe4 254
Ni67Cr16Mo16Fe1 254
Ni99.2 99
Ni99.6 99
NiBe2 427
Nichrome® 430
nickel 98, 203, 412
nickel 200 99
nickel alloys 412, 419
nickel-based alloys 87
nickel-beryllium alloys 427
nickel-chromium alloys 99, 414
nickel-chromium-aluminium alloy 428
nickel-chromium-iron alloys 99, 289, 418
nickel-chromium-iron-molybdenum-cobalt alloys 426
nickel-chromium-molybdenum alloys 99, 423
nickel-chromium-tungsten-molybdenum alloys 425
nickel-cobalt-chromium-molybdenum alloys 426
nickel-copper alloys 117, 425
nickel-molybdenum alloys 117, 425
nickel phosphide 473
NiCr16Fe 99
NiCr21Fe 99
NiCr21Mo 99
NiCr22Mo7Cu 99
NiCr22Mo9Nb 99
NiCr23Mo16Al 99
NiCrMo alloys 100, 254
Nicrofer® 4221 99
Nicrofer® 4823 hMo 99
Nicrofer® 5716 hMoW 99
Nicrofer® 5923 hMo 99
Nicrofer® 6020 hMo 99
Nicrofer® 6616 hMo 99
NiMo16Cr15W 99
NiMo16Cr16Ti 99
NiMo28 252

NiMo alloys 121, 124
Nimonic® alloy 80 A 418
niobium 130, 135, 138, 141, 239, 289, 432–434, 450
niobium alloys 434
niobium carbide 468
Ni-Resist® 67
$Ni_{12}Ge_3B_5$ 472
nitrides 198–199, 468
nitrile rubber 244
nitrogen-chromium-manganese steels 340
Nitronic® 50 438
Noricid® 9.4306 349
Noridur® 1.4593 396
Novolak® 251
NR 209, 220, 243–244, 261
NR/SBR 245
Nylon® 231

o
Oilon® PV-80 484
osmium 129, 169, 431
OT 4-0 449
OT 4-1 154–155, 440–441
oxide ceramic 197
oxide ceramic materials 467

p
P-105 60
PA 231
PA 6 209, 231
PA 66 209, 231–232
packings 503
Palatal® 243
Pallacid® 12
palladium 129, 144, 431
PAN 209
PAS 209, 231
passivated iron 335
PB 209, 259
PbSn alloys 430
PBT 209
PC 209, 232, 234
PCTFE 243, 494
$Pd_{50}Sn_{50}$ 431
$Pd_{66.7}Sn_{33.3}$ 431
$Pd_{75}Sn_{25}$ 431
PE 209, 221–222, 240, 243, 245, 261–262, 479
pearlitic gray cast iron 335
PE-C 209, 241
PEEK 231
PE-GF 213
PE-HD 207, 209, 213, 219–222, 259, 265

Index of materials

PEI 209, 231
PEK 209, 231
PE-LD 209, 222, 259
PE-LLD 209
Pentaplast® 497–498
Pentaplast® PL-2 498
Perbunan N® 243
perfluoroalkoxy copolymer 226, 229
perfluoroalkoxy polymers 243
perfluoroelastomers 263
perfluoromethyl vinyl ether 238
Perlon® 231
Permalloy® 52N 428
PES 209, 231
PET 210, 232, 234
PE-VLD 209
PF 210, 233, 243
PFA 210, 219, 226, 229, 240, 243, 258, 260–263, 265
PFA/TFA 494
PF-GF 210, 219, 264
phenol 248
phenol-formaldehyde 210
phenol-formaldehyde resins 7, 178, 208, 210, 216, 223, 233, 241, 243, 290
phenolic resins 482
phenol resins 207, 223, 258
phenoplasts 233
phthalate resin 234
PI 210
PIB 210, 479
plastics 80, 199
platinum 129, 148, 288, 430
platinum alloys 129, 430
platinum metals 431
plutonium 458
PMMA 210, 230
polyacetal 210, 484
polyacrylonitrile 209, 216
polyamides 207, 209, 214–216, 231, 290, 483
polyarylethersulfone 209, 231
polyarylsulfone 209, 231
polybutene 209
polybutylenterephthalat 209
polycarbonate 209, 215, 216, 232, 486
polychlorotrifluoroethylene 243
polyester 214–215, 232, 248
polyester resins 207, 216, 232, 240–241
polyester resins (glass fiber-reinforced) 258
polyesters 214, 290, 484
polyetheramide 210
polyetherimide 209, 231

polyetherketone 209, 231
polyethylene 209, 213, 221, 243, 257
polyethylene terephthalate 210, 232
polyimide 210, 248
polyisobutylene 210
polyisoprene 209, 243
polymer-cement mixture 183
polymer concrete 181, 183, 235
polymethylmethacrylate 210, 230
polyolefins 207, 210, 216, 221, 228, 243, 258, 261, 475
polyoxymethylene 210
polyphenylene sulphide 210, 231
polyphenylene sulphone 210
polyphenylene terephthalamide 180
polypropylene 208, 210, 213, 222–223, 242–243, 257
polystyrene 207, 210, 230
polysulphide rubber 238
polysulphone 210, 216
polytetrafluoroethylene 180, 210, 226–227, 243, 289
polyurethane resins 240
polyurethanes 210, 214–216, 235, 462, 487
polyvinyl acetate 210
polyvinyl alcohol 210, 214
polyvinyl chloride 14, 207, 210, 224, 243, 258, 480
polyvinyl esters 482
polyvinylidene fluoride 210, 226–227, 243
POM 210
porcelain 289, 466–467
Portland cement 181–182, 462
powder metallurgical materials 508
powder metallurgy zirconium 455
PP 207, 210, 213, 219–223, 243–245, 251, 259, 261–262, 264–265, 479
PPS 210, 231
PPSU 210
PRD®-49 484
PrGe 473
PrGe$_2$ 473
Pr$_5$Ge$_3$ 473
PrN 472
propeller bronze 50
PS 210, 230
PSU 210, 231
PT-3V 435–436, 440, 445
PTE 265
PTFE 7, 178, 180–181, 188, 210, 219, 223–224, 226–227, 227, 238, 243, 245, 247, 256, 258, 260, 262–265, 348, 409, 463, 493–494
PTFE compound 247

PTFE-perfluoroalkoxy copolymer   210
PTFE yarn   247
PUR   210, 235
pure aluminium   19–20, 22, 40, 294, 300–302
pure chromium   43–44, 173
pure cobalt   41
pure nickel   99
pure niobium   138
pure tantalum   137, 139, 435, 457
pure titanium   144–146, 149, 156–158, 166, 254, 435, 438, 446–447, 449
pure titanium sponge   150
pure zirconium   167
putties   463
PVAC   210
PVAL   210, 214
PVC   14, 40, 210, 224, 243, 245, 251, 258, 261–262, 480
PVC-C   207, 210, 219, 221, 225, 259, 262, 265
PVC hard   216
PVC-P   210, 224
PVC soft   216, 242
PVC-U   207, 210, 219–221, 224–225, 257–258, 264
PVDF   207, 210, 213, 219–221, 223, 238, 240, 242–243, 245–246, 251, 258–259, 262–263, 264–265, 494–495

q
QuaCorr®   243
Quacorr® 1200 FR   475
Quacorr®-RP 100A/RP 104B   475
quartz   464
quartz glass   185, 464
quartz sand   184

r
R30605   42
R50400   149, 156–157, 437, 440, 443–445, 448
R52400   149, 437, 448
R53400   157
R54520   438, 440
R60702   167, 451
R60704   451
R60705   451
R60802   454
R60804   452–454
Rapal®   467
reaction-sintered SiC   201
red bronze   54
refined aluminium   294

refractory materials   467
ReGe$_2$ 473
Remanit® 4306   355, 365
Remanit® 4335-So   355
Remanit® 4362   394
Remanit® 4429   396
Remanit® 4438   396
Remanit® 4462   393–394, 396
Remanit® 4465   397
Remanit® 4565   396
Remanit® 4575   397
ReSi   473
Re$_3$Si   473
ReSi$_2$ 473
resin   248
R glass   250, 258
rhenium   457
rhodium   129, 291
Rigidon®-4889   475
RSSiC   201
rubber   80, 236, 257
rubber coating   244
ruthenium   129, 431
ruthenium alloys   431
Ru$_{33.3}$Ta$_{66.7}$ 431
Ru$_{36}$Cr$_{64}$ 431
Ru$_{50}$Al$_{50}$ 431
Ru$_{50}$Ti$_{50}$ 431
Ru$_{50}$Zr$_{50}$ 431

s
S-5A   303
S 45 C   197–198
S30600   407–408
S31002   386, 407
S31254   87, 253, 407
S31500   406
S31603   106, 407
S31803   345
S32304   394
S32550   106
S32654   97, 253
S32750   253
S32760   253
S32803   339
S34565   396
S39209   384, 390–391, 393–394
S41500   346
S43000   73
S43025   73
S44400   73
S44626   73
S44627   73
S44635   397

Index of materials

S44700   73
S44800   73
S-AlMg5   307
SAN   210
sand   181–183
Sandvik® 2R12   359, 438
Sandvik® 2RE10   374, 386
Sandvik® 2RE-69   347
Sandvik® 2RK65   400
Sandvik® 3RE60   406
Sanicro® 28   385, 396
S/B   210
SBR   210, 237, 243–244
SBS   210
seals   503
selenides   474
semi-austentic steel   83
SG-70A   303
SI   210, 235
SiC   198–199, 201
α-SiC   201–202
Sigraflex®   460, 503
silicate glass   184
siliceous stones   467
silicon   473
silicon bronzes   56, 253
silicon cast iron   67
silicon dioxide   189
silicone rubber   208, 238
silicones   210, 235, 496
silicon monoxide   181
silicon nitride   470
Silumin®   309
silver   12, 288, 291
silver alloys   12, 291
silver-gold alloy   291
Si_3N_4 470
SiO_2 185
SiSiC   204–205
smelting slags   369
Sn-28Ni   432
sodium silicate glass   184
soft iron   330
soft rubber   236, 244
soft rubber CR   213
Solef®   243, 493, 495
SP-FE-8   496
SP-FP-7   496
ST   182
St 3   331, 331–332
St 35   92–93
St 50   501
stainless austenitic steels   338

stainless steel   51, 58, 223, 350, 372, 385, 390, 430
steel   51, 64, 189
steel 3   96
Stellite® alloy 6 B   312
Stellite® alloy 25   312
stone materials   181
stones   463
stoneware   467
stove enamel   242
structural steels   336
styrene   182
styrene/acryl copolymer   210
styrene/butadiene copolymer   210
styrene/butadiene rubber   210, 236–237, 243
styrene/butadiene/styrene block copolymer   210
styrene polymers   230
superaustenite   96
superferrite   72, 74
SUS 304   197–198, 350, 358
SUS 304 L   364
SUS 310 Nb   382
SUS 316   197–198
SUS 321   352
SUS 430   197–198
synthetic diamond powder   462
synthetic elastomers   507
synthetic fiber   245
synthetic rubber   208, 236, 246, 257

t

T   405
T20811   345
Ta-2.5W-0.15Nb   434
Ta-7.5W   434
Ta-40Nb   432–433
TaB_2 472
TaC   469
TaHf   143
TaMo   143
TaNb   143
TaNb alloys   136
TaNi   143
Tantalloy® 63   434
tantalum   7, 130–131, 134, 138, 166, 177, 190, 227, 239, 255, 289, 432, 433–435, 450, 463–464
tantalum carbide   134, 468–469
tantalum-tungsten alloy   433
Tantiron® E   336, 410
Tantiron® N   336, 410
TaRe   143

Index of materials

Ta$_5$Ge$_2$B 472
Ta$_5$Ge$_3$B 472
TaV 143
TaW 143
TaWHf 143
TaWMo 143
TaWNb 143
TaWRe 143
TaZr 143
TBA-26 461
TBA-700 461
TCR 10 504
TCR 30 504
technical-grade aluminium 301
Tedlar® 493
Teflon® 156, 188, 243, 289, 409, 474, 493, 496
Teflon AF® 229
Teflon FEP® 243, 493
Teflon PFA® 243, 263, 493
Tefzel® 243, 493, 495
tempering steel 74
terracotta 467
tetrafluoroethylene 229, 243
tetrafluoroethylene/hexafluoropropylene copolymer 209
TFE 134
thermoelastomers 206, 208
thermoplastic elastomers 208, 210
thermoplasts 206–207, 207, 213, 257
Ti-0.2Pd-15Mo 440
Ti-1Ni 438
Ti-1.5Ni 438
Ti-2Cu 444
Ti2Pd 149, 437–438, 440, 443
Ti-5Ta 440, 443
Ti-6Si 443
Ti-8Ta 437
Ti-10Mo 439
Ti-10Ta 437
Ti-15Mo 150–151, 440
Ti-15Ta 437
Ti-20Mo 439
Ti-27Ni-10P 435
Ti-27Ni-13P 435
Ti-30Mo 438–440
Ti-33Mo 440
Ti-35Nb 443
Ti-50Ta 440
TiAl2Mn1.5 440–441
TiAl4V2 435, 440, 445
TiAl6V4 239, 441, 443–444
Ti-Al-Zr-Mo cast alloy 152
Ti-Al-Zr-W cast alloy 152

TiB$_2$ 472
TiC 468–469
Ti-Code 12 149
TiCr alloy 174
Ti Grade 2 149, 156–157, 435, 437, 440, 443–445, 448
Ti Grade 5 441, 443
Ti Grade 6 438, 440
Ti Grade 7 149, 437, 448
Ti Grade 12 157
tin 129, 431
TiN 201, 203, 470–471
TiN-15Ni 203
TiN-30Cr 203
tin alloys 129, 431
tin-lead alloys 431
tin-nickel alloy 431–432
titanium 51, 73, 143, 147–148, 153, 156–157, 159, 166, 199, 227, 239, 254, 288, 435, 438, 440–441, 443–444, 446–448, 450
titanium alloys 143, 156, 254, 435, 438, 440–441, 444
titanium-aluminium alloy 153
titanium carbide 468
titanium material 159
titanium-molybdenum alloys 439
titanium-molybdenum-zirconium-niobium alloys 439
titanium-nickel-phosphorus alloys 435
titanium nitride 201, 470
titanium-palladium alloy 437
titanium-tantalum alloys 435, 437
titanium-vanadium alloy 445
titanium VT1-0 435, 437
TMPTMA 182
TPE-A 210
TPE-O/V 210
TPE-U 210
TPU 210
Tribocor®-532 N 434
trimethylolpropane trimethacrylate 182
Trovidur® 243
Trovidur PP® 243
tungsten 130, 456
tungsten carbide 468–469
tungsten carbide-cobalt alloys 469
TWF 18 504
TZM 412

u

U-0.75Ti 457–458
U-2.3Nb 457–458
U-3Fs 457–458

Index of materials

U7A 332–333
U10A 332–333
U-10Fs 457–458
Ucrete® 462
UF 235
UHB 25 L 365–366
UHB-724L 403
UHB-725LN 403
ultra high-strength steel 345
unalloyed cast iron 66, 335
unalloyed steels 56, 329, 331
unalloyed titanium 149, 441, 443
unsaturated polyester resins 210, 234, 240
UP 210, 219, 232, 234–235, 240, 243, 247, 259, 264
UP-GF 210, 219, 234–235, 251
uranium 457–458
uranium alloys 457
uranium-niobium alloys 458
Uranus® 625 99
Uranus® 825 99
urea-formaldehyde resins 235
urethane concrete 462

V

V-1Zr-20Cr 456
V2A 382
V-9Ta 457
V-10Cr-10Al 456
V-12Nb 457
V-13.3Cr-6.7Ti 456
V-13Ta 457
V-16.6Cr-33.4Ti 456
V-20Cr 456
V-21Nb 457
V-26Nb 457
V-30Cr 456
V-30Ta 457
V-31Nb 457
V-40Nb-10Ti 457
V-40Ta 457
V-50Ti 456
vanadium 456
vanadium alloys 456
vanadium-niobium alloys 457
vanadium-tantalum alloys 457
VB_2 472
VE 210, 240, 243
VE-GF 213, 265
Vestolen® 243
Vestolit® 243
Vetrodur-N® 243
VEW A-963 388
vinyl acetate 222

vinyl ester resin 207, 234, 241, 249
vinyl esters 210, 235, 240, 248, 251
vinyl polymers 261
Vitallium® 313–314
Viton® 156, 226, 238, 243, 263
vitreous carbon 369, 462
VLX 562® 345
VN 471
Voltalef® 243, 493
VT 1-0 154, 158, 435–436, 440, 443–446, 471
VT 1-1 154–155, 438, 446–447, 449
VT-14 438
VT-15 438, 444
VT-20 442
VT 5-1 154–155, 438, 440
VT 6-S 441

W

W1 332
W 4027 341
W 4059 341
Waterborne paints 242
WC 199, 469
WC6Ni 504
WC9Ni 504
wood 475
WSi_2 473

X

X1CrNi25-21 359
X1CrNiMoCuN20-18-7 87
X1CrNiMoN25-22-2 368
X1CrNiMoN25-25-2 349, 383, 396–397
X1CrNiMoNb28-4-2 339
X1CrNiSi18-15-4 348–350, 359
X1CrNiSi18-15-4 see X2CrNiSi18-15 353
X1NiCrMoCu25-20-5 84, 87
X1NiCrMoCu31-27-4 406
X1NiCrMoCuN25-20-7 87
X2CrMnNiMoN26-5-4 346
X2CrMoNb18-2 84
X2CrNi19-11 88, 96, 349, 355, 365
X 2 CrNi 25 20 359
X2CrNiMnMoNbN23-17-6-3 84
X2CrNiMnMoNbN25-28-5-4 87
X2CrNiMo17-12-2 110
X 2 CrNiMo 18 12 386
X2CrNiMo18-14-3 84
X2CrNiMoN17-11-2 88
X2CrNiMoN17-13-3 395
X2CrNiMoN17-13-5 84, 87, 90, 393–394, 397

Index of materials

X2CrNiMoN22-5-3   77, 84, 87, 390–391, 393
X2CrNiN23-4   394
X2CrNiSi18-15   353
X 2 CrNiSi 18 15   359–360, 362
X2NiCrAlTi32-20   90
X3 CrNi 18 10   88
X 3 CrNi 18 10   359
X3 CrNiMoN 17 12 2   88
X41CrMoV5-1   345
X4CrNi18-12   84
X4CrNiMo17-12-2   88
X5CrNi18-10   72, 80, 83–84, 87–88, 90, 110, 177, 352–353, 368
X5CrNiMo17-12-2   72, 80, 134, 177, 397
X5CrNiN19-9   352
X5NiCrMoCuTi20-18   400–401
X6Cr17   92
X6 CrNi 18 10   88
X6 CrNiMo 17 12 2   88
X6CrNiMoTi17-12-2   84, 90, 396
X6CrNiTi18-10   92, 95, 367–368, 409
X 8 CrNiMoTi 18 11   400–401
X8CrNiTi18-10   352–353, 383
X 12 CrNi18 8   87
X 12 CrNi 18 8   382
X17CrNi16-2   74, 76
XM-8   73
XM-27   73–74
XM-33   73
Xydar®   485

z

Zeron 100®   254
Zl-52   363
zinc   89, 159, 162–163, 450
zinc alloy   450
zinc brass   450
zinc layer   160
Zircaloy® 2   168–169, 450, 454
Zircaloy® 4   452–454

zirconium   147, 166, 177, 227, 239, 450, 451, 452, 454–456
zirconium alloys   166, 450, 454–455
zirconium carbide   469
zirconium grade 702   167
zirconium grade 705   166
zirconium (highly pure)   169
zirconium-titanium alloys   451
Zn-Al alloys   160
Zn/Al-eutectic   160
Zr-1.5Sn   451
Zr-2.5Nb   451, 455
Zr-5Ti-0.1Fe-0.3Cr-0.1Ni   452
Zr-5Ti-0.5Cr   452
Zr-5Ti-10Mo   452
Zr5Ti10Nb   451
Zr-5Ti-10Nb   452
Zr-7.5Ti-15Nb   452
Zr10MoTi   451
Zr-10Nb   452
Zr10NbTi   451
Zr-10Ti   452
Zr10Ti10Ta   451
Zr-10Ti-10Ta   452
Zr-10Ti-15Nb   452
Zr-10Ti-5Nb   452
Zr-10Ti-5Ta-5Nb   452
Zr-10Ti-7.5Nb   452
Zr-10Ti-7.5Ta   452
Zr-15Ti-15Nb   452
Zr-15Ti-5Nb   452
Zr-15Ti-7.5Nb   452
Zr-702   451, 453
Zr-704   451, 453
Zr-705   451, 453
ZrB_2 472
Zr-Ir alloys   170
ZrN   471
Zr-Os alloys   170
ZrTi   451
ZrTi alloys   451–452

Subject index

a

abrasion 197, 261, 264
abrasion corrosion 11
abrasion resistance 41, 262
absorbers 178
absorption 63, 226
absorption column 348
absorption columns 7, 260
absorption test 229
absorption towers 179
accessories 179, 189
accumulators 247
acetic acid (as inhibitor) 30, 33
acid corrosion 74
acid mixture 46, 81, 105, 106
acid pickling 69
acid pump 396
acid technology 214
acid wash solution 51
acid washing 51–52, 55, 67
acid-resistant coatings 233
acid-resistant filler 233
acoustic emission 307, 327
acridine (as inhibitor) 324
acridine orange (as inhibitor) 24
action of chemicals 211
activated carbon 226
activated diffusion 213
active corrosion 9
active stress corrosion cracking 83, 85
additives **10**
adhesion behavior 241
adhesive bond 245
adhesive bonding 225, 258
adhesive joint 225
adhesively bonded linings 245
adsorption process 249
adsorption vessel 223
aerospace applications 21

aftercondenser 368
after-etching 382
ageing **207**, 354
ageing resistance 249
ageing temperature 354
air contamination 181
air pollutants 14, 181
air temperature 14
air valves 226
air-conditioning unit 15
aircraft 14
aircraft industry 231
alizarin red S (as inhibitor) 24
alkaline pickling solutions 299
alkylamines (as inhibitor) 431
allyl thiourea (as inhibitor) 322
allylamine (as inhibitor) 26–27
allylpyridine thiocyanate (as inhibitor) 331
Alniflex process 454
aluchromizing 321
aluminium ions in nitric acid 383
aluminium oxide layer 18
2-amino-5-chloropyridine (as inhibitor) 34
3-amino-1,2,4-triazole (as inhibitor) 63, 65
p-aminobenzoic acid (as inhibitor) 299, 324
m-aminobenzoic acid (as inhibitor) 323
p-aminobenzoic acid (as inhibitor) 323, 325
2-aminopyridine (as inhibitor) 34
2-aminothiazole (as inhibitor) 321
2-amino-thiazole (as inhibitor) 63, 65
ammonia 63
ammonium nitrate production 439
amphoteric metal 18
anatase layer 445
aniline (as inhibitor) 78, 317
m-anisidine (as inhibitor) 57
p-anisidine (as inhibitor) 57, 299, 324–325

Subject index

o-anisidine (as inhibitor)   57, 325
anisidine isomers (as inhibitor)   57
annealing colors   382
annealing scale   56
anode material   156
anodic crack initiation   169
anodic dissolution   61
anodic polarisation   131, 144, 159, 170
anodic protection   **159**
anodised layer   21
anthranilic acid (as inhibitor)   30
anti-ageing agents   236
antipromotor effect   11
APF   103
apparatus engineering   227
apparatus material   235
approval procedures   252
aqua regia   201, 291, 311, 329, 429–430, 435, 462, 470, 473–474
aqua regia test   429
arc melting   133
arc welding   296
articulated lever specimens   250
1-aryl-3-formamidine-thiocarbamides (as inhibitor)   299
ASTM A 262 – Practice C   72, 392
ASTM A 262-C   341
ASTM A 262-D   399
ASTM G 3   126
ASTM G 30 test   121
ASTM 262 – Practice D   342
atmospheric corrosion   164
atomic percent factor   103
attack of the weld seam   116
autocatalytic effect   214
autoclave   191, 466
azeotropic concentration   6
azeotropic nitric acid   336, 406, 433, 454
azeotropic state   4
azo dye manufacturing   166

b

ball valve   43, 134
barrier   237
barrier layer   251
bellows   247, 259
bending strength   180, 204, 215–216, 233–234
benzaldehyde (as inhibitor)   78
benzidine (as inhibitor)   317
benzoic acid (as inhibitor)   30, 298, 299, 308
benzotriazole (as inhibitor)   317, 322, 332

benzylamine (as inhibitor)   26–27, 78
benzyl-2-methyl pyridine rhodanide (as inhibitor)   331
benzylpyridine thiocyanate (as inhibitor)   331
Berlin process   8
binder   178, 181, 201, 230, 233, 240
binder metal   469
biological attack   214
blister   237–238
blistering   234, 240, 250–251
block heat exchangers   180
Boehmite layer   300
boiler scale   57
boiler tubes   93
boiler water   199
boiling curve   5
boiling nitric acid   396, 423, 437, 455
bonding agent   236, 250
Box-Wilson   310
brass component   55
brass sheets   327
brightening baths   299
brittle fracture   168
brittleness   208, 215
bromoacetic acid (as inhibitor)   322–324
n-butylamine (as inhibitor)   24
butylamine (as inhibitor)   26–27

c

calomel electrode   113
carbide precipitation   123
carbon activity   124
carbon fiber   233, 248
carbon steel   61
carbon yarns   181
carboxylic acids   30
3-carboxylpyridine (as inhibitor)   34
4-carboxylpyridine (as inhibitor)   34
cases of damage   156
cast arc-melted tantalum   133
cast iron tanks   335
casting sand   201
catalytic activity   41
catechin violet (as inhibitor)   24
cathode   55
cathodic loading   132
cathodic polarisation   23, 131, 170, 175, 330
cation exchangers   288
CCT test   90
centrifugal pump   134, 181, 349
centrifugal separator   158

Subject index | 615

CERT test   90, 157
chain cleavage   214–215
chain degradation   214
characteristic mechanical values   216
chemical apparatus engineering   214
chemical cleaning   57, 66, 93
chemical degradation   211
chemical degradation processes   213
chemical descaling   444
chemical hoses   247
chemical industry   228
chemical loading   207, 211
chemical milling   444
chemical plant   252
chemical polishing   299, 302, 319, 382, 434
chemical process engineering   99
chemical process industry   149
chemical pumps   229, 348
chemical resistance   211, 215
chemical wastewaters   181
chemical-resistant coating   234, 249–250, 258
chemisorption   461
chimneys   235, 463
chip industry   262
chlor-alkali electrolysis plants   249
chlor-alkali industry   11, 227, 234, 251
3-chloro-2-hydroxypropanoic acid   447
3-chloro-2-hydroxypropanoic acid production   447
chloride corrosion   74, 77
chlorinating plant   227
chlorination unit   260
chlorine gas production plants   153
chloroacetic acid (as inhibitor)   322–324
p-chloroaniline (as inhibitor)   299
o-chloroaniline (as inhibitor)   317
m-chloroaniline (as inhibitor)   321
o-chloroaniline (as inhibitor)   323
p-chloroaniline (as inhibitor)   323–325
m-chlorobenzoic acid (as inhibitor)   30
p-chlorophenylthiourea (as inhibitor)   298
chromation   302
chromium carbide   198
chromium carbide coating   198
chromium carbide precipitation   75, 123
chromium coating   45
chromium depletion   198
chromization   450
circulation apparatus   463
city atmosphere   161–163
cladding devices   414
clean   58

clean filters   96
clean gas pipeline   223
clean iron surface   56
cleaning copper   318
coal ash particles   163
coating   69, 161, 212, 229, 236, 256
coating concrete floors   462
coating materials   240, 462
coating system   16
coefficient of friction   454
cohesion   211
cold deformation   112
cold vulcanisation   236
cold working   350
cold hardening   201
columns   62, 132, 178, 189, 233, 256, 258, 448
combination meshes   181
combustion gas   8, 80, 197
combustion of coal   163
combustion of polyvinylchloride   64
combustion plant   80
combustion products   14
compact seals   247
compacts   43
component   225
composites   248, 257
compressive strength   180, 182, 258
compressor   7
concentrated nitric acid   303
concentrating plants for nitric acid   463
concrete foundation   230
condensate   249
condensate region   194
condensate valves   226
condensation column   447
condensation of liquid   212
condensation water   63
condenser   62, 166, 223, 226, 240, 432
condenser pipe   51, 434
conductor   12
conductor material   14
construction material   258
contact behavior   12
contact resistance   14
contamination   91, 164
continuous temperature   207
continuous-cast material   294
conversion of residual austenite   74
cooler   223, 226, 406
cooling finger   368
cooling plants   388
copper wires   318

corporate publications   226
corrosion-current density   19, 61, 91, 201
corrosion inhibition   41
corrosion initiation methods   15
corrosion pit   85
corrosion potential   43
corrosion rate   183
corrosion resistance   48, 121, 130, 158, 184
corrosion underneath the coating   241
corrosion-protection coating   41
corrosion-protection lining   242
cost comparison   241, 264
cost considerations   221
crack propagation rate   250
cracking   69, 72, 120–121, 157, 170
crazes   212
creep behavior   69, 71
creep rupture test   217, 222
creep strength   214, 218, 221–222, 227
creep tests   216
crevice corrosion (in HCl)   72–73, 77, 112–113, 149, 156
crevice corrosion resistance   99, 149
critical pitting corrosion potential   112
cross-flow heat exchangers   180
crude oil   62–63
crude oil processing plant   63
cryotechnical applications   229
crystal violet (as inhibitor)   24
crystallite melting temperature   207
curing agent   233
current density/potential curve   19, 22, 25, 38, 41, 48, 49, 61, 77, 90–91, 93, 151, 164, 169, 198–199
cutting   225
cutting tools   468
cyanide electrolyte   292
2-Cyanopyridine (as inhibitor)   34
cyclic temperature change   258

d

decomposition   225
decomposition of residual austenite   76
decomposition temperature   227
decopperizing   300
decorative effects   302
degradation of plastics   11
dehydrochlorination   214
delamination   240–241
denitration plants   463
dental applications   239
n-denzylquinoline chloride (as inhibitor)   78
deposits   51, 57
desalting   63
descaling   410
desiccator   431
desilverized   292
detinning   432
dew point   350
dew point corrosion   243
dew point of water   63
dezincification   450
dezincing   55
2,4-diamino-6-phenyl-1,3,5-triazine (as inhibitor)   302
2,6-diamino-pyridine (as inhibitor)   63, 65
diaphragms   247
dibutylthiourea (as inhibitor)   51–52
dichloroacetic acid (as inhibitor)   322–324
1,5-di-p-chlorophenyl-2,4-dithiomalonamide (as inhibitor)   91
diethylamine (as inhibitor)   24
diffusion   9, 211, 225, 240, 251, 258
diffusion barrier   244
2,4-dihydroxybenzoic acid (as inhibitor)   298
dilute aqua regia   429
dilute nitric acid   429
1,5-dimethoxyphenyl-2,4-dithiomalonamide (as inhibitor)   91
dimethyl thiourea (as inhibitor)   322
dimethyl yellow (as inhibitor)   24
dimethylamine (as inhibitor)   24
dimethylol thiourea (as inhibitor)   316
1,5-di-p-methylphenyl-2,4-dithiomalonamide (as inhibitor)   91
diphenic acid (as inhibitor)   30
diphenyl thiourea (as inhibitor)   298, 308, 322
1,5-diphenyl-2,4-dithiomalonamide (as inhibitor)   91
dislocation etching   309
dislocation structure   358
dispersion annealing   410
dissolution potential   199
distillation   62
distillation columns   289, 369
distillation units   260
dithioglycolic acid (as inhibitor)   298, 308, 322
n-dodecylpyridine chloride (as inhibitor)   78

n-dodecylquinoline bromide (as inhibitor)   78
downstream evaporators   463
drop in hardness   74
droplet separator   223
dry blower dampers   224
Duranglas® apparatus   463
dyes   23

e

effusion   211
effusion barrier   72
elasticity   222
elasticity modulus   21, 211
electrical conductivity   12, 15, 133, 193
electrical equipment   225
electrical resistance   132
electron-beam melting   133
electrochemical chlorine production   156
electrochemical etching   381, 428
electrode   293
electrolysis   9
electrolysis cell   61
electrolysis plants   225
electrolyte   319
electrolyte cell   458
electrolytic conductivity   6
electrolytic polishing   310, 382
electrolytic processing of nickel   414
electronic assemblies   12
electronic equipment   15, 40, 225
electroplating industry   225
electroplating process   299
electropolishing   46, 48, 309, 318, 328
elongation   133
embrittlement coefficient   326
embrittlement (in HCl)   72, 120–121, 132–133, 143, 146, 167
α-emitting waste   435
emulsification   64
enamel corrosion   191
enamelled apparatus   464–465
enamelled container   190
enamelled vessels   191
enamel-lined pulsed absorption tray columns   466
enamelling furnace   241
endurance tests   **216**
equipment   225, 233
equipment made of tantalum   132
eriochrome black T (as inhibitor)   24
erosion-corrosion   11, 76, 261
erosion-corrosion behavior   74

erosion   197
erosion resistance   197
erosion resistant   153
etching   8, 292, 382
etching agent   302, 309–310
etching operation   429
etching rate   457
etching residues   301
etching solutions   315, 382, 417, 429, 464
ethylamine (as inhibitor)   24, 26–27
evaporation of concentrated nitric   466
evaporator system   51
evaporator units   91
evaporators   178, 189, 256, 386, 434, 438
exchange current density   156
explosively applied cladding   441
explosives industry   374, 400, 439
extinguishing agents   64
extraction agent   435
extrusion   229

f

factories processing nitric acid   348
$FeCl_3$ test   104
δ-ferrite   96
δ-ferrite attack   96
δ-ferrite formation   96
fertilizer factories   396
fertilizer industry   367, 468
fertilizer production plant   442
fiber content   248
fiber fractures   249
fiber orientation   248
fiber spraying   248
fiber-reinforced composites   248
fiber-reinforced products   258
filiform corrosion   15–16, 18
fillers   207, 236
filtration properties   197
first ply failure   249
fittings   179, 189, 233, 256, 259, 261, 262, 265
fixed-cover tank   348
flame spraying   241
flame-pyrolysis decomposition   242
flat seals   246–247, 263
flexural loading   237
flow of cooling water   66
flow rate   158
flue gas desulphurisation plant   260
flue gases   7, 161, 235
fluidised bed sintering   241
fluoropolymer lining   246, 258

fly ash   163
foil lining   244
food industry   9, 302
food processing   11
forced circulation evaporator   435
formation of blisters   212
fractionating column   62–63, 63–64
fracture strength   197
fracture stress   233
free corrosion potential   113
free of pores   257
freezing curve   5
freezing point   5
fructose (as inhibitor)   317
fuchsine acid (as inhibitor)   24
fuchsine (basic) (as inhibitor)   24
fuel element sheaths   436
fuel elements   385, 421, 424
fuel rod   454, 467
fume hoods   235
fuming hydrochloric acid   4
fuming nitric acid   288, 307, 433, 435, 452, 459, 461, 470
2-furancarboxyaldehyde-(2'-pyridylhydrazone) (as inhibitor)   37

g

galactose (as inhibitor)   317
gallic acid (as inhibitor)   318
galvanic coating   69
galvanic corrosion   451
galvanic current   144
galvanic element   55
galvanic zinc coating   72
gas filter element   197
gas nitriding   158
gas preheater   42
gas turbine industry   110
gaseous nitric acid   406, 461
gaskets   134
gasometer system   42
glass apparatus   463
glass coating   189, 256
glass corrosion   185
glass corrosion mechanism   184
glass equipment   187
glass fiber   248, 249
glass fiber-reinforced bellows   348
glass pipe system   134
glass structure   175
glass temperature   208, 215
glass-fiber reinforcement   224
glass-forming metal oxides   176

glow discharge   158
glucose (as inhibitor)   317
grain boundaries   147
graphite columns   7
graphite film   460
graphite heat exchangers   180
graphite products   178
graphite pumps   201
graphite seals   461
graphited cord   461
green chromating   16
GRP construction   258
GRP liner   258

h

haloacetic acids (as inhibitor)   299
hand-laid lamination   241
hard chromium coating   43
hard gold layer   311
hard rubber coating   236, 244
hard rubber lining   265
hardenable spring alloy   427
hazardous materials ordinance   4, 252
HCC   258
HCl absorption   180, 255
HCl concentration   180, 255
HCl diffusion   240, 250
HCl gas permeability   255
HCl initiation   18
HCl permeation   237, 260
HCl production   180
HCl recovery   201, 255
HCl solubility   9
HCl source   7
HCl stripping   180, 255
HCl synthesis plant   179
HCl test   15
HCl vapor pressure   257
health risk   258
heat exchanger jackets   348
heat exchanger tubes   67, 72, 149
heat exchangers   7, 54, 66, 68, 74, 80, 132, 153, 157, 166, 178, 189, 240, 241, 256, 260, 432, 434, 439, 448, 460, 461
heat generation method   34, 35
heat transfer   12, 80
heat transfer agent   349
heat-affected zone   103, 358
heated steam   63
heater   223
heating steam coils   351, 386
heat-treated state   49
heavy oil   14

Subject index | 619

hexamine (as inhibitor)   95, 317
high-energy fuels   456
highly concentrated nitric acid   360
high-performance alloys   110
high-performance plastics   230
high-speed steels   411
high-temperature creep   309
high-temperature reactors   462
high-temperature resistance   130
high-temperature water   135
high-temperature-β-mixed crystal   149
HOKO process   348
homogenisation treatment   97
hose engineering   247
hoses   226, 259
hot extraction   55
hot gas welding process   245
hot pressing   198
hot vulcanisation   236, 257
hot-pressing process   248
housing   348
Huey test   72, **289**, 338, 340–341, 346, 353–355, 357–359, 368, 379, 386–396, 399, 403, 405–406, 408, 418
humidity   14, 64
hydration   184
hydrazine sulfate (as inhibitor)   324, 470
hydroabrasion apparatus   158
hydrochloric acid engineering   256, 260
hydrochloric acid equipment   226, 258
hydrochloric acid pickling baths   56
hydrochloric acid production   7
hydrochloric acid pumps   230
hydrochloric acid recovery   225
hydrochloric acid separation   227
hydrochloric acid separators   260
hydrochloric acid solution   206
hydrochloric acid storage tanks   244
hydrochloric and chromic acid   46, 48
hydrochloric and nitric acid   46, 48, 80–83, 105–106, 109
hydrochloric and phosphoric acid   91
hydrochloric, nitric and sulfuric acid   89
hydrofluoric acid   212
hydrofluorocarbons   64
hydrogen-containing atmosphere   350
hydrogen-induced crack formation   61
hydrogen-induced cracking   69, 72
hydrogen absorption   11, 56–57, 72, 131–133, 143
hydrogen adsorption   169
hydrogen attack   189
hydrogen chloride gas   4
hydrogen chloride vapor   241, 244
hydrogen content   132
hydrogen effusion   69
hydrogen embrittlement   7, 126, 130–132, 143, 147, 167, 170, 255
hydrogen embrittlement fractures   170
hydrogen evolution   42, 88
hydrogen generation   34
hydrogen ion activity   193, 201
hydrogen liberation   35–36, 38
hydrogen overpotential   27, 131
hydrogen permeation   11
hydrogen source   132
hydrogen uptake   444–445
hydrolysis   59, 62, 157, 184, 214–215, 223, 226, 234
hydrolysis resistance   59, 215–216, 231
hydrolytic degradation   9, 214, **215**, 234, 240, 249
hydrophilic medium   234
m-hydroxybenzoic acid (as inhibitor)   30
p-hydroxybenzoic acid (as inhibitor)   30

i

immersion test   23
impact toughness   216, 233
impedance spectroscopy   164
impeller   201
implant steels   401
impregnating agent   178
impregnation   179
impregnation with synthetic resin   180
impurities (in HCl)   **11**, 124, 253, 257
incineration of waste   161
indentation hardness   216
indole (as inhibitor)   332
industrial atmosphere   15, 161, 163
industrial nitric acid   303
influence of hydrogen peroxide   445
inhibiting effect   470
inhibition   30, 91, 190–191, 194, 298, 441
inhibition effect   23, 27, 30, 93
inhibition efficiency (in HCl)   27, 30, 34, 39, 52, 57, 64, 65, 78–79, 91, 93, 165, 299–300, 300, 308, 316–317, 317–318, 321, 328, 331, 334, 429, 441
inhibitive action   23
inhibitor   10, 27, 34–35, 37, 39–40, 42, 51–52, 56–57, 63–64, 77, 92–93, 95, 148, 164–165, 298–299, 300, 308, 316–317, 317–318, 321–325, 328, 331–332, 334
inhibitor addition   43
inhibitor concentration   27, 34, 41

inhibitors for carbon steel 63
injection moulding 229
intensity/depth profile 136
intercrystaline corrosion 297
intercrystalline corrosion in boiling nitric acid 354
intercrystalline corrosion (in HNO_3) 302, 307–308, 336, 338–340, 342, 346, 348, 352, 358, 360, 363–367, 374, 381, 388, 393, 396, 417, 419, 427
intercrystalline corrosion on weld seams 353
intergranular attack 168
intergranular brittle fractures 146
intergranular corrosion (in HCl) 103, 116, 123, 168
intergranular cracking (in HCl) 147, 169
intergranular fracture 147
intergranular hydrogen embrittlement **146**
intergranular stress corrosion cracking 157
intergranular susceptibility 123, 198
intermetallic phase 72, 103, 120–121, 124, 199
internal liner 251, 258
internal pipe coating 241
iodoacetic acid (as inhibitor) 299
ion-exchange phase 190
ion exchange 251
iron inclusions 157
isatin (as inhibitor) 308
isatin-3-(3-thiosemi-carbazone) (as inhibitor) 308
isocorrosion curve 79
isocorrosion diagram 100, 144

k
Keller's solution 307
Keramchemie-Lurgi process 201
kerogen 379
kerosine-water mixture 63
Kesternich test 160
knife-line corrosion (in HNO_3) 123, 352, 359, 373

l
lack of pores 241
lamellar graphite 336
laminate coatings 240–241
laminate composites 248
laminate structure 251
last ply failure 249
lauryl acid (as inhibitor) 30, 33

lead coatings 300
leakage current 132
level of curing 251
liberation of hydrogen 27, 159
light water reactor 447
lime sludge 96
liner 242, 257
lining material 228
linings 80, 227, 236, 242
lip seals 247
litharge lead plate 156
LITHSOLVENT® 92–93
local element 19, 23, 157
longitudinal welding seam 157
long-term behavior 215–216
loop samples 327
loose linings 245
low-temperature properties 222

m
machine factories 308
machine part 225
machining 258
macromolecule 206, 213
maize extract 316
malachite green (as inhibitor) 24
Mannheim process 8
mannose (as inhibitor) 317
manual arc weldi 362
marine atmosphere 15
martensite decomposition 74
mat laminate 241
matrix fractures 249
mechanical polishing 382
mechanical properties 48, 180
medical implant 239
mediterranean countries 181
melt-spin process 170
mercaptobenzimidazole (as inhibitor) 321
2-mercaptobenzothiazole (as inhibitor) 321
methanol-hydrochloric acid solution 146
methanol/nitric acid 293
1-p-methoxyphenyl-3-formamidine-thiocarbamide (as inhibitor) 299
methyl red (as inhibitor) 24
methyl violet 6B (as inhibitor) 24
methylamine (as inhibitor) 24, 26–27
n-methylaniline (as inhibitor) 325
microcrack 179
microorganisms 211
microsealing 246
β-microstructural state 150

Subject index

mineral acid   288
mineral digestion industry   11
miscibility gap   6
mixed acids   81
mixed city-industrial atmosphere   15
mixed metal oxide   20
mixers   153
modified Huey test   417
molecular chains   206
molecular structure   206
molybdenum carbide   123
molybdenum depletion   123
$MoSi_2$ layers   412
Müller-Kühne process   401
multilayer composite   249
multilayer lining   245, 246
multilayer rubber coating   261
muriatic acid   4

n

NaCl initiation   18
NaCl test   15
1.8-naphthalic acid (as inhibitor)   30
naphthotriazole (as inhibitor)   322
α-naphthylamine (as inhibitor)   317, 325
neutralisation   8, 63, 228
neutraliser   63
neutralising agent   63
nickel layer   414
3-nicotinamide (as inhibitor)   34
nicotinic acid   292
nitration of organic substances   463
nitric acid   442
nitric acid + ammonia   456
nitric acid + gasoline   456
nitric acid and hydrogen fluoride   432
nitric acid (as inhibitor)   442
nitric acid condensate   368
nitric acid factories   440
nitric acid industry   289, 336
nitric acid plant   348
nitric acid plants   463
nitric acid pumps   348
nitric acid vapor test   431
nitric acid with hydrochloric acid   429
nitric and glacial acetic acid   459
nitric and hydrochloric acid   307, 310, 329, 333, 382, 402, 411, 414, 424, 428–430, 434, 442–443, 448, 462, 466, 469–470, 473–474
nitric and hydrofluoric acid   105, 109–110, 307, 313–315, 342–343, 371–372, 381–382, 397–399, 414–417, 421, 424–425, 428, 434, 443–445, 454–456, 460, 467, 469–470, 473–474
nitric and phosphoric acid   46–47, 310, 319, 403, 428, 463, 468
nitric and sulfuric acid   105, 108–109
nitric and sulphuric acid   307, 334, 336, 343–345, 400–401, 424, 428, 435, 442–443, 461, 463, 468–469, 475
nitric, hydrochloric and phosphoric acid   403
nitric, hydrochloric and sulphuric acid   382
nitric, hydrofluoric and hydrochloric acid   310
nitric, hydrofluoric and sulphuric acid   434, 464
nitriding   158–159
nitriding temperature   159
p-nitroaniline (as inhibitor)   323–325
nitrogen atmosphere   466
nitrogen compounds   63
p-nitrohydroxy-benzoic acid (as inhibitor)   308
nitrous gases   375
noble metal salts   148
nodular graphite   336
notch impact toughness   216
notched impact strength   360
nuclear energy plants   349
nuclear fuel reprocessing   450
nuclear fuel rods   454
nuclear fuels   288, 372, 447–448
nuclear power station   97
nuclear reactor construction   450
nuclear reactor industry   355
nuclear reactors   382, 421, 441

o

offset yield strength   346
oil   77
oil ash particles   163
oil distillation   62
oil industry   63–64
oil production   77
oil refineries   63
oil refinery   66, 72
open-air weathering test   15
operating temperature   227–228, 231
organic matrix material   248
organic waste   435
O-rings   247, 263
orthopaedics   314
oscillation load   214
overhead streams   62

Subject index

overheating 246
overpickling 10
overvoltage 156
over-vulcanisation 257
oxidation resistance 215
oxidative degradation 9, 215, **215**, 234
oxide layer 143, 153
oxidising agent 148
oxidizing ions 448
ozone 380

p

packaging material 302
packaging plastic 232
packings 247
paper industry 228
paper mill 96
particle-reinforced composites 248
passivation 148, 157, 198, 288
passivation of chromium steel 345
passivator 335
passive layer 87, 130, 136, 143, 175
passivity 154
paste extrusion 227
penetration composites 248
penetration rates 216
peristaltic pump 261
permanent die casting 303
permeability 9, 180
permeation 212, **212**, 241
permeation ability 6
permeation coefficient 212, 227
permeation effect 9, 212
petroleum distillation units 74
petroleum production 406
pharmaceutical industry 256
α-phase 149
μ-phase 103
ω-phase 149
p-phenetidine (as inhibitor) 299
o-phenetidine (as inhibitor) 317, 325
phenyl thiourea (as inhibitor) 322
m-phenylenediamine (as inhibitor) 323
o-phenylenediamine (as inhibitor) 323, 328
phenylhydrazine (as inhibitor) 324–325
phenylthiourea (as inhibitor) 298, 325
phosphate chromation 302
phosphate fertiliser 90
phosphate rock 90
phosphoric acid (as inhibitor) 470
photochemical degradation 229
photographic etching 301

photographic industry 380
photooxidative resistance 225
phthalic acid (as inhibitor) 30
physical vapor deposition 177
pickle 48, 56, 105
pickling 8, 46, 69, 381, 401, 444
pickling agent 10
pickling attack 45
pickling bath monitoring 56
pickling of stainless steels 110
pickling paste 382
pickling process 10–11, 55
pickling rate 445
pickling resistance 44
pickling solution 43, 299, 450
pickling time 69, 71
pickling towers 235
piezoelectric resonators 464
pigments 207, 223
pilot plant 134
pipe connections 421
pipe material 235
pipeline 57, 187, 259, 360
pipeline construction 259
pipes 179, 218, 264
piping for solvents 226
pitting 14, 73, 91
pitting corrosion (in HCl) 14, 19, 72, 77, 80, 87, 96, 99, 103, 108, 112–113, 156, 167, 175, 255
pitting corrosion (in HNO_3) 307, 367, 381, 401
pitting holes 157
pitting potential 169
pitting resistance 169
plant components 256
plant pickling 44
plastic composite 207
plastic deformation 132
plastic pump 262, 264
plastic sealing discs 350
plasticizer 207, 225, 236
plate heat exchanger 156, 260
platings 12
platinum 450
platinum layers 430
plutonium 372
plutonium oxide 467
polarisation curves 39, 67–68
polarisation measurements 25, 27, 39
polarisation resistance 40, 43, 77
polishing 310, 411
polishing agent 292

polishing bath   300
polishing copper   318
polyaddition   208
polycondensation   208
polyester barrier layer   250
polyethylene-coated carbon fibers   461
polymer cement   230
polymer concrete   235
polymer lining   260
polymer matrix   251
polymer-cement mixture   183
porcelain pipe system   134
pore diffusion   178
pore structure   180
potassium bichromate (as inhibitor)   154
potassium dichromate (as inhibitor)   441
potassium iodide (as inhibitor)   78
potential-pH diagram   12
potentiodynamic quick test   113
powder   198
powder coating   229, 240, 241
powder extrusion   227
powder sintering   241
power station systems   44
power stations   8
precipitation kinetics   121
preheater   63
preheater circuit   63
preheater stage   63
prepregs   248
pressure die casting   303
pressure sintering   227
pressure/temperature diagram   228
pressure-temperature-concentration diagram   5
pressurised reactor   135
pressurized water reactor   421, 424, 454
printed circuit   15
printing plates   450
process air dampers   224
process engineering   166
processing fuel rods   421
processing of nitric acid   441
production of ammonium nitrate   380, 408
production of carbide   450
production of fertilizers   441
production of nitric acid   408
production of nitroglycerine   374
production of Nitrophoska   403
production of TNT   374, 400
profile packings   247
propeller   396
propeller bronze   50

n-propylamine (as inhibitor)   24
propylamine (as inhibitor)   26–27
pT-diagram   247
PTFE coating   263
PTFE rings   368
PTFE yarns   181
pump   153, 179, 189, 197, 228, 233, 261, 265, 360, 439
pump components   261
pump industry   228
pump material   262
pure gas pipeline   223
pure nitric acid   382
Purex process   438
Purex-IWW solution   418
PVC burning   225
PVC waste   257
PVD process   174, 177
pyrolysis   178, 227
pyromellitic acid (as inhibitor)   30
pyrophoric reaction course   435
2-pyrrolcarboxaldehyde-(2'-pyridylhydrazone) (as inhibitor)   37

q
quartz glass   184
quartz sand   158
quartz sheets   464
quencher tower   7
quinoline (as inhibitor)   95

r
R glass   250
radical chain reaction   215
radioactive material   365
radioactive substances   379
radioactive waste   379, 438
radioactive waste solutions   115
radioindicator measurements   351
radionuclide method   59
radiotracer method   432, 454
rain water   181
reaction boilers   464
reaction tanks   374
reactor   189
reactor material   135
reactor vessels   134
recovery of the nitric acid   372
recovery plants   435
red fuming nitric acid   297
reduction factors   214, 217–218, 220, 222, 228
refineries   62

Subject index

regenerate dampers   224
regenerate pipeline   223
regeneration of ion-exchange resins   8
reinforcing materials   248
relaxation   246
release of hydrochloric acid   214
remediation measures   225
repassivating potential   167
reprocessing nuclear fuels   365, 456
resistance against stress corrosion cracking   97, 151
resistance criteria   216
resistance factor   217
resting potential   159
rhodamine B (as inhibitor)   24
risk of explosion   458
riveted connection   257
road tankers   246
rocket components   456
rocket fuels   307
rocket technology   288
Rodine 213   42
roll bonding cladding   441
rolling scale   56
rotary pump   231, 261
rotating mechanical seals   263
round seals   247
rovings   248
rubber coatings   236, 244–245, **257**, 261
rust   57
rust formation   59

S

safety valve   223
salicylaldehyde (as inhibitor)   78
salicylic acid (as inhibitor)   30
salt deposit   54
salt spray test   15
sand casting   303
sand-slurry pumps   222
sandwich samples   156
screws   69
scrubber   112
sealing disc   350
sealing layer   241
sealing material   156, 460
sealing technology   247
seals   226, 236, 246, 256, 263
seawater   49, 51–52, 97
seawater desalination plant   54, 153
seawater flash distillation plant   51
secondary hardening   76
self-regenerating oxide layer   432

semiconductor components   473
semiconductor technology   301
sensitization   354
service life   160
sewage filter   448
shaft   348
shaft cover   348
shaft seals   247
shell and tube heat exchanger   180, 227, 260
shrinkage   258, 260
shut-off gates   263
shut-off valve   96
SiC coating   198
SiC layer   181
sigma phase   199
sigma phase formation   338
silane binder   182
silicates   58
silicic acid   58, 194, 196
silicic acid saturation effect   190
silicoater process   242
silicon carbide material   204
silicone oil   194
silver chloride   291
silver solder   293
Simmerings®   247
sizing process   250
sliding gate housings   406
sliding gates   406
slow tensile tester   146
sodium bichromate   345
sodium hydroxide melt   302
sodium petroleum sulfonate (as inhibitor)   331
sodium phosphate (as inhibitor)   470
soft rubber coating   236, 244
soft rubber lining   265
solder   457
soldering   291
solution analysis   67–68
solution annealing   103
solution annealing temperature   150
solvent pump   224
solvent recovery   226, 246, 258
solvent separator   224
solvents   220
solvolysis   215
source of the fire   64
spark testing   257
spent fuel rods   468
spin-coating process   249
spinning process   248

Subject index | 625

spontaneous passivation 175
spot welding 359
stabiliser 207
stainless steel component 58
standard hydrogen electrode 46
statistical chain cleavage 214
steam 132
steam dampers 224
steam generators 67
steel industry 201
steel surface 56
steel tanks 441
steel vessel 257
steelworks 197
Stern-Geary equation 61
stirred autoclave 153
stirred tank reactor 264
stirrer 166
stirring blade 261
storage of highly concentrated nitric acid 348
storage tank 348, 374
storage tests **216**
stove enamel 240–241
strain loading 147
strain rate 90, 146
stratosphere 4
stress characteristics 246
stress corrosion cracking 313, 319, 326
stress corrosion cracking behavior 150
stress corrosion cracking (in HCl) 59, 74, 77, 80, 83, 85, 87–88, 90, 96, 112, 121, 149, 149–150, 167–168, 170, 250, 255
stress corrosion cracking (in HNO_3) 307
stress corrosion cracking (in methanol-HCl) 157
stress corrosion cracking tests (in HCl) 122
stress crack **212**, 214, 224
stress cracking 213, 227, 230, 232, 246
stress cracking (in HCl) 228–231, 234
stress intensity factor 250
stress-crack corrosion behavior 167
stringent Huey test 365
stripper 223
stripper column 7, 223, 261
stripper plant 180
stripper pumps 223
stripper reservoir 223
stripper towers 260
submersible pump 261
sugar industry 91, 93
sulfamic acid (as inhibitor) 470

sulphamic acid 54–55
sulphur bridges 236
sulphuric acid pickling baths 56
sulphuric and hydrochloric acid 442
supercritical water oxidation 116, 135
superficial electrolyte layer 164
superheated steam 421
supporting construction 244
surface corrosion (in HCl) 99
surface damage 132
surface layer 237
surface pretreatment 242
surface protection processes 242
surface treatment 450
surfactant production plant 225
susceptibility to cracking 249
susceptibility to intercrystalline corrosion 359
swell 238
swellability 221
swelling effects 9
swelling (in HCl) 9, 11, **211**, 213–214, 216, 225, 234, 240, 249, 258, 264
swimming pool hall 83, 87
synthesis reactor 447
synthetic diamond powder 462
synthetic fiber industry 9

t

Taber abraser 262
Tafel constant 61
Tafel line 20, 40, 95, 201
Tafel method 164
tank 256, 360
tank material 296, 303, 417, 435
tank wagons 289, 341
tankers 256
tannic acid (as inhibitor) 318
tanning 9
tantalum-lined autoclaves 465
tantalum tubes 132
tar pitch 178
Teflon® lining 143
Teflon® stoppers 144
Teflon® tape 144
temperature gradient 212
temperature limit 226, 238, 243, 256
temperature loading 207
tensile strength 69, 133, 168, 180, 216, 258
tensile stress 69, 214
tensile test 167
terephthalic acid (as inhibitor) 30

thermal conductivity 12, 72, 432
thermal decomposition 64
thermal degradation **214**
thermal equipment 260, 265
thermal expansion 72, 259
thermal shocks 197
thermal stability 117
thermally curable coating materials 241
thermite process 198
thermoforming 225, 258
thermolysis 214, 225, 258
thermometrics 27, 46
thermoplast lining 244, 245, 249
thermoplastic inner liner 257
thick nickel layers 413
thiocyanate (as inhibitor) 332
2-thiophencarboxaldehyde-(2'-pyridylhydrazone) (as inhibitor) 37
thiosemicarbamide (as inhibitor) 308
thiourea (as inhibitor) 95, 317, 324–325, 332, 334, 380, 382, 429
thorium-containing fuel elements 372
three-layer rubber coating 244
titanium electrodes 441, 448
titanium material 157
titanium mononitride 158
titanium pipes 439
α-titanium precipitation 149
titanium pumps 445
titanium tank 351, 385–386
m-toluidine (as inhibitor) 323
o-toluidine (as inhibitor) 317, 324–325
p-toluidine (as inhibitor) 299, 323
o-tolylthiourea (as inhibitor) 299
p-tolylthiourea (as inhibitor) 298
1-m-tolyl-3-tolyl-formamidine-thiocarbamide (as inhibitor) 299
tolyltriazole (as inhibitor) 317
torsional loading 169
torsional oscillation test 206
trans-cinnamaldehyde (as inhibitor) 78
transgranular cracking (in HCl) 169
transgranular cracks 90
transgranular stress corrosion cracking 72
transport of chemicals 246
transport tank 246
transportation of dilute nitric acid 341
triazole (as inhibitor) 322
trichloroacetic acid (as inhibitor) 322–325
triethylamine (as inhibitor) 24
3,4,5-tri-hydroxybenzoic acid (as inhibitor) 298
trimellitic acid (as inhibitor) 30
trimethylamine (as inhibitor) 24
trinitrotoluene 374
triple-point pitting corrosion 307
turbine pipes 421

u

ultimate elongation 216
uniform oxide film formation 299
Union Carbide process 134
uranium dioxide 467
uranium processing 385
uranyl nitrate 379
urea industry 407
urea plants 403
urea production 403
urea solution 405
urea synthesis 441
urea synthesis plant 405
urea synthesis solution 405
US Federal Specification DD-G-541B 187

v

valves 43, 227, 261, 434
vapor phase 190
vaporizer 352
ventilator casings 235
venturi scrubber 128
vessel 218, 224, 226, 235, 255–256, 265
vibration cavitation test 76–77
vickers hardness 41, 133
viscoelastic behavior 214
vitamin manufacture 158
Viton® sealing rings 156
volcanic vapors 4
vulcanisates 236
vulcanisation 236
vulcanisation accelerator 244
vulcanisation catalyst 236
vulcanised 244

w

washers 463
waste acid tanks 235
waste incineration 14, 225, 258
waste pickling bath 197
waste pickling solutions 201
waste sulphuric acid 335
waste tanks 355
waste water lines 463
water absorption 182
water absorption capacities 232
water activity 224, 233

water separator   226
water vapor permeability   241
waterborne paints   242
waveguides   291
wear behavior   198
wear regard   222
wear resistance   158, 181, 198, 201
weld filler   96
welded joint   298, 358
welded seam   246, 257
welding   112, 225, 258
welding material   308, 399
wet combustion   435
wet strength   242
wetting agent   314, 382
white fuming nitric acid   297
white liquor   96
white zone   307
white-liquor clarifiers   96
worm wheel shafts   348
woven filters   229

y
yellow chromating   16
yield point   133
0.2 % yield strength   69, 77, 87

z
Zn-Al coatings   160
Zn-Al protective coating   160
zinc anode   80
zinc coating   160
zinc layer   160
Zn/Al eutectic   160